W0079564

THE FRONTIERS COLLECTION

Series Editors

Avshalom C. Elitzur, Iyar, Israel Institute of Advanced Research, Rehovot, Israel

Zeeya Merali, Foundational Questions Institute, Decatur, GA, USA

Thanu Padmanabhan, Inter-University Centre for Astronomy and Astrophysics (IUCAA), Pune, India

Maximilian Schlosshauer, Department of Physics, University of Portland, Portland, OR, USA

Mark P. Silverman, Department of Physics, Trinity College, Hartford, CT, USA

Jack A. Tuszynski, Department of Physics, University of Alberta, Edmonton, AB, Canada

Rüdiger Vaas, Redaktion Astronomie, Physik, bild der wissenschaft, Leinfelden-Echterdingen, Germany

The books in this collection are devoted to challenging and open problems at the forefront of modern science and scholarship, including related philosophical debates. In contrast to typical research monographs, however, they strive to present their topics in a manner accessible also to scientifically literate non-specialists wishing to gain insight into the deeper implications and fascinating questions involved. Taken as a whole, the series reflects the need for a fundamental and interdisciplinary approach to modern science and research. Furthermore, it is intended to encourage active academics in all fields to ponder over important and perhaps controversial issues beyond their own speciality. Extending from quantum physics and relativity to entropy, consciousness, language and complex systems—the Frontiers Collection will inspire readers to push back the frontiers of their own knowledge.

More information about this series at http://www.springer.com/series/5342

Filipe Duarte Santos

Time, Progress, Growth and Technology

How Humans and the Earth are Responding

 Springer

Filipe Duarte Santos
Department of Physics
CCIAM-cE3c. Faculdade de Ciências da
Universidade de Lisboa
Lisbon, Portugal

ISSN 1612-3018   ISSN 2197-6619   (electronic)
THE FRONTIERS COLLECTION
ISBN 978-3-030-55332-6   ISBN 978-3-030-55334-0   (eBook)
https://doi.org/10.1007/978-3-030-55334-0

© Springer Nature Switzerland AG 2021
This work is subject to copyright. All rights are reserved by the Publisher, whether the whole or part
of the material is concerned, specifically the rights of translation, reprinting, reuse of illustrations,
recitation, broadcasting, reproduction on microfilms or in any other physical way, and transmission
or information storage and retrieval, electronic adaptation, computer software, or by similar or dissimilar
methodology now known or hereafter developed.
The use of general descriptive names, registered names, trademarks, service marks, etc. in this
publication does not imply, even in the absence of a specific statement, that such names are exempt from
the relevant protective laws and regulations and therefore free for general use.
The publisher, the authors and the editors are safe to assume that the advice and information in this
book are believed to be true and accurate at the date of publication. Neither the publisher nor the
authors or the editors give a warranty, expressed or implied, with respect to the material contained
herein or for any errors or omissions that may have been made. The publisher remains neutral with regard
to jurisdictional claims in published maps and institutional affiliations.

This Springer imprint is published by the registered company Springer Nature Switzerland AG
The registered company address is: Gewerbestrasse 11, 6330 Cham, Switzerland

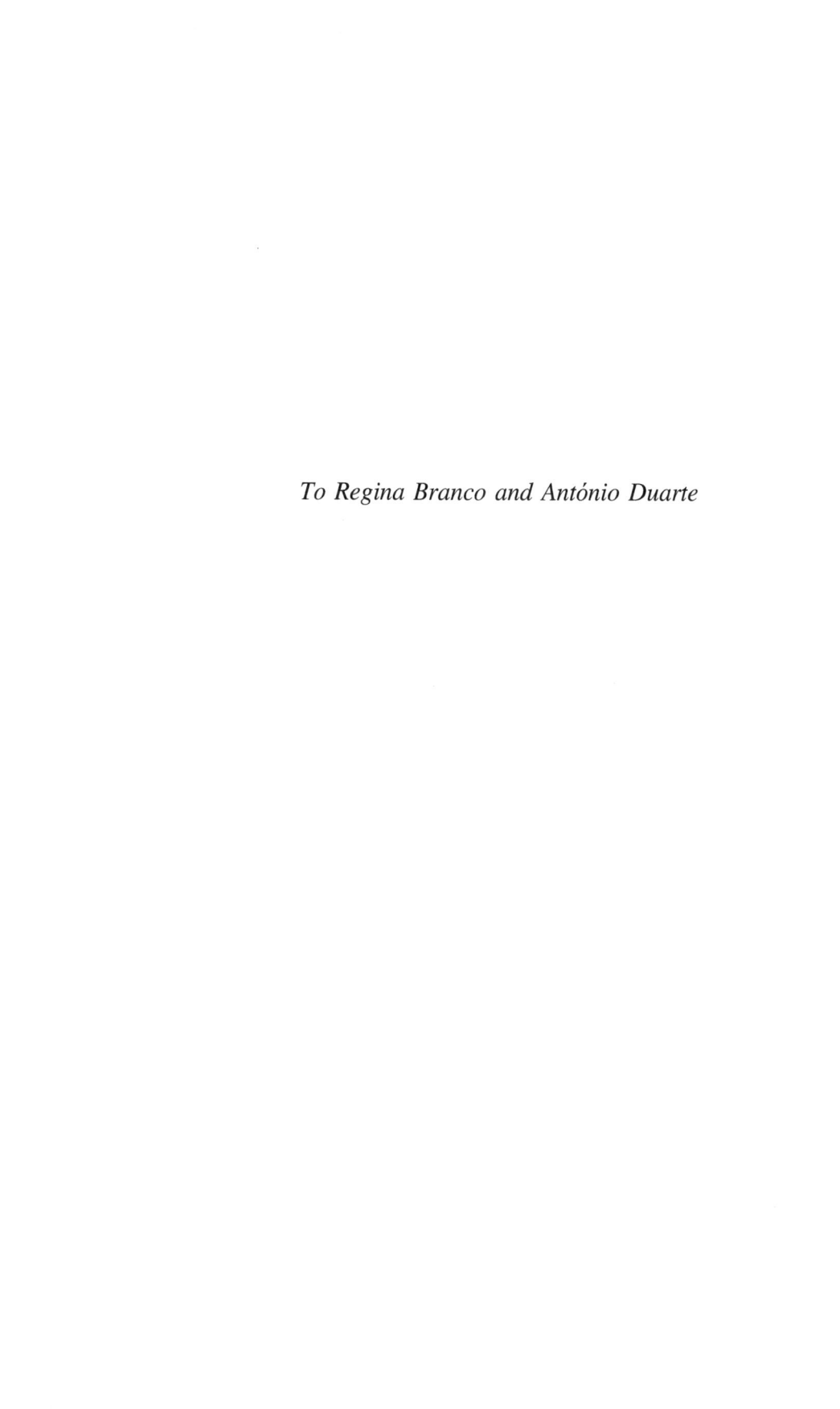

To Regina Branco and António Duarte

Preface

The guiding thread of this book is time and how *Homo sapiens* has been dealing with it during his journey. Time is present in all human actions, experiences, emotions, and expectations. We do our utmost to keep it as long as possible so that we can continue to enjoy living. Various concepts of time have emerged in the course of history that continue to be developed, showing their inexhaustible potential creativity. Our relation with time has been deeply changed by modernity and we now live under the constant pressure of time acceleration. There is an irresistible need to fill up time completely and permanently with everything planned and needed for self-interest and everything possible or enjoyable, as if we were afraid to face it alone. Concurrently, future time has become increasingly dominant, menacing, and uncertain.

The main aim of the book is to try to understand how the "new time" emerged by identifying the main concepts, ideas, hopes, processes, and systems that throughout history have shaped the present world with all its diversity and human achievements, challenges, problems, and crises. The underlying motivation is to understand the current global civilization essentially with the tools offered by science and to explore the possibility that this exercise may also shed light on the future.

Homo sapiens is the only surviving species of the genus *Homo* that emerged in Africa around 2.6 million years BP at a time of a global climate change that transformed a large part of the African tropical forests into savannahs. Natural selection in the new environment led to a strong encephalization that enabled the development of increasingly complex social behaviour and eventually to the development of symbolic thought, the capacity for spoken language and symbolic representation. The biological evolutionary process leading to the new species was completed between 300 000 and 200 000 years BP. After that time *Homo sapiens* underwent an extraordinary cultural evolution.

The Agricultural Revolution was the first major advance in that evolution and occurred very recently relative to the beginning of human time, just 10 000 years BP. It deeply transformed our relation with nature through the domestication of plants and animals. It also contributed to the development of new technologies and to increased social stratification, bringing about large scale violent conflicts and wars, densely populated cities, a much more diversified list of goods and services, consumer excesses, and a remarkable development and diversification in the visual arts, literature, and architecture. However, energy consumption growth and economic growth continued to be small, almost non-existent during the next 97.5 centuries. That situation changed radically just a quarter of a millennium ago during the Industrial Revolution, with the beginning of the high-energy-consumption-per-capita era, based on coal and later on oil and natural gas, and the proliferation of technological inventions based on the practical applications of modern science. Almost simultaneously with the invention of an improved steam engine by James Watt, Adam Smith proposed the concept of absolute advantage that played a crucial role in the development of capitalism and the instauration of the strong-economic-growth era. Until the mid-eighteenth century, average global growth in the production of goods and services—what is usually called global GDP— was very small and was driven almost exclusively by demographics and the colonisation of new lands by Europeans. This situation changed drastically with the advances of the Industrial Revolution and the development of capitalism. Economic growth gained pace first in some European countries and then progressively spread to a large part of the world, although with temporary slowdowns during major wars and crises.

Enlightenment and rationalism introduced the ideology that reason aided by the methodology of science guarantees that "the perfectibility of man is absolutely indefinite" (Condorcet 1795) and will lead the whole of humankind to a world of individual freedom, human rights, equality, solidarity, well-being, and material prosperity. There have been many social, economic, and political developments since that time, but we continue to be immersed in a culture that assumes the inevitability of progress, although recognising that there have been and will be periods of decline, deep disappointments, and dramatic intervals.

The French Revolution represented the first major burst of time acceleration but it also established a new relationship between the past and the future characterized by opposing ideas: on one hand human progress and on the other the uncertainty of a future without reference points. The future became more uncertain but in return it became filled with the promise of progress through the use of reason. Trying to unveil the nature and characteristics of our common immediate and middle term future, for each community, each country, and the entire globe is a fundamental and indispensable everyday task for politicians, business people, financiers, economists, stock-market analysts, writers, and political commentators around the world. But although these predictions and projections are all systematically used to construct a supposedly more secure and better future, its uncertainty is increasing.

Technology or "the useful arts", as they were called in the 1787 US Constitution, underwent an accelerated progress from the first quarter of the nineteenth century onwards, which revealed its power to influence and characterize the evolution of societies in Western Europe and North America. The establishment of science as the most powerful driver of technology happened in the second half of the nineteenth century through various inventions based on modern science and in particular with the generation and detection of electromagnetic waves by the German physicist Heinrich Rudolf Hertz (1857–1894) in 1887, following their theoretical prediction by the Scottish physicist James Clerk Maxwell (1831–1879) in an 1864 paper entitled *A Dynamical Theory of the Electromagnetic Field*, a discovery that revolutionized the way humans communicate. The twentieth century saw the establishment of another very powerful link, this time between technological progress and economic growth through the concept of total factor productivity, considered to be the residual attributable to innovation and technological improvement, which cannot be explained by changes in inputs.

These two links created an oversimplified causal chain between scientific research, technological innovation, and economic growth that erroneously convinced humankind that it had assured the perpetual economic growth needed to guarantee an everlasting and unlimited increase in human well-being and economic prosperity, at least for part of the human population. There is no doubt that science and technology provide the essential support for better human health and living standards, economic prosperity, and increasing availability and diversification of goods and services. However, science and technology constitute a necessary but not sufficient condition for the viability of perpetual improvement. There are other factors on which it depends.

After two devastating world wars in the first half of the twentieth century, the warring countries engaged in a period of reconstruction and development, starting in about 1950, in which the USA assumed the role of leader of the free world in opposition to the communist Soviet Union and its satellite countries. This period, usually called the Great Acceleration, witnessed a remarkable increase in the global population, urbanization, global GDP, land use change, exploitation of natural resources, communications and mobility, scientific research, and technological innovation, and it set the stage for two waves of democracy that went round the world. In the first three decades of the Great Acceleration, known as *Les Trente Glorieuses*, middle class incomes in countries with advanced economies grew strongly, while socioeconomic inequalities were reduced. This growth resulted mostly from extensive industrial development, which created a high number of jobs in the manufacturing, construction, and service sectors for workers with middle level qualifications.

From 1978 onwards, China started to open up its economy to the rest of the world, following the economic reforms introduced by President Deng Xiaoping, and it joined the World Trade Organization in 2001. India and other emerging countries were also integrated into the global economy and began providing goods and services at low prices to consumers in advanced economies, thanks to their cheap workforce. This process of globalization, together with the increase in the price of oil and other commodities, exported by developing countries, increased the speed of economic convergence between the two groups of countries, and it freed hundreds of millions of people from poverty in Asia, Latin America, and certain regions of the Middle East and Africa.

The Great Acceleration also led to the global economic and financial system becoming the prime driver of planetary-scale changes, usually referred to as global changes to the Earth system. Currently, we are observing global anthropogenic changes in atmospheric composition and circulation, ocean circulation, the climate, the polar regions, the global mean sea level, the water, carbon, nitrogen, and phosphorous cycles, pollution, land use and land cover, biodiversity, and the availability of natural resources, while we continue to produce various kinds of solid and liquid wastes and emissions at a rate at which they cannot be renewed or sequestered. In other words, humankind is overexploiting natural resources, polluting and degrading the environment, reducing biodiversity, and changing the climate at the planetary scale. These unintended but destructive changes, sometimes irreversible, are starting to backfire since they are affecting the well-being and economic prosperity of a large, indeed increasingly large, part of humankind, especially in the developing economies. Humans are intimately interconnected and dependent on the environment, much more than is usually admitted. If we interfere in this dangerous way with the various biogeophysical subsystems that compose the Earth system, we are in fact doing ourselves harm.

The solution to such global anthropogenic interference that one might expect from a rational and ethical point of view would be a progressive change in behaviour and a reform of the economic and financial system so that the relation with the environment, natural resources, and climate could reach a new and lasting equilibrium beneficial to the whole of humankind. The required changes are feasible with the help of the science and technology that is currently available. However, such changes will negatively affect a minority that has powerful vested interests in the current system, interests that are shared with most governments. That minority enjoys extreme economic prosperity and holds a disproportionate power over humankind. On the other hand, the required changes would allow the development of a sustainable model that would benefit the majority of people around the world.

The Earth system is like a large shared home to a group of people with very pronounced social and economic inequalities that consume more than is actually available and renewable, while more are arriving all the time. And the economic and financial system controlled by a powerful minority has succeeded in addicting people to a consumerism that is destroying this common home. In the absence of an effective response, the most easily available narrative takes the form of negating the science that would justify a transformational transition to a sustainable model, and of waging campaigns of disinformation and misinformation aimed at confusing and distracting people from the dangers of the developing situation. In the limit, an erroneous alternative is entertained, which would consist of terraforming Mars at a cost of many hundreds of trillions of dollars.

Attacks on what is considered to be politicized science or science classified by the state as "secret", to prevent it from being shared by citizens, are proliferating. Furthermore, widespread information noise, misinformation, disinformation, and mass manipulation in the social media is increasing distrust in and scepticism about science, promoting charlatanism, conspiracy theories, and political polarization in democracies. On the other hand, the incapacity of democracies to deliver the increasing well-being and economic prosperity that their political party leaders systematically promise to all citizens is weakening them.

There is a disturbing tendency for democracies to acquire some of the characteristics of oligarchies and for their governments to look more favourably on authoritarian governments, favour mercantilist and protectionist economic policies, dismiss the importance of human rights, and promote nationalist and nativist policies. Moreover, the malfunction and decline of some Western democracies is opening new spaces for the advancement of authoritarian regimes in emerging and developing countries, inspired by China's state capitalism.

All these developments represent a partial regression in some of the remarkable civilizational achievements initiated by the Industrial Revolution, by the worldwide development that occurred subsequently, and in particular by the past 70 years of progress in human cooperation, science, and technology, without any major world war.

What are the origins of the present situation and how is it likely to evolve in the future? What are the main obstacles in the way of reaching some form of sustainability? To what extent do the present and future developments stem from essential features of human nature and what are the chances of somehow circumventing that condition? These are some of the questions that this book tries to answer, using the knowledge and understanding provided by the physical, natural, and social sciences. Science provides an understanding of the physical and natural world in the past and present, and it projects the future by means of scenarios that portray a range of possible future outcomes. It helps to discover and develop technologies that have revolutionised our ways of life. It has extraordinarily improved human health and it is used in association with technology to fabricate a great variety of products and devices with extremely diversified potential uses, such as helping to reach greater economic prosperity, better well-being, and quality of life, while providing ways to produce powerful and destructive weapons and increasingly efficient addictive drugs. Human curiosity continues to be the main driving force of science and the same curiosity can also lead some to use the scientific method to reflect on the future of the human endeavour in the medium and long term and its effects on the Earth system, and to warn that the risks due to unsustainability of the current model are rising dangerously.

The present book pays special attention to the subject of climate change. Anthropogenic climate change is one of the more paradigmatic challenges facing humankind in the twenty-first and following centuries since it is an inadvertent collateral effect of the intensive use of fossil fuels. These have supplied abundant energy since the Industrial Revolution and were essential to build our present modern world. The analysis of the way humankind has reacted to this challenge reveals the inner workings of our relation with the environment and is particularly instructive about how the future is likely to unfold. Throughout the book and wherever appropriate, the specific characteristics and challenges created by climate change are addressed and compared with other issues. Particular attention is paid to the similarities between the present anthropogenic interference with the climate system and natural intensive-flux carbon injections into the atmosphere in the form of CO_2 and CH_4 that occurred in the geological past.

The following paragraphs attempt to guide the reader through the book by explaining the rationale for the sequence of chapters and sections in each chapter. In most cases, the reader can find here a very short description of the contents and main results of each section. This synthesis should help to localize the sections that are more relevant to following the main argument of the book and those that may be of particular interest to the reader.

Chapter 1 is about time, the unifying theme of the book. The first three sections play a central role since they contain the definition of the concepts of operative social time and operative time structure that are used throughout the book. Sections 1.4–1.7 deal with intertemporal choices, meaning those that involve decisions with consequences that have repercussions in the future. These are choices that lead us to feel the challenge of time and its all-embracing power over our plans and expectations and help us become acquainted with the uncertainty of the future. They are frequent and essential choices in everybody's personal life, but also choices that must be made at the individual level to construct collective decisions with implications for the present and future, and regarding institutions and all types of social groups from the local football team to professional groups and political parties, from the local community, to the city, the region, the country, or even the whole of humankind. Sections 1.4 and 1.5 deal with the main concepts of intertemporal choices, while Sects. 1.6 and 1.7 are more technical and address various ways of determining the optimum social discount rate and how they are crucial for intergenerational justice.

Section 1.8 introduces the analysis of the evolution of operational social time and operational time structure and the contribution to this process made by the advances in the measurement of physical time. Section 1.9 also deals with physical time and addresses the way the observation of cyclical phenomena, those defining time cycles, influenced its conceptual development. This relation provides the opportunity to introduce the concept of element of operative social time and in particular the element of operative time corresponding to a human lifetime, which plays a crucial role in our relation to time. Section 1.10 introduces a central notion of the book, namely the operative time of a social generation. Section 1.11 also deals with the deeply challenging meeting point between the ascendancy of time over humans and their essential eusociality. Sections 1.12 and 1.13 deal with multigenerational operative social time and historical time. They also offer a brief overview of our knowledge of ancient civilizations and introduce the idea of accelerated operative social time. Section 1.14 addresses the relevance of the human lifetime as an element of operative time in our horizon of expectancy and explores its implications in terms of time discounting. Section 1.15 discusses the time of the *Homo* genus and also *Homo sapiens* time and different forms of their termination. Sections 1.16–1.18 analyse the first records of human awareness of the solar and lunar cycles registered in European Upper Palaeolithic cave paintings and discuss how the Egyptian and Maya civilizations dealt with time. Sections 1.19–1.23 deal with physical time and how it is conceptually evolving with the theory of relativity and quantum mechanics.

Chapter 2 addresses the question of how humans became conscious of their lifetime as an element of operative time and how they dealt with death throughout historical time. Sections 2.3 and 2.4 analyse the concepts of mortality and amortality in evolutionary biology and the different strategies that living organisms have developed to extend their lifetime. The last two sections return to the subject of human amortality and death, but from a contemporaneous viewpoint.·

Chapters 3–5 present a narrative of *Homo sapiens* time that began with our primate ancestors in the Cretaceous, followed their evolution until the emergence of the *Homo* genus and of humans, and continues up to the present time. This narrative claims to identify the main driving forces of our success as a biological species in the Earth system and also the major successes of our cultural evolution. Naturally, not everything has turned out to be a success. There are sustainability shortcomings that stem either from the hardware of our biological heritage or from our cultural evolution during the past 100 centuries. The fact is that in the first quarter of the twenty-first-century humankind contemplates the future with increasing uncertainty and assailed by recurrent crises. The future was always uncertain, but over the past 250 years most people have always entertained a deep hope that progress is inevitable.

The three chapters do not follow a strict chronological order of events. Sections with conceptual discussions on topics that are considered especially relevant were included as appropriate, such as a section on altruism and cooperation at the end of Chap. 3. Chapter 4 deals with some of the fundamental conceptual pillars of modernity up to the contemporary operative social time, including economic growth, the irreplaceable role played by energy in making economic growth viable, and the origins and overpowering development of the digital information age. However, Chap. 4 also has three sections dedicated to the question of time that are relevant to understanding the advances of modernity. Section 4.3 addresses the acceleration of operative social time induced by modernity and Sect. 4.6 describes how humankind started to forecast the future with the help of probability theory and how the concept of risk was born and developed. Section 4.10 describes the various worldviews of human evolution in cyclical or linear time through successive ages provided by different religions. Finally, Sect. 4.11 deals specifically with the breakthrough of imagining a better future for humankind by means of the audacity and radicalism of the first utopias.

Chapter 5 plays a crucial role in the book since it describes the origin and development of the ideology of progress in its various forms of social, economic, political, ethical, and moral progress, as well as the overarching form of human progress. Special attention is given to scientific and technological progress in Sects. 5.8–5.11. Section 5.12 addresses the troubled encounter between the concepts of human economic progress and the environmental crisis and various solutions that have been proposed to relieve the stress. It is complemented by Sect. 5.13, which deals with the violent clash between indigenous environmentalism and economic progress. Sections 5.14–5.16 return to the issues of technological progress, but now applied to the military domain and the development of addictive drugs. Sections 5.17 and 5.18 consider briefly how the long awaited Fourth Industrial Revolution promotes the human ego and analyses some of its possible consequences. Section 5.19 deals with some of the present difficulties involved in making technological and economic progress compatible with reducing socioeconomic inequalities and preserving Western ethical and political values. The three following sections are centred on the way the writers of dystopian novels at the beginning of the twentieth century imagined what future societies dominated by technology would look like and how these prescient societies conform to what is witnessed nowadays. Finally, Sect. 5.23 is dedicated to ecotopia, a farfetched US utopia.

Chapter 6 attempts to provide a framework for addressing the problems of the contemporary globalized civilization by analysing a variety of human experiences, emotions, conditions, and values, discussing the way they are usually appraised, and how preferences and ratings can evolve in order to "solve" those problems. The analysis is partly based on Amartya Sen's different lifestyles, characterized by the words "opulence" and "utility", and by new ways of flourishing, which were later used by Tim Jackson to define three concepts of prosperity. Sections 6.2–6.4 describe how humankind, through the current economic and financial system, is forced to implicitly, passively, or unwillingly accept and support the extraordinary progress of ultra high-net-worth individuals, the perpetuation and proliferation of tax havens, and increasing wealth inequalities. Section 6.5 delves into the origins of the concepts of individualism and egoism and Sect. 6.6 describes the success of ethical and rational egoism in the twentieth and twenty-first centuries, and the way it is influencing politics in some countries, in particular the USA.

Sections 6.7–6.9 present a contrasting point of view to the preceding sections by proposing different concepts of prosperity, and in particular sustainable prosperity, which privileges social capital and environmental factors, and by revisiting voluntary simplicity. Section 6.10 introduces post-cooperation, or cooperation between the members of the contemporaneous social generation and the members of coming social generations, as a necessary condition to achieve strong sustainability. Such a goal imposes costs on the current social generation in exchange for a benefit that, in most cases, will only be tangible for future generations. A simple model shows that, once sustainability has been reached, the post-cooperation required to maintain it is reduced to a minimum. However, if post-cooperation is not implemented, the sustainability deficit will increase for each future generation as time unfolds. Sections 6.11 and 6.12 describe how science and medicine are doing their utmost to extend human lifespan in good health conditions and to achieve human enhancement, superintelligence, and finally transhumanism.

Chapter 7 uses the physical and natural sciences to address in more detail the impacts of the Great Acceleration on the Earth system and the risks that global changes are imposing on humanity in the twenty-first century. Section 7.2 shows that anthropogenic climate change is rapidly driving the Earth system into a non-analogue state, which should be taken as an ominous sign that humankind should reverse the course of action. Section 7.3 describes the evolving controversy in defining the Anthropocene as a geological epoch. Section 7.4 gives a succinct description of the most important materials that are used in the technosphere and are essential to maintain vigorous global economic growth, as well as the ways they may become scarce.

Section 7.5 deals with the biosphere and emphasizes that ecological systems are an example of complex dynamical systems. This condition implies that they can undergo critical transitions in which the system shifts abruptly and irreversibly from one stable dynamical regime to another at a critical threshold called a tipping point. Since the Cambrian explosion, the biosphere has undergone five critical transitions known as mass extinction events due to various types of global changes with either internal or external origin. The current pace of anthropogenic biodiversity loss is likely to transform itself into the 6th mass extinction if there is no significant change in our present relationship with the biosphere. If this mass extinction is allowed to proceed, it will take about 10 million years to return to the levels of biodiversity at the beginning of the Holocene. Section 7.6 discusses the various energy fluxes in the Earth system that are critical to humankind and sets the scene for analysing their future sustainability. In particular, it addresses the increase in the human appropriation of net primary productivity and the effect it has on biodiversity, the water and carbon cycles, and the capacity of ecosystems to provide essential services to humans. Section 7.7 displays the evidence for biodiversity loss and shows that healthy ecosystems supply services that underpin all aspects of human life.

Section 7.8 takes a different approach from the preceding sections by exposing the ecomodernist ideology, in which the deep anthropogenic transformation of the biosphere is assumed and promoted. From this point of view, the only solution now is to build an excellent Anthropocene supported by a technobiosphere that is also expected to be excellent. A better technobiosphere would involve direct human interference in the populations, genetic characteristics, and evolution of living organisms that are useful as sources of food and the evolution of other essential commodities, especially through genetic engineering and synthetic biology. Section 7.9 follows a closely related discourse and debates how genetic engineering can assure global food security up to 2100 and beyond. Finally, Sect. 7.10 readdresses the question of the human relationship with the biosphere and the technosphere.

Section 7.11 is dedicated to the question of energy, which plays a central role in the current model of perpetual global economic growth. It considers the energy fluxes, specifically in the technosphere and biosphere, and the concepts of exergy efficiency and energy sufficiency. Section 7.12 provides evidence that the biosphere and the technosphere have very different metabolic mechanisms and capacities. Furthermore, it shows that humankind has been using natural capital faster than the Earth system is able to renew it. The role of the circular economy, which tries to answer that challenge, is analysed.

Section 7.13 addresses one of the critical points of the present book. It starts by revisiting the conceptual debate about perpetual economic growth and moves on to the relationship between growth and debt. It then considers the likely possibility that current overexploitation of natural resources, environmental degradation, and climate change is already giving signs that perpetual economic growth will not be possible. It also presents some possible responses such as the circular economy, green growth, the dematerialisation of the economy, and the decoupling of economic growth from the use of energy and natural resources. It revisits the up-to-dateness of ideas, analysis, and recommendations put forward by the Club of Rome and ends by illustrating the regenerative capacity of the biosphere.

Sections 7.14–7.16 discuss three of the currently most important secondary global processes with strong social, economic, financial, political, cultural, and environmental impacts, namely globalisation, urbanisation, and migration. Section 7.14 presents a brief history of globalisation processes, including the ascent of the second modern globalisation process, now under way, describing the resulting benefits and shortcomings and the recent tendency to discredit its value. Section 7.15 deals with urbanisation, the present trends, the growing number of megacities and the challenges they generate, the urban environment, pollution and climate, urban health, and the relations between cities, use of natural resources, and the global environment. Section 7.16 begins by revisiting humans' remarkable capacity for endurance running and briefly reviews *Homo sapiens*' migratory movements out of Africa, which populated all the continents in a relatively short period of time. The contemporary problems regarding voluntary migration and forced displacement are analysed, along with their political implications. This section ends with a brief review of the travel and tourism sector, which has had the highest economic growth at global level in recent years, up to 2019.

The next seven sections all relate to the challenge of anthropogenic climate change. Section 7.17 deals specifically with the connection between the use of fossil fuels and climate change, while Sect. 7.18 addresses the feasibility of a global energy transition from fossil fuels to renewable energies. Section 7.19 analyses the problem of an ever-increasing global energy demand. Section 7.20 describes the extraordinary history of ideas to modify weather and climate, especially for military purposes and more recently to countervail anthropogenic climate change, a technology called geoengineering. It also provides an instructive opportunity to analyse the conspiracy theory surrounding aircraft contrails. Sections 7.21–7.23 address the emerging field of geoengineering research. Section 7.21 deals with the techniques of carbon dioxide removal and Sect. 7.22 with solar radiation management, a more promising technology to decrease global warming. Section 7.23 discusses the future of geoengineering and in particular its governance in the multilateral framework of the United Nations.

The last stages of writing this book coincided with the beginning of the COVID-19 pandemic. This dramatic event has shown once again how fragile and globalized is humankind's civilizational model and the economic and financial system that sustains it. Section 7.24 presents a brief analysis of the pandemic and its possible outcomes as regards the main sustainability issues.

Lisbon, Portugal   Filipe Duarte Santos

Acknowledgements

This book would not have been possible without the enlightenment and help provided by a large number of people. It reflects my many experiences and the knowledge I gained around the world on frequent professional trips, especially in Europe, Asia, and Central and South America. It benefitted from fruitful dialogues and debates with colleagues and friends in various institutions, especially in academia.

It would be impossible to mention them all so I will select a few. I would like to thank Rob Swart and António Telo for reading early versions of the book and for their valuable comments. I would also like to thank the illuminating discussions I had with Tim O'Riordan, Lusa Schmidt, Viriato Soromenho Marques, Rui Perdigão, António Amorim, and Gil Penha-Lopes. I am very grateful for the frequent discussions and sharing of information and references provided by Marta Sequeira. I also wish to thank the sociability, sharing of views, debates, and research collaborations during 11 years with lecturers and students of the Ph.D. Program on Climate Change and Sustainable Development Policies of the Universities of Lisbon, New University of Lisbon, University of East Anglia, and other participating universities.

I would like to deeply thank my family, especially my wife Amparo, for bearing with me during the process of writing.

Finally, I would like to thank Adlia Lopes and Raquel Brito for administrative and secretarial support, Alexandre Pereira for his professional collaboration regarding the figures, and Lus Dias and Stephen Lyle for their excellent work in the preparation of the final manuscript.

Lisbon, Portugal   Filipe Duarte Santos
June 2020

Acknowledgements

Contents

Chapter 1
Time

Truth was the only daughter of Time

Leonardo da Vinci, *The Notebooks of Leonardo da Vinci* (started in 1508)

1.1 Human Evolution and Time

We are the only biological species that has the ability for self-awareness and self-knowledge through thought and symbolic language. We do not know what it is like not to have these abilities but we understand that they make interaction and relationships with others and the environment around us much more complex. In this context, the environment is understood as all the biotic and abiotic conditions, factors, and elements that involve and affect ecosystems, humankind, and all living things in the Earth system. We also know that self-awareness and self-knowledge have created apparently endless opportunities for us, some of which have been used to transform and extraordinarily improve the quality of human life since prehistory, throughout the last great civilisations of the past and, above all, over the last four centuries.

We do not know the future of humanity in the medium and long term but it is possible to imagine several scenarios for how it will evolve. The problem is to identify the essential elements that define *Homo sapiens* and the elements that enable us to look into our future as a group. The search for the direction to take in the future is one of the most fundamental forms of human rationality and will potentially contribute the most to the sustainability of our civilisation. It enables us to discover the challenges and risks that we may encounter, and it allows us to attempt to minimise them using practical reason.

The essence of a complex evolutionary system can be defined as a set of intrinsic properties and fundamental functions that uncover and encapsulate its past and present behaviour and build scenarios for the future. It is the set of properties that characterise the system's evolutionary process and are central to its complexity. It is the profound and determining nature of the direction its transformation takes.

© Springer Nature Switzerland AG 2021
F. Duarte Santos, *Time, Progress, Growth and Technology*, The Frontiers Collection,
https://doi.org/10.1007/978-3-030-55334-0_1

The evolutionary process discussed here is the evolution of humanity, although we must recognise that it is a process that cannot be separated from the natural evolution of the Earth system and, ultimately, the evolution of the Universe. The Earth system is a complex one that comprises, as well as the planet's core, the human subsystem and five subsystems—the lithosphere, the hydrosphere, the cryosphere, the atmosphere, and the biosphere—which form the ecosphere. The human subsystem is the functional system that groups together all our activities, processes, and products, including complex and diverse social, cultural, economic, financial, scientific, and technological structures, the renewable and non-renewable natural resources used, farming and forestry, material and energy flows, and all the systems and physical infrastructures we build. In other words, the human subsystem is the whole human enterprise recognised us a functional system anchored and integrated in the Earth system. This subsystem includes the technosphere, a concept defined by Haff (2014b) as part of the Earth system that has been transformed by different human technologies, including the means and infrastructures involved in those processes, such as the interconnected energy systems, systems for the production and consumption of goods and services, including industry, agriculture and food, communication systems, and transport and housing infrastructures, as well as services at the global level. The technosphere goes beyond the ecosphere, occupying part of outer space where spacecraft, satellites, and space debris can be found. The technosphere is gradually expanding into the Solar System and some space probes have even gone beyond. Furthermore, the exploration of mines has extended the technosphere to the greatest depths of the lithosphere. Despite the enormous complexity of the *Homo sapiens* population, its many actions and social relationships, its interactions with the environment, and the various successive stages in its journey, we may still reflect on the essence of the evolutionary process of humanity. This may lead us to discover the fundamental processes and essential features of our lineage that have determined our evolution from primordial times to the *Homo* genus.

The origin of the concept of essence can be found in Aristotle's (384–322 BC) *Metaphysics*. While trying to answer the question of what forms a thing's substance, Aristotle introduced the concept of essence through the expression *to ti ên einai*, which is, translated literally, "the what it was to be", although it is arguable whether this really gives the original meaning of the expression (Owens 1978). The translators who translated the text into Latin used the expression *esse*, which means to be and later gave rise to the word *essentia*. Since then, essence has been a concept that is frequently used in everyday speech, in literature, and in philosophy.

In philosophy, the concept of essence became the cornerstone of metaphysics and rationalism. Martin Heidegger (1889–1976) and Jean-Paul Sartre (1905–1980) stated that existence predated essence, to highlight that ways of being and acting are just choices that each person can make. Based on those choices, people acquire and project an identity that they claim is their essence. The essence that is sought here is not the essence of individual existentialism but the essence of the evolution of the *Homo sapiens* population as a biological species. Personal, subjective, and identity-based aspects of each human life are a few major manifestations of the complexity that surrounds the essence of evolution.

If we delve into the evolutionary side of the human process, we find one form of essence that is truly fundamental: time. Of course, everything related to humankind has happened and will happen in time. But the relationship goes deeper than that. Time influences and limits all human actions, each one of our lives and, ultimately, the evolution of humanity itself. We must find out how man has dealt with time and its relentless command to understand the history and future of the human condition. One of the central aspects of modern times is the acceleration in the evolution of technology and social and cultural changes which impose a different pace on life. This acceleration has generated time structures that restrict our time and force a larger proportion of it to be occupied, which leaves us in the perplexing situation of not having time. To analyse the origins and future consequences of this accelerated pace of time, we must start by reflecting on the notion of time and the different concepts of time that we have built throughout our history.

The word "time" has a clear operative sense but causes bafflement when we try to explain or define its meaning. In other words, we are able to use the word without hesitating but we find it extremely difficult to illustrate what time is from an ontological perspective. This was acknowledged by Augustine of Hippo (354–430) when he wrote in Book XI of *Confessions*: "What then is time? If no one asks me, I know: if I wish to explain it to one that asketh, I know not." We are aware that time is not something that we can escape from in order to analyse it from the outside, as we would an object. We also have no way of perceiving time itself independently from the perceptions that constantly allow us to be conscious and interpret the environment that surrounds us. However, time is real and can be understood in the intuitive sense defended by Henri Bergson (1859–1941) and Martin Heidegger. Time in an operative sense is an essential, unavoidable reality, the most important in human experience and evolution. Moreover, in its different forms and appearances, it is always a human construct, as recognised by the Dutch–American philosopher of science Bas van Fraassen in the context of his constructive empiricism, when he said "there would be no time were there no beings capable of reason" (Dowden 2009).

Time is a recurring and incredibly rich theme in philosophy. It was first explored by pre-Socratic philosophers in the 6th century BC in their search for the essence of things. Parmenides, who was active in the first half of the 5th century BC, was the first to conclude that time had something unreal about it, as Plato (c. 428–c. 346 BC), Immanuel Kant (1724–1804), and J.M. McTaggart (1866–1925) would later maintain (Hoy 1994). Discussion about the philosophy of time has been kept alive over the last 25 centuries and new life was breathed into it at the start of the 20th century.

We intuitively acknowledge that there is a flow of time that produces a past, a present, and a future. Here, the past is considered to be unchanging and the future undefined and uncertain. The passage of time consists of a constant transmutation of the present into a near past, accompanied by the transmutation of a near future into the present. Analysing in this intuitive, common way of interpreting time leads to the philosophical theory of presentism, according to which only sensations, perceptions, objects, and events that are temporally present exist. An analysis of time may also lead to the philosophical theory of eternalism, which argues that both past and future events exist, even if we cannot experience them. In accordance with this point of view, the flow of time is an illusion of our consciousness, in other words, time does not flow because it is always present everywhere. A version of eternalism that comes closer to perceptions connected to psychological time is the theory of time called the growing block universe theory of time (Tooley 2000), according to which the past and present exist but the future does not, and is gradually uncovered by the flow of time. In this situation, the past block and the present block gradually increase, revealing the evolution of the Universe.

We can clarify, or even attempt to resolve, the dilemma of the flow of time by using an alternative form of analysis. The central idea involves focusing attention on ordering events by time, in other words, with approaches like event X happened two days before event Y or event X and event Y took place at the same time. In this case, we do not need to refer to the flow of time. Sentences that include references to the past, present, and future can be reformulated using relations of order in time. The two points of view—placing events in a past, present, or future that are constantly transmuting, or relating pairs of events using temporal relations, before, after, or simultaneously—correspond to McTaggart's A-theory and B-theory, respectively (McTaggart 1908). A-theory, which corresponds to a description of events as existing in absolute time and belonging to the past or the future, leads to contradictions that McTaggart interprets as being the result of an illusory conception of time. He therefore defends B-theory, according to which reality is non-temporal. Curiously, these two theories are relevant to relativistic theories of physical time, as we will see later. The concept of time can also be approached from a phenomenological perspective in which the consciousness and perception of time and use of the word "time" are analysed. The latter path is particularly relevant to the reflections proposed in this book.

1.2 Physical, Psychological, and Biological Time

Time is a central, constant theme in physics, psychology, philosophy, religion, and the social sciences but it is not possible to find a sufficiently broad-reaching and consensual definition that applies to all the fields in which it is involved. To start the analysis, we may use the perspective of the psychology of time. The perception of time fundamentally involves the concept of succession, the result of perceiving a relationship of order between two or more successive events, and the concept of duration, the result of perceiving two successive events (which could be two successive transitions, for example the moment when a sound is first heard and then the end of that sound) (Fraisse 1984). Duration does not exist in itself, i.e., independently of perception of the events that define it. The concept of duration requires the concept of succession. Succession, moreover, contrasts with simultaneity, and duration is extinguished when simultaneity occurs.

A full understanding of the concepts of succession and duration is only acquired at 7–8 years of age when children achieve the ability for abstract reasoning (Friedman 1982). It is also during that period that children begin to have an abstract awareness of a certain type of time, namely that personal, private, non-transferable, and unyielding time that is normally called psychological, phenomenological, or mental time. It is important to remember that strictly speaking we do not perceive psychological time itself but only the changes and successive events that we later process in abstract forms of successions and durations until we build an awareness of psychological time.

The concept of psychological time is clearly distinguished from the concept of physical time. This is essentially the result of finding that duration can be measured consistently and universally. Physical time was constructed by measuring durations, first using calendars and later using different types of clocks, which are becoming more and more precise. Newton, in the *Scholium*, an introductory essay to his renowned book setting out the laws of motion and gravity (Newton 1687), begins by establishing the conceptual basis of the distinction he proposes between absolute and relative motion. That is where he introduces "relative, apparent, and common" time and space and "absolute, true, and mathematical" time and space. Relative time is time measured in cyclical durations linked to the uniform motions of objects, particularly stars such as the Sun and the Moon.

The problem was to know whether a motion was really uniform. Newton intuited that it would never be possible to prove such a property beyond all doubt. This underlies Newton's postulate: "Absolute, true, and mathematical time, of it self and from its own nature, flows equably without relation to anything external, and by another name is called 'duration'" (Newton 1687). Finding a reliable measure for time was actually a highly debated issue in the 17th century, especially among astronomers. It was known that the mean solar day was around four minutes longer than a sidereal day, and this led to diverging measurements of time. And it was not certain that the rotation of the Earth was uniform. Newton said that it was, but he considered that it was a contingent assertion that needed to be proved. Newton's model of physical time became deeply rooted in people's minds because it has a structure compatible with psychological time, for which it began to function as a support and reference.

The emergence of Einstein's theory of relativity (Einstein 1905a, b) broke that relationship and definitively shifted the concept of physical time away from the concept of psychological time. In the end, there is no absolute time and space in physics, and the spacetime of relativity theory is a structure that cannot be reached merely by directly processing the perception of successions, durations, and distances. It becomes necessary to use a physical theory expressed through mathematics and examine whether or not measurements of the spatial and temporal coordinates of events obey the equations that define the theory.

Perception of psychological time is not associated with any particular sensory organ. However, our brain gives us that ability via a complex, distributed system that involves the cerebral cortex, the cerebellum, and the basal ganglia. Using this system, which includes an internal clock based on time defined by regular neural impulses, we are able to place past events in time and assess and compare time intervals with durations that vary between milliseconds and several decades. Psychological time includes perception of what we call the present which, in terms of measuring physical time, is a relatively short time interval. E. Robert Kelly (1883), who used the pseudonym E.R. Clay, called this limited and somewhat mistaken perception "specious present". William James described it as the short duration to which we are immediately and incessantly sensitive (Andersen and Grush 2009).

The limits of our ability to perceive duration, for instance the duration defined by two successive events, have been established by psychologists and neurologists. We are unable to perceive durations of less than 0.1 seconds for visual stimuli or durations under 0.01–0.02 seconds for audio stimuli. Below these limits, we interpret events as simultaneous even though from a physical perspective they are not. On the other hand, when perceptions of durations are compared with measurements obtained from clocks, we conclude that perceptions are subjective and vary according to emotional state and age, although they are on average reliable enough for most of our activities.

Psychological time can dilate and contract in relation to physical time. A day can seem longer to a young child than it does to an adult. In dangerous situations, when we are afraid, suffering, or under intense stress, time appears to run faster, while the outside world runs at a slower pace. In contrast, when we are absorbed in a rewarding or routine activity, time seems to run more slowly, while the outside world advances at a faster pace. Perception of the duration of past time intervals is also influenced by what we did at the time. Generally speaking, when a period of time is filled with varied, pleasant activities, experiences, or events, we feel like it went past quickly; when we assess its duration later, it feels like it was long. In contrast, a period of time with little to fill it takes longer to pass but in retrospect appears short (James 1918).

Michael Flaherty (1991) attempted to explain the variation in perception of the passing of psychological time compared with physical time. This variation is generally related to the density of experiences and conscious activities that require or force our cognitive attention, emotional involvement, or decision-making actions per unit of physical time, as a result of the circumstances the person finds themselves in. Problematic or difficult circumstances demand greater awareness and cognitive, emotional, or decision-making intervention, and this increases the density of experiences. As a result, time appears to pass more slowly and there is time dilation. Examples of this include intense pleasure, for example an orgasm, intense suffering, torture, violent combat, and situations of great danger. Psychological time also dilates during discovery, wonder, and pleasure at a feeling, sight, experience, or new knowledge, as well as during deep meditation or mental bliss. Forced inactivity, for example in prison or as a result of unemployment, can also cause time to dilate due to forced, concerned, and emotional introspection which fills time unhealthily. Tedium is the most benign way of perceiving time dilation.

On the other hand, if the density of experience and intense, interesting, or new activities per unit of physical time is low, time seems to go by more quickly, in other words it seems to contract. This is the case when we are doing something we like which occupies us without involving a high density of emotional states or different activities. It is also the case if we are involved in a routine, relatively pleasant activity that we are able to perform almost unconsciously.

The contraction of time is also found in our perception of the duration of past time intervals and its intensity depends on the density of significant, varied, and relevant cognitive, emotional, and decision-making experiences. We have an extraordinary ability to selectively forget most of our past actions and experiences. This forget-fulness concerns above all things that are part of our everyday routines and that we consider to a greater or lesser extent irrelevant. Only the most important and relevant experiences, actions, and events remain in our memories. Due to this restructuring of time caused by such forgetfulness, past time tends to contract and that contraction tends to increase as the density of important memories decreases.

Despite the subjectivity of our perception of time, we frequently find ourselves in situations in which we feel like time does not go faster or slower than the physical time shown by clocks and watches. Under these circumstances, psychological and physical time are synchronised and we are able to assess the physical duration of the time periods we perceive relatively accurately. This ability is acquired by the practice of establishing equivalence between the units of physical time, particularly minutes and hours, and the duration of cyclical personal and professional actions, as well as environmental and social cycles and the physical infrastructures where we live.

Drugs also interfere with our perception of duration. Some, such as cocaine, interfere with the dopamine regulation system. When the dopamine concentration increases, our internal clock speeds up, psychological time moves faster and external events appear to last longer. Dopamine is the main neurotransmitter involved in processing psychological time in the brain. It is associated with pleasure, motivation processes, and learning from reward mechanisms, the inclination towards addiction to gambling, sex, and drugs, and certain neurological diseases, including schizophrenia, Parkinson's disease, and attention deficit disorder with hyperactivity.

As well as the ability to be aware of psychological time, our organism has invol-untary cyclical behaviour lasting around 24 hours that includes biochemical, phys-iological and behavioural elements. This is known as the circadian rhythm, a name which comes from the Latin *circa*, meaning "near", and *diem*, meaning "day". It is a rhythm that is endogenous to our organism, regardless of the environment we find ourselves in, but it is relatively sensitive to some kinds of external stimuli, known as *zeitgeber* ("synchronisers" in German), of which the most important is sunlight. Circadian rhythms can be observed in most living organisms, including some bac-teria, fungi, plants, and animals. There are other biological rhythms linked to the tides, the different seasons, and the year. During the evolution of life on Earth, the adaptation of living beings to some cyclical events in the external environment, such as the alternation between day and night, has generated cyclical behaviour equivalent to the biological clocks that have created biological time.

Circadian rhythms originated in the evolutionary process of adapting internal cell physiology to the diurnal cycle. Little is yet known about the mechanisms that first triggered this adaptation process. They may have evolved to stop the complex process of DNA replication from taking place in the daytime to protect it from solar ultraviolet radiation, which was initially much more intense at the surface, before the stratospheric ozone layer was completely formed. Later, when the concentration of oxygen in the air increased, the evolution of biological clocks probably resulted from mechanisms for protecting oxygen at the molecular level, involving oxidation–reduction cycles by way of antioxidant enzymes (Edgar et al. 2012).

In mammals, the primary centre for regulating circadian rhythms is the suprachi-asmatic nucleus in the hypothalamus, in the central area of the base of the brain. If it is removed, circadian rhythms are completely destroyed in the organism, in particular the regularity of sleep and wakefulness periods. The genes that regulate circadian cell mechanisms in the human species are present not only in the suprachiasmatic nucleus but also in several organs and tissues, including the skin (Zanello et al. 2000). There are minuscule cellular clocks spread throughout our bodies that together consistently regulate our biological time.

1.3   Operative Social Time

The notion of time is not limited to the concepts of psychological time, physical time, and biological time. Time also has a social dimension that has been recog-nised by the founders of modern sociology. Max Weber (1864–1920) (Weber 1905) called attention to the importance of considering awareness of time when studying the evolution of mentalities and the development of the modern era. He defended the theory that the "spirit of capitalism" requires a special relationship with time. Emile Durkheim (1858–1917) (Durkheim 1912) went even further in his analysis by postulating a social time that transcends individual experiences of time. According to Durkheim, time is a social institution, i.e., an essentially collective phenomenon, a product of collective consciousness.

In other words, we are dealing with a structural construct that arises from the awareness of and ability to process psychological time together with the use of language and the fact we live in society. Pitirim Sorokin (1889–1968) and Robert Merton (1910–2003) (Sorokin and Merton 1937) conclude, in their methodological and functional analysis of social time and the relationship with physical time, that social time reflects the pace of activity in different societies. Calendars, for example, are systems for organising social time that express the pace of collective activities and at the same time have the role of ensuring that activities are regular. Hassan (2009) considers time to be social and believes that it fundamentally exists in the social sphere, although he emphasizes that, in this field, it does not in any way constitute an absolute and universal structure similar to the one Newton suggested for physical time.

The diversity and complexity of human perception, processing, analysis, knowl-edge, action, and communication capabilities in social settings have generated dif-ferent concepts of time that are compatible with each other but impossible to place in a hierarchy. Physical time therefore essentially represents the ability to measure and compare the duration of various cyclical phenomena and the fact that the results of the measurements form a consistent set of data that makes it possible to structure and communicate intelligible, universal, falsifiable physical laws (Popper 1959) that can be expressed using the language of mathematics. That ability and the applica-tions it produces do not necessarily imply the existence of an external physical time independent of humankind, to which all other concepts of time would be subordinate.

As well as the concept of social time introduced by Durkheim, analysis of the social nature of time reveals another conceptual form of time. The word "time" is frequently used in our everyday speech to communicate to others the time relationship between experiences, activities, situations, and events, as well as their temporal locations in the past, in the future or even in the different stages of human life. It also serves to identify, designate, compare, reflect, and transmit an assessment, related to scheduled objectives, on durations defined by a wide range of activities, events, and personal or collective experiences. This extremely frequent practice of using the word "time" and others directly related to temporality expresses and characterises consciousness and the operational use of a concept of time in interaction and communication within society, which gives it an eminently social dimension.

For example, when we see someone we met when they were young after a long lapse of time, we sometimes mention how time has changed them. We also say that we don't have time, or that there is or isn't time to carry out a specific action or project. Other common expressions include how I have used, use, or will use my time. We also say that something was a waste of time, that we have gained time or that we have to bide our time. We also use the expression that there is "a time for everything in life", as in verses 1 to 8 of *Ecclesiastes 3*.

The idea that there is a right time for a specific act was thoroughly explored by the Greeks and brought about the distinction between numerical or chronological time called *Chronos*, which is close to physical time, and divine, metaphysical time called *Kairos*, which literally means "the right or opportune moment".

In conclusion, the way of dealing with the framework of time for the entire experience of life in society is reflected in the use of expressions of temporality in different forms of human communication. It corresponds to a concept of operative time in social contexts that is not identified in any of the concepts of time mentioned above: physical, psychological, biological, or social, in the sense of the concept introduced by Durkheim and developed by other authors.

Operative social time, or simply operative time, is an operational form of temporality that is structured by events, activities, experiences, and expectations experienced, shared, and communicated in the social context. It is the most important form of temporality from the perspective of social behaviour and interactions, and it is the form that has the closest relationship with human life. This form of social time is described as operative because its origin and function are found in the operational framework of events, actions, activities, experiences, and expectations, and in the perception of time regarding all forms of change. It is not an abstract and static temporality but is instead functional and dynamic. It reflects the pace of change in various aspects of our surroundings that we seek constantly to adapt to. These rates of change help determine operative social time and generate an operative social time structure, or simply an operative time structure, that conditions our actions, activities, experiences, and expectations; in other words, it conditions the way we use the time we have.

Operative social time is different from Durkheim's social time because it is an analytical concept that describes a capacity, a practice, and an experience at the level of the individual. Operative social time is created, processed, and experienced by each member of a certain society in a broad sense, which may be a local community, a country, a group of countries, or a civilisation. The range of operative social times of the members of a society defines that society's operative social metatime, which will be called simply operative social time. Operative social metatime is more abstract and describes a characteristic of the society in question. In this respect it comes close to Durkheim's concept of social time.

It is also important to define the operative time structure of a society as the structure emerging from the experience of its operative social time. It reflects the dynamics, in other words the pace of change affecting its social, cultural, religious, economic, political, scientific, technological, and environmental aspects, and conveys the structural time constraints generated and imposed by that society. Regarding this last aspect, the operative time structure of every society is and has been constrained throughout history by the way that society measures and uses physical time in organising its activities. Operative time structure reflexively affects the way of life, activities, and expectations of the relevant society and all its members. Both operative social time and an operative time structure are specific to the society to which they refer. They are transformed with the flow of physical time because of society's evolution, especially regarding its social, political, economic, and technological aspects.

Let us consider some extreme examples. The operative social time of a hunter–gatherer who lived in the Dordogne, in the south of France, during the Upper Palaeolithic roughly 20 000 years ago, was different from the operative social time of a worker who helped build the Great Pyramid of Giza in Egypt between 2580 and 2560 BC and different from the operative social time of a modern-day New York stockbroker. Throughout human history, each society with a well-defined identity has had its own pace of activities, experiences, ways, and expectations for the occupation of time, which generates an operative social time and an operative time structure that are particular to that society.

Due to the coexistence of human societies at different stages of development, which currently range from indigenous peoples in some tropical regions to more varied societies of advanced economies and even to an emerging global society, their operative social times and time structures coexist but are different. The globalisation process, defined as the tendency towards the free flow of people, trade, capital, knowledge, and technology throughout the world, which we have witnessed since the mid-20th century, is associated with a time and a specific operative time structure that tends to progressively overlap with times and operative time structures on smaller spatial scales.

Operative social time is built around ordering and assessing personal activities, experiences, and expectations in time and sharing and communicating them in the social context, as well as knowing and evaluating the ordering and assessment made by others. It involves psychological time but goes further by including our ability to reflect on, analyse, assess, evaluate, manage, and communicate our experience within a specific temporal framework, and this framework is influenced by and influences those of others. Ultimately, it is a time that reflects the flow of human, individual, and collective experience, shared in a certain social context. Operative social time identifies, absorbs, and integrates the past, the present, and the future in the sense that these three dimensions interact and connect to each other in different ways.

As well as relationships of causality between past and future events, whether internal or external to human beings, the present connects to the past through the imagination, in which it was constructed, of an anticipated future. The future connects to the present and the past through an anticipated future that was imagined in them, especially by way of expectations and desires regarding that future. This type of unbreakable relationship between the three dimensions of time has an important role in the hermeneutics developed by Heidegger in his well-known book *Being and Time* (Heidegger 1927). In operative terms, the memories and imagination of events and future experiences play a central role in life, particularly in planning a direction and future actions.

It is today possible to locate and study these mental processes at the level of the brain. Nuclear magnetic resonance imaging has made it possible to identify areas of the brain where a network of connections is established, known as the default mode network, which becomes active when recalling memories of past experiences and when imagining and simulating possible future experiences (Addis 2007; Szpunar 2007). It is therefore possible to study the role of our recollections of past events when imagining the future in terms of brain activity. One important conclusion of these analyses is confirmation of the conjecture that memory provides a basis for building scenarios that makes it possible for us to imagine the future (Boyer 2008; Schacter 2012). In this prospective mode, the brain selectively stores information regarding the past for later use when imagining, simulating, and foreseeing future events. Comparing and including elements of past experience when simulating future events generates a vital adaptation process. These results reveal our difficulty at cerebral level in dealing with events that we can imagine in abstract terms but do not correspond to past activities, experiences, or events.

1.4 Intertemporal Choices

One of the actions that has contributed the most to shaping personal operative social time is the performance of intertemporal choices, i.e., choices that involve decisions with consequences that have repercussions in the future. These types of decisions, which are very common in everyday life, force us to imagine the future and make comparisons between the costs and benefits that take place at different times. They are personal decisions in which time, the perception of time, and the perception of the finiteness of a lifetime play an essential role. When they relate to the person himself or herself, they help determine the future of his or her health, eating habits, education, profession, personal relations, sexuality, family, consumption habits, economy, assets, physical exercise, well-being, and quality of life, among many other things. When they are decisions made at a personal level but have implications for the society or country in which the person lives, or even some with a scope extending to humanity as a whole, they contribute to determining the future of those groups in all their different aspects: social, political, economic, and environmental.

Adam Smith (1723–1790) (Smith 1759) was one of the first people to analyse the issue of intertemporal choices, since he considered that our behaviour is determined by a struggle between two processes that he associated with "passions" and the "impartial spectator". The latter plays the role of an imaginary moral guardian. When dealing with a decision between immediate gratification with high future costs or delayed gratification with lower future costs, the impartial spectator "serves as the source of self-denial, of self-government, of that command of the passions which subjects all the movements of our nature to what our own dignity and honour, and the propriety of our own conduct, require" (Smith 1759). It is this spectator that provides us with the ability to restrain the impulses of immediacy, to control impatience, to scrutinise and assess the consequences of current decisions in the future, and to have as passionless and balanced a view in the medium and long term as possible so that we are able to plan our future and the future of coming social generations prudently and safely.

Less than a century later, the importance of intertemporal choices for the economy was solidly recognised and studied by John Rae (1796–1872), who believed the a nation's wealth to be dependent on its ability to save and invest (Rae et al. 1906). Studying his narrative reveals the transformation in the organisation of societies and their values between that time and the present. Rae believed that the two main factors that contribute to saving money are philanthropic motivations resulting from "the prevalence throughout the society of the social and benevolent affections" and the propensity to exercise self-restraint, i.e., "the extension of the intellectual powers, and the consequent prevalence of habits of reflection, and prudence, in the minds of the members of society" (Rae et al. 1906). In Rae's view, there are two factors that counteract those trends. The first is somewhat surprising in today's age: "When engaged in safe occupations, and living in healthy countries, men are much more apt to be frugal, than in unhealthy, or hazardous occupations, and in climates pernicious to human life" (Rae et al. 1906). The second, more universal and timeless factor is the desire generated by the perspective of immediate consumption together with the simultaneous feeling of frustration or privation associated with the possibility of delaying that immediate gratification. We prefer to enjoy something pleasurable or a benefit now rather than postponing it. Choosing to plan and obtain results that are distant in time and having the patience to wait for them rather than gain immediate, unsustainable results is one of the hardest exercises for human willpower.

This time preference plays a central role in the characterisation of operative social time and the determination of how we manage the future. In general, the immediate and the short term interest us much more than the medium and the long term. Time preference is a human trend that has been acknowledged since time immemorial. One of the most famous and oldest illustrations is Aesop's (620 BC–564 BC) fable in which the grasshopper, who prefers to spend the hot days of summer singing, is unconcerned about its future the following winter, while the ant works to save food to feed itself during the winter.

Time preference can be defined as a preference for satisfaction or happiness from consuming something or using a service now rather than postponing it, and it manifests itself in several ways. One of the most widely studied in both people and animals, since it is relatively easy to quantify, is the greater value assigned to an immediate reward over the value assigned to the same reward postponed (Kalenscher 2008; Hayden 2016). This devaluation caused by the postponement time is called time discounting and the way the value of the reward falls over the time of postponement is called the discount function. To choose a reward with a long delay over one that has a shorter delay or is immediate, its value must be considerably higher.

Several empirical methods of applied psychology have been used to study the time discount function in humans and other animals. The conclusions reached are affected by significant uncertainty and are sometimes conflicting (Frederick et al. 2003). There are a great many factors that make it difficult to measure a time discount function on the basis of laboratory studies (Chabris et al. 2008). Despite this problem, it is safe to conclude that the time discount function does not follow an exponential law. Generally speaking, the time discount rate, or rate of time preference $s(t)$ of a differentiable time discount function $D(t)$ that decreases over time is defined by

$$s(t) = -D'(t)/D(t) .$$

For around 80 years, most analyses of intertemporal choice in the context of economics were based on the model of discounted utility, introduced by American economist Paul Samuelson (1915–2009) (Samuelson 1937). In this model, which is still systematically used, it is assumed that people assess the gains and losses resulting from intertemporal choices in accordance with exponential time discounting, just as the financial markets assess gains and losses over time. Deviations in relation to the discounted utility model are often interpreted in neoclassical economics as anomalous patterns of economic behaviour.

The use of an exponential function

$$D(t) = d^t ,$$

with a value of d between 0 and 1 to represent the time discounting function implies a time discounting rate that is fixed and independent of time equal to

$$s = -\ln(d) ,$$

where d is the discount factor. The higher the time discounting rate, the greater the depreciation of gains or losses over time. The constant value s ensures dynamic consistency of the discount function since in this model time preference is the same at all times, i.e., it is independent of time. However, the rate at which people discount future rewards decreases as the length of the delay, so the personal time discounting function does not show that dynamic consistency. The empirically deduced behaviour is not exponential but closer to a hyperbolic or quasi-hyperbolic function (Ainslie 1975, 1992; Doyle 2013).

In addition to psychological analyses, neuroscience can also provide a very important contribution to understanding intertemporal choice processes by locating them in the brain and establishing how they relate to different types of mental processes, particularly representation, anticipation, willpower, and self-control (Berns et al. 2007). McClure et al. (2004) used magnetic resonance imaging to show that there are two systems in the brain involved in choices between immediate monetary rewards and those that involve a delay.

When the decision is made to choose an immediate reward, the region that tends to be activated is the limbic system, also known as the paleomammalian cortex, linked to the dopaminergic system. The regions of the dorsolateral prefrontal cortex and lateral parietal cortex are activated in intertemporal choices, regardless of delays in rewards. The relative involvement of the two systems is related to the delay linked to each choice, with greater frontal–parietal activity when long-term options are chosen, i.e., those with the greatest delay (McClure 2004). These results suggest that there is greater involvement of the prefrontal and parietal cortex when assessing rewards in a more distant future, as well as when making decisions and planning that involve the medium and long term, due to the greater abstraction required by time periods that are further away. It is likely that the prefrontal cortex, as well as codifying time discounted values to make it possible to make intertemporal choices, also intervenes in the abstract assessment of the consequences of those choices over future time (Kim et al. 2008). To conclude, the part of the brain linked to the most recent evolution of the phylogenetic lineage of *Homo sapiens* is generally speaking more involved in mental processes relating to the long term than the parts that evolved long before, which are generally more involved in immediacy and the short term.

1.5   The Increase in Rate of Time Preference

Although all of us regularly make intertemporal choices, our rates of time preference or time discounting rates are different and variable. Rate of time preference is an important characteristic of operative social time. Each person uses his or her own rate of time preference in his or her operative social time. The rate of time preference of a community, social group, or country, or indeed the global population, reflects the average rate of time preference of individual members of the relevant groups. People who have a high rate of time preference tend to favour their interests and well-being in the present or very short term more than the average amount chosen as standard. People who have a low rate of time preference assign greater importance to their interests and well-being in a more distant future compared to the average amount. The personal rate of time preference is determined by a range of factors of different kinds, which include social, psychological, professional, cultural, and economic factors, and also age. Moreover, empirical studies reveal that the rate of time discounting varies depending on the nature of the object of discounting and is higher for basic goods, such as food and water, than for money (Odum 2011). There is intensive research to explore the relationship between empirically estimated time discounting rates and a wide range of behaviours and personality traits. Several authors have connected time discounting rates to intelligence levels (Shamosh et al. 2008) and have linked the time preference related to self-control in children to their future health and prosperity (Moffitt et al. 2011). Population groups affected by various types of clinical problems, such as smoking, alcoholism, obesity, or drug and gambling addictions (Chabris et al. 2008), have been the subject of special attention.

Obesity is a particularly interesting case because one of the main contributing factors to being able to control weight is making the decision not to eat more now to enjoy better health in the future. The rate at which future benefits of not harming health are discounted has a direct impact on decisions regarding the quantity and type of food eaten now. It is also an important example because obesity and excess weight are growing around the world, with obesity doubling between 1980 and 2016 in 70 countries and growing continually in many others (GDB et al. 2017). The World Health Organisation estimated that in 2014 more than 1.9 billion adults over the age of 17 were overweight, while 600 million adults and roughly 107 million children were obese. At global level, these conditions are linked to more deaths than malnutrition and famine. In the USA, obesity is one of the main factors in premature death and is believed to be responsible for around 18% of deaths in people aged between 40 and 85 (Masters et al. 2013).

From a biological perspective, the cause of obesity or just being overweight is the consumption of more calories than the organism expends. Many reasons are given for the recent increase in these conditions around the world. One of the main reasons is the unhealthy but tempting types of food that the food industry promotes using sophisticated marketing techniques. Other reasons are related to the various technological transformations that have modified food products, food preparation, and eating habits. Another relevant technological transformation is the one that drives children and adults to spend, on average, several hours a day interacting with televisions, mobile phones, computers, and other electronic devices. This new type of enticing and intensive occupation favours mindless eating habits which encourage weight gain. On the other hand, several authors, using different methods and indicators, have argued in favour of the theory that an increase in the rate of time preference has been observed in recent decades, claiming that this has contributed to the current epidemic (Blaylock 1999; Komlos 2004; Smith 2005; Zhang 2008; Cavaliere 2014). For many people, healthy eating is not worth the effort or sacrifice. According to Blaylock et al. (1999), the challenge is particularly difficult in the USA because Americans seem to discount time heavily, as can be seen in the extraordinarily low levels of savings and high levels of credit card debt. These results are particularly interesting because they identify a more fundamental cause that may influence other behavioural deviations as well as obesity, such as a tendency to get into debt, a reluctance to save, and addiction to drugs or gambling.

The main obstacle to implementing this approach is to find an objective enough way to evaluate the rate of time preference so that it can be used to assess any trend towards an increase in its average value over the last few decades. There are only three methods for measuring the rate of time preference: estimates made on the basis of simulations of consumption and saving using the Euler equations (Lawrance 1991), using proxies such as personal savings rates and personal debt levels (Komlos et al. 2004), and interviews with questions that aim to reveal time preference, for example, asking whether a lack of willpower is the greatest obstacle to weight loss, in the case of obesity (Zhang and Rashad 2008). Jonathan Parker (1999) analysed the decline in individual and private savings, as well as the increase in personal consumption as a percentage of the US GDP since the 1980s and concluded that one of the most likely causes is the increase in the population's average time discounting rate. This increase may in part be the result of technological advances that provide faster and faster access to a growing range of goods and services and faster and more efficient channels of communication and marketing. New facilities and opportunities stimulate our desire for rapid consumption and almost instant gratification in the use of a growing number and variety of goods and services supplied by the market. Becoming used to such intensive and accelerated consumption processes tends to increase the rate of time preference.

However, identifying the current economic model and the technological progress that supports it as one of the possible origins of this increase in the rate of time preference, as manifested in behavioural changes that have negative connotations such as less patience, less willpower, and self-control, is an embarrassing and inconvenient conclusion that will need to be rigorously scrutinised by science. Could Adam Smith's (1759) "impartial spectator" be becoming less attentive and intervening less or, using John Rae's (1834) slightly more modern narrative, are "habits of reflection, and prudence, in the minds of the members of the society" becoming less prevalent?

1.6   Social Discounting of Future Time in Economics

Intertemporal choices are crucial in economics. Consumers are always choosing between spending money and consuming goods and services now or not spending and saving that money instead. The decision is subjective and complex due to the many motivations that may be involved. The neoclassical theory of economics aims to model the consumer's decision-making process by weighing up market interest rates and the consumer's own time discounting rate. A high time discounting rate or a low interest rate tends to increase immediate consumption, which is considered ideal for economic growth. In this context, the time discounting rate is the characteristic of the personal utility function that determines the choice between consuming more now or pushing that consumption into the future. The economic concept of utility is defined as the measure of preference or satisfaction that a consumer feels about using the goods and services available. The utility function measures an individual's relative degree of preference or satisfaction regarding the different goods and services he or she consumes.

The steady development of the economic system that emerged in the Industrial Revolution generated a new domain for making intertemporal choices using specific methods and values. It is therefore important to distinguish between personal intertemporal choices that are constantly made in everyday life, known as general choices, and personal intertemporal choices that are made at personal or institutional level but follow methods and values specific to the economic system prevailing in the world, known as economic choices. It should be stressed that the two types of intertemporal choices interact and influence each other, and it is sometimes difficult to distinguish between the types in concrete situations. The advantage of establishing this distinction comes from the fact that general choices are universal and timeless while economic choices are inextricably linked to an economic system that is located in a certain time in history, even if it is constantly evolving.

All humans, regardless of the era or place in which they lived or live and whatever their social, economic, political, religious, cultural, and environmental condition, make general intertemporal choices. Economic intertemporal choices began to come into their own in the 19th century. The analysis and optimisation of such choices began to flourish from that time on, requiring specific learning and experience. They are made most frequently and efficiently by people at the forefront of humankind, characterised by having the greatest economic power. They have a dominating impact at global and national levels, essentially due to the benefits arising from economic growth. On the other hand, general intertemporal choices include approaching and analysing decisions made by the contemporary generation regarding the consumption of natural resources, environmental degradation, and climate change, which will have an impact on future generations. These topics are not the focus or main objective of economic intertemporal choices, although they may receive some attention when such choices are made.

Economic intertemporal choices are largely determined by the time discounting rate that makes it possible to convert cost and benefit flows with a future economic value into the equivalent value in the present. There is an enormous range of economically important projects for which the most suitable time discounting rate needs to be determined. On the one hand, we have private, corporate projects that are generally short in duration, lasting less than one social generation, and social projects, which are often run by governments and frequently have time frames spanning several social generations. Ethical principles come into the calculation of the corporate and social rate of time discounting. However, ethics has a special relevance in calculating social time discount rates since they involve matters of intergenerational solidarity. In general, corporate rates are higher than social rates because people are most interested in their prosperity, well-being, and quality of life in the short term and are more averse to risk during their lifetimes, while society as a whole tends to have a view that reaches further into the future, encompassing the medium and long term.

One of the most important issues about social discounting rates is the way they deal with and try to resolve matters of fairness and intergenerational equity and justice. The social discounting rate is systematically used in public policy. It is that rate which is used to calculate the amount of future investment in health, education, research, and development. It is also used to evaluate investment in protecting the environment and combating global changes in the medium and long term. In the specific case of climate change, the value of the social time discounting rate to be used in climate change mitigation has been the subject of a long controversy (Goulder and Williams

2012). To do cost–benefit analysis of climate change mitigation we need to estimate the current value of future impacts of climate change. This is done by calculating the social cost of carbon, which measures the marginal costs of impacts caused by emitting to the atmosphere one extra unit of carbon as carbon dioxide, usually chosen as one tonne of carbon (NASEM 2017a).

Before carrying on, it is important to ponder how the concept of time preference enters into economic intertemporal choices. In economics, there are two reasons for discounting future time. The first comes from what is known as "spure time preference" and the second originates from the rate of consumption growth per capita, i.e., it is a direct consequence of economic growth. To make the distinction between these two elements clear, it is worth discussing a highly influential and well-known article by the British mathematician and philosopher Frank P. Ramsey (1903–1930) (Ramsey 1928), a close friend of Ludwig Wittgenstein (1889–1951), whose interest in economics was encouraged by John Maynard Keynes.

In that article, which aims to describe optimal economic growth in an intergenerational context, Ramsey begins by accepting that "we do not discount later enjoyments in comparison with earlier ones, a practice which is ethically indefensible and arises merely from the weakness of the imagination" (Ramsey 1928). This expression represents an apology for a pure rate of time preference, or utility discount rate, of zero and has been defended by several economists (Dasgupta et al. 1999). However, this possibility could never be applied to the concept of time discounting that comes into general intertemporal choices. Time discounting is unavoidable when we make such choices.

However, immediately afterwards, Ramsey tells us that he will introduce time discounting for utilities in the future later in the article. According to Ramsey, due to the accumulation of wealth by successive, later generations, future utilities are not equal to those in the present but should be discounted at a constant social time discount rate s. For example, if we choose "a rate of ρ per cent, utility at any time would be regarded as twice as desirable as that a hundred years later, four times as valuable as that two hundred years later and so on" (Ramsey 1928). This type of time discounting is the result of expectations inherent to unlimited economic growth encouraged primarily by investment and scientific and technological progress, which continuously improves and renews goods and services. In a world where future generations will certainly benefit from growing economic prosperity, current utilities will gradually lose their capacity to attract and satisfy. In neoclassical economics, imagining a different world is generally considered to be outside its area of interest.

How to combine the two components identified by Ramsey in order to establish the social time discount rate s that is used to calculate future benefits or losses was a question that dragged on for years. Lind (1982; Hepburn 2007) suggested that the optimal social discount rate should be equal to the time preference rate determined by interest rates relating to consumption, in turn estimated based on the return obtained from financial investment available in the market. So, for a particular utility function, the social discount rate is given by the formula

$$s = \rho + \mu g ,$$

known as the Ramsey rule, where ρ is the pure rate of time preference, μ the elasticity of the marginal utility, and g the rate of consumption growth per capita, used as a proxy for the economic growth rate. Using this formula means admitting that the social time discount rate is partly determined by the interest rates available to investors in the market. Even if ρ is zero, as Ramsey supposed, since μ and g are positive except in highly anomalous situations, the value of the social discount rate is positive.

The greater the rate of consumption growth per capita, i.e., the greater the economic growth rate that stimulates that consumption, the greater the social temporal discount rate and, as a result, the greater the devaluation of the future. The above formula for s has been used on a systematic basis in cost–benefit analyses for a huge range of social projects. This practice has been useful in formulating public policy around the world in the short and medium term, typically up to 30 years.

1.7 Intergenerational Justice and Long-Term Economic Intertemporal Choices

Nevertheless, in the long term, when several successive social generations are involved, some conceptual difficulties emerge. The most important is knowing to what extent we should be interested in the well-being of future generations. Many of the decisions taken today regarding politics, the economy, and the environment around the world have long-lasting implications for future generations. This is the case, for example, for choices concerning the way we extract and use natural resources, energy options for controlling climate change, the control and reduction of pollution, environmental degradation, and the loss of biodiversity. Natural resources and environmental assets have a long natural duration on the order of billions to tens of millions of years, unlike the assets created by humans. Looking after the environmental assets we depend upon is therefore a challenge that spans hundreds or thousands of years at least and must therefore involve many human generations.

This issue is directly linked to the matter of intergenerational justice, to which the usual principles of justice do not apply because there is no direct reciprocity between generations that are not contemporaneous. On the other hand, intergenerational justice is forced to deal with a certain type of specific uncertainty that arises because our knowledge becomes gradually more limited the further into the future we look. A systematic analysis of the concept of intergenerational justice was made by John Rawls (1921–2002), an American moral and political philosopher, in which he proposed a principle of "just savings" (Rawls 1971, 2001). In his radical egalitarian interpretation of liberal democracy, Rawls believes that the practice of intergenerational justice obliges the present generation to ensure that future generations have "the conditions needed to establish and to preserve a just basic structure over time" (Rawls 2001).

The conceptual compatibility of the principles of intergenerational justice with economic intertemporal choices is not easy to achieve and has led to several models and utility optimisation criteria for a set of future generations. In the frequently used model of discounted utilitarianism, it is accepted that total utility, in other words, utility for the whole group of generations, may be broken down into a sum (or an integral, if time is considered as a continuous variable) of flows of utility relating to each successive time interval, weighted by a time discount function, viz.,

$$U(t) = \sum_{\tau=0}^{T-t} D(\tau)u\big(c(t+\tau)\big) , \qquad (1.1)$$

where T is the total period of time under consideration, D is the time discount function, u describes the utility path, and c is the consumption factor. Having adopted utilitarianism, u is a function of consumption growth c. The goal is to maximise U for different consumption currents $c(t)$. Applying (1.1) to the intergenerational case means choosing infinite T.

If, according to the principles of intergenerational justice, we decide to treat all generations equally, the values of the utility u will be independent of time. In this model, when time ceases to be discounted, i.e., if it is assumed that D(t) = 1 in (1.1), the series diverges and U does not have a finite limit because all periods of time have positive and comparable levels of utility. It is therefore concluded that it is impossible to select the consumption current that maximises U if we demand full intergenerational equity over an infinite period of time. Introducing the time discount function is a way of devaluing the future in relation to the present, which allows for a convergence of the series and the calculation of consumption functions that lead to optimal economic growth.

As well as the discounted utilitarianism model outlined briefly above, there are others that seek to find compatibility between intergenerational equity and the optimisation of economic growth for an infinite period of time (Heal 2005; Gosseries 2008). One of the most interesting is the model due to Chichilnisky (1996), developing an economic theory that formalises the goal of not devaluing the interests and utilities of future generations. She uses neoclassical economic growth theory, the only one the author believes has the clarity and substance needed for the task at hand. The model is based on two axioms designed to define sustainable development. The axioms are used to order utility paths—enabling an optimisation of economic growth—becoming sensitive to consumption in both the immediate future and the long term. Sensitivity to the present is obtained through the condition that ordering is not only determined by the "tails" of the utility paths, i.e., by long-term consumption. Sensitivity to the future is translated by the condition that there is no date after which consumption becomes irrelevant for determining the ordering of the different utility paths. Using these axioms, it is only possible to order and therefore optimise consumption currents when the total utility is given by the following functional, in which time is considered as a continuous variable:

$$U = \alpha \int_0^\infty u\big(c(t)\big)D(t)\mathrm{d}t + (1 - \alpha) \lim_{t\to\infty} u\big(c(t)\big). \tag{1.2}$$

In this equation, $D(t)$ is a social temporal discount function and the parameter α, with a value between 0 and 1, determines the relative weight of the integral in relation to the asymptotic component, i.e., the short and medium term in relation to the long term. The first term in (1.2) is a discounted integral of utilities and therefore a generalisation of the discounted utilitarianism mode. The second term represents the level of sustainability of utility in the long term. There is no solution to the equation if the social discount function is exponential, but solutions can be found if the social discount rate falls over time, coming asymptotically closer to zero at infinity.

Notwithstanding the theoretical interest of these developments in neoclassical economics, it would probably be impossible for contemporary governments and societies to put a system of intergenerational justice into practice effectively within the current economic model, based as it is on the very same neoclassical economic framework. The essential issue is knowing to what extent the present generation is willing to change its behaviour, deny itself some benefits, and invest what is needed to minimise or avoid adverse effects for future generations in its actions and in the actions of the generations that came before it, especially regarding the environment, natural resources, and climate change.

It is hard to gain political support from those alive today to change their behaviour, lose benefits, and invest in the lives of future generations whose members are not yet born, given the urgency and wide range of very real problems already faced by the present generation. In practice, an investment of this type will have to be made following the rationale and rules of the free market. Choosing to invest to avoid or make up for future harm assumes that the alternative investments available would have an interest rate at least equivalent to the social discount rate used to be able to overcome the opportunity costs. If the discount rate were much lower, it would be preferable to make alternative investments. In other words, in our present economic system, to decide upon the value of the social discount rate that should be applied, the interest rates practised in the financial market should be taken into consideration. If we use social discount rates of 3–5%, values that are often used in the short term, the value of future benefits in 100 years' time becomes very small. For example, a 100-euro investment now with a constant social discount rate of 4% would be worth 1.9 euros after 100 years and only 0.03 euros after 200 years.

The conclusion within the current economic system is that it is not worth investing heavily in solving long-term issues regarding environmental degradation, including climate change. The reason for this is the assumption that perpetual economic growth is guaranteed, so future generations will surely be richer and will have other, currently unknown, scientific and technological means to solve their problems. In practice, this attitude has prevailed and has created positive feedback because, by heavily discounting the future and investing little in long-term matters, the present generation always favours the practice of carrying out actions that have adverse impacts on future generations.

Despite this situation, there are numerous studies that attempt to adapt the current economic system to deal with long-term issues by way of models for calculating social discount rates in the intergenerational context (Hepburn 2007). The aim of these studies is to make the processes involved in making general intertemporal choices compatible with the processes involved in making economic intertemporal choices. Some authors argue that utility discount rates of zero should be used in long-term social cost–benefit analyses (Ramsey 1928; Pigou 1932; Broome 1992). This decision means adopting a position of impartiality among generations in terms of their well-being, based on the principle of intergenerational justice. Interestingly, the contemporary generation ends up being penalised in this model because, however low its consumption may be, a reduction is always justified in light of an unlimited number of future generations to whose well-being it should contribute. The same situation would be repeated with subsequent generations. However, the problem may be considered entirely theoretical because the risk of humanity's extinction is likely to be greater than zero. Menahem Yaari (1965) demonstrated that, if that risk follows a Poisson distribution, a positive utility discount rate is obtained that is constant over time.

Using a positive utility discount rate is not necessarily unfair to future generations because economic models assume that they will have greater wealth and will enjoy more knowledge and technology. This was the reasoning behind including the rate of consumption growth per capita g when defining the social discount rate. While social discount rates for the short term are fixed, the rate g varies in time and there is great uncertainty about its behaviour over time intervals that include several social generations. Discount rates higher than the economic growth rate cannot be sustained over long periods of time. The value of g may become negative if there is prolonged economic slowdown in which capital ceases to be productive, for example, due to highly adverse environmental situations (Weitzman 1994). In the long term, it is natural for social time discount rates to fall over time, in part because, as already discussed, that is the empirically determined human perception of time preference (Pearce et al. 2003; Hepburn 2007; Doyle 2013).

The main reason, however, is that, from a human standpoint, the constant social discount rate goes too far in devaluing what will happen in the distant future. Charles Harvey (1994) was one of the first economists to defend the importance of what he called "soft discounting" in long-term public policy, meaning the use of social discount rates that fall over time. But there are other reasons that would recommend using social discount rates that fall in the long term. When the uncertainty over the economic conditions of future generations is considered, specifically uncertainty about the value of the future economic growth rate, it can be concluded that optimising intergenerational efficiency requires the use of discounting rates that fall over time in social cost–benefit analyses (Weitzman 1998; Gollier 2002; Philibert 2006). Social discount rates that fall are also, as already discussed, a necessary condition for applying Chichilnisky's sustainable development model (Chichilnisky 1996).

In the context of ecological economics, Thomas Sterner (1994) studied the consequences of calculating the social discount rate when limits exogenous to the economy are reached for its growth. He accepted that economic growth, rather than being exponential, follows a logistic function determined by the Earth's carrying capacity and concluded that in that case the social discount rate would fall over time. However, the point of view that the planet's physical limits impose limits on economic growth is highly controversial and actively rejected by economists who theorise and support the economic system that currently prevails at global level.

The great, long-term environmental challenges of the 21st century have not been faced consistently, so they are not being resolved. The conceptual framework of neoclassical economics is broad and robust enough to incorporate the issue of the long term and sustainability in a formal manner, but that advantage is only of theoretical value. In practice, society has been unable to convert those advances in economic theory into action. The problems of unsustainable use of natural resources, environmental degradation, pollution, and climate change are getting worse at global level and the current political, financial, and economic system continues to focus on the challenges of the present and the short term and to significantly discount the time of future generations. According to Pascal Boyer (2008), one way of reducing the effects of temporal discount is to focus attention on simulating the emotional effects of future rewards, shifting the focus away from time factors. This practice may help adapt our behaviour to favour decisions with medium and long-term—and not just short-term—implications. From an emotional perspective, we find it perfectly natural to value future scenarios that are on the whole more positive than those that are negative. In fact, this tendency is evidenced in brain activity, too, which increases when we imagine a positive scenario rather than a negative one (Sharot et al. 2011). The strong tendency to engage in unrealistic optimism is an essential component of our behaviour.

1.8 The Evolution of Operative Social Time and Operative Time Structures

While operative social time is built in the context of social relations, in the many different forms and content of communication, information, dialogue, and debate, physical time is an abstract construction of the human mind that is, in essence, independent of those relations. However, the growing ease with which time could be measured and the increasing precision of the measurements themselves over history have influenced operative social time and the resulting operative time structures. Furthermore, while psychological time represents a basic personal ability to fully process perception of successions and durations, operative time represents the ability to form a collective framework to organise, communicate, and share the temporal side of experience and the human condition. In turn, these are conditioned by operative social time and by the operative time structure associated with it.

Time, in the sense of operative social time, became increasingly important during the first period of modernity, from the beginning of the 16th century until the start of the French Revolution in 1789. The first watches, worn as pendants or on suits, date back to the middle of the 16th century and were made in Germany. It became fashionable to remain aware of the passage of time in this way and this awareness began to regulate day-to-day activities. Time became so important that the Polish–British philosopher Zygmunt Bauman (1925–2017) says that the history of time began in modernity and that modernity is, above all, the history of time (Bauman 2000).

One of the aspects that contributed most to the value assigned to time was the extraordinary evolution of faster means of transport. Reducing journey times between towns or countries became a matter of pride and a sign of progress and belief in technology. The higher speed of transport links helped "disconnect" space from time by making the time needed to travel between two points less dependent on distance. It also made it possible to conquer new spaces, such as the space around the Earth, the Moon, and the planets in the Solar System.

This "disconnection" of space from time highlighted the inequalities between people who could afford to use faster forms of transport and those who could not, which is a trend that has accompanied most technological innovations. The reduction in distance provided by new modes of transport and communication and their social influence is also known as time–space compression (Harvey 1990). In the modern era, time has become increasingly important for the economy, which is very clear in statements made by two American politicians, Benjamin Franklin (1705–1790), who made the celebrated remark that "time is money" (Houston 1748), and John Fitzgerald Kennedy (1917–1963), who said "we must use time as a tool, not as a couch" (Kennedy 1961). Operative social time is currently monetised in a highly unequal way throughout society. Each person has acquired an economic and financial value that is measured by the monetary value of his or her minutes, hours, days, or months of presence, intervention, or work.

In the modern era, not only has social time gained more economic and financial value, but the temporal resolution of the operative time structure has become more "refined"; in other words, the resolution of this structure has increased. This is exemplified by the performance of athletes. In the Olympic Games in Ancient Greece, they ran as fast as they could to arrive first, but performance in different races could not be compared and time itself was practically irrelevant. Today, we can measure tenths, hundredths, and thousandths of a second, and even much smaller time intervals, so the main goal is now to reduce the time spent by athletes to cover the requisite 100, 200, or 400 metres and establish records in registers that include the altitude of the track and impose restrictions on wind speed.

The operative time structure of modern societies has made minutes and sometimes even seconds relevant. Greater complexity and social interdependence, a higher density of events, commitments, and requests (especially in urban life), the frequent need to do everything more quickly and with higher time resolution, which makes it possible to record changes and the pace of change in greater detail, mean that we have the impression that time is speeding up. Making use of time within the new, faster time has become a necessary condition for success. The acceleration of operative social time is different from the contraction of psychological time mentioned previously. They are phenomena with some similarities, but relate to different concepts of time.

Despite the overwhelming presence of time in life, we also try to deny it, something humans have probably done ever since they became aware of time. To do this, they created the concept of eternity. The attraction of eternity reflects the relentless fight for a full, intense life that postpones death for as long as possible. It reveals the difficulty that humans have always had in accepting and living with the finiteness of their own lives. The solution proposed by most religions to help deal with this ancestral form of depression is to interpret death as a transition to another form of existence.

Eternity is a concept that is intimately linked to religion and the nature of God, rather than a physical concept. It can be understood as the quality of that which is outside time (timelessness) or the property of lasting forever (everlastingness). Nonetheless, the latter meaning is not a "duration" in the physical sense, because physical time has a beginning and very probably an end.

1.9   Time Cycles and Elements of Operative Social Time

Physical time, unlike psychological time, biological time, and operative social time, is external to the human organism and is not limited to complex biochemical processes in the human organism. Instead, it is an abstract concept that has been built over millennia. Although there is a great deal of doubt surrounding the stages and mechanisms involved in that long process, we can say based on studies of the first civilisations that the observation of cyclical phenomena was a determining factor in starting that building process. This conclusion is bolstered by anthropological studies carried out over the last two centuries on the ways indigenous populations assess and measure the passage of time (Nilsson and Fielden 1920). Human time, represented by the cyclical regularity of certain natural phenomena, contrasted with unlimited divine time. The Celts in Ireland symbolically represented eternity by a period of a year and a day, i.e., a cycle representing time was broken by adding one unit (Roux and Guyonvarc'h 1969).

We find external time cycles in the environment around us and internal time cycles linked to human life itself. The process of adapting to external cycles and a constantly changing environment certainly played a decisive role in the biological evolution of hominids that led to the emergence of the *Homo* genus in the *Hominidae* family. Recall that this family whose members are the hominids includes four genera of species currently in existence: *Pan* (chimpanzees), *Gorilla* (gorillas), *Homo*, and *Pongo* (orangutans). Later, the observation and interpretation of external cycles played a fundamental role in the cultural evolution of *Homo sapiens*.

The most important external cycle is the alternation between day and night, which rules most of our behaviour and habits and is linked to the apparent motion of a star, the Sun. Another extremely valuable cycle in life is the year, responsible for the changing seasons and linked, although more subtly, to the apparent motion of the Sun on the celestial sphere. We also have a lunar cycle that does not have a clear relationship with life, something which made it more mysterious and seductive to our ancestors.

The day is the most important element of operative time and we use it constantly in our communication and dialogue with others. For example, we might say that yesterday was a productive day or that tomorrow will be a rest day. The term "element" of operative time is used instead of "interval" because content is more important in this context than the points that delimit the time period. A day is also a period of physical time that has been used to measure operative time, using calendars, since very early in human history. The day also underpins the meticulous measurement of physical time, which serves to control the less meticulous measurement of time performed by our clocks. By using the day to measure physical time, it became necessary to measure it more and more precisely and universally as physical knowledge and technology advanced, as will be seen below.

The day as a period of physical time is an entirely abstract time interval with no kind of connection to everyday routines and events. But the day is also an element of operative social time that is inextricably linked to each of our personal experiences as members of the society we are part of. Every human society has its own day-element of operative time, defined by the specific characteristics of the set of day-elements of operative time of its members. A society's day-element of operative time helps build that society's operative time structure.

A human society's day-element of operative time has different structures, event densities, types of activity, and expectations depending on the particular society's cultural characteristics and level of development. For example, the day-element of operative time of a country in Africa's Sahel region is very different from the day-element of operative time in Europe, despite their relative geographical proximity.

We also have yearly cyclical behaviour that follows the cycle of the seasons and is reflected above all at the psychological and occupational levels. For most of the human population, the cycle of the seasons is nowadays fairly irrelevant, although it affects a few professions, such as farmers or those in the tourism sector. However, the annual cycle affects us through the climate of the seasons and we normally have a longer period of annual holiday that often helps define the year. The day and the year are the most important elements of operative time. They are the basis for the calendar on which all events are recorded, from religious celebrations to birthdays, commemorations, and important historical events.

There is yet another element of operative time that is extremely relevant but has characteristics that are very different from the day and the year. That element is the period of life, from birth to death, whose passage we observe in ourselves and in our companions and all living organisms. Curiously, at first glance, we are all tempted to describe life as a cycle, when really it is the opposite.

This ambiguity has a very revealing nature and background from a conceptual standpoint. Life is not a cycle because its end does not lead to a beginning, as the phases of the Moon do, for example. After the new Moon, the sliver of the quarter crescent returns, advances to the full Moon, and eventually goes back to the new Moon. In life, on the other hand, every moment is truly unique, different from all the others. And that is why we so much enjoy looking for apparent repetitions in situations, projects, feelings, and moods.

Life only has some of the characteristics of a cycle if we consider it in a social context, that is, if we take into account the successive repetition of periods of life that arise, at the most basic level, from reproduction within a family or population. Regardless, we can be convinced that life is not finite if, even though it has cyclical characteristics, we believe that life continues after death for an essential form of the being, whether the soul, Hinduism's Atman, or some other form.

One of the contrasts that must have left a great impression on humankind since prehistoric times was the contrast between the surprise and uncertainty of everyday human life and the irregular length thereof on one side and the regularity of the apparent daily motion of the Sun on the other. The greatest mystery was this surely superhuman regularity, which had to be preserved at any cost, given its vital importance. Nothing in human life, in the volatile relationships between humans or the constant, unpredictable struggle for survival, would lead us to believe that the regularity of the apparent daily motion of the Sun was guaranteed. It took *Homo sapiens* millennia to become sure that it would be guaranteed for at least the next 5 to 7 billion years. Only after this enormous period of time will the Sun lose its present appearance and become a giant red star that will occupy most of the sky and even envelop Earth itself (Schröder and Connon 2008). The regularity of the lunar cycle and the cyclical motion of Venus and the other planets in the celestial sphere must have caused the same kind of bewilderment.

1.10 Operative Time of a Social Generation

Operative social time forms the conceptual framework that lets us deal with and communicate to others durations defined by events, situations, actions, activities, and personal or collective experiences, located and established in time. These are elements of operative time that play an important role because they let us structure our reflection on and communication about the past, the present, and the future. In our personal experiences, we often use expressions like "when I was young", "when I was at secondary school", "when I started working", "when I go on my next journey", or "when I go and live in my new house".

These are examples of elements of operative social time that are defined by specific parts of our personal experience but have a clear and transferable social value for others, especially those in the same generation, friends and neighbours and, in general, all those we can usually communicate with in some way. An element of operative time corresponds roughly to a period of physical time, but its limits do not necessarily have a precise definition in terms of physical time. They are defined by activities, events, and personal or collective experiences that often do not have a precise position in physical time. Consequently, an element of operative time does not correspond exactly to a period of physical time. Moreover, each element of operative time is unique and cannot be repeated, while periods of physical time are abstract and independent of the human experiences with which they may be associated. Periods of physical time of the same length do not necessarily correspond to operative times of the same length for a given person. Nevertheless, the elements of operative time may also be used to describe intervals of physical time by using a process of abstraction in order to measure them approximately and use them in communication.

In the absence of clocks, indigenous peoples use everyday tasks to identify and communicate different intervals of time lasting less than a day. Anthropological reports from Madagascar in the 19th century, for example, indicate that "a rice-cooking" often meant half an hour and that "the frying of a locust" meant a moment (Sibree 1896). Many other peoples use the cooking times of different types of food or the time needed to follow certain common land or river routes to describe intervals of time of different durations (Nilsson and Fielden 1920).

The complexity of operative social time is even greater when we use expressions like "in our grandparents' day", "at the time of the Second World War", "at the time of the financial and economic crisis in the West at the start of the 21st century", "in our time", or "in our generation's time". We need to distinguish between an element of operative time and the social time that may be connected to it if it is long enough, for example, when discussing our generation's element of operative time and the operative social time itself of that generation.

The example of the operative social time of the contemporary generation is particularly interesting because it creates the possibility of essentially describing the individual and collective life of our generation in all its different aspects: social, economic, political, military, cultural, religious, recreational, scientific, technological, and environmental. It even makes it possible to describe its specific operative time structure at a spatial scale that can vary from local to global and also to describe the constant transformation of the characteristics of human life over time. The word "generation" is used here to mean "social generation" (Pilcher 1994) and it should be clearly distinguished from the concept of generation used in biology or family genealogy, which could be called the "family generation".

A social generation is a group of people born within a specific period of time and who share similar social, economic, and cultural experiences. A good example is the Lost Generation, first defined by Gertrude Stein as the generation who fought in the First World War in Europe, which corresponds roughly to those born between 1883 and 1900. The Baby Boomer Generation comprises those born after the Second World War, between 1946 and 1964, a period that was marked by a significantly increased birth rate in those countries involved in the war. This generation lived through a time of extensive social, economic, scientific, and technological development. This period is also characterised from a political standpoint by the shift of a large number of autocracies to democratic regimes.

Sociologists, historians, and demographers have been studying and naming the generations that have followed in countries in the West. The Baby Boomers were followed by Generation X, roughly corresponding to those born between 1965 and 1979, and they in turn were followed by the Millennials and Generation Why or Generation Y, which corresponds to the following period, between 1980 and 2000. The latter name refers to the inquisitive nature of its members, the first to be born and grow up in the Information Age, defined by its intensive use of new information and communications technologies (ICT).

Finally, we get to the second generation of the Information Age, the Post-Millennial Generation, also known as Generation Z, who were born after 2001. It is the first affected at the outset by the financial and economic crisis in the West of 2008–2009, the volatile nature of commodity prices that began at the start of the new millennium, the threat of terrorism, especially in Western countries, stagnation or weak economic growth in advanced economies, and the worsening problem of anthropogenic climate change.

Generally speaking, each person shares their time with three or four successive generations: their own, their parents' generation, sometimes their grandparents' generation, and the two after their own (their children's and grandchildren's, if they have them). We do not share any other generations' times and we only have indirect, fractured knowledge about them based on reports, reading, conjecture, and interpretations that are naturally influenced and distorted by the particular view of our own generation. Despite the limitations of personal experience, the operative times of generations succeed each other with a dynamic of their own that reveals the nature and characteristics of social, economic, political, military, religious, cultural, scientific, technological, and environmental changes at local, national, and global level.

1.11 The Cyclical Nature of Generational Time and the Uniqueness of a Lifetime

Here, we find one of the most remarkable properties of the concept of time. The time of a social generation represents an element of operative time with cyclical characteristics, in the light of the succession of social generations, although each period has relatively flexible boundaries. On the other hand, as an element of operative time, a human lifetime is well-defined, unique, personal, and cannot be repeated. Knowing how to use and deal with this particular element of operative time is an enormous—perhaps the greatest—challenge that faces *Homo sapiens*.

At a certain point in life, we begin to become fully aware of that element of time and realise that it is bounded. We normally take steps to make sure it is as long as possible, enabling us to fulfil as many of our expectations as possible and achieve some of our ambitions. Having full awareness and accepting the opportunities and restrictions of this element of operative time is therefore a recurring motive for anxiety, hope, responsibility, and sometimes disappointment and anger. The way we are and the way we live are largely shaped and conditioned by that awareness and above all by the relationship between our identity, our experience and interaction with the society in which we live, and our awareness of the various physical and mental transformations we undergo as time goes by, which inevitably lead us to death.

The self-portraits of Rembrandt van Rijn (1606–1669) are a remarkable example of the search for personal identity, independent of those transformations, through dialogue with an accurate visual representation. Rembrandt painted, engraved, and drew around 85 self-portraits over 41 years, from the age of 22 until a few months before he died. This suggests his need to reflect on himself over time and to understand the identity of others through the visual aspect of his own identity, and so be able to depict them faithfully without betraying their identity. His extraordinary capacity for psychological analysis is revealed in this series of self-portraits, in which time gradually transforms the expression of the strength and pride of youth and his years of great notoriety, into an expression of the frailty, nostalgia, and introspection of old age. He sought to capture the idea that Heidegger rather negatively called "being-towards-death", which he believed to be one of the fundamental aspects of the human concept of time (Heidegger 1927).

The cyclical nature of a generation's element of operative time is social and ultimately arises because we are able to reproduce and because we are primarily social beings who are normally in constant communication and interaction with others. When social interaction is limited, the finiteness of a human lifetime as an element of operative time seems more marked.

The seeming contrast between the cyclical nature of the element of generational time, or simply generational time, and the uniqueness of the lifetime as an element of operative time is one of the essential challenges of the human condition. To what extent should we, on the one hand, appreciate and enjoy something that is unique and personal and cannot be repeated and, on the other, value our active participation in the social generational cycle by helping it to perdure and contributing to its sustainability? How can we resolve the apparent conflict between these two lines of thought? To what extent are personal interests compatible or incompatible with the interests of future generations? These seemingly abstract issues have major implications for the analysis and resolution of many of the challenges that we meet in life.

All social generations are unique although there is some underlying continuity and consistency in the evolution between successive generations. Each social generation naturally has a value to its members that is very special compared to others. Our lifetimes are part of our generation. It is within our generation that we establish and develop relationships, friendships, camaraderie, projects, and expectations. It is within our generation that we experience individual and shared victories and defeats, suspicion, envy, enmity, hatred, and conflicts. This enormous network of different kinds of human relations, with connections that reach wider and wider throughout the world, creates unique cohesion and identity that help define our generation, its operative social time, and the operative time structure that is associated with it.

All of us live in the operative time of our social generation and as we get older we find it increasingly difficult to belong to the time of younger social generations. As difficult and incomprehensible as the the element of operative social time of a forthcoming generation may seem, it is always an opportunity and a form of rebirth, building new solutions full of hope and creating an alternative view of the world. It is above all an opportunity to demonstrate that the new social generation has been able to overcome the problems and limitations of the previous generation. The operative social time of the contemporary generation of a local, national, and global population focuses on the present and is primarily characterised by the major events that have affected or affect it through activities, behaviour, reflections, concerns, assessments, expectations, and trends in the various fields of human activity, especially those of the age groups that are most active and most involved in that generation.

Each generation not only makes efforts to develop and assert a collective aware-ness, a vision, and an intervention in the world that is different from previous gener-ations, often reactively, but also believes that it will make a decisive contribution to what it believes to be world progress or, more precisely, the progress of its generation in relation to previous generations. This positioning is the one made by most of the population that forms the social generation and does not stop there being differing positions at individual level, varying in number and diversity, although their social manifestation is negligible. Each social generation reacts and corrects, in its own way, the legacy of those that came before, and believes it makes an innovative and determining contribution to the progress of its community, its country, and the world. The process is repeated with every new generation, but the practical implementation is always different, adaptive, and unique.

If a particular generation was especially disruptive and divisive or if it was marked by conflict, violence, or war, subsequent generations will seek to oppose this and reverse the trend. These responses and counter-responses form an intergenerational dialogue that has marked a large part of the history of nations and civilisations.

The contrast, tension, and dialectic between the cyclical nature of social gen-erations and the uniqueness of the social time element of a lifetime contributes to maintaining the prevailing opinion that, apart from some rare exceptions, such as the two World Wars, each generation lives in a much better world than those that came before it and that future worlds will be even better. This opinion stems from human nature and it translates into the deeply held belief that social generations sail in a never-ending current of progress driven by science and technology. The current is never-ending because, even if humanity is in a difficult situation that shows a trend towards getting worse, most people will tend to deny the evidence, believing that progress will somehow continue. This is the point of view of progressionism, which has originated from a progressionist evaluation of modern history (Marx and Mazlish 1998). As regards the future, most transhumanists defend progressionism (Verdoux 2009).

Each new social generation invests in technological progress generated by a suc-cession of industrial revolutions as its main instrument, the trophy and banner it carries to stand out from the backwards, retrograde view held by the previous social generation. It does not much matter that in the meantime the environment is dete-riorating at several spatial scales, including globally, and that natural resources are being overexploited, because these problems will eventually be resolved by tech-nology, replacing natural capital with human capital. It is likely that in the medium and long term, areas rich in biodiversity will be almost exclusively confined to con-servation areas, mostly in remote regions where the land has no other profitable uses. What is truly important is captivating the attention of consumers, surpassing supposed technological impossibilities to create new opportunities for services, con-sumption, and experiences that are as dazzling as possible, and allowing consumers to stand out from each other. The cutting edge of each generation's progress is led by an elite whose intelligence and technological prowess must also win over the majority of consumers at national and global level by identifying and satisfying their most immediate, emotional, and primal desires and expectations, anaesthetising as far as possible their ability to realise the dependence that may entrap them. Naturally, the progress brought by each generation creates winners and losers. Some of the losers are those who are least apt at joining and benefiting from the new social and technological changes of our generation. There are many dissidents in each new social generation, but a one-dimensional view of technological progress, together with subordination of the environment and natural resources against which they fight, is what prevails and is currently the driving force of historical time.

There has been a long historical path leading to this simplification. One important metaphoric milestone on this journey was the book *Candide, ou l'Optimisme*, signed by a German called Doctor Ralph. The real author, who sought to remain anonymous for his personal safety until 1768, was Voltaire (1759), the literary pseudonym of François-Marie Arouet (1694–1778). In *Candide*, Voltaire ridiculed the philosophy of optimism of Gottfried Wilhelm Leibniz (1646–1716), who saw God as perfect and, although recognising that the world was not, argued that God created it to be as good as possible. The ideas of Pangloss, Candide's teacher and tutor, are a satire of Leibniz's philosophy and can be summed up in the expression "all is for the best in the best of all possible worlds" ("tout est au mieux dans le meilleur des mondes") (Voltaire 1759).

Voltaire also mentioned Leibniz's principle of sufficient reason when Candide, on his travels, arrived in Lisbon on the eve of 1 November 1755, All Saints' Day, and sees the impact of the earthquake that caused some 70 000 victims and destroyed the city. Pangloss asks him: "Is there anything to gain here?—What could be the sufficient reason of this phenomenon?" ("Il y aura quelque chose à gagner ici?—Quelle peut-être la raison suffisante de ce phénomène?"). In *Candide*, Voltaire also ridiculed religion, the Catholic Church, theologians, governments, armies, and wars. The book was immediately denounced as going against religion and being an attempt on the state, and was therefore banned. Nonetheless, this did not stop more than 20 editions from being published, and more than 20 000 copies were sold by the end of 1759, the year in which it was published for the first time. The book was also banned in the US and even in 1929 a worker at the Boston customs took the initiative of confiscating several copies that were to be used in French classes at Harvard University because it was considered obscene (Haight 1935). At the end of the book, Candide eventually rejects Pangloss' insistence on an optimistic view that God made the best world possible, concluding that "we have to cultivate our garden". This simple sentence reflects a profound change in stance that opens a path to the never-ending current of progress or progressionism. It is noteworthy because it also metaphorically dissuades us from merely looking forward to and waiting to enjoy the perfection of the garden of paradise. We continue to "nurture our garden" with great enthusiasm. It is now more and more artificial, complex, and contingent, but it is so dazzling, especially for the most credulous and short-sighted, that it is worth the perfection of which we dreamed in bygone times, but which few now believe in.

A large number of recent books, reports, and conferences tend to emphasize that humankind has never had a world better than today's and that we have more than enough reasons to be optimistic about the future. One of the defenders of this current of positive and optimistic thought is Steven Pinker, who considers that we are living through the least violent moment in the history of humanity (Pinker 2011). In his latest book (Pinker 2018), he points out that we are in a new enlightenment period based on reason, science, humanism, and progress and that all indicators of human progress have improved and have never been better than now. Microsoft founder Bill Gates, one of the iconic figures of our time, called it "my new favorite book of all time". Unfortunately, the same cannot be said about environmental indicators at the local, regional, and global level. The new wave of optimism that adds to and enhances the never-ending current of progress suggests that there is a need to defend it from some discordant voices that are worried about some emerging and growing trends towards violence and the impacts of humanity on the Earth system and the negative consequences that they are beginning to have on human indicators. The former include terrorism, civil wars, and armed violence that are common in some fragile states (OECD 2018) and the challenge for human health posed for some countries by the number of non-combat-related gun deaths, particularly the USA and Brazil (Murray 2006) and countries in Asia and Africa for which it is harder to obtain data. The latter include the impacts of pollution in the air, water resources, soils, coastal areas, and the ocean on human health and well-being. Furthermore, it includes the impacts of climate change on health and well-being and on food security, especially in some more vulnerable regions.

1.12 Multigenerational Operative Social Times and Memory of the Past

We could be more abstract and consider elements of operative social time that go beyond one social generation and include a succession of social generations. This is true for the time of the Renaissance, the time of the Middle Ages, and even the time of the ancient Egyptian civilisation, the ancient Chinese civilisation, or the Maya civilisation. These elements of multigenerational operative time involve the lives of communities that have lived in successive social generations over several decades, centuries, or even millennia. What defines each of these elements of time is the fact that a great number of generations underwent and shared a set of experiences and values that have a specific, recognisable identity. They are also characterised by the operative social times of each of the successive generations that make up the element of multigenerational time, which have their own evolution, as we have seen. Each civilisation during the cultural evolution of *Homo sapiens* has produced an operative social time with specific characteristics. For the longest civilisations, it may be possible to identify and describe the evolution of operative social time throughout the history of that civilisation.

Multigenerational operative times can be characterised by analysing historical, cultural, and archaeological records. Doing so requires a highly abstract approach and involves significant uncertainty, since we naturally cannot witness what happened at those times. However, multigenerational operative times play a fundamental role in reassembling and understanding our common past at the national, regional, and global levels. Analysing them reveals an identity and consistency that likely extends beyond the present and makes it possible to peer into the future. When we talk about them, we seek to discover and interpret the different kinds of perceptions, notions, outlooks, expectations, fears, values, activities, experiences, and social relations of the people who lived at those times. This is possible because we share the same impulses, essential modes of behaviour, and capacity for knowledge that are characteristic of our species. The final goal is to reconstruct and describe the successive operative generational times that form, as far as possible, part of an element of multigenerational operative time, including the rationale behind how they developed and their *raison d'être*, i.e., the justification for establishing an element of multigenerational operative time with specific, defined boundaries.

These boundaries correspond to transitions that involve highly diverse factors, such as the economy, the environment, natural resources, periods of peace, conflict, and war, migrations, politics, religion, and culture. Ultimately, it involves rebuilding the past of the different human societies.

Building multigenerational operative times is a relatively recent process in the course of cultural evolution, and analysing it reminds us of the importance that memories of past events have acquired throughout that evolution. It is likely that notable events regarding the social group were first memorised and mentioned in oral communication and transmitted from generation to generation, often in the form of stories, fables, and myths. Later, they were represented in the form of engravings and paintings. This transition constituted an adaptation in the evolution towards increasingly complex social behaviour. Symbolic representations, such as artefacts with engravings or figurative representations and the rock paintings of the Upper Palaeolithic, go a lot further than the spoken language because they last; they can be seen and their messages can be communicated over time to several social generations. It is likely therefore that the process of using symbolic representation in physical forms played an important role in the evolution of multigenerational operative social time.

At the start of the historical period, the Narmer Palette, found in Hierakonpolis and sculpted between 3200 and 3100 BC, is one of the oldest representations of a past event. Its figures represent the conquest of Lower Egypt by king Narmer of Upper Egypt who led the unification of the nation. It was also at the end of the 4th millennium that the first writing systems emerged in Sumer, using cuneiform characters, and in Egypt, using hieroglyphs. The oldest clay tablets containing cuneiform pictographs show that their main function was as an *aide-mémoire*, recording agricultural accounts, such as the number of bags of grain or the number of head of cattle. Later, cuneiform characters were also used to record the king's activities, especially his military campaigns, as well as myths and early forms of literature. From the very beginning, the writing system developed in Egypt, using pictographs, phonograms, and determinatives, enabled writing that was close to spoken language and more appropriate for dealing with abstract concepts. The hieroglyphs were widely used to report past events, especially those relating to the pharaohs' lives and military campaigns. However, in both the first civilisations of Mesopotamia and in Egypt, there were no systematic, chronologically ordered records of the past, of the kind that would later be identified as history.

One important characteristic of human societies and civilisations of all times is the cultivation and transmission of the memory of some past events in the form of stories, fables, allegories, myths, and epics. Each generation believes that some past stories have great social, moral, or environmental educational value and so should be passed on to new social generations. During this long transmission process, narratives are transformed and the most outstanding persist, travelling through and influencing a large portion of the world. Examples include the *Epic of Gilgamesh*, an epic poem from Sumer, the oldest texts of which date from 1800 BC, the *Mahabharata* and the *Ramayana*, two epics originated in the Vedic civilisation in India, around 800 BC, and the *Iliad* and the *Odyssey*, two epic poems about the Greeks, attributed to Homer and written in the same period. Despite their otherness, these founding texts deal with themes, feelings, emotions, and existential issues that remain alive and we continue to share.

The difference from history is that this memory of the past does not seek thoroughness and an exact chronology of the facts but rather a summary of a succession of events, their importance, and their message in terms of human values. In the West, history emerged later with Herodotus (c. 484–425 BC), born in Halicarnassus, now Bodrum in Turkey. Herodotus was the first to attempt to reconstruct the past systematically from critically analysed observations and reports. His main goal, set out at the start of his book *Histories*, was to avoid the memory of human events being lost in time, especially the notable achievements of Greeks and non-Greeks in the wars in which they were involved. It is at times impossible to distinguish between fiction and historical description when reading *Histories*, although it is known that he directly reported what he saw and heard, including myths and legends, on journeys to Egypt, Mesopotamia, and probably the south of Italy.

In the Chinese civilisation, the oldest historical narrative is *Zuozhuan*, or *Commentary of Zuo*, probably written in the 4th century BC. It is a chronicle of life in the State of Lu, where Confucius was born, for the period from 722 to 468 BC, focusing on military actions, diplomacy, and politics. The goal of historical veracity in the narrative is confirmed by the mention of several eclipses, the dates of which have been proven, and the first recorded passage of Comet Halley in 611 BC (Pimpaneau 1989).

In Egypt, the first known record of an intention to systematically reconstruct the past dates from the Ptolemaic period, when the Hellenisation process had already begun, following the arrival of Alexander the Great in 332 BC. Pharaoh Ptolemy II Philadelphus, who reigned from 283 to 246 BC, asked the priest Manetho to write the history of Egypt, apparently with the aim of preserving the memory of the Egyptian civilisation and bringing it closer to the Hellenic civilisation. This resulted in the book *Aegyptiaca*, which contained the first list of the Egyptian dynasties.

The book has been lost, however, and we only have quotations from it and synopses drawn up by later Roman historians. The first was Flavius Josephus, a Romanised

Jewish historian who lived in the Christian era roughly between the years 37 and 100. He cited Manetho's work in his book entitled *Jewish Antiquities* (Josephus, c. 93–94), where he criticises Greek historians for distorting the Jewish history and not recognizing the antiquity and universal significance of the Jewish people, especially Apion, an Egyptian writer born in Alexandria and a contemporary of Josephus. The controversy between Josephus and Apion took place at a time of conflict between the Greek and Jewish communities of Alexandria. It is interesting to note that the long Egyptian civilisation, despite its omnipresent concern with prolonging human life after death, made little effort to systematically record the history of the living. It only began to do so during its decline and against a backdrop of interaction and tension between different cultures and religions that struggled to mark their ancestry and precedence.

The construction of the history of the past has evolved a great deal since the first civilisations. The goal is to build a consistent, credible narrative of the past based on all available sources, such as oral records, written texts and documents, archaeological and monumental heritage, weapons, artefacts and the different forms of visual art. It is nonetheless impossible to remove the temporal relativism of history given that the interpretation of historical sources is influenced by a historian's vision and approach, unavoidably conditioned by a very different spatial and temporal framework. This difficulty is acute, for example, when attempting to perform ethnographic comparisons to study the behaviour of prehistoric social groups.

As well as temporal relativism, there has been a spatial relativism arising from history having tended towards choosing topics, frameworks, hypotheses, and methods centred on Europe, since the Greek civilisation. The current and future challenges of globalisation demand a global history that makes it possible to share multiple global narratives and include historical perspectives and traditions from non-Western countries (Sachsenmaier 2011).

There is even a third, particularly important form of relativism related to the future. History is not just a narrative of the past but also a "check on the future" since, as the historian Reinhart Koselleck has pointed out, it includes a timeline of the past and the conceptions about the future that were made in that past (Koselleck and Tribe 2004). Each generation carries out a check on the imagined future by way of a critical comparison between its "space of experience" and its "horizon of expectation". Here, the concept of horizon of expectation is used beyond its specific meaning in literary history in the context of reception theory (Jauss and Benzinger 1970).

This clash between possibilities and perspectives about the future, expressed throughout the past, with the experience of the present gives us greater awareness and knowledge of ourselves and the world to which we belong. It also offers us a measure of relativism of our ability to plan the future and the uncertainty involved in doing so.

1.13   Historical Time

Koselleck was one of the first to develop a theory of historical time essentially based on the idea that time has a history and that history is time (Koselleck 2002). The evolution of the most significant, intense or long-lasting events over physical time that establish a dialectic between the horizons of expectation of successive social generations and those generations' spaces of experience generates a specific time structure and pace. This multigenerational pace and structure, which relates to a society in a broad sense (i.e., it can include one country or a group of countries), defines what we shall call the historical time of that society. With this form of timing, accelerations and decelerations in the pace of change of both multigenerational operative social time and historical time can be identified.

The definition of a community or country's historical time is inevitably affected by the subjectivity involved in identifying which spaces of experience and horizons of expectation prevailed. The exercise becomes even harder if we wish to define the historical time of humanity, which inevitably involves including and summarising the times of history of at least those countries that made decisive contributions to characterising it. Generally, there are periods in which the historical time speeds up because expectations for the future and the experiences encountered acquire a new, special value that causes operative social time to accelerate. It is common for these elements of accelerated operative social time to eventually characterise the history and culture of a country or group of countries. Historical time is built through the sequencing of and dialectic between (generational or multigenerational) elements of operative social time, characterised by revolutions and corresponding reactions, by peace or by war, by the stability of populations or migration, by wealth or poverty, abundance or misery, hope or disillusionment.

The flourishing of the modern era is the most eloquent example of that acceleration in the historical time. Until the end of the Middle Ages, time was practically absent from history. In other words, major historical events, some separated by long periods of time, had a compatibility that gave them a timeless nature. To illustrate this, Koselleck and Tribe (2004) cites the famous 1529 painting by Albrecht Altdorfer depicting the Battle of Issus in 333 BC between the Greeks, led by Alexander the Great, and the Persians, led by Darius III. The Persian fighters are shown as the Ottoman Turks in the army of Suleiman the Magnificent who carried out the Siege of Vienna, also in 1529. The depiction of the Battle of Issus, around 1800 years beforehand, was a credible metaphor for the contemporary situation. There is nothing in the picture to point out that time had meanwhile consumed eighteen centuries.

In that era, historical time was still relatively slow because horizons of expectation were very limited. The main expectation about the future was eschatological (from *eschatos*, meaning "last" in Greek) and it consisted of an apocalyptic prediction of the Last Judgment. In Judaeo-Christian doctrines, historical time is linear, irreversible, and finite because there will be a final day after which the chosen will receive the eternal joy for which man was created.

With the arrival of modernity, the emergence of the idea of progress and hope for an eternally better future deeply changed historical time. We became convinced that historical time is changeable and that an evolution driven by progress can be observed. As well as evolution, acceleration also emerged. After the Industrial Revolution, the acceleration intensified, pushed forward by economic growth based on the growing supply of readily available energy, greater scientific knowledge and growing technological innovation. This cultural evolution invites us to reflect on the growing acceleration of the historical time and its relationship with progress and sustainability.

1.14   Horizons of Expectation Beyond the Operative Social Time of the Contemporary Generation

As well as their role in defining historical time, space of experience and horizon of expectation also play a determining role in characterising operative social time. A person's horizon of expectation is predominantly located in the physical time period of his or her life. His or her space of experience is necessarily situated within that period. Given the great importance of past experiences and future expectations in our lives, that location tends to focus the framework in which operative social time is inserted within the element of operative time of a life. The events, activities, experiences, and expectations undergone, shared, and communicated in the social context, which structure the operative social time of a person, are mostly located within the element of operative time of his or her life.

The central nature of the operative time of someone's life is, in the end, one form of time discounting, and is characterised, as we have seen, by the adopted rate of time discount. What someone thinks will happen to human societies and the environment in which they are going to live after their own death, whether they are expectations, forecasts, or projections, or merely assumptions about natural disasters, has a much lower value than what someone imagines will happen during their own life. In the latter case, events and situations have a direct personal value, whereas in the former they have only an indirect value based on different forms of family and social solidarity, involving descendants, other members of the family, friends, people who are from the same area or nationality, and maybe even humanity as a whole.

Solidarity with descendants and family members is clearly much greater than with other communities and tends to fall as the size increases. Naturally, all forms of solidarity tend to disappear as we imagine a more and more distant future. The value of future events and situations, imagined, foreseen, or planned through scenarios obtained using models, gradually falls with the rate of time discount as we move away from the contemporary social generation until we reach complete social, political, and economic irrelevance. Beyond a certain time, this has a purely academic value fed by curiosity and recognised only by a small community of scientists specialised in the matter.

Similarly, the space of experience of a social generation is located exclusively, and the corresponding horizon of expectation located primarily, in the element of operative time of that generation due to the great importance of past experiences and future expectations of a social generation. This means that the operative social time of the contemporary generation is focused on the element of operative time of that generation. The spaces of experience and horizons of expectation of future generations tend not to be part of the horizons of expectation of the contemporary generation, which does not encourage intergenerational solidarity, equity, or justice.

For these goals to develop, willingness needs to be created to obtain information, mobilise interest and knowledge about the social, political, economic, and environmental problems affecting the contemporary social generation and about scenarios and projections for the evolution of those problems for physical times that comprise the social times of several future generations. That willingness, interest, and knowledge are only accessible to a very small part of humanity that is not confronted with the urgent need to guarantee its survival every day and has received the necessary training for this. However, those who belong to this small part may not be curious, or they may not be interested in the matter. This attitude is likely because they are convinced that economic progress will inevitably give future generations growing economic prosperity, well-being, quality of life, and a capacity for increasing consumption of goods and services which will make our intergenerational solidarity with those generations practically unnecessary. We are dealing with a conviction we have inherited from modernity and which has since had a deep and lasting influence on our relationship with time.

Each social generation is fully convinced that it can in some way surpass the previous generation. The desire to achieve this dream is unquestionable and uncontrollable. Each generation feels an obligation to demonstrate to itself and the generations that came before it, particularly the last one, that it has brought new momentum to progress. There is no place for steps backwards in the succession of social times of future generations. In this context, talking about the future with pessimism will always lead, sooner or later, to an impasse, rejection, and isolation. For the same reasons, suppositions about the lack of sustainability of the current global development paradigm do not gain popularity and are considered to be groundless and catastrophic. With the removal of a final eschatalogical ending, it has become much more difficult to assume any kind of future instability or decadence. This would implicitly acknowledge an inability to guarantee the benefit of the continuous progress of civilisation for successive future generations. It would mean challenging the arrow of progress, or in other words, the point of view of progressionism.

If we are convinced that the model and global system that currently supports civilisation lack sustainability, that the lack of sustainability is a risk for future social generations, and furthermore, that we have the moral duty to help reduce that risk, the road we have ahead of us is to adapt our relationship with time to today's challenges. In other words, it means beginning to transform the operative social time of the contemporary generation in order to make it compatible with sustainability. If we are not convinced about the lack of sustainability, we follow the mainstream path in which future generations will react to try and overcome recurrent crises.

1.15   *Homo sapiens'* **Time**

The evolution of hominids led to the emergence of the *Homo* genus and later to the *Homo sapiens* species, which prevailed as the only species of that genus after a slow process of natural selection. We do not know how long the *Homo sapiens* species will live but the probability of it not coming to an end is practically non-existent. There is, then, an element of operative social time that we can call *Homo sapiens'* time that includes and surpasses all the other elements of operative social time. Although its duration is unknown, it is an element that is well defined by the emergence of our species on Earth and its future extinction. We have no knowledge of how that extinction will come about and whether or not it will give rise to other *Homo* species. In any event, it is practically certain that there will be an end. It is also a safe bet to say that *Homo sapiens* will use all its ingenuity and determination to extend its time as much as possible. The duration of this time is determined primarily by *Homo sapiens'* current and future cultural evolution and it encapsulates essential aspects of our species' behaviour and values and its relationship with the environment, considered from the perspective that it may extend beyond planet Earth.

We can reflect on and seek to determine *Homo sapiens'* time. It is a time that deeply characterises us. It begins with the emergence of the species in Africa, extended by a spectacular cultural evolution that led to a position of dominance and command over the biosphere and a great capacity to benefit from and interfere with the Earth system. Eventually, that time will reach an end and we can only imagine the reasons for which that will happen, but they are intimately linked to essential aspects of the process that directs our evolution.

To what extent did the evolution that led from the *Australopithecus* to the *Homo* genus and from the early days of the *Homo* genus to *Homo sapiens* contain the seeds for future success and the opportunities and risks that the species faces today? Is it possible to discern in this process whether or not our species will end because of natural environmental transformations or whether we will in some way bring about our final evolution and our own end? Will there be Darwinian biological evolution to form a new species or has *Homo sapiens* made that type of evolution impossible? Will the power of science and technology produce a form of transhumanism that enables the post-Darwinian evolution of *Homo sapiens*, driven with the aim of improving human capacities in accordance with pre-defined standards? Although it is highly unlikely (Sandberg et al. 2018), we may still ask whether we will find other intelligent beings in the Universe that also have self-awareness and self-knowledge during *Homo sapiens'* time. All these questions fit into that time, and the answers to them, although sometimes only in the form of conjectures, must be consistent with each other and with our past. The integration of time and evolutionary consistency is one of the main goals of reflecting on *Homo sapiens'* time.

Biological evolution took about 4 billion years to produce humans starting from the first living organisms on Earth (Bell et al. 2015) and we are the only human species currently alive on Earth. We know from palaeoanthropology that there were many species close to us at the phylogenetic level, some of which lived alongside us but have since become extinct. Fossils demonstrate that, as we go back in time, those species become less and less similar to humans. The closest ones that survive are chimpanzees (*Pan troglodytes*) and bonobos (*Pan paniscus*) of the *Pan* genus. The split into the Panina subtribe, which includes the *Pan* genus, and the Hominins of the Homininae subtribe, which includes the *Homo* and *Paranthropus* genera, was a complex process that probably took place over a long period of time from 10 to 7 million years ago (White et al. 2009). It is thus natural to ask when, how, and why our ancestors became human. What were the natural selection drivers that led the *Homo* genus to emerge in the Homininae subtribe and for *Homo sapiens* to emerge in the *Homo* genus, whose members are referred to generically as humans?

Homo sapiens' time began when it appeared in Africa around 300 000 years ago (Hublin et al. 2017). It is likely that *Homo sapiens'* ancestors in the *Homo* genus already had some human characteristics. In other words, the oldest representatives of the *Homo* genus in fossils were possibly the first beings with human attributes. The evolution in the phylogenetic line that gave rise to *Homo sapiens* was propelled by natural selection mechanisms on an evolutionary journey with its own characteristics over a period of survival that has lasted for 2.5 million years. We can also talk about the time of the *Homo* genus as the time at the start of which humans' first characteristic signs emerged. Its end will coincide with the end of *Homo sapiens'* time.

Defining and characterising *Homo sapiens'* time is not the same as performing an eschatological analysis of humanity's final destiny and the final events of history, a reflection which often lies within the scope of theology. *Homo sapiens'* end may be the result of a natural or anthropological global catastrophe that quickly wipes humans out. That is the only case in which its end will be relatively well defined in physical time. We could imagine a transition from *Homo sapiens* to another Darwinian or post-Darwinian species but that process would be longer, which makes the end of the time indistinct and harder to locate in physical time. It is likely that, in the event of a post-Darwinian evolution, some of the post-human population will live alongside a population of *Homo sapiens*.

Within the physical time flow of the Universe, it will be easier to identify the end of *Homo sapiens'* time *a posteriori* in temporal scales with low resolution. After the end, physical time will continue to flow unstoppably until the heat death of the Universe. As for psychological time, human biological time, and operative social time, we do not know how they will evolve but they will be dependent on *Homo sapiens'* time.

One of the main difficulties of the *Homo* and *Homo sapiens* time analysis is our inability to imagine periods of time that last hundreds of thousands of years and the impossibility of obtaining and analysing all the information about the relevant events and processes that took place within them. In fact, we have only extremely limited and fragmented access to them. We have built up a reasonably complete knowledge of the time that separates us from the discovery of agriculture around 10 000 years ago. But if we continue into the past, archaeological records and fossils become increasingly scarce. The operative inability for comparison is due to the fact that the time of the *Homo* genus is much longer than 10 000 years. There are almost 250 consecutive periods of 10 000 years separating us from the possible origins of the *Homo* genus, or around 125 000 biological generations if we assume that each generation corresponds to an average of 20 years.

Each of the successive changes determining the evolutionary process that led to *Homo sapiens* was important. The success of each had a reason and left a mark on mankind but we cannot reconstruct all the small steps that have been taken place over at least 2.5 million years. Although it is extremely difficult to reconstruct the past time of *Homo sapiens*, it is possible to reflect on and look for the essence of our species' time that determines its future evolution.

1.16 The First Human Records of the Cycles of the Sun and the Moon

Homo sapiens' behaviour and process of adaptation to the surrounding environment were clearly influenced by external time cycles, some with surprising regularity. Furthermore, the relationships among external cycles, and between them and the human lifetime, caused great curiosity and sparked an enormous effort to decipher and interpret them throughout the history of civilisations. It is very likely that *Homo sapiens* recognised at an early stage that those relationships were of vital importance and contained essential messages that humans would need to understand. Let us look at the oldest human records identifying external cycles that have survived to the present day. The records possibly reflect some of the first steps in the long process of constructing physical time.

It is likely, although impossible to prove, that *Homo sapiens* was aware of the crucial relationship between the Sun and life from very early on, whence, thanks to its cognitive abilities, an appetite for observation and a desire to interpret their cyclical movements would have developed. We are a long way from being able to imagine what the concept of time was, and how it evolved, for our prehistoric ancestors. There probably was an operative social time connected to the first forms of acknowledging external time, evidenced by the cyclical patterns of some particularly important natural phenomena. We only have some signs of the attention aroused by those cycles.

One of the oldest known signs dates from 17 000 years ago. At the Lascaux cave in the Dordogne, in France, the gallery at the entrance is aligned with the Sun's rays at twilight on the summer solstice. It is unlikely to be a coincidence that the twilight Sun lights up the wall of the great Hall of the Bulls at the back of the cave for around an hour for a few days of the year around the summer solstice. The full Moon can also be seen from inside the cave before the early hours of the morning on the winter solstice.

An alignment of the entrance to the cave with the Sun at dawn and twilight on the solstices or equinoxes can be seen in a hundred other caves that are known to have been inhabited by prehistoric humans in the south of France, including Combarelles, Font-de-Gaume, and Bernifal (Jègues-Wolkiewiez 2000). It is plausible that our ancestors of the Upper Palaeolithic, especially from 35 000 years ago onwards, were well aware of the annual cyclical movements of the Sun and their relationship with the seasons. It is not hard to imagine that they performed some form of Sun worship, but we do not have any remnants of those activities.

There are also signs that they observed the cycle of the Moon in detail, as the most conspicuous and easily observed cycle in the celestial sphere, which coincided approximately with the menstrual cycle, although entirely fortuitously. Alexander Marshak (1964) was the first researcher to suggest that the series of marks found on small stone, bone, and horn artefacts from the Upper Palaeolithic were symbolic records of the lunar cycle. The best known example is a small plate of bone from around 32 000 years ago found in the Abri Blanchard cave, also in the Dordogne, which has 69 incisions that appear to describe the apparent motion and phases of the Moon. There are several dozen objects with similar marks, although it is hard to make any scientifically reliable conclusions about what they mean. Recently, Michael Rappenglueck (2004) of the University of Munich interpreted several series of marks painted on the walls of the Lascaux cave as symbolic representations of the Moon and the constellations of stars. Sets of 13 and 29 small squares and circles among the remarkable paintings of bulls, horses, and antelope may possibly represent the number of days on which the Moon is visible during the lunar cycle.

Later, in Neolithic Europe, we find several clusters of megaliths, like those at Stonehenge, Carnac, and Évora, in which the latitude at the location, the positioning and alignment of the megaliths, and their position in relation to the different clusters are related to the cycles of the Sun and the Moon in the celestial sphere. The positioning of the clusters and the geometry within them can be easily linked to the solstices and equinoxes. It becomes clear that our ancestors knew the periodic movements of the Sun and Moon in detail and sought to interpret their meaning and the relationship between them.

The concept of a year naturally originated with the cycle of the seasons and gave rise to the tropical year or seasonal year, defined as the period of time between two successive vernal equinoxes. In other words, it is the period between two successive passages of the Sun past the vernal point, defined as one of the two points where the ecliptic and the celestial equator cross. The vernal point is also known as the first point of Aries because it serves as the starting point for the right ascension, the azimuthal coordinate that together with declination forms the equatorial coordinate system of astronomy. The average duration of a tropical year is roughly 365.24 days or, more precisely, 365 days, 5 hours, 28 minutes, and 45 seconds.

The lunar calendar was one of the first to appear with 12 or 13 months, corresponding to years of 354 or 384 days, since the synodic period of the Moon is roughly 29.5 days. The problem with those calendars is that, without inserting years with more or fewer months, the calendar shifts away from the cycle of the seasons that define the tropical year and is no longer synchronised with agricultural cycles. It should be noted that, despite this inconvenience, lunar calendars are still used. The Islamic calendar, called *Hijri*, is one example. A year has 12 months, corresponding to the lunar synodic periods, and shifts against the seasons by 11 or 12 days every tropical year. Synchronisation with the seasons occurs roughly every 33 Islamic years. The Roman calendar that came before the Julian calendar was also lunar.

Other cultures adopted lunisolar calendars that recorded the motions of both the Sun and the Moon. Synchronisation between the solar and lunar cycles could be achieved by inserting a 13-month year between two consecutive 12-month years. The lunisolar calendar gives a more precise indication of the seasons, in other words, it is more compatible with the tropical year.

However, if the year is defined as the sidereal year, i.e., the period of time between two successive identical positions of the Sun in relation to the stars, the lunisolar calendar forecasts the position of the full moon in the celestial sphere in relation to the stars and their constellations more precisely. Due to the precession of the Earth's axis, which has a period of roughly 26000 years, the vernal point moves to the west in the celestial sphere, along the ecliptic, by approximately one degree every 72 years. This means that the vernal point moves through all the constellations of the Zodiac. The first person to realise that the equinoxes and solstices moved through the Zodiac towards the west was probably Hipparchus (190–120 BC). He was also the first to define the vernal point, which, in his day, was located at the westernmost point of the Aries constellation. It is now found in the western part of the Pisces constellation, and in the 26th century it will move into the Aquarius constellation. Another consequence of the precession of the Earth's axis is that a tropical year is roughly 20 minutes shorter than the sidereal year.

Many civilisations have used lunisolar calendars established in relation to the tropical year or the sidereal year. They were used by the Sumerians, Chinese, Hindus, Hebrews, Celts, and other peoples of northern Europe. The great importance given to the numbers 12 and 13 in many cultures is probably derived from the relationship between the duration of the solar and lunar cycles.

1.17   The Influence of External Cycles in the Egyptian Civilisation

The Egyptians also started off using a lunisolar calendar of 12 months that lasted for the duration of a lunar cycle, grouped into three seasons of four months. Each lunar month began in the early hours of the first morning on which the waning crescent of the moon could no longer be seen. Given that the lunar calendar was 10 or 11 days shorter than the solar calendar, a year with a thirteenth month called Thoth, the Egyptian god associated with the Moon, arts, writing, and magic, was inserted periodically. At the start of the third millennium BC, a 365-day year began to be used for the first time in the history of civilisations, probably for administrative and tax reasons. It was known as the civil year, lasting 12 months of 30 days and five additional days or epagomenal days which coincided with the birthdays of Osiris, Isis, Horus, Seth, and Nephthys.

In Egypt, the most important external cycle, further to the apparent daily motion of the Sun, was the annual flooding of the Nile which fertilised the fields and naturally prepared them for the crop farming that was essential to people's lives. The three seasons of the year lasted 120 days each and corresponded to the three fundamental phases of that cycle. The first, called Akhet or inundation, began with the cyclical flooding of the Nile valley at the start of summer, caused by monsoon rains in the region of Ethiopia and the upper Nile, and lasted from mid-July to mid-November. The second, called Peret or the season of emergence, was the time for planting and the resurgence of plant and animal life; it lasted from mid-November to mid-March. Finally, Shemu or the season of low water was the time for harvests, from February to July.

The start of the civil year in Egypt was ingeniously tied to the start of the rising waters of the Nile's flooding cycle. Over many years, the Egyptians observed a curious synchronisation between the start of the floods in Thebes and the heliacal rising, just before sunrise, of the Sirius star (*Alpha Canis Majoris*) in the Great Dog constellation, the brightest in the sky. Sirius (a name which comes from the Ancient Greek *seirios*, for "glowing") was invisible in the sky for around seventy days and reappeared close to the summer solstice to announce the floods. For the Romans, the reappearance of Sirius announced the hottest days, in other words, the dog days.

In Egypt, the day on which Sirius rose again just before dawn was chosen as the start of the civil year. Sirius was personified by the goddess Sopdet, also called Sothis in the Greco-Roman period from 332 BC to 395 AD, and was represented in sculptures by the body of a woman with a five-pointed star on her head. Sopdet formed a triad with her husband Sah, the divine personification of the Orion constellation, and her son Soped or Sopdu, the god of the eastern border of Egypt. These three gods form a replica of the divine family of Isis, Osiris, and Horus. The association between Sirius and the constellation of Orion came from the proximity between the constellations and, as a result, the fact that the heliacal rising of Orion took place close to that of Sirius.

The period of seventy days during which Sirius could not be seen in Egypt was another external cycle to which the Egyptians assigned great symbolic value. Sirius, as we have seen, was the star of Isis and its disappearance was associated with the time when Isis hid, until her son Horus was born, exactly when Sirius reappeared. This was the resurrection that gave new and abundant life to the fields on the banks of the Nile. It was a cycle of death and resurrection that the Egyptians associated with funeral ceremonies. The mummification process ideally lasted seventy days and was used by all social classes, although with very different levels of sophistication. At the end of those days, the mummy was placed in its final resting place, whether a great pyramid or a simple tomb. Respecting the period of seventy days allowed the deceased to be reborn.

We have many signs that the Egyptians paid close attention to the night sky and the daily and annual cyclical apparent motions of the stars. From the Ninth or Tenth Dynasties, around 2100 BC, paintings in sarcophagi lead us to conclude that 36 stars or small groups of stars had been identified, located in a line along the ecliptic, which they called *bakiu* and were later known as the Decans. The remarkable property of the Decans is that their heliacal rising was separated by ten days, running through a total of 360 days, around five days less than the tropical year. The name of the Decans comes from the Greek *dekanoi*, which means "tens". There were tables to calculate divergences in the appearance of the Decans in relation to the seasons of the year due to the difference of roughly five days. The motion of the Decans in the sky served as a night clock, and the rising of each Decan coincided with the start of a decanal hour. Furthermore, the successive heliacal risings of the Decan stars after the end of the cycle in which they are not visible (due to their position in relation to the Sun) formed a calendar. By closely observing the night sky, the Egyptians of more than 4 000 years ago were able to tell not only the decanal time at night but also the time of year they were in. We would not know how to make those measurements now because clocks, watches, and today's other accurate devices for measuring physical time have made it unnecessary to do so.

Patient observations of the sky over many centuries showed the Egyptians another remarkable aspect of the celestial sphere. There are stars that do not have a cycle that includes setting and can be seen all night long and all year round. They are circumpolar stars, said to be indestructible or untiring by the Egyptians, and they symbolised eternity. Two particularly important stars were Kochab (*Beta Ursae Minoris*) in the *Ursa Minor* constellation and Mizar (*Zeta Ursae Majoris*) in the *Ursa Major* constellation. At night, they followed a circular arc around the north celestial pole with a radius of around 10 degrees. Inside that imaginary circle was paradise, the eternal place for the Egyptians. We should remember that 4 500 years ago, when the pyramids of Giza were built, the pole star of the *Ursa Minor* constellation was further away from the North Pole in the celestial sphere because of the precession of the Earth's axis, the period of which varies but is currently around 24 600 years. At the time, it was the Thuban star (*Alpha Draconis*), in the Draco constellation, that was closest to the North Pole.

The Egyptians assigned great importance to the north direction and knew that the North Pole was roughly halfway between Kochab and Mizar. The Great Pyramid of Giza, built by Pharaoh Khufu, known as Cheops by the Greeks, is aligned with the north celestial pole with a precision of around three minutes of arc, which is a truly remarkably thing for the architects of the time. Its base is an almost perfect square, in which its north face measures 230.45 m and its south face 230.25 m, with a height of 147 m. It is formed by around 2.6 million blocks of limestone or granite weighing a total of 5.9 million tonnes. The construction of this grave is one of the most surprising achievements in the history of humankind, carried out by a nation state that, at the time, had a population of only around one million. It is not so surprising that many have tried to ascribe its construction to extra-terrestrials!

The construction of the pyramids of Giza, using the engineering means and technology available at the time, displays an extraordinary stability and social cohesion, capacity for planning, clarity of objectives, ingenuity, intelligence, determination, and persistence. It also reveals that human societies are able to mobilise capacities for cooperation and coordination that are apparently limitless in times of peace. The process of building the Khufu pyramid has fascinated successive generations of Egyptologists, although it is impossible to fully decipher it (Lehner 1997; Smith 2004; Houdin 2008).

Egypt has a mild climate, free from violent or random storms like the tropical cyclones or severe extratropical cyclones that occur in Mesoamerica and China, regions where noteworthy civilisations also flourished. Nature was relatively predictable and included many cycles that needed to be understood, such was their vital importance. The astral cycles were the most fascinating. Only the indestructible stars were almost stationary and represented a contrast with the fragility and finiteness of human life.

According to the Egyptians, life continued after death. They had to preserve the body of the deceased by using mummification and providing food and everyday objects. The chaos of death was deeply feared. The pharaoh, who represented divine order in the world of the living, was "the one who had no end"; in other words, he did not plunge into the underworld, called Duat. He was identified with the indestructible stars that never disappeared into the Duat.

The step pyramid of Saqqara was the first large pyramid of its kind and was built thanks to the brilliant Imhotep, the first great architect of which there is memory in the world and who lived under the reign of Pharaoh Djoser in the 27th century BC. On the north side of the pyramid, there is a separate building in stone called a *serdab*. Two orifices were found on the sloped north face of the *serdab*. Their function was to allow Pharaoh Djoser, represented inside the *serdab* by a seated statue (in Saqqara there is a copy of the original, which is currently at the Museum of Cairo) to see the indestructible stars he would join after death with his own eyes. This constant view gave his Ka access to eternal life.

The Egyptians observed, recorded, and sought to interpret external cycles, their duration, the relationship between them, and their importance for life, but they also envisaged the absence of cycles, unlimited time, i.e., eternity. The contradiction developed through the opposing positions of an internal life cycle and the deep-seated belief in another life after death. They found an inspiring analogy in the celestial sphere by observing the stars that underwent cycles of visibility and invisibility and circumpolar stars that were always visible.

1.18   Cyclical Time in the Maya Civilisation

In Mesoamerica, several civilisations, including the Zapotec, Olmec, and Maya civilisations, developed complex systems for referring to events in time that demonstrate a deeply original concept of time. We have the most detailed knowledge about the Maya concepts. The defining feature was the construction of multiple cycles of time that were designed to situate great historical events and also the events of day-to-day life. Some were based on the cyclical apparent motions of the Sun, Moon, and Venus, while others had no correlation with external cycles and their origins are not fully understood. The main idea seems to have been that history followed the cyclical nature of time. The Maya found an intrinsic symmetry in time that made it inevitable for events to repeat themselves. Everything that happened was considered as part of a cycle of time and would grow, flower, decay, and die unless it was renewed at the end of the cycle. The repetition of a certain notable configuration of positions in the different cycles also meant that it was possible to foresee what would happen in subsequent days with the same configuration.

The main cycles are the Tzolk'in, or sacred count of days, the Haab', the calendar cycle, the rulers of the night, the lunar cycle, and the cycle of Venus. The sacred cycle is a cycle of 260 days according to which each day is described using a combination of a number between 1 and 13 and a name for the day from a set of 20 names. Each day name gave it intrinsic, determining properties to be respected, but the details of them have been lost. The same combination is only repeated after 260 days. Two independent cycles were combined to create a new cycle. It is not known precisely what the origin of the sacred cycle was, nor the possible meaning of the number 260. However, it is known that the number 13 was particularly important for the Maya and that both the Maya and the Aztecs used the vigesimal or base 20 numeral system.

The Haab' is a cycle of 365 days equivalent to the civil year in the ancient Egyptian civilisation. It is divided into 18 months of 20 days and an additional cycle of five days at the end of the year called Wayeb, comparable to the five epagomenal days of the Egyptian calendar. The Wayeb was believed to be a dangerous time, when the barriers were removed between the intermediate world of the living and the underworld, or Xibalba, the source of sickness and death. The Haab' began on the summer solstice and it is estimated that it was first used around 550 BC (Bricker 1982). There was a convergence and similarity between the solar calendars of the Egyptian and Maya civilisations, despite the distance between them in space and time. However, the Maya's sensitivity to the passing of time and the search to decipher the future led them to build and imagine and interpret other more complex cycles.

The Tzolk'in and Haab' cycles run independently of one another. A certain combination of dates in the two cycles will be repeated after a cycle with a number of days equal to the least common multiple of 260 and 365, or 18 980 days, equivalent to 52 years. This new cycle, called the Calendar, was extremely important to the Maya and later also to the Aztecs. The end of the Calendar was considered to be a time of renewal in which houses, buildings, and sculptures would be destroyed and rebuilt. There were other notable cycles.

Nights were organised into a cycle of nine in which each night was governed by one of nine gods of the underworld. We know the name that the Aztecs gave them, but in the Maya civilisation only the glyphs used to represent them are known. The Maya also observed the motions of the Moon and recorded the lunar cycles using a series of glyphs on stelae and slabs and in cave paintings.

We also know that they systematically observed the motions of some planets, especially Venus, which they noticed alternated between being the morning and the evening star. The Dresden Codex is one of only four codices to have been preserved from the late Classic period to the end of the Postclassic period (14th to 16th centuries). It contains detailed calculations regarding the Venus cycle and solar eclipses. There was much more information about the Maya's astronomical knowledge in other codices but they were systematically burned by the Spanish clergy after their arrival in Yucatan in the 16th century because they were believed to be damaging to their campaigns of religious conversion.

The heliacal rising of Venus was believed to be an extremely important event but a bad omen, especially prone to the outbreak of war. They calculated that the cycle lasted 584 days, which is a very close approximation to the 583.92 days of Venus' synodic period. The great Venus cycle is the result of synchronisation of Venus' synodic period, the Tzolk'in cycle, and the Haab' cycle, and corresponds to 37 960 days, i.e., 104 Haab' or 2 Calendar cycles. At the Xultun ruins in Guatemala, wall paintings have been discovered that were made in the 9th century, in the late Classic period, long before the codices. They include hieroglyphs about the Maya calendars and tables of the motions of the Moon and possibly Mars and Venus (Saturno et al. 2012).

In spite of their focus on the cyclical nature of time, the Maya also recognised that it had a linear dimension, which they recorded using a long count, a calendar which probably originated before the Maya civilisation. The long count had the great practical advantage that it could be used to relate events that were very far apart in time, further to the periods of the different calendars. Its fundamental unit was a year of 360 days, formed of 18 sets of 20 days called Tuns, a word derived from "rock", possibly due to the practice of piling stones to signal the passage of the Tuns. Each day was named K'in, which also means "Sun", and each set of 20 days was a Winal. A set of 20 Tuns made up one K'atun, which literally means "twenty stones", and 20 K'atuns were one Baktun, i.e., 400 stones. This system of counting time was constructed using successive powers of twenty—400, 8 000, 160 000, 3 200 000, etc.—and could be projected into the past and future, quickly reaching spans of time greater than what we believe to be the age of the Universe and its foreseeable future duration.

The start of the long count, i.e., year zero for the Maya, is 13 August 3114 BC, a date which is long before the pre-Classic period of Mesoamerican civilisations on which, in accordance with the Maya's creation myth, the current world inhabited by humans appeared. Since then, 13 Baktuns have passed, and the start of the 14th took place on 21 December 2012, a date which would probably have been celebrated by the Maya for leading to the start of a new cycle. This date was erroneously linked to a non-existent eschatological Maya prophecy.

In the last two centuries, there have been great advances in knowledge about the Maya civilisation. We know that there is still a great deal yet to be deciphered, although this is largely impossible because of the destruction of the codices and the pillaging of temples over the centuries. The civilisation was organised into highly aggressive rival city states, and none could effectively dominate the others. They were violent theocracies in which human sacrifices were commonly practiced as a way of offering nourishment to the gods. The Maya believed that without this nourishment the Sun would cease to rise and the world would come to an end. With their sacrifice the victims earned honour in the afterlife. One of the most prized offerings was the decapitation of a captive king (Sharer and Traxler 2005). Heart extraction became a more frequent form of human sacrifice in the Postclassic Maya period, probably influenced by the Aztecs who carried human sacrifices to an extreme.

Excavations that started in 2015 revealed for the first time the emplacement of the *tzompantlis* or skull racks, referred to by the Spanish conquistadors, on a street behind the cathedral in Mexico City, where the temple complex of Tenochtitlan was located and destroyed in 1521 (Wade 2018). The human sacrifices were performed on top of the pyramid-temple in front of the two small temples it had on top, one dedicated to Huitzilopochtli, the war god, and the other to Tlaloc, the rain god. The victims, believed to be mainly prisoners from wars that later lived for some time in Tenochtitlan with Aztec families, were sacrificed by priests on a stone slab. The abdomen and diaphragm were sliced open with a sharp obsidian blade and one of the priests would grab and extract the hearth which was considered to be the seat of the human soul where god was present. Divinity was associated with the physical hearth and in particular the Sun was imagined as a hearth, round, hot, and pulsating, that went about delivering hearth-souls. After death, the corpses were decapitated and the skulls placed on the *tzompantli*. When they decayed the priests fashioned the skulls into masks or placed them in two towers of skulls built with mortar that flanked the *tzompantli*. The large scale sacrificial deaths in the Aztec empire have been interpreted as a functional form of adaptation that helped control the growth of the populations in the surrounding territories (Wade 2018).

It is very likely that most of the sacrificial victims were willing to die because it was considered to be a glorious end. We know that from the written reports of the Spanish, in particular from the Dominican friar Diego Durán (1537–1588) who wrote one of the first books on the Aztecs and was fluent in Nahuatl, the Aztec language (Durán 1581). When the Spanish criticized the human sacrifices, the Aztecs made sarcastic remarks and ridiculed the weakness of the Christians. If they freed the victims to be sacrificed the offer was "indignantly rejected and they demanded to be sacrificed" (Diaz del Castillo 1632). Some did not want to die by the obsidian blade and they wept and faltered when they performed the long rituals before the sacrifice (Kerkhove 2006). However, they were a minority and their behaviour was viewed as a bad omen and an insult to the gods. The Aztecs probably had such a strong belief in the afterlife that sacrificial death was an opportunity for a particularly glorious afterlife. According to Diego Durán, the victims sang and danced as they marched to their deaths "with great joy and gladness" (Kerkhove 2006; Durán 1581).

The Mayas deified blood. It was a precious liquid, awesome and valuable. Thus, the kings and queens, since they descended from gods, had the duty of providing their blood at important ritual occasions in the presence of noblemen and the people. Spilling blood guaranteed continued favour with the gods.

Maya art and architecture show a deep concern with life, death, and Xibalba, the place of fear. Most pyramid-temples of the Classic period were built to be used as tombs for kings, as in the Egyptian civilisation.

But the world of Mesoamerica, with its tropical forests and relatively frequent natural disasters, volcanic eruptions, tropical cyclones, torrential rains, and droughts, was a more complex, mysterious, and violent environment than Egypt. The Maya believed that the world had a precarious existence that depended on periodic pleas, offerings, and human sacrifices. Probably the relatively more peaceful environment in Egypt made the Egyptian gods less demanding when it came to blood and human lives.

The Maya believed that they were ruled by sacred powers that controlled the structure of space and time, rain, wind, storms, earth, water, the sky, and all the other forces of nature. Through ritual acts, the king sought to convey and control the immense supernatural powers in order to guarantee the life and work of mortals. The cyclical nature of time was a reflection of the cycles they observed in life and in the terrestrial and cosmic environment. Using the cycles of time enabled them to peer into the future and decide which occasions were ideal for different types of actions, like deciding the good moment or *Kairos* in the Greek civilisation. The creation myths themselves had a cyclical nature because the current world had been preceded by others created by gods of different types but eventually destroyed. The sky and the cyclical motion of the planets were certainly a source of great fascination, a mystery that needed to be observed attentively over a long time in order to be deciphered and properly interpreted. Linking these cycles to the cycles of life and nature and building the relevant calendars was a way of establishing harmony between celestial events and sacred rituals so that they could influence the future.

1.19 The Measurement of Physical Time

In the ancient civilisations, the systematic observation of external cycles related to the apparent motion of celestial objects, especially focusing on the Sun and the Moon, led to the concept of physical time and later to the science and technology of chronometry. As an abstract form of organising and counting days, calendars differ from clocks, which are physical instruments that can measure intervals of time that last less than a day.

We have seen how the different civilisations developed several types of calendars all based around the cycles of the seasons and the apparent motions of the Sun and the Moon. Another common characteristic is that they were used essentially for religious purposes, although they were also used for civil ends. A religious origin can also be found in the Gregorian calendar currently used around the world. It was introduced by Pope Gregory XIII in the bull *Inter gravissimas* on 24 February 1582 to replace the Julian calendar that had been implemented by Julius Caesar in 46 BC. In the Julian calendar, the year is split into 12 months and has an average length of 365.25 days to bring it closer to the length of the tropical year. However, this amount is not exact and the difference causes a delay of around three days in four centuries of the Julian calendar in relation to the equinoxes and the seasons. Easter, instead of falling close to the spring equinox on 21 March, as decided at the first Council of Nicaea in 325 AD, shifted in relation to the equinox. In 1582, the spring equinox took place on 11 March in the Julian calendar instead of on 21 March. The Gregorian calendar corrected the delay and altered the length of the average year to 365.2425 days in order to bring it closer to the tropical year. However, this approximation still contains a very small error of around one day every 3300 years in relation to the average tropical year. Currently, the vast majority of countries use the Gregorian calendar although some, including India, Bangladesh, Israel, and Myanmar, also use different, religious calendars.

As for clocks, it is likely that the first were the obelisks built in Egypt since around 3500 BC. The movement of the shadow of the obelisk on the ground made it possible to distinguish between the morning and the afternoon and also served to identify the solstices and equinoxes. The longest and the shortest days of the year corresponded to the shortest and longest length of the obelisk's shadow at midday, respectively.

Physical time is measured by clocks, using a time pattern, i.e., specifying time intervals of equal length and reference points in time. In fact, we immediately find two possible definitions for those time intervals: the solar day and the sidereal day. The alternation between day and night is the result of the Earth's rotation, but to establish the rotation period there needs to be a reference point in space. It should be noted that by using a cycle to measure physical time, we have to be sure that those cycles correspond to equal lengths of time. For the Earth, this condition requires its angular velocity to be constant. But is it really constant?

The physical day is defined as the time interval between two successive upper transits of a certain reference point in the sky. Upper transit occurs when the point of reference crosses the celestial meridian of the place where we are standing on the surface of the Earth, in an east–west direction. The celestial meridian is an imaginary circle drawn through the north point of the horizon at that place, the zenith, and the south point of the horizon.

The most obvious and natural choice was to use the Sun as a celestial reference point because of the crucial importance of its apparent daily motion to human life. With this choice, we define the solar day. But we can also choose a star, such as Sirius, instead of the Sun. In that case, we have the definition of the sidereal day. Thinking a little more about this, we reach the conclusion that the solar day and the sidereal day are not the same length. This happens because the Earth, due to its rotation around the Sun, has to turn slightly more each day for the Sun to cross the celestial meridian. In other words, the star crosses the celestial meridian roughly 4 minutes before the Sun. It is very easy to observe this phenomenon. According to our clocks, stars rise and set on the horizon around 4 minutes earlier every day. After fifteen days, the difference is an hour, which can be observed easily.

Greenwich Mean Time (GMT), used throughout the world, is the mean solar time at 0° longitude, conventionally located on the meridian that runs through the Royal Observatory, Greenwich, in London. It is based on the concept of the mean solar day, linked by an equation to the solar day in any place in the world, and is subdivided into 24 hours, 1 440 minutes, or 86 400 seconds. The surface of the Earth is divided into 24 time zones, within which the same time is used on clocks. The time difference between adjacent zones is normally an hour, although in some cases there are differences of 30 or 45 minutes.

In 1935, the International Astronomical Union recommended using Universal Time (UT), which has several versions (UT0, UT1, UT2), and, in 1967, Coordinated Universal Time (UTC) was adopted (Guinot 2011). These standards for measuring physical time are based on the period of the Earth's rotation relative to distant celestial bodies, such as some remote but very bright stars and quasars.

The first atomic clocks were built in the 1940s, based on the hyperfine transitions of hydrogen (^{1}H), caesium (^{133}Cs) and rubidium (^{87}Rb). These are transitions between quantum states of atomic electrons, in which electromagnetic radiation is emitted at a particular frequency and therefore corresponds to a period that we take to be constant. This assertion can be checked using experiments based on current knowledge of physics. After the first precision atomic clock was built in 1955, using ^{133}Cs (Essen and Parry 1955), atomic clocks were used more and more, and this led to the introduction of International Atomic Time (TAI) in 1971. This is standard coordinated time set by roughly 400 atomic clocks located in more than 50 national laboratories of different countries. These clocks are synchronised to an accuracy of about 10^{-9} per day. TAI is used to define the second in the International System of Units (SI) used in modern physics, based on the electromagnetic radiation frequency of the transition between two specific states of ^{133}Cs.

The SI second was therefore defined as 9 192 631 770 periods corresponding to that frequency. This multiple was chosen so that it would be as close as possible to a fraction 1/86 400 of the mean solar day. However, this unit of physical time is slightly variable in its duration because it is known that the duration of the mean solar day is not constant. The Earth's rotation is constantly slowing down because of the resistance to tidal movements caused by differential gravitational forces that the Moon and the Sun exert on the Earth. The mean solar day is getting longer, although this variation is very small. At the end of the 20th century, its duration in SI seconds was 86 400.0013 s. The accumulated difference corresponds to 1 s roughly every 800 days. To ensure that the TAI day, which is much more precise, does not diverge from the mean solar day, it was decided to add leap seconds periodically to TAI. In 1972, before leap seconds were added, UTC was 10 SI seconds behind TAI. Since 1972, 27 SI seconds have been added to UTC, the last of these on 31 December 2016 at 23:59:60 UTC.

This is a sign of more recent efforts made by humanity to bring its calendars and physical time measurement standards closer, with ever increasing accuracy, to the fundamental cycles of the solar day and the seasons of the year. One of the first notable examples of advances in this direction was made around 5 000 years ago by the Ancient Egyptians, as we have already seen, when they introduced epagomenal days in addition to the 12 months of 30 days so that the year would last 365 days.

Our ability to measure shorter and shorter time intervals has increased extraordinarily. It is now possible to issue pulses of electromagnetic radiation lasting only 100 attoseconds (1 attosecond = 10^{-18} s) and stabilise the position of the maximum oscillation of the related electric field to an accuracy of 12 attoseconds (Koke et al. 2010). This time interval is roughly half the atomic unit of time, equal to 24.188 843 265 09 attoseconds, which is the fastest timescale for physical processes in the outer electron shells of an atom. The shortest interval of time to have been measured is 0.85 attoseconds in the observation of the photoelectric effect (Ossiander et al. 2018).

As very short time intervals can now be measured, the duration of a day can also be determined very precisely to see whether it is constant or not. We must remember that Newton accepted the Earth's rotational motion as uniform, but believed it to be a statement that needed to be proved. The current answer to that question demonstrates the great complexity of phenomena in the Earth system that influence the duration of a day. According to the laws of physics, the vector sum of all torques acting on a body is equal to the time rate of change of the angular momentum of that body. If no external torque acts on a system of interacting objects, then their total angular momentum is constant. In the case of the Earth, its angular velocity is decreasing due to the slowing down of rotation caused by the tides, as already mentioned. Analysis of fossilised sedimentary deposits marked by tidal movements in estuaries called "tidal rhythmites" shows that, around 620 million years ago, a day lasted only 21.9 hours (Williams 2000).

Given that the angular momentum of the Earth–Moon system is roughly constant, the decrease in the Earth's angular momentum is made up for by the increase in the angular momentum of the Moon's rotation around Earth, which leads to an increase in the distance of the Moon from Earth of around 3.8 cm per year. Moreover, there are factors that cause changes in the moments of inertia of the Earth on different timescales, and these also produce variations in the length of a day. One example is the glacial isostatic rebound of continental regions at high latitudes, for example, the very slow rise of Scandinavia due to the melting of ice that covered it during the last glacial period. Combined with the tidal effect, this adjustment of the Earth's crust has produced an increase in the length of a day of roughly 1.7 ms per century over the last 2 700 years (Stephenson et al. 1995).

On a much shorter timescale, the periodic accumulation of snow in winter in the Northern Hemisphere and some earthquakes and tsunamis also cause small changes in the length of a day. Finally, there is the influence of the atmosphere on the rotation period of the solid part of the Earth system. The angular momentum of the atmosphere varies due to the general circulation of the atmosphere and, in particular, due to certain phenomena that alter the intensity and dominant direction of zonal winds. This variation is compensated by a variation in the angular momentum of the solid part of the Earth, so the angular momentum of the Earth system remains constant. Measurements made between 1981 and 1983 using beams of laser radiation reflected by the Moon and satellites, together with observations based on extragalactic radio sources, have led to the conclusion that the general circulation of the atmosphere causes variations in the length of the day (Morgan et al. 1985). In an El Niño event, the angular momentum of the atmosphere slightly increases due to the change in the general circulation of the atmosphere in the equatorial region of the Pacific. This causes persistent westerly winds instead of trade winds, which have a significant easterly component. This change involves a slowdown in the rotation of the Earth, leading to an increase in the length of a day by 0.8 ms when El Niño is at its most intense (Pavlis et al. 2017). In the El Niño–Southern Oscillation (ENSO), the angular momentum of the body of water of the ocean also varies, and as for the atmosphere, this influences the Earth's rotation vector (i.e., the vector in the direction of the Earth's rotational axis and with magnitude equal to the angular velocity), and hence also the length of the day. In conclusion, most variations in the length of a day over timescales of days, weeks, and a few years are rooted in phenomena connected to the atmosphere, hydrosphere, and cryosphere.

Newton's question can now be answered in the negative. The Earth's rotation is not uniform. There is an enormous range of anomalies in this motion, and their explanation requires detailed knowledge of the Earth system and increasingly precise ways of measuring physical time over ever shorter timescales.

1.20 Physical Time in the Theory of Relativity

Einstein's theory of relativity (Einstein 1905a, b) transformed our understanding of physical time and irreversibly moved it away from psychological time. Between the absolute physical time of Newton's *Principia* and relativistic physical time there is a deep change in paradigm in the fundamental physical concepts of space and time. The theory of relativity focuses on recognising that throughout the Universe the speed of light, or more precisely, the speed of all forms of electromagnetic radiation, is invariable. When this speed is measured, the value obtained is always the same, regardless of the observer's state of motion and the speed of the light source. Einstein came to this conclusion by applying a principle of "democracy" or equivalence that is often used in physics. In the case of special relativity, it consists in postulating that observers in inertial reference frames, i.e., non-accelerating frames, arrive at the same laws of physics. In other words, the laws of physics are the same for two inertial observers in uniform relative motion. Specifically, this means that Maxwell's equations, which describe electromagnetic phenomena, are invariable, and this is only possible if the speed of light is also invariable.

This statement is counterintuitive, since it means that two observers that move in relation to each other in the direction of a light ray record the same value for the speed of that light. In relativistic physics, the speed of light becomes a "higher" concept, that is, more important and determining than the concepts of space and time. The latter transform and become flexible to guarantee the invariance of the fundamental laws of physics. In relativity, space and time are inextricably merged in a four-dimensional spacetime. An instantaneous physical situation or occurrence that happens at a specific place and time is an event represented by a point in space-time, in other words, by a set of four numbers, three of which represent the x, y, and z coordinates that define the position in three-dimensional space and a fourth, temporal coordinate t. The entire special theory of relativity essentially flows from the invariance of the spacetime metric

$$ds^2 = -c^2 dt^2 + dx^2 + dy^2 + dz^2 , \qquad (1.3)$$

in other words, the differential ds of the generalized distance s in four-dimensional space is the same in any inertial reference frame. In the equation, dt, dx, dy, and dz are differentials of the four coordinates and c is the speed of light. We can replace those differentials with finite intervals $\Delta t, \Delta x, \Delta y, \Delta z$, corresponding to the differences in coordinates between two events. While it lasts, each event describes a path in spacetime called a world line. An object describes a set of world lines.

With the loss of the absolute character of space and time, the absolute character of the simultaneity of two events is also lost. Consider two events, A and B in spacetime, and suppose the interval between them is space-like, i.e., their spatial separation in some inertial frame is greater than the distance through which light travels in the time interval separating the two events. This means that Δs^2 in (1.3) is positive. In this situation, there are moving observers that see A and B as simultaneous, while others say A came before B and yet others that B came before A. Despite this surprising result, which goes against our intuition of time flowing in one direction, the principle of causality is upheld. A cannot be the cause of B, nor B the cause of A, because it is not possible to connect A and B by way of a light signal. Since no information in the physical universe is propagated at a speed greater than the speed of light, there is no way to establish a physical cause–effect relationship between A and B.

If, on the other hand, the spacetime interval between A and B is time-like, i.e., their spatial separation in some inertial frame is less than the distance travelled by light in the time interval between the two events, then Δs^2 is negative in (1.3) and a moving observer can be found for whom A and B occur at the same point in space. However, if A precedes B for one observer then A precedes B for all observers in the Universe. A may be the cause of B because, in this case, A and B can be connected by a light signal. Thus the fact that in the theory of relativity simultaneity of events is not absolute but depends on the observer does not call into question the principle of causality. There is no absolute physical present, since the simultaneity of events depends on the observer. There is also no flow of relativistic physical time because, as we have seen, the ordering of some events in time depends on the observer. This absence of a flow of physical time means that future events for some observers are present events or past events for others.

Unlike the already mentioned growing block universe theory, in which the flow of time generates an expansion of the block of the past and of the present, in the theory of relativity, past, present and future are amalgamated in the four-dimensional block of spacetime that constitutes the framework for the evolution of the Universe. The lack of a flow of relativistic physical time, as well as being compatible with causality, does not hinder the direction of time, i.e., the asymmetry between the past and the future that is demonstrated in most physical processes at macroscopic level. The omnipresence of this asymmetry in phenomena and processes that we directly perceive with our senses produces the relentless human awareness of an irreversible direction for time, that there is an "arrow of time", a concept developed in 1927 by British astronomer and physicist Arthur Eddington (1882–1944).

One of the clearest demonstrations of the arrow of time for humans is ageing. Nonetheless, the laws of physics, in the field of classical mechanics and in quantum mechanics, at the microscopic level of atomic and particle physics, exhibit time reversal symmetry, except for a small asymmetry in the decay of some elementary particles through the weak nuclear force. This means that, when run backwards in time, the films of most physical phenomena are likely to represent another possible physical situation. The arrow of time is a property of the behaviour of macroscopic systems, i.e., those that contain a large number of molecules or atoms, and stems from the second law of thermodynamics, which says that, in an isolated physical system, entropy tends to increase with time.

The absence of a flow of relativistic physical time and the invariance of "before" and "after" relations between events related by causal relationships has similarities with McTaggart's B-theory. However, these are only apparent similarities between profoundly different concepts. Time is not something beyond, or independent of, the way we approach it. Time depends on a purpose, it depends on what we are looking for: if we are interested in measuring time very precisely and using it to build physics, then we are dealing with physical time and with relativistic physical time in particular. When using the concept of time operationally in interaction and communication within the society around us, we are in the realm of operative social time. If we are analysing and testing the ability and limitations of our perception of physical time, we are dealing with psychological time, which in turn underpins the construction and intercommunication of operative social time. If our goal is to reflect on time entirely freely and abstractly, without conditions or limitations on uses or specific functions of the concept of time, but paying attention to perspectives and issues continually created by those uses or functions and by the dynamics and content of the reflections themselves, we are in the realm of the philosophical analysis of time, or, if we prefer, philosophical time. McTaggart's reflections concern the latter concept of time.

Physical time reveals even more surprises. According to the special theory of relativity, observer A, sitting at a fixed point in some inertial reference frame, observes the watch of an observer B moving in that frame to be running more slowly relative to the time shown on his own watch. This time, measured by the watch travelling with him, is called the proper time of that observer. Something similar happens with space. Observer A, when measuring the length of a one-metre bar in B's frame of reference aligned with the direction of the relative motion of that frame of reference, finds a value of less than a metre. This contraction, known as the Lorentz contraction, increases as the relative velocity increases.

In terms of time, A says that B's watch is going more slowly than his own, i.e., it is undergoing time dilation. This dilation increases as the relative velocity between A and B approaches the speed of light. At its most extreme, when the relative velocity approaches the speed of light, B's time, as observed by A, stops flowing altogether. At first glance, this situation seems as though it might satisfy our deep desire for conditional immortality.

We must remember, however, that B observes the same phenomenon in A. The time dilation seen by each observer is reciprocal. At first glance there appears to be an irreconcilable contradiction, but that is not the case. Let us imagine that observers A and B are the same age and that at the beginning they are in the same inertial reference frame and same point in spacetime. Let us also suppose that B begins a journey on a spacecraft, where he reaches speeds approaching the speed of light, which takes him away from A. In the end, he decides to come back and meet with A, who has remained in the inertial reference frame in the meantime. When A and B meet again at the same point in spacetime, B sees that A is much older than he is and that the age differences depend on the speeds at which he travelled in relation to A. The symmetry between A and B has been broken due to the three stages of B's journey when movement was accelerated, i.e., following departure, when accelerating to reach cruising speed, when changing the direction of motion to be able to return, and when slowing down to meet with *A* again.

This surprising result has not been tested with people because we have not yet managed to build spacecraft that travel at speeds close enough to the speed of light for the difference in age between those who travel and those who stay on Earth to become noticeable. When we have spacecraft that reach such speeds we will be able to go to some of the exoplanets orbiting 3152 stars (Schneider 2020) that are already known in May 2020 to have planetary systems with one or more planets and return to Earth alive. When returning, we would find that time on Earth had passed much more quickly and that it would likely be difficult or even impossible to understand the transformations in the people, societies, countries, and civilisations that we left behind at the beginning. If the speeds of our spacecraft were much higher, close to the speed of light, most of our friends and family would already have died. The news is not good for immortality. Our biological time is also subject to the laws of physics, so the observer that travelled at speeds approaching the speed of light to an exoplanet and returned to Earth to find that his family and friends are no longer alive has the same life expectancy as he would have had if he had stayed on Earth.

Although we are still far from these remarkable experiences, it has already been possible to confirm this prediction of relativity in the Hafele–Keating experiment (Hafele and Keating 1972) in which four atomic clocks were transported around the world on commercial jet flights, once travelling east and once west. When they returned, the time shown on the clocks was compared with the proper time on atomic clocks that remained stationary at the US Naval Observatory. The clocks that travelled east lost 40 ± 23 nanoseconds (10^{-9} s) and those that travelled west gained $275 \pm 21 \times 10^{-9}$ s, which is close to what is expected from relativity. Later, the Hafele–Keating experiment was repeated with greater accuracy, confirming the predictions of relativity (NPL 2005).

To interpret these results, we need to turn to the general theory of relativity because the reference frame at the Earth's surface is not an inertial reference frame, for two reasons. The first is that a fixed frame on the Earth's surface is an accelerated frame because of the Earth's rotation. The other is that, on the Earth's surface, we are in the gravitational field generated by the Earth.

Let us briefly look at how times are calculated in the Hafele–Keating experiment. The general theory of relativity developed by Einstein in 1915 (Einstein 1915, 1916), is today's theory of gravity, replacing Newton's law of gravity, and generalising the special theory of relativity. Together with quantum mechanics and the Standard Model of fundamental interactions— electromagnetic, weak nuclear, and strong nuclear—it establishes the most complete and most successful model we have for describing the physical forces and fields in the Universe. We do not yet know, however, how to combine and perhaps unify general relativity with quantum mechanics.

In the general theory of relativity, gravity, rather than being a force field in space, becomes a property of spacetime in neighbouring regions of matter or energy, due to the equivalence between the two. In other words, the presence of matter or energy curves spacetime. In the presence of large masses such as stars, black holes, or galaxies, spacetime becomes curved and rays of light follow the geodesics of that spacetime, which are curved trajectories instead of straight lines. Equation (1.3) is no longer valid, being replaced by another with a spacetime-dependent metric tensor $g_{\mu\nu}$, which defines the geometric and causal structure of spacetime. Einstein's gravitational field equations (Einstein 1915, 1916) relate the metric tensor with the energy–momentum tensor, which describes the distribution of mass and energy in spacetime.

Just as a time dilation is indicated by a clock in one inertial reference frame that moves in relation to a clock situated in another inertial reference frame, there is also time dilation for a clock located in a region of spacetime deformed by the presence of mass or energy when compared with a clock situated in an inertial frame far away from the influence of that amount of mass or energy. To give a concrete example, a person who lives on the ground floor of an apartment building ages more slowly than someone who lives on a higher floor of the same building because the latter is subject to less intense effects of the Earth's gravity on average over his or her lifetime. The age difference is very small and in total reaches only values of the order of microseconds (10^{-6} s) over the average human lifetime.

The time dilation at a specific point located within the gravitational field of a large mass is the same as the time dilation caused by the escape velocity at that point, in other words, the velocity needed to escape that gravitational field. At the Earth's surface, the escape velocity is approximately 11.2 km s^{-1}. As a result, if we ignore the Earth's rotation to simplify the argument, the time dilation recorded by the watch of an observer located on the Earth's surface and observed by a distant observer located in an inertial reference frame that is static in relation to the first observer, is the time dilation corresponding to the speed 11.2 km s^{-1}.

When someone falls into the gravitational field of a black hole, an observer outside and very far away from the action will observe that person's watch to slow down before stopping when it reaches the event horizon, i.e., an imaginary boundary that surrounds the black hole, where the escape velocity is the same as the speed of light. As this is a speed that cannot be surpassed, nothing can escape the black hole beyond the event horizon, even light. As for what happens physically inside a black hole, beyond the event horizon, we do not yet know the laws of physics needed to find out. This does not stop us from exercising our freedom of thought and reflecting on what time is like inside a black hole, but such situations are in the realm of philosophical time and not physical time. This is similar to the situation regarding time before the Big Bang.

1.21 What Is the Relevance of Relativistic Physical Time in Our Day-to-Day Lives?

Physical time dilation in the theory of relativity may seem to be irrelevant to our lives and our day-to-day existence in today's civilisation because its effects are only significant when speeds are very high and close to the speed of light. However, this is not actually the case, because the technology on which we depend today requires the use of clocks that are able to measure very small time intervals with great accuracy and the effects of the theory of relativity can be plainly seen. Our dependence is well exemplified by the Global Navigation Satellite Systems (GNSS). These include the NAVSTAR Global Positioning System, or simply GPS, which was the first, developed by the USA since 1973 to provide its armed forces with a satellite navigation system, Russia's GLONASS, the EU's Galileo, and China's BeiDou-2, which will go global in 2020. GPS currently comprises a constellation of 33 satellites in orbit around the Earth at an altitude of 20 180 km, with an orbital velocity of 14 000 km/h and an orbital period of 12 hours.

Every satellite has an atomic clock on board, synchronised with the clocks on other satellites in the constellation and atomic clocks on the ground, with a nominal precision of 10^{-9} s. Each satellite broadcasts its time measurement and other relevant data by broadcasting electromagnetic signals. The constellation of GPS satellites is organised so that, for each point on or close to the surface of the Earth, at least four satellites are "visible". This visibility of four synchronised signals allows a GPS receiver to know not only the GPS time but also its position in space with an accuracy that can be as high as a centimetre for the most advanced military applications.

When we travel by plane, our position is continuously determined by detecting and processing GPS signals. Any failure in this aircraft positioning system may be fatal. However, the applications of GNSS go much further than air transport and have become essential in practically all fields of human activity. GNSS are indispensable in the military domain and are a crucial element in the search for military dominance. If GNSS were to become inoperable for a few hours, it would quickly cause chaos in human activities. Mobile phones, most of the internet, air, maritime, and most terrestrial traffic control, the financial markets, and most weather forecasting systems would come to a standstill.

In the civil sphere, one of the most important functions of GNSS is to provide a standard time that can be accessed using a detector of its electromagnetic signals. GPS provides GPS Time (GPST) continuously, with a nominal precision of 14 nanoseconds. GPST is synchronised with TAI but maintains a 19 second difference from it, and the information contained in signals sent by GPS satellites includes the difference between GPST and UTC. Access to GPST has become indispensable to the running of the different infrastructures and services that support most human activities, such as telecommunications, electricity grids, financial transactions, commerce, transport services, several activities related to the use and exploration of outer space, and scientific research, especially in physics, geophysics, and astrophysics.

Let us now see how the theory of relativity comes into the functioning of GNSS. According to special relativity, a fixed observer on the Earth's surface with a GPS detector will find that the atomic clocks on GPS satellites run slow, losing 7×10^{-6} s per day in relation to his proper time recorded by an atomic clock (Ashby 2002) due to the relative velocity of the satellite. On the other hand, the satellite clocks are further from the Earth's center, at a location where the curvature of spacetime caused by the presence of the Earth's mass is lower than at the surface, so they run faster, gaining 45×10^{-6} s per day in relation to the observer's clock at the surface (Ashby

2002). In conclusion, GPS satellite clocks gain on average roughly 38×10^{-6} s per day, i.e., 38 000 nanoseconds, which is incompatible with the accuracy required by a GPS positioning system. If this relativistic divergence is not corrected, it can cause positioning errors on the order of 10 km per day. This problem is resolved by putting the GPS atomic clocks back to make up for the gravitational time dilation and including a microcomputer in GPS receivers that, based on the data received from the satellite, compensates for other relativistic effects.

Although nearly everyone in the world directly or indirectly uses GPS or other GNSS, very few realise that its functioning relies on the theory of relativity. This may be thought of as just a "fun fact", something curious but entirely irrelevant. Such an attitude is part of a widespread trend that devalues the complexity and fragility of contemporary civilisation. It is just one example, among many others, of how the background and contingencies inherent to providing services and supplying products that are essential to today's lifestyle are dismissed and ignored. More and more products and services are being created by technological advances. Nonetheless, most consumers are less and less interested in understanding the underlying science and technology and the implications that their production and consumption have for human life, particularly health, and for natural resources.

Even more seriously, some do not know or understand, and may even deny that the science they enjoy and upon which they depend is a coherent structure built using the scientific method, the conclusions and predictions of which are universal and potentially falsifiable (Popper 1959). It thus makes no sense to accept only those that are useful to us and deny those that are inconvenient. This emotive and selfish differentiation is equivalent to denying the scientific method and the entire scientific body of knowledge built on observation and experimentation. The consistency of the system that made it possible to provide the goods and services that support our well-being and prosperity is beginning to be at risk and this makes it less safe and potentially dangerous.

1.22 Physical Time in Quantum Mechanics

There is much more to learn about physical time beyond the theory of relativity, where time plays a role similar to the position variables, since the four form the coordinates of spacetime. In non-relativistic quantum mechanics, the dynamic variables are represented by operators, but time is not. Time therefore does not have eigenvalues, in contrast to dynamic variables such as position, linear momentum, and energy, and is instead simply a parameter with a continuous domain (Hilgevoord and Atkinson 2011).

In relativistic quantum mechanics, it is possible to define an operator corresponding to time but the problem remains of knowing whether time can be divided indefinitely or whether there is a lower limit for the duration of a time interval. Combining Newton's gravitational constant G, Planck's constant of quantum mechanics h, and the speed of light c, we can construct a quantity t_P with the dimensions of time, called the Planck time, through the relation

$$t_P = \left(\frac{hG}{2\pi c^5}\right)^{1/2} = 5.4 \times 10^{-44} \text{ s}.$$

This very short timescale is likely to be where quantum gravitational effects become important. However, we cannot yet study physical events occurring on time scales of the order of the Planck time since the time intervals we are currently able to measure are many orders of magnitude greater (Ossiander et al. 2018).

To solve these issues, we would need to develop a quantum theory of gravitation: we would need to be able to quantify the field of the relativistic theory of gravitation which, as already mentioned, we have not yet managed to do. This new theory will probably involve a granular structure for spacetime in which the granules would play a quantum role in spacetime. There are many attempts to build a quantum theory of gravity that is falsifiable, i.e., can be cross-checked with experimental or observational tests. One of the most promising attempts is string theory. Nevertheless, we are still far from solving these fundamental problems.

There are yet other very important aspects of quantum mechanics in which time plays a decisive role and that relate to the process of measuring dynamic variables, in other words, the interaction involved between the macroscopic scale and the microscopic scale in the act of measuring.

Let us imagine that a high energy photon goes through a process called pair production, in which it creates an electron and a positron that fly apart. In accordance with quantum mechanics, the two particles thereby created are said to be in an entangled state, that is, a state which is the superposition of several possible states. More generally we find that that a pair of particles (or a group of particles) are in an entangled state when they are generated or interact in ways such that the quantum state of each particle cannot be described independently of the quantum state of the other (others), even when the particles are separated by a large distance. The entanglement of states is a fundamental feature of quantum mechanics with extraordinary properties and consequences.

Suppose that the pair-production experiment is prepared in such a way that the two particles are in a state with spin zero called a singlet state. We do not know the spin state of each particle but if we measure the spin state of one of the particles we immediately know the spin state of the other, simply by applying the conservation of angular momentum.

What is surprising, and was unacceptable for Einstein, is that quantum mechanics only gives us the probability distribution of the possible results obtained from measuring a dynamical variable when repeating the same measurement, always prepared in the same way, many times. It does not predict the result that is obtained in any single measurement. It may seem amazing that we do not know that. To explain this observational result, we have only two hypotheses: either after the pair production each of the two particles has a well-defined spin state (but quantum mechanics as we know it today is unable to allow us to foresee what value will be obtained in the measurement) or the measurement of the spin state of one of the particles reveals a form of remote communication with the other particle that fixes its spin state, thereby ensuring the conservation of spin in the system's final state. The second hypothesis, known as the non-locality hypothesis, is profoundly different from the first because it corresponds to an essential, unbreakable uncertainty rather than the lack of knowledge of a physical mechanism that may or may not exist.

John Stewart Bell (1928–1990) showed that the two hypotheses could be distinguished by experiments (Bell 1964) and, in 2015, three independent experiments were carried out that unequivocally demonstrated the non-locality of quantum mechanics (Hensen 2015; Shalm 2015; Giustina 2015). The physicists who coordinated these three remarkable experiments, Ronald Hanson, Sae-Woo Nam, and Anton Zeilinger, were awarded the *John Stewart Bell Prize for Research on Fundamental Issues in Quantum Mechanics and their Applications* in 2017 by the University of Toronto.

The problem of non-locality is that it apparently clashes with the theory of relativity because the instantaneous connectivity between particles separated in space is incompatible with the speed of light being the maximum speed at which matter and all known forms of information in the Universe can travel. One way of seeking to make the two theories compatible with one another is to postulate retrocausality, i.e., assuming that quantum communication between particles can be made in the direction of time as we perceive it, from the present to the future, and in the reverse direction of time, from the present back to the past, which we are unable to perceive. In the case of pair production, measuring the spin state of one of the particles would mean that it sends a signal at that moment, at a speed that is not above the speed of light, towards the past up to the moment of pair production, and then continues towards the future before arriving at the other particle exactly at the moment of the measurement. Retrocausality means we avoid the connectivity between two particles at the moment of measurement having to be instantaneous. A different way of attempting to explain this is to say that, after pair production, the two particles opened an immediate line of communication between one another, regardless of the distance between them. How spacetime "opens" to allow this line of communication, we do not know.

One of the fundamental characteristics of quantum mechanics is its time reversal symmetry. Put another way, the equations of quantum mechanics remain valid whether time t develops forward or backward, whether we are moving from the present to the future or from the present to the past. It has recently been possible to demonstrate, through certain reasonable hypotheses, that the symmetry of quantum mechanics in relation to time reversal must involve retrocausality (Leifer and Pusey 2017).

We are reaching a situation in our knowledge of the physics of the Universe, and physical time in particular, in which the incompatibilities between quantum mechanics and the theory of relativity are becoming clearer and require a solution. Up until now, it has been possible to include the theory of relativity in quantum mechanics, i.e., make quantum mechanics covariant, but what probably needs to be done is to start with relativistic spacetime and find a way to include quantum mechanics in it. The concept of time involved in retrocausality obviously cannot be understood through psychological time or operative social time. But we should not be surprised, because relativistic time also has incomprehensible features.

In 2009, a group of theoretical physicists (Linden et al. 2009) proposed a new relationship between quantum mechanics and time that may contribute to explain further the arrow of time. They argue that the fundamental mechanism by which a macroscopic quantum system reaches thermal equilibrium is its increasing quantum entanglement with the external environment surrounding it. The idea is to research the thermalisation process of an isolated system using only the dynamical laws of quantum mechanics. In this approach, the reason for which our cup of coffee cools down rather than warming up or remaining at its initial temperature is ultimately the quantum entanglement of states as the molecules of the hot liquid collide with the molecules of the cup and form a growing number of entangled quantum states until thermal equilibrium is reached. In conclusion, it is clear that physical time, just like all the other incarnations of the concept of time, continues to be a highly fertile area for thought and research. Physical time holds enigmas that are still far from being deciphered. Will we be able to decipher them in *Homo sapiens*' time?

1.23   The Beginning and Likely End of Physical Time

Physical time certainly had a beginning and will very likely have an end. Space is expanding, causing the Universe also to expand, as can be seen by observing the systematic motion of galaxies away from the Milky Way, where the Solar System is located, at a velocity proportional to their distance from us. This relation, called Hubble's law, was proposed by the American astronomer Edwin Powell Hubble (1889–1953) in 1929 (Hubble 1929). Hubble's law implies that galaxies were closer in the past and if we go back in time we reach a point at which their total mass was concentrated in one object, the density of which was extremely high. That is the time of the Big Bang, a singularity that took place 13.799 ± 0.021 billion years ago (Ade et al. 2014), in which all the laws of physics we know emerged. The state of the Universe at the Big Bang did not depend in any way on what may have happened before it because the laws of physics that rule the current universe did not exist. Because events before the Big Bang cannot be observed, and are therefore inaccessible to science, we can say that physical time, which corresponds to one of the coordinates of the relativistic spacetime, arose with the Big Bang. This does not mean that we cannot think about a time before the Big Bang, but that time is not relativistic physical time but rather philosophical time.

At the Planck time, only 5.4×10^{-44} seconds after the Big Bang, the Universe was a huge concentration of energy at an extremely high temperature and in a small volume. In these extreme conditions, the theory of relativity and quantum mechanics forced a fusion between time and space that made it impossible to differentiate between them. In conclusion, physical time began at the Big Bang and immediately afterwards was irreversibly amalgamated with space. Only later, as the Universe cooled, did physical time take on the characteristics that we know it has today.

As for a possible end to physical time, let us start by noting that in 1998 Adam Guy Riess and his team (Riess et al. 1998) concluded, by observing very distant supernovae that the expansion of the Universe is accelerating, moved by what was called dark energy. This energy represents roughly 68% of the Universe's total energy (Ade et al. 2014), but the nature of it is still poorly understood. Roughly 5% of the energy in the Universe is in the form of baryonic matter, i.e., the matter that forms the mass of the atomic nuclei, electrons, atoms, and molecules that make up the stars, planets, and living things. Finally, the remaining 27% (Ade et al. 2014) is in the form of dark matter, the nature of which is also poorly understood.

Cosmologists believe that the future evolution of the Universe is determined by the dark energy that works against the force of gravity and causes a continual, exponential expansion of the Universe. In the absence of this dark energy, the force of gravity would dominate and the Universe would end up reversing the expansion process and collapsing to form a new singularity. It is unlikely that the singularity will happen. Due to the presence of dark energy, the expansion will continue indefinitely. In this theory, known as the Big Freeze, the temperature of the Universe will get asymptotically closer to absolute zero, due to the expansion of space (Adams and Laughlin 1997).

Eventually, after about $10^{1}4$ years, stars, planetary systems and planets will no longer form because the interstellar gas is not renewed and runs out. Forms of life as they exist now will be impossible. Black holes will begin to take over the Universe, but even they will end up being extinguished due to Hawking radiation emission (Page 1976). The Universe will become progressively darker and more inert, its entropy will reach a maximum, and information processing of any type will become impossible because there will no longer be any form of change. It will then be impossible to measure physical time because the concept will no longer make sense. The heat death of the Universe will involve the end of physical time. In conclusion, according to our current understanding of the evolution of the Universe, physical time began at the Big Bang and will probably end with the heat death of the Universe. These two events define an interval of physical time that we shall call the physical time of the Universe.

Curiously, in some mythologies and religions, humankind has created the idea of a time of the Universe, and used it through operative social time dating back long before the construction of physical time. This time of the Universe includes the creation of the world, eventually leading to an end that has several eschatological interpretations that often correspond to ways of uniting or reuniting the human and the divine. The main difference between the narrative that underlies the physical time of the Universe and the mythological and religious narratives that underlie the operative element of time of the Universe is that in the latter the centre stage is given to humans and their gods and in the former humans are starkly irrelevant.

Chapter 2
Human Life and Immortality

2.1 The Origins of Awareness of Death and the Notion of Life After Death

It is impossible to retrace the steps in the development of human awareness about lifetimes and, consequently, the end of life from the primordial times of *Homo sapiens*. It is also impossible to know when the development of language made it possible for lifetimes to become an element of operative social time. Nonetheless, it is very likely that during the long evolutionary journey from the early hominids to humans, a growing awareness about lifespans and the end of them developed. This was in part self-awareness and also awareness about the lifetimes of fellow human beings, especially close family members, in the sense of elements of operative time. As a result, an awareness of death also emerged. This awareness was probably distinctive, leading to specific forms of behaviour in our earliest primate ancestors. It is perhaps similar or comparable to what we see today among members of the Homininae subfamily, which includes humans, chimpanzees, and gorillas.

Specific behaviours can be seen in chimpanzees when they are faced with the death of a member of their group, and they can be interpreted as attempts to revive it or expressions of frustration and grief. In some communities, observations have been made of mothers carrying the bodies of their children around for several hours or even days after they had died from disease. They showed them affection, slept next to them and were visibly upset when they were separated from them (Biro et al. 2010). From a human point of view, it would seem as if the mother was trying to do everything to remain with her dead child. In other cases, chimpanzee mothers radically change their behaviour after the death of a child, shifting from intense protection and affection to indifference. In the case of bonobos in Ivory Coast, Christophe Boesch and Hedwige Boesch-Ammerman observed one case in which a member of the group died after being attacked by a leopard (Pettitt 2002). The adults in the group gathered together around the body of the dead bonobo while letting out cries, and then sat in silence. Afterwards, the males let some of the females inspect and caress the body. While they stood guard around their late companion, they sometimes let out laughs and

© Springer Nature Switzerland AG 2021
F. Duarte Santos, *Time, Progress, Growth and Technology*, The Frontiers Collection,
https://doi.org/10.1007/978-3-030-55334-0_2

grimaced, probably to relieve the shock and tension. These types of behaviour show a relatively evolved awareness of the transition from life to death. There are other animals that display specific and rather complex behaviours when one of their kind meets death, including elephants, whales, dolphins, and crows.

Somewhere in time, during its evolution, *Homo sapiens* became aware of its identity and of death, and acquired some ability to foresee the future. At the same time, it would have developed existential death anxiety (Langs 2004; Lehto and Stein 2009), an emotion that is likely specific to it. These processes may be linked to the emergence of symbolic thought and the capacity for spoken language, which may have been acquired around 100 000 years ago (Henshilwood et al. 2009a). The defence that humans built against existential death anxiety was to deny it, and this can produce surprising behaviour (Becker 1973). The denial of death is one of the most powerful expressions of the psychological defence mechanism of denial, which Sigmund Freud (1856–1939) was one of the first to study (Freud 1901). Recently, denialism has become a fairly common form of psychological behaviour whereby overwhelming scientific evidence is denied because it is considered too uncomfortable to accept (Specter 2009). This new tendency is likely to be related to the emergence of anthropogenic existential risks (Bostrom 2002; Bostrom 2013). The *Homo* genus species have successively survived natural existential risks, such as extreme natural climate changes, supervolcanic eruptions, and earthquakes over some 2.5 million years.

At the present time, we have to deal with global catastrophic risks, that can be natural, such as large asteroid collisions, supervolcanoes, and natural pandemic diseases, or anthropogenic, such as a generalized nuclear war followed by a nuclear winter, accidental or engineered misuse of emerging technologies, such as nanotechnology or artificial intelligence, engineered pandemic diseases, extreme anthropogenic climate change, failure of geoengineering to countervail anthropogenic climate change, undetermined forms of global ecological collapse, extreme consequences from overexploitation of natural resources coupled with increasing demand, worldwide social and political tyrannies and global economic severe contractions generated by the inner workings of the current economic and financial system, by an escalating trade war or by concurrence with the preceding risks. Some of these global catastrophic risks, because they can be perceived as having the potential to endanger humankind in a long-lasting or even irreversible way, are called existential risks (Leslie 1996; Posner 2004; Smil 2008; Bostrom 2002; Bostrom 2002; Bostrom 2013). An increasingly common way to react to the anxiety generated by these risks is to understate or deny them (Rees 2013; Spratt and Dunlop 2018).

The oldest known signs of awareness of the transition from life to death are the use of graves. The first were probably simple adaptations of natural pits. The oldest known remnants date from the Middle Palaeolithic period and reflect the importance assigned to the person who died. At the Qafzeh and Skhul caves in Israel, *Homo sapiens* fossils have been found that date back 80 000 to 100 000 years. They are linked to what are thought to be the first recorded funeral rites. In the human graves, there are bones of valuable game animals: in the Qafzeh cave, a child's skeleton was accompanied by the skull and antlers of a deer (McCown 1937), and in the Skhul cave one of the skeletons was laid on its back and had the jaw of a wild boar in its hands

(Vandermeersch 1970). The presence of game animal bones in significant positions beside human skeletons suggests that their role was to accompany the deceased into the afterlife. These primitive versions of funeral rites unmask the importance that the death of a member of the group held for the others, and may even have reflected belief in some form of existence beyond death. Although we cannot say more from the data available, it is important to note that the most ancient forms of funeral rites are easily understood by us because they are similar to later, more complex and developed rituals, some of which can be observed in contemporary indigenous peoples.

For *Homo neanderthalensis*, funeral rites would only appear to have emerged later in its existence. The first evidence that Neanderthals buried their dead was found in the Shanidar Cave in northern Iraq, where the remains of pollen from several plants of medicinal value were found (Leroi-Gourhan 1975; Kooijmans et al. 1989). Analysis of Neanderthal fossils found at the La Chapelle-aux-Saints cave in France and dating back roughly 50 000 years revealed the use of graves for the dead (Rendu et al. 2014; Basset 2015), although the evidence presented is not considered sound enough by some researchers (Dibble et al. 2015).

Interestingly, it was also around 50 000 years ago that, according to DNA analysis, miscegenation between Neanderthals and humans probably began (Fu et al. 2014). Although there are signs that Neanderthals buried their dead, no objects accompanying the deceased or signs of other rituals have yet been found.

In the *Homo* genus, only *Homo sapiens* evolved to the point of having more complex funeral practices. They were an important part of their cultural evolution, which became particularly remarkable and fast from around 45 000 years ago onwards in western Eurasia (Bar-Yosef 2002; Powell et al. 2009). During that transitional period, known as the Upper Palaeolithic Revolution, humans gained the cognitive ability to develop and integrate knowledge about nature, technical knowledge, (especially relating to hunting tools), and social intelligence (Mithen 1996). *Homo sapiens* possessing these new abilities are often referred to as having the behaviour of modern humans. Nonetheless, it is important to realise that human cultural evolution was not a continuous and uniform process. On the contrary, there were steps forward and steps backward in different regions of the world, in an unstable process of adaptation to constant change driven mainly by food availability, environmental and climate changes, extreme phenomena, migration, and demography.

2.2 Human Life, Time, and Immortality in Mythology and Religion

Is there some way to uncover how the concept of time has evolved throughout *Homo sapiens*' history? Studying and trying to build a picture of symbolic human thought through myths are good ways to look for an answer. Myths are symbolic narratives about the origins, nature, and future of the world and humans. They reflect the human propensity to find meaning in events, animals, plants, water sources, rivers,

Fig. 2.1 Paul Gauguin: *D'où venons-nous? Que sommes-nous? Où allons-nous?*, Tahiti, 1897, Museum of Fine Arts in Boston, Massachusetts, USA. The painting was made in the winter of 1897–1898 when Gauguin was in a time of personal crisis and represents his spiritual and pictorial legacy. Time flows from right to left and the various scenes representing the questions about the human condition posed in the title take place beside a forest stream with the sea and the mountains of an island on the horizon. *Photo credit* The Picture Art Collection/Alamy Stock Photo

mountains, and other features of the environment that help us build an interpretation of our relationship with the world. The tendency to over-interpret the meaningfulness of things and events has probably accompanied *Homo sapiens* since the very beginning and continues to be a subject of current interest (Eco et al. 1992) in semiotics, defined by John Locke (1632–1704) as the doctrine of signs (Locke 1690). Many myths try to answer our eternal questions about ourselves, our beginnings, and our future. Paul Gauguin summed these questions up in his famous work *D'où venons nous, que sommes nous, où allons nous?*, painted in Tahiti in 1897 and 1898, and which he considered to be his final masterpiece (see Fig. 2.1). Myths are also narratives that tell us about cosmology and its relationship with human society, and help us accept the human condition and social circumstances.

Recently, Witzel (2012) studied and compared the development of mythologies around the world using a method similar to that used by linguists to build evolutionary trees of languages (Gell-Mann and Ruhlen 2011) and by biologists to create phylogenetic trees. The result was to divide all the mythology systems studied into two main groups: the Laurasian group, which includes mythologies from North Africa, Eurasia, and the Americas, and the Gondwanan group, which brings together the mythologies from Sub-Saharan Africa, the Andaman Islands, Papua New Guinea, Australia, and some ethnicities from other areas in southern India and Malaysia. The terms were inspired by the names of the two subcontinents, Laurasia and Gondwana, that were formed when Pangaea broke up around 175 million years ago, although the geographical areas do not fully coincide.

The geographical distributions of the two groups of mythologies are related to the colonisation routes followed by *Homo sapiens* when they left Africa, reconstructed by comparing the DNA of people who currently live in different areas of the world. When leaving Africa for the east, over the Gulf of Aden or Sinai, the "Southern route", 125 000–70 000 years ago (Forster 2005; Macaulay 2005; Appenzeller 2012), humans took the Sub-Saharan mythologies from the Gondwanan group with them along the coast of the Indian Ocean to Papua New Guinea and Australia. According to Witzel, Laurasian mythology came later and originated in South-West Asia around 40 000 years ago. It spread first throughout Eurasia, around 20 000 years ago, and later through the Americas to Patagonia, over the Bering Strait. Both groups developed from a primitive group of mythologies that would have developed in Africa a long time before the remarkable migration outside the continent began. Witzel gives it the name Pangaean, like the continent Pangaea.

The structure of mythologies in the Laurasian group is characterised by a narrative that includes the creation of the world from chaos or darkness, the subsequent generations of gods, the time of the demigod heroes, the emergence of humans, and an eschatological view with a sometimes violent end to the current world, but with hope for later rebirth. The mythologies also often involve the creation of a Father Heaven and a Mother Earth, great floods to punish human arrogance, and magical, mischievous deities linked to the emergence of cultural expressions. The Laurasian narrative is structured by the time of the universe and includes the origins and development of the human presence within it. It is possibly the first model of history.

In Gondwanan mythologies, the existence of Heaven, the Earth and the Sea is taken for granted and their origins are not part of the narrative. There is no end of the world and therefore no future rebirth. Most Gondwanan mythologies describe the emergence of humans and their forms of cultural expression through a forest of tales in a universe that already existed, the origins of which are not addressed.

Over the last 3 000 years, many of the characteristics that distinguish Laurasian mythologies were adopted and reformulated by some of the major world religions, especially Zoroastrianism, Judaism, Christianity, and Islam. The Abrahamic religions replaced Laurasian polytheism with a monotheist framework but retained the narrative of a universe created by God and a final demise with the promise of a paradise, a word that comes from *pairi-daeza*, which means "walled garden" in Iranian. The concept of time that evolved to reach global acceptance and use was not the one inspired by the Gondwanan mythology in which humans emerge within a timeless, pre-existing framework. Instead, it is the Laurasian concept with a beginning and an end that has persisted, in which the element of operative time of the human life is like a metaphor for the operative time of the universe. Furthermore, it is curious to find parallels between the Laurasian concept and the scientific concept of the physical time of the universe.

The records left by the first great civilisations enable us to retrace the evolution and increasing complexity of the concepts of life and life after death. Although all discussed these issues, Egypt paid them particular attention. The Egyptians were fascinated by the enigma of the end of life from the very beginnings of their civilisation, and they constantly sought to decipher and reinterpret it. This essential concern can be seen in all aspects of social, political, and religious life and, above all, in the funeral rituals and monuments from the simplest mastaba to the Pyramid of Khufu.

In the first dynasties, only the elite in society could be initiated into the mysteries of death and have access to the set of rules that allowed them to achieve immortality.

Later, during the First Intermediate Period and following deep social, political, and religious unrest, access to eternal life became more democratic. As mentioned, all Egyptians could then enjoy the afterlife as long as they knew the requisite sacred formulas and "words of power". From the Eleventh Dynasty onwards, the use of magic inscriptions on sarcophagus walls became more common, as did placing rolls of papyrus next to the mummy containing essential texts to allow the deceased to achieve eternal life (Assmann 1984). The texts were selected by the relatives from a collection that was called the Book of the Dead. At the time, however, it was called "Book of Emerging Forth into the Light". The formulas described in the book allow the deceased person to move unimpeded from the world of the dead to the world of the living. They consist of long monologues in which the deceased speak to themselves, to the gods, and to the people of the other world. The Book of the Dead also contains visionary recitals, during which the deceased make use of worship, flattery, and magic. The emotional state is intense and ambivalent, oscillating between the arrogance of considering themselves gods and begging the gods to save them, while displaying impotence against demons. In the end, it is a pathetic fight for survival, to avoid a "second death" or falling into non-existence, using sacred formulas and calling on all physical, psychological, and moral forces. It is one of the earliest most vivid accounts of existential death anxiety.

The hearts of the deceased were judged in the underworld, called Duat, against the ethical and moral principles of Maat, the goddess of justice, truth, balance, and order that prevents the universe from falling into chaos. But it is not enough to have been pure and just in life, those who die must also be sanguine and have the esoteric knowledge needed to avoid the pitfalls and traps set by evil spirits. In some ways, the struggle for survival in life continues into the afterlife in the struggle to leave the Duat.

The relationship between the world of the living and the world of the dead is intimately linked to Osiris, the imprisoned, paralysed man-god and protagonist of the primordial cosmic tragedy (Griffiths 1980). Osiris was the first-born son of the god of the Earth Geb and the god of the sky Nut and the first pharaoh. According to his cult, he was killed by his brother Seth, who envied him. Seth, the god of chaos, storms, and the desert, was then able to usurp his brother's throne. Isis, Osiris' wife and sister, searched for his body with the help of Nephthys; she found him and buried his body in the desert. However, Seth discovered him, cut him into fourteen pieces and spread them around the world. Isis did not give up, and with the help of Thoth's magic and Anubis' art of embalming, she rebuilt his body and wrapped it in bandages. She prepared the funeral and was rewarded by the gods, who resurrected Osiris as the god of the dead. When he was revived, Isis became pregnant with Osiris' child, Horus, represented by a falcon, and he overcame Seth and restored order to Egypt. Osiris represents the internal cycle of life, death, and resurrection, the enigma of the duality of life and death, and the struggle between order and chaos. The myth of Osiris was also the mythological basis for mummifying the dead, which was designed to avoid the decomposition that follows death and prepare the bodies for the afterlife. The cyclical essence of Osiris led the Egyptians to associate him with the external cycles of floods of the Nile and the growth cycle of annual crops that begins with sowing seeds and ends with the production of new seeds. The death of plants was seen as the death of Osiris and germination as the resurrection.

The Egyptians believed in a form of existence after death in which the integrity and functions of the human body remained. That is why they filled up the tombs with objects and products for the deceased to use.

The Dharmic religions of the Indian subcontinent developed more complex concepts of life after death by introducing a cosmic dimension. In Hinduism, the concepts are based on Brahman–Atman, the reality and identity of the whole that transcends space, time, and causality. Brahman is not only all of the external world but also the essential identity or soul of each human being, of each animal or plant, of each stone, river or mountain. This soul of each being is its Atman and the Atman is the Brahman. Human experience can teach us of the ultimately identical nature of the Brahman and the Atman. According to the writings of the Chandogya Upanishad (Hume 1921), written before the 6th century BC, "just as honey is made from different flowers, so Brahman is the Atman of each living or non-living being". The inability to understand the absolute reality of the Brahman–Atman inevitably leads to continuous cycles of birth, life, death, and rebirth through reincarnation, called *samsara*. This does not have a beginning and, in most cases, does not have an end. The cycles of death and reincarnation take place on Earth, in the many different forms of human, animal, and plant life. They can end when, at death, the form cannot be distinguished from the Brahman.

The sequence of the cycles is determined by karma. According to its law, all the good and bad that we do in life—thoughts, words, and actions—determine the nature and events of future reincarnations. The idea introduced earlier by the Egyptians can be found here, in that the quality of existence after death is determined by what is done in life.

Buddhism makes a more advanced analysis of nature, psychology, and human conditions and projects their sublimation into something beyond death. Siddhārtha Guatama (Buddha), who lived in the sixth century BC, by recognising that everything in the Universe is transitory and constantly changing, concludes that a permanent reality cannot exist inside or outside the person and refutes the Brahman–Atman idea and the existence of the Atman or soul in Hinduism. The path and the obstacles to be released from the cycles of reincarnation become clearer.

Buddha also acknowledges the existence of desire or obsession by the possession of material things, the search for power, influence, and control over people, and the search for physical and intellectual gratification, while always remaining convinced that full satisfaction and permanent fulfilment are achievable states. Only extinguishing this vicious thirst, which he called Tanha, can lead to nirvana, where the cycles of reincarnation cease. The flow of consciousness that goes with the successive reincarnations begins to follow a well-established route to terminate the transitory states and achieve the peace of perennial and absolute happiness or nirvana.

The resurrection of Osiris, which made him the god of the dead, is one of the first well-documented examples of the resurrection concept in history. There are several later gods and goddesses that also rise from the dead, including Baal, Dionysus, and Persephone. The resurrection of Jesus, which demonstrates his divine nature and justifies his proclamation as the son of God, adds a new doctrinal dimension that plays a central role in Christianity. Believing in the resurrection of the dead and judgment by Jesus Christ is also part of the Christian belief system. For believers, the resurrection of Jesus and the resurrection of the dead definitively surpass death and offer eternal life.

2.3   Immortality and Mortality in Evolutionary Biology

Taking it for granted that living is something wondrous and unique, then living longer with quality of life and well-being and health, in full control over one's faculties and organic functions, is one of the major, if not the main, goals of life. From this point of view, it is hoped that, above all in countries with advanced economies (countries with a relatively high GDP per capita and high levels of industrialisation), science and technology are attempting to extend human life in the best health conditions and beyond all imaginable limits. The practical application of scientific and technological advances in life extension will initially tend to be very limited in the poorest countries because of their relatively high cost and the fact that in these countries the main concern for a large proportion of the population is survival. However, it is likely that those scientific and technological advances will end up spreading to all countries. Currently, for some social classes with more economic resources in countries with advanced economies, the quest to delay ageing and extend human life as long as possible has become a very important goal.

There are several limitations that make biological immortality impossible to achieve. Even if it were potentially possible to indefinitely extend human life, it would not be feasible to fully avoid fortuitous fatal accidents, including those that may result from situations linked to developments in social, economic, and environmental conditions on Earth or beyond Earth. In other words, immortality always ends up being interrupted randomly by matters of chance and, ultimately, by the long-term development of the environment on Earth and in the Solar System.

It is therefore important to distinguish between absolute immortality, which goes beyond the framework of biological life, and conditional immortality or ammortality, in which the organism is not limited by ageing or by disease but can potentially be destroyed sooner or later by unavoidable random causes.

The form of life that we have on Earth makes both conditional immortality and mortality as the result of ageing possible, although during evolution into progressively more complex organisms there has been a clear preference for the latter. The first traces of life on Earth may date back to between 4.1 and 3.8 billion years ago (Bell et al. 2015). Living organisms were unicellular and their cells, known as prokaryotic cells, did not have a nucleus. Only much later, between 2.1 and 1.6 billion years ago (Knoll et al. 2006) did eukaryotic cells with nuclei emerge, a remarkable form of adaptation to an atmosphere with a growing oxygen concentration resulting from photosynthesis. Unicellular living beings with prokaryotic cells still exist and constitute the domains of bacteria and archaea, which are abundant in the oceans and include extremophiles and also the organisms that produce methane (CH_4) in our digestive system. Reproduction in these organisms is asexual and consists mainly of a binary fission process in which the cell divides itself into two identical cells with the same genetic material as the parent, also known as clones. It involves only one parent individual and the parental identity is lost in the division process.

The first multicellular organisms originated from prokaryotic unicellular organisms between 3.5 and 3 billion years ago (Grosberg and Strathmann 2007), but most evolved later in different, independent ways from eukaryotic unicellular organisms (Parfrey and Lahr 2013). With multicellularity, cells acquired specific, differentiated functions and ended up constituting organs. Finally, multicellular organisms diversified profusely, creating three kingdoms: fungi, plants, and animals.

Unicellular organisms have a form of conditional immortality because the original organism reproduces through cell division to produce clones that only stop living if they are destroyed by an accidental external factor, a process known as necrosis. Generally, their immortality is only broken due to random, fatal conditions occurring in the environment. It should be noted, however, that cells produced in binary fission are not exact copies of the parent cell. The division process generates malformations, which accumulate in the proteins, leading to a lower cell reproduction rate (Lindner et al. 2008).

In multicellular organisms, the situation is different because the cells are subject to a process of programmed death or apoptosis. The term "apoptosis" is derived from the Greek word for petals falling from flowers when they wilt or the leaves falling from trees in the autumn and was used for the first time in a famous article by J. F. Kerr, A. H. Wyllie, and A. R. Currie, published in 1972 (Kerr et al. 1972). The dysfunctions of apoptosis are highly relevant in medicine. Excessive apoptosis leads to atrophy, while insufficient apoptosis causes an uncontrolled proliferation of cells, an example of which is cancer.

In general, multicellular organisms are also subject to an ageing process at the cell level, which eventually leads to death. The differentiation between germ cells and somatic cells is ultimately responsible for the ageing processes (Kirkwood and Austad 2000). The fact that the separation of these two types of cells takes place just when the embryo begins to grow ensures that genetic or regulatory modifications in somatic cells that occur during the development process do not have consequences for gamete formation. These are the cells for sexual reproduction, the sperm and the egg, which fuse at the moment of fertilisation, starting the development of the embryo of a new organism of the same species. Although there are exceptions, there is a strong correlation between differentiation of germ and somatic cells and sexual reproduction.

To have emerged and to have persisted and proliferated over hundreds of millions of years, sexual reproduction had to provide evolutionary advantages. It had to improve Darwinian fitness, which is essentially the ability to survive long enough to reproduce. One of its main advantages from an evolutionary standpoint is that it helped provide greater genetic diversity and therefore higher adaptive capacity in a continuously changing environment. It is as if the cost of the evolution of species, giving rise to an enormous range of increasingly complex organisms including intelligent beings, namely humans, is ageing followed by death. If we think about it, we are likely to acknowledge that it is a small price to pay for our existence, but also one that deeply perturbs and conditions us.

We are still a long way from fully understanding the cellular, genetic, and molecular processes that lead to ageing, and, in particular, the reasons why multicellular organisms have such different lifespans. We know that ageing results mainly from an accumulation of somatic damage caused by an organism's decreasing investment in maintaining and repairing DNA. Longevity is regulated by a range of genes that control maintenance and repair actions. Cells may not divide properly, proteins may not acquire the right functional configuration when they are formed, and this can lead to tissue deterioration and fatal diseases. Nature's strategy has been to allow this damage to accumulate without being duly repaired, which inexorably leads to the decline and death of the organism. On the other hand, it has invested in the renewed vigour and adaptation capacity brought by descendants generated through sexual reproduction.

However, there are no fundamental theoretical reasons why multicellular organisms should not have evolved towards conditional immortality. There are examples of some species that explore that path in different ways.

2.4 Survival and Life Extension Strategies

Let us start with a relatively simple but fascinating strategy for prolonging life under harsh conditions. Tardigrades or water bears are small animals measuring around 1 mm, closely related to arthropods. They can survive in extreme environments by entering a state of complete dormancy called cryptobiosis. They curl up, dehydrate, and turn off all metabolic processes, and are able to stay that way for many years. In this state, they can withstand temperatures, pressures, and doses of ionising radiation that would be fatal for almost all living organisms on Earth. Recent experiments have shown that they are able to wake up and return to life after spending several days in the very high vacuum of outer space (Jönsson et al. 2008).

A more complex and particularly interesting case is the hydra, a small cnidarian with a cylindrical body in the form of a polyp that can be found in unpolluted freshwater pools and streams. When there is abundant food available, the hydra reproduces asexually through gemmulation. In other words, it produces gemmules on the wall of the polyp that develop until they become miniature adults, detach themselves and move on to lead an independent life. However, when environmental and feeding conditions become adverse, some species of hydra develop sexual organs to reproduce sexually.

Laboratory observations of various groups of *Hydra vulgaris* over a four-year period in favourable environmental and feeding conditions showed no signs of senescence or decrease in the reproduction rate (Martínez 1998). Mortality appeared to only come from external causes. However, when the development of sexual reproduction is induced, several physiological functions begin to deteriorate over time, and mortality increases with age. The lifespan reaches an average of four months. Depending on the environmental circumstances, hydra can attain conditional immortality or they can age and die, which makes them the ideal testing ground for research on the biological processes responsible for ageing.

Conditional immortality in hydra is the result of their extraordinary ability to reproduce asexually, because adults have stem cells that are able to proliferate continuously. Every adult is like a timeless embryo, in which the genes that regulate growth are constantly rejuvenating the body. In most multicellular organisms, including humans, stem cells become less active and numerous over time, which impairs their ability to maintain the rate of forming new cells. This loss causes a decline in tissue function and ageing of the organism. In the case of the hydra, the existence of a strong relationship between the FoxO gene and the lack of senescence was recently demonstrated (Boehm et al. 2012). A comparison between hydras with a normal FoxO gene and other hydras in which the gene has been deactivated shows that organisms without the gene have a significantly smaller number of stem cells and undergo disturbances in their immune systems. The same gene is also found in humans and there are reasons to believe that it plays a regulatory role similar to the one found in hydras. The study demonstrated that increased longevity is highly dependent on maintaining stem cells and a functioning immune system.

The *Turritopsis dohrnii* jellyfish is a hydrozoan measuring a few millimetres in diameter and found in the waters of the Mediterranean. It uses a different strategy to achieve conditional immortality. The life cycle of these hydrozoans begins in the form of small larvae called planula that start off swimming freely and then attach themselves to the seabed and produce a colony of polyps that are all genetically identical clones. Each of these polyps eventually detaches itself, becomes free, and evolves into an adult jellyfish that can reproduce sexually through fertilisation of eggs in the aquatic environment to generate larvae, thereby starting the cycle once again. But in its adult state, the jellyfish has created an alternative option to continue life. If the jellyfish faces adverse factors, including senescence, it can invert its development process and return to the polyp stage (Piraino et al. 1996).

What is surprising and apparently unique is that a sexually mature individual that is able to reproduce can reverse its ontogenic sequence using a process of cellular transdifferentiation, in which it forms a cyst on the seabed that develops into a colony of polyps. In theory, this process can be repeated indefinitely, which makes the jellyfish immortal, provided that it does not suffer a fatal accident, of course. The remarkable aspect of cellular transdifferentiation in *Turritopsis dohrnii* is the ability of some genes to turn off their activity while others become active in order to reactivate genetic programmes used in the initial stages of the jellyfish's life cycle. Although it is very difficult to keep a colony of *Turritopsis dohrnii* in captivity, Shin Kubota, of the University of Kyoto, saw his colony reborn eleven times over two years (Kubota 2011).

The regeneration capacity of planarians, a type of worm that belongs to the *Platyhelminthes* phylum, has been known since 1766, when the Prussian zoologist Peter Simon Pallas (1741–1811) described for the first time how a complete new organism could be generated by a small piece of the animal's head (Brøndsted 1969). Fragments equivalent to only 1/279th of the animal's body size can regenerate in a few weeks to form a new animal (Morgan 1898). This ability is due to the presence of stem cells in planarians that can differentiate into all the types of cells that make up the organism, including germinal cells (Handberg-Thorsager 2008). In addition to this, stem cells in planarians also have the unique characteristic of avoiding senescence and maintaining their ability to reproduce (Sánchez Alvarado 2012). In our bodies, stem cells progressively show signs of ageing as they divide to ensure the organism's growth when it is young, to heal a wound or replace cells that have died. Their ability to reproduce decreases and, as a result, they become unable to replace specialised organ cells and tissue when they grow old and die.

One of the signs of ageing cells at genetic level is the shortening of the structures that form the end of the chromosomes, called telomeres. In the cell division process, telomeres become shorter and at a certain point they prevent new divisions from taking place. The enzyme that controls the length of telomeres, called telomerase, is only active during the growth period, so telomeres begin to get shorter with age. We may thus conclude that, for an organism to be immortal, its cells should be able to maintain the length of its telomeres.

In planarians, reproduction can be sexual or asexual. Some species of planarians are only asexual, while others can reproduce either way. In sexual species, planarians are hermaphrodites, and have both testicles and ovaries. In asexual species, planarians divide into two parts: one with a tail and the other with a head, and both of the resulting pieces regenerate the remaining parts to form two new worms. Recent studies have concluded that, in asexual planarians, the gene responsible for forming telomerase drastically increases its action in the partition and regeneration process, enabling telomeres in stem cells to remain the same length as they divide to regenerate the required tissue and organs (Tan et al. 2012).

Planarians' ability to maintain the length of the telomeres in the asexual reproduction regeneration process potentially enables conditional immortality. Curiously, the process for maintaining the length of telomeres in sexual planarians is different, although they have similar capacities for regeneration (Tan et al. 2012).

Several types of life forms give us clues on how to construct organisms with conditional immortality. Nonetheless, it is quite clear that the choice made on the evolutionary journey of natural selection to reach more complex organisms was mortality. The form of life found on Earth allows for conditional immortality to be chosen but its Darwinian fitness in the evolution of species has revealed itself to be lower than the fitness of mortality.

2.5   Human Life and Immortality

With the emergence of modern science in the 16th century and its subsequent development, especially in the fields of biology and medicine, the concept of death has taken on new dimensions. In the world today, there are many concepts regarding the end of life and a possible life after death with associated interpretations: some are religious and involve a wide range of forms of worship, others are hybrids in which religion and science seek harmony, and some are simply scientific.

Death is the subject of reflection at different depths and levels of resistance and anxiety, and probably some fear. It is an especially private topic about which we generally talk little. It is therefore difficult to understand the evolution of the concept and understanding of death in human history, just as it is difficult to know how the contemporary generation perceives and deals with the end of life.

One common feature in all these reflections is the fact that death is inevitable. This is a certainty that counterbalances all the uncertainties we find in life and in day-to-day existence, and it has an irrepressible influence on the operative social time, partly because science and technology have accustomed us to breaking almost every barrier, some of which had previously seemed insurmountable. Another common trait is that death represents a definitive leap into the unknown. It is a definitive transition, but the time and circumstances in which it occurs remain unpredictable.

Death is the most eloquent proof of our irrelevance in time and space. It is a moment that instantly establishes complete equality among all those who share it. We are entirely equal in birth and in death, although they occur in different ways. Naturally, once our lives have ended, the memory of our lives, of our greater or lesser influence and power, our experience, work, and knowledge, however remarkable, influential, humble, or insignificant they may have been, begins to fade and is inexorably lost. The memory may fade quickly or slowly, but eventually there will be no memory left. The places where we liked to spend our time, those that we visited, those that we did not visit, and in fact the Earth and the Universe as a whole will continue to exist and evolve, entirely indifferent to our short lives. Everything that we build during our lifetimes, including friendships, cooperation, power networks, wealth, and possessions, becomes detached and meaningless. The seemingly enormous relevance of all that construction abruptly vanishes. The conviction held by some that our consciousness, identity, and memories of our existence are suddenly lost forever may be painful. Living with this conviction is difficult and gradually, throughout the history of civilisations, some have acknowledged and commented on the absence of meaning in human life. Through the voice of Macbeth, William Shakespeare (1564–1616) gave us one of the first and most eloquent descriptions of that point of view (Shakespeare 1623):

> And all our yesterdays have lighted fools
> The way to dusty death. Out, out, brief candle!
> Life's but a walking shadow, a poor player
> That struts and frets his hour upon the stage
> And then is heard no more. It is a tale
> Told by an idiot, full of sound and fury,
> Signifying nothing (Macbeth, Act 5, Scene 5).

Most of human history can be interpreted as an instinctive, irrepressible, and committed denial of that literal insignificance. Really, life is what is truly important. Furthermore, the force that determines and motivates human behaviour, which justifies our human ambitions, expectations, and achievements and renders them comprehensible, from prehistory to the present day and in the foreseeable future, is the drive to fully enjoy life, to enjoy well-being and prosperity, to delay its end in the best possible health, regarding both physical and mental conditions and, for many, to believe that there is some form of life after death. This is the fundamental guideline for our behaviour, which prevails over all others. For the vast majority of people, their greatest ambition is to satisfy the unstoppable desire to live longer, with greater well-being, and to fulfil all the expectations they build up throughout their lives.

There are certainly exceptions, diversions, and pathologies, but they are far from being a significant number within the human population. The incalculable value of life should not stop us from accepting its end as a natural and harmonious transformation that continues our journey in nature and in the cosmos under a different form.

2.6 The End of Life

Although human death is something very clear and obvious to us, how is it defined scientifically? Biological death is the irreversible termination of all the organic functions that keep the organism alive. One inevitable consequence of death is that the human body begins to decompose. Although the concept of death is clearly established, it is hard to ascertain the exact moment when it occurs, in other words, it is hard to know how to determine if and when it has taken place.

Historically, death was first believed to coincide with the termination of cardiac and respiratory activity. In 1740, however, the Paris Academy of Sciences officially recommended mouth-to-mouth resuscitation for drowning victims. From the middle of the 20th century, cardiopulmonary resuscitation evolved extensively and was widely used to save many lives. Using cardiopulmonary function to define the moment of death was therefore no longer quite adequate. A more recent alternative to the cardiopulmonary criterion is brain death, i.e., the termination of all brain functions, but even here there is some ambiguity (Cranford 1995). Brain death is sometimes also called cerebral death, although that is imprecise, since the cerebrum makes up only part of the brain. The brain is formed of the cerebrum, the main organ responsible for consciousness, cognition, and thought, the cerebellum, responsible for coordinating and controlling voluntary muscle movements, and the brainstem, responsible for the most basic homoeostatic mechanisms, such as heart rate, breathing, and indicating pain. A person can now be dead according to the brain death criterion while their metabolic processes are kept running by means of life support equipment. The former cardiopulmonary criterion would determine that the person was still alive. a whole-brain concept of death is now adopted as the foundation for legally declaring death in many countries, although the specific criteria for brain death vary for legal and cultural reasons (Bernat 2006).

The whole-brain death criterion leads to an organic definition in which death is identified with the organism irreversibly losing its functioning as a whole. The prevailing trend is towards considering that death only occurs when all brain and brainstem functions, breathing, and blood circulation cease. When that termination occurs, all perception, cognition, action, memory, emotion, and consciousness are irreparably lost.

Nonetheless, the question of knowing how biological death relates as a concept with the essence and identity of the human person still stands. The meaning of essence in this context is strictly ontological and refers to the property or properties that make us people with individual identities. We are essentially beings with the capacity for consciousness and that is the condition needed for the existence of an individual, human identity. Following this line of argument, death is the irreversible loss of consciousness or, in other words, the irreparable loss of our unique human identity. That which some call eternal oblivion.

This transition from life to death may be relatively fast or slow. In the extreme case of a coma, the person enters a state of unconsciousness in which there is no reaction to light or sound stimuli and the daily sleep cycle is lost. A coma leads to death in many cases, but, in some, consciousness is recovered. It can also lead to a vegetative state that may last more than a decade. An extreme example of this is Elaine Esposito, who was born in Chicago in 1934. She went into a coma after surgery on her appendix, when she was 6 years old, and she remained in a vegetative state for 37 years until she died.

In the Greek and Roman civilisations, it was common to practise euthanasia, from *euthanatos* or "good death" in Greek. This is a quicker death to avoid prolonging pain and suffering in a terminal situation, although Hippocrates (c. 460–c. 370 BC) apparently opposed it (Mystakidou et al. 2005). Christianity fights against euthanasia and suicide because it asserts that human life is an unalienable gift from God.

At end of the 19th century, the issue of euthanasia returned, starting a long, complex, and intense debate that is still ongoing, intensifying, and expanding throughout the world to this day. The definitions and boundaries between euthanasia, suicide, and murder are actively discussed in the fields of religion, law, and above all bioethics. In 2018, the practice of euthanasia is considered legal in Belgium, Canada, Colombia, the Netherlands, and Luxembourg. Assisted suicide, when the person takes the initiative him or herself, is currently legal in Germany, Switzerland, the Netherlands, and some US states. Most countries in the world are reviewing their legislation on the topic. The first private institution for euthanasia opened in the Netherlands in 2012 and is called Levenseindekliniek or "clinic for dying". Most patients have neurological or cardiac diseases or cancer. In the period of one year, it received 714 requests and performed 104 euthanasias. It is likely that these clinics will start to spread to different countries in the future.

Suicide has been a constant theme throughout history in religions, philosophy, social sciences, and medical science. According to Christian doctrine, suicide breaks our relationship with God since the control over our body is limited to its employment while God retains the dominion or authority over it (Durkheim 1897). In the Middle Ages, those who committed suicide lost the right to Christian burials and graves. Their corpses were often hung, burned, or dragged through the streets and dumped into rivers. With the Enlightenment, the traditional Christian attitude toward suicide was questioned and finally it was legalized in much of the Western world during the XXth century. Nevertheless, suicide continues to generate a range of points of view. Some continue to believe it to be objectionable on religious or moral grounds, while others defend it as legitimate in certain extreme circumstances or as a way of imposing specific objectives on other people. Although there have been records of suicide attacks throughout history, they have become more notorious in recent times with the kamikaze pilots of the Second World War and especially, since the 1980s, with attacks in Lebanon, Sri Lanka, Israel, Iraq, Afghanistan, Pakistan, India, Syria, Yemen, and many other countries.

The annual number of suicide attacks of this type carried out by members of terrorist groups has risen from less than five in the 1980s to more than 180 in the first half of the 2000s (Atran 2006). A decade later, and the annual number of suicide bomb blasts has increased to several hundred: 452 in 2015 and 469 in 2016 (Schweitzer et al. 2017). The 469 attacks in 2016 were carried out by around 800 terrorists in 28 countries, causing 5 650 fatalities and wounding 9 480 people. Around 70% of attacks were attributed to the Islamic State, followed by Boko Haram with 11% (Schweitzer et al. 2017).

Suicide attacks are a new way of systematically using terrorist violence in the battle against enemies that have great superiority in a conflict. This type of extremely asymmetric war, the success of which depends mostly on the enemy's reaction and media coverage, has shown itself to be highly effective in boosting and expanding political and military movements. It requires enormous spending on prevention and security measures in the affected states and is likely to create a climate of insecurity. Its extent and persistence, especially in the Middle East, is a modern phenomenon, the evolution of which is unpredictable. One of its characteristics is the practically complete incapacity or impossibility for human dialogue between the opposing parts. It is likely that many of those who kill themselves in terrorist attacks do not have suicidal psychopathologies, although it is very difficult to validate this statement scientifically (Sheehan 2014). For those who do not suffer from such pathologies, life may perhaps have a value lower than the value attributed to the expected glory arising from the successful suicide attack.

In dysthanasia, from the Greek for "bad death", the life of a patient with an incurable, terminal disease is prolonged artificially, supported by medicines and advanced medical technology such as defibrillators, ventricular assist devices, extracorporeal membrane oxygenation, and so on. This type of life extension beyond anything that would be expected in the past has been made possible by the remarkable development of scientific knowledge and technological means. Modern medicine has enormously increased the ability to keep patients alive in terminal stages (Warraich 2017). The reason for doing so, despite the high costs involved, is to allow patients to live to the very limit of the possibilities offered by current science and technology. In these cases, quality of life is something relative and subjective, and is believed to be less important than the supreme value of life. Often, patients are no longer able to properly communicate their wishes and feelings to those who assist or accompany them. Eventually, the limits of life support treatments are reached. Defining those limits from a practical point of view involves several types of medical, technical, and economic criteria, and is an ethically delicate matter. In these types of procedure, death is far from being a natural process. Prolonging life artificially, seeking to go beyond all limits, is a way of temporarily denying something that is inevitable. For the patient, the cost of that denial is likely to be an artificial death, with prolonged suffering, attached to machines. Studies carried out through interviews demonstrate that there is a great divide between what people prefer during the end stage of their lives and what really happens to them (The Economist 2017). Nevertheless, in many countries, individuals are given the opportunity to decide by means of a living will whether or not they intend to be kept alive artificially in the event of not being able to express their wishes. A growing number of people are using it.

The costs of life support in the last two months are rapidly increasing, especially in countries with advanced economies, and comprise a high proportion of total health care costs; they are currently responsible for around 30% of health care costs in the United States of America (Neuberg 2009). In 2009, medical and hospital costs for the last two months of life paid for by the US government through the Medicare system totalled $ 55 billion (CBS 2010), more than the annual budget of the US Department of Education. Nevertheless, it is estimated that in 20–30% of cases treatment does not have a significant effect on the progression of the disease. The enormous costs of extending life in its final stages, supported by public funds, is inevitably in competition with health expenses for young people and for adults, whose quality of life benefits from that investment for a longer period of time because of a longer average life expectancy of those age groups. The fact that this distribution of public money is relatively well accepted socially and politically is probably rooted in existential death anxiety. Although it is a universal issue, it takes on very different forms in different countries, depending on their economic power and religious, cultural, and social characteristics.

Unlike dysthanasia, orthothanasia is natural death, in which patients follow their own paths, without prolonged life-support treatment, but in the best possible conditions for relieving physical and psychological symptoms that cause suffering.

Chapter 3
The Initial Part of *Homo Sapiens'* Time

3.1 Flowering Trees and the Evolution That Led to Humans

Flowering trees took part in some of the most decisive steps in the evolutionary journey that led to the emergence of hominids. During the Devonian geological period, between 419 and 359 million years ago, vascular plants began to diversify and the first trees and forests appeared. At the end of this period, the first plants with seeds emerged, an innovation that was revolutionary because it allowed a species to expand its territory quickly. Thanks to the invention of seeds, forests of conifers spread around Earth and began to dominate the flora until they reached their peak at the end of the Jurassic, which took place between 201 and 145 million years ago. It was at that time, 160 million years ago, that the evolution of plants led to another remarkable innovation that would have profound consequences for the plant and animal kingdoms: the appearance of plants with flowers and fruits, called angiosperms, from the Greek *angeion* for "receptacle" or "vase" and *sperma* for "seed".

In these plants, the ovules that contain the female cells and later transform into seeds after fertilisation are wrapped in and protected by an ovary that is transformed into a fruit. Often, the fruit is edible, which means animals become agents of dispersion and colonisation, along with wind and water. The key idea was to protect female cells, as happened in other episodes of the evolution of species. Angiosperms dispersed quickly across all the continents during the Lower Cretaceous and ended up overtaking forests of conifers and dominating global flora, becoming a great evolutionary success. Their flowers also have the crucial advantage of allowing animals to participate in and facilitate the reproduction process, which is the central aim of Darwinian evolution.

This participation includes the pollination of flowers carried out by a wide range of animals, including insects, birds, and bats. This means that genes can be transferred between isolated plants of the same species for which pollination by wind would be highly unlikely and even impossible. Thanks to this interaction, plants and their pollinating animals began to co-evolve. Co-evolution is the simultaneous

© Springer Nature Switzerland AG 2021
F. Duarte Santos, *Time, Progress, Growth and Technology*, The Frontiers Collection,
https://doi.org/10.1007/978-3-030-55334-0_3

evolution of two species in which a genetic change in one can be triggered by a genetic change in the other (Vermeij 1994). In short, angiosperms allowed for the emergence of a magnificent biodiversity and now form the energy base and structure for the vast majority of the Earth's ecosystems. We know of around 300 000 species of angiosperm (Christenhusz and Byng 2016), which represents approximately 80% of the total number of plant species.

What is the connection between the success of flowering plants and primates? The answer is that the latter emerged in the Cretaceous period at a time of expansion and diversification for flowering trees (see Fig. 3.1). Estimates based on fossil records indicate that primates began to diverge from other mammals around 81.5 million years ago (Tavaré et al. 2002). The fossils of the oldest known primates have been found in Eurasia and North America. The oldest common ancestor of primates was a relatively small, tree-dwelling animal that had three major characteristics: hands and feet adapted to life in the trees, with nails instead of claws; front-facing stereoscopic vision in which the images provided by each eye overlap, allowing depth perception; and a higher level of encephalisation (the relationship between brain mass and the animal's body mass) compared to other families of animals. It is very likely that these characteristics were the result of adaptation to the increased diversity of forests of angiosperms during the Upper Cretaceous. Front-facing vision made it easier to locate and consume small plant elements, particularly the fruits which were becoming abundant and diverse at the time.

Perhaps the move towards encephalisation arose from natural selection caused by adaptation to an increasingly complex forest habitat. These animals had to master the forest's three-dimensional structure and know how to seek out and remember the location of edible fruits. Life among the trees caused longer and stronger lower limbs to develop, with prehensile feet to make it easier to support the body when the upper limbs collected fruit and insects. As the arms and shoulders became more flexible, the hands adapted to catch, hold, and examine them. Bringing food to the mouth by hand, after careful inspection, is a specific characteristic of primates.

The primate branch that led to the *Homo* genus separated from the orangutan branch roughly 14 million years ago and from the gorilla branch roughly 9 million years ago. Finally, around 6–7 million years ago, there was another split: one branch led to chimpanzees, the primates closest to our species, and the other led to the *Homo* genus, in which humans emerged (Ravosa and Dagosto 2007).

3.2 The Influence of Climate Change on the Emergence and Evolution of the *Homo* Genus

Transformations in the environment caused by climate change are one of the most important factors in the evolution of species throughout the history of life on Earth. It is likely that this cause and effect relationship has played an important role in the biological evolution of the *Homo sapiens* phylogenetic line, although it is very difficult to decipher what really happened in detail. Let us briefly analyse the climate change that has taken place since the beginning of the Cenozoic, the geological era

Fig. 3.1 Silk floss tree (*Ceiba speciosa*) in flower, photographed in Mato Grosso, Brazil, where it is called *paineira-rosa*. *Ceiba speciosa* is a deciduous tree of the Malvaceae family, the same family as the baobab, native to the tropical and subtropical forests of South America. The fruits are ovoid capsules that contain black seeds surrounded by a fibrous and fluffy matter like cotton. The trees of the genus *Ceiba* have many applications, including medicinal uses. *Ceiba pectranda*, is a giant tree of the tropical forests in Central and South America. For the Maya it was sacred, a symbol of the Universe, representing a pathway between its three levels: the underworld, the middle world where humans live, and the upper world with the thirteen levels of the Maya heaven. Indigenous peoples of Brazil also consider it sacred. It produces large quantities of nectar, consumed by bats as well as moths and other insects in the process of pollination. Photograph by the author

that began 66 million years ago and extends to the present day, for two main reasons. The first is that it was during this era that the average global atmospheric temperature reached a high point, between 54 and 48 million years ago, after which it has tended to fall until recently. The other reason is that it was during the Cenozoic that the first radiation of the Primate order into several families took place, including the ancestors of the Hominidae family.

In the Cretaceous, before the beginning of the Cenozoic, the Earth's climate was in a stage of global warming, in part due to the evolution in the configuration of the continents and ocean currents determined by plate tectonics. The continents were still relatively close together following the break-up of the supercontinent Pangaea which began at the beginning of the Jurassic, around 200 million years ago, and there were no polar ice caps. At the end of the Cretaceous, a period of extensive volcanic activity occurred in the Deccan region of India, with considerable eruptions of basaltic lava. The enormous volume of lava released solidified, eventually forming extensive stepped mountains called Traps which initially covered roughly 1.5 million km^2 and whose remnants can still be seen in the Deccan Plateau in central western India. This intensive volcanic activity released large amounts of CO_2 into the atmosphere, and this contributed to increasing the average global atmospheric temperature.

During the Cenozoic, the Earth's climate system saw a complex evolution with exceptionally hot periods, without polar ice caps, and cold periods with large masses of polar ice (Fig. 3.2). Shortly before this period began (Renne et al. 2013), a meteor measuring about 15 km in diameter collided with the Earth, in Chicxulub, in the region that is now the Yucatan Peninsula in Mexico. The collision caused catastrophic environmental changes that led to the fifth mass extinction of species. Due to the huge temperatures caused in the impact, large amounts of ash, dust, and aerosols were launched into the atmosphere. There were widespread fires, which in turn released a lot of soot. For around a year, all these particles, particularly sulphate aerosols, obstructed sunlight, stopping photosynthesis, and lowering the average global atmospheric temperature by several degrees Celsius. The stratospheric ozone layer was temporarily destroyed and there was intense acid rain that was highly destructive to terrestrial and marine ecosystems.

Since the collision took place in a region that is high in carbonate rock, its instant vaporisation after the impact caused large amounts of CO_2 to be released into the atmosphere. The amount of carbon launched into the atmosphere in the form of carbon dioxide, estimated at 4 600 GtC (gigatonnes or 10^9 tonnes of carbon) generated a radiative forcing of 12 Wm^{-2}, enough to increase the average global atmospheric temperature by 7.5°C (Beerling et al. 2002). This positive radiative forcing was initially counteracted by the negative radiative forcing caused by sulphate aerosol emissions. However, as the residence time of aerosols in the atmosphere is on the order of months to a few years, much shorter than the residence time of CO_2, on the order of hundreds of years, global warming eventually prevailed. Recent isotopic analyses of fossilised fish remains suggest that the global warming caused by the Chicxulub impact would have been on the order of 5°C and lasted 100 000 years (MacLeod et al. 2018). The environmental changes that took place immediately after the collision, as well as the global warming it caused for tens of thousands of years, were highly destructive for the biosphere.

We know about the climate of the Cenozoic largely from analyses of the relative variation of stable isotopes of oxygen ($\delta^{18}O$) and carbon ($\delta^{13}C$) in living organisms, such as foraminifera collected from sediment at the bottom of the ocean (Douglas and Woodruff 1981). These studies show that 9.7 million years after the violent start to the Cenozoic, there was another extreme climate event in the transition from the Palaeocene geological epoch to the Eocene. For a relatively short time, around 56 million years ago (Turner 2018), the average global atmospheric temperature was very high (Kennett and Stott 1991). It is estimated that during that period, called

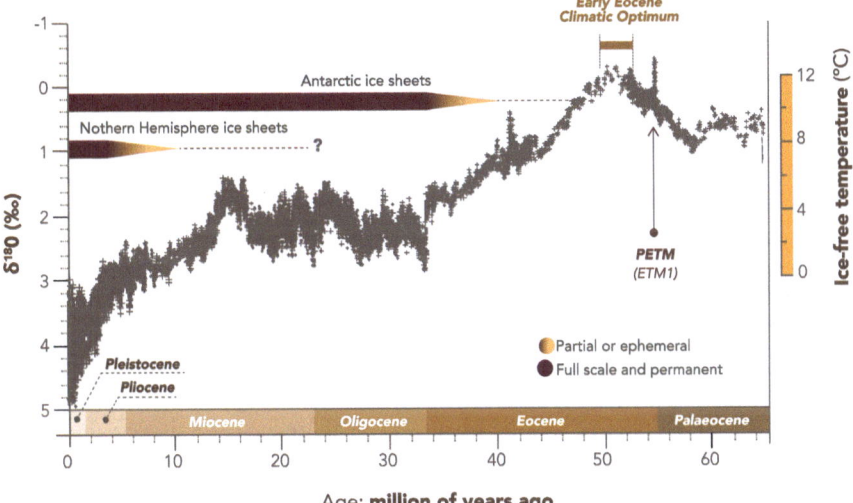

Fig. 3.2 Global climate over the past 65 million years. The climate curve is a stacked deep-sea benthic foraminiferal oxygen-isotope curve. The $\delta^{18}O$ temperature scale on the *right-hand axis*, was computed on the assumption of an ice-free ocean; thus it only applies to the time preceding the onset of large-scale glaciation on Antarctica, about 34 million years ago. The $\delta^{18}O$ measurement in the biogenic calcite of foraminifera provides a proxy for the global ocean temperature. It is defined as the ratio of the heavier ^{18}O isotope and the lighter ^{16}O, which is more easily evaporated from the ocean and sequestered in glacial ice on land. The figure clearly shows the Early Eocene Climatic Optimum and the very short-lived early Eocene high-temperature event called the Paleocene–Eocene Thermal Maximum, or PETM (also known as Eocene Thermal Maximum 1, ETM1). Large-scale glaciation on the Arctic started only about 3 million years ago. Adapted from Zachos (2001)

the Paleocene–Eocene Thermal Maximum, or PETM, the global mean temperature initially rose 5–8°C in around 20000 years and then remained largely stable for the approximately 100000–200000 years of the PETM (see Fig. 3.2) (McInerney and Wing 2011).

The PETM is an example of a natural global change in the Earth system caused by high emissions of carbon into the atmosphere. Studying it helps us foresee the response by the climate system, and the biosphere in particular, to current anthropogenic interference with the climate. We know that a large amount of carbon was released into the atmosphere during the PETM because it was recorded in variations in the isotopic composition of that element. Carbon has two stable isotopes, ^{12}C, which has an abundance of 98.93%, and ^{13}C, which has a much lower abundance of 1.07%. When incorporating carbon from atmospheric CO_2 into plants, photosynthesis favours the ^{12}C isotope over the ^{13}C isotope because its mass is lower. As a result, the carbon in plants has a different isotopic signature from the signature observed in CO_2 in the atmosphere or minerals. The signature varies depending on whether the plants are C_3 or C_4 or if they use crassulacean acid metabolism (CAM) in photosynthesis.

Throughout the world, the strata corresponding to the PETM show a marked reduction in $^{13}C/^{12}C$ ratios, which indicates that a large amount of carbon was released into the atmosphere, very likely in the form of CO_2 and CH_4. There are several examples of this type of event in the Earth's history, known as carbon isotope excursion, but the PETM is the one that involved the highest carbon emissions.

Richard Zeebe and his collaborators (Zeebe et al. 2016) calculated that the initial carbon release measured 2 500–4 500 GtC and lasted at least 4 000 years, which means an emission rate of 0.6–1.1 GtC per year. Note that this method for estimating the amount of CO_2 launched into the atmosphere cannot be used for the Chicxulub collision because in that case the carbon was lithic rather than organic in origin.

The PETM left us another significant record. In marine sediments deposited thousands of metres into the depths of the ocean, corresponding to the PETM, a sharp fall can be observed in the amount of carbonates as a consequence of the substantial reduction in the number of marine organisms. This reduction was very likely the result of the acidification of the ocean caused by an increase in atmospheric CO_2.

Little is yet known about the causes of the PETM. The warming at the beginning of the Cenozoic culminated in the Early Eocene Climatic Optimum, around 51 million years ago, one of the times when the average global atmospheric temperature reached a maximum (see Fig. 3.2). Volcanic activity probably intensified during this unstable period, following the basaltic lava emissions of the Traps, which launched large amounts of CO_2 into the atmosphere. A significant part of global warming at the start of the Eocene was due to increased CO_2 concentrations in the atmosphere (Anagnostou et al. 2016) caused by geological transformations. However, an additional, biological process is needed to explain the carbon isotope excursion during the PETM. This may have been the emission of large amounts of CH_4 into the atmosphere from the dissociation of methane hydrates of biological origin stored in the continental shelf of the Arctic regions, caused by ocean surface waters becoming warmer (Gu et al. 2011).

At the beginning of the Eocene, which began 56 million years ago when the Earth's atmosphere reached high temperatures, mammals radiated into two new orders that would become particularly important in the future: Artiodactyla, even-toed ungulates, which currently include pigs, oxen, deer, sheep, and camels and Perissodactyla, odd-toed ungulates, which include horses, rhinoceroses, and tapirs. It was during that same period that there was a marked adaptive radiation of the Primate order into 60 genera, which represents much greater diversity than at present. Our prosimian ancestors were close to the lemur, the tarsier, and the loris. The three radiations emerged consecutively, separated by short time intervals of around 10 000 years, the last being the primates. These events represent a remarkable example of acceleration in the evolution of species (Gingerich 2006). It was probably caused by exceptionally favourable environmental conditions, including the climate and a higher atmospheric oxygen concentration. The PETM period was notable for geographical redistribution in marine and terrestrial ecosystems, rapid evolution in some species, and the extinction of others, such as some benthic foraminifera.

Later, the evolution of the three orders was influenced by changes in the climate and terrestrial ecosystems that involved the development of grasslands where vegetation was dominated by Gramineae, a family of plants that emerged at the end of the Cretaceous (Prasad 2005). The expansion of the grasslands came at the cost of a reduction in forested areas and was largely the result of the progressive transition to a cooler, drier climate in the last 52 million years (Retallack 2001). This marked shift in the environment caused an adaptive radiation in some mammals especially horses, antelope, and other mammal species. The grasslands and savannahs were where our *Homo* genus ancestors evolved, as did the animals on which they fed, some of which ended up later being domesticated. Around 7 million years ago, adaptation to a climate that was drier and had a lower concentration of CO_2 in the atmosphere led to an expansion of Gramineae coverage; these plants use C4-type photosynthesis in which carbon fixation requires less water consumption. CO_2 concentrations in the atmosphere fell during the Cenozoic from levels on the order of 1400 ppmv (parts per million by volume), in the Early Eocene Climatic Optimum (Anagnostou et al. 2016) to levels below 400 ppmv in the last 5 million years (Zhang et al. 2013).

Climate change in the Cenozoic resulted mostly from two types of forcing with different origins and timescales (Zachos 2001). The climate forcing that caused the cooling over periods of time on the order of tens of millions of years has its origin in the lithosphere and is connected to a reduction in magmatic activity and in tectonic plate movements. These movements produced changes in the relative position, geography, and orogeny of the continents, in ocean currents, and in variations in greenhouse gas concentrations, particularly CO_2, connected to transformations in the lithosphere and the responses of the biosphere. During the most recent part of the Cenozoic, particularly when the Northern Hemisphere ice sheets started to form, climate variability increased over time intervals of tens to hundreds of thousands of years, becoming heavily influenced by the astronomical forcings that generate the Milankovitch cycles (Milankovitch 1930). Naturally, we know in greater detail the climate evolution as we approach the present because we can use more climate proxies, which reduces the uncertainty.

Roughly 52 million years ago, the average global atmospheric temperature, which was at least 10°C higher than pre-industrial temperatures (Anagnostou et al. 2016), began to fall (see Fig. 3.2) and the temperature gradient between the equatorial regions and the poles increased, essentially due to the increased difficulty in transporting thermal energy from the equator to higher latitudes. The decrease in heat flow was the result of changes in ocean currents caused by the movement of the continents that led to the opening of the Drake Passage between South America and Antarctica and the closing of the connection between the Atlantic and Pacific Oceans in what is now Panama. The first half of the Cenozoic was also a period of intense orogeny in which the formation of the Himalayas, the Tibetan Plateau, and the Rocky Mountains in North America perturbed the general circulation of the atmosphere and reduced the atmospheric CO_2 concentration by intensifying the weathering of rocks in the new mountains. This orogeny contributed to reducing the average global atmospheric temperature and helped begin glaciation.

Roughly 34 million years ago, an ice sheet formed in Antarctica due to a fall in atmospheric CO_2 levels to below 750 ppmv (Pagani et al. 2011). It went on to melt to some extent in the Oligocene and eventually extended again in the Miocene, around

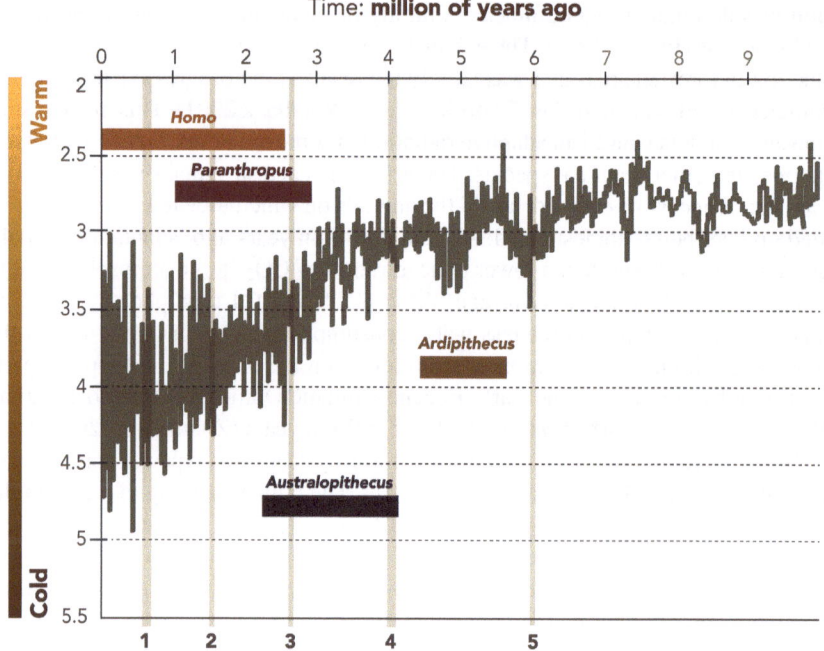

Fig. 3.3 A close-up of the last 10 million years of the global climate curve of Fig. 3.2. The numbers *n* ranging from 1 to 5 the *lower horizontal scale* represent the times of important events in the history of the Hominini sub-tribe: (5) represents the approximate time when the Hominins emerged, (4) was the time when bipedality became well established, (3) is the time of the first sharp-edged stone tools of the Oldowan technology and also probably the time when the *Homo* genus appeared, (2) was the onset of long-endurance mobility, and (1) marks the beginning of an intense encephalization process that culminated with *Homo neanderthalensis* and *Homo sapiens*. Notice that, since (5), climate variability has increased, which implies a greater environmental uncertainty, and that, since about 4 million years ago, the global mean temperature has decreased more quickly, causing significant environmental change across the world and in particular in Africa, where it has transformed tropical forests into savannahs. Adapted from Zachos (2001)

12 million years ago. In the Northern Hemisphere, the formation of ice sheets started much later, around 3 million years ago (see Fig. 3.3) (Zachos 2001).

As we approached the present day, the Earth's climate became not only cooler but also more variable, with relatively fast and large oscillations in the average global atmospheric temperature (see Fig. 3.3). At the same time as the polar ice cap in the Northern Hemisphere was being formed, palaeoclimatic records of the global mean surface temperature reveal the emergence of Milankovitch climate cycles, initially with a period of 41 000 years and, later, around 1 million years ago, with a period of 100 000 years (see Fig. 3.4). The main cause of these climate cycles, identified by Milutin Milankovitch in the first half of the 20th century (Milankovitch 1930), is climate forcing arising from small variations in the orbital eccentricity of the Earth

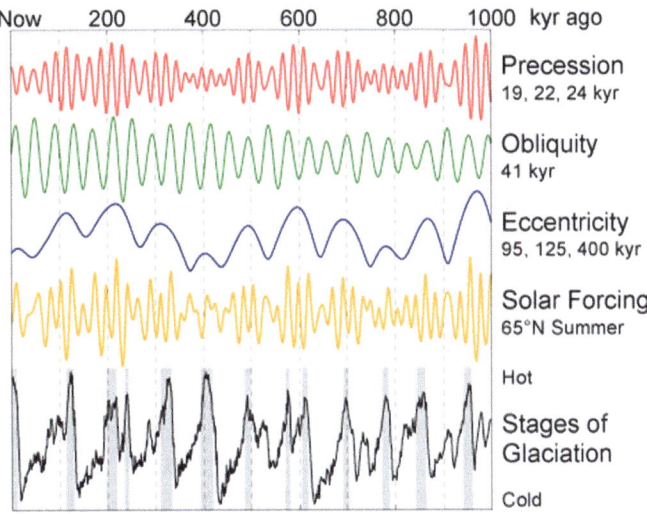

Fig. 3.4 The Milankovitch climate cycles result from climate forcing associated with small variations in the orbital eccentricity of the Earth around the Sun, in the obliquity of the Earth's rotation axis, and in the precession of that axis. These cyclical changes cause variations in the amount of solar radiation received at different latitudes (solar forcing) generating climate cycles with periods lasting 100 000, 41 000, and roughly 26 000 years, respectively. The figure shows the variations of the solar forcing at 65°N in the summer and their correlation with the observed glacial cycles, which have a mean period of about 100 000 years. The orbital data shown are from Quinn et al. (1991). Principal frequencies for each of the three kinds of variations are labelled. The glacial data are from Lisiecki et al. (2005). The *grey vertical bars* in the lower curve indicate interglacial periods, defined here as deviations in the 5 kyr average of at least 0.8 standard deviations above the mean, and coincide with relatively high values of the solar forcing. Image produced by Robert A. Rohde licensed in Wikimedia Commons under the Creative Commons Attribution

around the Sun, in the tilt of the Earth's rotation axis, and in the precession of that axis, which cause variations in the amount of solar radiation received at different latitudes, generating climate cycles with periods lasting 100 000, 41 000, and roughly 26 000 years, respectively.

The climate in Africa, where the *Homo* genus emerged, is largely determined by the monsoon, the intensity of which varies with the 26 000-year cycle connected to precession (Larrasoaña et al. 2003). When this movement causes very warm summers in the Northern Hemisphere, the monsoon in Africa becomes more intense and the continent becomes wetter above the equator, as happened over a period of 5 000 years between 11 000 and 6 000 years ago. During this time interval, the Sahara was covered in savannah. However, 5 500–6 000 years ago, the wet period ended and the Sahara became a desert over a period of roughly 1 000 years. The formation of the polar ice cap in the Arctic produced colder climates at high latitudes and a cooler, dry, and windy climate in Africa. The African continent was affected by the 41 000 and 100 000 year Milankovitch cycles of the global climate, together with the 26 000 year wet–dry cycle associated with the precessional motion of the Earth rotation axis.

We know with some confidence that the marked transition in Africa to a cooler and drier climate between 3 and 2.5 million years ago coincides with a period when the *Australopithecus* branch split into the *Homo* genus and the *Paranthropus* genus (Behrensmeyer 2006). *Australopithecus* was adapted to a habitat of tropical rain forests which, with climate change, were partially transformed into sparser forests and savannah. This environmental change forced a new adaptive evolution that produced the *Paranthropus* genus with a robust anatomy specialised for a herbivorous diet. The latter became extinct around 1.2 million years ago. It also produced the *Homo* genus, with a more generalised anatomy, a cranium with a greater volume, around 650 cm^3, a flatter face, and smaller molars than the *Paranthropus*.The only surviving species is *Homo sapiens*.

It is likely that the first systematic users of stone instruments belonged to the *Homo* genus. The oldest known instruments emerged around 2.6 million years ago (see Fig. 3.3) and belong to the stone tool technology known as Oldowan because many examples were found at the palaeoarchaeological site at Olduvai Gorge in Tanzania. Encephalisation, one of the most distinctive features in the evolution of the *Homo* genus, is likely related to the adaptation process of a relatively recent biped to a drier climate that produced grassland and savannah with open spaces where it was difficult to compete in the search for food and fend off predators (see Fig. 3.3). Encephalisation is also related to the development of complex social behaviour in small groups that developed protection and cooperative intragroup strategies and highly competitive and aggressive intergroup strategies. With the fragmented and limited knowledge available to us, it is hard to establish a relationship of cause and effect between climate change and the emergence of the *Homo* genus, but it seems likely that it played a relevant role (Behrensmeyer 2006).

Almost half a million years after the *Homo* genus emerged in Africa, *Homo erectus*, sometimes also called *Homo ergaster*, migrated to Eurasia roughly 1.9 million years ago. From there it quickly spread through India, China, and the island of Java, but also through Georgia and to the west until it reached the Iberian Peninsula, all regions where fossils of the species or closely related species have been found. Fossils of *Homo heidelbergensis*, which may have evolved from *Homo erectus*, were found in various parts of Europe, same dating to about 600000 years ago. This species is closely related to the Neanderthals (*Homo neanderthalensis*), who appeared about 400000 years ago and occupied a large part of Europe and western Asia, and to the Denisovans, whose fossils have been discovered in Siberia. Fossils have been found in China from the period between 900000 and 125000 years ago bearing characteristics between *Homo erectus* and *Homo sapiens* (Gao et al. 2010), demonstrating that there was a very significant evolution in the *Homo* genus in that region of the world. These results are the basis for the multiregional theory of the evolution of *Homo sapiens* as an alternative to the out-of-Africa theory, which has a solid genetic foundation in the genetic analysis of human populations around the globe.

It seems likely that the extraordinary migration of *Homo erectus* was caused or encouraged by the effects of the change to a drier climate that transformed the flora and fauna of Africa. The change in habitats and vegetation made *Homo erectus* more carnivorous than its ancestors and this drove it to greater mobility to increase opportunities for finding food by hunting or necrophagy.

Roughly 2.6 million years ago, a significant speciation began in Africa in the *Bovidae* family, which led to the emergence of species better adapted to savannah and grassland, particularly antelope (Vrba 1995). Some of these bovines migrated from Africa to Asia, taking with them the *Homo erectus* that hunted them.

It was probably at this time that the ability of *Homo* species to adapt to significantly different environmental conditions began to develop. They spread throughout a vast area that included Africa and much of Eurasia, but the population density was probably very low. Information about demographics can be obtained by interpreting certain differences in the comparative analysis of human genome sequencing using demographic models. This type of analysis suggests that the total population of our ancestors up to 1.2 million years ago would have been very small, less than 26 000 individuals (Huff et al. 2010). The problems associated with the very low population density, which certainly made conditions for reproduction more difficult, were later successfully overcome by *Homo sapiens*. One of the likely causes for the extinction of *Homo neanderthalensis*, probably a descendent of *Homo erectus*, around 40 000 years ago, was having a population spread out over a vast area of Europe and the westernmost part of Asia with very low population densities. It lived in small, isolated groups that practised endogamy (Sánchez-Quinto and Lalueza-Fox 2015).

Fossil records indicate that *Homo sapiens* emerged in Africa, and until recently it was thought that it emerged in East Africa around 200 000 years ago. However, fossils of a primitive version very close to *Homo sapiens* were recently discovered in North Africa, in Morocco, that were around 315 000 years old (Hublin et al. 2017). It is most likely that *Homo sapiens*, or very close ancestors of *Homo sapiens*, occupied most of Africa 300 000 years ago and reached the northernmost region of Africa at a time when the Sahara had rivers, lakes, and savannah.

The global climate was warmer and wetter then. It was in the final stage of a warm interglacial of one of the Milankovitch cycles that lasted approximately 100 000 years (see Fig. 3.4). These cycles have a relatively warm interglacial period lasting roughly 10 000–30 000 years, followed by a rather longer glacial period, with an average duration of roughly 80 000 years, when the average global atmospheric temperature is approximately 6–8°C cooler than during the interglacial period (Berger et al. 2016).

After the interglacial period that culminated around 325 000 years ago, a glacial period began during which climate variability increased. The effects of this variability can be seen in the Olorgesailie Basin in Kenya, where a transition from Acheulian stone technology to another more evolved technology can also be observed 320 000 to 295 000 years ago. The latter technology often used obsidian from sites more than 25 km away, and iron-rich rocks were exploited to extract red pigments used in artefacts (Brooks et al. 2018). It seems likely that climate change and the new environmental conditions it created forced the *Homo* population that lived in the Olorgesailie Basin to develop new technologies and perhaps a symbolic culture to survive (Brooks et al. 2018; Potts et al. 2018). It could therefore be presumed that this transition was at the root of the emergence of *Homo sapiens*, although fossils of the *Homo* population that produced that transition have not yet been found.

The glacial period that followed the interglacial 325 000 years ago ended with another short interglacial roughly 240 000 years ago (see Fig. 3.4). A new glacial period followed, this time with average global temperatures that reached very low values in the period between 195 000 years ago and the beginning of the last interglacial period around 123 000 years ago, called the Eemian (see Fig. 3.4).

During this very cold period, known as "Marine Isotope Stage 6" in palaeoclimatic studies based on analyses of deep sea sediments (Emiliani 1955), Africa became a cooler, drier continent with extensive desert, arid, and semi-arid areas, except for some small regions, particularly in the south, with relatively abundant vegetation.

This difference in climate was very likely at the root of a reduction and a geographical and genetic split in the *Homo sapiens* population in Africa. Although it is very difficult to trace the migratory movements and demographics of that population, some information can be obtained by analysing the genome of individuals of different ancestries (Gronau et al. 2011). Kim et al. (2014) analysed the complete genome of five individuals belonging to the Khoisan people who inhabit the south of Africa and have retained the practice of shepherding and a hunter–gatherer culture until the present day, and one individual of the Bantu people, dominant in the west of Africa.

A coalescence analysis of genomic sequence data demonstrated that the Khoisan and their ancestors formed the group with the greatest population since the populations split in Africa between 150 000 and 100 000 years ago (Kim et al. 2014). The non-Khoisan group, including the Bantu, saw a marked decrease in population after the split, probably due to the change to a more arid climate in the central and northern regions of Africa, and it lost roughly half its genetic diversity.

Later, there was a new split in the non-Khoisan *Homo sapiens* population with migrations from Africa to Eurasia, which probably began roughly 120 000 years ago (Bae et al. 2017). Initially, there were many relatively small migrations, first through the southern tip of the Arabian Peninsula, which explains the appearance of *Homo sapiens* fossils in southern and central China dating from 70 000–120 000 years ago. Later, roughly 60 000 years ago, there was a more significant migration responsible for the subsequent presence in inland Europe. It was this migration that would have heavily reduced genetic diversity in most contemporary non-Africans (Bae et al. 2017).

It should be noted that *Homo sapiens* fossils were found recently in the Misliya cave at Mount Carmel, in Israel, which have been dated as being more than 177 000 years old (Hershkovitz et al. 2018). This first departure from Africa through the Middle East was not favourable to *Homo sapiens* penetrating inland Europe and Asia, perhaps because at the time the climate was moving into a progressively cooler, drier stage until it reached the lowest global average temperature of the last glacial period, i.e., Marine Isotope Stage 6.

If we compare these migrations by *Homo sapiens* outside Africa with the previous migration of *Homo erectus*, we can conclude that they are very different in terms of both the timescale and the driving forces behind them. For the latter, the main impetus probably came from the speciation and migration of the fauna *Homo erectus* depended on.

In conclusion, most of today's human population finds its ancestors in the *Homo sapiens* population that migrated from Africa to Eurasia roughly 60 000 years ago and then to other continents, so its genetic biodiversity is relatively small when compared to the Khoisan. One of the most remarkable aspects of this population of *Homo sapiens* was that it gradually occupied the whole world, including regions with very different climates and biomes, from the steppes of northern Eurasia to the arid regions which border deserts and even tropical rain forests. The group that remained in Europe and the Middle East led the Upper Palaeolithic Cultural Revolution around 40 000 years ago, during a glacial period.

3.3 Climate Change Events in the Holocene Which Influenced Civilisations

The Holocene geological epoch, defined as the end of the last glacial period, began around 11 700 years ago. The climate become warmer and wetter and stabilised during the current interglacial period. This stabilisation to a milder climate was probably one of the drivers that contributed to the emergence of agriculture roughly 11 000 years ago in the Fertile Crescent in the Middle East and, a little later and independently, in other areas of the world, such as China and South America. Until then, humans were hunter–gatherers and population densities were very low. With the emergence of agriculture, populations grew significantly and the first city-states were formed in Sumer. The transition from the last glacial period to the current interglacial started the emergence of the great civilisations of the last 6 000 years.

During this period, it has been possible to use palaeoclimatic data to identify several climate change events and some of the related impacts on those civilisations. Prolonged periods of droughts are generally much more destructive than wet periods because they cause a scarceness of water and foodstuffs essential to human life. There are well-documented examples of shifts to drier climates that led to the weakening and eventual collapse of a culture or civilisation and its replacement by another.

One of the most intense cases of aridification in the Northern Hemisphere during the Holocene happened in the Middle East and part of northern Africa and was the result of the Milankovitch cycle (see Fig. 3.4) lasting roughly 26 000 years that, as already mentioned, began the desertification of the Sahara 6 000 to 5 500 years ago. This climate change caused people who lived there to migrate to the Nile valley, where they would later build the Egyptian civilisation (Parker et al. 2006; Brook and Barnosky 2012). The same climate event probably also contributed to the collapse of the Ubaid culture, the first stage of the Sumer civilisation, which lasted from c. 8 500 to 5 800 years ago in Mesopotamia and was replaced by the Uruk culture, a more evolved stage of the same civilisation. Later, around 4 200 years ago, there was a new climate event characterized by droughts that had relatively well-documented impacts on several regions of the world.

Let us begin with the Egyptian civilisation, where the cyclical floods of the Nile had vital importance for people's lives. Although, as we have seen, the start of the annual flooding could be foreseen, the maximum height reached by the waters varied a great deal and was unpredictable. Very high flooding would be destructive and could devastate settlements and riverside infrastructures. On the other hand, weak floods reduced agricultural production and caused widespread famine. The progress of the flooding was essential to planning the new year and probably to calculating the annual amounts of tax.

Priests in the temples measured the height of the Nile's waters using nilometers and encouraged their reputation among the people that they could foresee the time of the annual flooding (Bell 1970). Some of these nilometers have lasted to the present day, such as those at Elephantine, Edfu, Esna, Kom Ombo, Dendera, and Thmuis. They are formed of corridors and stairs that lead to the river, and whose steps are gradually submerged as the floods advance, or they are formed of wells, also accessible by stairs that are connected to the river through tunnels. Nilometers were used for more than 5000 years and there are written records of the Nile's water level for most of the last fourteen centuries. Analysis of this data shows that the variability in the Nile's flooding is related to the El Niño–Southern Oscillation (ENSO) phenomenon (Quinn 1992). The East African Monsoon is the main source of precipitation that feeds the Nile via the waters of the Blue Nile. During El Niño events, the waters of the Eastern Equatorial Pacific are abnormally warm, and this produces upward movements in the atmosphere. On the other hand, in more westerly regions, including the Indian Ocean, there are anomalous downward motions which weaken the monsoon and cause droughts on the Ethiopian plateau, where the source of the Blue Nile is located, and very low water levels in the Blue Nile and the Nile.

With the construction of dams, the water levels of the Nile downstream were no longer related to El Niño. However, an analysis of the historical records of floods from 622 to 1522 and from 1871 to 1997 (Eltahir and Wang 1999) shows that the higher frequency of El Niño events observed since the end of the 1970s, for roughly four consecutive decades, in comparison with previous periods, is very likely an anomaly caused by anthropogenic climate change (Trenberth and Hoar 1996). As well as the higher frequency, projections based on climate scenarios indicate that extreme El Niño and La Niña events will gradually become more frequent in the future as a result of climate change (Cai et al. 2014; 2015; Wang et al. 2017).

The reign of Pharaoh Pepi II, who took the name Neferkare, began around 2246 BC and was one of the longest in Egyptian history, perhaps more than 90 years. After his kingdom, historical records become scarce, but it is known that in a short period of time of around 20 years, 18 kings and probably one queen rose to the throne. Egypt went into a deep crisis. It is also known that around the year 2150 BC and for two or three decades, the Nile floods fell drastically, sands invaded part of the river valley, Faiyum Oasis dried up, the soils of the delta deteriorated, and famine spread throughout Egypt, paralysing the political institutions and sowing chaos (Hassan 1997; Stanley et al. 2003).

The biographical part of the inscriptions on the tomb of Ankhtifi, the governor of Edfu and Hierakonpolis in the 9th dynasty, includes the passage "the whole country has become like locusts in search of food". People were driven to carrying out dreadful atrocities due to the famine, including, very probably, cannibalism. Temples were vandalised and looted and statues were destroyed. At the end of Pepi II's reign, the Pharaoh's central government collapsed and the governors of the different regions, called nomarchs, assumed power at local level and fought each other. The so-called First Intermediate Period in Egypt's history began. Nevertheless, around 100 years later, central governance returned with the reunification of Egypt achieved by Pharaoh Mentuhotep II. His reign began the Middle Kingdom that started around 2055 BC and lasted until 1650 BC.

The deep crisis that affected Egypt produced a new political framework charac-terised by greater sensitivity to social issues, mercy, and compassion. This would likely have been the first time in the history of civilisations that a government, based on a strong, centralised hierarchy, adopted, albeit in an embryonic way, social con-cepts of equality that involved the pharaoh protecting the weakest and poorest in society, especially at times of adversity. Later, these concepts and practices flour-ished in different ways with Christianity and Islam. One of the clearest signs of the transition to new forms of equality was making immortality accessible to all and not only to the pharaoh and ruling elites. The formula found for that accessibility was to consider that, for the purposes of access to immortality, everyone is a king, thereby creating equality among all.

Palaeoclimatic data reveals that the period of abnormally dry climate that began around 2200 BC and caused the very low levels of the Nile was also experienced in other regions of the world, from North America to the Mediterranean, the Middle East, East Africa, India, and China. It is very likely that this was the main cause of the collapse of the Akkadian Empire in Mesopotamia, which took place around 4170 years ago. Palaeoclimatic analyses of sea sediments deposited in the Persian Gulf, transported partly by the prevailing winds of the north quadrant, show that there was an increase in the amount of dust at the time because of the more arid climate in Mesopotamia (Cullen et al. 2000).

The shift to a drier climate also had an impact on the Indus valley civilisation in the Bronze Age, which flourished between 4600 and 3900 years ago and whose main cities were Harappa and Mohenjo-daro. Roughly 4200 years ago, a depopu-lation of the great urban centres began, and commerce dropped off with Egypt and Mesopotamia which were also impacted by droughts, starting a process of decline for the Harappan civilisation that eventually led to its extinction approximately 1700 years ago (Staubwasser et al. 2003). In India, in the state of Gujarat, in a periph-eral area of the Indus Valley civilisation, recent archaeological studies show that, at that time, people were changing their agricultural crops as a way of adapting to the weakening monsoon that made the region more arid (Pokharia et al. 2017). It is also likely that, in China, the same climate change contributed to the disappearance of the Liangzhu culture, the last of the jade cultures in the Yangtze River Delta.

In the Iberian Peninsula, during the Bronze Age, the drier climate 4200 years ago was at the root of intriguing stone constructions called Motillas that can be found in Castilla La Mancha, in Spain, near Ciudad Real. Archaeological research in recent years has led to the conclusion that Motillas were buildings designed to make use of groundwater and to store water and cereals at a time of severe drought (Mejías Moreno 2014). A particularly deep well dating back roughly 4 000 years, probably the oldest one in the Peninsula, has been found at Motilla de Azuer. It would have been used to get water from a deep groundwater level. The construction of wells in the Motillas Culture was successful in dealing with the drought, and this helped boost the transition to a more complex, structured society.

Concerning the origins of the drought climate event in the Holocene 4 200 years ago, little is yet known. It may have been related to variations in the surface temperatures of the West Pacific, the Indian Ocean, and the North Atlantic.

There are other examples of droughts that caused collapses or transitions in civilisations. In the period between 750 and 900 AD, the Classic Maya civilisation collapsed, in part due to prolonged periods of drought (Kennett et al. 2012; Evans et al. 2018a). A similar situation occurred in the Tiwanaku Empire between 1000 and 1100 AD and in the Khmer Empire based in Angkor in Cambodia in the 14th and 15th centuries (Buckley et al. 2010).

Before modern meteorology and climatology had developed in the 20th century, droughts were considered to have a supernatural origin connected to divine plans, or else they were simply mysteries. In many different cultures, there are gods and goddesses associated with rain, weather, water, fertility, and the abundance of crops. Examples include the god Baal in the Middle East, Chaac in the Maya civilisation, and Tlaloc in the Aztec civilisation. There are historical records of a wide variety of rituals and sacrifices to bring rain in different cultures around the world. Some of these rituals are still used at times of drought.

Humanity is currently facing anthropogenic climate change that is intensifying droughts and making them more frequent in many regions of the world (Dai 2013; Schwalm et al. 2017). Unlike in the past, we now know the cause of the trend towards aridification that can be seen around the globe, but we have been unable to address effectively the causes behind it. The persistence and greater frequency of these new droughts depends solely on our behaviour, but we have not been willing to change it. In countries with advanced economies, we do not need to perform rituals or sacrifices to the gods to stop the new trend towards drought. Now, we appear to be the gods and goddesses, but we consider that we do not need to eradicate the causes of droughts because we suppose that we will always be able to overcome and cancel out its serious effects.

There was recently a drought in the Syrian region that lasted 15 years, from 1998 to 2012, and it was particularly intense from 2006 to 2009. It was probably the most intense drought of the last 500 years. Some researchers consider that its consequences helped form the conditions leading to the civil war that began in March 2011 (Kelley et al. 2015), which has killed between 311 000 and 475 000 people and has driven roughly 5.1 million refugees out of Syria.

Due to anthropogenic climate change, decadal average precipitation has been falling in the Mediterranean, particularly in the Iberian Peninsula, the Balkan Peninsula, and the area of the Middle East that is home to Israel, Jordan, Palestine, and Syria. Droughts are becoming more common and longer, and the drought in Syria is part of that trend, one that will certainly worsen in the future. Anthropogenic climate change will have profound social and economic impacts around the globe, including increases in the number of conflicts over the coming decades and centuries (Hsiang et al. 2013; Carleton and Hsiang 2016).

3.4   Symbolic Representation, Time, and Art

The oldest known artefact that is likely to have been worked by the *Homo* genus is a fossil *Pseudodon* mussel shell with geometric engravings found in the collection from excavations made by Dutch palaeoanthropologist and geologist Eugene Dubois (1859–1940) in 1891 at a riverside settlement inhabited by *Homo erectus* in Trinil, Java (Joordens et al. 2015). Studies carried out on the finds show that *Pseudodon* fresh water mussels were used for food, pierced at the point that helped open them most easily, and also used for making tools. The dating of sediments and pollen grains found in the shells at the archaeological site gives an age between 430 000 and 540 000 years old.

One of the shells has a geometric engraving with several very pronounced straight lines, including a zigzag in a continuous line with a design, part of which looks like the letter M. When they were alive, this species of mussel had a black film, so the incision made in the shell is clearly visible because of its light colour on the dark background. During an attempt to reproduce a similar engraving, including its microstructure, it was concluded that it could have been made by a piece of sharp flint or a shark tooth. The second possibility is more likely, given the number of such teeth found at the site. No similar records are known, but it seems likely that the zigzag from Trinil was engraved by a *Homo erectus* (Joordens et al. 2015). What is most fascinating in this find, which is almost half a million years old, is that it is similar to something anyone might draw on a shell in a moment of leisure. It shows that the mind of *Homo erectus* was already able to free itself of the immediate demands of the present and hence digress, producing a careful engraving with a design that appears abstract to us.

The reason for discussing art at this point, especially the first artefacts and rock paintings made by humans, is its profound relationship with operative social time. This relationship has probably been the most essential characteristic of art since its most distant beginnings. The geometric design on the Trinil shell is something with a human value, an ability to interest and captivate humankind, that remained for roughly half a million years until it was found again in 2014. Art, particularly the visual arts, often tends towards conditional immortality in *Homo sapiens'* time. It is conditional because any work of art may be destroyed accidentally or sometimes intentionally; immortality in *Homo sapiens'* time because there are many works of art that hold human value in this element of operative social time, and for this reason they are carefully preserved for future social generations. Works of art are one of the most important ways of building multigenerational operative social times, and they are powerful vehicles of intergenerational communication.

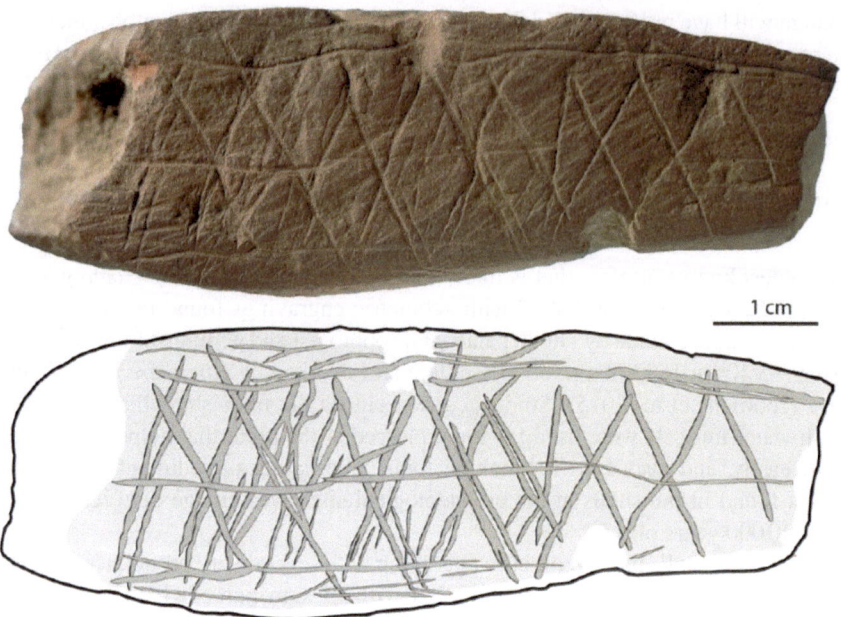

Fig. 3.5 Ochre piece with a deliberately engraved cross-hatched design found among other engraved ochres from c. 75 000 year old levels of Blombos Cave, Western Cape, South Africa. Additional pieces of incised ochre recovered from c. 75 000–100 000 year old levels at Blombos Cave have also been found (Henshilwood et al. 2009). These artefacts represent the oldest known examples of graphical representation in modern human culture.   *Credit* Fig. 3.9 of Henshilwood et al. (2009)

The next oldest known artefacts with geometric designs were found in the Blombos Cave in South Africa (Henshilwood 2002; Henshilwood et al. 2009) and were made by *Homo sapiens* 100 000–70 000 years ago (see Fig. 3.5). The collection of human artefacts found in the cave is very rich and includes two engraved ochre stones, a fragment of bone that is also engraved, beads made from shells, bone instruments, and stone tools that are very advanced for the time (Mourre et al. 2010). It is believed that these items, considering their complexity, diversity, and age, are the first visible signs of symbolic thought and a proxy for the emergence of the capacity for spoken language in humans (Huybregts 2017). Up to now the inhabitants of the Blombos Cave represent the start of modern human behaviour in Africa roughly 77 000 years ago (Henshilwood et al. 2009), long before its relatively fast blossoming in Eurasia around 40 000 years ago in the Upper Palaeolithic Revolution mentioned earlier.

Perhaps prior to this revolution modern human behaviour had evolved more slowly in Africa for roughly 250 000 years, but was now manifested, with greater exuberance, in the Blombos Cave. It seems highly likely that the capacity for spoken language developed during a period before the separation of *Homo sapiens* populations at least 100 000 years ago, long before it fully emerged and became externalised (Huybregts

2017). This means that the arrival of more fluent spoken language took place independently in the separate populations. We do not have records of those developments, but the first forms of symbolic representation on the Blombos artefacts mean we can guess that the cognitive and symbolic thought capacities needed to make them were also essential to advanced oral communication with our peers, i.e., oral language (Miyagawa et al. 2018). Symbolic representations on material media or visual arts, using the current form of expression, due to their tendency towards timelessness, help us understand the essence of our evolution. These were the representations that later, after the advent of agriculture, served as a conduit for written language.

On one of the Blombos ochres, intentional engravings were found of series of parallel lines in two directions, framed by another two lines made using an engraving technique involving several stages, in which the lines were etched first in one direction and then in another. This is a small abstract design that we are able to consider like any other, whatever age it may come from, and whose manufacture we understand perfectly (see Fig. 3.5).

Interestingly, there is another species of the *Homo* genus, *Homo neanderthalensis*, which we know also made abstract representations in the Middle Palaeolithic at least 64 800 years ago, in three caves in the Iberian Peninsula (Hoffmann et al. 2018b). Uranium–thorium dating of these paintings (Pike et al. 2017) shows that the Neanderthals were painting on the walls of caves in Eurasia at least 20 000 years before *Homo sapiens*. It was known that Neanderthals used pigments to paint their bodies and adorned themselves with shells, eagle claws, and feathers (Zilhão et al. 2010; Hoffmann et al. 2018a), but there was no conclusive proof that they also did rock art. The paintings found are geometric elements such as dots, lines, discs, geometric designs in the form of stairs, and hand stencils, all done using red pigment. Most of the paintings are located at places that are hard to reach, in the depths of caves and where the walls form special shapes. These characteristics demonstrate the intentionality of placing what were very likely symbolic signs in places that also have an unknown symbolic value.

We know very little about the behaviour and symbolic thought of the Neanderthals, but the discovery of rock paintings shows that they had cognitive and symbolic abilities that were very similar to those of humans, going against what had so long been thought. Abstract drawings etched on flint stones have also been found in Crimea, dating from 37 000–35 000 years ago, and they were very likely the work of Neanderthals towards the end of their existence (Majkić et al. 2018).

The less advanced culture of the Neanderthals had so frequently been cited to explain their extinction shortly after *Homo sapiens* arrived in inland Eurasia. But with the discovery of rock paintings, their extinction gained another dimension that may also be interpreted as a sign of our own vulnerability to extinction. The great cultural similarity between Neanderthals and the first humans makes our later cultural evolution less comprehensible and more random and contingent. Some authors, using demographic models (Powell et al. 2009), attribute the rapid cultural evolution and its maintenance in modern human behaviour around 45 000 years ago to higher population density, which would have driven greater development and complexity in social relations. However, there are likely to have been many causes, most of which remain unknown (Vaesen et al. 2016).

The oldest figurative artefacts made by *Homo sapiens* date from the beginning of the Upper Palaeolithic, around 40 000 years ago, and are part of the Aurignacian period. The first to be found, in 1939, at the entrance to a cave in Germany, in the Swabian Jura, was the Hohlenstein-Stadel Lion Man (Lobell 2012), made of mammoth ivory. It depicts a standing man or woman with a lion head and is part of a vast collection dating from 35 000–40 000 years ago, indicating that the characteristic behaviour of modern man had already begun.

Later, in 2008, close to the entrance to Hohler Fels cave in the same region of Germany, a mammoth ivory sculpture of a woman, known as the Venus of Hohler Fels (Conard 2009) dating back 35 000 years was found, together with other Aurignacian human artefacts. The figurine powerfully highlights the feminine attributes related to sex and reproduction. Bone and mammoth ivory flutes were also found in the cave, and are some of the first known musical instruments, demonstrating the extraordinary blossoming of human culture at the time (Wilford 2009).

Curiously, by mere chance, the two oldest figurative sculptures depict very different facets of human behaviour: one shows a mysterious and notably symbolic man–animal hybrid and the other depicts our pervasive sexual and reproductive impulses in an artistic language approaching what we would today call pornography. Both carry messages from our distant past that are somewhat enigmatic but intelligible, and in that sense they have become timeless in *Homo sapiens*' operative social time.

It was also in the Upper Palaeolithic periods that parietal art with a well defined style was developed inside some Eurasian caves. They are particularly concentrated in Europe, from the Urals to the Iberian Peninsula, but there are also examples in Indonesia of caves with animal paintings in the district of Maros. One of the most remarkable, thanks to what we today believe to be the aesthetic quality of the paintings, is the Chauvet cave in southern France (see Fig. 3.6), where the oldest paintings date back 32 000 years (Quiles et al. 2016). There are many other examples of caves decorated with paintings, including those of Lascaux and Pech Merle in France and Altamira in northern Spain (see Fig. 3.7).

This art is mostly animalistic, depicting animals of several species, predominantly ungulates that are visually well depicted and generally viewed in profile. It was not decorative art for spaces normally inhabited by humans. Humans intentionally ventured inside caves, often reaching the deepest areas that were hardest to reach, and there, by torchlight, they covered the walls with symbolic depictions in the form of paintings and drawings. It is common to find paintings that appear to have been inspired by the zoomorphism or anthropomorphism of the shapes of the walls, thereby creating something approaching three-dimensional representations. People are not widely represented, and when they do appear, they are mostly human fragments: hands, vaginas, phalluses, eyes, noses, deformed faces, and incomplete human beings or human–animal hybrids. People are not drawn with the same intention of achieving faithfulness or accuracy in representing form and essence that is present in the depictions of animals.

Fig. 3.6 Palaeolithic mural depicting horses, aurochs, and woolly rhinoceroses in the Chauvet cave situated above the former location of the Ardèche River, near Vallon-Pont-d'Arc in Southern France. The paintings have been dated between 32 000 and 30 000 years BP in the Aurignacian period. The Chauvet cave was occupied by humans during two periods from 37 000 to 33 500 years BP and from 32 000 to 28 000 years BP. About 21 000 years ago, the entrance collapsed and the cave was completely sealed off until its rediscovery on 18 December 1994 by Jean-Marie Chauvet, Éliette Brunel, and Christian Hillaire (Chauvet et al. 1995). *Photo credit* Heritage Images, Hulton Fine Art Collection, Getty Images

Animals are much more prominent and appear to emerge as metaphors for humans, since the much deeper *raison d'être* of the depictions was certainly people and their social life and groups. Full-body human depictions appear almost exclusively on mobile art that can be moved to the open air, generally found close to the entrances to caves where humans could live, such as the numerous sculptures like the Venus of Hohler Fels, including those from Dolni Vestonice, Lespugne, Willendorf, Moravany, Mezine, and Monruz. Another important point is that inside caves there is an iconographic organisation of symbolic depictions as if they were shrines. It is therefore possible to rationalise the layout of systems of abstract signs and animal panels, including their internal composition, on the walls, ceilings, and topography of caves (Leroi-Gourhan 1975). This set of characteristics of rock art suggests that we are dealing with human behaviour that expressed symbolic thoughts and could be used to develop, transmit, and share a worldview based on fundamental principles. And even though we are still a long way from fully deciphering those principles, they probably form something like a cultural or religious canon.

Fig. 3.7 Palaeolithic painting made with ochre and charcoal of the now extinct steppe bison (*Bison priscus*) on the Polychrome Ceiling of the Altamira cave at Santillana del Mar, Spain. The multi-coloured cave paintings, rock engravings, and drawings found in the cave have been dated from 36 000 to 15 000 BP, which includes the Aurignacian, Gravettian, and the Solutrean periods. *Photo credit* Jesus de Fuensanta, Stock by Getty Images

It is remarkable that this canon lasted roughly 22 000 years over a vast area of Europe (Testart 2012), much longer than the current civilisation, which originated in the Agricultural Revolution beginning roughly 10 000 years ago. The disparity between these two time intervals effectively reveals the extraordinary acceleration in human cultural evolution starting with the transition from groups of hunter–gatherers to the emergence of agrarian societies. During those 22 000 years, there were evolutions in stone tools, hunting and fishing implements, and adornments, and probably in many other human activities, while the essential characteristics of rock art remained the same. This art has the virtue of bringing us closer to the humans of that time and especially their forms of symbolic thought. We will never fully know their beliefs and views or their explanation for humankind and the world, but we can sense that they already thought in a very similar way to us.

The cave paintings at Chauvet, Lascaux, Altamira, and many others reveal a capacity for representation and synthesis, an expressiveness, a balance, a style, and a beauty that rival the drawings and paintings that we most appreciate in historical and modern art. Picasso, when he saw the Altamira paintings for the first time, belonging to the Magdalenian culture that flourished 17 000–12 000 years ago, exclaimed "after

Altamira, all art seems decadent" (see Fig. 3.7). The concepts of art and aesthetics obviously emerged much later in cultural evolution. Nonetheless, we are profoundly sensitive to what we now say to be the beauty and aesthetic value of the Altamira paintings, just as we are with the paintings at the Sistine Chapel, for example. They were done with very different purposes that we don't fully grasp now, particularly in the case of the former, but both share an extraordinary capacity to attract our attention and thought and to delight a large number of people. Works of art, as well as their aesthetic characteristics and subjective beauty, always have a human meaning that comes from the fact that they are man-made. Nature may be beautiful or sublime, but those characteristics do not have the value inherent in a human creation.

The time that separates us from those who made the Altamira paintings dissolves when we try to understand and interpret their visual representations. They are an unequivocal sign that one of the most important characteristics of human behaviour is the use of systems of symbolic visual representation shared not only by the communities in which their creators lived but also by their future social generations.

The importance of visual representation continues to grow in our age through a wide variety of sophisticated digital devices for visual representation and communication. Throughout prehistory and history, it has become unthinkable to imagine modern man without those systems of representation. The fact that they are excellent forms of communication, interaction, and social cohesion mean these systems play an essential role in building operative social time. It is likely that the development of visual symbolic representation during the Palaeolithic made a decisive contribution to developing and establishing the concept of operative time of a social generation or generational time.

We know that the decorated caves were frequented by humans for long periods of time, indeed over thousands of years. Chauvet cave, for example, was inhabited 37 000–33 500 years ago and later 32 000–28 000 years ago (Quiles et al. 2016). During each of those long periods, lasting roughly 3 000 years, more than 150 social generations passed through the caves, probably sharing the social and intergenerational value of the paintings (see Fig. 3.6). We also know that the caves were visited by humans of different ages. Fossilised footprints belonging to adults and children are found on the floor of the Niaux cave in the Vicdessos valley in the French Pyrenees, where the paintings were made during the Magdalenian period. It is highly likely that the symbolic depictions in those caves were among the most important media for a multigenerational cultural system that helped define the essential tension between the cyclic nature of generational time and the uniqueness of a lifetime.

What would those humans have intended, and what would they have wanted to externalise, express, and communicate to one another, when they covered the deepest areas of caves with drawings and pictures in the Upper Palaeolithic? This is a fascinating topic that has been, and will continue to be, extensively researched and discussed. Many advances have been made, especially using methods of iconographic analysis, but it will never be possible to fully decipher the conceptual processes and systems of ideas and convictions that inspired them. What we do know with some certainty can be sorted into several theoretical interpretive models. The first was proposed by Henri Breuil (1877–1961) in the 1950s and asserts that rock art has its origins in the magic of hunting (Breuil 1952). Animal paintings, particularly those that are traversed by lines that can be interpreted as arrows, would be to promote a good hunt. This model has a limitation in that it cannot explain the many seemingly abstract symbols that appear systematically in caves located over broad geographical areas. This is the case, for example, of a vertical P or a club-shaped symbol called a claviform, whose shape may have come from the schematic representation of a woman seen in profile, and so could have been a symbol of fertility.

The second model, developed by Jean Clottes and Davis Lewis-Williams (Clottes and Lewis-Williams 2001), is based on shamanism and establishes a link between visual representations and shamanic trances. Although forms of shamanism may have been present in human behaviour at that time, the explanation for Palaeolithic rock art culture is likely to be much more complex.

The third model, created by Alain Testart (2016), is broader and based on the already mentioned recognition of the iconographic organisation of the caves, interpreted as shrines, and proposes a formal analysis of all rock art involving abstract or anthropomorphic signs, animals, and spatial organisation. This analysis led to the conclusion that the canon of Palaeolithic hunter–gatherers that painted caves in Eurasia reflected a totemic view of the world, perhaps similar to the one found today among Aboriginal Australians. This totemism would have been a coherent way to institute a relation and to classify society and nature by establishing a kinship type of relationship between the multiple different clans of a social group and the animal species. Rock art is an indirect, symbolic way of structuring social life through those relationships. The cave represents the world in its original, chaotic state, but with a principle of underlying order. Painters sought to free and recreate what was already latent on the cave walls. Humans, unlike animals, were still in an undifferentiated, fragmented, amorphous, unformed, and sometimes human–animal hybrid state. Most abstract symbols suggest fragmented feminine forms inspired by sexual and reproductive attributes found in Venus-type figurines, which suggests that the fundamental principles of life and reproduction prevailed in the iconographic design of these caves. The superposition of female symbols over groups of animals may be interpreted as a way of propitiating reproduction and the continuation of life. These are probably the first visual depictions of myths, primordial cosmogonies, and religions, but we will never be able to decipher their stories. We cannot read or hear them, but we understand the essential meaning of their messages. In that sense we share the same social time, *Homo sapiens'* operative social time.

At the end of the Upper Palaeolithic and in the Mesolithic, open-air rock art began to develop in the form of engravings on rocky surfaces or isolated rocks, called petroglyphs. Unlike cave art, the figurative depictions became more anthropomorphic and less animalistic although both humans and animals are present. The human figures are schematic and do not have the realism of the drawings and paintings of animals in Palaeolithic caves. Man and woman are no longer represented in an unformed, fragmented way, and attention is now focused on describing human activities. Depictions frequently show scenes of fighting, hunting, dances, rituals, and food harvests and, in general, the figures became less enigmatic.

3.5   The Agricultural Revolution, Surpluses, and Excesses

At the beginning of the Neolithic, the practice of agriculture, the use of new methods to polish stone tools, and the emergence of ceramics, weaving, and other technologies developed and profoundly diversified human activities, social organisation, and religions. Agriculture and other drivers of change, such as the development of trade and metal-working, led to new weapons, new types of violent conflict, and new cultures. Architecture, that is, built heritage, and the arts, including literature, the visual arts, and music, flourished in various regions of the world. The Agricultural Revolution began in the Fertile Crescent in the Middle East roughly 10 000 years ago, at the end of the transition from the last glacial period to the current interglacial period when the average global atmospheric temperature stabilised close to 15°C and the climate became wetter. A little later, the Agricultural Revolution emerged independently in China and Central America.

From the Fertile Crescent, the practice of agriculture spread to the Balkan Peninsula and the Nile valley, where it gave rise to the Egyptian civilisation, ending up covering most of Europe from roughly 4 000 years ago onwards. Studies of human DNA found at archaeological sites now show us how the great human migrations of prehistory took place, and in particular, how agrarian migrant populations expanded and assimilated throughout Europe (Reich 2018; Guilaine 2017). One notable example of these migrations was the expansion of the bell beaker culture throughout Europe, which began roughly 4 900 years ago. Some of the oldest tall bell beakers of the so-called maritime style were found in Iberia. The culture lasted around 1 000 years and stretched over most of Europe including Portugal, Spain, France, Italy, the United Kingdom, Ireland, Denmark, Germany, Poland, Hungary, and Romania (Guilaine 2017). The transition from groups of hunter–gatherers to an agricultural society was not always continuous or irreversible. There have been cases where, due to several forms of adversity, such as overexploitation of natural resources or epidemics, there has been a return to the previous way of life. The city-states of Sumer give us the first examples of the consequences of deforestation and the salinization of soils caused by intensive agriculture (Scott 2017).

With the development of agriculture based on the domestication and use of some plants and animals, food surpluses were generated, part of the population was freed from strictly agricultural activity, and a social process of division of labour began, which created an increasing specialisation of human activities, new occupations and professions, and eventually new forms of social stratification. Division of labour increased efficiency in the production of goods and services in the agricultural society and created more opportunities to develop new technologies. In turn, these technologies helped boost efficiency and diversify the production of goods and services. Such processes, many of which were amplified by positive feedback, created surpluses of different types—not only food—and therefore greater wealth. Society appropriated that wealth in a very unequal way. In the social stratification process driven by the Agricultural Revolution, surpluses generated more wealth and this was distributed in society according to the relative power of different social actors and groups, which accentuated and created new inequalities of power and wealth (Lenski 1966; Boix 2010).

It is enlightening to review how the concept of the origins and reasons for social stratification in human societies evolved in the West. In the first period of modernity, the theme of inequalities particularly interested Jean-Jacques Rosseau (1712–1778). In his famous book *Discours sur l'origine et les fondements de l'inégalité parmi les hommes* (Rousseau 1755), Rosseau argues that humankind first lived in an egalitarian "natural state" and that it was the later development of civil society that eventually destroyed what he called moral and political equality. The first major description and classification of social stratification was produced by Max Weber and published posthumously in 1922 (Weber 1922). Weber believed that three distinct elements contribute to social stratification. The first is social class, which reflects economic situation and wealth accumulated in buildings, land, factories, financial assets, and other property. The second is status, which reflects the prestige that others ascribe to a certain person and is connected to ethnic and religious groups, titles of nobility, ancientness of families, and sporting, business, scientific, and artistic elites. Finally, there is the power that people or groups of people possess to achieve their goals through pressure groups and political parties, especially by appropriating the means of control and coercion at the disposal of the state.

In the mid-20th century, very special attention was paid to the functional aspects of social stratification, highlighting that economic inequalities play an important role in society as an incentive to acquire skills and perform activities that lead to an increase in wealth (Davis and Moore 1945). In 1966, Gerhard Lenski (1924–2015) (Lenski 1966; Collins 2004) went further and sought to identify the reasons for social stratification and its variations in the history of human societies. From his perspective, hunter–gatherer societies are very close to complete equality, while economic and social inequalities emerged with the development of what he called primitive agrarian societies. These inequalities became more intense in the transition to advanced agrarian societies, when they peaked, before falling to moderate levels after the transition to industrial societies. With agricultural societies, city-states emerged that developed a class of trained warriors to wage war and protect them from other city-states. Society adopts a political organisation focusing on military aristocracy that holds a significant portion of political power and land ownership. Profound economic inequality is generated between this political-military class and the social class of farmers who do not belong to the aristocracy, formed of peasants, servants, and slaves. The concept and practice of war emerges as an organised form of violent conflict between city-states.

One of Lenski's most important contributions was to highlight that inequality operates only in the economic production that is surplus, going beyond the level of production needed to guarantee everyone's survival. This observation implies that the potential for inequality increases with increased productivity. However, according to some authors, analysis of the theme of inequalities, why they get worse, and the consequences of that worsening has been scarce (Boix 2010). Later, at the end of the 20th century, science, particularly by way of ethnography, anthropology, and archaeology, enabled advances in the reconstitution of the evolution of human societies since prehistoric times. Alain Testart revisited the theme of the transition from hunter–gatherer to agricultural societies and concluded that there are economic and social inequalities comparable to those in agricultural societies in more than half of today's hunter–gatherer societies studied by him (Testart 1982). More recent studies have also revealed that there was no radical transformation in human behaviour in that transition and that a capacity for social stratification and the development and institutionalisation of social and economic inequalities were always in some way explicit in human societies, and perhaps in all species of the *Homo* genus (Smith et al. 2010). Social stratification was likely not particularly pronounced because surpluses were generally rare. This is latent behaviour that waits for the opportunity to show itself. If surpluses emerge and increase, humans make keen use of them and, by way of proactive intervention in the different mechanisms of production and consumption of those at the top of the hierarchy, they reinforce and amplify social stratification. If they fall, inequality and social stratification tend to decrease to a minimum. A society that is already used to surpluses and marked social stratification generally finds it more difficult to adapt to a reduction or scarcity of natural resources and food.

Recently, several authors have stressed that the Agricultural Revolution is at the heart of many of our current problems because they result from the long and irreversible path travelled since we stopped being hunter–gatherers (Demoule 2017; Scott 2017). We certainly still have a lot to learn about the circumstances and causes of the agricultural revolutions that took place independently in several regions of the world, but there have clearly been positive and negative aspects. The initial motivation was a result of the important discovery that some plants and animals could be domesticated and that their persistent and continued use would tend to improve food security and, as a result, produce reproductive advantages for populations of hunter–gatherers. The increase in human population was the main reason *Homo sapiens* followed the path of the Agricultural Revolution. It also provided more well-being because homes in settlements offered better comfort and protection against wild animals and bad weather. The new system also revealed negative features from the very beginning. Sometimes, agricultural collapses due to droughts, floods, or plagues led to famine and misery. Agricultural communities are sometimes invaded, attacked, sacked, and pillaged by armed invaders. Living in sedentary, sometimes overloaded communities close to domestic animals gave rise to new lethal diseases and epidemics. Although food security was generally greater, the diet was probably less nutritious and varied than hunter–gatherers diets, at least for some of the population (Schoeninger 1981; Crittenden and Schnorr 2017).

However, the Agricultural Revolution brought the promise of abundant agricultural production, able to satisfy appetites and bring prosperity. Later, it brought the experience of increasingly different and luxurious lifestyles, increased opportunities for social mobility, professional diversification, enrichment, ascent to power, domination over peers, military victories, land conquests, and lucrative pillaging. The Agricultural Revolution represented the beginning of a process of civilisation which the vast majority of people in the world joined freely over thousands of years. It was the first step enabling the emergence of the great ancient civilisations, in which the most diverse technologies were discovered, where mathematics, philosophy, and primordial forms of natural and social sciences began to flourish. It was the abundance produced by the Agricultural Revolution that enabled the remarkable development of the arts: architecture, visual arts, theatre, dance, music, literature, and poetry. The Agricultural Revolution was the cradle of the development of the different contemporary cultures, including Western culture. This process of civilisation also created an almost immediate acceleration of the exponential increase in human population. After a while, it led to the militarisation of societies, to wars and large scale battles, to deep and growing social and economic inequalities and crises, and later on to widespread degradation of the environment and the overexploitation of natural resources. After the Industrial Revolution, the latter trends became more serious and began to contribute to climate change.

These negative trends, which have manifested themselves throughout the process of civilisation, are not grounded in the Agricultural Revolution itself, but have a complex origin involving some essential characteristics of human social behaviour which enabled and developed those trends, when *Homo sapiens* took advantage of the opportunities offered by the Agricultural Revolution. Put another way, human nature was not such that it could avoid the negative outcomes. Furthermore, the essential characteristics of the human biological heritage that condition individual and collective behaviour do not change over periods of centuries or millennia, as experience has demonstrated. It is likely that they will not change in *Homo sapiens*' time. The Agricultural Revolution actually represented the start of a process of a human enterprise that in its present form is likely to be unsustainable in the medium and long term, but the seeds responsible for the behaviour that produced that lack of sustainability were already present in *Homo sapiens* long before the Agricultural Revolution. Is it really the case that the "Agricultural Revolution was one of the biggest frauds in human history" (Harari 2014)? Would it have been better for us to remain hunter–gatherers? How can it be a fraud if the current improvements in global indicators of human well-being can be traced back to the civilizations that emerged from the Agricultural Revolution, as is so keenly and frequently pointed out? Or is blaming the Agricultural Revolution for some of our current problems just finding a scapegoat for the difficulty we continue to have in evolving culturally towards sustainability, freedom, human rights enforcement, solidarity, and equity? It will very likely be impossible to complete this cultural evolution, but we should always try as hard as possible. This apparent contradiction is part of the human condition. It is preferable to acknowledge our past inability and endeavour to overcome it than to point to a false scapegoat. Such an acknowledgement will be essential if we are to change our individual and collective behaviour through small but definite steps.

Since the Agricultural Revolution, humanity has grown accustomed to surpluses in the production of goods and has been able to use them to increase the power of a few groups of people in society and then indirectly, slowly, and differentially to increase economic prosperity, well-being, and quality of life for the whole of the human population. There have been abuses and deviations in this process with dramatic consequences, but on average the outcomes have been positive for roughly 100 centuries. The way in which the economic and political power generated by surpluses is distributed has varied over history. However, surpluses are vital for the current economic system to keep functioning. Without the constant increase in productivity that produces surpluses, the system tends towards dangerous dysfunctionalities. At the most basic level of impulses, desires, fears, and ambitions, what drives this system is the fact that the proliferation of surpluses makes it possible to have a practically unlimited variety of behaviours that often tend to become extreme, and whose meaning for those who engage in such behaviour and for society ranges from very positive to very negative. Examples include cooperation, philanthropy, generosity, profligacy, intemperance, extravagance, luxury, greed, egoism, aggression, and violence. Our attraction to extreme behaviours, i.e., excesses, is part of our unalienable biological legacy. We can curb the impulses that drive us to engage in them, but we cannot suppress them completely. Probably we will have to live with them until the end of *Homo sapiens'* time.

Individual and collective extreme behaviours are largely responsible for the heritage of civilisation and culture that we benefit from today. This includes works carried out by various civilisations such as the Egyptian pyramids, the Great Wall of China, the Maya temples, the medieval cathedrals in Europe, and the wide variety of Renaissance works of art. It also includes the great voyages of discovery and exploration throughout the entire Earth system and outer space. At the individual level, we find remarkable examples of philanthropy, cooperation, protection, and love for the poorest and most vulnerable, full and tireless dedication, often in highly challenging conditions, to creation in mathematics, philosophy, science, technology, and art, to the cause of medicine and public health, to the cause of businesses producing goods and services, and to public service and governance. At the other end of the spectrum of extreme behaviour, we find cruelty, murder, torture, persecution, pillaging, terrorism, the systematic aerial bombing of populations, targeted killings especially by drones, genocides, the Holocaust of the Second World War, the dropping of atomic bombs on cities and even, more generally and commonly throughout the world, a huge range of crimes, thefts, and especially corruption.

3.6 A Contemporary Excess in the Distribution of Surpluses

The recent evolution in wealth distribution in advanced economies gives us a current example of surprising collective extreme behaviour. The case of the USA is paradigmatic and was recently studied by Piketty et al. (2016). When the distribution of annual increases in post-inflation, after-tax income is compared for two consecutive 34-year periods, one ending in 1980 and the other in 2014, two very different curves emerge (see Fig. 3.8) (Leonhardt 2017; Piketty et al. 2016). In the 1946–1980 period, the greatest increases are found in the middle classes and lowest incomes. The income increase of the middle class is roughly 2%, representing a doubling of post-inflation after-tax income in 34 years. In the 1980–2014 period, the curve is reversed and now the smallest annual income increases are found in the lowest incomes and the greatest increases are among the richest segments of the population. The middle class, rather than annual increases in post-inflation after-tax income of 2%, saw increases of 1% in the more recent period. In this period, 2% increases are only found among the 3% richest people. At the far end of the graph showing the highest incomes, the curve of annual income increases has a very obvious peak that reaches increases of more than 6% (see Fig. 3.8). This means that it is the richest and only the richest who have had high rises in income over the last decades. Wealth inequalities are becoming more pronounced and this has social and political consequences. The strength of the wealth increase for the 1% richest is a surprising excess for many but quite natural for some. There are multiple causes for it. An important one can be summed up in the observation that in the last three to four decades more of the income generated by the economy has flowed into profits rather than salaries, compared to the period immediately after the Second World War. Another important aspect is that the American economy is growing but the growth rate in the last decades has been on average lower than the average in the post-war period. In other words, the surpluses are decreasing instead of increasing as the large majority of the population unflaggingly demands.

It is also important to highlight that while the average external trade surplus for the 1946–1980 period represented 0.5% of the USA's GDP, there was an external trade deficit of 2.7% of GDP in the 1980–2014 period (Piketty et al. 2016). In the last decades, not only has economic growth been weaker, but the economy has become less competitive in the global market. American society's reaction to these adversities has come through a complex network of technological, political, economic, and social events that have redistributed the now more limited surpluses, favouring the rich over the middle class and poorest. All these developments demonstrate a strong and unfailing faith in the capitalist and liberal free enterprise economic model that most Americans believe to be not only the driving force behind the USA's ascension to becoming the world undisputed superpower, with the largest economy and the strongest military power, but also more generally the origin and essence of the USA's identity and exceptionalism (Lipset 1997). A closer analysis identifies a range of technologies, economic and political processes, legislations, and regulations that have helped to cause the recent increase in economic and social inequalities in the USA and in other countries with advanced economies. These include a trend towards globalisation that has lowered salaries in the industries most affected by external competition and technological advances that have further increased differences in salaries according to qualifications and skills, particularly in the field of digital technologies. Most of the increases in earnings inequality results from the increased difference of the earnings among companies, especially between high and low technology companies (Barth et al. 2014). ICT and the systematic use of outsourcing have allowed companies to concentrate on their core business and to reduce the number of employees to a minimum of highly paid jobs. Furthermore, in the new business models, ICT has transformed social operative time into a continuum of activity with no idle times in order to increase productivity. Other important factors include lower salaries caused by high and persistent unemployment rates in some sectors and by the

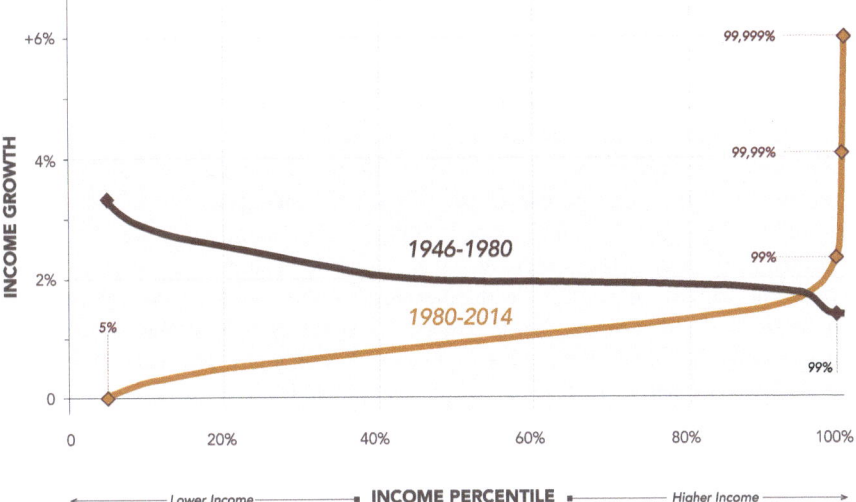

Fig. 3.8 Inflation-adjusted annual average income growth using income after taxes, transfers, and non-cash benefits, as a function of income, in the USA for two consecutive periods of 34 years after the Second World War. The *upper black line* corresponds to the first 34-year period 1946–1980, immediately after the war, at the beginning of the Great Acceleration. Middle class growth was typically 2% per year, which implies that the household's income almost doubles every 34 years. While the low income families had an annual income growth above 2%, the 99th percentile had an income growth below 1.5%. This situation is reversed in the following 34-year period 1980–2014, represented by the *ochre line*. The annual income growth now increases with income starting from no-growth for the very low income household's to growths of 6% for the 99.999th percentile. Figure adapted from Leonhardt (2017) and data from Piketty et al. (2016)

concentration of economic power in large companies. These trends lead to a loss of representativeness and bargaining power by the USA's trade unions. At government level, policies tend to counter progressive taxes, i.e., reduce the tax rate on the highest incomes and create highly complex special regimes that lower their tax burden. It is also important to mention the financialisation and increasing deregulation of the American economy, which leads to the Minsky cycle or the so-called "shampoo economy" characterised by bubble, bust, repeat. However, the attractiveness of this model, based on the accumulation of public and private debt, is that it is likely to achieve annual GDP growth rates of 4% or more in the cycle's growing phase. At the end of each cycle the collapse of the financial system is avoided using massive injections of taxpayers' money, primarily coming from those with average and low incomes. This cycle is an efficient way of transferring financial and economic power to the highest incomes and also to large multinational corporations.

The response of the contemporary economic system to the perceived failure of slow growth in most advanced economies is to strengthen its essential driving mechanisms. One of the most important is the elite of the top investors and CEOs of the major corporations. Their vision, intelligence, experience, combativeness, and endurance is crucial to create new jobs and to achieve economic growth. The economic system rewards their performance by differentially increasing their wealth up to unforeseeable limits, well above all segments of society. It is assumed that this incentive salience mechanism contributes to maintaining the country's exceptionalism and economic and military hegemony in the world.

Economic inequalities are also getting worse in other countries with advanced economies, where young people in the Millennial generation will be among the most affected. Countries with advanced economies are finding it harder to generate surpluses, and the ones they do create are accentuating wealth inequalities although in different degrees. In countries with emerging economies and others, the situation is more diverse and complex, but in general all have a greater potential capacity to generate surpluses than countries with advanced economies. However, it is likely that this capacity will progressively decrease at global level.

According to the 2018 World Inequality Report, the percentage of total income in 2016 shared by the 10% highest earners was 37% in Europe, 41% in China, 46% in Russia, 47% in the USA–Canada, around 55% in Sub-Saharan Africa, Brazil, and India, and 61% in the Middle East (Alvaredo et al. 2018). In the 1980–2016 period, inequalities grew sharply in North America, China, and India and moderately in Europe. In Russia there was steep growth during the years that followed the collapse of the Soviet Union, and then stabilisation followed by decline from 2008 to 2016 (Alvaredo et al. 2018). According to a report by the Swiss bank UBS (2018) the total billionaire wealth in the world increased by 19% in 2017, reaching $ 8.9 trillion. The combined wealth of the 2158 billionaires grew $ 1.4 trillion in 2017, which is considered the greatest absolute growth ever (Stadler et al. 2018). The wealth of the global uber-rich is now increasing much faster than at the end of the XIX century and beginning of the XX century when multi-generational families in Europe and the USA, such as the Carnegie, Ford, Guggenheim, Rockefeller, Rothschild, Vanderbilt, Wallenberg, and Wendel among many others, started to control vast wealth. China, a country where the conditions for economic growth are particularly favourable, is the one where the number of billionaires is growing fastest, at a rate in 2017 of two new billionaires per week. The progress of wealth inequalities throughout the world reveals that at the personal level rational egoism and ethical egoism are convictions that are increasingly accepted, defended, and diffused in human societies.

Globally, in the medium to long term, the capacity to generate surpluses is likely to fall in every country although at different rates. World annual real GDP growth per capita has been falling since the 1960s (see Fig. 3.9) (Maddison 2006; Jancovici 2012a) and will likely continue to fall due to the difficulties in satisfying global primary energy demand, overexploitation of natural resources, demographics, degradation and destruction of the environment, and climate change. We can solve the problem if we change our behaviour and use technology wisely to do so. Without a change in human behaviour, technology will only mitigate the problem and further degrade the Earth system. It is therefore likely that in the future the practice, invented in the Neolithic, of systematically generating increasing surpluses that are as valuable and abundant as possible, will become more and more difficult and eventually impossible. The growing scarcity of surpluses will make it hopeless for the majority

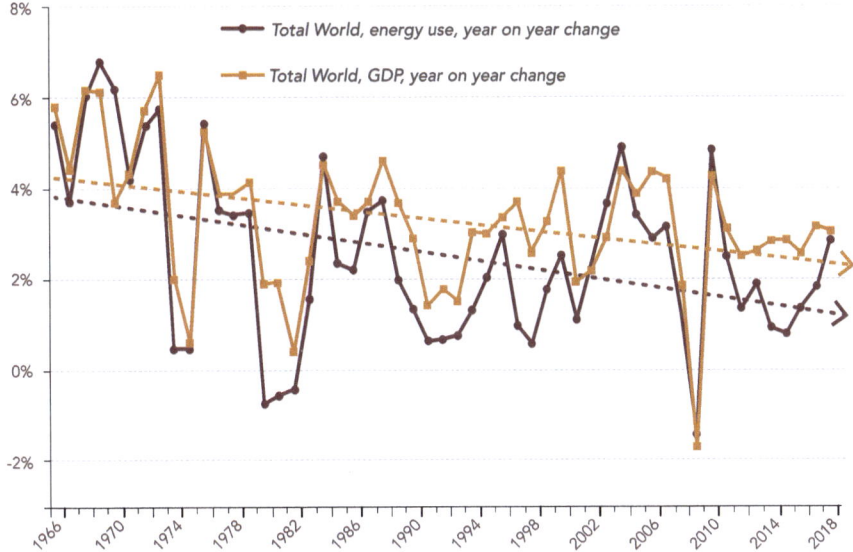

Fig. 3.9 The two curves represent the annual percentage increase in global real GDP per capita (*ochre*) and in global energy consumption per capita (*brown*) for the period 1965–2018. The two curves appear to be strongly correlated and in most cases the drop in energy precedes the drop in GDP, as in 1980, 1989, 1997, and 2005. The linear regressions show that both energy growth per capita and real GDP growth per capita have tended to decrease since 1965. Data from the BP Statistical Review and the World Bank. Figure adapted from Jancovici (2012b)

of human beings to profit from our ancestral ability to enjoy all imaginable kinds of extreme behaviour. A minority is likely to continue lavishly enjoying the excesses allowed by their extreme wealth. This situation, if it comes to pass, will mean we have reached the limits of the highly successful model that has made it possible to increase economic prosperity, well-being, and quality of life for most of humanity. To many this will seem an unhelpful alarmist or catastrophist vision of the future. The different perceived visions are points of view based on specific narratives, including the narrative of science, and they will all meet the opportunity to be confronted with the unfolding evolution of *Homo sapiens*' time in the Earth system during the coming decades and centuries.

3.7   A Brief Introduction to Altruism and Cooperation

Altruism is a form of behaviour characterised by a willingness to do things that brings advantages to others, even if it results in a disadvantage for those that perform them. The word has its origins in the French *autrui*, for "other people", which came from the Latin *alter*, and was introduced by Auguste Comte in 1852 (Comte 1852) as an antonym to the French word *égoïsme* in his defence of a "religion of humanity", also known as religious positivism. For him, altruism was a type of behaviour that formed emotional bonds, promoted goodness, protection, even reverence. Initially, the concept of altruism was applied exclusively to humans to describe acting out of concern for another's well-being, although with some likely disadvantages for those that act in that way. Altruistic acts can be observed in human relations within families, especially between parents and children, in charitable actions, humanitarian aid in many different situations of need or emergency, and in many other circumstances. In this common and intuitive sense, altruism therefore counters selfishness. In English being selfish means acting by placing one's own interests before those of others, while egoism is the tendency to think of self and of self-interest. Egoism can also be a rational or ethical conviction. It should be noted, however, that the distinction between altruism and selfishness is hard to establish from a conceptual standpoint, and it depends on adopting either a purely psychological, descriptive approach or an ethical, normative point of view. In the case of the former, the issue may arise of whether there are truly altruistic acts that bring no personal benefit or gratification to those who perform them. In the case of the latter, altruism is an integral part of the ethical doctrine that argues for the moral duty to help, protect, and defend others, most of all family members, the needy, and those who are in difficult situations, even if it involves personal sacrifice. In this normative view, the practice of altruism is formally different from selfishness and opposes the rational egoism and ethical egoism that will be discussed below.

Normative altruism is practised in individual and group behaviour and also institutionally in the social and governing structures of today's civilisation. There are many challenges and problems to overcome at the individual, collective, and institutional levels, such as social exclusion, violent conflict, war, and migrations in many parts of the world. Furthermore, extreme and moderate poverty, malnutrition, hunger, and lack of access to health services continue to severely affect many hundreds of millions of people all over the world. To what extent and under what circumstances the current trend and future evolution of ethical altruism at personal, collective, and institutional level is able to prevail over the current trend toward ethical and rational egoism at regional and global level is an open question.

Darwin was the first to acknowledge that examples of altruism in animal behaviour are surprising and appear to challenge the mechanism of natural selection. If natural selection acts exclusively at the scale of the individual living organism, then it does not appear to be possible for altruistic behaviour to become established, develop, and prevail since, by definition, altruistic behaviour is detrimental to the organism itself and is therefore likely to reduce its capacity to reproduce. Evolution is based on ferocious competition between individuals and should therefore reward selfish behaviour. However, altruistic behaviour is frequent in the animal kingdom, especially among animals that live in complex social structures, such as ants, bees, and termites. In 1975, altruism was considered by E.O. Wilson to be the central theoretical problem of sociobiology (Wilson 1975).

Altruistic acts in animals have to some extent a different nature, because there is no reason to believe that they are performed consciously, as they are in humans. Altruistic behaviour is very varied in the animal kingdom. For example, vervet monkeys in the *Chlorocebus pygerythrus* species give alarm cries to warn other members of the group about the presence of predators, although doing so normally attracts attention to them, thereby increasing the likelihood of their being attacked. Leafcutter ants of the *Atta* and *Acromyrmex* genera, found in the tropical and subtropical regions of the Americas, are considered to be among the animals with the most complex social behaviour, apart from humans. The workers are sterile and dedicate their lives to caring for the queen, building and protecting the nest, and seeking food to feed the larvae. They also have activities analogous to agriculture and build structures like cities (Wilson 2012).

The first proposal to solve the problem of knowing how natural selection created altruistic behaviour was ventured by Darwin right at the start (Darwin 1871). He considered the case of primitive tribes in which, at times, some individuals would sacrifice their own lives to avoid betraying their companions or to protect them. A group with several members that are able to warn, aid, and defend the other members of the group, often with personal sacrifice, would, according to him, have a better chance of doing better than or conquering other groups that do not have, or have less, altruistic behaviour. From this perspective, altruistic behaviour would have evolved through a mechanism of between-group selection, or intergroup selection, called group selection. However, the relationship between this social mechanism and the genetic evolution of the group's members remained obscure for many years.

Several mathematical models have shown that group selection is a relatively weak evolutionary force compared with selection within the group, or intragroup selection, and has difficulty explaining the variety and strength of altruistic behaviour observed in animals (Maynard Smith, 1964; 1998). One of the problems identified with the group selection process is "internal subversion" (Dawkins 1976) in which "free riders" within the group benefit from the altruism of others but are not altruistic themselves. Free riders have clear advantages for survival and reproduction, so the emergence of a genetic mutation relating to this selfish behaviour should proliferate quickly. By applying the concept of free riders to the human management of limited shared and unregulated resources, we arrive at the well-known problem of the "tragedy of the commons", studied by Garrett Hardin (1968), which reveals the conflict between individual and collective interests. Thus there is a tendency for intragroup selection and intergroup selection to produce diverging outcomes regarding the development of altruism.

New explanations for the biological evolution of altruistic behaviour began to appear in the 1960s. The evolution of altruism is determined by its Darwinian fitness, a central concept in the theory of evolution of species that measures the reproductive success of an organism in passing its genes to the next generation. It is defined as the average contribution to the following generation's gene pool made by an average individual of a specified genotype or phenotype. If the differences between the alleles of a certain gene affect fitness, then the frequency with which they appear in the succession of future generations tends to vary, favouring alleles that have better fitness. Altruistic acts reduce the donor's fitness but increase the fitness of one or more other individuals of the population (Trivers 1971). This is the definition of absolute or strong altruism in which the fitness is absolute because it refers to the individuals who practice altruism themselves. In relative or weak altruism the fitness reduction of the donor is measured in relation to the recipient. The word "cooperation" is also often used to describe social behaviour that is close to altruism but is more general in nature (West et al. 2007).

The word "cooperation" is often used to describe a more general social behaviour in nature than altruism, but which nevertheless carries a cost to the individual, even if relatively small (West et al. 2007). Some authors use the term "cooperators" for the individuals in a group that perform altruistic actions. Individuals whose actions do not reduce their own fitness or increase the fitness of other individuals in the group are called defectors. In a group with a mixed population, defectors have, by definition, an average fitness that is higher than the fitness of cooperators. As a result, natural selection tends to work towards increasing the population of defectors and, after some time, cooperators disappear. The problem is to find out what mechanism causes natural selection to encourage cooperation. Why should cooperation exist in a biological world dominated by Darwinian competition for reproductive success?

Let us consider the gene responsible for a certain kind of altruistic behaviour in its carrier, such as sharing food with others. Those who do not have this gene are selfish because they do not share the food they gather but, occasionally, they effortlessly receive food that is given to them by altruists. We would expect, then, that this gene would disappear from the population because it diminishes the fitness of those who carry it. However, if altruists share food only with their family members, there is a strong likelihood that those who benefit also have the altruism gene. This means that the gene can spread to the population through natural selection, which means that altruistic behaviour will prevail. This mechanism, known as kin selection, was first suggested by William Hamilton (Hamilton 1964a, b), who expressed in mathematical form the condition for the altruist gene to be favoured by natural selection. In its simplest form, the condition is given by the formula

$$c < br ,$$

where c is the cost to the individual who performs the altruistic act, or donor, b is the additional benefit to the beneficiary, and r is the coefficient of relatedness, which is determined by the genetic relationship between donor and beneficiary. Costs and benefits are measured in terms of Darwinian fitness. The coefficient of relatedness is defined as the probability of the donor and beneficiary sharing, in the same locus (the place in the chromosome where the gene is located), identical genes by descent, i.e., the likelihood that they are copies of genes that belonged to common ancestors. In this theory, altruistic behaviour spreads thanks to the kinship evolution mechanism; although altruistic acts reduce the fitness of the individual organism that performs them, they increase the organism's inclusive fitness. This concept is defined as the sum of the individual's own fitness and the fitness of relatives, weighted by the coefficients of relatedness between each relative and the individual. According to this model, natural selection acts to maximise the inclusive fitness of individuals in the population. The crucial point here is that the selection of genes for social behaviour is assumed to be dependent on the social effects on genetic co-bearers (Bourke 2011).

Since the middle of the 1990s, several authors have highlighted the importance of group selection in explaining the evolution of advanced forms of social behaviour that are particularly altruistic. One of the most important and controversial articles (Nowak et al. 2010) in this field attempts to demonstrate the limitations of using the concept of inclusive fitness to explain the evolution of eusociality, the most evolved form of social life in animals, which can be seen in ants, bees, wasps, termites, and the naked mole-rat (*Heterocephalus glaber*). Eusociality is characterised by the presence of several generations in the same nest and an altruistic division of labour based on reproductive and non-reproductive groups. Altruistic behaviour is also found in all eusocial animals in the way they protect the nest from enemies, whether they are predators, parasites, or competitors. According to Nowak, the evolution of eusociality is a two-stage process (Nowak et al. 2010). First, through genetic mutation, an allele is acquired that favours the continued presence of a mother and her descendants in the same nest rather than future generations dispersing to form new nests. Once this stage is complete, natural selection begins to work on the group, leading to more complex levels of eusociality that include the division of labour, harming the individual interests of some members of the group. According to Wilson (2012), the very low probability of pre-adaptation processes that favour the construction and defence of a nest and the continued overlapping of several generations in that nest helps explain why the phenomenon of eusociality is so rare in the animal kingdom.

The first ideas on group selection, which were heavily criticised from the 1960s onwards, evolved to give rise to the theory of multilevel natural selection. This new approach acknowledges the existence of selective forces within groups and between groups in the evolution of species that have social behaviour. An important point in sociobiology is the fact that each level of the natural selection hierarchy favours different adaptations, which sometimes compete with each other. As we have seen, selection among members of the same group may favour the appearance of free riders, who have selfish behaviour that works towards individual benefit and harms the group as a whole. Conversely, selection among groups of the same species favours behaviour that benefits the fitness of a particular group or set of groups, although it may have a negative effect on the fitness of the species as a whole. Generally, adaptation at a certain level of the hierarchy requires a process of selection that tends to harm adaptation at lower levels.

The theory of multilevel selection identifies how each of those levels of the biological hierarchy participates in the evolutionary process and concludes that selection at the intragroup level does not necessarily prevail over between-group selection. Regardless of the hierarchical selection level at which the adaptation takes place, it must always ultimately be reflected in the genes.

All primates are sociable animals, but hominids developed sociability to a very advanced form in the groups they lived in. The evolutionary process that led to the *Homo* genus, and later to the appearance of *Homo sapiens*, very likely produced selection mechanisms at two levels of biological organisation. At the upper level,

groups compete among each other, favouring cooperative behaviour and social characteristics of a cooperative nature among the members of the group. At the lower
level, members of the same group compete among each other to gain personal benefit.
The opposing nature of the selection processes at two hierarchical levels favoured
dual behaviour in humans: one part hero, defending the interests of the group, the
other part villain or anti-hero, defending their individual interests within the group, in
other words, having selfish behaviour. It also favoured conflicting motivations. Each
of us is aware of the tension involved in choosing between heroism and cowardice,
between truth and lies, between generosity and greed, between aiding and abandoning, between getting involved and running away, between altruism and egoism. It is
likely that a key event in the evolutionary process of the *Homo* genus was a "guarded
form of within the group egalitarianism" (Wilson and Wilson 2007). The transition to
a suppression of differences in fitness within groups allowed intergroup selection to
become a powerful evolutionary driver. The success of the group crucially depends
on its internal cohesion and the efficiency of cooperative work within it. However,
within-group selection has only been supressed but not eliminated. D.S. Wilson and
E.O. Wilson sum it up by saying that (Wilson and Wilson 2007): "Selfishness beats
altruism within groups. Altruistic groups beat selfish groups."

In some species, cooperative behaviour evolved through genetic selection and
cultural selection. Culture is a complex and wide-reaching concept used in both
anthropology and in biology. In the latter context, it can be defined as the range
of behavioural characteristics that distinguish different groups of the same species.
According to this definition, the concept of culture applies as much to animals as to
humans, and consequently to the evolutionary process of hominids. The occurrence
of different forms of cultural behaviour is well documented in chimpanzees, bonobos, orangutans, and dolphins (Whiten 2009). To develop cultural behaviour, it is
very important to have a powerful medium and long-term memory, which requires
advanced cognitive ability. In this regard, the *Homo* genus is in a privileged position
because of the extensive encephalisation (by a factor of more than 2.5) that occurred
from *Australopithecus afarensis* to *Homo sapiens* in less than 4 million years (see
Fig. 3.3). This is one of the most significant and surprising anatomical characteristics of our evolutionary process, and one of the likely reasons for our perceived
success in the process of the evolution of species on Earth. We know that human cultural evolution became more distinctive and faster around 40 000 years ago, although
complex cultural behaviour, like abstract thought and syntactical language, appeared
earlier (Mcbrearty and Brooks 2000; Wynn 2009; Henshilwood et al. 2009; Huybregts 2017). It was through the interdependent processes of genetic evolution and
cultural evolution that *Homo sapiens* became the champion of cooperation, surpassing all other species in the complexity of its cooperation and defection. Both genetic
and cultural between-group selection have been very important drivers of human
social evolution.

Presently, a high level of cooperation and division of labour can be seen in small
contemporary groups of hunter–gatherers, much more so than in other primates.
These types of behaviour are now manifested and developed with greater complexity
and on much larger social scales, including the level of nation states, through acts as

varied as warfare, help for refugees, natural disaster assistance, security systems for people and property, social security, tax systems, public health services, and many forms of humanitarian aid. The costs of cooperative behaviour in today's societies are high, although many believe that the corresponding benefits are still not enough to achieve the objectives that have been set. In any case, we are considering only cooperative behaviour among people living at the same time, which corresponds approximately to the coexisting members of three consecutive social generations. Our cultural evolution has as yet been unable to develop systematic and durable forms of cooperation with future social generations. This cooperation is necessary in view of our unprecedented capacity to interfere with and transform our "home", that is, the Earth system. By actively carrying out that transformation solely for the well-being of current generations we will very likely leave our "home" in a state that will be much less friendly and favourable for future generations. We are extremely proud of our capacity for transformation and it is generally believed that there are no limits to the way it can be used for our benefit. Interestingly, it has been developed in a species with a highly evolved and complex social behaviour, where cooperation within contemporary generations is strong. But the difficulty in developing cooperation with future social generations has now become a crucial factor for the well-being of our species in the medium and long term. It is probably impossible to develop a higher hierarchical level of natural selection involving a set of successive social generations because Darwinian fitness is unable to overcome the gap between temporally separated generations. The solution of the problem will thus depend almost exclusively on cultural evolution.

Chapter 4
Modernity and the Acceleration of Time

4.1 A Brief History of Economic Growth

Two stages can be clearly distinguished in the history of economics. Until the mid-18th century, average global growth in the production of goods and services—what is normally called global economic growth—was very small and was driven almost exclusively by demographics and the colonisation of new lands. This situation changed radically with the Industrial Revolution. Firstly, economic growth gained pace in some European countries and then progressively spread to a large portion of the world, although with temporary slowdowns in some regions. For a quantitative analysis of this crucial historical phenomenon, we must begin by defining a measure of economic growth.

One of the most commonly used indicators to evaluate a country's economic development and standard of living is its gross domestic product (GDP), which measures the monetary value of all goods and services produced in that country in a certain period of time, normally a year. It was created immediately after the Great Depression by the economist Simon Kuznets (1901–1985), following a request by the United States Congress for a report published in 1934. It purported to be a production indicator and not a welfare indicator. Later, John Maynard Keynes (1883–1946) perfected the concept, and since the Bretton Woods Conference in 1944, it has been used more and more often around the world as the main indicator of a country's economic performance and living standards. To receive aid from the Marshall Plan for Europe after the Second World War, countries first had to calculate their GDP.

An indicator that comes closer to people's everyday lives is GDP per capita, although it does not reflect income distribution among the population or the inequalities found in that distribution. Nonetheless, GDP per capita gives us a first rough approximation of the average standard of living in that country, which includes prosperity, well-being, life expectancy, level of education, and income and consumption levels. To compare the GDP per capita of different countries, we must also be aware that the market value of goods and services varies from country to country. To compare economic productivity and standards of living between countries and across

© Springer Nature Switzerland AG 2021
F. Duarte Santos, *Time, Progress, Growth and Technology*, The Frontiers Collection,
https://doi.org/10.1007/978-3-030-55334-0_4

time, economists introduced the concept of purchasing power parity (PPP). This provides a way to compare different countries' currencies through a basket of goods approach. But there is another challenge. What matters most in terms of economic development is the production of goods and services, and not how much they cost. The price of a particular product, for example a tonne of wheat, may increase without their being any increase in the total amount of wheat produced by the country, which would lead to an artificial increase in the GDP. To overcome this difficulty, a real GDP is defined as an inflation-adjusted measure that reflects the value of all goods and services produced by a country in a given year, expressed in base-year prices. This is also often referred to as constant-price or inflation-corrected GDP. Nominal GDP is calculated at current market prices and therefore includes the changes in prices due to inflation. Financial assets retain their initial nominal value but their market value is affected by inflation and by other market forces. On the other hand, tangible goods such as real estate, gold bullion, art, antiques, jewellery, and other collectibles have a physical form and natural value and often constitute a powerful inflation hedge. In short, the indicator that matters most here for a brief reconstruction of the history of the world economy is constant-price GDP per capita.

There is a current of opinion considering that the present economic and financial system overestimates the importance of GDP, and there has been growing criticism and efforts to go against that trend (Costanza et al. 2009; Van den Bergh and Antal 2014). GDP is an economic indicator that does not reflect levels of health, education, and other aspects of social development, environmental quality, and questions regarding sustainability. It does not take into account the relationship between time spent on production and the quantity of goods and services produced, nor what that means for working conditions. It does not take into account the evolution and possible distortions in the distribution of the wealth generated by an economy. It does not account for the value of the negative externalities of production, such as pollution and environmental damage, although market transactions made to correct those effects are included. GDP excludes goods and services traded outside the market, or non-market production, such as voluntary work carried out individually or as a group, food products made for personal consumption, and goods that can be accessed free of charge, such as digital platforms and services. Moreover, GDP does not reflect the advantages involved in the growing diversity of products accessible to consumers. It excludes the underground economy, which incorporates non-recorded activities for which tax is not paid, and other inherently illegal activities, such as arms or drug trafficking. Despite these limitations, GDP continues to be the preferred and most widely used indicator of living standards around the world. The reason for this is that the current economic and financial system tends to value money and the capacity it provides to consume products and services as the most important measure of well-being.

Pioneering studies by the economic historian Angus Maddison enabled an estimation of the evolution of GDP per capita at constant 1990 international dollars (an international dollar would buy in each country a comparable amount of goods and services a US dollar would buy in the USA) for the different countries in the world since the start of the Common Era in the year 1 AD (Maddison 2006, 2007a). The results of these studies can be used to estimate that the world's constant-price GDP per capita in 1990 international dollars was 467, 615, 666 and 7215 for the years 1, 1700, 1820, and 2006, respectively. The world's population for the same years is estimated at 226 million, 603 million, 1.041 billion, and 6.56 billion, respectively. For the first 1 700 years, the variation in world GDP per capita was very small. Most of humanity lived a life of subsistence agriculture in rural areas. There were occasional periods of misery and famine caused by wars, epidemics, and extreme meteorological events, such as droughts, very heavy rain, and cold or heat waves. The world population grew relatively more than the GDP per capita between the beginning of the Common Era and 1700, but at a much lower rate than in the centuries that followed.

Although the period was a time of extraordinary social, scientific, and cultural progress, human societies lived in what is known as the Malthusian trap, identified clearly for the first time by Thomas Robert Malthus (1766–1834), an English cleric and economist. In his well-known book *An Essay on the Principle of Population* (Malthus 1798), written in 1798, he started with the idea that humanity essentially has two basic needs—subsistence and sex—and concluded that the natural trend of human population growth would always be greater than the agricultural production capacity needed to provide subsistence for the population. Malthus believed that human population growth would be exponential, doubling every 25 years, as long as unlimited resources were available. However, the arithmetic progression of agricultural production limits human population to increasing by only the same amount every 25 years. Supposing that human population was 1 billion in 1800, it could potentially reach 256 billion in 2000, but according to Malthus it could not be greater than 9 billion at that date (see Fig. 4.1), which agrees with what has happened. According to his analysis, if a new invention can double agricultural productivity to allow, on average, double the food available per person, then this increase contributes to "the yearly increase of an unrestricted population" (Malthus 1798). He goes on to say that "the happiness of a country does not depend, absolutely, upon its poverty, or its riches, upon its youth, or its age, upon its being thinly, or fully inhabited, but upon the rapidity with which it is increasing". This growth causes a reduction in food consumption per capita and the situation returns to the previous balance of subsistence. The Malthusian trap strikes once again. In this situation, GDP per capita, despite some oscillations, did not have any long-lasting tendency to grow.

Not long after 1750, at the time when Malthus was born, the Industrial Revolution started to get under way in England. However, its more lasting social and economic effects were only felt from about 1820 onwards (Hobsbawm 1968). As it spread to other countries, it has provided the world with extraordinary economic growth for roughly 200 years, thereby freeing it from the Malthusian trap (see Fig. 4.1). Curiously, Malthus published his book as a reaction to the belief that was taking hold at the time about the possibility of unlimited social, moral, and material progress for humanity, as Nicolas de Condorcet (1743–1794) and William Godwin (1756–1836) had argued in different ways. Malthus' book was widely criticised when it was published and is still a controversial reference today.

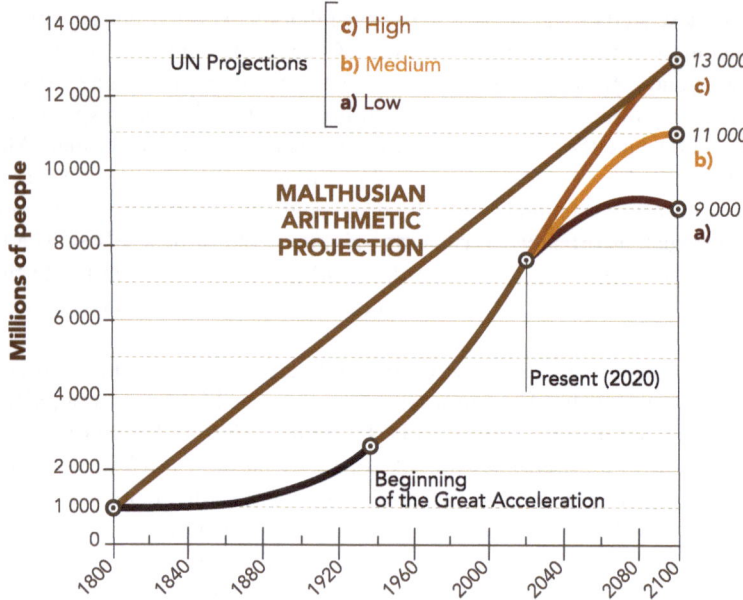

Fig. 4.1 Observed evolution of the global population since 1800 up to the present and UN projections up to 2100 according to three scenarios: low, medium, and high (UNDESAPD 2019a). The *straight line* represents the population growth limit proposed by Thomas Malthus, corresponding to adding up the population in 1800, estimated at one billion, every 25 years, reaching 13 billion in 2100. Interestingly, humankind has managed to experience exponential global population growth since about 1850, but according to the medium scenario, is likely to reach a maximum near the end on the 21st century or sometime after, without ever crossing the arithmetic projection relative to 1800, describing a "Malthusian oscillation". Figure by the author

The Industrial Revolution led to the transition between two very different economic and energy systems and two periods of human history that are profoundly distinct; some economic historians call the latter "the era of modern economic growth" (Mokyr 1990). Historical reconstitutions of daily life before the Industrial Revolution indicate that most people in Europe lived in conditions that we now define as poverty or extreme poverty, with localised exceptions of rich merchants and people linked to the governing or military power. Since the Industrial Revolution, there have been three critical transformations: energy use per capita, global GDP per capita, and global population have all increased enormously. At the same time, the growth in GDP per capita has seen highly unequal behaviour between regions. This differentiation has generated countries that, despite their domestic economic and social inequalities, are richer globally than other countries that are much poorer globally. World energy consumption per year increased from 20.35 GJ (10^9 J) in 1800 to 565.43 GJ in 2018 (BP 2019), which corresponds to a factor of 27.8 (see Fig. 4.2). During the same period, world energy consumption per capita per year rose from 20.35 to 74.45 J, which corresponds to a factor of 3.66 (Tverberg 2018; BP 2019).

Global primary energy consumption

Global primary energy consumption, measured in terawatt-hours (TWh) per year. Here 'other renewables' are
renewable technologies not including solar, wind, hydropower and traditional biofuels.

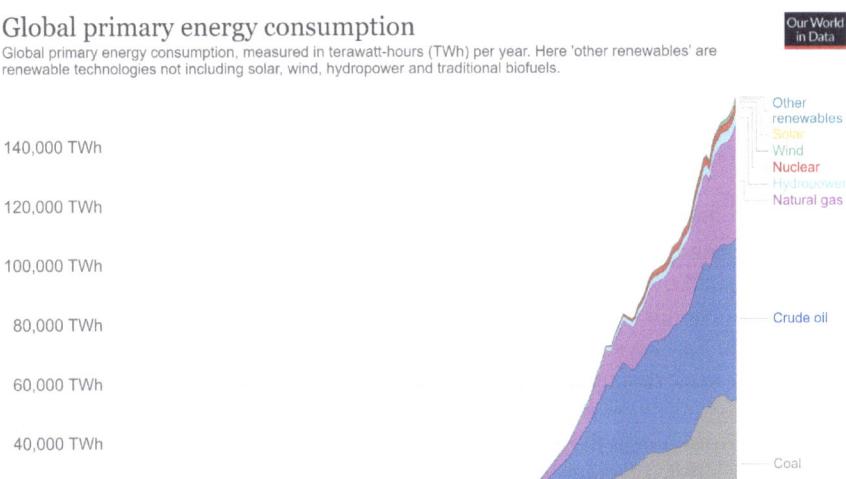

Source: Vaclav Smil (2017) and BP Statistical Review of World Energy   CC BY

Fig. 4.2 Global primary energy consumption in terawatt-hour (TWh) (1 TWh $= 3.6 \times 10^6$ GJ)
for the period 1800–2018. The figure shows that the global primary energy consumption increased
very slowly during the 19th century, somewhat more steeply in the first half of the 20th century,
and started to increase much faster after the end of the Second World War, at the beginning of
the Great Acceleration. Renewable primary energy sources include the traditional biofuels, which
provided almost all of the primary energy up to 1850, and the recent modern renewables. Notice the
importance of fossil fuels that account for about 80% of the primary energy consumption. *Source*
Ritchie and Roser (2015)

From 1820 to 2006, the world population increased by a factor of 6.3 and world GDP
per capita increased by a factor of 10.8. However, there are large differences in this
economic growth indicator between countries. For example, it grew by a factor of
24.7 in the USA, 7.1 in Egypt, and 4.6 in India. Another way of uncovering this deep
differentiation is by comparing GDP distribution with world population distribution.
Before 1820, the two distributions converged because a country's GDP was roughly
proportional to its population. Since 1820, the two distributions have diverged. The
USA, for example, currently has 5% of the world's population and 21% of its GDP,
while Asia, excluding Japan, has around 60% of the population but only 30% of the
GDP.

The economic growth that followed the Industrial Revolution was associated with
profound social changes. The wealth that was primarily held by the merchant, governing and military elites until 1780 is now concentrated in the hands of the economic
and financial elites, which possess most of the wealth and usually prefer to be involved
indirectly in governance.

Another crucial aspect is the impact that the growth in the world constant-price GDP has had on natural resources and the environment. It is estimated that the world GDP was worth 105, 694, and 47 267 billion 1990 international dollars in the years 1, 1820, and 2006. Note the very high growth by a factor of 68 in the last 186 years. World economic growth—which has been boosted by several factors, in particular by an abundant supply of energy in the form of fossil fuels, modern science, and technology—has created a model of safe, abundant food supply, qualified, accessible education and increasing life expectancy and health, well-being, material wealth, and mobility, all of which is in theory within everyone's reach. It has also created increasing social and economic inequalities and has deeply transformed the functioning of the Earth system. However, these disadvantages are generally considered to be mostly irrelevant in comparison with the benefits of enjoying that standard of living or the expectation of enjoying it at some point.

What reasons led to the Industrial Revolution in England in the middle of the 18th century? This is a complex and highly debated question (Deane 1965) that will not be addressed here. However, it is important to mention that the process has its roots in the Middle Ages and involved many favourable social, political, economic, scientific, and technological factors (Mumford 1934). At the time, England benefited from a period of social and political stability following the unification of England and Scotland, with the creation of the United Kingdom and reliable legislative, judicial, and administrative structures and procedures. Two other crucial factors were James Watt's (1736–1819) discovery of an efficient steam engine, following the first attempts by Thomas Newcomen (1664–1729), and the relative abundance of coal and iron in the British Isles. The steam engine was an extremely important device to convert the chemical energy of coal into kinetic energy. It was used initially to pump water out of mines, and later in rail and maritime transport, and to mechanise several industries, including the textile industry.

The main feature of the Industrial Revolution was increased efficiency in the production of goods by mechanisation in the new factory system that gradually replaced the domestic system. A machine could do the work of many men in much less time. The power of the machine to accelerate operative social time was one of the most important drivers of the Industrial Revolution. This power had already been felt in the domain of timekeeping. Throughout the Middle Ages, the need arose to ensure that the use of time in the different everyday religious activities was kept regular. The Benedictines in particular were very concerned with making use of the entire day and dedicating it to God, in accordance with their motto *Ora, lege, et labora*, meaning 'pray, read, and work'. Prayers throughout the day were held at the canonical hours. There were initially seven, but an eighth was added by Saint Benedict of Nursia (480–547). Mechanical clocks were used to announce these hours from the 14th century onwards. The regular ringing of the bells in bell towers provided a way to ensure that time was fully used in monasteries and later in towns, profoundly changing the operative time structure. The emergence of mechanical clocks and the new operative time structure that it created started to influence all fields of human activity, but especially work in factories. The temporal resolution of the operative time structure increased as the efficiency of industrial production began to be monitored by clocks, which made them the key machines in the modern industrial era (Mumford 1934).

4.2 Origins and Expansion of Capitalism

Until the mid-18th century, the interventionist economic theory of mercantilism dominated England. In this system, the world economy is supposed to work as a zero-sum game in which one state's increase in wealth involves a decrease in wealth for another state or states. It was at that time that a group of economists led by Adam Smith presented an alternative view that was gradually adopted and revolutionised the world economy. This theory was summarised in the famous book *The Wealth of Nations* written in 1776 by Adam Smith (Smith 1776), a Scottish moral philosopher and economist who was a great friend of the philosopher and historian David Hume (1711–1776), a fellow Scotsman. The proposal made in this book claimed that, when a landowner or trader had more income than needed to sustain a family, he could use the surplus to hire more help to expand the business and therefore make even more profit. When the additional profit is reinvested in production, it drives the growth of wealth, prosperity, and well-being in the surrounding society. Adam Smith identified and defended the idea that the division of labour and expansion of the markets by reusing profit offered unending opportunities to increase wealth and well-being. Smith argued that economic agents should act in their own interest, which together with competition and consumer preference produces dynamics that lead to economic prosperity. It should be noted, however, that recognising the central role of one's own interests in the economy did not lead him to explicitly defend selfishness. The underlying revolutionary idea is to place the impulse of greed that encourages an increase in personal income at the heart of wealth-production for the community. Smith thereby overcame the contradiction that existed at his social time between morality and the accumulation of personal wealth. Being rich could be morally recommended if the wealth obtained was reinvested in production to increase it, generating more wealth for society as a whole in the process. The revolutionary character of these ideas is clear if we remember that in the Bible, it is practically impossible for a rich person to enter the kingdom of God (Matthew 19:24).

From an economic perspective, Smith's ideas, together with modern science, technological innovations, maritime discoveries, and the expansion of the West's power throughout the world, opened the doors to the extraordinary growth of the economies and prosperity of Western countries. This growth then progressively extended to other countries and become global. There is currently a global economy that is governed by the standards of the capitalist system and has a determining influence on national economies. Our common future is therefore unavoidably dependent on the performance and evolution of the current economic system. In the last three centuries, most of humanity has become firmly convinced that it is possible to put the idea of progress into practice and, more importantly still, it has grown accustomed to having increasing and unyielding confidence in the future. It is this that allows us to believe in using credit, based on the premise that future resources and opportunities will certainly be more abundant and better than at present. Using credit makes economic growth possible, which boosts confidence in the future and makes it easier to provide more credit; this creates a virtuous circle of growth.

It has been a long journey to reach the current economic system since interest-yielding money lending was considered a sin in the Christian Europe of the Middle Ages, when it was known as usury. From a theological point of view, mankind cannot benefit from the passage of time because time is governed by God. Furthermore, charging interest on a loan to someone in a difficult situation was interpreted as a form of exploitation that contradicted Christian values, which include helping the poor without expecting a material reward. In Luke's Gospel, Jesus says "But love your enemies, do good to them, and lend to them without expecting to get anything back. Then your reward will be great" (Luke 6:35), which in our time is likely to sound to most people quite extraordinary and impracticable. However, the Old Testament had already opened a way to monetary loans in Deuteronomy 23:20, which reads "You may charge a foreigner interest, but not your brother, so that the Lord your God may bless you in everything to which you put your hand in the land you are entering to possess." The distinction between foreigner and brother is left open to several interpretations. For Saint Ambrose (340–397), an influential Bishop of Milan, those who are not brothers do not share the Christian faith, so they could potentially be enemies that are fought and killed. The Saracens were clearly included in this category, but in relation to the Jews who lived close to Christians in towns the situation was more complex. In practice, the Jews, who were subject to the same verse by way of the Torah, played the role of loaners or usurers, often forcibly, in an economy that needed credit for trade and war, and sometimes this led to injustice and terrible persecutions. In the Early Middle Ages, the evolution of trade and the financial system that supported it imposed the growing use of credit and at the end of the 15th century the first tables limiting the amount of interest appeared. Usury began to mean charging interest over the maximum permitted limit.

Adam Smith is considered the first author who explained and defended the economic principles that about 80 years later became known as capitalism. The surprising etymology and evolution of the word "capitalism" reveals the tensions and controversies inherent in the economic system devised by Adam Smith. From the end of the 12th century onwards, use of the word capital became common. It was derived from the Latin *caput*, for head of cattle, as a synonym for wealth or an amount of money that could grow (Braudel 1985). Much later, in the XVIII century, Anne Robert Jacques Turgot (1727–1781), baron de l'Aulne, known as Jacques Turgot, a French statesman and economist, was probably the first to use the word "capitalist" in French in his book *Réflexions sur la formation et la distribution des richesses* published in 1766. Later, the word was used in English by the writer and agriculturalist Arthur Young (1741–1820) in his work *Travels in France*, published in 1792. A distinction was gradually made between capital in the form of money or assets with monetary value that are invested in production and non-productive wealth that remains untouchable or is spent on luxuries and unproductive activities.

The word "capitalism" was first used pejoratively in France by the historian and politician Louis Blanc (1811–1882), in his book *L'Organisation du travail* published in 1851, and by the politician Pierre-Joseph Proudhon (1809–1865) in the book *Les confessions d'un révolutionnaire*, published in the same year. Proudhon rejected both capitalism and communism and defended an economic theory of mutualism which involves the control of the means of production by the workers and usufruct property norms. In England the word "capitalism" was first used in a more prosaic way by the novelist William Makepeace Thackeray (1811–1863) in his novel *The Newcomes*.

Curiously, Friedrich Engels (1820–1895) and Karl Marx (1818–1883), who both defended different forms of socialism, preferred to use in their writings the expression "capitalist system" and "capitalist mode of production" rather than "capitalism". The term "capitalist system" appears often in Karl Marx's *Das Kapital* (Marx 1867, 1885). According to Marx, capitalists pay their workers less than the value added to the goods produced by their labour, in other words, only enough to keep the worker at a level of mere subsistence. The difference is the surplus value which allows capital to grow and which, according to Marx, lies at the heart of the system's instability.

It is important to distinguish between the ideal economic model proposed by Adam Smith and the development of what we now call a capitalist system which established its roots in the Middle Ages. Long before the Industrial Revolution and the economic theories of capitalism, there were already bills of exchange, stock markets, securitisation, brokerage firms, credit institutions, and central banks. What happened at the beginning of the 18th century was that a large-scale credit system was formed that made it possible to finance the production of a growing quantity of goods and services by benefiting from the expansion and exploitation of the colonies. The extraordinary growth of the economy was not based only on free labour but also on many forms of forced labour and the reinvention of slavery, mostly aimed at exploiting the plantations and mines in Africa, Asia, and the Americas (Boutang 1998).

It is estimated that, from the 16th to the 19th centuries, around 12.5 million people from the west coast of Africa were forced to cross the Atlantic to the Americas as slaves in inhuman conditions. A large number of slaves—it is hard to estimate how many—died being captured and transported from the inland to the African coast or during the sea journey, so only approximately 10.7 million actually arrived in the Americas (Eltis and Richardson 2008). Ultimately, by 1820, there were around four times as many Africans crossing the Atlantic as Europeans. The migration of slaves was part of a highly lucrative triangular trade between Europe, Africa, and the Americas, in which the European states were not directly involved. It was a typically capitalist undertaking: the companies performing the slave trafficking obtained financing from the market in accordance with the laws of supply and demand, and issued shares that could be traded in the London, Amsterdam, and Paris stock markets. The triangular trade in the Atlantic was what generated the most profit in Great Britain in the 18th century and played a central role in financing the Industrial Revolution. A more detailed analysis shows that one of the most important driving forces for that trade and profit was the desire of Britons and other Europeans to consume sugar-sweetened drinks, sweets, chocolates, cakes, and other treats that made sugar production in the Caribbean an outstanding business. The sugarcane plantations and sugar production were based on forced labour from slaves who were the private property of their owners and were bought, sold, and treated like merchandise. Slavery has existed throughout the history of humanity and, although it is now forbidden by law in every country, forced labour can still be found in several forms all over the world. After a long and often violent process, slavery was abolished by law in the European colonial powers from the middle of the XVIII century onwards, and in the USA in 1865 (Blackburn 1988). The success of capitalism in the modern age results to a large extent from its ability to combine competitive capital accumulation with scientific advances and technological innovation. Furthermore it constitutes a fertile ground to put to use some of the most innate, irrepressible, and unquenchable impulses of human nature such as egoism and greed. Capitalism promotes consumerism in a way that allows greed to reach unlimited private profits using wage labour. Through its success, the capitalist system also encourages the combination of naivety, credulity, and fantasy with egoism and greed. This association is not ingenuous and constitutes the heart of the recurrent bubbles produced by capitalism. Economic bubbles occur when the value of some types of assets grows on the basis of belief in an implausible or inconsistent view of the future. The increasing value attracts more investors; demand becomes contagious and causes a self-sustained growth in value that ends up collapsing, sometimes causing serious financial crises.

The first economic bubble that caused a financial crisis occurred in England and resulted from the formation of the South Sea Company in 1711, the year in which Adam Smith was born, by Robert Harley (1661–1724), a politician who became Lord High Treasurer, equivalent to prime minister of the government. It was a joint-stock company that aimed to decrease the huge public debt caused by successive wars by selling merchandise to the Spanish colonies with the objective of finally monopolising trade with South America. However, this monopoly was impossible to achieve, despite Spain's weakness in the midst of the War of the Spanish Succession. With the treaties of Utrecht (1713) and Rastatt (1714), which put an end to the war, England was able to supply 4 800 slaves per year from Africa to the Spanish colonies in South America, via the island of Jamaica. From 1719, persistent rumours about the enormous potential value of trade with the Americas made stock prices soar. From lords to peasants, everyone wanted to buy shares and benefit from the promised wealth. But in 1720 the company collapsed and many investors were ruined. England's economy was also deeply affected. Some investors who had access to insider trading and political connections to the people who set up the scheme leading to the bubble had sold their shares just before the collapse and profited greatly. While the value of its shares had been growing, the South Sea Company had become the model for hundreds of other joint-stock companies over a broad range of trades. They issued shares that were eagerly traded in the hope of increasing their value and producing an amazing profit for those who managed to sell them before the collapse. Mackay (1854)) has left us unforgettable descriptions of the history of those bubbles. They formed because some people were convinced that others were credulous or naive enough to believe that money could be made from nothing and that they could earn unlimited money purely from the process of speculation.

Almost three centuries later, economic bubbles are still around but they now often have a global impact. Moreover, they are much more complex, involve a range of different financial instruments, and are more difficult to predict. They constitute a well-accepted risk of the capitalist system since their origin stems from the unchangeable human impulses of self-interest, egoism, greed, naivety, and credulity. The advantages of the system are generally considered to largely outbalance the disadvantages. The last economic bubble burst in September 2008 and produced a global financial and economic crisis so serious that it can only be compared to the Great Depression of the 1930s. The bubble originated in the USA when subprime mortgages were granted for the purchase of real estate. The mortgages were often granted to people with no job, income, or known assets and became known as NINJA (no income, no jobs and no assets) or NINA (no income and no assets) mortgages. As subprime mortgages were hard to pay off, a highly sophisticated system needed to be developed to securitise the credit. The risk was diluted by many banks using derivatives that could be traded on international financial markets, such as mortgage-backed security and collateralised debt obligations, the value of which depended on the increasing market value of the real estate. The average value of houses in the USA rose by a factor of six between 1976 and 2006 but it peaked that year and then began to fall. The collapse in real estate market value ended up causing the collapse of the Lehman Brothers investment bank and started the West's 2008–2009 financial and economic crisis, which became global and affected every country in the world, albeit in different ways and in some cases for long periods of time.

The dependence of the current economic system on financial institutions guaranteed the heads of such institutions that, when it became clear that the debts of irresponsibly granted mortgages could not be paid, governments would eventually reimburse them for the losses, disregarding the often dramatic consequences that this had on taxpayers, companies, and a country's economy. An audit by the USA's Government Accountability Office revealed in July 2011 that the Federal Reserve Bank discreetly spent a total of $16 trillion between December 2007 and July 2010 to control the financial and economic crisis triggered by the subprime bubble, by bailing out many of the world's banks, corporations, and governments, an amount that is greater than the USA's $15 trillion GDP in 2010. The presidents of the Federal Reserve Bank, Ben Bernanke and Alan Greenspan, along with other important bankers, tried to oppose the audit by arguing that it would be damaging to the financial markets.

These events reveal that the stability of the global financial and economic system is an unquestionable supreme value. However, on the other hand, bubbles continue to be the likely result of a gigantic game built up over the years, a game that eventually generates a small number of big winners and an enormous number of losers, some victims of their credulity and naivety and many others just victims, without any involvement in or connection with the process itself. Since the 2008–2009 crisis, the constant risk of new bubbles has been fought with attempts to correct the system through greater regulation and supervision of financial institutions and transactions. The reaction from the system has been to increase the development of shadow banking, an informal parallel financial system that runs outside the regulation and supervision rules and seeks to offer better credit conditions and profit than the formal financial system. Shadow banks are characterized by the scarcity or absence of information on the value and nature of their assets, with non-transparent governance, maintaining more or less structured connections to banks within the formal system, but without having access to the liquidity of central banks or public-sector credit guarantees. In general, the greater the pressure of regulation and supervision on the formal financial system, the more highly developed the informal system becomes. The two are intimately linked through complex, constantly changing connections, so instability travels between them. The global assets of shadow banking grew by 7% in 2013 and were valued at $75 trillion that year, which represents around 50% of the other banks' assets (FSB 2014). The countries with advanced economies are those that have relatively large informal financial systems, but emerging economies, particularly China, also have a prosperous shadow banking system.

As at the start of the 18th century, the financial market continues to be full of analyses and rumours about the most lucrative investments, unmissable opportunities to become rich, and the risks of a huge range of different kinds of economic bubbles. The main difference is the enormous complexity and size of the current financial system, the immense variety of financial products on offer, and the dizzying speed of trading. According to a recent report by the IMF, the world economy is under an increasing risk of another financial crash mainly because global debt levels are well above those at the time of the last crash in 2008, and governments, regulators, and corporations have been unable to regulate the financial system (IMF 2018). Today's system is based on perpetual economic growth, which, is unlikely to be indefinitely sustainable. Nonetheless, in spite of various warnings (Meadows et al. 1972, 2004; Daly 1993), continuous growth has been achieved since the 18th century, with the exception of a few periods of crisis. The best way to guarantee economic growth has been by increasing investment and encouraging people to consume as much as possible. Such consumerism is made possible mainly through loans, following the "borrow and spend" motto. Over the last few decades, a parallel trend has appeared in which individual and institutional investors heavily invest in shares and financial products using successive loans in a process that was partially responsible for the 2008–2009 crisis. This tendency to maximise profit in the financial sector by reducing the use of capital to stimulate the productive economy, where profitability is likely to be lower, tends to have a negative effect on economic productivity. "Consumer capitalism" is now in an ongoing competition with the emerging, aggressive "financial capitalism".

The main factors that led to the success of maintaining economic growth for more than two centuries were the increase in the global population, the geographical expansion of economic activity and of the exploitation of natural resources throughout the world, and the advance of science and technological innovation. The last two driving forces create new products and services, industries, labour markets, and consumer patterns that can guarantee capital accumulation and further investment. The current overriding belief in most financial, economic, and corporate environments is that scientific research and technology will be able to renew and enable economic growth with no end in sight. However, as already mentioned, there are signs of difficulties in achieving a robust and sustained economic growth, especially in countries with advanced economies, and at the same time service an increasing public and private debt at the global level (Keen 2011). The global economy follows modern mainstream economics, which is a version of neoclassical economics based on individualism and more specifically on the assumption that the economy must be guided by the optimization of individual self-interest that follows directly from rational egoism.

Despite the recent difficulties mentioned above, the current economic system remains as powerful and immovable as agriculture since its invention in the Fertile Crescent roughly 10000 years ago. There are innumerable examples that this robustness is generally taken for granted and that the current economic system will likely follow the same pathway up to its inevitable conclusion. Alternative economic systems are considered unrealistic and impractical. The well-known research journals in economics do not generally publish papers that deviate in any fundamental way from mainstream neoclassical economics (Stockhammer and Yilmaz 2015).

Communism, the political system that emerged as an alternative to capitalism, could not fully adapt to the social, economic, and technological transformations of the last few decades and was unsuccessful in a significant number of countries. It retains a limited expression in the contemporary world as a form of governance. There are currently five countries that are one-party states ruled by communist parties that follow different varieties of Marxist–Leninist ideologies, namely the People's Republic of China, the Republic of Cuba, the Lao People's Democratic Republic, the Socialist Republic of Vietnam, and the Democratic People's Republic of Korea. In the latter the reference to Marxism-Leninism was replaced in the Constitution of 2009 by the Juche ideology, meaning "self-reliance", developed by Kim Il-sung. In China the communist party adopted a form of state capitalism that is having remarkable success in promoting economic growth and benefits from the support of most of the population. China is not a collectivised economy, as in the first communist countries. However, governance is assured through the rule of the communist party, disregarding Western-style democracy. This new mix is the legacy of Deng Xiaoping, who is likely to be the politician who has lifted the most people out of poverty in the history of the world so far.

Currently, in 2018, there are more than ten multi-party states where the communist party rules or is part of the ruling coalition. The number of political parties that declare themselves as communist or anti-capitalistic and have parliamentary representation at national level is greater than 100, but there many more political movements of that nature throughout the world without parliamentary representation. Capitalism, although dominant at the global level, continues to be vehemently criticised by some, and generates an intense search for credible alternatives, which up to now have been unable to attract enough popular support to lead them to success. Not all countries of comparable development have benefited from capitalism to the same degree. The differences are probably linked to social, religious, and cultural reasons. Max Weber (1905) argued that the development of the capitalist system in north and central Europe was boosted by the Protestant work ethic. The crucial innovation was to bring together incentives to accumulate capital with disincentives to spend it on mundane pleasures or store it unproductively. The Protestant Reformation introduced a moral obligation for each individual to serve their secular calling as well as possible for the glory of God and to achieve salvation. In Catholicism, salvation was guaranteed to all those who followed the sacraments of the church and submitted to the authority of the clergy. These guarantees are not found in Protestantism, so there is a need to assert self-esteem and self-confidence. Connecting these values to a simple life, where luxury is often rejected, encouraged the accumulation of capital. The historical origins of the Industrial Revolution in Great Britain probably helped the practice of capitalism to be generally more efficient and beneficial to society in terms of human development indicators in Anglo-Saxon countries. These are also the countries where the virtues of the capitalist economic system and its ability to adapt to new circumstances and challenges are most often highlighted and defended. The main Anglo-Saxon countries—Australia, Canada, the USA, the UK, and New Zealand—constitute a family, with a rather unusual unity based on friendship relations, that has led the world since 1815, the beginning of Britain's Imperial Century (Hyam 2002), engendering the Pax Britannica, and later the Pax Americana (Porter 2006; Kiernan 2005).

Initially, capitalism was primarily an economic theory, but thanks to its expansion and success, it has gradually become normative in practically all areas of human activity. It influences individual and collective behaviour, education, justice, politics, science, and technology. Two supreme, salvific values have emerged: economic growth, a relatively abstract, collective value to which all other social, political, and environmental values must be subordinated; and money, an omnipresent, dominating, and powerful medium and measure of value at the individual and collective levels, that can be exchanged for the acquisition and enjoyment of every imaginable kind of goods and services provided by the unlimited bounty of contemporary capitalist societies.

The Industrial Revolution, modern science and technology, and the extraordinary growth of economies all over the world, promoted by capitalism over the last three centuries, have produced a vast range of social, political, ethical, and cultural change. Family relations, class structure, and society have been restructured, gender equality has become accessible, a large part of the rural population has moved to cities, leading to the global phenomenon of urbanisation, and mobility has increased vigorously throughout the world, facilitating contact between people with different cultural, political, and religious backgrounds. Migration, segregation, and integration issues have become increasingly important. A growing number of countries have achieved democracy, and human rights have been proclaimed and applied to a growing part of humanity. Demographics and economics are core issues now, and will remain so in the future. The environment, climate change, and sustainability have become emerging issues that are both embarrassing and controversial.

4.3 Acceleration of Operative Social Time and Modernity

Besides the transformations directly related to the strengthening and expansion of capitalism, there has been another deeper and more essential change that affects all aspects of human life and is linked to the concept of time. In the last 300 years, through an accelerated process of temporal restructuring, time has begun to have a dominating presence in our lives. Today, all our actions are referred to and conditioned by time, and the performance of those actions is constantly monitored by frequently checking clocks and other electronic devices that measure time.

The prototypes of current operative time structures were the assembly lines of the first factories of the Industrial Revolution. The division of labour in production forced the introduction of strict working hours that all workers had to follow. If some did not comply with them, it had a multiplicative effect on the assembly line and reduced production per unit time, i.e., it reduced productivity. First factories and then schools, hospitals, public administration, and trade all adopted working hours that had to be followed rigorously. The town itself began to operate as a metaphor for the public clock installed in the church's bell tower or in the clock tower. The idea emerged that one should not "waste time" but instead make full use of every minute, otherwise an irretrievable opportunity would be lost.

The 18th century contributed decisively to highlighting the relationship between time and space and opened the way to defining a universal standard time. Improvements to England's roads at the time enabled a mail coach service to be introduced in 1784, subject to a timetable that was followed scrupulously. The mail coach was founded by John Palmer, owner of the Theatre Royal of Bath, mayor of that town, and later Member of Parliament for that constituency. The coach left Bristol at 4pm and arrived at London's central post office at 8am the following day. The time in each town was set by the local sundial, so the time in towns to the west of London was behind the time in the capital. Bristol, for example, was 10 minutes behind. The time in towns to the east was ahead. Although these were short periods of time, they caused confusion and potentially accidents. The problem was overcome by carrying a mechanical timepiece in the mail coach with hands that were adjusted during the journey. When commercial rail transport began in 1830, the need to harmonise the time between towns became vital because journey times were now much shorter. In 1847, the rail companies agreed to match the timetables of all lines to the GMT of the Royal Observatory, Greenwich, rather than using the local time in Glasgow, Liverpool, or Manchester. GMT was used as the reference time to set the marine chronometers on boats berthed at the port of London that served to determine longitude on ocean voyages. Setting was performed by means of a "time ball" installed in 1833 in the Observatory tower. This was a large, painted wooden ball that fell, guided by a vertical axis, at precisely 1pm. When time signals were first broadcast over the radio at the beginning of the 20th century, time balls were no longer used, although the Greenwich time ball continues to fall every day, to tourists' delight, as a reminder of the past.

While he was director of the Royal Observatory, Greenwich, the Astronomer Royal George Airy (1801–1892) spread GMT throughout Great Britain by transmitting the time using "galvanic" signals carried by cables installed alongside railway lines. This was the name then used for electrical signals. The new time became known as "railway time". Clocks in the clock towers of some British towns had two minute hands, one showing railway time and the other local time. In one of his reports, Airy said that he was pleased that the "Royal Observatory is thus quietly contributing to the punctuality of business throughout a large portion of this busy country" (Whitrow 1989). Finally, in 1880, legislation drawn up by the British government established the use of GMT throughout the country. For the first time in the history of the world, time was no longer determined locally by the apparent daily motion of the Sun, but was replaced by a single, national standard time. This form of unification extended to the entire globe with the adoption of UTC and the different time zones in 1960. Since then the world has had a single standard time, the hours, minutes, and seconds of which serve as the operative time structure for all human activities around the globe.

In all countries, but especially the most industrialised among them, there is an increase in the complexity of social, political, economic, industrial and technological activities, and supporting structures, and also in the chains of dependence between those different fields of activity. Societies operate increasingly as highly complex, dynamic, and unstable mechanisms with chaotic, non-linear behaviour. Small disturbances or localised events can lead to regional, national, or global changes and crises. The greater complexity of human activities and the growing scope of the chains of dependence between them require much more efficient and precise planning, regulation, and management of time.

There is an increasing density of experiences, perceptions, actions, activities, events, and changes per unit time, which amounts to an acceleration of operative social time. In other words, the pace of life is increasing (Levine and Norenzayan 1999; Wiseman 2007). This acceleration is closely linked to increasing speeds: in human performance and activity, in the production, distribution and consumption of goods, in making services available, in the communication and spread of all kinds of news through news media and social media, in the access to knowledge, and in land, sea, and air transportation. On larger scales of time and space, it relates to a faster pace of social, cultural, political, economic, and technological changes, and finally, the acceleration of historical time.

In the past, the scale of time for decisive changes was generally longer than human lives, so adaptation to change often took place over several generations. Currently, the scale of time for change is frequently shorter than human lives, which requires training for constant adaptation throughout one's lifetime.

The tempo of life, in the musical sense of the word "tempo", as beats per minute, has been getting faster. When the number of events, pieces of information, news, experiences, decisions, actions, responsibilities, and projects per unit time increases, the tempo of life gets faster and we are faced with an acceleration of operative social time. Acceleration in a linear motion is the rate of change of the speed with respect to time or the time derivative of the speed. Continuing with the analogy, it is as if the tempo of life is changing from lento to andante, to allegro, to vivace, and ever closer to vivacissimo and prestissimo. Today's societies generate a great deal of their disciplinary and planning power by establishing operative time structures that strictly regulate the use of time. Examples of this include timetables, regulations, planning, and operating hours for businesses, public institutions and services, schools, hospitals, and transport. At a personal level, time structures condition the duration, sequence, speed, and pace of our work activities through work schedules, time-framed performance goals, meetings, presentations, and deadlines for completing all sorts of tasks and projects. Further to these activities, often planned down to the minute, there is leisure time, which includes what has been called since the 1970s "quality time": time entirely dedicated to families and partners, friends, and favourite pastimes. In many countries, and especially in urban areas, the increasing complexity of society and human life overloads time dedicated to compulsory activities while squeezing leisure time or quality time, placing them in a secondary position. Clock time is now the ever present and relentless disciplining mechanism of all human activities.

The acceleration of operative social time varies across countries and is very likely to be higher in the industrialized countries with advanced economies. To measure it we would have to start by measuring the tempo of life or the pace of life, which requires the identification of proxies. As indicators of the pace of life, Levine and Naranzayan (Levine and Norenzayan 1999) used the average walking speed in downtown locations, the speed with which postal clerks completed a simple request, and the accuracy of public clocks. By measuring these indicators, they compared the pace of life in large cities of 31 countries around the world, finding that the pace of life was fastest in Japan and the countries of Western Europe, and was slowest in economically undeveloped countries. A later study using identical methods to those employed in the previous work, found that about 10 years later the pace of life in the same countries and cities was 10% faster (Wiseman 2007). The biggest changes were found in East Asia, where the pace of life had increased by 20% in Guangzhou, China, and 30% in Singapore, which was considered the "fastest moving city in the world".

The tempo of life also varies within a society. For some, the overloading of time with all kinds of events and activities regarding family, work, and leisure, is greater than the average for the society in which they live, while for others it is less. Some manage to feel less time-stressed, with frequent periods of leisure time or longer periods where the pace of life slows down. Various studies have tried to correlate these different groups of people with their income level, finding that people who were always or often stressed had the highest earnings (Hamermesh 2019). However, most people would choose to feel time-stressed rather than income-poor. In any case we are all influenced by the acceleration of operative social time in the society in which we live. Furthermore, this acceleration is rooted in factors that determine the collective behaviour of its members. There are social, economic, political, and technological factors, but the main guiding force is the current global model of development, which has its origins in modernity.

The world 'modern' has its roots in the Latin *modernus*, which in turn comes from the Latin *modo* for 'recently', 'just now', or 'presently'. *Modernus* was used by the senator Cassiodorus, who lived in Italy in the 4th century, in the kingdom of the Ostrogoths, to refer to his time as opposed to *antiquitas*, the ancient times or the time of the ancients. The word 'modern' has been used since the XVI century as a way to divide operative social time into opposing halves (Perrault 1688). Modernity is the condition of being modern, but the word has also been used frequently since the 19th century in the sense of an intellectual, cultural, socioeconomic, political, governmental, scientific, and technological set of ideas, norms, attitudes, knowledge, and practices that replace and oppose a previous, old, traditional, and obsolete set. With this meaning, it has become a topic that is often discussed in the social sciences, and there are various definitions for the historical periods of modernity, including their beginnings and ends. Modernism was a philosophical and artistic movement that flourished in Western society during the late 19th and early 20th centuries and rejected traditional values and techniques, while emphasizing the importance of individual experience. However, the word 'modernism' is also used today to mean the adherence to some form of modernity.

Charles Baudelaire (1821–1867) used the word "modernity" in 1863 in the book *Le Peintre de la Vie Moderne* (Baudelaire 1863) to define life in a time of change and rupture with the past in the great city of Paris, and to describe art's ability to capture and convey the singular nature of experiencing that time. Baudelaire lived during much of the Second French Empire led by Louis-Napoleon Bonaparte (1808–1873), Napoleon III, a Caesarist political regime in which the emperor was elected by universal male suffrage. The Second Empire was founded on what we may describe today as triumphant modernity, in which everything connected to the world of science, technology, industry, and politics was valued and developed with an unquestionable belief in progress. Baudelaire tried to find new forms and models of artistic expression within that triumphant movement. For him, "modernity is the transient, the fleeting, the contingent; it is one half of art, the other being the eternal and the immovable" ("*La modernité, c'est le transitoire, le fugitif, le contingent, la moitié de l'art, dont l'autre moitié est l'éternel et l'immuable*") (Baudelaire 1863). He characterized modernity in art by the importance given to the transient and the ephemeral, without losing the content of a work of art that should be timeless. The new value of the fleeting moment of the present is clearly expressed when Baudelaire writes that "the pleasure we derive from the representation of the present is due, not only to the beauty it can be clothed in, but also to its essential quality of being the present" ("*Le plaisir que nous retirons de la représentation du présent tient non seulement à la beauté dont il peut être revêtu, mais aussi à sa qualité essentielle de présent*") (Baudelaire 1863). In this book, published in 1863, Baudelaire defends painting en plein air since it is a way to capture the setting, landscape, and light of the present directly in the open air and in situ. This form of painting became possible with the emergence of tubes of oil paint and was widely practised by impressionists. With modernity, the present became sufficiently valuable, diverse, and rich to develop its own identity and distinction. The new role of the present reflects the new operative social time of modernity, more focused on full enjoyment just now, on the ephemeral, fashion, outfits, and the art of appearance and visual impact. It also reflects acknowledgement that the present contains, within its complexity, a message that lingers in the past and acquires historical value. Modernity has been able to use the present to enrich historical time. An amazing journey has been made in just a century and a half, from the time of Baudelaire's precursory ideas about modernity and the richness of the present to the overvaluation of instant gratification that modern technologies and especially the ICT constantly give us, in the form of mobile digital devices, for instance. Instant gratification, one of the most basic drives in humans, is related to the tendency to seek pleasure and avoid pain and to seize any opportunity for pleasure as it comes. Modernity has proven itself extremely capable of satisfying our craving for instant gratification. However, by enjoying such short term activities, we may be distracted from meeting longer term goals.

Although Baudelaire was a very innovative writer for his time and sought to assimilate modernity into his creative work, he was eventually overtaken by the dynamics of progress. Throughout his life and particularly in his book *Fleurs du Mal* (Baudelaire 1857), he condemned and firmly opposed the urban plans and projects of Georges-Eugène Haussmann (1809–1891), known as Baron Haussmann, because the plans disrespected the city's historical heritage. Baudelaire rebelled against the destruction of most of the labyrinthine medieval city in Paris and could not have imagined the success of the Second Republic's modernity in urban planning. The future revealed that Napoleon III's gamble on setting Paris against London was a winning one and transformed Paris into a prototype modern city admired by successive generations until the present day. Urban models respond to the aesthetic, social, economic, and environmental demands of their time. None is perfect at the time, and they gradually become unsuitable, but some last an impressively long time. Haussmann's model for Paris was enormously successful and was the inspiration for many modern urban areas all over the world. It is interesting to note that some authors have acknowledged recently that Haussmann's model had, as well as aesthetic, historical, and symbolic value, certain characteristics that respond well to the current sustainability concerns faced by large urban areas (Jallon et al. 2017).

From a formal perspective, modernity is an intrinsically temporal concept rooted in a comparison between the view of the world associated with the operative social time of one or several successive social generations and the view that the same generations associate with a previous operative social time interval. It requires an awareness of history and involves an underlying expectation or belief in an evolution or some form of progress. From a historiographical perspective, it is commonly considered, as already mentioned, that the first stage of modernity, or the modern era, began at the start of the 16th century, during the Renaissance, just after the beginning of the Age of Discovery, and ended with the French Revolution in 1789 (Berman 1982). It was followed by a period up to the 20th century, which Anglo-Saxon historians call classical modernity (Osborne 1992), during which the Industrial Revolution and the capitalist economic system flourished, science and technology caused relatively significant economic growth, and the British Empire began to prevail at world level. From the 20th century onwards, the terminology began to diversify: some authors claim that a new stage of modernity began, while others conclude that modernity was replaced by postmodernity (Lyotard 1979). Furthermore, some French historians prefer to call the period from the French Revolution to the present day the Contemporary Age.

The close relationship between modernity and the acceleration of time has been analysed by many authors, including Marshall Berman (1982), Peter Conrad (Conrad 1999), Reinhart Koselleck (2004) and Hartmut Rosa (2013). Modernity is ultimately the acceleration of time that manifests itself in several ways in our everyday lives and also at the economic, scientific, technological, social, political, and cultural level, and in the arts. In this set of sectors, some operate mainly as driving forces for the acceleration of time and, as a consequence, for the advancement of modernity, especially the economy and science and technology. Others primarily reflect the social response or reaction to the specific and changing characteristics of the acceleration process, and they also contribute to the characterization of modernity, from the point of view of social behaviour, politics, culture, and the arts.

4.4 The Role of Technology in the Acceleration of Operative Social Time and Economic Growth

The development of science and technology has played a central role in the acceleration of operative social time (Rosa 2013). The two processes are closely related and strengthen one another. In Leslie White's determinist perspective, "technology is the independent variable, the social system the dependent variable" (White 1949). Later, Gerhart Lenski, in the context of a theory of social development, suggested that technological advances are the fundamental factor for the evolution of civilisations and cultures and that the greater a society's technological transformation, the faster its rate of change will be (Lenski 2005).

One of the most eloquent expressions of the influence of new technologies on the acceleration of operative social time is the concept of technological singularity, first proposed by the mathematician Stanislaw Ulam (1909–1984) in 1958 and later widely developed by several scientists, including Vernor Vinge (1993) and Ray Kurzweil (1999). The singularity is the purported climax of continued increase in the speed with which technological innovations emerge. Believers in this forthcoming event foresee it happening in the middle of the 21st century, although the nature of the singularity is still ill-defined. It is expected to lead to the transformation of humans into "post-humans", giving them superhuman intelligence and prolonging their lifespan indefinitely. The use of these new extreme abilities will increase the diversity and the rate of production of goods and services, prolong working lives, and enhance our physical and cognitive capabilities, thereby profoundly changing social life, the economy, and operative social time.

At present, there is already a time crisis caused by a scarcity of time in countries with advanced economies and in many other countries, especially in their urban areas (Rosa 2013). The belief that "time is money" continues to become entrenched more deeply and more globally, although people think about time and money in very different ways. Both time and money became scarce and precious resources and the future is likely to accentuate these tendencies. A guaranteed way to transmit an image of success is to look time-stressed with a very full agenda and little time available. A professional with ample time available would be incomprehensible, especially for the governing economic and financial elites. The time crisis is related to technological advances that are designed to give us more time, in a seemingly paradoxical way. A significant number of technological innovations make it possible to save time by carrying out specific functions, activities, or journeys in less time. The paradox lies in the fact that the more appliances, devices, processes, infrastructures, and means of mobility we invent and produce to save time, the less spare time we have. According to Rosa (2013), "we don't have any time although we've gained far more than we needed". However, there is no paradox here for most people because what they really want is to gain more time to increase its loading capacity, not just to gain time. People may be lured by the perspective of more leisure time but they often do not use the extra time in that way. There are many different motivations for overloading time, although they all tend to share the property of satisfying some irrepressible need. One of them stems from the deep-rooted and ever expanding relationship that has been established between time and money. The more time we gain, the more money we are likely to make with that extra time. The relation between time, money, and subjective well-being has been analysed by many authors and there is empirical evidence that prioritizing time over money can be a stable preference that leads to greater subjective well-being (Whillans et al. 2016).

Contemporary workplaces are frequently characterized by long working hours per week, constant availability, and ever increasing speed of work, although time overloading has detrimental effects on individuals, families, and organizations (Blagoev et al. 2018). Furthermore there is evidence that time stress increases the risk of several diseases of the nervous, endocrine, and cardiovascular systems. A study by Pfeffer and Carney (2018) showed that people who are keenly aware of the economic value of their time, or in other words, tend to think of time as money, are generally more psychologically stressed and exhibit higher levels of the stress hormone cortisol than do people for whom the economic value of time is less important. In two experiments, people nudged into an economic mindset self-reported higher levels of psychological stress compared with participants whose monetary value of time was not emphasized (Pfeffer and Carney 2018). Time scarcity is one of the main drivers of current modernity. Many of those who are at the cutting edge of their profession and have the ability to transform society fill their time to the limit and often complain about not having time. The speed at which significant events fill their personal operative social time is a measure of success. A break or delay in professional activity may compromise an opportunity or give competitors an advantage. The overloading of time is presently an agent of socio-economic distinction. Generally, the greater the professional and economic success, the greater the risk of time stress and acceleration of personal operative social time. This strong correlation assigns a pioneering, avant-garde, and progressive value to the acceleration of operative social time.

Technology also influences economic growth, which is a dominant concept in the social, economic, and political fields. It is therefore important to analyse this relation. In order to do so, we start by considering the concept of production function, defined as the relationship between the quantities of the factors of production (inputs), such as labour and capital, and the amount of product obtained in the form of goods and services (outputs), assuming that the most efficient available methods of production are used.

From an analytical perspective, economic growth is only possible if the inputs that support the production process grow or if new ways of achieving greater outputs from the same inputs are used. The productivity of an economy describes and evaluates the efficiency of production, which can be measured by the ratio between the outputs and inputs used in the production process. Several theories emerge at this point to identify the factors of production. In the classical economic theory, developed by Adam Smith and David Ricardo, the factors of production or "component parts of the price of commodities" (Smith 1776) are land, labour, and capital. The neoclassical theory of economics re-analysed the production function and identified a new factor of production. Let us look at how that conclusion was reached.

The first quantitative evaluation of the production function was only made after the Second World War in the USA, a country that was already highly industrialised, and it was carried out by the economist Moses Abramovitz (1912–2000) (Abramovitz 1956). He calculated the growth in labour and capital as factors of production and the growth in output in the American economy between 1870 and 1950 and concluded that the growth in the inputs accounted for only 15% of the growth in output. In other words, the latter was much greater than would be expected solely on the basis of the growth in capital and labour. In a statistical sense, there was a residue of 85%, which is now called the total factor productivity (TFP) or multi-factor productivity. TFP is a measure of the efficiency of all the inputs to a production process, and its increases generally result from technological innovations or improvements, or in other words from technological progress.

The economists Robert Solow (1956) and Trevor Swan (1918–1989) (1956) studied the same topic, with an economic model established within the framework of the neoclassical theory of long-term economic growth and based on capital accumulation and growth in the economically active population. The model provides a way to identify the contribution of technology to growth and evaluate the importance that it has had in boosting it. Solow (1957) demonstrated that about 85% of the growth in output in the American economy per worker in the period 1870–1950 can be attributed to technological innovation, which agreed with the residual growth obtained by Abramovitz. These results marked an important turning point since most economists since Adam Smith had been using economic growth models in which the growth was primarily obtained by adding inputs to production, especially inputs of capital.

As a result, since the middle of the 20th century, investment in scientific research and technological innovation has been considered essential by the governments of countries with advanced economies—later gradually extending to all countries— as a necessary condition to ensure economic growth. In conclusion, in mainstream neoclassical economic theory, the factors of production are labour, capital, including fixed capital (land, factories, buildings, machinery, laboratories, computers, and other physical assets essential to production), financial capital, and what is often called technological progress, which is expressed quantitatively by the concept of TFP.

In the 1990s, several economists (Romer 1990; Grossman and Helpman 1991; Aghion and Howitt 1992; Jones 1995a, b) used different neoclassical economic models to demonstrate once again that investment in developing new technologies and promoting technological innovation in a country would cause its economy to grow. Nonetheless, this correlation, which was considered uncontroversial and vital, was short-lived. It was called into question in the specific case of a range of digital technologies and more generally in ICT (Brynjolfsson 1993). In the USA, TFP grew remarkably between 1891 and 1972, with an average annual growth rate of 2.33% (Gordon 2000, 2016), peaking between 1940 and 1965. This wave of TFP growth was mainly the result of technological innovations in the second half of the 19th century and the beginning of the 20th century, which spread throughout the economy and revolutionised industry, spurred by their remarkable positive impact on people's ways of life. In Europe and Japan, the same wave of growth arrived later, due in part to the effect of the two World Wars. In the 1970s, the annual TFP growth values began to fall in the USA and remained close to 1%, except between 1996 and 2004 (Gordon 2015, 2016).

The remarkable growth in ICTs between 1980 and 2014, and increasing access to computers and the internet, did not lead to growth in TFP in the USA, which remained practically the same for the whole period. This discrepancy with the expectations created was highlighted by Solow in 1987, six years after mass production of personal computers began, and became known as Solow's Paradox or the Productivity Paradox. In an article in the New York Times Book Review, Robert Solow summarized his thoughts stating that "You can see the computer age everywhere but in the productivity statistics" (Solow 1987). The extraordinary scientific and technological development of the last four decades, especially in ICT, the physical sciences, biotechnology, and health sciences and technology has not had a significant impact on economic growth in countries with advanced economies. In the USA, real GDP grew an average of only 1.55% per year in the period 2005–2014, which is half the annual average of 3.12% for the period 1974–2004 and less than half the annual average of 3.62% for the period 1929–1974 (Gordon 2015).

The decline in economic productivity in the USA, and also in most other countries with advanced economies, since the 1970s has been widely analysed and debated from the standpoint of the neoclassical economic theory and from other points of view. In the first case the only agreement is that there is no single clearly identifiable cause, but rather several types of cause that interact with each other. The causes most often given are connected to social factors, such as an ageing population, especially in countries with advanced economies, which reduces the labour input into the economy, the increase in wealth inequalities, which impair increases in consumption, decreasing average levels of education in countries with advanced economies, and the unsuitability of vocational training. Financial causes are also invoked, such as an increase in public and private debt in relation to GDP in almost all countries in the world, the scarcity of capital to invest in productive activities, which reduces the capital input into the economy, inadequate investment in the production of goods and services, in part due to a preference for investment in financial products, unsuitable allocation of public and private investment and governance factors, for example, unsuitable legislation and policy. In the opinion of the economist Lawrence Summers, some of these causes have led to the recent anæmic economic growth in the USA, which he calls secular stagnation (Summers 2014), a concept first used by Alvin A. Hansen at the time of the Great Depression (Hansen 1938). However, he believes that the downturn is temporary and that we have to be patient, since with the right policies high economic growth will return. Other neoclassical economists believe that countries with advanced economies are on the verge of a new technological revolution that will certainly produce robust growth.

The point of view of ecological economics, while agreeing with some of the above drivers, also emphasizes the importance of physical and environmental causes, such as increasing global demand for water and energy, overexploitation of natural resources, environmental degradation, and anthropogenic climate change. The overuse of natural resources creates an "ecological credit crunch" that has similarities with the financial credit crunch that happens when an economic bubble bursts. According to this standpoint, the question of the environment and natural resources are critical, and generate long term trends that make it unlikely that robust growth will pick up again soon.

On the other hand, from the standpoint of neoclassical economics, there is an emerging tendency to argue that it is precisely overly restrictive environmental regulations that are harming economic growth. As regards climate change, it is forcibly argued by some that an energy transition away from fossil fuels into renewable energies would harm economic growth and employment. There is clearly an increasing polarization as regards the attribution of causes to the ongoing decrease in the average global growth of GDP.

Although advances in science and technology in the past few decades do not appear to have been able to free economic growth from a period of relative stagnation, especially in countries with advanced economies, due to a wide range of reasons, it is widely recognised that they continue to have a highly beneficial effect on well-being and the quality of life.

4.5 The Role of Energy in Economic Growth

The likely causes of slow economic growth can be further illuminated by considering the global picture and in particular the relation between global GDP growth and global energy consumption. The average annual growth of real world GDP rose from 2.1% in the period 1900–1940 to a maximum of about 3.8% in the period 1940–1970 and has fallen since then, with average values for the periods 1970–2000 and 2000–2014 of approximately 3.3% and 2.7%, respectively (Maddison 2006; USDA 2018). If we consider real world GDP per capita, an indicator that comes closer to the experience that each person has of the behaviour of the economy and how it influences individual lives, we can conclude from World Bank data that it fell from 3.5% in the 1960s to 2% in the 1970s, and stagnated below 1.5% in the period 1980–2017. In the OECD countries, the fall in real GDP per capita has been more marked, starting with figures above 4% in the 1960s and hitting values below 1.2% in the period 2000–2017.

What are the causes of this systemic behaviour that has been going on for more than half a century? A relevant observation is that over the same period of time the annual growth of real global GDP per capita tends to follow the growth in the world's energy consumption per capita (see Fig. 3.9) (Jancovici 2012b). Remembering that energy is not considered to be a factor of production in the production function makes this correlation surprising. When primary energy consumption per capita rises or falls, real GDP per capita tends to follow a similar behaviour. A clear example was the 1979 shutdown in Iranian oil exports caused by the Iranian Revolution, which reduced global oil supply by only 4%, but caused oil prices to more than double in the following year and subsequently led to economic recessions in Europe and in many other regions of the world (Hénin 1983). The primary goal of the economic system is the production and consumption of goods and services, and all these processes require energy, which is a physical concept with a precise definition that does not always correspond to its meaning in everyday life. Energy in physics is the capacity to do work and it exists in various forms, which can be classified into two groups, kinetic and potential. Work is done whenever a force causes an object to be displaced and is calculated by multiplying the length of the path followed by the component of the force along that path. Mechanical, thermal, and electrical energy and electromagnetic radiation are four different forms of kinetic energy, while gravitational, electromagnetic, weak nuclear, and strong nuclear are the four different types of force fields present in the universe, each generating a type of potential energy. One of the most important points about energy is that it cannot be created out of nothing. Energy can only be converted from one form to another. This is a law of physics that human ingeniousness cannot change. Furthermore we cannot use all the energy input, but only a fraction. The ratio between the useful output of converted energy and the energy input is the energy conversion factor, which is about 1–2% in the photosynthesis of most crop plants and about 20% in internal combustion engines.

For instance, strong nuclear potential energy is converted into thermal energy in the Sun's interior through nuclear reactions. The thermal energy is converted into electromagnetic radiation emitted by the Sun's surface and this solar electromagnetic radiation, when it reaches the Earth, is converted through photosynthesis into the chemical energy stored in plants and in the animals that feed on them. Finally, the conversion of the chemical energy stored in humans and animals into mechanical energy allows them to move and live. There is no energy production in any of these stages, just a chain of energy conversion processes with a significant fraction of "wasted" energy in the end.

The production of any good or service requires the use of natural resources that must be collected, prepared, transformed, and transported. All these processes require the work to be done by the forces involved, and therefore require the consumption of energy. Energy is necessary to produce and cook food, heat and cool homes, operate transport systems, and run all sorts of businesses, schools, hospitals, and other kinds of institution. From a historical standpoint, the intensive use of energy has been a necessary but not sufficient condition to explain the success of the Industrial Revolution, and the scientific, technological, and economic progress that followed. When, by whatever cause, the supply of primary energy is reduced, productivity falls, making it much harder, or even impossible, for the economy to grow.

The reliance of an economy's growth on an energy consumption that is also growing is the main reason for promoting energy efficiency. Its ultimate goal is to disconnect economic growth from the growth in energy use as far as possible, or in other words, to decrease the energy intensity of the economy, defined as the average amount of energy required to generate a unit of GDP in that economy. In the period 2000–2016, global energy intensity fell by an annual average of 1.6% (Enerdata 2017). The value and behaviour of this energy indicator in different regions and countries is highly variable and depends on the energy mix, the level of dependence on energy imports and exports, and social, economic, technological, and environmental factors. According to the World Resources Institute's databases, Saudi Arabia and Russia are examples of countries with high energy intensity, while countries with advanced economies have generally moderate energy intensities, and some developing countries, such as Bangladesh and Costa Rica, have low energy intensity.

Despite the reduction in the global energy intensity, it is safe to say that the global demand for energy will continue to rise due to demographic pressures and the expected growth of real global GDP per capita. Scenarios for the future growth of the global energy demand are many and varied. According to the International Energy Agency, a likely scenario is for the global energy consumption to rise by 30% by 2040 in comparison with 2017 (IEA 2017a), which is an enormous challenge for world industry and economy.

In prehistoric times, the average annual energy consumed per capita was roughly 0.36 GJ per year, corresponding to the average work carried out by an adult human body. That average value has now increased to 75 GJ, as already mentioned, but the differences between countries are considerable. For instance, in Afghanistan people use on average 7 GJ per capita and per year, while in Iceland that value increases to 709 GJ. The main reasons for such differences are related to the level of economic development, energy availability, and climate.

The global consumption of primary energy per capita has increased in the period 1860–1980 at an annual average rate of 2.5%, although with a slowdown and greater variability in the period of the two World Wars. Then since 1980 annual growth has decreased to an average of 0.4% in the last 30 years (Jancovici 2012b). In the period 1980–2000 the world energy consumption per capita stagnated at about 64 GJ, then increased to about 78 GJ in the period 2000–2011, before increasing more slowly in the period 2011–2017 (BP 2018). The decrease in energy intensity was insufficient to compensate for the slow increase in world energy use per capita since 1980. Furthermore, the slow decrease in global energy consumption growth per capita became associated with a decrease in the annual growth of real world GDP per capita (see Fig. 3.9). The main response to this unwelcome situation has been to assume that the deceleration in the annual growth of real world GDP per capita can be inverted by enhancing the productivity of the global economy with increasing credit. This response provokes an accumulation of public and private debt, which enhances the risk of debt defaults.

The slow increase in annual global energy consumption per capita since 1980 has been attributed to a decrease in the supply of fossil fuels. Coal production has been generally lower and the cost of fossil fuel exploitation, particularly coal and oil, is increasing (Jancovici 2012b; Tverberg 2012). The energy return on energy invested (EROEI) of those fuels, defined as the ratio between the amount of energy produced from a certain primary source and the amount of energy consumed to obtain that energy production, has been systematically falling, which is to be expected for a non-renewable natural resource.

The idea that energy supply plays a crucial role in the economic growth of all countries, especially those with advanced economies is not a new one. However, energy is not recognised as a factor of production in the production function of neoclassical economic theory. This has led to a long and probably irresolvable debate between neoclassical economists and ecological economists (Illge and Schwarze 2006). The latter argue for a view of economics in which natural inputs, essentially energy and natural resources, are centrally important and place limits on economic growth (Georgescu-Roegen 1971; Costanza 1980; Daly 1997a). In between the extreme positions of the two camps, one finds economists close to the neoclassical view that have recently highlighted the indispensable role of energy in economic growth using the framework of the neoclassical formalism (Ayres and Warr 2009a; Stern 2011). The relationship between energy use and economic output has been analysed with neoclassical economic methods using the Granger causality concept, but according to the meta-analysis of scientific literature carried out by Bruns et al. (Bruns et al. 2013), there is no consensus on how to model that relationship, so the causality cannot be firmly established.

Another way of dealing with the issue is to use the cost-share theorem from neoclassical equilibrium theory, according to which costs are shared in proportion to the elasticity of the factors of production in the production function. In an economy in equilibrium, the cost share of labour is about ten times that of energy, so the latter is not considered as a primary factor of production. In conclusion, neoclassical economics only recognises labour and capital as inputs, although it acknowledges that there are relationships between energy and production. However, it is assumed that these relationships can always be controlled by prices, capital, and technology. The neoclassical point of view establishes a formalism in which energy is relevant for production but plays a secondary role relative to the accepted factors of production.

The crucial point is that the factors of production are solely dependent on human activities. In other words it is postulated that productivity is independent of non-human factors. Neoclassical theory has the reputation of being one of the most powerful and successful human theoretical constructions, credited with creating economies where, on average, the value of indicators of human consumption and economic prosperity continue to grow throughout the world. This success was created by people and it is assumed that they will have the necessary creativity and ingenuity to be able to prolong it indefinitely. Nonetheless, it is not entirely clear how perpetual economic growth can be achieved, given its dependence on energy and natural resources and the laws of physics, and in particular thermodynamics.

The solution to this problem can be found in the operative social time that is privileged by neoclassical economists, which focuses on the short and medium term of 50–60 years rather than on the long term (Stiglitz 1997). They are primarily concerned with immediately providing robust economic growth because that is the way to meet the expectations of a large majority of populations. They promote the acceleration of operative social time by implementing multiple strategies to lure consumers to consume and thereby enhance economic growth as quickly as possible. The long term and the contradictions to which perpetual growth can lead do not concern them, or may concern them, but only from a purely academic and theoretical point of view. The operative social time of the neoclassical culture of economics favours the present and the short term, and in the medium term, only deals with the element of operative social time of the contemporary generation.

4.6 Predicting the Future. Probability Theory and Evolution of the Concept of Risk

The relationship between modernity and time goes well beyond the acceleration of operative social time. There is another dimension in this relationship that specifically concerns the future. In the first stage of modernity, the development of mathematics made it possible to launch the foundations of probability theory and, with it, formulate the concept of risk as it is currently used (Bernstein 1998). Defining risk in terms of the mathematical probability and the value of the expected loss has changed our relationship with the future. It has become possible to evaluate quantitatively the likelihood of danger, harm, or loss in the future as a result of a given human action or inaction or of a natural unprovoked disaster. The future has stopped being an unknown entity, only assessed and commented by oracles, prophets, and clairvoyants. With the modern age and the help of science and technology, it has become possible to assign likelihoods of occurrence in terms of probability to some kinds of hazardous future events and situations.

Risk assessment and management have become common practice in many different fields of human activity and have in particular led to the development of the insurance industry. Specific and powerful methods have been developed to make continual risk assessment and management in almost all domains of human activity and enterprise. We now deal with risks in health, human factors, security, military operations, social, recreational, and cultural activities, information systems, transport, business, economic and financial activities, technology, and the environment. Besides, as already mentioned, we are now starting to use the emerging concept of existential risk (Bostrom 2002).

Probability theory was born, like many other ideas and theories, from the impulses that characterise human nature, in this case the attraction to gambling. Antoine Gombaud (1607–1648), a French writer who called himself Chevalier de Méré, had an intense social life and devoted himself to gambling at the social salons he frequented. By betting that he would get a 6 at least once if he threw a die four times in a row, he managed to win money. As he was an amateur mathematician, he calculated that he would have the same chance of winning if he bet that he would get a double 6 by rolling two dice 24 times. He was surprised to find that he lost money when he did this.

A second problem that Méré was unable to solve was the "problem of the division of the stakes" or "problem of points", that had been recognised since the Middle Ages. Two players play a series of rounds that each one has an equal chance of winning, such as flipping a coin to get heads or tails, with both players betting the same amount of money on each round until one of them has a total of an agreed number n of wins, at which point the game ends and the winner gets all the money. The problem is knowing how to divide all the money bet by the two players if the game is interrupted by unexpected circumstances.

Méré sent these two problems to the *Academia Parisiensis*, an informal forerunner of the *Académie Royale des Sciences*, founded in 1666 by Jean-Baptiste Colbert (1619–1683). The former was set up in 1635 by Marin Mersenne (1588–1618), a French philosopher, Catholic theologian, physicist, and mathematician who discovered the laws of acoustics and published them in his book on Universal Harmony (Mersenne 1636). He was one of the most learned people of his time and he kept up correspondence and shared knowledge with the great figures of mathematics and science in Europe. The austere and brilliant French mathematician, physicist, and philosopher Blaise Pascal (1623–1662), who corresponded with the Academy, decided to try and find a solution to Méré's problems in 1652. The first was easy: Méré's way of calculating probabilities was incorrect, because he added probabilities instead of multiplying them. By multiplying, one actually concludes that the probability of winning the game with one die is $1 - (5/6)^4 = 0.51774\ldots$ instead of $(1/6) \times 4 = 0.6666\ldots$, both numbers larger than 0.5. When throwing two dice, the correct answer is $1 - (35/36)^{24} = 0.49140\ldots$, which is smaller than 0.5.

The second problem was more difficult. Pascal began to correspond with the magistrate, mathematician, and physicist Pierre de Fermat (1607–1665) in 1654 and they both came up with the same solution by following different paths. In Pascal's letter to Fermat of 29 July 1654, he revelled in the universal nature of the conclusions when he wrote: "I can clearly see that the truth is the same in Toulouse and in Paris" ("Je vois bien que la vérité est la même à Toulouse et à Paris"). Pascal's proposed solution used, for the first time, the concept of mathematical expectation or expected value of a random variable, conditional expectation, and even the martingale, which is a sequence of random variables where the conditional expectation of x_{n+1}, given $x_0, x_1, x_2, \ldots, x_n$, is always just x_n. The martingale is a very rich concept in probability theory, and one that is used extensively in gambling strategies and in the fields of economics and finance.

During a stay in Paris, the Dutch mathematician and physicist Christian Huygens (1629–1695) heard about the debate between Pascal and Fermat, and in 1657 wrote the first treatise on probability theory (Huygens 1657). This was followed by contributions from Jakob Bernouilli (1655–1705), Thomas Bayes (1701–1761), and Joseph-Louis Lagrange (1736–1827). In the 1780s, Nicolas de Condorcet, Permanent Secretary of the Paris Academy of Sciences at the time, argued for the importance of applying mathematics to social sciences and realised the power of probability for predicting the future and therefore preparing for a better future. One of his main objectives was to construct a unified project conjugating politics, education, and science. In 1792, he wrote a book on probability calculus (Condorcet 1805), published after his death, where he strongly advocates the teaching of this subject area: "Since there is nothing more suitable (than calculating probabilities) for destroying speculative and practical errors that impede progress and work against the happiness of the human race") (*Il n'en est point de plus propres (que les connaissances du calcul de probabilités) à détruire les erreurs spéculatives ou pratiques qui arrêtent les progrês et s'opposent au bonheur de l'espèce humaine*). He concludes that we must not ignore any means of making such knowledge widely available. The omnipresence of the future and a constant preoccupation with controlling it and filling it with hope, doubt, and fear was born with probability theory and its applications, and it coevolved with modernity until we reached the "risk society".

Probability theory was essential to the creation of mathematical statistics, which was first developed at the beginning of the 19th century. The method of least squares was first described by Adrien-Marie Legendre (1752–1833) in 1805. Various branches of science, in particular astronomy, physics, and biology, started to use probability models and statistics. The foundations of statistical mechanics, one of the fundamental domains of modern physics, were established by Ludwig Boltzmann (1844–1906) in 1872. During the 19th and 20th centuries, statistics and probability theory were increasingly used by social scientists in many areas, such as experimental psychology and sociology. Let us now go back in time and explore the etymology of the word "risk", which has uncertain origins somewhere in the Mediterranean region. It probably comes from *rhizikon, rhiza*, a Greek navigation term that means "root", "rock immersed in the sea", or "underwater obstacle". The word was a metaphor for the dangers that seafarers had to avoid at sea. In Latin, we find the words *resicum, risicum*, and *riscus* meaning "cliff" or "reef", which led to the words *risico, risco*, and *rischio* in Italian, *risque* in French, *riesgo* in Spanish, and *risco* in Portuguese.

Another possibility is that the origin of the meaning that we now give to the word "risk" comes from the Arabic word *rizq*, which represents the means of livelihood that Allah provides everyone with. This influence may have come into effect when the Arabs started to trade and do business with the Europeans around the Mediterranean Sea. The underlying idea behind the word *rizq* was that, although Allah would always provide sustenance for everyone, believers could not know for sure what profit they would have at the end of the day. It was this notion of uncertainty that was probably transmitted to non-Muslims with a somewhat inverted meaning. Risk became an awareness of the danger involved in an enterprise, for instance travelling to unknown lands to seek God's bounty.

In any case, the word "risk" spread around Europe from Italy in the 12th century and gradually came to be used to mean the challenge and danger of going forward into the unknown to explore, become wealthy, and achieve success, or the danger associated with many kinds of endeavour.

The richness of human feelings, conditions, and ideas involved in the use of the word "risk" goes well beyond its etymology. Risk is essentially linked to uncertainty and therefore to time. The way that people have interacted with uncertainty has evolved profoundly through history, and most likely through prehistory, since the early stages of consciousness. The modern technical use of the word "risk" is strongly anchored in the principles of probability theory, but the ability to evaluate a future danger and think up ways to deal with it and ways to conceptualize this type of permanent condition of uncertainty about the future is very likely to have preceded *Homo sapiens* time.

In historic times we know the importance of the idea of fate in many mythologies. In Greek mythology the Fates or Moirai are three weaving goddesses in charge of the destiny of both men and gods (Grimal 1985). Their names are Clotho, the spinner, Lachesis, the allotter, and Atropos, the inflexible or inevitable, associated with death. According to Plato's Republic (Plato, circa 380 BC), the Fates are the daughters of Ananke, the personification of necessity and inevitability. In those times it was recognized that people could make decisions that would interfere with their future, but it was believed that they could not escape a pre-ordained fate. Even the gods could not change their fate. The conclusion was therefore that people should not tempt fate through foolhardiness, foolishness, or vanity because it would anger the gods and therefore be harmful and dangerous for them. It is interesting to note that this concept, which is the one having the most similarities with our current concept of risk, still exists and is still used in a very similar form. Currently, the expression "to tempt fate" means to do something that one knows is dangerous or likely to have negative effects. Furthermore, we continue to say that there is a risk in going against our most deep-rooted dispositions and propensities, which has the same functionality as advising us not to go against fate and destiny. So initially, there was a tendency to take a fatalistic view of the future, which was to a large extent pre-ordained.

However, an important transition occurred at the beginning of the 16th century, when we started to convince ourselves that we had some significant ability to influence our own future. The need emerged to analyse and assess the benefits that could be gained from certain actions and projects, and also the dangers that they implied. The future became more complex and fragmented by different degrees of opportunity and danger. The first dictionary to mention the word *riscare* in Italian defines it as: to hazard, to jeopardize, to endanger (Florio 1598). The concept of risk became closely linked to what we currently refer to as risk-taking, which is a characteristically human predisposition. It is part of human instinct to embark on relatively high risk activities if we perceive that it may lead to valuable rewards. To achieve their adult identity, most adolescents need to practice different sorts of high risk activities.

There was no methodology available for systematically evaluating the dangers associated with risk-taking decisions or activities up to the time when modern science started to flourish and probability theory was developed. Time was considered to be an important player in dealing with that uncertainty. As already mentioned, *kairos* in Ancient Greek was the right time to say a particular thing or the opportune time for action. To decide when to initiate an action or start executing a plan which was likely to be dangerous, such as a voyage or a military attack, people asked for the advice of oracles and clairvoyants, and looked for all sorts of signs. Notice that this way of dealing with the uncertainties involved in risk-taking activities still persists nowadays. It is not uncommon throughout the world to be superstitious about selecting the best time to start an initiative with potentially beneficial and dangerous outcomes. In conclusion, there are many meanings for risk in contemporary societies that coexist and interact, and also many different definitions. Furthermore, risk is a rapidly evolving concept because it is strongly related to time, uncertainty, and technology.

The basic model to measure risk is to assume that it is given by the product of the probability that the event occurs and the severity of the consequence, usually expressed in terms of the value of the resulting loss. Diversified and specific methodologies have been developed to quantify the risk involved in different areas of risk management, in particular in projects relating to or in the operation of security, health, information, energy, industry, transport, economic, and financial systems. The estimated costs of the risks involved, also called the known unknowns, define the cost contingency of the project or operating system. Cost contingency is calculated using heuristic methods based on experience-based techniques, empirical methods based on the historical factors that drive risk, probability distribution methods based, for instance, on Monte Carlo simulations, mathematical modelling methods based on artificial neural networks and fuzzy sets, and many other methodologies.

With modernity, it became possible to make quantified estimates of a wide variety of risks using probability theory, statistics, and science. But modernity has also generated a huge range of new risks. This aspect was emphasised by Ulrich Beck (1944–2015), when in the 1980s he considered that we were living in a "risk society", characterised by "a systematic way of dealing with hazards and insecurities induced and introduced by modernization itself" (Beck 1986). Beck believed that a "reflexive modernisation" had emerged in the 1970s that reflected a growing concern with issues related to the political and economic management of current, emerging, and future risks, particularly those arising from new technologies. Constant, instant access to news and information causes us to be aware of an increasingly broad range of individual and collective health, security, terrorism, war, migratory, political, economic, financial, and environmental risks. There are also technological risks connected to modernity which Anthony Giddens calls manufactured risks (Giddens 1999), such as the use of genetically modified foods and nuclear accidents, and emerging technological risks such as those arising from genetic engineering and human cloning and the widespread use of synthetic biology, robotics, and artificial intelligence.

Most people, particularly in countries with advanced economies, have an education and academic training that enables them to understand at least superficially the enormous variety of risks, but they generally do not have enough knowledge to fully comprehend and assess each of them, nor the time and predisposition to acquire such an ability. Nevertheless, most people are sensitive to the problem and using the information available build their own very diverse opinions, which are unevenly reasoned and often contradictory, although all with the same face value. This huge cacophony of viewpoints on risks makes a decisive contribution to reflexive modernisation. Society has become trapped in the snare of risk. We are forced to decide on risk while aware that we ourselves are unable to assess it objectively, so it has become necessary to trust in the assessments made by experts who use probabilistic and statistical methods and complex, abstract algorithms that we do not fully understand. They are the ones who warn us about risk levels for terrorist attacks, armed conflicts, financial collapses, natural disasters, climate change impacts, and current and future technologies. Generally, the more familiar and understandable a technology is, the lower the risk we assign to it, even if that goes against statistics. It is relatively common to have a fear of flying, but the risk of dying on a civil aviation flight is around 100 times less than the risk of dying in a car accident, according to a study carried out in the USA (Savage 2013). Travelling by motorbike involves a risk 3 000 times greater than travelling by plane. Our reaction to risk currently depends on our perception of it, based on how much we trust abstract risk assessments, on our knowledge of the most serious events that have actually taken place, on the media we regularly use, and especially on the opinions expressed by the identity group to which we belong. This reflexiveness of risk may have a greater or lesser ability to influence the political, social, economic, and technological decisions of a country depending on the nature of its political regime and historical, religious, and cultural characteristics. The current reflexiveness of risk has implications not only on the contemporary generation, but also on future generations.

One clear example of this multigenerational effect is the choice of primary energy sources and corresponding energy conversion technologies. The continued intensive use of fossil fuels involves potentially serious risks for the contemporary generation and especially for future generations due to anthropogenic climate change, which is mainly caused by carbon dioxide emissions into the atmosphere. There are other primary energy sources, all of which carry some risks and associated contingency costs, besides different investment and operating costs. Choosing between the different primary sources and choosing an energy mix appropriate for a country is a complex process that involves weighing up a range of social, economic, financial, political, and environmental factors, as well as people's perception of risk in the short, medium, and long term. The importance of the latter factor heavily depends on the system of government, the level of industrialization, and the level of socioeconomic development. For example, in countries with advanced economies, public opinion favours discontinuing use of both fossil fuels and nuclear energy and replacing them with renewable energies. In some large emerging economies, such as China and India, that energy transition is considered to be incompatible with the economic development needed to reach the status of an advanced economy. Thus, they prefer to keep and develop nuclear energy. In any event, the assessment of the risk involved in choosing the best primary energy mix for a given country, necessarily involves the short, medium, and long term. One of the most challenging and interesting analyses to be pursued within the next 100 years or so is to interpret and explain the motivations and rationality for the comparative risk assessment between fossil fuels and nuclear energy as primary energy sources that is currently made at global level.

Besides the risks that can be modelled and reasonably incorporated in probabilistic predictions, we are also confronted with damaging natural or man-made occurrences that are very difficult or almost impossible to envision, because they are very rare or entirely new as regards the human experience. The Deepwater Horizon oil spill in 2010, the Fukushima nuclear accident in 2011, the extremely intense tropical cyclones, such as the Hayan in 2013, and certain large scale terrorist attacks are examples of such occurrences, often called the unknown unknowns. In the case of natural disasters we may be able to forecast the occurrence, but unable to estimate the full extent of the damage because of the complexity of the affected systems and knock-on effects. The question arises as to whether there are risks inherently unknowable to us or whether we potentially have the ability to envision and assess all kinds of risk if we devote enough attention to the problem.

The concept of threat in contemporaneous societies is intimately related with security. As a political value of a given society, security is defined by the individual and collective value systems of that society. Arnold Wolfers (1962) considers that "security, in an objective sense, measures the absence of threats to acquired values, in the subjective sense, the absence of fear that such values will be attacked". Threat is the potential occurrence, the individual, entity, or action that has the power to harm life, property, all kinds of human systems, or the environment. Risk can be interpreted as including threat as a component, but goes a step forward, because it defines the unwanted outcome resulting from the occurrence, which exposes the threat, and determines its likelihood and associated consequences. Thus the use of the word "threat" by itself is usually interpreted as an admission that we cannot entirely control or prevent a future event with harmful effects, or that we are unable to fully determine those effects. In this context it is significant that in the 1980s the concept of threat was extended from the military to the environmental realm and various authors recognized the emergence of new security threats that included population growth, resource scarcity, and environmental degradation (Mathews 1989). The Brundtland Commission, a report published in 1991 (WCED 1987), refers repeatedly to environmental threats and states that: "The whole notion of security as traditionally understood in terms of political and military threats to national sovereignty must be expanded to include the growing impacts of environmental stress locally, nationally, regionally, and globally. There are no military solutions to 'environmental insecurity'."

4.7   Origins and Development of the Digital Information Age

The advances of science and technology have changed the operative time structures, as well as contributing to the acceleration of social time. This power to transform has become particularly effective in the Digital Information Age, or simply the Information Age, the most recent period of modernity, characterised by the intensive development of ICTs (Gleick 2011). The word 'inform' has its roots in the Latin verb *informare*, which means to give form, give form in the mind, and instruct, and was used by Cicero (106–43 BC), among others. In the Middle Ages, the verb 'inform' became associated with obtaining information in prosecution and criminal cases. It later lost this negative connotation and began to represent what can be conveyed or communicated by any means of transmission or obtained from direct or indirect observation and learning, such as facts, news, and more generally data.

Information, when received and processed by the intellect, promotes awareness and consciousness, leading to knowledge and enabling the acquisition of new abilities and ideas. It is not an inert element that is possessed like an object, but rather something that becomes part of ourselves, encouraging and enriching thoughts and actions. It is something that has the potential to transform. The expression 'Information Age' conveys the idea that easy access to information and the recognition of the empowerment that it provides constitutes a good way of characterizing current social time.

The development of science in the 20th century created the conditions for a profound evolution in the concept of information. The discovery of electromagnetic phenomena and the theory of electromagnetism led to the possibility of propagating electrical signals along cables, and later electromagnetic waves in space. These remarkable scientific advances triggered the development of technologies to communicate information using electromagnetic waves over long distances.

Telecommunication is the transmission of information of any nature, such as signals, words, writings, images, and sounds, in the form of electromagnetic signals as by telegraph, telephone, radio, television, and internet. The medium through which the signals are transmitted can be a wire, cable, optical fibre, or just free space in the case of wireless transmission. A telecommunications system must have a transmitter and a receiver, which can be combined in a single device called a transceiver. The electromagnetic waves that carry the signal can be radio waves, microwaves, infrared radiation, or visible light, the only difference being their wavelength or frequency. Furthermore, the signal sent between the transceivers is codified, and there are two ways of encoding it: analogue or digital.

In the former, the amplitude of the wave varies continuously in time and it is this variation that contains the information in coded form, whether it be music, speech, signs, or images. In digital signals, the coding has a binary format in which the amplitude varies discontinuously between just two values, 0 and 1. In the transmission process, the signals inevitably become distorted by interference from other signals, and this creates noise and errors in the information transmission. Digital signals are more reliable because errors transmitted are more easily removed than in analogue signals. Moreover, digital signals are easier to reproduce and process, have better scalability with technology, and carry more information per second than analog signals.

The development of digitization, along with digitised information communication and the expansion of its uses has been so extraordinary since the middle of the 20th century that it has become known as the Digital Revolution. This success is largely due to the mathematical analysis of the conditions that optimise digitised information transmission processes, carried out by Claude Elwood Shannon (1916–2001) in his landmark article written in 1948 (Shannon 1948; Shannon and Weaver 1949). Shannon applied the binary algebra developed by George Boole (1815–1864), in which real numbers are replaced by a variable with only two values—true or false—and the prime operations of elementary algebra are replaced by the operations of conjunction, disjunction, and negation, to develop an algebra of electronic circuits in which the two values correspond to two possible positions of a switch: open or closed. He also proved that any type of information—document, book, image, or music—could be coded and transmitted electronically using a base 2 number system, in other words using a binary digit that only assumes two values, commonly represented as either 0 or 1. A bit, short for 'binary digit', is the unit of information in computing and digital communications, and is also called a shannon (Sh), named after Claude Shannon. Bits are usually grouped in 8-bit clusters called bytes. Thus a single byte may have 2^8 or 256 different values.

Based on the realisation that information is also a way of resolving uncertainty, and considering that this can be achieved with varying degrees of efficiency, Shannon defined the entropy of information using an equation equivalent to the one used in thermodynamics. To simplify matters let us suppose that someone sends a message to another person. If the message tells the receiver something that they already know, the information is pointless and its entropy is very low. If the message conveys new information, it has high entropy. Trying to quantify these concepts one finds that the entropy of a message is the expected number of bits of information contained in the message, taking into consideration all the possibilities that may be inferred from the message transmitted. If the message halves the number of possibilities, it transmits one bit of information. For example, if a sender intends to use some transmission channel to send a receiver the result of a random choice of a positive integer, and if the message states that the integer is even, it contains one bit of information. If the message does not reduce the number of possibilities, it does not contain information and its entropy is zero.

The entropy of a message is the smallest number of bits needed to codify it so that it can be transmitted. If an encoding scheme manages to reach this lowest bound it is said to be lossless. However, a message may include some redundancy, in other words it may have more bits than necessary to transmit the information it contains. Redundancy may be useful for eliminating errors introduced in transmission channels that involve a lot of noise. This usefulness can also be seen in human communication. To send a message, it is sometimes appropriate to use expressions that are redundant with regard to their essential content to make communication easier and avoid interpretation errors. Shannon further demonstrated that it is only possible to transmit information without errors in a communication channel if the transmission rate, expressed in bits per second, is less than a certain amount, now called the bandwidth. In a channel with a small bandwidth and a lot of noise, it is always possible to transmit information without errors if the transmission rate lies within the bandwidth and if the redundancy is increased so that errors introduced by the noise can be identified and removed in the decoding process.

Information theory, initially developed by Shannon, unified the issue of communication and storage of digitised information within the increasingly diverse range of devices and technological media—personal computers, tablets, email, internet, mobile phones, GPS satellites, high definition television, and digital data storage devices, including CDs, DVDs, USB drives, and memory cards, among many others.

The process of digitising cultural heritage information has been explosive. It is calculated that in 1986 only 0.8% of all the information stored in the world was digitised and that the rest was conserved in analogue formats, such as documents, books, works of art, photographs, and films (Hilbert and Lopez 2011). In 2000, that percentage had risen to 25% and in 2007 to 94%. A 2012 IBM report (IBM 2012) shows that every day around 2.5 exabytes (10^{18} bytes) of information are created, including shopper information, posts in social media, digital pictures and videos, cell phone and internet signals, and many other digital signals. To get a better idea of the immensity of this number, a character in a digitised text, such as a letter or a number takes one byte. The complete works of Shakespeare contain roughly 5 megabytes and the complete works of Beethoven around 20 gigabytes.

It is calculated that the digital universe, in other words, the total amount of accessible digital data, doubles at least every two years and will increase from 4.4 zettabytes (10^{21} bytes) in 2013 to 44 zettabytes in 2020 (IDC 2014). The largest part of the data generated up to 2020 will be produced by machine sensors and smart devices communicating between themselves over data networks, and not by humans. A significant proportion of this data universe is transient, and will not be stored in a lasting way. This is true for mobile phone calls and messages, and a significant percentage of television footage and surveillance system images.

Data banks are growing in a wide range of fields, from the natural and social sciences, particularly medicine, the environment, economics, and sociology, to transport, and civil and military security and spying systems. This is due to the increasing use of networks of sensors on the Earth and in space on satellites, radiofrequency identifiers, and other data detection and recording systems. As ICTs have developed, the growing information flows have changed the way that economics, finance, governance, and social relations operate. It has become possible to store, analyse, and process large-scale, complex data banks known as "big data", which are used globally for a wide range of purposes. Big data evolves through growth in data acquisition and storage, increased processing speed, and the ever more varied types and origins of data and related acquisition technology. Furthermore, it is now possible to analyse these giant data banks for a huge range of purposes, thanks to the increase in computer memory, faster processors, more intelligent software, and more evolved algorithms.

The phenomenon of big data has been characterized by the five Vs: Volume, Velocity, Variety, Veracity, and Value. It is significantly changing how information is used in three main ways. In the past, to carry out a statistical study on a system with many elements, a representative sample was sought. Now, the main goal is to obtain data on the entire system. It has become feasible to store, process, and analyse that data, which makes it possible to explore and find out about the characteristics and behaviour of the whole system and its subsystems. Secondly, in the past, there was great concern with the reliability and veracity of all data, whereas it is now possible to deal with the uncertainty that arises from some data not having the desired quality. Thirdly, the use of big data allows us to discover many characteristics and statistical correlations between data relating to the whole system, and this helps us to find out how it behaves, even if we do not fully understand how it works. In strictly operational terms, knowing about the existence of an important correlation may matter more than discovering its cause. Hilbert and López (2011) carried out a pioneering study on the evolution of the world's technological capacity in digital information communication and storage. They also evaluated the computing capacity on computing devices, including personal computers, supercomputers, and other digital devices. Between 1986 and 2007, the world's per capita effective capacity for bidirectional telecommunication and for installed storage information capacity grew with average annual growth rates of 28% and 23%, respectively. In the same period, growth in the global per capita general purpose computation was even higher, with an average annual growth rate of 58%.

Despite these new extraordinary capabilities, most of the digital information that is exchanged and stored globally has an ephemeral value and a large part of it is used just once or a few times. Nevertheless, a large part of the human-generated information is considered to be very important or even essential at the moment when it is produced or received, although its value collapses rapidly with time. Social time is becoming increasingly populated by ephemeral happenings and experiences.

In 2013, only 22% of the information in the digital universe would be likely to be considered useful if it were previously analysed, characterised, and tagged, and less than 5% was actually analysed (IDC 2014). It is estimated that in 2020 the useful percentage will grow to more than 35%. The global information data storage capacity is growing more slowly than the production of this kind of data. In 2013, there was capacity for 33%, but it is estimated that this capacity will fall to 15% in 2020 (IDC 2014). The percentage of data processed by cloud computing is also predicted to decrease from 20% in 2013 to 15% in 2020.

The growth of digital information data is far beyond the capacity and the time needed to revisit it and decide about what to be stored. This feature reflects and amplifies our natural tendency to possess and store. In everyday life, we feel a growing digital information overload—frequent communication through smartphones and computers, news intermingled with advertisements, some specifically directed to our preferences, all sorts of data from photographs, to articles, reports, and books—which tend to monopolise our ability to use time if we do not remain constantly vigilant and keep making choices. A large part of this information data remains available and accumulates in our electronic devices without being used.

The current digital world reflects the growing security issues and concerns. There are several levels of security that range from protecting privacy and confidentiality and protecting financial and commercial transactions to protecting the security of people, their assets, communities, and countries. Around 20% of the digital universe currently has some form of protection (IDC 2014), but that percentage is likely to increase. The quantity of digital information generated by every person in the world, including phone calls, messages, emails, documents, photographs, videos, music, articles and books, and other products, is much smaller than the average amount of information created about that person, including the received phone calls, messages, emails, registered movements, security and surveillance system files, records and documents regarding citizenship, institutions and associations, among a great deal of other data. This disparity is partly a consequence of our condition as citizens, voters, and taxpayers. It is also a consequence of the tremendous importance that the economic system assigns to each individual as a consumer or a potential consumer for an increasing range of goods and services, and of its attempts to motivate people to consume. Finally, it is also rooted in the huge effort to identify, monitor, and combat all those who in some way put the security of people and assets at risk.

One of the main features that define the Information Age is access to connectivity. A high and rapidly growing percentage of the world's population has access to digital information and communication devices: it is estimated that in October 2018 the world had 4.176 billion active internet users, 3.397 billion active social media users, and more than 5.1 billion mobile phone users (Statista 2018). The initial ideas that led to the internet originated in France, the United Kingdom, and the USA with the development of electronic computers. In the 1980s the English engineer Tim Berners-Lee while doing computer research at CERN, invented the World Wide Web allowing hypertext documents to be linked into an information system and accessible from any node in the network. In 2015, approximately 3.3 billion people were connected to the internet.

Today, access to the internet is often considered a human right because it serves as a platform that allows democracy to develop, enhances education and vocational training and services, creates economic opportunities, protects populations in natural disasters and conflicts, and enhances healthcare services, particularly through telemedicine. Target 9.C of the United Nations Sustainable Development Goals is to "Significantly increase access to information and communications technology and strive to provide universal and affordable access to the Internet in least developed countries by 2020".

The financial and economic activities related to the Information Age have become very significant and attractive. In accordance with a World Bank report, the global number of mobile phone contracts rose from less than 1 billion in 2000 to more than 6 billion in 2016 (WB 2012). The global economic sector linked to mobile phones, known as the mobile economy, was worth US\$ 3.1 trillion in 2015 and generated 4.2% of the world's GDP in 236 countries and territories (GSMA 2016). It is calculated that the sector will hit an economic value of US\$ 3.7 trillion in 2020, with a higher average growth than the rest of the world's economy. The most successful device at the moment is the smartphone, which combines the functionalities of a personal computer operating system with the specific functions of a mobile phone. Production of smartphones is growing much more than the production of personal computers and traditional mobile phones, sometimes known as dumbphones, and their technological evolution has been very fast. According to the Boston Consulting Group, data download speeds increased by a factor of 12 000 between 2009 and 2013, and the cost of data transmission has fallen in recent years to a few cents per megabyte. There are currently more than 2 billion people who have smartphones, but this number will likely double by 2020.

Smartphones have revolutionised the mode and form of communication as well as our relationship with the world around us. Using different types of sensors, smartphones know their exact location and the speed at which they are moving. They make it possible to navigate the physical world as well as the virtual digital world. Smartphones are the key to the success of the internet of things, a system of inter-related computing devices, mechanical and digital machines, sensors, objects, or people that are provided with unique identifiers and the ability to transfer data over a network without requiring human-to-human or human-to-computer interaction. The internet of things is designed to intervene and help in our personal and professional lives, including our homes, property, vehicles, shopping, and security, production, and consumption systems. All these objects and systems are expected to maintain a permanent operating connection by receiving data and sending remotely controlled instructions by way of smartphones. In the future, we are likely to become "smart patients" by voluntarily collecting physiological data through sensors implanted in the body, including temperature and blood sugar level, thereby increasing the amount of information and the quality and efficiency of dialogue with our doctor.

Humanity is now navigating in two parallel worlds, the physical world of the Earth system, where we live in a biological sense, and the virtual digital world of the ICT, where our minds spend an increasing amount of time navigating. The way towards the future is to integrate these parallel realms as far as possible with a wide variety of devices and structures located in different points of physical space and virtually mapped in digital space so that we can extend our presence and power over the physical world. Our direct perception of the physical world is conditioned by our physical presence and therefore depends on our movements through space, and ultimately on the speed of the transport system that is used. In the virtual digital world the relationship between space and time is universally regulated by the speed of light in vacuum, i.e., $299\,792\,458$ m/s. With digital information transmitted at this speed, the transmission times in the region of the Earth system are very small, which makes distances irrelevant. Thus the virtual digital space of the Earth system, and the surrounding outer space where satellites operate, is populated by information that is almost in real time. There is actually a delay for a signal to be received from very far away, which is about a quarter of a second for each satellite (located at an altitude of about $38\,000$ km) that is used in the communication channel. The time that the electromagnetic waves take to travel explains the delay between answer and reply in TV interviews to the other side of the world.

When our common social space extends beyond the Earth system and embraces the entire Solar System, then distances will effectively be relevant to ICT communication. Remember, for example, that an electromagnetic signal takes between 3 and 21 minutes to propagate between Earth and Mars, depending on the relative position of the two planets in their orbits.

Until that time distances will be voided, space will be compressed according to our wishes, and time will become timeless and compliant because virtual access to events previously recorded is independent on their temporal order. The fact that digital devices are being designed to satisfy and develop a whole range of innate and gratifying impulses and desires without time or space limitations encourages adaptation to a world that is constantly changing. It also makes us feel like we are in a fluid world in which practically all rigid structures can be bypassed.

4.8 How the Digital Age Is Changing Social Life, Cognitive Processes, and Operative Time Structures

Initially, digital ICT media devices had very specific, differentiated functions but they have gradually begun to share many of their functions. Today, we can access the news, social media, online lectures and courses, music, games, videos, and films on our televisions, personal computers, and mobile phones. All this bounty is instantaneously and constantly accessible through today's digital ICT media devices, which now form an inseparable part of billions of people's everyday lives. The internet platforms are now the map and the clock, the printing press and typewriter, the film and photography camera, the calculator, telephone, music player, radio, and TV for billions of people (Carr 2008). Since the 1980s, we have been witness to the transition from the role of being a citizen to the role of being a digital citizen. It is probably impossible to overemphasize the importance of this transition and the implications it has on life, especially social and political life, and on our common future. There is a genuine effort to promote responsible and ethical digital citizens. It is fully recognized that parents, teachers, and peers share the responsibility to advance digital citizenship. To be safe and responsible in online digital environments, people need training and advice, especially young people.

There is also an unbreakable belief in the immense benefits and enlightenment to be gained from the ICT Path, to use an analogy with the Eightfold Path. To find an historical analogue to the attachment and insuppressible dependency that billions of people now have on their digital communication devices, we only have to consider the religious faiths and their sacred books, cult objects, and cult images. The two types of relation have a strong commonality in the belief that both are paths to enlightenment, which although different are considered to be perfectly compatible. The main difference between them concerns the way time is used and valued. The first case focuses on the "specious present", utterly filled with the thrill, enjoyment, and utility derived from social interactions, messaging, news updates, information, music, games, online shopping, and many other everyday activities. The second fills time with various forms of spiritual enjoyment provided by faith, and the time horizon goes far beyond the "specious present" to encompass prayer, religious rituals, observance of the moral codes and laws of the scriptures, and many other activities specific to each religion. In the first case, we tend to use the word addiction to refer to excesses, while in the second, devotion is the expression usually employed.

Social media platforms are one of the areas where growth is stronger. It includes Facebook, Google+, LinkedIn, Instagram, Pinterest and WeChat, cross-platform messaging, such as WhatsApp, blogs and micro-blogging services, such as Twitter, Tumblr and Weibo, video-sharing services, like YouTube, and photo-sharing services, like Flickr. The initial development of social media can be traced back to 1997, when the social network SixDegrees.com created CompuServe. Shortly afterwards, in 2004, Facebook emerged as a Harvard student's extension of the printed journal "Facebook" for undergraduates, and became available to the general public in 2007. YouTube and Twitter were launched in 2005 and 2006. The essential transformation was that many people could easily and gratuitously produce and share information with many people, while in the past, through printed books, only very few shared information with many people.

The use of the internet, in particular Wikipedia and other online encyclopaedias, and social media has a very positive impact on education, and especially on out-of-school learning and research. It allows children and adolescents to interact with their peers and adults, and also to learn about all sorts of events, objects, geographical sites, subject matters, and devices in a way that is closely related to their interests and natural tendencies. This form of learning is particularly effective, because it is co-created and unintentional to a large degree. The use of the internet and social media has generated a "participatory culture", in which students create, connect, and collaborate with each other and with a global audience (Greenhow et al. 2017).

The social interactions provided by social media act as mediators to the surrounding cultural, political, economic, and financial environment of the digital citizen. Furthermore the internet and social media are increasingly used for proselytizing and also for mobilizing and enlisting people into campaigns at local, national, and international level. Smartphones and social media have been used to develop, convene, and coordinate all sorts of social, political, and environmental movements, and in particular protest movements with a level of success that would have been unimaginable in the past (Kidd and McIntosh 2016). Examples can be found in what happened at the start of the Arab Spring in Tunisia in 2010, in Cairo from 2011 to 2013, in Zuccotti Park in New York in 2011, in Hong Kong in 2014, in Ferguson in Missouri in 2014 and 2015, and in France with the *Gilets jaunes* uprising in 2018, to mention just a few cases.

The internet and the social media platforms companies are fully integrated into the current globalized economic system and constitute one of its driving forces. They started to operate in a very ingenious and paradoxical way. Customers of the internet and social media companies are not their users but their advertisers. The users use the utility as much as they wish freely, without any form of payment. However, there is a non-monetary price to be paid. Advertising in social media is nowadays one of the most powerful ways for a business to succeed, by reaching an increasing number of prospects and customers. To increase their profitability, social media companies manipulate the attention of their platform users and they promote addiction to the services they provide. Their aim is to encourage users to spend as much time as possible on the platform so as to get to know their interests and preferences, sell that knowledge to the advertisers, and then use that time to send commercial and political messages, often guided by algorithms.

They thrive on the "network effect" that arises from being able to attract an increasing proportion of potential users. After reaching the status of leaders in the market, they are chosen and used because nearly everyone else is already there. In this way they enjoy rising marginal returns and benefit from the economy of scale, which explains their extraordinary profitability and growth. They function in the winner-takes-all logic. It is estimated that Facebook and Google control over half of all digital advertising revenue. Facebook took about eight years to reach one billion users, but just half that time to double that number. Since the number of potential users is growing more slowly than what is required to sustain the high profitability of the process, social media platforms are inventing new business models where, besides advertising, they sell products and services directly to their platform users.

The price that users of internet and social media platforms pay is also political. The news content provided by social media is often generated by algorithms designed to match the political preferences of the receivers. This manipulation contributes to favour information which supports the beliefs and opinions that the receivers already hold, and avoids information that contradicts them, generating a self-perpetuating selective exposure which promotes political polarization. Such polarization is a way to confine people's freedom of mind and autonomy to think and act freely. The internet and social media are increasingly used to distribute fake news and information, which is also deplorable. Furthermore, it is becoming increasingly evident that governments collect large amounts of information on their citizens, allegedly for security reasons. People cannot fully realize the scale of these activities and exactly what personal data is actually obtained by governments through internet platforms and digital communication devices.

It has been argued that such new developments can eventually lead to the transformation of some of the present-day advanced economies, which increasingly share the characteristics of oligarchies, to become authoritarian states supported in part by large global ICT corporations. There are many uncertainties about this scenario. However, it may have a relatively high probability because it is likely to optimize economic growth and consumerism, while at the same time impairing human rights.

What is becoming increasingly evident is that ICT digital communication devices are changing the way humans interact socially, and the processes and structure of their mental activities. Social relationships are increasingly technologically mediated. We no longer need physical presence to talk, discuss, debate ideas, or express love and anger and many other feelings. And neither is physical presence required to exchange images and videos of ourselves, others, and the world. Inevitably our empathy and our ability to socialize and enjoy the physical presence of fellow human beings is likely to be devalued. Along the ICT Path, slowly but inexorably, social interactions acquire the characteristics of interactions between technological beings or the like of robots. The model of technologically-mediated relations favours ephemeral relations and a swiping behaviour.

Face-to-face encounters are often time consuming and inefficient for many objectives. It is much easier to use the internet and social media to select friends or lovers, supposedly compatible and physically perfect, and to find most of the relevant information about them, than to rely on face-to-face dating. Cyber selection, including the huge variety of online dating applications, is becoming a strong competitor to old-fashioned natural non-cyber selection, at least in some countries and circumstances. Younger generations, particularly millennials, consider that digital communication has removed a barrier in social relationships, made communication easier and quicker, and created new ways to resolve conflicts.

One of the cognitive processes that is most affected by the internet and social media platforms is attention (James 1922). Attention is a state of arousal that enables selective concentration on a discrete aspect of information or the environment around us. However, by attention we usually mean an on and off activity which has a relatively short duration. Concentration of the mind on some activity or object usually involves longer periods of time and can be exercised with different levels or degrees. It is considered odd or inadequate to say that a smartphone user is not attentive to what he is doing with his or her device, but often it is said that students are not attentive in their classes. The reason is that mobile digital information devices are designed to promote attention to an atomized stream of information that is entirely chosen by the user, nearly always on the basis of the user's personal interests and enjoyment. In class, attention and concentration of the mind on the same subject matter is required, even though it may not interest the student. One of the reasons why digital devices are so attractive is that they specialize in providing information and various forms of enjoyment like particles flowing in a current. Attention is happily provided to each particle since each of them was deliberately chosen to satisfy our enjoyment and interest. This process is self-perpetuating, but concentration of the mind, especially at a high level, is largely unnecessary.

These relatively long periods in which attention is fragmented and scattered are transforming our operative time structures at a global level. According to a recent report from the UK (Ofcom 2018), people are online for an average of 24 hours a week, which is twice as long as 10 years ago, but some reach an average of 40 hours per week. If this progression were sustained there would be no time available to do anything else by 2050. Obviously, the average percentage of available time that is used on the internet will reach a saturation point, but we don't know what that value is going to be. Studies carried out by operators and marketing companies show that people look at their mobile phones an average of 150 times a day (KPCB 2013), mostly to send messages, emails, and tweets, and make calls, listen to music, play games, and access the internet to check social networks and news.

Most users cannot be separated from their mobile phones. They carry them around from the moment they wake up until the moment they fall asleep, when they are normally left on their beds, on bedside tables, or at least nearby. In the future, it will be possible to remain permanently connected through devices known as wearables, drivables, flyables, and scannables (KPCB 2013). Wearables are hands-free devices. Some already exist, such as Google Glass—glasses that contain a small, voice-activated smartphone—but most are still in the development stage.

The compulsive and irresistible use of digital information devices is essentially driven by a desire for gratification as often and intensely as possible. At the neurological level, this gratification corresponds to the release of dopamine in the human brain. Dopamine can also lead to a false sense of accomplishment. Nevertheless, users seek the immediate satisfaction of receiving expected and unexpected text messages, social media posts, instagram photos, and emails, reading a new piece of news that interests them, and especially viewing videos and photographs, listening to music, and playing games. The use of digital information devices has provided a way to dispel the tedium and loneliness that plagued past generations. This use is also a way to dilate psychological time, filling it with a high density of experiences and cyclical actions that force our attention and emotional involvement. For generations Y and Z, time is filled as much as possible by smartphones, especially when travelling on public transport, during breaks at work, at meals, when waiting for something, or when moving around in the street. In such moments, attention is focused on digital activity and the external environment becomes remote.

This form of exclusive attention was tragically demonstrated on 23 September 2013, in the subway in San Francisco, USA, when surveillance cameras in one of the carriages caught footage of a man waving a gun and repeatedly making intimidatory gestures with it. He ended up killing a passenger he did not know, Justin Valdez, a student at San Francisco State University, by shooting him in the back. The video also shows that there were many passengers in the same carriage who did not notice his aggressive behaviour and did not try to avoid what happened because they were so absorbed in their digital devices.

We are together in the same space but in fact alone. Paradoxically, we are focused on an activity that includes interaction and intense sharing with others, but outside that space (Turkle 2011). The compulsive need for continuous surveillance of what is going on around us is written into our genetic code and is shared with our ancestors and today's primates, among many other animals. In evolutionary terms, it can be explained because, in order to survive when searching for sustenance by hunting or gathering food, people had to be as vigilant and alert as possible to what was happening around them, to identify dangers, such as threats from competing groups or predatory animals. Nowadays the focus of that attention has profoundly changed, especially regarding location in space.

Immediate surroundings are no longer as important as they once were. We remain deeply interested in finding out as quickly as possible what is happening, especially things that may affect us or be of special interest, but all such relevant events are now spread throughout space and are incomparably more abstract.

Idleness, tedium, and solitude continue but we can now chase them away by overloading time. Time can easily be filled with digital communication activities that attract, distract, please, and alienate. In the end, more diverse experiences have been created. All were potentially possible and just needed technology's capacity for invention and adaptation for them to appear, flourish, and transform us culturally. Digital information devices have broadened human social abilities and experiences in a socially acceptable way.

An essential characteristic of human behaviour that is fully exploited by the digital communication technologies is the propensity to act and try out new personalities and identities that partly free us from our own. A lack of physical closeness in social interactions using digital networks and the intensity of these connections tends to change the nature of interpersonal relationships. The intensive use of connections to develop these relationships encourages people to create an electronic personality or e-personality that represents a version that is generally less inhibited, more impulsive, and more narcissistic than the parent personality (Aboujaoude 2011). The internet tends to remove the psychological barrier to impulsiveness and creates an anonymity that encourages people to exhibit a personality with bolder, coarser, and more aggressive behaviour, sometimes reflected by disloyalty, intimidation, and persecution. Furthermore, the intensive use of the potential offered by the internet gives us a misleading feeling of being at the centre of the world, because all our desires are potentially at our fingertips.

Finally, the ICT Path attends to another human passion: games. It provides the possibility of playing at any time and in any place, to kill time, relax after a day at work, or simply satisfy the desire to play. It has been estimated that in 2011 gamers spent on average 3.7, 3.2, and 1 hour playing on games consoles, computers, and smartphones, respectively (PwC 2012). Several studies have shown that playing games for long periods of time each day has negative consequences for young people's development. According to Przybylski and Mishkin (2016), teenagers who spend more than three hours a day playing games tend to be hyperactive, get involved in fights, and lose interest in school. There are about 150 million active gamers online every day around the world. People can gamble online by playing roulette, blackjack, pachinko, or baccarat, for example, which means that they do not need to go to casinos or arcades. This represents a highly lucrative global market that was worth 35 billion dollars in 2014. It is forecast to have a compound annual growth rate of 10.6% between 2014 and 2018 (Stocks 2015).

Use of digital communication devices is highly psychoactive. This use causes a transition in our mental processes which, because it generally leads to more gratifying feelings, tends to make the repeated checking of these devices addictive, and to generate anxiety when it is not possible to do so. Several studies have shown the personal and professional advantages that arise from voluntarily restricting time spent on digital information devices, for example by creating "predictable time off" (Perlow 2012).

In the Information Age, time has lost some of its viscosity. It is now easier to take advantage of every minute and make it useful and gratifying. Thus there is no time to lose; we can always be doing something more than simply being. Time flows without the effort linked to boredom or undesirable, recurring memories that in the past used to infest our idle moments. The giddiness of exhaustive use of the "specious present" tends to diminish the past and the opportunities to analyse it. We experience the present as intensely as possible, with less time left for concerns about the past and the future. In contrast, there is a very strong tendency towards ensuring that the "real" good times, as opposed to the "digital" times, are recorded in the virtual digital universe, stored in the memories of our communication devices. The digital cameras on smartphones and tablets are now widely used for selfies with friends and to take photos of celebrations, gatherings, meetings, trips, cities, monuments, art works in museums, and many other things, some of which are very private. Most of this gigantic archive immediately becomes available on social networks, building and structuring a timeless and compliant time and a dimensionless space. When in front of a monument, landscape, picture, or sculpture, people often spend longer trying to capture a digital image of it than they do actually observing and appreciating it. Each person's digital universe has become a refuge in which time and space are firmly under their control.

How the internet is changing our memory and thinking habits, and probably also the way we think, has been a subject of considerable interest, but there is a notorious lack of scientific consensus. John Brockman writes that the internet "is the infinite oscillation of our collective consciousness interacting with itself" and that ultimately "it is about thinking" (Brockman 2011). In the collection of essays entitled *How Is the Internet Changing the Way You Think?* (Brockman 2011), one finds a wide variety of points of view and some violent disagreements about the value of the internet for our civilization and its future. Some consider that the internet is one of the highest achievements of the human species, while others write that it is the greatest detractor to serious thinking. There is no strong endorsement of the view that the internet is changing the way we think. It is often conceded that the internet is increasing our multitasking skills but diminishing the ability for mental concentration that was enhanced by a print culture.

Nicolas Carr, in particular, has consistently argued that the internet is negatively changing the way we think. His article entitled "Is Google making us stupid?" (Carr 2018) was heavily criticised by various authors, such as Jamais Cascio, who replied that "Google isn't the problem; it's the beginning of a solution" (Cascio 2009). He goes on to summarize the predominant point of view of the mainstream techno-optimists: "In any case, there's no going back. The information sea isn't going to dry up, and relying on cognitive habits evolved and perfected in an era of limited information flow—and limited information access—is futile. Strengthening our fluid intelligence is the only viable approach to navigating the age of constant connectivity" (Cascio 2009). In the following year a report by the Pew Research Center stated that, in a survey in the USA, "76% of those surveyed agreed with the statement: 'By 2020, people's use of the Internet has enhanced human intelligence; as people are allowed unprecedented access to more information they become smarter and make better choices. Nicholas Carr was wrong: Google does not make us stupid' " (PRC 2010).

Carr emphasizes that the internet is a medium characterized by repeated interruptions in time, whence its intensive use fragments information processing and thought to a stream of particles (Carr 2010). He then invokes the neuroplasticity of the brain to argue that the more time we spend in this particulate mode of thinking, which involves surfing, skimming, and scanning, the better we become in that mode. Different modes of thinking may tend to be atrophied. In the end we will gradually move towards a stereotype whose technological existence we have always dreamed of.

The Information Age and its ICT devices have produced a deep transformation in societies and in the political, economic, and financial systems of all countries significantly contributing to the globalisation process. The new digital world, and the connectivity that provides access to it, have made it possible to improve health conditions in many regions of the world, prevent and better manage natural disasters, boost social inclusion and cohesion, improve access to education and professional training, promote access to information and active citizenship, and enable civil society movements and organisations in social, political, economic, and environmental fields.

On the other hand, the increasing use of the virtual digital world to explore and enjoy its psychoactive capabilities, leading to dopamine surges in the brain, produces a resonance effect that gradually turns it into a habit and a need, and if this need is denied, it produces stress and withdrawal. Internet and social media companies know very well how to entangle their products with hijacking techniques that create compulsion loops liable to lead users into addiction (Alter 2017). Tech addiction exists and is a serious problem for many people. However, calls for moderation and containment in the development of addictive technologies have very little or no effect on the internet and social media companies or on the governmental institutions that regulate them. Users are expected to go through a natural selection process, where they are expected to avoid the addictive traps and elect the positive outcomes of the internet and social media platforms. The prevailing rationale is that the ICT Path is unquestionably one of the main avenues to the future. People should just adapt and take advantage of the rapidly changing ICT. The adventure pressed upon us by technology leaders is supposed to be irresistible and inevitable and we must all believe in it.

4.9   The Surveillance Era, Cyberattacks, and Cyberwarfare

The compression of the dimensionless virtual space, the timelessness of virtual time, the universal access to digital communications, and the abundant, diverse, and instantaneous gratification that it provides, have elicited a reaction that is problematic but quite predictable and natural in human terms. Personal movements, journeys, tastes, habits, trends, relationships, problems, plans, conversations, and private data that go through personal digital information devices can be detected, identified, recorded, and used by others for all purposes. One way to collect sensitive information is through spam, which can be sent through email, instant messaging, chat rooms, and voice over Internet Protocol (IP) conversation systems, such as Skype and Google Talk. The IP of a personal computer constitutes its digital signature on the internet. IP spoofing is used by hackers to steal passwords and bank details, to hold customer information to ransom, and sell personal data. Their main motivations are to obtain cash, to address the challenge of breaking the unbreakable protection systems and gain the recognition of their peers, for revenge and to boast about their exploits in the global digital space. Another risk is to shame and hurt selected people on social media through an identity theft of their personal information data. Computers and smartphones are constantly targeted by attacks made using many types of malware and spyware, but the vast majority of users learn to defend themselves with firewalls and antivirus software.

Personal activities and private information flowing through personal digital information devices are also avidly recorded and scrutinized by governmental institutions.

Allegedly, they do this exclusively for the country's security or, using an alternative formulation, for what is interpreted as the well-being and security of its citizens. In the USA, the National Security Agency (NSA) is an intelligence agency responsible for global surveillance, which has entire departments dedicated to accessing, collecting, checking, and evaluating the data on billions of mobile phones and other digital devices around the world. The NSA seizes the opportunity to intercept most of the world's digital communications because it flows physically through the USA, where a large part of the internet infrastructure is located.

The NSA's clandestine PRISM project operated secretly from 2007 until the Guardian and the Washington Post revealed its activities on 6 June 2013. These revelations were based on statements made by Edward Snowden, a former analyst hired by the NSA who left the USA, took refuge in Hong Kong, and finally obtained political asylum in the Russian Federation. Although there had been strong suspicions about the extent and depth of electronic spying carried out by the USA, Snowden's revelations unleashed astonishment, concern, and agitation throughout the world, including among USA politicians and government officials. Robert Mueller, then FBI Director, told Congress on 13 June 2013 that Snowden's leaks had caused "significant harm to our nation and to our safety" and that the USA would hunt down and prosecute leaker Edward Snowden. PRISM is able to intercept and record content from all forms of electronic communication around the world, including emails, mobile phone calls, internet browsing, images and data transmitted by satellite, and all kinds of personal, business, institutional, and government data found in digital format on telecommunications networks both inside and outside the USA.

The documents revealed by Snowden in 2013 showed that the major internet and social media platforms participated in the PRISM program as data providers, including Google, Facebook, YouTube, and Skype (Gellman and Poitras 2013). Immediately afterwards the corporate executives denied that they had knowledge of the PRISM program and stated that they did not provide information directly to the program. Furthermore, they all emphasized that they took user privacy very seriously.

The governments of various countries, in particular those of the Anglophone Five Eyes intelligence alliance—Australia, Canada, New Zealand, the UK, and the USA—acknowledged that they knew about the PRISM program and that they were actively collaborating with it, especially in security and anti-terrorist operations. However, it has been proved through the leaked documents that PRISM's objectives were much broader and included gathering information from other countries on their political and economic activities, businesses, international tenders, contracts, competitiveness, energy, especially fossil fuels and in particular oil, commodity deals, and many other issues related to politics, economics, and finance. Other countries, in particular Germany, France, and Brazil, consider that secret spying activities are inacceptable among partner countries and allies.

American intelligence institutions intercept electronic traffic abroad mostly under the rules established directly by the President of the USA under Executive Order 12333, dating back to President Ronald Reagan (Goldberg 2017). Global surveillance programs under EO 12333 are largely unchecked by either the legislative or the judicial branch, and conducted entirely under the authority of the president. Americans have showed some concern about their communication records being analysed as part of bulk surveillance programs conducted abroad under EO 12333, since it offers less privacy protection than surveillance programs under the Foreign Intelligence Surveillance Act (FISA), since the latter are regularly reviewed by Congress (Goldberg 2017). Part of the FISA legislation is also controversial in the USA because its surveillance programs, although aimed at foreigners, have resulted in the USA government monitoring the digital communications of Americans without a warrant from a judge. Section 702 of the FISA Amendments Act forces internet and social media platform companies, such as Google and Facebook, to share data with the intelligence agencies, in particular the NSA and the FBI, and let them carry out searches on massive databases containing billions of personal records. Non-Americans are the prime targets of the surveillance programs under EO 12333 and FISA, but naturally they don't have any way to express any concerns they may have.

Many other countries, including the United Kingdom, Germany, France, Canada, Australia, the Russian Federation, China, India, and Iran have telecommunications interception, recording and search systems, and programs similar to those in the USA. Some countries share the data and the results of their analyses according to alliances and common interests. These surveillance programs are used to find suspect communications, especially those that may involve the planning for acts of terrorism. In principle, the communications and data that are considered to be trivial or irrelevant have their privacy protected through a process known in the USA as minimisation. However, as already mentioned, there are well-founded suspicions that the global surveillance programs are also used for other objectives, such as to obtain information on competing companies and international tenders in order to give advantages to national companies.

The preoccupations with privacy invasions through computers and smartphones are very unlikely to disappear in the age of surveillance. Our digital devices function like a two-way mirror, reflective on one side and transparent on the other, similar to those used in interrogation rooms, police stations, prisons, and reality television. Digital platforms have a built-in automatic capability to infer the interests, intentions, feelings, and thoughts of their users from their behaviour on the reflective side of the two-way mirror. Using sophisticated algorithms on the transparent side, they can then predict in a statistical way the future evolution of the collective preferences, trends, and behaviours of their users. These predictions have huge economic value in the advertising market and constitute the foundations of their profitability. Douglas Edwards, who was Google's first director of marketing and brand management in 2001, wrote in 2011 that: "Everything you've heard or seen or experienced will become searchable. Your whole life will be searchable" (Edwards 2011).

Our human experiences and activities in the digital virtual space are fully surveyed, recorded, and statistically transformed into behaviours that become valuable and eagerly traded assets. The economic system has found a new and highly profitable kind of asset consisting in human behaviour databases that compete with other more common types of goods and services. According to Soshana Zuboff, we have entered the age of surveillance capitalism (Zuboff 2019). The internet of things presupposes a massive program of surveillance of human actions, habits, movements, personal relations, and physiological systems and functions. With data obtained through surveillance programs and processed with the help of artificial intelligence, people will know better in the future how to optimize their use of time, how to improve their health, how to live longer, and how to diversify, enhance, and create new human experiences. One finds here another pathway to the technological existence we are supposed to dream of. Losing control over our privacy is a negligible disadvantage compared with the benefits of the surveillance era. In short, the smartphone, one of the main paradigms of the digital era, is the most successful consumer product ever made, and presently about four billion adults out of a total of 5.5 billion actually own one. The promise it brings for an increasingly technological future is unquestionable.

Cyberattacks ranging from installing spyware on personal computers to destroying the infrastructure of entire nations have increased considerably and are now more sophisticated and dangerous. Cybersecurity issues have become a day-to-day struggle for governments, public institutions, banks, and all kinds of businesses, private institutions, and organizations. Some cyberattacks, in particular those coming from China, target technology firms, with a particular emphasis in recent years on attacks on biotechnology companies aimed at stealing their research secrets and intellectual property. Pharmaceutical, defence, mining, and transport companies are also hit.

The cost of cyberattacks in 2017 has been estimated at USD 600 billion, or 0.8% of the global GDP (Lewis 2018). This cost represents about 1/3 of the cost of all international crime or the costs related to counterfeiting and piracy, which are estimated by the Global Financial Integrity and the World Trade Organization at USD 1.6–2.2 trillion and USD 1.8 trillion annually. The highest costs of cyberattacks may have been incurred in East Asia and the Pacific Region, where they are estimated to represent about USD 200 billion, or 0.89 per cent of the region's GDP. In Europe, Central Asia, and North America the costs represent about the same percentage of the region's GDP (Lewis 2018). In Africa, the costs of cyberattacks are still relatively small.

Some of the most notorious cyberattacks happened in waves starting apparently in 2007 and continuing into 2010, when the control systems of 14 industrial facilities in Iran were infected with a powerful computer worm: a type of malware that spreads copies of itself from computer to computer, changes the systems it passes through, and in particular can destroy operating systems (Kushner 2013). The common feature of the 14 facilities was that they all had supervisory control and data acquisition systems made by Siemens. One of the systems was installed in the uranium hexafluoride centrifuges that were used to separate ^{235}U from ^{238}U, located at the Natanz underground nuclear facility.

Initially, nobody officially claimed responsibility for the cyberattack, but analysts agreed that the sophistication of the worm pointed to the involvement of government organisations. It is now implicitly acknowledged that the worm was produced in a collaboration between the USA and Israel (Sanger 2012), in part at Dimona, the Israeli nuclear complex in the Negev desert where Israeli nuclear bombs have been made. The aim was to slow down and eventually incapacitate the Iranian nuclear programme and therefore dissuade the Israelis from making a pre-emptive conventional air strike against Iran. The worm, called "Olympic Games" by its creators, but known as

"Stuxnet" by analysts, forced the Natanz centrifuges to spin too fast and ended up destroying them. The worm is also able to display fake information about what is happening on the control panels, so that operators are not alerted.

Stuxnet was taken into Iran on an infected USB drive carried by a double agent, probably from the People's Mojahedin of Iran or Mojahedin-e-Khalq, who oppose the regime in Iran and are supported by the Mossad, the Israeli secret service. It is estimated that almost 20% of Iran's centrifuges were destroyed in 2010, and this would have delayed the Iranian nuclear programme by several months. However, some research has indicated that the evidence of the worm's impact on Iran's enrichment program is circumstantial and inconclusive, and that after the attacks the Natanz enrichment capacity increased significantly and Iranians became more cautious about protecting their nuclear facilities (Barzashka 2013).

Stuxnet was the first cyberweapon with major destructive power to be used operationally, and it opened the way to cyberwarfare. Interestingly, use of this new type of weapon was initially linked to the fear of proliferation of nuclear weapons, which in 1945 revolutionised wars and geopolitical and military strategies. Progress in the destructive power of weapons depends exclusively on human will and on the development and cutting edge of science and technology. As in the case of nuclear weapons, the success of cyberweapons has created new capabilities and new risks. They have legitimated a new form of war to which all countries are now vulnerable, including the USA, although only a few have the capacity to develop and apply this.

The creators of Stuxnet were convinced that the worm would be confined to Natanz, which was air-gapped from outside networks, but in 2010, Stuxnet escaped from Natanz, probably on someone's laptop and it propagated stealthily between computers that use the Windows operating system or through the use of contaminated USB drives. As a result, after apparently fulfilling its mission at Natanz, Stuxnet propagated throughout the entire world and ended up infecting control systems in many countries, including the USA, as in the case of the Chevron oil company, although without serious consequences. Because it spread through telecommunications networks, Stuxnet could be observed, studied, and possibly replicated, perfected, and adapted for other specific purposes. Several pieces of malware similar to Stuxnet are known, including Duqu, Flame, and Gauss. Some of them were probably made by the same people who made Stuxnet, with goals similar to those in the Iran attack.

The first large-scale cyberattack on critical infrastructures took place on a power grid in Ukraine on 23 December 2015, leaving 225 000 users in eastern Ukraine without electricity for almost six hours. It was done using the malware Black Energy 3 and started through the infection of an office PC using a USB. As well as cutting off electricity, the attack also blocked the electricity company's telephones so that users could not inform it about the situation. The Ukranian government and Western countries attributed the attack to hackers working for the Russian Federation, and the fact that it did not affect the whole of Ukraine was taken to imply that it was just a warning. In 2016, there was another large-scale attack on the Ukrainian electricity power supply systems in Kiev.

The vulnerability of critical infrastructures to cyberattacks, including electricity, gas, and water supply systems, results mainly from the low security of the ICT components used to improve the infrastructure maintenance efficiency at reduced costs. These components allow the intrusion of malware infections that end up disabling the control systems of the infrastructures. Countries around the world are forced to live with the risks that come from potential cyberattacks on all kinds of military and industrial facilities and public services. Some countries have higher risks than others, due to their geostrategic relevance or to ongoing conflicts, and some are better prepared than others to face those risks. The challenge is to protect the most critical institutions and infrastructures by being able to detect cyberattacks and neutralise them as quickly as possible.

4.10   Time Narratives of Human Experiences in Past and Future Ages

The idea that human existence progresses in eras or ages with different trends towards improvement or deterioration is present in many cultures and civilizations. In the Ancient Greek civilization, Hesiod, one of the first Greek poets, who lived around 700 BC and was a contemporary of Homer, described in his poem *Works and Days* how humanity had passed through five ages: the Golden, Silver, Bronze, Heroic, and Iron Ages. The Golden Age was spent under the reign of Cronus, son of Gaia, the Mother Earth, and Uranus, the father of the sky and life, and the youngest of the first generation of Titans. Men lived among the gods at Olympus in harmony, peace, and abundance and neither aged nor needed to work. Death was like gently falling asleep. The Golden Age ended when Zeus, the son of Cronus and Rhea, overcame the Titans.

In later ages, under the reign of Zeus, men started to have much shorter lifespans and spent a great deal of time at war. In the Bronze Age, Zeus decided to destroy the corrupted generation of mankind by a flood, in the days of Prometheus' son Deucalion and Pyrrha, and when the Bronze Age men died, they went to the Underworld. The Heroic Age was the first one in which mankind's situation improved in comparison with the previous two. This was the age of the warriors of Thebes and Troy. During the Iron Age, which Hesiod describes as his own, the situation got worse again. The gods abandoned mankind, and moral degradation, suffering, and misery took hold. Instead of a path of systematic progress and enlightenment, one finds a complex story with ups and downs.

Interestingly, Cronus had a close relationship with time because his name is homophonous with Chronos, a god in Greek mythology who personifies time and destiny and was more closely connected to Orphism. This relationship produced a notable evolution in the mythological symbolism of time. According to Plutarch, the Greeks frequently confused Cronus and Chronos. Later, the Romans fused the two together into a single god of time called Saturn, who had greater influence on Western culture. Father Time, a mythological figure inspired by Saturn and therefore by Cronus, is depicted in paintings and sculptures with a scythe and an hourglass, a symbolism that is also often linked with death.

Saturn inherited from Cronus the story that he castrated his father Uranus on Gaia's orders using an enormous stone scythe and devoured the children he had with Rhea to stop the prophecy that one of them would overthrow him. To keep Cronus from devouring his sixth son, Zeus, Rhea hid the child on the island of Crete and replaced him with a stone wrapped in swaddling clothes. She gave the stone to Cronus and he swallowed it. Symbolically this stone, called *omphalos*—navel in Greek— became important because it represented not only Zeus' birth and power, but also, metaphorically, the centre of the world. There are several examples of *omphaloi* in the Mediterranean region. In Delphi, an astounding sanctuary located on the slopes of Mount Parnassus, a marble *omphalos* was found in the sacred part of the temple near the place where Pythia, the priestess to Apollo, delivered the oracles. In Jerusalem, there is one in the form of a marble goblet in the Greek Orthodox worship area of the Church of the Holy Sepulchre. This *omphalos* represents the centre of the Christian spiritual world.

The Roman poet Ovid, who lived between the 1st century BC and the 1st century AD, also developed the theme of human ages in his poem *Metamorphoses*. Of Hesiod's Five Ages, he removed the Heroic Age, but retained the others. The narrative acquires a historical character but the trend towards deterioration over time remains. In the Bronze Age, men began to engage in wars, but respected the gods. In the Iron Age, truth, faith, and loyalty left the world and tricks, traps, and greed emerged. The Earth, which had been shared by all until that point, along with sunlight and the air, was fragmented into nations, marked out by borders. Besides exploring the Earth for food, men began to explore the planet's insides to extract gold and iron, and this generated wealth, weapons, and crimes. Human relationships became more difficult and more aggressive. Astraea, the virgin goddess of innocence and purity and a symbol of goodness and justice, was the last to leave Earth, soaked in blood.

Ovid doesn't tell us that he lived during the Iron Age. He says that that age belongs to the past and that other events and peoples have emerged in the meantime, while decadence has remained. Neither Hesiod nor Ovid talk about a future time, although the former writes that he would have preferred to be born after the Iron Age, which leads us to think that he believed in the possibility of a better future.

In the Abrahamic religions, future time is dominated by an eschatological view in which the world ends with the resurrection of the dead, the Last Judgement, and the emergence of a much better world to replace the current imperfect and unrecoverable one. Several Christian theologians foresaw a year for the Second Advent or Parousia of Christ and the end of the Earthly world, including Hippolytus of Rome, who mentioned the year 500 AD, and Pope Sylvester II, who predicted that it would be at the end of the first millennium. During the remarkable expansion of Christianity throughout the Mediterranean and later, during most of the Middle Ages, it was often thought that the world was getting older and that the time left in the future until the end of time, expected to bring with it a lot of misery and pain, was much shorter than the time of the past. Many looked forward to the world to come as a form of liberation from deprivation and misery. After so much darkness, the light would finally shine forth and the Last Judgement would take place. In Christianity, it was unthinkable that the world could somehow transform the life of the majority of people in the population so that it would become significantly better in terms of health, well-being, and wealth. The only promise of improvement was eschatological, in the form of the final destiny, and was dependent on faith and living in accordance with the scriptures.

The Roman–African theologian and philosopher Saint Augustine (354–430), bishop of Hippo Regius, now Annaba in Algeria, established a Christian view of history divided into seven ages, similar to the seven days of the creation of the Universe according to Genesis, with each of the first six lasting roughly 1000 years. This view was prevalent throughout the Middle Ages and later until the Enlightenment. The first age started with Adam, the four following ages included the events of the Old Testament, and the sixth or current age began with the birth of Jesus Christ. The seventh age is not an age of this world, but runs parallel to the six previous ages, and includes the eternal rest that comes after the Second Coming of Christ to oversee the Last Judgement as described in the Apocalypse of John, also called the Book of Revelation. It is the age of the eternal kingdom of God. Isaiah records the Parousia of Christ by saying that "I saw 'a new heaven and a new earth,' for the first heaven and the first earth had passed away, and there was no longer any sea" (Revelation 21:1; Isaiah 65:17). God himself will be with the faithful and "He will wipe every tear from their eyes. There will be no more death or mourning or crying or pain, for the old order of things has passed away" (Revelation 21:4; Isaiah 25:8).

With this division of history into periods, Augustine went against the beliefs of millennialism, largely based on verse 20:1–6 of the Apocalypse of John, which emerged during the first centuries of Christianity, as an expression of a messianic hope that life on Earth could be improved. According to the doctrines of millennialism Christ would return to build his kingdom on Earth for 1000 years, before the Last Judgement and eternal peace. The central idea was to create, within history, the promise of a supernatural world of peace, well-being, and abundance on Earth for the human community, which was considered impossible in Christianity. Salvation and eternal peace could only be reached at the individual level and beyond history through the Last Judgement.

Although rejected, millennialism had great influence on many religious movements and confessions, especially on Protestantism. Furthermore, it was also a source of inspiration for many resistance movements, since the millennial belief in a better Earthly world with more peace and prosperity infused people with a revolutionary spirit that threatened authority. The repeated failure to establish God's kingdom on Earth have been unable to dissuade believers, such as Jehovah's Witnesses, from renewed expectations, and likewise in secular areas, this has not discouraged attempts to construct utopias. Around seven centuries after Saint Augustine's death, Joachim of Fiore (1135–1202), a mystic and theologian from Calabria, argued in favour of a type of millennialism in which history was divided into three ages by analogy with the Trinity: the Age of the Father, corresponding to the Old Testament; the Age of the Son, from the Advent of Christ until 1260; and the Age of the Holy Spirit, which would begin that same year, a date suggested by verses 11:3 and 12:6 of the Book of

Apocalypse, with the arrival of the Antichrist. The last age would be the Kingdom of the Holy Spirit, a time of peace and harmony, when the words of God transmitted by the Gospel would finally be understood and put into practice, society would become free, and the hierarchy of the Church would be replaced by the Order of the Just. Joachim of Fiore believed in a better world within humanity's grasp, a world that would resolve the violent conflicts of his social time and, in particular, reconcile and unite the Muslims and the Christians who were fighting each other in the crusades. That new world would begin in 1260, with the era of the Holy Spirit.

Despite the great popularity Joachim of Fiore enjoyed in his time, the doctrines that he defended were regarded with considerable suspicion by the Church. They were declared heretical at the Synod of Arles in 1263 after Pope Alexander IV had condemned them. In accordance with the doctrine defended by the Church, the promise of a future that really was better for humanity had to be part of an eschatological destiny, and any attempt to predict a better future in historical and political terms was rejected. However, Joachim of Fiore's ideas and the example he set with his life continued to inspire many Christians during the Middle Ages, and influence later religious and secular movements. One curious example is the analogy that the German nationalist historian Arthur Moeller van den Bruck draws in his book *Das Dritte Reich*, published in 1923, between Joachim of Fiore's three ages and the three reichs that he identifies in German history. The third of these—triumphant and prospective—was adopted by the Nazis, who assigned it a duration of a thousand years.

The idea that human history is a succession of ages, starting with a state close to perfection and moving gradually towards spiritual, moral, and physical deterioration does not appear only in the West with the Greeks, but also in Eastern civilizations. Some assume that the flow of time in human experience reveals growing and ever deeper imperfections, drawing it relentlessly towards decline, finally halted somehow through rebirth or return. In the Hindu texts of the *Mahabharata* and the *Manusmriti*, or the *Laws of Manu*, the cycle of the world is split into four ages or Yugas called: Satya Yuga, Treta Yuga, Dvapara Yuga, and Kali Yuga, where we find ourselves today.

Each age heralds a different stage of civilization. The first Yuga, as in Greek mythology, was the Golden Age, an age of justice, perfection, oneness with one god, one religious text or Veda, and one ritual. Humans were giant, strong, virtuous, and learned. They led very long lives, on the order of 100 000 years, and had hundreds of children. It was during this age that the caste system emerged, initially playing its role without oppression or envy. During the second Yuga, the first signs of a gradual decline in spiritual values appeared, with the emergence of cupidity, wealth, and envy. Sedentary, agricultural, and urban societies formed. The natural oneness of men with the divine world began to be lost, and the Dharma bull lost the leg of austerity. Men now had to work, suffer, and beg the gods. Life expectancy became much shorter, between 1 000 and 10 000 years. In the third Yuga, humans acquired more tamasic qualities, reflected in greater inertia, apathy, and lethargy; they became weaker and more vulnerable to disease, succumbing to avarice, vice, and dishonesty. There was still no ageing, but human life was shorter than during the previous Yuga. The Dharma bull lost its cleanliness and mercy legs and by now had only the leg of truth, although it would become weaker as the Kali Yuga progressed. This was the Dark Age, the age of conflict, also called the Iron Age, marked by a severe decline in human biological, intellectual, ethical, and social standards. Disease, desperation, and armed confrontation dominated. Humans were more ignorant, followed the scriptures less closely, and were wracked with imperfection and vice, while their life expectancy continued to fall. Weak, unreasonable rulers ceased to defend people and became dangerous. Water and food were scarce and there was large-scale migration. In the Hindu scriptures, the four Yugas are linked to gold, silver, bronze, and iron, just like the four ages in Greek mythology and the four kingdoms of Nebuchadnezzar's dream in Chap. 2 of the Book of Daniel, which may suggest a common origin.

The Kali Yuga started with the death of Krishna, probably at the beginning of the fourth millennium BC, and would end when Kalki, the tenth and last avatar of Vishnu, known as the destroyer of foulness, darkness, and ignorance, returned to Earth. He would come on a white horse with a blazing sword to fight the Danavas, which could be compared with Christianity's demons, and start the new Satya Yuga or Golden Age. This would in turn lead to a new Mahayuga, a complete cycle of four Yugas.

Time in Hinduism is cyclic, and it relentlessly causes deterioration and death. Kalki therefore represents eternity embodied by the unending recommencement of new cycles. The concept of time is linked to the word *kalka*, which initially meant the right moment for a sacrificial ritual. During the Vedic civilisation, *kalka* began to be used to mean time in a more abstract sense, considered to be the fundamental principle of the Universe. Its power for deterioration, destruction, and death linked it to the goddess Kali, "the black one" or "the force of time", one of Shiva's wives.

Buddhism also identified successive ages of decline in human history which, unlike Hinduism's Yugas, valued the present more highly than the future. In the Pali Canon in the first branch of Buddhism, Buddha mentioned that Dharma was already in decline in his time. He foresaw that his teachings would gradually be forgotten until a period of greed, lust, violence, and death was reached, after which the memory of Buddha would disappear altogether. There would then be a future messianic Buddha, called Maitreya or Ajita in other texts, who would restart the cycle. This doctrine had a great influence on the Mahayana branch, which emerged later when Buddhism spread to the Far East through Tibet, China, and Korea, as far as Japan.

A three-age theory was developed in China and Japan in the Middle Ages— the true Dharma, the imitation of Dharma, and the end of Dharma. This reflected Buddhism's growing inability to influence human behaviour. The last age, known as Mappo in Japanese, started a period of conflicts, violence, natural disasters, and famine that Buddhism's protective capacities became powerless to stop. In Japan, the social conflicts at the end of the 10th century were interpreted as a sign of Mappo's approach.

In contrast with Hinduism, where the past and future are subject to a superstructure of unending cycles, eschatology in the Abrahamic religions defines a single, redeeming end. The nature of the world, of life, and of time radically changes in that last deliverance from the physical world. The drama of existence is strengthened by the linear nature of time and the definitive singularity of the final conclusion.

In Hinduism, the Earthly cyclical nature of natural phenomena and of life is extended by an abstraction process to embrace the cosmic and divine order. The concept of the eternal return, in which the universe passes through similar successive cycles, present in the Indian subcontinent—in Hinduism, Buddhism, Jainism, and Sikhism—devalues the awareness of the irrecoverable nature of past time. Western culture adopts a quite distinct point of view. As time flows, the focus is on its uniqueness and unrepeatability, which implies that time is precious and must be used to the full and with eagerness.

However, the idea of eternal return has also been explored in Western philosophy by Heinrich Heine (1797–1856), a German poet, essayist, and journalist of the Romantic movement, and by the German philosopher, philologist, and scholar Friedrich Nietzsche (1844–1900), as a physical concept related to predeterminism and not as part of a supernatural cycle that includes reincarnation. Heine followed a mechanistic view in which time is infinite but the particles that make up concrete bodies are finite. According to his analysis, this implies that the number of combinations that they assume is also finite and determinate, leading to an eternal play of repetition (Heine 1869). Nietzsche went further and concluded that human existence is repetitive; he considered it a very heavy burden to realize that one would have to live, not just once more, but innumerably many times (Nietzsche 1882). In the narrative of modern physics, the eternal return of physical states in our Universe is most unlikely, due to the probabilistic nature of quantum theory.

The principle of a dependent arising or dependent origination plays a central role in Buddhism, and is closely related to the concepts of causality and time. It is a way of explaining the origin of *dukkha*, that is, dissatisfaction and human suffering, and its name in Sanskrit is *pratityasamutpada*, which means "that which arises" (*samutpada*) "while depending" (*pratitya*) on conditions, or due to conditions. According to the Pali Canon, the principle states that: "When this exists, that comes to be; with the arising of this, that arises. When this does not exist, that does not come to be; with the cessation of this, that ceases". Dependent origination is an ontological relationship of conditionality between events, objects, or human experiences within a particular time frame, and not a connection that can be analysed by the principle of conditionality. With this system of interpretation, the phenomena of human experience, sensations, perceptions, thoughts, and awareness and the physical phenomena themselves arise as composite entities formed by interdependent components. In dependent origination, rather than highlighting the idea that cause and effect are separate and presupposing a specific temporal relationship, it is said that they form an entity that emerges in the human mind in composite form, or that cause and effect emerge together. This emphasis dismisses temporal order and is at odds with the view of Newtonian mechanics built around the search for causal relationships between phenomena, conditional on a direction for time.

4.11 The First Utopias and Increasing Freedom to Conceptualize a Better Future for Humanity

Let us now return to the West and the time of the Renaissance in Italy when humanism, the arts, and the origins of modern science were developed. These highly significant cultural advances in terms of their future impact on civilization occurred in Italy in the 15th century, a time marked by an increasing awareness of crises, impending disasters, and the prospect of apocalypse. For people at the time, the future did not seem benign or promising for humanity. On the contrary, many believed that the end of the world was nigh. Leonardo da Vinci (1452–1519), in his *Codex Atlanticus*, left us magnificent drawings of maelstroms of water from catastrophic floods together with prophecies that reveal his concern for the end of the world and the terrible suffering that it would cause humankind (Gantner 1958). These fears were not foreign to the feeling of crisis that was felt at the time, particularly the denunciation of corruption, despotism, and exploitation of the poor made by Girolamo Savonarola (1452–1498) in Florence.

In the same social time as the Renaissance, Thomas More (1478–1535) recognised that the society in which he lived had many imperfections, especially poverty, injustice, greed, and staggering inequality. He decided to respond to the challenge with a fictional narrative. The idea was to describe a world with a system of government which, to his eyes, would decisively improve human living conditions and well-being. The book, called *Utopia* (More 1516), was ambiguous and contradictory, probably to ensure that his more heterodox ideas would not prevent the book's acceptance by the Church. The title was coined by More himself from the Greek "ou" for "not" and "topos" for "place", and literally means "nowhere". The place that didn't exist was located on an island called Utopia somewhere in the Atlantic, and the perfectly ordered society that lived there is described to Thomas More, who writes in the first person, by a Portuguese adventurer and seafarer called Raphael Hythloday, also a bibliophile and philosopher. The invented name, as with other names in Utopia, is a kind of a pun with a clear intentionality. It combines the Hebrew biblical name of Raphael, a famous archangel and messenger of God, with a Greek compound word Hythloday derived from *hythlos* meaning absurd or meaningless and probably *hodaia* meaning merchandise. Combining these meanings we obtain the idea of a "dispenser of nonsense", which reveals More's message.

Raphael lived on Utopia for more than five years and only decided to return to England, the contrasting real society where good governance was prevented by pride, private interests, and flattery, to spread the good news about the new world. The opinions of the two main characters, More and Raphael, are constantly contrasted and negotiated without the book pointing to any definitive solution. More manages to interest and inspire his readers and create a space for them in which to stretch their imaginations and get them to ask themselves whether it would really be possible to build a better world than the one they knew at the time. This pioneering approach created a new style of literature and a new conceptual framework for reflecting on and idealizing the future that have since been widely developed.

Interestingly, the prevailing meaning of the new word "utopia" is that of an impracticable social and political system or more generally something ideally perfect but impossible to achieve, which often has a negative connotation. Plato's *Republic* (Plato, circa 380 BC) is not a utopia, although it is sometimes suggested that it was the first. Plato was convinced that the laws of the New Republic could be put into practice in the real world, but that the road to achieving that would be long and arduous. The first utopias emerge when the reality of the recent past becomes sufficiently problematic or unbearable to drive people to imagine a better future. This would be built by conceiving of a consistent political and social system in the hope that it could one day be implemented. Later utopias also use an *argumentum ad absurdum* when reacting to the shortcomings and adversities of the recent past by showing that it inevitably leads to an absurd, dysfunctional, or altered future. Periods when the space of experience is blocked, there is little hope, and the future becomes bleak tend to create a horizon of expectation centred on a utopia. Many utopias have been developed throughout history with different focuses: on religious, economic, political, scientific, or technological aspects. All reflect their social time together with the corresponding space of experience and horizon of expectation.

Two utopias that were particularly important thanks to the change they announced in the relationship between science and technology and the construction of societies with greater well-being, abundance, and economic prosperity were developed by Tommaso Campanella (1568–1639) and Francis Bacon (1561–1626). Campanella was a Dominican friar from Calabria, the son of a cobbler, and a philosopher, theologian, and astrologer who opposed Aristotle's teachings. He supported the empiricism of Bernardino Telesio (1509–1588), the Renaissance philosopher whose book *De Rerum Natura Iuxta Propria Principia* (On the Nature of Things according to their Own Principles) helped launch the bases of the scientific method. Both Telesio and Campanella professed the idea that to know the nature and the workings of the world it was not enough to exercise abstract reason. The data received directly by our senses also had to be analysed. This was thus an early and pioneering form of empiricism.

Campanella defended the common ownership of material goods, like Thomas More in *Utopia*, and believed in the imminent arrival of the age of the Holy Spirit prophesied by Joachim of Fiore, which he predicted would happen in 1600. Due to his heterodox views, he was accused of heresy and sentenced to death by the inquisition but, by feigning dementia, he ended up being imprisoned for 27 years. Freed in 1629, he spent his last years in France, protected by Cardinal Richelieu, and died in Paris at the Dominican convent. While in prison, he wrote many works, including *The City of the Sun*, written in Italian in 1602, and *The Defence of Galileo*, a courageous defence of Galileo Galilei (1564–1642) against the persecution by the Church, published in 1616.

The City of the Sun is a theocratic urban utopia, partly inspired by Thomas More's book, which is described to us in the form of a poetic dialogue between the Grandmaster of the Knights Hospitallers and a Genoese sea captain, his guest and a former companion of Christopher Columbus. The City of the Sun is located in Taprobana and is organised into seven concentric rings corresponding to the seven planets of medieval cosmology. The centre contains the temple. All knowledge of mathematics and the various sciences, from the physical sciences to medicine, is presented and described on the city's walls. The city's perfection is founded on the exclusive use of reason and is demonstrated in the regularity and high quality of spiritual and social activities and architectural structures that seek to reproduce on Earth the divine order and cycles of the cosmos.

Everything in life is regulated down to the smallest detail, including clothing, food, working hours, reduced to only four per day, and sexual relations. There are no slaves but only citizens who work cooperatively. Love born of desire has ceased to exist and rules for procreation are strictly obeyed. The great ruler, an elected priest named Hoh, also known as the Metaphysic, governs the city with the help of a triumvirate of princes called Pon, Sin, and Mor, meaning power, wisdom, and love. There is no private property. All material goods, houses, rooms, beds, and even women and children are common property. As the sea captain who has been in the City of the Sun tells us, the residents have put an end to selfishness by removing the motives that underlie them. The rhythms and rituals of the City generate a cyclical time free from the contingency of linear historical time, considered as a synonym of degradation and physical and moral imperfection.

With his utopia, Campanella seemingly sought to demonstrate that it was possible to live on Earth in a perfectly harmonious way by developing and applying knowledge based on the laws of nature in full harmony with the Gospels. Curiously, this was the first attempt to build a better world based to a large extent on what today we would call technology. Campanella went beyond utopia and sought to implement it by building a world monarchy that would bring together all peoples converted to Christianity. He chose Spain and its young empire under the reign of Philip II as the prototype in his book *The Monarchy in Spain*, published in 1599, and later turned to France.

Like other great utopias, *Utopia* and *The City of the Sun* have the power to fascinate. They also contain ideas that were later adopted by most modern societies, like abolishing the death penalty, which Thomas More defended for robbers, and abolishing slavery, defended unconditionally by Tommaso Campanella. However, utopias also inspired many of the features that characterized failed social and political movements and systems on the extreme left and right of the political spectrum. Utopias are the most eloquent proof that we are unable to foresee the future in an organic and coherent way with all its complexity, but can only glimpse some of its fragmented forms. Although people today are likely to consider that implementing *The City of the Sun* would be an unbearable nightmare, the book was very popular in its time, and this helps us to understand Campanella's operative social time.

Francis Bacon was a contemporary of Tommaso Campanella but had very different social origins and upbringing. Campanella was a religious man from very humble beginnings, who led a turbulent and difficult life and remained faithful to the Catholic Church, despite his criticisms of the institution itself. Francis Bacon was the son of Nicholas Bacon, a minister in the court of Queen Elizabeth I, and Anne, a highly cultured mother, the daughter of King Edward VI's tutor. He had a highly privileged upbringing and education and was an active politician as well as a philosopher and jurist.

There are similarities between their ideas on natural philosophy and religion, but Francis Bacon went much further than Campanella in defining a new path for science and the potential benefits it could bring to society. Bacon was one of the most important figures of the transition period from the Renaissance to the modern era and had a considerable influence on the history of humanity. He also used utopian fiction to communicate his ideas in his book *New Atlantis*, published posthumously in 1627 (Bacon 1627).

The book describes the island of Bensalem, in the Pacific to the west of Peru, which is discovered by the crew of a European ship after it is wrecked there. The residents live under a form of Christianity that accepts the freedom to profess other religious beliefs and encourages respect for wiser, morally exemplary members of society, as well as a deep sense of order, discipline, peace, and harmony. Bensalem society's most important institution is Solomon's House, which in Bacon's own words is "the noblest foundation (as we think) that ever was upon the earth; and the lanthorn of this kingdom. It is dedicated to the study of the Works and Creatures of God" (Bacon 1627). It represents a metaphorical reconstruction of Solomon's Temple, symbolic of the restoration of true religion but now also applied to the recovery of natural philosophy. Bacon had a dual program of restoring the human relationship with God and at the same time restoring and developing the human dominance of nature through natural philosophy (McKnight 2006).

In modern language Solomon's House is conceived as a research institute that accords priority to experimental science run by 36 wise men. Twelve of these are "merchants of light" who organise scientific expeditions to other parts of the world, while the others devote their time to fundamental and applied research. The main goal of Solomon's House "is the knowledge of causes, and secret motions of things; and the enlarging of the bounds of human empire, to the effecting of all things possible" (Bacon 1627). The wise men have observatories in caves, mountains, and coastal areas, there are artificial wells and fountains, including one called *Water of Paradise* "for health and the prolongation of life" (Bacon 1627). There are also pharmacies, centres for mechanical arts, furnaces for metallurgy, laboratories to study optics and acoustics, and soils to improve agricultural productivity as well as gardens and orchards where medicinal plants are cultivated and where bigger and sweeter fruit can be grown. These activities help increase the knowledge of nature and produce potentially interesting and useful new phenomena and products with the ultimate aim of improving living conditions for Bensalem's residents.

Bacon was firmly convinced that the observation and study of nature would bring practical benefits. He considered that such a project was also acceptable from a religious point of view, because studying God's creation meant knowing God better. Furthermore he was convinced that scientific endeavour was also compatible with social stability. The model provided by Solomon's House helped to inspire the foundation of the Royal Society in 1660 and ended up becoming one of the ideal prototypes for today's university research centres and state-run or private research laboratories.

Francis Bacon had the opportunity to experiment some of his ideas when he played an important role in setting up several English colonies in North America, Virginia, the Carolinas, and Newfoundland. He was the guiding spirit behind the colonization schemes and responsible for drafting the charter of government of the colony in Virginia. The imprint of his foundational vision has probably contributed to the predominance in the United States, since its creation, of the idea that science and technology drive human progress. After more than 350 years we are witnessing the first cracks in this system, in particularly because some corporations no longer view science as a provider of truth, but rather as one of the many inputs into production that can hinder profitability, a standpoint which has led to the manipulation or negation of specific scientific knowledge in some cases.

Bacon, however, did not use the word "progress" in the way it is commonly used today. Its stems from the Latin *progressus*, which means advancement compared to a certain object, such as a place. In the Middle Ages, the word was used to describe the forward motion of the king or queen with his or her retinue on an official trip across the land. In England, "progress" only began to be used in the figurative sense of development and advancement towards better human living conditions and economic prosperity in the 17th century, but then it fell into disuse. Curiously, the figurative use of the word was recovered in the USA, while in England it was considered an Americanism. It eventually became more widespread in the 19th century, when the effect of the Industrial Revolution and modern science on a large part of the population began to make itself felt, in particular as regards public health care, transport, and housing facilities.

In many ways, contemporary societies keep trying to achieve the technological utopia that Bacon imagined. At the end of *New Atlantis*, Bacon lists some objectives in the appendix *Magnalia Naturae*, the Wonders of Nature, which are still pursued actively by science and technology. It starts with "the prolongation of life", one of his greatest concerns, which he believed could be resolved by science, and continues with "the restitution of youth in some degree", "the retardation of age", and "the curing of diseases counted incurable". But he goes much further than medicine and mentions "the increasing or the exalting of the intellectual parts", one of the current ambitions of transhumanism, the "making of new species", pursued by synthetic biology, "drawing up new foods out of substances not now in use" and "instruments of destruction, as of war and poison", which is still one of the foremost fixations of humanity.

The new empirical method created by Bacon replaced Aristotelian logic and established an analytical system that made it possible to observe, experiment, find out about, and exploit new domains of nature without limit, to the benefit of humans. Symbolically, on the cover of the book *Novum Organum* (Bacon 1620), the flagship of the new "discovery machine" makes for unknown seas, passing between the Pillars of Hercules, which traditionally carry the inscription *Non plus ultra* (nothing beyond), leaving behind it the old world of the Mediterranean, and setting out to explore the new world. Bacon was firmly convinced that the application of scientific knowledge in conjunction with a reformed religion based on a pure form of gospel Christianity would be able to transform and improve human living conditions and lead to peace, harmony, and prosperity. He believed that a better world was within reach of future generations and that the most important thing to achieve it was to develop the sciences and their applications through cooperation and organisation. He had a well-defined plan for the future that would make it very different from the past. His horizon of expectation was profoundly different from his space of experience. Bacon made it possible to believe that historical time was changeable and "progress" was possible for humanity by following a path supported by science and a reformed Christianity. The emphasis on this complementarity faded away rapidly with time after Bacon and may even be surprising nowadays. Most people would now agree with Bacon when he writes that "knowledge and human power are synonymous" (Bacon 1620). But in his social time, this was a bold and revolutionary statement. Nonetheless, Bacon was particularly careful to claim that his progressive views about the benefits of exploring, studying, and understanding nature and putting scientific knowledge to practical use were compatible with Christian teachings.

In one of his first books, *Valerius Terminus* (Bacon, c. 1603), the meaning of whose title remains obscure, he argues that knowledge of nature is legitimate and cannot be confused with something that God does not want humans to investigate. The research and free exploration of nature, including human nature itself, a new secular field that Bacon called "natural history", was considered compatible with the non-violation of the two trees of Eden, the tree of knowledge of good and evil and the tree of life. Eating the forbidden fruit of the latter led to eternal life (Genesis 3:22), but Bacon sought only to prolong human life.

Although some historians would say that Bacon manipulated religious ideas in order to provide authority for his political agenda (White 1968), it is more likely that he was genuinely convinced about the compatibility of the dual endeavour of fulfilling God's law and developing natural philosophy to improve human well-being.

In any case *New Atlantis* represents a watershed in Western culture where well-being and material prosperity accomplished by humans starts to overshadow spiritual salvation enabled by God. David Innes, a minister of the Orthodox Presbyterian Church, considers that Bacon opened the way to progressively substitute the path towards religious salvation by the secular path of material comfort and economic prosperity, and states that the Christianity of Bensalem is a "fundamental assault upon, transformation of and ultimate displacement of Christianity" (Innes 1992; McKnight 2006).

Francis Bacon was the first to state unequivocally that science and technology would be able to prolong life expectancy, one of the greatest human ambitions, and perhaps the one that in the end is most highly valued. Interestingly, this was the first time that religious narratives were reversed; most of them postulate that, in the distant past, the duration of human life was generally much longer, but that human behaviour over time led to a spiritual, moral, and physical deterioration which irredeemably shortened lifespans. Bacon restored hope for a longer life through human ingenuity and determination, without questioning the inaccessibility of eternal life, which can only be granted by God.

To a large extent modernity began in harmony with Christianity, or at least not in a position of open antagonism. It brought in the idea that humans would always have the means and ability to make their future better than their past. There are diverging and frequently opposing views on the most appropriate political, socio-economic, and environmental models and solutions, but the need to develop scientific research and technological innovation is agreed upon unanimously. Since Francis Bacon's time, science and technology have remained the human activities that provide the essential support for better living conditions, better health and well-being, greater economic prosperity, and an increasing availability of goods and services to consume. The acknowledgement that technology is the vehicle for unlimited progress has the characteristics of a utopian conviction and may be the last myth in *Homo sapiens*' time.

Chapter 5
The Triumph and Challenges Created by the Idea of Progress

5.1   The First Proponents of the Idea of Progress

The idea of progress was born, is deeply rooted and has played a central role in the development of Western civilization. Robert Nisbet even believes that "No single idea has been more important than … the Idea of Progress in Western civilization for three thousand years" (Nisbet 1980). Let us consider then how it has evolved since Francis Bacon's time. The French writer and natural philosopher Bernard de Bovier de Fontenelle (1657–1757) established a link between the 17th century, when France was arguably the major power of Europe, and the Enlightenment of the 18th century.

Influenced by the theories of René Descartes (1596–1650) and the emerging impact of science on technology, Fontenelle was one of the first to defend the idea of inevitable, universal, and linear human progress, and was a forerunner of the Enlightenment, although without using the word "progress". He stressed the fact that this upward movement, based on science and the new technologies, was the result of a consistent and enduring construction made over time and built upon the advances and discoveries achieved in previous periods. He was a distinguished academic, author of works of literature, history, and science, and one of the first writers to popularize science. In his essay on astronomy "Entretien sur la pluralité des mondes" (Fontenelle 1686), he presents an easily understandable and pleasurable exposition of Descartes' physics, Copernicus heliocentric system, and Galileo's view of the Earth, Sun, and planets. Fontenelle writes in the preface: "I wished to represent philosophy in a way that was not philosophical" (Fontenelle 1686). A prodigy with compliments, he avoided controversies and when he was asked one day why he had so many friends and no enemies throughout his life, he answered that he had managed it "using two axioms; that everything is possible and everyone is right".

The advances of science and the ideas of tolerance, rationality, and openness of spirit to observation and experimentation received a new and important boost from the Encyclopaedists, the authors of the famous *Encyclopaedia* (Diderot and D'Alembert 1751) called *Encyclopédie, ou dictionnaire raisonné des sciences, des arts et des*

© Springer Nature Switzerland AG 2021
F. Duarte Santos, *Time, Progress, Growth and Technology*, The Frontiers Collection,
https://doi.org/10.1007/978-3-030-55334-0_5

métiers, drawn up and published between 1751 and 1772 under the leadership of Denis Diderot (17031784) and Jean le Rond d'Alembert (1717–1783).

Anne Robert Jacques Turgot (1727–1781), known as Jacques Turgot, and Marie Jean Antoine Nicolas de Caritat, Marquis of Condorcet, known as Nicolas de Condorcet, both of whom were Encyclopaedists, were the first to analyse and explicitly advocate the idea of progress. Turgot did so in a remarkable speech in Latin entitled *On the Successive Progress of the Human Spirit*, delivered on 11 December 1750 at the Sorbonne, where he had completed his degree in theology. The text of this lecture, which was released after his death, contains the first complete, coherent statement of the perfectibility of human society: "The total mass of the human race, by alternating between calm and agitation, good and bad, marches always, however slowly, towards greater perfection" (Turgot 1750). The fundamental idea, which is still valid today, is that progress is not constant, or the same for all, but is statistical in nature. In other words, it is a property of a population that manifests itself over periods of time that are relatively long compared with a social generation.

Another crucial aspect of Turgot's speech is his concept of a historical time (*temps de l'histoire*) that contrasts with a time of nature (*temps de la nature*). The time of nature is determined by the phenomena of nature which, "governed as they are by constant laws, are confined within a circle of revolutions which are always the same" (Turgot 1750). Historical time is different, since "reason, the passions and liberty ceaselessly give rise to new events" (Turgot 1750). The overall view of a historical time marked by the march of progress becomes even clearer when he writes "the human race, considered over the period since its origin, appears to the eye of a philosopher as one vast whole, which itself, like each individual, has its infancy and its advancement" (Turgot 1750).

Jacques Turgot shared with the Encyclopaedists the conviction, revolutionary at the time, that man learned through his senses, reason, and experience and not under the influence of any religious authority. In the *Encyclopaedia* (Diderot and D'Alembert 1751), man is defined as a being that "feels, reflects and thinks" outside of any kind of relationship or dependency on the divine world. For Turgot, Christianity is a generous and charitable deism, the greatest benefit of which "has been in having enlightened and propagated natural religion", in which, as opposed to revealed religion, the existence of God is achieved by reason and is therefore accessible to all men.

Turgot gave up being ordained as a priest and pursued a career in law, telling his friend Pierre Samuel du Pont de Nemours (1739–1817) that "it is impossible for me to give myself up, all my life, wearing a mask". Pierre du Pont de Nemours, was a writer and economist who moved to the USA, after being arrested during the Reign of Terror of the French Revolution, and his son Éleuthère Irénée du Pont de Nemours (1771–1834) set up the DuPont Company, which is still one of the largest chemical multinationals in the world.

Turgot became interested in economics and was one of the greatest apologists of a liberal economic model based on an advanced form of physiocracy, an economic theory characterized by the assumption that the wealth of nations is derived from land agriculture. He was the first to enunciate the law of diminishing Marginal Returns in agriculture and his economic ideas constitute an agrarian capitalist model adapted to the framework of absolute monarchy. The succeeding economic model was the classical capitalist model developed by Adam Smith, which was strongly influenced by Turgot's theories. Many of the essential ideas that he developed in the *Wealth of Nations* (Smith 1776) were already present in the writings of Turgot, such as the "necessary price" that included a "normal" return on capital, part of which was to be accumulated (Meek 1963). In 1774 he was appointed minister of the kingdom and a month later Controller-General of Finances, in which position he attempted to carry out reforms to France's financial and economic policies, but was forced to resign in 1776.

David Hume (1711–1776) died in the same year. The Scottish philosopher, historian, and economist shared ideas on progress similar to those expressed by Turgot and Condorcet, but with less passion and more caution. In his book *On the rise and progress of the arts and sciences* (Hume 1777), he considered that progress required an environment of political stability, something that Condorcet was far from experiencing towards the later part of his life.

Nicolas de Condorcet was educated at a Jesuit school in Reims, where he revealed his ability for mathematics. He then studied at the Collége de Navarre in Paris and his thesis, written in 1765 and entitled *Essai sur le calcul intégral*, was presented to the Academy of Sciences. He became a student of the mathematician, physicist, and philosopher Jean le Rond d'Alembert, turned down a military career to devote himself to mathematics, and at the age of 26, became an elected member of the Academy. It was d'Alembert who introduced him to the salon of Melle de Lespinasse, where he met the circle of philosophers of that time and in particular Voltaire and Turgot. Both became friends, turning his attention to the moral and political questions of their social time and influencing his work and public career.

Condorcet's life changed abruptly when he became Permanent Secretary of the Academy of Sciences and accepted the invitation by Turgot to become inspector at the National Mint in Paris in 1774. From then on, he focused his attention more on philosophy, economics, and politics, while pioneering the application of mathematics to the social sciences (Granger 1956), especially the use of probability theory and statistics. The expression "social mathematics" was first used and defined in an article in a revolutionary journal entitled *Journal d'Instruction Sociale* that went into publication in 1793 at the beginning of the Reign of Terror (Condorcet et al. 1793).

Condorcet became a keen defender of human rights, particularly gender equality and the rights of minorities, Jews, and black people, and fought to abolish the death penalty. His works characterise the ideals of the Enlightenment and rationalism, and caused controversy at the time but ended up having a significant influence in France and beyond. With the downfall of Jacques Turgot in 1776, after he tried to implement several liberal reforms, Condorcet was disheartened, but became better aware of the obstacles facing the implementation of his ideals.

In 1786, he married Sophie de Grouchy, a beautiful and intelligent women twenty-one years his junior, who had an exceptional humanist education, spoke good English, and shared the same interests and ideals as Condorcet. She was a writer and translated some of the major works of Adam Smith and Thomas Paine (1737–1809) into French.

Sophie de Condorcet received guests at the salon of her home, the Hôtel des Monnaies in Quai Conti, attended by the Marquis de Lafayette (1757–1834), Thomas Jefferson (1743–1826), Thomas Paine, Adam Smith, Honoré-Gabriel Riquetti de Mirabeau (1749–1791), Germaine de Stael (1766–1817), André Marie Chénier (1762–1794), Pierre Cabanis (1757–1808), and many other illustrious figures of the Revolution and the Enlightenment.

Unlike other philosophers of his time, Condorcet decided to play an active role in the Revolution, which he believed could lead to a rationalist reform of society. He was elected to the Legislative Assembly in Paris in 1791 and, thanks to his impartial approach to the affairs at hand, he became Speaker. He wrote a report on the development of public education and suggested establishing the metric system, together with Jean-Charles Borda (1733–1799), Pierre-Simon de Laplace (1749–1827), Joseph-Louis Lagrange (1736–1827), and Gaspar Monge (1746–1818). He ended up siding with the Girondins against the Montagnards when he voted against the execution of Louis XVI because he opposed the death penalty, and instead suggested sentencing him to a life of forced labour on the galleys.

In 1793, the Girondins lost control of the National Convention, and Condorcet, a member of the committee appointed to draw up the constitution, criticised the Montagnards' newly proposed constitution. On 8 July of that year, the "theorist of Gironde", as they called him, was accused of treason and the National Convention voted in favour of an arrest warrant. Alerted to this, he took refuge at the very modest home of Rose Marie Vernet, at 21 rue Servandoni in Paris, where some of his friends had lived as students. Rose Marie was the widow of sculptor Louis-François Vernet, a close relative of the renowned Vernet painters. Condorcet did not know her but she nevertheless stepped forward to shelter the fugitive. Condorcet and his wife Sophie de Grouchy, who visited him and supported him in hiding, and who was now working in order to survive, suffered deeply from the situation in which the country found itself. Many of their closest friends were summarily accused and guillotined without appeal. Neither had imagined this end when they had shared such great enthusiasm for the Revolution at its beginnings in 1789.

It was under these circumstances that Condorcet, in July 1793, with Sophie's support and encouragement, started to write his well-known book *L'Esquisse d'un tableau historique des progrès de l'esprit humain* (Condorcet 1795). This in turn was only a sketch of a much longer work he planned to write, which would be called *Tableau historique des progrès successifs de l'esprit humain*. The book constitutes a formulation of the ideology of progress within the framework of an historical analysis and reveals Condorcet's unbreakable belief in the perfectibility of man. The message becomes pathetic and even paradoxical if we consider the time and the dramatic circumstances under which it was written, precisely when history started to put seriously into question the hopes of the Encyclopaedists. Condorcet revealed an extraordinary attachment to his principles precisely when his everyday experience contradicted them most cruelly.

The book is organised into ten epochs, in which he highlights the invention of writing, the progress of the human spirit in Greece, the revival of the sciences in the West, the invention of printing, the time of Descartes, and the formation of the French

Republic, while the tenth epoch is about the future progress of mankind. Condorcet's aim was to discover in historical time the signs pointing to the relentless progress of the human spirit. His underlying belief was that this progress was the determining thread that had always guided history. Progress in the future could be secured and planned by developing the sciences, perfecting moral ideas, and implementing human rights. Turgot had never gone so far; he had simply admitted that progress was possible, although neither constant in time nor egalitarian.

For Condorcet, human perfectibility could always be surpassed and had only physical limits. In fact he wrote that "the perfectibility of man is absolutely indefinite; that the progress of this perfectibility, henceforth above the control of every power that would impede it, has no other limit than the duration of the globe upon which nature has placed us. The course of this progress may doubtless be more or less rapid, but it can never be retrograde; at least while the earth retains its situation in the system of the universe, and the laws of this system shall neither effect upon the globe a general overthrow, nor introduce such changes as would no longer permit the human race to preserve and exercise therein the same faculties, and find the same resources" (Condorcet 1795).

This statement is curious and premonitory because it clearly has a long reach that the author was far from being able to fully grasp in his time. Some saw in it the first expression of a technological singularity. The constraints expressed by Condorcet are surprisingly prophetic of contemporary challenges due to global change. It was obviously impossible for him to have imagined that human overexploitation of natural resources and environmental degradation could undermine sustainability and eventually limit human progress. However, he admitted that human progress would be impaired if the "laws of this [Earth] system" could produce a "general overthrow" and "no longer permit the human race [...] to find the same resources".

Condorcet predicted that the march of human progress would eventually lead to some form of globalisation. According to him progress would benefit first some men and then spread to others until it reached several countries, until finally the "the moment knowledge shall have arrived at a certain pitch in a great number of nations at once, the moment it shall have penetrated the whole mass of a great people, whose language shall have become universal, and whose commercial intercourse shall embrace the whole extent of the globe" (Condorcet 1795). Condorcet left us with the most influential definition of the idea of progress, based on the conviction that the natural and social sciences would lead us to a world of individual freedom, human rights, equality, justice, moral solidarity, well-being, and material abundance. In his own words: "Then will arrive the moment in which the sun will observe in its course free nations only, acknowledging no other master than their reason" (Condorcet 1795).

L'Esquisse is not a utopia. Condorcet believed that it was possible to establish a new political and moral order. He referred in his writings to Bacon's *New Atlantis*, but did not share his vision of a perfect kingdom illuminated by science and Christianity. His vision was rather that reason applied to the social and natural sciences could establish a new secular society that ensured freedom, human rights, well-being, and prosperity.

There were many voices in the XVIII century that did not share the historical optimism of Condorcet, even among the Encyclopaedists, such as Diderot and d'Alembert. The latter, a mentor and friend of Condorcet, stated in his *Discours préliminaire à l'Encyclopédie* (Diderot and D'Alembert 1751) that "Barbarism lasts for centuries; it seems that it is our natural element; reason and good taste are only passing". Instead of linear progress based on a gradual enlightenment of the horizon of expectation propelled by reason and the desire for human freedom defended by Condorcet, Georg Wilhelm Friedrich Hegel (1770–1831) and Karl Marx defend a dialectic approach that integrates the contrasting and often cyclical periods of history, where progress is subdued by human aggression and violence, war, class struggles, and ideological conflicts.

After nine months at Rose Marie Vernet's house, Condorcet became convinced that his enemies had discovered his hideaway and that his protector was in danger and he decided to run away and fool those keeping watch on him. He went to Fontenay-aux-Roses to the family house of the Suards, his former close friends, but they refused to give him asylum even for one night. Injured and starving, he went into a tavern at Clamart and ordered an omelette with many eggs. The combination of his extreme hunger and visible exhaustion with the politeness and good manners distinctive of the upper social classes raised the suspicions of the innkeeper, Louis Crépine, and a group of boors sitting at a table nearby. Without hesitation, they grabbed him and took him forcefully and triumphantly to the nearest *Comité de Salut Publique* at Bourg-l'Egalité, which was the name of Bourg-la-Reine during the Revolution. There it was found that he had no valid documents and he was handed over to the authorities, who locked him up in a miserable cell at the local prison. After two days, on 29 March 1794, he was found dead on the floor of the cell floor, but the reasons behind his death remain mysterious. At the bicentennial of the French Revolution on 12 December 1989, President François Mitterrand symbolically placed the ashes of Nicolas Condorcet in the Pantheon in Paris, but his tomb is actually empty because his remains were never found in the mass grave of the cemetery at Bourg-la-Reine.

5.2 Evolution of Progress

Condorcet's ideas on progress were very fertile from the point of view of ideological debate and became an indispensable reference for the future. The concept of progress evolved and became a powerful indicator of social, political, economic, scientific, and technological transformations and expectations. *L'Esquisse* became a famous book in part because of the tragic circumstances under which it was written. Curiously, on 2 April 1795, the National Convention voted in favour of the proposal by Pierre Daunou (1761–1840), a French historian and politician, that the State should pay for the publication of 3000 copies of the book, less than two years after it had imposed the death sentence on the author, now referred to as the "unfortunate philosopher". The Conventionnels agreed with Condorcet's ideas on progress and they wanted the book to become a classic.

Nonetheless, criticisms of Condorcet appeared immediately after his death. Critics included Louis de Bonald (1754–1840), a Catholic royalist counter-revolutionary who, a year after *L'Esquisse* was published in March 1795, released a text in which he stated that "the fanatical picture that this philosopher gives of his hypothetical society can explain to us the inconceivable phenomenon exhibited by revolutionary France" (Bonald 1796). Bonald continued to defend the apocalyptic salvation announced by Catholicism and believed that science was usurping it, forgetting the brutal, everlasting realities of passion, conflict, and human violence. Henri Saint-Simon (1760–1825), one of the first ideologues of socialism, positivism, and industrialisation, had a different opinion. He praised *L'Esquisse* for "demonstrating that the progress of civilisation has consistently tended towards the establishment of the industrial system" (Saint-Simon 1824).

Thomas Malthus made a deeper, well-founded criticism of Condorcet and Godwin's progressive outlook. For him, misery was an integral and unstoppable part of societies; he stated that "no possible form of society could prevent the almost constant action of misery upon a great part of mankind, if in a state of inequality, and upon all, if all were equal" (Malthus 1798). A statement that is at once wise, wistful, disheartening, and, for that reason, dangerous. Condorcet and Godwin had a different opinion and argued that the way to resolve misery and reform social structure was to use reason, follow the Enlightenment principles of freedom, equality, justice, and progress, and work out ways to make the most for society's well-being and prosperity from the benefits of scientific and technological progress. These ideals were the driving force behind major advances in the last two centuries, but new forms of misery and poverty persist, along with deep inequalities. It is as if we are forever forced to pursue something which experience shows to be unreachable, despite our enthusiasm.

The French Revolution opened a new horizon of expectation that had little in common with past space of experience. Historical time sped up and Maximilien de Robespierre (1758–1794) himself invited his fellow citizens to accelerate it further. When speaking at the Constituent Assembly on 10 May 1793 he said: "The time has come to call upon each to realise his own destiny. The progress of human reason laid the basis for this great Revolution, and you shall now assume the particular duty of hastening its pace". Little more than a year later, on 28 July, Robespierre was guillotined in the *Place de la Révolution*. The pace of operative social time reached lightning speed in the final years of the Revolution until the beginning of the Napoleonic regime on 9 November 1799.

As well as this acceleration of time, the French Revolution established a new relationship between the past and future characterised by two opposing ideas: progress and unpredictability. The expectation of human progress was intimately linked to the Revolution, but it created a completely unknown future without reference points, as Diderot recognised when he wrote: "What will follow this revolution? We don't know" (Dieckmann 1951). With the French Revolution, the future became more obscure and unpredictable but in return it became filled with the promise of progress through the use of reason. These two new attributes continue to be strengthened by the development of modernity and place enormous and increasing relevance on the future. Revealing the nature of our common future for each community, each country, and the entire globe has become a fundamental, indispensable, and everyday task for politicians, businesspeople, financiers, economists, writers, and political commentators around the world.

An historical perspective of over two hundred years reveals that Condorcet naively believed the thesis that reason could be used to promote social and political advances, just like when it is applied in science and technology, and that reason could work to produce reliable and everlasting progress in both cases. Nevertheless, the ideology he set out for human progress in *L'Esquisse* is still the politically dominant conceptual framework to address the future, in which progress can always be achieved. From the 19th to 21st centuries in Europe and the USA, the future had unconditionally embraced the promise of human progress, and this conviction has generated a feeling of confidence and hope, a clear direction and perspective for the horizon of expectation, and the further entrenchment of the linear nature of time. We are completely immersed in a culture that assumes the inevitability of progress, although recognising that there have been and will be periods of decline, deep disappointments, and dramatic interludes. However, looking at the big picture of historical time, our dominant culture leads us to believe that progress is guaranteed, although in different and changing forms.

The Founding Fathers of the United States, who led the Thirteen Colonies to the American Revolution and helped write the 1787 Constitution, considered that they were realizing a practical application of Enlightenment ideals, and they believed that the idea of progress could be put into practice through a political system organised specifically towards improving human well-being and economic prosperity. The Constitution explicitly mentions the progress of science and technology, the latter expressed using the term "useful arts". In Article 1, Section 8, Clause 8, known as the copyright clause, it is stated that one of Congress' responsibilities is: "To promote the progress of science and useful arts, by securing for limited times to authors and inventors the exclusive right to their respective writings and discoveries".

Thomas Jefferson (1743–1826) argued that the "Empire of Liberty" created in the USA should extend to the north, now Canada, the rest of the Americas, and eventually the whole world. This view of reforming the world has served as the model for the USA's international policy until recently and was used consistently to justify intervention in the Spanish–American War, the First World War, the end of the Second World War, and the war on terror which began in 2001.

The idea of progress has not been abandoned and continues to be the only available framework for the future. While the French and Scottish thinkers of the Enlightenment had a dominantly empiricist idea of progress, Immanuel Kant (1724–1804) reasoned in an a priori way to the conclusion that humanity progresses. Kant considered that, if nature was not to be in vain, or in other words, if human faculties were not to be considered useless, then over time mankind would develop all the human faculties. Progress measures the development of such faculties over successive social generations. However, for progress to happen, it requires an environment of freedom and peace that, taken as far as possible, will lead globally to a harmonious federation of republics. Kant believed that ultimately mankind's progress was an immanent feature of human nature.

Georg Hegel had a less peaceful view of progress, asserting that human aggression and war are an inevitable part of the process of human development. He placed universal history at the heart of his metaphysics and believed that over historical time there is progress in the world, considered as a complex, diverse, and unstable whole (Hegel 1830). Hegel's view becomes clearer when it is remembered that he recognised only three "great men" in history: one was Napoleon, whom he saw on the eve of the Battle of Jena, 13 October 1806, and the other two were Alexander the Great and Julius Caesar.

Charles Darwin (1809–1882) opened up a new dimension for the idea of progress with his theory of evolution. Could evolution through natural selection be a form of progress? Darwin seemingly thought so when he wrote at the end of his book *On the Origin of Species* (Darwin 1859) that "As natural selection works solely by and for the good of each being, all corporeal and mental endowments will tend to progress towards perfection", a view which is heavily contested by contemporary biology and, in particular, by Stephen Jay Gould (1941–2002). In the context of evolutionary biology, adaptation can be qualified as a form of progress since it is defined as the accumulation of genetic characteristics that, in a certain evolutionary lineage, contribute to a specific adaptation. In other words, adaptation is the process that adjusts biological organisms to their environment by enhancing their evolutionary fitness. However, humans no longer have a fixed environment since they have become able to profoundly modify their natural environment to best serve their interests, and most of us now live in artificial environments. The concept of human progress cannot be simply reduced to a form of biological adaptation.

Nonetheless, the philosopher Herbert Spencer (1820–1903) used the evolutionary biology of Jean-Baptiste de Lamarck (1774–1829) and Darwin as a starting point to argue that progress essentially consists of biological evolution from roughly homogeneous structures to increasingly complex heterogeneous structures. By applying this theory to man, he went on to defend a form of scientific racism saying that "We may infer that the civilised man has also a more complex form of heterogeneous nervous system than the uncivilised man" (Spencer 1857). Spencer was one of the promoters of Social Darwinism, a pseudo-scientific construction which rejected the principle of equality between all human beings. The theory resulted from the application of evolutionary biology, and in particular the concepts of natural selection, to human society. It assumed that some human beings are biologically superior to others and that only the fittest should survive in society. "Social Darwinism" was a term rarely used by its proponents, although often in a pejorative sense by its opponents.

Various social and political movements during the 19th and the first half of the 20th centuries in Great Britain, Germany, the USA, Japan, and other countries were influenced by Social Darwinism. The theory provided a justification for the supposition that Westerners were intellectually, emotionally, and physically more advanced than the other peoples of the world. This form of racial hierarchy was often used as an ideological foundation for the colonisation of Third World countries up to the 20th century. The British sociologist Benjamin Kidd (1858–1916) interpreted the remarkable expansion of the British Empire as a vindication of Social Darwinism writing: "We watch the Anglo-Saxon overflowing his boundaries, going forth to take possession of new territories, and establishing himself like his ancestors in many lands" (Kidd 1894). He goes on to say that "He [the Anglo-Saxon] has been deeply affected, more deeply than many others, by the altruistic influences of the ethical system upon which our Western civilization is founded", acknowledging the progress relative to the colonization of the other "races like the ancient Peruvians, the Aztecs, and the Caribs". For Benjamin Kidd progress is imperative and ineluctable when he writes that "Progress is a necessity from which there is simply no escape and from which there has never been any escape since the beginning of life" (Kidd 1894). These views on progress constitute a notorious example of the human incapacity to perceive and integrate historical time so that they may realize the ephemeral and contingent character of their operative social time.

Ten years after the publication of Spencer's book, Karl Marx's *Capital* came out (Marx 1867, 1885). Historical materialism was born from the belief that it is possible to solve the paradox of the Industrial Revolution. According to Marx, the paradox consisted in simultaneously generating a rich capitalist social class and a relatively poor proletariat who frequently lived in degrading conditions. This situation urgently needed to be overcome while maintaining the impetus of industrialisation and economic growth. He was inspired by the socialist ideas of Henry Saint-Simon, Robert Owen (1771–1858), and Charles Fourier (1772–1837), whom Karl Marx and Friedrich Engels called "utopian socialists" in the Communist Manifesto published in London at the end of February 1848. In the same year the so-called 1848 Revolutions began to emerge in several European countries. In historical materialism, human progress in historical time is manifested in the successive stages of primitive communism, barbarism, slavery, feudalism, capitalism, socialism, and communism. The idea of progress therefore became an inherent part of this form of historical determinism.

The 1848 Revolutions in Europe, sometimes known as "The People's Spring", represented the first great crisis for capitalism and challenged mainstream ideas about the function and power of money in society. The geographer David Harvey considers that the social and economic upheaval that hit most of Europe produced a new social time that was very different from the social time of the Enlightenment, and that it was the acceleration of this latter time that culminated in the French Revolution (Harvey 1990). In the interpretation of historical materialism, a cyclical or alternating social time emerged, propelled by the recurring crises of capitalism. The question "What time are we in?" acquired greater social, political, cultural, and philosophical importance.

It was in the 18th century that the idea of progress began to be put into perspective and criticised. The first to do this was Jean-Jacques Rousseau, a pre-Romantic who denounced the role of progress in society because it generated injustice, violence, and inequality. He believed in perfectibility only for individuals, achieved by improving behaviour, although always subject to steps forward and steps backwards. At the end of the century, the first of the Romantic movements began to emerge. They considered society at the time to be sophisticated but corrupt and in moral decline, despite economic, scientific, and technological progress. The heart of this contradiction lay in the disconnection between the emergence of the concept of a progressive society, which is an abstract entity that includes all its members, and the enormous diversity of individual situations and experiences ranging from the most abject misery and poverty to material luxury, extreme wealth, frivolity, and strange perversions.

This is where we again revisit the challenge and dangers of elaborating on "the total mass of the human race", as Turgot wrote, and setting that discourse against each person's unique, individual experience. The comparative and quantified analysis of human well-being and health conditions and the characteristics of human lifestyles and communities became possible through conceptualisation and measurement when indicators were designed to hand such abstract entities, and later with the application of mathematics, especially statistics, to the social sciences, which began in the 18th century. These analyses benefited from extraordinary developments based on the use of ICT and they are now essential to political activity and to social, economic, and environmental planning. Nonetheless, we must not forget that we are dealing with human beings who have specific, non-transferable experiences, each with their own spaces of experience, horizons of expectation, and versions of operative social time.

Romanticism praises authenticity, the power of the emotions, and artistic creativity, and it values the past, particularly medieval times, and pristine nature free from what it views as human adulteration. Defenders of Romanticism, including Friedrich Schlegel (1772–1829), Friedrich Schelling (1775–1854), and Samuel Taylor Coleridge (1772–1834), believed that the movement towards mass society dominated by trade, the economy, and profit would damage individual freedom and social order. They resisted the idea of progress because it sought to impose a linear, simplistic narrative for the development of society, similar to the development of scientific knowledge and technology. This imposed linearity is contradictory with the changing values of the arts, the creativity of artistic expression, and the advancements and setbacks in people's lives as individuals, as a community, or as a nation.

In the 19th century, while societies were starting to profit from extraordinary progress in modern science and technological innovations that revolutionised the way of life around the world, a group of philosophers and historians emerged, including Arthur Schopenhauer (1788–1860), Alexis de Tocqueville (1805–1859), Jacob Burckhardt (1818–1897), and Friedrich Nietzsche, among others, who manifested scepticism about the possibility of implementing the idea of progress in historical time. Many of the concepts that they developed still form the foundations of most criticisms that appeared later in the 20th and 21st centuries. All these writers, thinkers, and philosophers either identified with the philosophical movement of pessimism, particularly Schopenhauer and Nietzsche, or were very close to it, and all foresaw several features of the future that really happened afterwards.

Schopenhauer found no signs of progress in history and wrote in one of his later works that "History shows us the life of nations and can find nothing to relate except wars and insurrections; the years of peace appear here and there only as short pauses, as intervals between the acts" (Schopenhauer 1851). This pessimism may shock us after 74 years without a world war but during the social time of Schopenhauer, wars and insurrections were indeed very frequent. Schopenhauer believed that man could only progress in terms of behaviour and morals through religion, philosophy, and the arts but, unlike the philosophers of the Enlightenment and Georg Hegel, he did not believe that man's rational nature would necessarily help that process.

Tocqueville is often presented as a defender of an optimistic vision of the future in which the unbreakable confidence in the USA's progress would eventually spread and benefit the whole world. However, in the second volume of his book *On Democracy in America* (Tocqueville 1840), he wonders prophetically about the future of democracy. In Chap. 6 of this volume, which has the intriguing title *What Sort of Despotism Democratic Nations Have to Fear*, Tocqueville argues that democracies run the risk of walking towards mild forms of despotism in which leaders behave like schoolteachers rather than tyrants. In these situations, citizens are so absorbed in providing their own comfort and well-being that they become uninterested in the surrounding society. Above them, there is a protective power that guarantees them security and the opportunity to consume goods and services.

Democratic despotism gives them the illusion that they are free and control their own future, but in fact they have very limited influence over governance. Tocqueville gives us a summary of his concerns when he writes (Tocqueville 1840): "Having thus taken each citizen in turn in its powerful grasp and shaped men to its will, government then extends its embrace to include the whole of society. It covers the whole of social life with a network of petty, complicated rules that are both minute and uniform, through which even men of the greatest originality and the most vigorous temperament cannot force their heads above the crowd. It does not break men's will, but softens, bends, and guides it; it seldom enjoins, but often inhibits, action; it does not destroy anything, but prevents much being born; it is not at all tyrannical, but it hinders, restrains, enervates, stifles, and stultifies so much that in the end each nation is no more than a flock of timid and hardworking animals with the government as its shepherd."

Jacob Burckhardt (1818–1897) was a Swiss historian with a Calvinist education who was greatly attracted by art. He devoted most of his time to studying the history of art and culture, and he was one of the first to describe the time of the Renaissance. He was also interested in his own social time and was highly critical of the developing nationalisms and militarisms in Europe at the time. On 24 July 1889, in a letter to his friend Friedrich von Preen, he wrote that his "mental picture of those terrible simplificateurs who will one day descend upon our old Europe is not an agreeable one". In his view, they would be relentless demagogues, able to build a tyranny armed with the tools of industrial capitalism, science, and technology. Burckhardt was conservative and averse to extremisms, and he did not embrace the teachings of democratic liberalism or share the modern belief in progress. Just like Tocqueville, he had concerns about the evolution of democracy, since he considered that it promoted vulgarity, the devaluation of culture, and the corruption of politics.

His criticism was also directed towards the financial system, when he wrote (Burckhardt 1929): "The state incurs debts for politics, war, and other higher causes and 'progress' […] The assumption is that the future will honour this relationship in perpetuity. The state has learned from the merchants and industrialists how to exploit credit; it defies the nation ever to let it go into bankruptcy. Alongside all swindlers the state now stands there as swindler-in-chief." Burckhardt was also one of the first thinkers to point out nostalgically that, since the 17th century, Western civilization had deeply transformed the forests, rivers, and cities of Europe with modern technology, industrialisation, and economic growth. According to him, the new environment was being gradually dehumanised.

Nietzsche, like Schopenhauer, believed in the possibility of progress strictly in an individual sense, in that individuals can improve their moral behaviour and develop culturally. However, he questioned the possibility of realizing the idea of progress in human society that had been inherited from the Enlightenment. Nietzsche used the doctrine of eternal return (Nietzsche 1883) to refute the linear nature of the idea of progress. He criticised Darwin harshly, especially the statement quoted above in which he acknowledged that species progress towards perfection through natural selection. When looking at *Homo sapiens*, Nietzsche saw the opposite of what he considered to be progress. He expressed this by writing: "What surprises me most when surveying the great destinies of man is always seeing before me the opposite of what Darwin and his school see or want to see today: selection in favour of the stronger, in favour of those who have come off better, the progress of the species. The very opposite is quite palpably the case: the elimination of the strokes of luck, the uselessness of the better-constituted types, the inevitable domination achieved by the average, even below-average types" (Nietzsche 2003). Nietzsche did not believe that science would be able to resolve all the world's problems. The only possible paths to reach such an objective were the will to power, which he believed to be the main driving force in humans, and overcoming the human condition, which at its most extreme leads to the "overman" or "superman" (*Ubermensch*).

On the cultural side, closer to art and aesthetics, several movements emerged at the end of the 19th century following Romanticism: Aestheticism, Symbolism, the Decadent Movement, and Decadentism. These sought to free themselves of the materialistic concerns of industrial society which were becoming more and more dominant and widespread. They argued that art's purpose was art itself and that it should not be used for religious, political, or social ends. They reacted to the philistinism of the industrial age, which was focused on materialistic values and despised art and beauty, intellect, and spiritualty. Charles Baudelaire, in his *Fleurs du Mal* (Baudelaire 1857), which was considered "an outrage against public morality" and "an offence against religious morals", was one of the leading figures of the Decadent Movement. He cultivated sophisticated eroticism and sensuality and an extraordinary imagination for the grotesque and the perverse. Other "Decadents" appeared later, such as Arthur Rimbaud (1854–1891), Stéphane Mallarmé (1842–1898), Jovis-Karl Huysmans (1848–1907), and Oscar Wilde (1854–1900).

The concept of decadence is ancient and appears in some of the first myths about the history and future of humans. Both progress and decadence are relative, complementary concepts with a dialectic relationship over historical time. One of the first metaphorical allusions to the alternation of progress and decadence was made by Bernard de Chartres, who was born in the 11th century and died around 1130. He compared his contemporaries to dwarfs standing on the shoulders of giants and said that this was the only thing that allowed them to see further. The metaphor was likely inspired by the recognition that the scholars of his time had acquired valuable knowledge by reading the classical philosophers and writers of Greece and Rome, whose books were preserved at the cathedral schools. In 1676, Newton used the same idea in a letter to Robert Hooke (1635–1703), who was one of the central figures of experimental science in the 17th century, in which he said "If I have seen further it is by standing on the shoulders of Giants".

The natural cycles of biology suggest an alternation between progress and decadence, but this analogy is a long way from reflecting the complex relationship that has been established between the two concepts throughout history. The concept of progress that emerged in the period of the Enlightenment was continuous and cumulative. These attributes are closely linked to the characteristics of scientific and technological advancements. Nonetheless, from the end of the 19th century onwards, the concept of human society's progress has been criticised by several movements, particularly anti-science and anti-rational ones. Scientific and technological progress is acknowledged, but the perception and experience of the effects of this type of progress in everyday life and lifestyles create a feeling of restriction, anguish, and alienation that some see as a form of decline or human decadence. Society is experiencing various forms of progress and decadence that coexist and were created by different schools of thought and visions of the world. These visions are embraced by different people according to their identities and allegiances to social and political causes. However, the globalized economic and financial system and national governments officially defend an unquestionably optimistic view of sustained progress.

On 1 January 1901, the New York Times published an article saying that: "The World is optimistic enough to believe that the twentieth century [...] will meet and overcome all perils and prove to be the best this steadily improving planet has ever seen" (Colton 1982). The optimism held by many at the start of the 20th century developed into a much more reserved and cautious assessment of the future at the beginning of the 21st century. Only 13 and 38 years after 1901, the First and Second World Wars began, bringing horrors and serious setbacks to civilisation. The Second World War ended with two atomic bombs being dropped on cities in Japan, and for the first time *Homo sapiens* realised that he had more than enough power to provoke the end of *Homo sapiens* time. This was a wake-up call and he has since avoided another world war.

5.3   The Great Acceleration

Nonetheless, the period that followed the last world war generated greater confidence and hope in the future. It was a period characterized by strong economic growth. Poverty was significantly reduced in many parts of the world and several countries, especially the emerging economies, became fully involved in world trade. It was also marked by the emergence of democratic governments in many nations, by an extraordinary expansion in the applications of science and technology in most human activities at the global scale, by a remarkable increase in mobility and flows of communication and information, and by an acceleration in global population growth. The thrust from the strong economic growth and the spread of the benefits of scientific and technological progress across the world translated into an acceleration in operative social time for a large portion of humanity. Time, for the hundreds of millions of people who benefited from this process, began fill up with more activities, new opportunities, new challenges, greater communicability and mobility, and an outstanding access to news and information from everywhere in the world.

This remarkable acceleration in population, the economy, the production and consumption of goods and services, and the use of natural resources meant that the impact of human civilisation on the Earth system at the global, regional, and local scales became conspicuous. It was nicknamed the "Great Acceleration" at the 2005 Dahlem Workshop (see Figs. 5.1, 5.2 and 5.3) (Hibbard et al. 2006). The Earth system is, after all, our planet and our common home, as stressed in Laudato Si', the second encyclical of Pope Francis (Francis 2015).

The part of the system that interests us the most, since it is where we find ourselves most frequently, is the one closest to the surface and includes the five biogeophysical subsystems of the ecosphere that have already been introduced: the lithosphere, hydrosphere, cryosphere, atmosphere, and biosphere. From the standpoint of systems science, the dynamics of the Earth system are characterised by the set of material and energy flows, by the physical, chemical, and biological processes, and by the cycles that interact with each other and occur at various temporal and spatial scales, including the global scale (Schellnhuber et al. 2005). The Earth system, at a certain point in its history, began to provide for the presence and evolution of living organisms and for the environment of the ecosphere that allows them to function on the surface and adjoining regions. Later, human impacts began to interfere intensively and extensively with the ecosphere and particularly with the biosphere. Thus, to analyse global dynamics, it became useful to consider that the Earth system also comprises a human subsystem whose dynamics is mainly determined by social, political, and economic drivers. This organisation of concepts reflects the idea that humans, although they are themselves part of the biosphere, have acquired a capacity for transformation and disruption of the other subsystems. It acknowledges the fact that some parts of the dynamics of the biosphere are currently being disturbed and sometimes dominated by the dynamics created by human interference.

When analysing and modelling the interactions between the human subsystem and subsystems of the ecosphere, the new entities created are often qualified as social-ecological systems (Binder et al. 2013). These systems with a dual nature undergo changes at the global, regional, and local levels.

Changes induced by the human subsystem have only become global since the Age of Discovery, which began in the 15th century, developing further with the Industrial Revolution to become more intense as globalisation began to thrive in the 20th and 21st centuries. Global changes driven by the human subsystem, called global anthropogenic changes, include global population growth, urbanization, global production and consumption of goods and services and changes in the global economy, global energy use, global use of renewable and non-renewable natural resources, and land use changes. The new situation that emerged in the 20th century was that many

Socio-economic trends

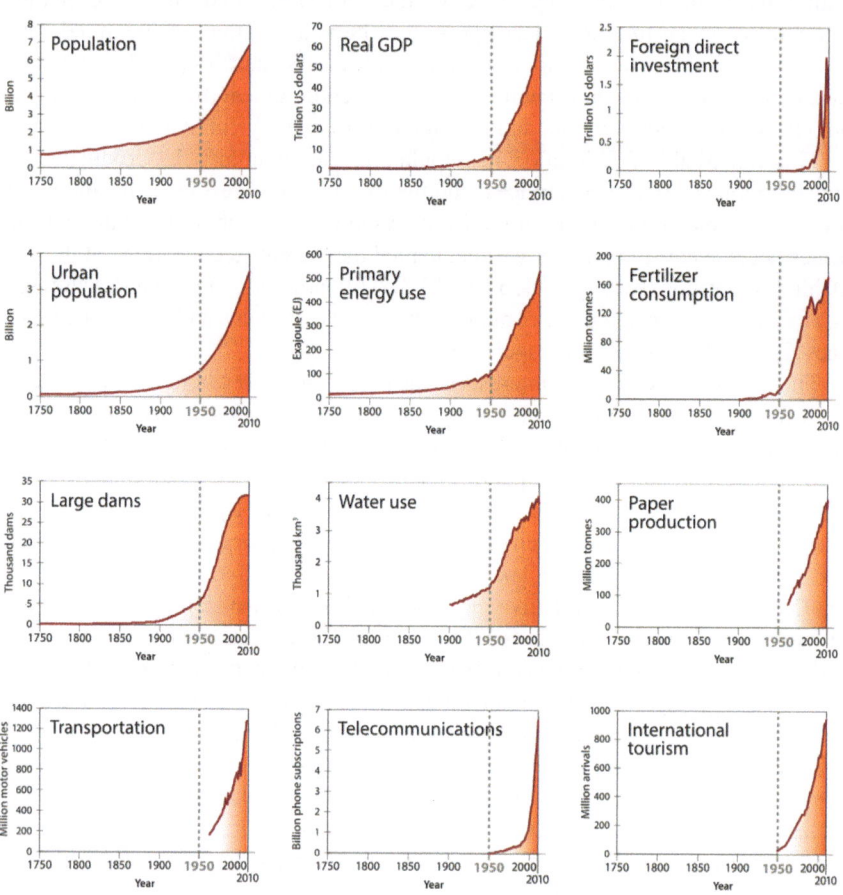

Fig. 5.1 Great Acceleration graphs showing socio-economic trends from 1750 to 2010 (Steffen et al. 2005, 2015). **a** Twelve major socioeconomic indicators, including real GDP, population, foreign direct investment, energy consumption, water use, fertilizer consumption, telecommunications, transportation, and international tourism. The *vertical dashed line* in 1950 indicates what is usually considered to be the beginning of the Great Acceleration. Data compiled by the International Geosphere–Biosphere Programme (IGBP) and by the Stockholm Resilience Centre. Published online at http://www.igbp.net/

human activities began to cause global changes in biogeophysical subsystems (see Fig. 5.3).

Some of these changes are much faster and more intense than the global changes that result from the natural evolution of Earth's biogeophysical subsystems, referred to as global natural changes. The impact of human civilisation on biological, chemi-

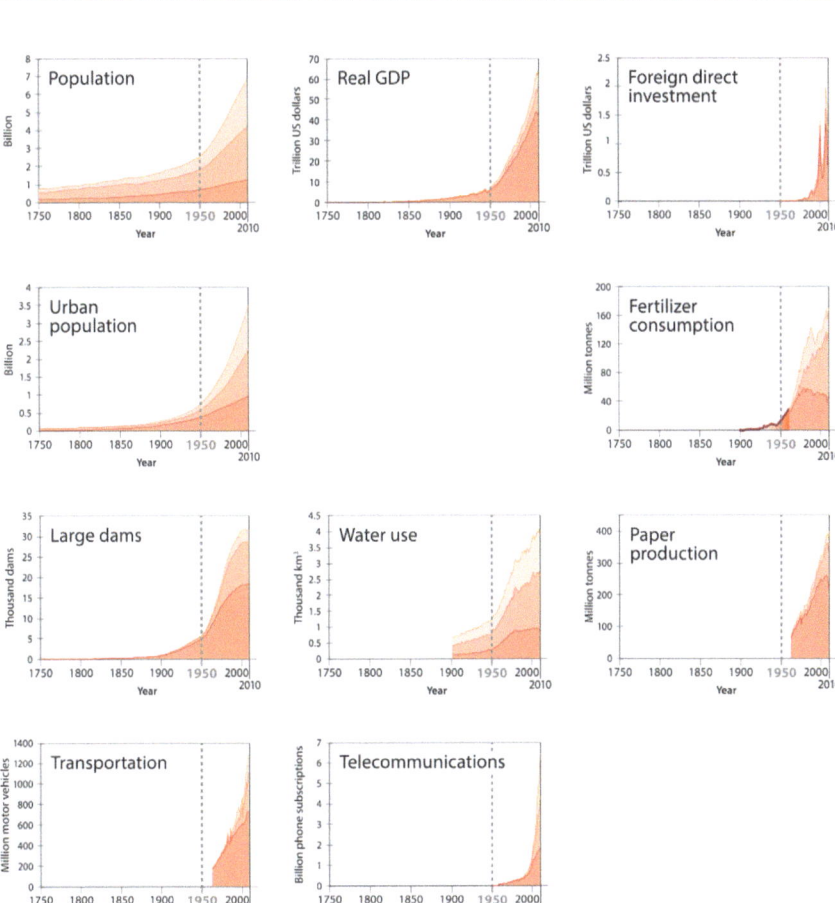

Fig. 5.2 Great Acceleration graphs showing socio-economic trends from 1750 to 2010 (Steffen et al. 2005, 2015). Ten of the previous socioeconomic trends split into the contributions from the relatively wealthy OECD countries, the BRICS nations—Brazil, Russia, India, China, and South Africa—and the other countries. The differentiation reveals the underlying socioeconomic inequalities that are not apparent in the global aggregates. Most of the economic activity and consumption of goods and services take place in the OECD countries, which in 2010 accounted for 74% of the world GDP, but only 18% of the world population. Data compiled by the International Geosphere–Biosphere Programme (IGBP) and by the Stockholm Resilience Centre. Published online at http://www.igbp. net/

cal, physical, and geological processes in the Earth system has become so extensive and intensive that several authors, including Eugene F. Stoermer and Paul Crutzen, believe it necessary to recognise that a new geological epoch has begun. It follows the Holocene, whose name comes from a combination of the Ancient Greek *holos*,

Earth system trends

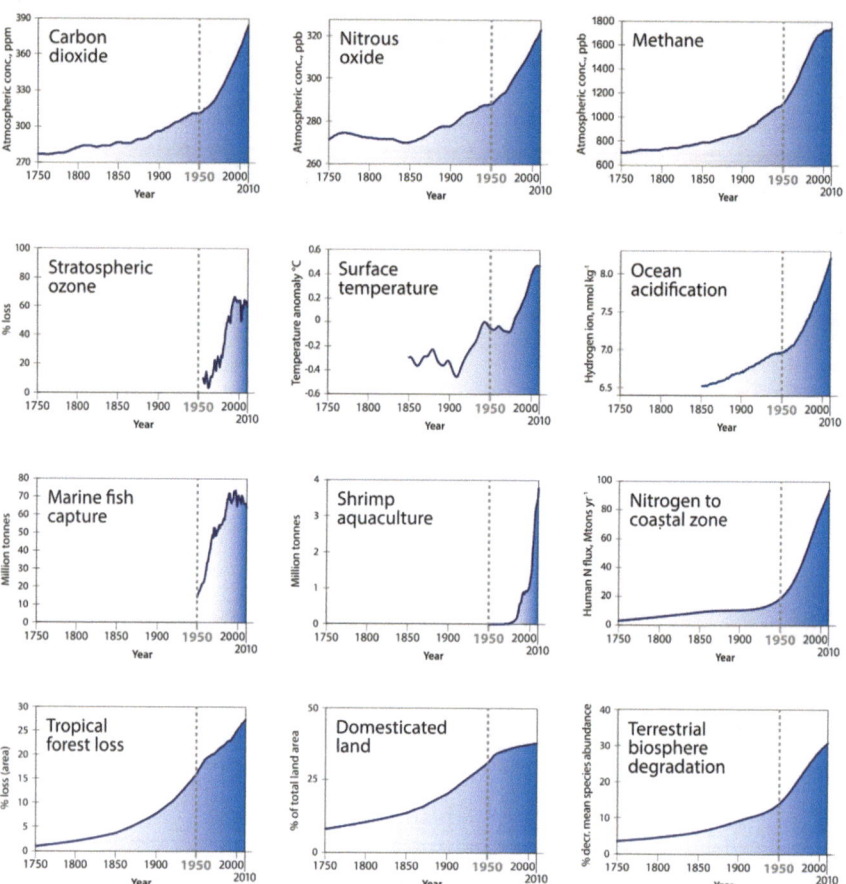

Fig. 5.3 Great Acceleration graphs showing twelve Earth system trends from 1750 to 2010 (Steffen et al. 2005, 2015). Most of the observed global changes are directly linked to the global economic and financial system. The *first row* shows the increase in atmospheric concentration of the main greenhouse gases with large anthropogenic emissions—CO_2, CH_4, and N_2O. The post-1950 acceleration of the Earth system indicators is clear. Data compiled by the International Geosphere–Biosphere programme (IGBP) and by the Stockholm Resilience Centre. Published online at http://www.igbp. net/

meaning "whole" or "wholly", and *kainos*, meaning "recent". The Holocene followed the Pleistocene and began at the end of the last glacial period 11 700 years ago. Stoermer and Crutzen have called the proposed new geological epoch the Anthropocene (Crutzen and Stoermer 2000), formed from the Ancient Greek with *anthropos* meaning "human being" and *kainos* as above.

Global anthropogenic changes include climate change, average global sea level rise and ocean acidification, stratospheric ozone depletion, changes in the carbon, water, nitrogen, and phosphorus cycles, changes in food chains, global freshwater use, biodiversity loss, desertification, and pollution of the atmosphere, oceans, rivers, and soils. All this is happening often rapidly and without control. "Pristine, wild nature" is a thing of the past. That "nature" has been transformed by global anthropogenic changes. Nature has become a changeable and uncertain concept, constantly being transformed by the interference of the human subsystem. Humans are now in a position to decide what nature is going to be in the future in order to better satisfy their interests.

While in the 20th century man tried to test the limits of aggression against himself, in the 21st century, he is trying to test the limits of aggression against the Earth system to which he belongs and upon which he depends.

5.4   Contemporary Views of Progress

Where does the idea of progress stand in relation to sustainability and environmental sustainability? This question is particularly interesting because it focuses on progress, one of the key ideas of Western civilization, whose influence is found all over the world. The answer would implicitly characterise the Western view of the current state of the world. However, instead of an answer we find an immensely complex, diverse range of answers that have no clear resultant direction and include openly contradictory messages. Complexity also emerges in terms of semantics.

It has already been noted that the original meaning of progress is associated with the Latin root of moving forward towards a goal anchored in space. Nowadays, the word progress is used to describe movements, developments, or processes which may have a positive or negative human connotation. Thus it generally describes a transformation, change, or process that follows a discernible evolutionary rationale. It may be said that there is progress in a human illness, in a political crisis, or in the environmental deterioration of a country, region, or the world. There is a tendency to use the word "progress" in a negative context when dealing with a process that is considered inevitable, which reveals a deep, underlying meaning of progress as something guaranteed deterministically over a long period of time, in spite of interruptions and temporary setbacks.

The use of the word "progress" in a context involving human values and activities has been extended to a wide variety of domains. Important topics for discussion are social, economic, political, ethical, and moral progress, as well as the progress of science and the progress of technology (Almond et al. 1981). Further to this diversity of specific applications, there is also an older idea of human progress applied generally for humanity, or for the population of a certain region or country, which has its roots in the Enlightenment.

The recognition of advances in each of these forms of progress is a long way from being consensual. For each type of progress there are people who argue that there is progress and others that deny it. The debate is important because it reveals the horizon of expectation and shows how segmented and diverse that horizon is, with expectations that are often contradictory. The belief in or approval of progress in its various forms and applications is usually called progressionism, while progressivism is a specific political theory that advocates social reform, a theory that had a significant influence in the government of the USA in the two first decades of the 20th century.

In the analysis of contemporary views on progress, let us begin with human progress. The idea of a sustained human progress, which originated in Western civilization, is far from being shared throughout the world, in part because it assumes a linear concept of time different from the cyclical time of other civilisations. Nonetheless, although the idea of progress varies considerably, there is a remarkable convergence about how to characterize human progress in terms of indicators of well-being, health, and economic prosperity. It is agreed that a country progresses if poverty is reduced, economic prosperity increases, social and economic inequalities are reduced, life expectancy increases, infant mortality and morbidity decreases, education, training, and development levels go up, and employment opportunities increase. Progress also depends crucially on freedom and justice, on respecting, protecting, and fulfilling human rights and on a less degraded and polluted environment. When making quantitative assessments of these features, one finds larges differences between countries with advanced economies, which can be defined as those belonging to the OECD, and the rest of the world. In other words, human progress in this general form has very different expressions depending on the country's state of development, although they all tend to coincide in their standards for long-term expectations.

What can be expected for the future evolution of human progress? The answer to this question is complex because we are dealing with two large groups of countries in very different situations and a wide variety of different dominions where the concept of progress can be applied. One should consider how progress is advancing in the social, economic, political, scientific, technological, and environmental domains. A brief sector-by-sector analysis is presented in the next section.

However, before that analysis, it is relevant to recognize that the rather consensual contemporary approach to assessing human progress through indicators of well-being, health, and economic prosperity, as briefly described, is far from the approaches suggested by Kant, Schopenhauer, Nietzsche, and other thinkers of the 18th to 20th centuries. They emphasized that human progress depends primarily on the development of the human faculties, in particular on the personal development of higher standards of moral behaviour and of cultural interests. Implicitly it is acknowledged that human progress depends on a voluntary individual attitude. The current materialistic narrative devalues individual responsibility and reduces human progress to advances in the public domain and to sets of quantifiable indicators. For example, who would dare to insist that human progress depends on people assuming the responsibility to repudiate corruption in politics, economics, and finance? Have we lost any hope that the "human race as a whole, over time, will develop all the human faculties" (Kant 1784)? If one accepts that human progress depends primarily on a personal effort to improve moral and cultural behaviour, then one should question how to make that goal practicable in a global economic system based on optimization of individual self-interest, or in other words on rational egoism. In the long term the role of personal moral self-development in a transhumanism based on technology becomes an interesting subject of discussion.

5.5 Social and Economic Progress

Social Progress Imperative, an organisation that was founded in the USA in 2012, has created the concept of a Social Progress Index, which includes roughly 50 indicators grouped into three categories: basic human needs, well-being, and opportunities. The index is now used in several regions and countries of the world, including the EU, where it is currently being tested. The indicators have already been applied to 150 countries, and the aim is to use the index regularly as a complement to GDP. As would be expected, applications of the Social Progress Index reveal big differences between OECD countries and the others, especially the least developed nations.

As for the economy its progress is currently measured using GDP. In the last 50 years, the world's real GDP has grown continuously, except in 2009 due to the West's financial and economic crisis. The growth rate was far from uniform across the globe and there was one outstanding success. From 1978 to 2010, with economic reforms introduced by Deng Xiaoping (1904–1997), China's GDP grew vigorously, without significant periods of recession, and with an average real annual growth rate close to 10% (CEIC 2016). This was a unique event in the history of humanity, taking into account the huge number of people that benefited from it. In contrast, the USA had 11 business cycles between 1945 and 2007, each with a growth period lasting an average of 59 months, alternating with shorter periods of recession lasting about 11 months (NBER 2018). The average annual GDP growth figures for those periods was much lower than 10% and reached a high of 5.6% between October 1945 and November 1948.

The "Chinese economic growth miracle" once again placed the country in the position it held from the year 1 AD until the 19th century, when its GDP accounted, on average, for 25% of the world's GDP (CEIC 2016). In 1820, it reached the maximum level of 33% of the world's GDP, and from then on it systematically fell, reaching its lowest point of 4.6% in 1950. During the period of the economic miracle, it grew from 5.2% in 1980 to 17.5% in 2008 (CEIC 2016). Despite the financial and economic crisis of 2008–2009, China continued to assure robust economic growth, with an average growth rate between 2008 and 2015 of around 8% (CEIC 2016). China's continued economic growth is seen with mistrust and as a potential threat to the West, especially in the USA.

In spite of all the good news, there are also some surprising aspects of GDP growth. As already mentioned, real world GDP is still growing, but the average of the annual percentage growth reached a maximum of 3.8% in the period 1940–1970 and then started to decrease systematically, reaching 2.7% in 2000–2014. The deceleration of real GDP growth has been more pronounced in OECD countries, especially since the 2008–2009 crisis. The average growth of real GDP in the USA in the last 20 years has been close to 2% and close to 1% in the EU and Japan. The paradox of the global slowdown of real GDP growth is a vindication of Solow's Paradox, given that science and technological innovations have been flourishing since the end of the Second World War up to the present. Most probably the reason for the continuing paradox is the negative effect of overexploitation of natural resources, and the increasing effects of environmental degradation and anthropogenic climate change, on the global economy.

Among the various reasons presented to explain why the paradox of the slowdown in real GDP growth is more striking in OECD countries, one of the most frequently mentioned is the ageing population, which makes it more difficult to revive the economy using monetary stimuli. Another likely cause is the increase in total debt, including public debt, household debt, and company debt, which is in part the result of those stimuli and gradually becomes unsustainable if economic growth is weak or non-existent.

According to the Bank of International Settlements, in the third quarter of 2015, the total debts of the USA, the EU, and Japan were 248%, 269%, and 387% of their respective GDPs. Meanwhile, China's total debt at current prices has grown around 160% since 2008, reaching 248.6% of GDP in the same quarter of 2015. Although monetary stimuli have been unable to generate robust economic growth in Western countries and Japan, they have had more success in China. Such a large accumulation of debt in these four regions and in many other countries, dangerously increases the risk of creating new financial bubbles that could lead to serious global economic crises.

Total global public and private debt has systematically grown in the last decades and, according to the International Monetary Fund, reached $164 trillion in 2016, which represents 225% of global GDP. In the capitalist economic system, an absence of growth is devastating because it leads to a vicious circle in which weak growth or economic contraction forces companies to reduce their activity to be able to maintain their profits, and this in turn tends to cause an economic slowdown and widespread unemployment.

The future of economic progress, understood specifically as growth in GDP, is undergoing a phase of some uncertainty. If economic growth becomes harder to achieve, or even impossible, in countries with advanced economies and gradually all over the world, this may imply the beginning of a difficult transition to a new economic and political paradigm. According to Piketty (2013) and other economists, robust economic growth tends to be an incidental event in economic history, so its continuation in the future is not guaranteed. Despite its accidental nature, it is viewed politically around the world as an indispensable objective for development that we must constantly strive to achieve. In Europe, high annual growth in GDP, of the order of 5–10%, was only achieved in *Les Trente Glorieuses* from 1946 to 1975 (Fourastié 1979), during the period of reconstruction after the Second World War. A similar phenomenon occurred in some emerging economies, starting in the 1970s, accelerating the process of economic convergence with the economies of advanced countries; this was especially remarkable in China during the period of the "economic growth miracle".

Unlike the variability of the rate of economic growth, the rate of return on capital has been stable, generally between 4 and 5%. This difference, in Thomas Piketty's view, is increasing the accumulation of wealth in the hands of one very small section of society, thereby causing increasing inequality and social and economic instability (Piketty 2013). Both income from capital and inequality in wealth distribution have been increasing since the 1980s. Since 1960, roughly 60% of economic growth has gone to the world's richest 1% (Piketty 2013). This concentration of wealth undermines the meritocracy that supports democratic societies and produces a political system that serves the interests of the few who concentrate that wealth and ignore the interests of the majority. There are signs that the world is again moving into a new form of "patrimonial capitalism" of the kind that flourished during the *Belle époque* in France between the Franco-Prussian War in 1870 and the First World War in 1914.

The US economy has been the main driving force of the world economy since the 20th century. Furthermore the USA has defended the Western world's main principles and values and, with its enormous military power, has played a controlling role in world geopolitics. Average growth in real GDP per capita in the USA between 1948 and 2000 was 2.3%, but it fell to an average of 1% for the period between 2000 and 2016 (Eberstadt 2017).

Since the end of the 20th century, the US labour market has gone through significant transformations. The number of civilian workers aged between 25 and 55 who are not registered as unemployed but are not working or looking for employment has increased substantially. Currently, there are three in this situation for every unemployed worker. The number of adult civilian workers in the economy is currently the lowest in the last 30 years. According to the Bureau of Labor Statistics, the percentage of the USA population over 20 years of age that had civilian jobs fell between 2000 and 2016 from 64.5 to 59.7%, the fastest reduction since the end of the Second World War. This decline in the labour force affects women as much as men and has dramatic social consequences.

At the same time, wealth creation in the USA has been growing rapidly in the 21st century. Between 2000 and 2016, it is estimated that American families and non-profit organisations have increased their wealth in financial assets and all other kinds of capital assets from 44 to 90 trillion dollars. This spectacular increase took place despite the 2008–2009 financial crisis. The growth in wealth was accompanied by a pre-eminent increase in inequality, as already mentioned.

However, what concerns the middle and lower classes is above all its current economic insecurity and not the increase in inequality. This latter aspect is viewed as an abstract feature of an immovable economic system that apparently has no direct impact on people's everyday life. For them inequality is the concern of sterile intellectual elites moved by covert envy.

In short the American economy has managed to produce a growing number of capital goods, but it has not been able to create growing work opportunities. The working middle classes of the USA feel like victims of this imbalance and are beginning to feel unsure about the future. This in turn fuels a lack of belief in government institutions and politicians. The promise of economic growth that dominated the culture of American society in the 20th century, particularly its white elites, is not fully meeting expectations at the beginning of the 21st century.

There are several signs that the lack of robust economic progress is slowing down social progress. A recent study (Case and Deaton 2015) demonstrates that mortality and morbidity among non-Hispanic white middle-aged people grew between 1999 and 2013 and shows suicide and drug and alcohol abuse as the most likely causes. Data from the Centers for Disease Control and Prevention (CDC/NCHS) show that the mortality rate due to excessive drug use grew between 1999 and 2014 for all population groups—non-Hispanic whites, Hispanic whites, and non-Hispanic blacks. The highest rate affects non-Hispanic whites and grew over that period from 6 to 18 per 100000. There are other indicators of social decline, such as stagnation or a slight reduction in life expectancy, reduction in health conditions, and less social mobility, which is one of the brand images of the American socio-economic model (Eberstadt 2017).

North Dakota, the state with the lowest mortality rate for excessive drug use, was the state that had the greatest economic growth across the whole of the USA in the period 1999–2014, thanks to the economic returns from a surge in oil and natural gas production using hydraulic fracturing or fracking, particularly in the tight oil deposits of Bakken Shale in the north-western part of the state. In 2012, North Dakota's GDP grew by 13.4%, around five times the national average, which led the state governor, Jack Dalrymple, to tell CNN (Hargreaves 2013) in June 2013 that "There's nothing like an oil boom to get things rolling".

The correlation between economic growth, social decline, and the evolution of fossil fuel exploitation can also be seen in the mortality epidemic caused by drug use that began in the Appalachian Mountains, which cover a large part of West Virginia and Kentucky, a region whose past wealth came from farming and the exploitation of coal mines and forests. In West Virginia, where there was a marked decline in jobs in the coal mines, mortality caused by drug use grew by a factor of eight between 1999 and 2014, from 4 to more than 35 per 100000 (CDC 2019). The surge in unemployment is in part the result of the automation of mining operations, competition with lower sulphur coal mines in the west, particularly those of the Powder River Basin in Wyoming, and coal price volatility.

The fact that the largest economy in the world is unable to maintain the economic growth it had in the three decades after the Second World War is a worrying sign, not only for countries with advanced economies but also for emerging economies and for the whole world. The reasons behind this situation are debated extensively in various forums and by many experts. Of the selection of reasons already mentioned, those cited most frequently are slow recovery from the 2008–2009 crisis, high public and private debt, companies' lower propensity to invest, and structural problems in the economy, such as the ageing population, the increase in inequality, and the greater difficulty in generating economic growth from technological progress. According to some analysts, these factors as a whole could help to create a vicious circle of secular stagnation.

Many believe that there is a political answer to secular stagnation that can get the US economy back to its robust growth of the last century. The current administration under President Donald Trump is trying to develop that policy based on growth incentives such as lowering taxes for the wealthiest Americans and big corporations, financial deregulation, environmental deregulation, and the introduction of protectionist measures to reduce its bilateral or multilateral trade deficits with its trading partners. To boost economic growth, the US government is also creating support and incentives for the exploitation and consumption of fossil fuels, especially coal, and cancelling previous policies for reducing greenhouse gas emissions, particularly those of CO_2.

Probably the most dangerous aspect of the current US policy is the assault on the world trade system through the imposition of import tariffs on China and other countries. Such a policy reveals both the awareness of a decline and the determination to invert it, albeit in an erroneous and counterproductive way. The main problem is the current account deficit resulting from savings falling short of the required investment. Although tariffs can change the composition of trade flows, they only have a weak and indirect effect on the savings–investment balance. The national security rationale for starting trade wars is steering the USA away from the path of statecraft that it enjoyed previously.

Since the end of the Second World War, global trade has grown 50% faster than GDP, mainly due to several rounds of liberalization at the WTO. Increasing world trade has been essential to global development and it has doubtless enriched developing countries, especially those that complement their trade policies with higher investment for education, capacity building, and the construction of infrastructures.

The signs in the USA after more than two years of these policies are apparently encouraging: the US stock market indices are increasing, the economy is seeing more robust growth, the deficit is rising, inflation is rising moderately, and the Federal Bank considers that the economy is improving enough to allow it to increase interest rates. The issue of how long the current economic growth will last and to what extent it is sustainable or, on the other hand, whether it will increase the risk of a serious economic and financial crisis, is to a large extent considered to be an irrelevant question at the moment. Right now, all imaginable methods must be used to encourage economic growth and applaud its immediate benefits. A vast majority of people have unwavering faith that this is the safe path to a future that will be able to meet the growth expectations that have been created and have so far always been met. The fact that this path increases the risk of social, economic, and environmental crises in the USA, in several regions of the world and even at global level, over medium and longer time frames, is believed to be practically irrelevant, since it is only an uncertain supposition. If these things should ever become a reality, others will solve the problems using the inexhaustible human ingenuity. This excessive focus on short-term results at the expense of long-term interests is a form of extreme short-termism.

On the side of long-termism several authors have been arguing since the 1970s (Meadows et al. 1972, 2004) that economic growth as it was understood up to the beginning of the 21st century has ceased to be a viable goal for several reasons that operate in the medium and long term and can be put down to limited natural resources and negative externalities in the biogeophysical subsystems of the Earth system. More recently, Richard Heinberg (2011) has alleged that robust, sustained economic growth has become impossible because the expansionist course of our industrial, technological, and globalising civilisation is colliding with insurmountable natural limits. Specifically, he believes that economic growth is being slowed down by the depletion of natural resources, the negative impacts of growing pollution and environmental degradation, and also by the dependence of the world's economies on rising levels of debt.

Such long-term concerns are currently gaining support from the frequent publication of scientific papers and reports that emphasize the urgency to address environmental problems, the overexploitation of natural resources, and climate change. One recent example is the report published by the Institute for Public Policy Research, a UK centre-left think tank, entitled *This is a crisis. Facing up to the age of environmental breakdown* (Laybourn-Langton et al. 2019), based on a meta-study of academic papers, government documents, and reports prepared by non-governmental organizations. The authors argue that politicians and policymakers have failed to grasp the gravity of the environmental crisis facing the Earth. According to the views expressed in the report, human impacts have reached a critical stage and threaten to destabilise societies and the global economy, making it impossible to achieve sustained social and economic progress.

5.6   Political Progress

Political progress can be understood as a movement, in periods of one or more successive social generations, towards ideal systems of government characterised by the satisfaction of various types of values, beliefs, standards, and expectations that individuals hold and that can be organized into four groups. Political progress is assessed by the degree to which people consider that their government succeeds in fully respecting and satisfying those values, beliefs, standards, and expectations. The main set of values that a given society chooses to judge government performance depends on the country's state of socioeconomic development and the dominant religion and culture, and it evolves with time.

The first group of values can be referred to as freedom and human rights, and it also includes justice, equality, cooperation, participation, accountability, compliance, order, security, peace, stability, and other values of the same nature. The second group is the economic group and includes as values and expectations salary increases, tax reductions, employment opportunities, economic growth, and higher purchasing power. This group of values has acquired overwhelming significance due to the increasing importance attributed to economic prosperity and consumerism since the beginning of the Great Acceleration. Furthermore, in their assessment of government actions and policies people include the protection, preservation, and reinforcement of their national identity and self-esteem, the control of immigration to ensure the preservation of national identity, customs, the way of life, religious and cultural balance, security, and job security specifically related to immigration. The new tendency to stress the importance of this third group of identity values in advanced economies is in part the result of immigration pressure from fragile states affected by terrorism, civil wars, armed violence, droughts, and environmental degradation. Finally, there is a fourth group that will be referred to as environment and sustainability. This includes the preoccupation with environmental degradation, overexploitation of natural resources, climate change, and other sustainability issues.

Society includes groups of people with different political views that attribute different degrees of importance to each of the four groups of values. For instance one group gives an overwhelming importance to the economy and identity groups of values while another group does the same for the freedom and rights and the environment and sustainability groups of values. This characterisation does not mean that the first group does not recognize the importance of the freedom and human rights group or the latter the importance of the economy and identity groups. It a question of greater or lesser emphasis. The more different groups of people stress their fidelity and unwavering commitment to defend their preferred group of values, the stronger becomes the political polarization in society.

Democracy is arguably the system of government that best guarantees the fulfilment of the freedom and human rights group of values, far better than oligarchies and autocracies. However, democracy does not specialize in fulfilling the economic or the environmental and sustainability groups of values. The importance that a society attributes to each of the four groups of values and expectations, then uses to judge government policies, can change in such a way that democracy is forced to lose its essential characteristics. At the end of the 20th century the West cherished the idea that liberal parliamentary democracy, born in Great Britain at the start of the 18th century and later adopted in most countries with advanced economies, is the final stage of progress in systems of government. Francis Fukuyama, in his book *The End of History and the Last Man* (Fukuyama 1992), writes that the fall of the Berlin Wall in 1989, followed by the collapse of the Soviet Union in 1991, was "the end point of mankind's ideological evolution and the universalization of Western liberal democracy as the final form of human government". This statement reveals the ethnocentrism of the West, incorrectly convinced that the final battle was between liberal democracy and the Soviet version of communism. For Fukuyama, communism collapsed in Russia in 1991 because it was unable to provide economic prosperity in the medium and long term, and he implicitly assumes that democracy is immune to such a problem. The West believes erroneously that the applicability of the ideal of democracy is timeless and that it can be put into practice everywhere in the world and under all societal, economic, and environmental circumstances.

John Rawls has a broader, more conceptual, and less simplistic view of history based on the goal of a "well-ordered society" of free citizens that understand, approve, and act in accordance with the principles of justice and fairness within an egalitarian economic system. His theory of political liberalism legitimates the use of political power in a democracy, allows for freedom of religion, makes it possible to establish a shared view of morality and fairness, and hopefully leads to a permanently peaceful and tolerant international order.

Democracy requires relatively high average education levels and a generous availability of time for information exchange, knowledge, and reflection, and for the development of a culture of participation, dialogue, and negotiation. Experience has shown that the chaotic tendencies of a democracy can be better suppressed with a well-educated citizenry. Since these characteristics imply a significant dependence on time, the acceleration of operative social time tends to have negative effects for democracy. The Digital Revolution and the universal and continuous use of social media facilitates political propaganda, mobilisation, and organization of protest movements. It also produces never-ending opportunities to advertise, influence, and share political opinions all the time. There is an emerging need to continuously collect, monitor, analyse, summarize, and visualize politically relevant information from social media (Stieglitz and Dang-Xuan 2013). This acceleration in the pace of political activities and news favours emotional over rational responses in the decision-making process.

One example of this phenomenon was the extraordinary increase in Google searches in the UK to find out the consequences of Brexit the day after the referendum vote. Many people had formed their opinion through mechanisms driven by series of messages on Twitter, Facebook, or other social media platforms with content that reflected a fragmented, incomplete, partial, or even false view of the facts. Political leaders are subject to the same process of acceleration in time, which provides them with enormous quantities of information of very different and varying quality but leaves them with little time available to analyse the data, reflect, negotiate, and deliberate.

The systematic use of online political communication forces a simplification of the ideas expressed in social media messages, usually with no more than a few hundred characters, which contrasts with the growing complexity of the problems that governance has to face nowadays. We are confronted with a modern and extreme version of the "terrible simplification" that Burckhardt predicted and so much disliked.

On the scale of several successive social generations, time also has the effect of modulating the value assigned to democracy. In Europe, most members of the Baby Boomers and Millennial social generations do not remember and do not know what it was like to live under the totalitarian regimes of the 20th century. Only some of the youngest of those generations dwell enough on history to be able to appreciate the value of democracy and recognise the need to nurture it.

To assess political progress it is useful to review the recent evolution of representative democracies and other government systems throughout the world. Democracy, literally "government of the people", is a superior form of governance in that it supports the exercise of political freedom, human rights, and participation and co-responsibility at individual level in defending collective interests. Naturally, the practical applications of the model are far from perfect. Democracy has taken many different forms throughout history and has recently spread through the world in a series of steps forward and steps back. Huntington (1991) has identified "long waves" of democratisation in the 19th and 20th centuries. The first wave saw the formation of 29 democracies, but, with the emergence of totalitarian regimes in Europe, the number of democracies fell to 12 in 1942. The second wave started at the end of the Second World War and reached a peak of 36 democracies in 1962. Afterwards, there was a decrease that lowered the number to 30, but in 1974 the third and most powerful wave surged with the Carnation Revolution in Portugal, followed by Greece and Spain. The new wave then extended to several countries in South and Central America, South-East Asia, Eastern Europe after 1989, and finally Sub-Saharan Africa, raising the total number of democracies in the year 2000 to 120, according to the Freedom House classification criteria.

In the 21st century, there has been a new decline. Some democracies have become, in practice, disguised autocracies, because they do not fully respect political freedom and democratic rights and processes, as in the case of the Russian Federation. Flagrant dictatorships that openly adopt fascism, communism, or military rule have gone out of fashion and almost disappeared from the world. With new standards of political correctness resulting from an increasingly globalized world that values and promotes global communications and visibility, the new political and military leaders have acquired a sense of decorum and respectability. These leaders definitely prefer to give the appearance of a democratic regime that holds regular elections. Furthermore, democratic breakdowns tend to result more from the decisions and behaviour of the elected governments than from military coups and violent seizures of power. Since the beginning of the 1990s the majority of democratic failures have resulted from the subversion of democratic institutions carried out by elected political leaders, as in Georgia, Hungary, Nicaragua, Peru, the Philippines, Poland, Russia, Sri Lanka, Turkey, and Ukraine (Levitsky and Ziblatt 2018).

Many democratic governments feel the need to find ways to subvert democracy in order to conserve power. The challenge then is to make these new ways look like legal procedures, to make them acceptable by the courts and to portray them to voters as improvements in the way democracy functions. Courts are often manipulated to serve the interests of the government and people who oppose the regime are persecuted, for instance by facing employment, fiscal, and legal harassment. Newspapers who dare to be openly critical are bought off or forced to engage in self-censorship.

It is increasingly difficult for non-governmental organizations and human rights defenders around the world to denounce government laws and practices that work against freedom, human rights, and corruption. A recent report by Amnesty International (Amnesty International 2019) indicates that "groups working to promote or defend human rights are smeared, stigmatized, put under surveillance, harassed, threatened, prosecuted on spurious charges, arbitrarily detained, and physically attacked; some human rights defenders are even killed or forcibly disappeared simply for the work they do".

In the West, there are several countries where democracy is declining, mostly due to the electorate's reduced confidence in politicians and public institutions, more frequent obstructions to the normal functioning of government institutions, tribalism, and political apathy. Another sign of deterioration is the emergence of parties that defend extreme nationalism or ideologies that have an openly undemocratic past, and of political leaders who systematically use demagogy to win votes. They manipulate more naive, gullible, or ignorant voters' discontent to endorse policies that have attractive outcomes but are impossible to implement. One of the predominant concerns in contemporary societies is to seek effectiveness, increase the ability to focus and achieve goals, and ensure the efficiency of procedures. In the field of democracy, the application of this trend disproportionately favours the capture of votes in elections, using all legitimate and illegitimate means, which downplay the ethical, moral, social, and economic values of democracy.

This weakening of democracy can also be seen in public opinion. A recent study (Foa and Mounk 2016) in the USA, showed that 72% of respondents from the generation preceding the Second World War gave the highest score possible, 10, in response to the question of whether or not they thought it was essential to live in a democracy, whereas of those in the Millennial generation only 30% gave the same score. In Europe, there is a similar trend, although the difference between the two generations is less pronounced. The study also reveals that, on average, interest in politics has fallen among younger generations, although this does not mean that active membership of political parties is necessarily lower among a small part of the younger population.

Several causes can be identified for the decline of democracy in many parts of the world. A major driver of decline is the economy. Economic stagnation and economic crises tend to polarise society, maximising opposition between groups that have different political, social, and economic identities. Extreme polarization increases the risk of the democratic institutions not working properly and weakens democracy. The financial and economic crisis of 2008–2009 caused a global recession and continues to have negative effects in many countries because recovery has been very slow. It unveiled deep political, financial, and economic weaknesses in the West's most advanced democracies. The bailout of banks and other financial institutions carried out by governments using taxpayers' tax money, while the elite that controls them continued to increase their privileges and wealth, has reduced voters' trust in the economic system that supports democracy. In the West, especially in the USA, the government has shown little interest in creating the conditions needed to avoid future crises similar to the one that started in 2008. On the contrary, these crises are seen simply as a low price that has to be payed for an economic system that provides business cycles whose growth periods significantly increase economic prosperity and consumerism.

One of the most worrying problems for democracies over the last four decades is the decoupling between productivity growth and employment growth in countries with advanced economies. Work salaries no longer have the role of redistributing wealth that they had since the mid-18th century. With technological advances, the labour market has become polarised by favouring employment opportunities in positions for highly qualified people. This polarisation creates job insecurity and a stagnation or reduction in average salaries, which impairs the social contract upon which the West's liberal democracies have been based. This polarisation phenomenon is similar in companies. The increase in productivity is mostly concentrated on a relatively small group of technological companies that are at the forefront of developments in ICT and in the new technologies of the Fourth Industrial Revolution. The convergence of profit into such companies does not benefit the whole of the business landscape and contributes to making wealth inequalities worse. Current trends tend to create economic insecurity for the majority of the population, limiting their horizons of expectation under growing inequalities, and thereby encouraging political radicalism.

Economic and financial elites hold an increasing part of the country's political power in advanced economies. Through their economic power and influence on frequently corrupt politicians, they defend their own corporate and personal interests rather than voters' interests. The extraordinary power of the economic elites in the USA was recently analysed by Gilens and Page (2014) from the point of view of different political theories. Their best description of the contemporary situation is what they call "economic-elite domination" and "biased pluralism" rather than "majoritarian electoral democracy" or "majoritarian pluralism". In the USA and in other countries, there are unmistakeable signs that the social and economic insecurity among the lower and middle classes is in part the result of an offensive driven by political forces, but guided and financed by multinational corporations controlled by the wealthiest 1% of the population (Lafer 2017). Corporate lobbying and financing of political activities has become practically unlimited. Lobbying has been able to get legislation passed that keeps wages low, reduces sick leave, diminishes the power to protest against working conditions, and reduces funding for public services.

The corresponding reduction in the independent power of governments devalues democracy, until in the most extreme cases it becomes practically irrelevant. Awareness of this irrelevance stimulates the emergence of certain political movements, sometimes called populist movements, which challenge mainstream political views, the established order, and elite narratives, ending up polarising opinions and making them even more extreme.

The second kind of effect weakening the democratic movement throughout the world can be found outside the group of countries with advanced economies. The main example is China which, although not a representative democracy, has become a world-renowned success due to the "economic growth miracle". This success has encouraged the Communist Party of China to discreetly defend its governance model, known as the China Model, in which one can find some ideas from the doctrines of Confucius (551–479 BC). The essential elements of Confucianism are the unity of China, represented and led by the emperor, respect for the sovereign and for social hierarchy, the promotion of education and study, which leads to virtue and maintains order, and togetherness and harmony in the family. The current governance model aims to be a meritocracy in which political leaders are renewed every decade and there is a considerable effort to recruit new people to the highest positions based on merit, talent, experience, and motivation. However, there are signs that this may be changing. For instance, the National People's Congress recently passed a resolution that allows Chairman Xi Jinping to stay in power indefinitely.

The meritocracy system has its origins in the imperial examinations, a highly demanding and thorough selection process for mandarins, the Western name for the state's high-ranking civil servants, trained and educated according to the principles of Confucianism. High-ranking officials came from different regions in China, which had their own languages. To understand each other when working together in

Government institutions, they talked in a language from the north of China known in Chinese as *Guānhuà* or "language of the officials". Jesuit missionaries called it "mandarin", a word derived from the Portuguese word *mandarim* for "minister and councillor", which originated from the Sanskrit word *mantrin* for "commander". At the beginning of the 20th century, the dialect of mandarin spoken in the urban area of Beijing was adopted as the official language of China, and called standard Chinese. The institutionalisation of the principle of meritocracy dates from the Han dynasty (206 BC–220 AD) and the imperial examinations lasted 1300 years, from the year 605 until they were abolished in 1905, shortly before the end of the Qing dynasty.

The Chinese governance system has the advantage of avoiding the institutional obstructions and slow decision-making processes that frequently affect representative democracies. However, it has the disadvantage of curbing political freedom, controlling and restricting public opinion, repressing and imprisoning dissidents, and censoring publications and postings on social media. Although perverse, this repression allows the authorities to see the substantive reasons for protests, giving them the chance to eliminate them. As well as the economic growth miracle, the Chinese government has managed to reform the state and provide it with a legislative framework and functional institutions within a relatively short period of time compared to what would very likely be needed in a democracy.

The vast majority of Chinese people are comfortable with their political system, primarily because it has provided robust economic growth and therefore an increase in their economic prosperity, quality of life, and well-being. The Communist Party of China conveniently recovered Confucianism and now, when faced with Western criticisms, claims that China is not a dictatorship or an autocracy, but a Confucian regime.

According to a 2013 study by the Pew Research Center (PRC 2013) that was carried out in several countries, the question "How satisfied are you with the country's direction?" had the highest proportion of positive answers in China, at 83%. In Russia, positive answers represented 37%, while in the democracies of Germany, the USA, Great Britain, and France, those satisfied totalled 57%, 31%, 26%, and 19%, respectively.

Western countries, particularly the USA, tend to be very critical of the circumstances and future of the economic growth miracle, but what is important in geopolitical terms is that it has returned China to a position it had lost within the concert of nations. China is now on a path that may lead it once again to become the largest economy in the world. It was a singular process that demanded enormous sacrifices by the population. It is estimated that there were 287 million migrant rural workers in China in 2017 (CLB 2019), with about 60% coming from large distances. This is probably the largest migration in human history.

The Chinese are mostly proud of the remarkable transition they have made, and this is not down to democracy. It would not have been possible to achieve such a broad, large-scale process without a strong and pervasive empathy and understanding between the population and the governing elites. It is important to remember that the miracle of Chinese economic growth took place with labour and environmental protection laws that are inadequate compared to countries with advanced economies.

Would it have been possible to achieve the same levels of economic progress without those serious shortcomings? It is impossible to say. For most of the population it was the right path to follow because it returned China to the historical position it had lost, despite the numerous instances of forced displacement and degradation of the environment, many of which were violently opposed by the people. The dominant driving force was to rebuild the Chinese people's pride in their country and boost their self-esteem.

The China Model, also known as the Beijing Consensus (as opposed to the Washington Consensus, which has fallen into disuse) is beginning to influence several countries in Africa, the Middle East, and South-East Asia. Kenneth Lieberthal, emeritus professor at the University of Michigan, and Wang Jisi, Director of the School of International Studies at Peking University and one of the most important ideologues of China's ascension to power, believe that "many developing countries that have introduced Western values and political systems are experiencing disorder and chaos" (Lieberthal and Jisi 2012).

The 21st century has been especially prolific in disruptive events, such as the provocation of war in Iraq (whose public justification was shown to be non-existent), the great recession of 2008, the economic stagnation in countries with advanced economies, terrorism in various parts of the world, the civil war in Syria, the recent threats of trade wars, social and political upheavals in Europe, the rise in migration movements, and frequent climate-related natural disasters throughout the world, such as very intense tropical cyclones, forest fires, intense and frequent heat waves, prolonged droughts, and devastating floods. There is also the feeling of economic insecurity, rising income and wealth inequalities, and increasing poverty in most countries with advanced economies. The affected population feels a growing nostalgia for a lost identity and the greatness of days gone by, along with frustration, confusion, and bewilderment. In such countries democracies are having increasing difficulty satisfying people's expectations regarding the economy and identity groups of values, which constitute important gauges for measuring political progress.

The issue can be summed up by the idea that, in those countries, representative democracy is losing its ability to find solutions to resolve social, economic, financial, geopolitical, environmental, natural resource, and climate change problems at national, regional, or global level because these problems are becoming increasingly difficult and complex. A more analytical way of expressing this would be to say that the short, middle, and long term unsustainability of the current globalised social, economic, and financial system, particularly as regards its relation with the environment, natural resource exploitation, and climate change, is chipping away at the viability of representative democracies. Restricting the analysis to economic values, we may conclude that the unsustainabilities that result from environmental degradation, the overexploitation of natural resources, and climate change are making it increasingly difficult to sustain strong economic growth at the national, regional, and global level. Restricting further the analysis to natural resources, governments are likely to face an increasing risk of conflicts generated by competition to gain access and exploit valuable natural resources, especially energy resources and certain types of minerals.

However, we may predict that this lack of sustainability and the problems it is causing to the world economy and to democracies will continue to be denied by the great majority of political leaders, their governments, and the majority of people who support them, and by most of the major leaders of the multinational corporations that do everything they can to sustain the current system.

Trying to mitigate the effects of contemporary and future unsustainabilities without acknowledging and addressing their causes, but instead denying them, significantly reduces the number of political options that democratic governments have at their disposal to fulfil the increasingly difficult goal of assuring a robust economic growth in their countries. In this mystifying scenario, such difficulties became incomprehensible to the majority of the population and end up intensifying society's political polarization and devaluing democracy. The space and opportunity to dissent must not hinder the unquestionable political goal of assuring strong economic growth under very adverse circumstances. The world will very probably be confronted with an inevitable transition to other paradigms of development more compatible with the management of increasing unsustainabilities, and this transition may well require an adaptation of representative democracy to the new paradigms.

Is it possible to assess political progress for each group of values used to evaluate the performance of governments in different countries and in a given period of time? The answer to this question would require a full study using the methodologies of the social sciences. Each country has its own standard of progress for the four groups of values and these standards are constantly changing. It is likely that a majority of Chinese people consider that China has achieved considerable political progress in the last 40 years in the economy and identity groups of values. However, in the eyes of most people living in the West, there has been very little or no-progress in the freedom and human rights group of values during that period in China. The West is currently experiencing difficulties to make progress in the freedom and human rights group compatible with progress in the identity group. At the global level it is increasingly problematic to make progress in the economy group compatible with progress in the environment and sustainability group.

5.7 Post-truth Politics

Democracy also faces the risk of what is known today as post-truth politics, which exploits the tendency to believe in statements and the repeated assertion of talking points that appeal to emotion and meet expectations, but do not reflect the facts. The Oxford Dictionary, which chose the term "post-truth" as its word of the year in 2016, defines it as an adjective "relating to or denoting circumstances in which objective facts are less influential in shaping public opinion than appeals to emotion and personal belief". It has always been the case in politics, in both autocracies and democracies, that decision-makers omit, deny, manipulate, or fabricate the truth while at the same time accusing opponents of the doing the same. In post-truth politics, it is no longer necessary to carry out the "reality control" discussed by Orwell, because people are now willing and eager to deny the truth and believe in what is currently known as fake news. All that is needed is to appeal to their emotions, anxieties, concerns, identity ideals, and above all, their expectations.

The term "post-truth" was used for the first time by the Yugoslavian-born American writer Steve Tesich (1942–1996), in an article published in Nation magazine about the Watergate scandal, the Iran-Contra affair, and the First Gulf War. The accumulation of gaffes, blunders, failures, cover-ups, and lies proclaimed by the government led him to write: "We came to equate truth with bad news and we didn't want bad news anymore, no matter how true or vital to our health as a nation. We looked to our government to protect us from the truth." Further on, he wrote: "In a very fundamental way we, as a free people, have freely decided that we want to live in some post-truth world."

Current post-truth politics does not aim for or need to systematically confront the truth head-on. The main objective is to distort and manipulate the truth. The truth persists, but is often jealously protected from public view by the few. The main goal is to exploit and strengthen political biases and prejudices that shape and give rise to the complaints and expectations of the electorate. In this way different identities are accentuated and exaggerated, forcing an opposition between "us" and "them" in the conviction that it will help to win votes in elections and, in the end, power.

The post-truth phenomenon is favoured by the wide variety of sources of information and news available in the media. Newspapers, magazines, TV, internet, and social media abound with information, misinformation, disinformation, factual news, rumours, biased news, and fake news, all presented with the same value. One reason why the amount of fake news, misinformation, and disinformation circulating in cyberspace has increased is the rising power of dark money, funds of unknown origin and sponsor trolls and bots aimed at creating confusion and mistrust (Iyengar and Massey 2019). This tendency is especially strong in the USA and is often driven by political rightwing thinktanks such as Americans for Prosperity, an advocacy group funded by David and Charles Koch to promote lower taxes and less government regulation (Iyengar and Massey 2019).

Just as faithfully reproducing facts is not a protected value, falsity has become an accepted practice. A fake but minimally plausible news story, and especially one that is able to please a particular identity group, acquires the qualities of truth when it is shared on social media networks used by that group. Believers usually refuse to recognise that it is untrue even when it is proven to be so. The spreading of misinformation, disinformation, and fake news in the media is frequently done with the intent of deceiving or damaging individuals, groups, organizations, and governments, often with the intent to obtain political, financial, and economic gains. In controversial areas of politics, economics, finance, and social discourse, it is common for facts to be immediately disputed by contrarian made-up facts or counterfacts that are presented as having the same value. The competition between facts and the corresponding counterfacts extends to areas of science and technology, believed by many people to be liable to politicisation. Social media and the internet are the ideal way of spreading and lending credibility to counterfacts.

Another common practice that distances people from truth is the attribution of hidden, and often conspirative, motivations and causes to some facts and decisions, a practice called *dietrologia* in Italian, and translated as "dietrology" in English. Conspiracy theories are a more structured form of dietrology that result from constructing distorted or plainly false narratives to describe and explain historical, political, economic, and financial events, and also scientific or technological facts. Dietrology stems from the belief that there is always an explanation *dietro*, or behind, institutional analyses and explanations. People that practice dietrology tend to consider that nothing happens by accident, nothing is at it seems, and everything is somehow connected (Douglas and Sutton 2011). Furthermore they believe that evidence showing the falseness of conspiracy theories is further proof of their truth, making them unfalsifiable.

The vehicle for circulating misinformation, disinformation, and fake news is neutral and the receiver is free to assign it the credibility and veracity that he or she sees fit. Credibility mostly depends on the trust that the receiver has in the source. If the source is approved by or related to the receiver's identity group, the content has a high probability of becoming credible. The maintenance of such a stance tends to develop into an unwillingness to receive, value, analyse, and face up to information, news, and statements from sources outside the identity group to which the person belongs. Facts only become relevant if the identity group is receptive and interested in knowing them. These attitudes may provide a feeling of comfort and protection by strengthening the identity group, without having to confront opposing views, but it does not favour a capacity for dialogue, understanding, and coexistence between people belonging to different groups in society. They bring short-term benefits but are prejudicial and dangerous in the medium and long term because they tend to aggravate tensions. Time always ends up revealing the relativism and transience of identities and of supposed unyielding differences between "us" and "them". This personal or historical experience should be used as an argument to seek dialogue, understanding, and tolerance.

The post-truth period is also an opportunity to question whether or not there are truths in the social environment. This question has no direct relation with the philosophical debate about the nature of truth and contemporary discussions about the correspondence, coherence, and pragmatist theories of truth (Burgess and Burgess 2011). Truth is different from an empirical fact, which is an objective matter that can be observed directly or indirectly through the senses, and can therefore be verified. Empirical facts are independent of time and not a matter of opinion. Truths are constructed beliefs, grounded in observations and experience, which are in conformity with facts. Truths are more malleable than empirical facts and they usually evolve with time.

In the current post-truth era, there is also a trend to oppugn some empirical facts, such as selected scientific facts and visual evidence, in particular photographs and TV images. In the post-truth world some empirical facts become subjective and are negated because they are in conflict with one's personal opinion and views. If the number of those who question a given fact is high, and especially if the number grows on social media networks, the fact loses relevance and is finally negated. This type of social negation of an empirical fact creates what has been named an "alternative fact", which counters the empirical fact. The expression was introduced by Lellyanne Conway when she defended a false statement about the number of people attending the inauguration of President Donald Trump on 20 January 2017. Alternative facts have become a mainstay of popular culture.

Falsehood also reaches the very identities of Facebook, Twitter, and other social media users. It is relatively easy to create a fake account based on the account of a real user that duplicates, transforms, and distorts his or her personality. Some social media companies, such as Devumi, create fake accounts by copying real users account details and profile pictures to create realistic bots that are sold as Twitter followers to whoever would like to become more popular and influential on social networks. In 2018, there were an estimated one billion fake accounts on Facebook. It is calculated that, on average, user accounts of social media are 40% fake. Fake personal accounts serve as a virtual "passport" to make purchases, gain access to public and private services, and carry out political activities. The new armies of fake users can be used for commercial, economic, or political ends, in particular to spread fake news that becomes a weapon against people, political parties, governments, and foreign countries.

To prevent anyone creating an account and putting just anyone's name on it with no accountability, it is necessary to develop some form of identity check to guarantee that every account is linked to a real registered user with a name and an address that can be verified. However, joining a network has to be extremely simple and leaving it has to be as difficult as possible. What is particularly important is to increase the number of personal accounts, because that is the main indicator that ensures growth of profitability. Attempts at regulation and legislation by governments are largely inconsequential because of the economic and political power of the large social media corporations.

Social media developed and spread under the idea that they were ideal instruments for encouraging freedom, knowledge, and a better future for all of humanity. The great effectiveness of this ideal model led the digital technology giants—Amazon, Google, and Facebook—to acquire enormous power and influence over users. Amazon has specialised in finding out what and how people buy, Google in knowing what people like and want to find out about, and Facebook in knowing what and how they communicate. This knowledge has generated enormous profits and has transformed the three companies into powerful monopolies.

The fact that social media platforms can be used to misinform, disinform, spread false statements, and fake news is considered by society as an unfortunate but unavoidable side-effect that can be mitigated but not eliminated. Social media platforms can also be used to disseminate hate material driven by xenophobic, sexist, political, and other motivations. This opening is being systematically used at national and international levels, especially through the advertising and publishing systems that the platforms provide. Russia uses it regularly to influence the results of elections and referendums in its favour in several Western countries. In March 2018 it was revealed that Facebook allowed Cambridge Analytica, a political data-mining company founded by the conservative media executive Steve Bannon and the conservative billionaire Robert Mercer, to access personal information data of 87 million Facebook users without their knowledge or consent. Among the political events for which politicians paid Cambridge Analytica to use personal information extracted from the data breach were the 2015 and 2016 campaigns for Donald Trump and Ted Cruz in the USA, the 2016 Brexit vote in the UK, and the Mexican general election of 2018. Social media platforms are also vulnerable to the attacks of hackers who exploit security flaws of popular features in the networks to steal account credentials of tens of millions of users.

Privacy and identity loss are recurring risks in social media platforms but it is very difficult to find solutions because regulations and legislation addressing this problem would tend to harm the profits of these corporations. In a recent study carried out by the Pew Research Center (PRC 2017), experts were evenly split when asked the following questions: "In the next 10 years, will trusted methods emerge to block false narratives and allow the most accurate information to prevail in the overall informa- tion ecosystem? Or will the quality and veracity of information online deteriorate due to the spread of unreliable, sometimes even dangerous, socially destabilizing ideas?" 49% answered yes to the first option and 51% yes to the second. The experts of the latter group explain their opinion by saying that they believe that humans shape tech- nology advances and its pace of progress for their own, not-fully-noble purposes and that bad actors will thwart the best efforts of innovators to remedy today's problems. Some more pessimistic even foresaw a world in which widespread misinformation, disinformation, and mass manipulation would cause broad swathes of the public to give up trying to be informed participants in civic life (PRC 2017). In the unstruc- tured fluid environment of social media, a sociological phenomenon has taken hold in which the most popular opinions are usually those that promote self-interest, rational egoism, alienation, and short-termism, and strengthen fast growing emerging iden- tity groups. Although there are many exceptions, opinions that require self-control, rational analysis, tolerance, cooperation, social responsibility, the adherence to moral and ethical values, and the fulfillment of medium and long term objectives are likely to be less popular.

China adopts a different policy for controlling the use of the internet and social media networks by censoring rumours considered to be biased, disinformation, fake news, violence, and pornography. Anonymous social media users are blocked. Global platforms like Facebook and Twitter are banned and replaced by powerful Chinese platforms like Baidu and Sina Weibo. The websites of some Western newspapers of public record are also banned from the internet. The contents of Chinese digital platforms are overseen by supervisors who analyse it and report infractions directly to the police. Based on these complaints, the companies responsible for the platforms are forced to pay heavy fines. The state also influences the governance of some large ICT companies by taking management positions in return for buying shares. Restrictions and controls are primarily driven by political and social motivations. It is imperative to obstruct political dissent in social media, as well as what the governmental officials consider to be unacceptable social behaviour. Protecting the established model and collective interests of society is considered an unassailable obligation that requires crushing some of the individual political and social freedoms cherished by Western culture.

In line with the tendency to place collective values before individual values, the legislative framework and procedures for the protection of internet users' private information data is weaker in China than in the West. This defencelessness favours the development of artificial intelligence applications that use personal information data to optimize marketing and promote online shopping of goods and services. Such technological advances tend to reinforce consumerism and economic growth. Eventually the opportunity to advance ICT in a market environment with hundreds of million of consumers is likely to create an economic advantage relative to the West. On the other hand the Chinese system is particularly effective in the protection of state security regarding the risk of intrusions, external cyberattacks, and cyberwarfare. The system is also effective at controlling the risk of malevolent and ethically irresponsible use of the internet and social media platforms, and serves as a model for other countries.

Post-truth tendencies can be seen in almost all fields of human activity. Ethical relativism encourages modern societies to tolerate forms of dishonesty that involve limited, controllable risks with impunity. These practices are often considered to be positive since they reveal the ability, courage, and determination to defend forcibly personal interests or other interests, which is a valuable asset in a very competitive professional and marketing environment. The narratives presented in marketing and in personal CVs and the information available about companies and all sorts of institutions and organizations are often more favourable than the truth of the facts. But as the validity of documents is increasingly ephemeral, exaggerations quickly fade away. The receiver of the information already suspects it and makes allowances accordingly. The use of post-truth in personal, professional, and institutional relationships may create immediate, short-term advantages, but in the medium and long term it has a tendency to cause those relationships to deteriorate, breaking trust and solidarity and thereby undermining the soundness of the social fabric. This is a further example of the blossoming of a type of cunning that benefits from immediacy, from the acceleration of operative social time, and from the associated tendency to overvalue the short term.

The consequences of post-truth politics for democracy are negative. Post-truth exploits credulity, ignorance, resentment, antagonism, and hatred. As the different "truths" contradict each other more and more often, a divided society is formed, split into identity groups that antagonise each other, thereby increasing the risk of confrontation and authoritarianism. Representative democracy is based on the conviction that there is a language common to everyone based on the use of solidarity, reason, and knowledge. It presupposes that people and especially politicians agree on empirical facts. Having a diverse range of political opinions is perfectly natural, legitimate, and beneficial for democracy but those who defend them must communicate and engage with one another within the same framework of moral, ethical, and scientific values and principles. In the transition that some democracies have undergone throughout history in their transformation to become autocratic regimes, we can observe a stepping-up of post-truth politics and its growing success.

5.8   How Science and Technology Triumphed Through Wonderment, Adventure, and Utility

Before addressing the question of scientific and technological progress, it is important to analyse the way science captivated and endeared society through wonderment, recreation, and fun, as well as through its value for epistemology, utilitarianism, and its capacity to improve human health and well-being. The public spectacle of science began in the Age of Enlightenment, the 18th century European intellectual movement, which according to French historiography ran from the death of Louis XIV, the Sun King, in 1715 until the French Revolution in 1789. Inventors, science communicators, and enthusiasts offered demonstrations and experiments in the streets, squares, markets, and cafés that amazed an audience eager to experience new sensations, enjoy themselves, and be surprised and dazzled. In several European cities, especially London and Paris, observations and experiments with mechanics, optics, electricity, magnetism, and the chemistry of gases were carried out using a huge array of instruments, such as scales, densimeters, thermometers, batteries and electric machines, vacuum pumps and telescopes for observing the Moon, planets, and stars.

Some became very popular, such as the electric discharges of the Leyden jar performed by Abbé Jean-Antoine Nollet (1700–1770), a French physicist who devoted himself primarily to studying electricity and was a teacher of experimental physics at the Collège de Navarre in Paris. The Leyden jar was invented by Ewald Georg von Kleist (1700–1748) in Germany in 1745, and Pieter van Musschenbroek (1692–1761) explained how it worked at Leiden University in the Netherlands a few months later. The Leyden jar was the first means of accumulating electric charge that could then be discharged at the experimenter's will, and it was a very important instrument for the study of electrostatics.

The first Leyden jar, invented by von Kleist, was a primitive condenser comprising a glass jar filled with water, with a metal electrode inserted through the stopper at the mouth of the jar. By connecting the electrode to an electrostatic machine, electric charge accumulates in equal amounts but with opposite charges at the inner and outer surfaces of the jar. When the jar is held in one hand and the other hand is placed on the metal terminal, electricity passes through the human body, sometimes with some violence, depending on the amount of accumulated charge. Musschenbroek carried out this experiment without due care and received a strong electric shock that scared him so much that, in a letter to his friend René Antoine Ferchault de Réaumur (1683–1757), the well known French entomologist, he wrote (Heilbron 1999): "I advise you never to try yourself, nor would I, who have experienced it and survived by the grace God, do it again for all the kingdom of France."

Musschenbroek's advice was not followed, because humans dearly love new, exotic, and extreme experiences. Nollet carried out highly successful and popular public demonstrations in which the electric discharges from the Leyden jar ran through one, two, or more people. In the spring of 1746 he performed the experiment on 180 royal guards connected by metal rods held in their hands, for the King and

Queen and many courtiers to see, at the Galerie des Glaces in Versailles. To witness the 180 guards jumping all at the same time under the influence of the electric discharge must have been an extraordinary sight, demonstrative of science's enormous potential power. Later, Nollet did a similar experiment in Paris on 200 monks disposed in a circle. Given that the 180 guards or the 200 monks all jumped almost simultaneously, he concluded correctly that the speed of propagation of electricity had to be very high. There were long queues of people waiting to take part in Nollet's experiment and experience the electric discharge, which was called "sensation scientifique". Faced with the amazing proliferation of experiments that revealed and demonstrated physical forces, mysterious phenomena, and invisible fluids with extraordinary properties, the European Academies of Science sought to separate fantasy and esotericism from what they believed to be true science.

The Age of Enlightenment also saw the first automata and the first attempts to build an artificial man, or human robot. One well-known example is the wonderful *Joueuse de tympanon* (see Fig. 5.4), built in Germany by the clockmaker Peter Kinzing and the cabinetmaker David Roentgen, which played eight melodies, one composed by Christoph Willibald Gluck (1714–1787). The automated music performer represented the highest technology of the day and enchanted Queen Marie-Antoinette (1755–1793) so much that it was later given to her. Marie-Antoinette was known in her time as Madame Déficit owing to her lavish spending and her opposition to the social and financial reforms proposed by Jacques Turgot and Jacques Necker (1732–1804). She gave the *Joueuse de tympanon* to the Academy of Sciences before being imprisoned and guillotined on 16 October 1793.

Other notable examples were the "speaking machine", a manually operated speech synthesiser, built in 1769 by the Hungarian inventor and writer Wolfgang von Kempelen (1734–1804), and the "writer", a mechanical boy who could write with a quill, built by the Swiss watchmaker Pierre Jaquet-Droz (1721–1790) and his son Henri-Louis, exhibited in public in 1774. The writer dipped a goose feather into an inkwell and wrote a 40-character sentence.

In 1748, Julien Offray de La Mettrie (1709–1751), who was trained as a physician and made pioneering advances in neurosciences, published *L'Homme machine* (La 1747), in which he developed a radical, atheistic mechanicism of the human body. This displayed analogies with the automata of the French inventor Jacques de Vaucanson (1709–1782), whose devices included the famous *Canard digérateur*. The book, published in Leiden, was forbidden and burned the following year forcing him to leave the Netherlands. La Mettrie defended the idea that human beings had no immaterial soul, writing that "l'âme n'est q'un principe de mouvement, ou une partie matérielle sensible du cerveau qu'on peut, sans craindre l'erreur, regarder comme un ressort principal de toute machine" (La 1747). He was one of the first thinkers to conclude that there is an unbroken continuity between humans and the rest of nature and that humans are just complex animals. His materialistic and hedonistic views eventually forced him into exile in the court of Frederick the Great (1712–1786) and the circle of intellectuals at Sanssouci in Potsdam.

All these new ideas, experiments, and objects amazed people and represented the triumph of a new operative social time propelled by science and technological

Fig. 5.4 *La Joueuse de tympanon* built in Germany at the beginning of the 1870s was one of the more sophisticated of the early automatons. The robot actually plays music by striking the instrument's strings with sticks. It is said that about one hundred artisans from 26 different trades participated in its fabrication. In 1864, the automaton was ceded to the *Musée des Arts et Métiers* in Paris and was restored by the famous French master magician Jean-Eugène Robert-Houdin (1805–1871). *Left*: The automaton playing. Photograph by Jean-Pierre Dalbéra, Paris, France, licensed under the Creative Commons. *Right*: The inner workings. The device was wound up and the movement of the hands directed by a revolving cylinder hidden under the skirt. Photo credit: La historia de los autómatas—Baúl de Chity

inventions. Science and its applications, as well as being interesting, entertaining, and provocative, began to have unexpected, even surprising practical uses, and to generate new concepts and avenues of thought. Many people, especially in the elites, started to believe, admire, and promote science.

One of the most remarkable inventions of the Age of Enlightenment, thanks to its revolutionary nature, was finding a way to fly, thereby realising Icarus' dream. The first demonstrations of small balloons filled with hot air that rose up and floated in the air for some time were made by Bartolomeu de Gusmão (1685–1724), a Portuguese Jesuit priest born in Brazil. The demonstrations took place in Lisbon during the month of August, 1709, some in the presence of King João V (Gusmão 1709). However, the invention was only improved and applied on a bigger scale by the brothers Étienne Montgolfier (1745–1799) and Joseph Montgolfier (1740–1810), paper manufacturers in the Ardèche, in the south of France. The first successful ascent using hot air, or what was called *air raréfié*, was achieved in the town of Annonay on 4 June 1783.

News quickly reached Paris, where a crowd of Parisians gazed in awe at a balloon rising on 19 September of the same year in Versailles, carrying a duck, a chicken, and a sheep on board, in the presence of King Louis XVI. Finally, the first men to carry out a free flight, that is, without the hot air balloon being tethered by ropes, were François Pilâtre de Rozier, an ambitious young chemist, and the Marquis of Arlandes on 21 November 1783. At 1.55 pm, the ropes that tethered it to the ground were cut, the balloon rose into the air and the crowd fell completely silent, overcome by a mixture of wonder, anxiety, and concern for the two balloonists. However, when Rozier signalled unemotionally to the spectators with his hat and waved a handkerchief, the audience broke into applause and cheers. The balloon left the gardens of the Château de la Muette in the west of Paris, flew over the Seine, and landed around 20 minutes later in Buttes-aux-Cailles, in the south-east of the city.

In the meantime, an alternative technology was developed by the French physicist and mathematician Jacques Charles (1746–1823), using a balloon containing hydrogen, then known as "flammable air", which was obtained by pouring sulphuric acid onto iron. The project was supported by the Academy of Sciences and financed using a crowdfunding system organised by the geologist Barthélemy Faujas de Saint-Fond (1741–1819). The balloon was built with the help of two engineering brothers, Anne-Jean Robert (1758–1820) and Nicolas-Louis Robert (1760–1820). After solving several technical problems connected to producing the hydrogen and introducing it into the balloon, which was made of silk sealed using a rubber-based varnish, the flight took place on 27 August 1783 and was observed by thousands of Parisians, including Benjamin Franklin. At the time Franklin was the first Minister Plenipotentiary from the United States of America at the Court of Versailles, the name then given to the Ambassadors to France. The balloon rose over the Champ de Mars, where the Eiffel Tower now stands, and then moved north-east, followed by horsemen, covering a distance of almost 21 km in 45 minutes before landing in the village of Gonesse.

The villagers ran to the landing site *en masse*, convinced by two friars that it was a monster. Terrified, they destroyed what was left of the balloon using stones, pitchforks, scythes, knives, and blunderbusses. This incident led the government to release a warning to people, assuring them that the balloons were not only inoffensive to people but could also become useful for society.

Shortly afterwards, Jacques Charles himself boarded a new hydrogen balloon with his collaborator Nicolas-Louis Robert. On 1 December 1783, at 1.45 pm. in the Jardin des Tuileries, after hearing a cannon fired, the balloon rose smoothly into the air in the presence of 400 000 awed and enthusiastic spectators (see Fig. 5.5). It was equipped with a thermometer and a barometer, which were used to make the first observations of the physical state of the atmosphere above ground level. It flew southwest until, when it was nearly sunset, Charles decided to descend, having travelled around 36 km in 2 hours and 5 minutes. Excited by this good performance, he went up in the balloon again, this time alone, to make observations of the atmosphere at an even higher altitude. He very likely went higher than 3000 m, at which point he felt a severe pain in his ears. He saw the sun again, when the ground was already in darkness, and a second sunset. Charles became very well known in Paris and began to

Fig. 5.5 Print of the first hydrogen balloon flight with Jacques Alexandre César Charles and Marie-Noël Robert ascending from the Tuileries Garden, Paris, on 1 December 1783, as viewed from the Pont Royal. The print was made by the French etcher Benoît-Louis Prévost (1733–1816), who was a member of the first team of engravers of the *Encyclopaedia*, edited by Diderot and D'Alembert (1751). One of his best known works is the frontispiece of that *Encyclopaedia* depicting *la Raison et la Philosophie arrachant son voile à la Vérité rayonnante de lumière*. Photo credit: World History Archive/Alamy Stock Photo

have many admirers. In 1785 he was elected a member of the Academy of Sciences. The balloon in which he flew was very similar in design to today's gas balloons, and already had all their main elements, including a safety valve for releasing hydrogen, ballast for controlling altitude, and a barometer for measuring it.

As often happens with new technologies, airships were frequently used for military purposes in reconnaissance operations immediately after they were invented. In fact, on 2 April 1784, less than five months after the first free flight, the Company of Aeronauts (*Compagnie d'Aérostiers*) was formed, the first air force unit in history. On 26 June, the Company intervened in the Battle of Fleurus, which took place between Charleroi and Namur in what is now Belgium. The battle set the forces of revolutionary France and the powers of the First Coalition—which included the Habsburg Monarchy, the Dutch Republic, and Great Britain—against each other. Jean Coutelle and General Antoine Morelot stayed in the balloon *L'Entreprenant* for nine hours, observing and reporting on the movements of the enemy forces and throwing messages to the ground, which were promptly taken to the command of the French troops. This information helped position the troops and direct the artillery fire more effectively and they were important in ensuring the Coalition's defeat. The results of the battle led Austria to lose control of its territories, where Belgium is now located, and strengthened the French Revolution politically. Airships were then regularly used in several conflicts up until the First World War.

This is a remarkable example of the success of a new technology that was able to captivate and dazzle crowds, further scientific knowledge, interest and thrill balloonists, and finally find use in warfare. Furthermore, airship flights opened the way for the development of aircraft and spacecraft technologies and industry, creating limitless opportunities for business, profit, adventure, tourism, and new experiences and pleasures in the atmosphere and in the immensity of outer space.

Three hundred years after the beginning of the Age of Enlightenment, science and technology continue to excite, dazzle, and captivate, although in different and much more diversified ways. In contemporary social time, there are countless examples of human fascination with new technologies, especially in the field of ICT. An example of this fascination for new technologies is the yearly launch of new smartphones by the various competing companies. The launch of Apple's iPhone 7 in September 2016 was a global event, taking place almost simultaneously in 28 countries around the world, and attracted thousands of consumers eager to have the new and exciting gadget in their hands. Some had to wait more than two days to be among the first to enjoy such a privilege. With the pace of smartphone evolution moving so fast, there's always something even more exciting waiting for next year. Consumers buy a new handset when they see something new that is really exciting for them so the companies must keep producing all sorts of technological and marketing innovations to guarantee their continuous growth.

The enthusiasm for and attraction to new technologies and their new products extend beyond smartphones to other personal devices, such as tablets, computers, audio-visual systems, cameras, domestic appliances, cars, planes, boats, and a never-ending array of gadgets and machines. Some consumers buy to satisfy their basic needs, but others buy just to enjoy keeping up with the latest trends, fashions, and innovations. From the point of view of a large part of the manufacturing companies, satisfying the basic needs of people throughout the world, especially the poor, who can only spend small amounts of money, is not particularly attractive. It is much more lucrative to produce sophisticated, diversified, and expensive products using innovative new technologies, especially when they hit on highly successful products and devices.

The essential human characteristic of seeking adventure, discovering the unknown, going to the edge of emotion and survival, and surpassing all boundaries has found in science and technology a highly fertile ground in which to flourish. A very telling example of these unquenchable desires is the exploration of the Solar System and what lies beyond it. Only 12 years since the launch of Sputnik 1, on 4 October 1957, mankind went to the Moon on the Apollo 11 mission, and 43 years later in November 2000, the first long-term residents arrived at the International Space Station (ISS) (see Fig. 5.6), a remarkable example of international cooperation between countries that confront each other, sometimes dangerously, at lower altitudes and on Earth.

The fascination with space exploration and the development of space technologies overcomes economic, political, and military rivalries. The ISS completes an orbit around the Earth in around 93 minutes at altitudes varying between 330 and

Fig. 5.6 NASA Astronaut Robert L. Curbeam Jr. (*left*) and European Space Agency (ESA) astronaut Christer Fuglesang, both Space Shuttle STS-116 mission specialists, participate in the first of three planned sessions of extravehicular activity (EVA) for the construction on the International Space Station on 12 December 2006. Visible on Earth is the Cook Strait that separates New Zealand's North Island (*left*) from the South Island (*right*). STS-116 Shuttle Mission Imagery. Photo credit: NASA

435 km, and its management and scientific programme are coordinated by five space agencies: NASA (National Aeronautics and Space Administration, USA), Roscosmos (State Corporation for Space Activities, Russian Federation), JAXA (Japan Aerospace Exploration Agency), ESA (European Space Agency), and CSA (Canadian Space Agency). The ISS serves as a microgravity and low Earth orbit space environment research laboratory. One of the main goals is to have astronauts who spend long periods of time in space to study the effects on their health, with the aim of preparing for future journeys and colonisations in the Solar System and interstellar space. It also serves as an Earth environment and meteorology observation platform, an astronomical observation platform, a laboratory for biology and physics experiments, and a place for testing and improving space equipment to be used on voyages within the Solar System and in particular to Mars. Going to the ISS, floating weightlessly in space, seeing the Earth, the oceans, continents, mountains, ice sheets, forests, and deserts must be a unique and unforgettable experience that makes lives worth living and dreams come true.

From 2001 to 2009, Roscosmos enabled seven tourists to go to the ISS. They paid tens of millions of dollars for the trip. This space tourism was interrupted so that the ISS could be used more by astronauts from other partners to the project, but Roscosmos plans to resume the service. Roscosmos has signed a deal with Space

Adventures, a US-based space tourism company founded in 1998, to carry two "spaceflight participants", a euphemism for space tourists, to the International Space Station in late 2021. The prospective astronauts will fly to the ISS aboard a Russian Soyuz spacecraft, but their names haven't yet been revealed.

Meanwhile, in the USA, there is an emerging new space industry promoted by private companies investing in space tourism and space exploration, both orbital and suborbital. One of these companies is Virgin Galactic, a spinoff of Richard Branson's Virgin Group, whose VSS Enterprise was the first SpaceShip Two (SS2) spaceplane built by Scaled Composites for Virgin Galactic. VSS Enterprise suffered a catastrophic in-flight breakup during a test and crashed in the Mojave Desert killing one of the pilots. On 13 December 2018, VSS Unity reached an altitude of 82.7 km, crossing the Kármán line, which, according to some institutions, defines the limit between the atmosphere and outer space, although there is no agreed international law defining the edge of outer space. One of the main objectives of Virgin Galactic is to provide suborbital spaceflights to space tourists, taking them to an altitude of 100 km in outer space, where they will experience the absence of gravity for a few minutes and see spectacular views of the Earth.

Another company pursuing the vision of space exploration is Elon Musk's SpaceX, which successfully used its Falcon 9 rocket to launch the Dragon crew capsule from Florida's Kennedy Space Center and carry two NASA astronauts to the ISS on 30 May 2020. It was the first launch of NASA astronauts by a private company and also the first from US soil after the Space Shuttle Program ended in 2011.

SpaceX has very ambitious plans. One of its objectives is to establish a self-sustaining Mars colonisation. SpaceX has already developed the Falcon launch vehicle family and the Dragon spacecraft family, which can deliver payloads into Earth orbit. Currently, it is developing the Big Falcon Rocket, expected to be capable of launching 150 tons or more into low earth orbit in the early 2020s, more than any other launch vehicle under development at the present time. The first cargo mission is planned for 2022, to be followed by a crewed mission in 2024. Musk believes that humanity's future will include thriving self-sustaining cities in Mars. His strong convictions lead him to say that: "I'd like to die on Mars, just not on impact".

Mars fascinates because, relatively speaking, it is the most benign terrestrial planet for humans, but it would still be extremely difficult to establish our current civilization there. It is situated quite far away, at a distance that varies constantly between about 54 million km and 401 million km. It has an extremely rarefied atmosphere (with a total mass equal to only 6% of the mass of the Earth's atmosphere), comprising 96% CO_2, and it has an average surface temperature of $-63\,°C$, which varies between about $20\,°C$ at the equator at noon and $-153\,°C$ at the poles. All these problems for humans can be solved with current science and technology, but only at a high cost.

For more than 40 years, robotic explorers have been studying Mars, and much has been learned about its geology and atmosphere. More missions have been attempting to explore Mars than any other astronomical body in the Solar System, except the Moon. There have been many failures but the number of planned missions and countries engaged in the exploration is fast increasing. The space agencies of the

USA, Europe (ESA), Russia, China, and India are all planning to send humans to Mars in the current century.

To carry out these and many other space exploration programmes, science and technology must continue advancing. This progress is what will make it possible to follow up the human adventure that began on Earth at the limits of the Solar System and perhaps beyond it, despite the immense difficulties caused by the fact that these are such extremely inhospitable environments for humans. Such difficulties are considered to be almost irrelevant compared with the expected achievements of space exploration, which will fully satisfy our ego and therefore the most essential aspirations that characterize *Homo sapiens* as a species. Technological progress has the promise of continuously creating entertaining, thrilling, and dazzling new experiences and opportunities, and most importantly, of improving human heath, well-being, and economic prosperity, and solving many of the problems that humanity currently faces regarding the environment and natural resources. There are negative sides to technology and technological innovations, but without technological progress, people would think that time had stopped and that we will inevitably regress to the primitivism of the past.

5.9 Progress in Science

Science stands out from other areas of human activity, such as morality and politics, because it is cumulative construction, a characteristic clearly acknowledged by Fontenelle in the 17th century. The horizon of science broadens with new knowledge supported by theories built on previous knowledge. George Sarton (1884–1956), one of the first historians of science, identified this cumulative character as an unmistakeable form of progress and wrote that "progress has no definite and unquestionable meaning in other fields than the field of science" (Sarton 1936). William Henry Bragg (1862–1942), in the context of the physical sciences, also defended an epistemic point of view when he stated that "if we give to the term Progress in Science the meaning which is most simple and direct, we shall suppose it to refer to the growth of our knowledge of the world in which we live" (Bragg 1936). Progress is a normative concept that requires criteria to judge its performance. In the application to science the usual criteria to evaluate progress is the continuing cumulative epistemic legacy that science delivers.

In the latter half of the 20th century, a philosophical debate developed on scientific progress, which involved dealing with other aspects of science as well as the objective of broadening knowledge of the physical world and of all living organisms. This debate involved Karl Popper (1902–1994), Thomas Kuhn (1922–1996), Paul Feyerabend (1924–1994), and Larry Laudan. They focused on issues related to defining scientific truth and scientific progress, as well as on the methods, structure, and process of building scientific theories. Furthermore, it was recognised that the concept of progress can also be applied to the different aspects of scientific activity, notably aspects relating to epistemology, methodology, science education, the professional status of scientists, and the impacts of scientific knowledge on socio-economics. In the latter field, John Desmon Bernal (1901–1971), inspired by the ideals of Marxism, argued that the progress of science should be assessed in terms of its contribution to better serving the social and economic interests and needs of humanity (Bernal 1939).

In the epistemic context, science has made it possible to broaden and deepen our knowledge of the Universe, the Earth system and its subsystems, life on Earth, and living organisms including man. The fundamental laws of physics, which play a central role in understanding and predicting natural phenomena, from the largest to the smallest scales of time and space, are now much better known than in the previous three centuries. There is still no quantum theory of gravity, but the field of application of its unknown laws is found mostly in the phenomena that took place at or before the Planck time after the Big Bang. It is obviously impossible to retrodict to the Big Bang itself, or to reproduce it in a laboratory to test competing quantum theories of gravity. It is therefore an area in which research is almost entirely theoretical, without direct mechanisms to test predictions through experimental checks as should usually be done in the scientific methodology to distinguish a true theory from a false one. These trends, specific to some areas of physics, and the search to find a physical theory of everything, i.e., a theory that provides a single, consistent description of all the fundamental interactions and the way they have intervened since the Big Bang, have led the physicist David Lindley to believe that the end of physics is approaching (Lindley 1993). In epistemic terms, this would mean that progress in physics would be coming to an end.

Shortly afterwards, John Horgan went even further, arguing that we are moving towards the end of science, since the scientific theories that make it possible to explain and understand observed phenomena are relatively well established (Horgan 1996). The observed phenomena range from the spatial scale of the Standard Model of elementary particles—fermions (quarks and leptons) and gauge bosons, which are the field theoretical carrier particles for the four fundamental interactions of physics—to the scale of the stars, planetary systems, including life on Earth, up to the galaxies and the cosmological structure of the Universe as a whole. Major changes or revolutions in the edifice of physical knowledge are not expected, just gradual advances that will not significantly change the structure of the building or our current view of the Universe. In other words, the extraordinary progress of science in epistemic terms that has taken place in the last 300 years is not likely to continue indefinitely at the same pace. Of course, there may be surprises and the building will gradually become more complete, complex, and coherent.

In Horgan's view, the great barrier to the epistemic progress of science in the future is its past success. In more explicit terms, it is highly unlikely that a scientific revolution will lead to abandonment of the theory of electromagnetism, summarised in Maxwell's equations, or the special and general theory of relativity, or quantum theory, or Darwin's theory of evolution, or even the DNA based theory of genetics. What we can expect are adjustments and improvements that do not change the central structure of these theories. The hypothesis of the end to scientific progress expressed by Horgan has generated a great deal of interest and a highly negative reaction from scientists, which is significant in itself. This reaction is partly due to the impact that an end or slow-down in the epistemic progress would have on the professional aspects of scientific life, and therefore on the status of scientists and their institutions in society.

Science is not only about unlocking the fundamental laws that govern physical and biological phenomena, but also about understanding and forecasting the behaviour of systems in nature and social fields. A system is a group of interacting elements that form a whole with identifiable behaviour patterns. They range from the very simple to complex systems composed of many elements whose behaviour is difficult to model due to the nature and complexity of the interactions between elements or with the external environment.

Examples of complex systems go from condensed matter, to cells, to ecosystems, to the Earth system and its subsystems, and from the human body and the human brain to groups of people, communities, nations, and groups of nations, to the whole of humanity. The study of this wide variety of complex systems requires different methodologies that range from those used in the physical and natural sciences to those of the social sciences and humanities.

At a higher level of complexity, we have systems science (Mobus and Kalton 2015), which is an interdisciplinary field that studies the complexity of systems, in particular systems of systems or systems that involve subsystems of different kinds, as for instance socio-ecological systems. Systems science can be used in almost any field of human activities, from the study of the physical and living world to the study of human societies, psychology, medicine, engineering, technology, and business management. It uses specific methodologies, such as system dynamics, agent-based modelling, and big data techniques.

In contrast to the situation regarding the search for the fundamental laws, there is much more to know about complex systems and ways to predict their behaviour as reliably as possible. Such advances may be considered to provide a measure of the progress of science. In applied science and in technological innovation, the production of new materials, new processes, new products and services, new technologies, and new living organisms also constitute criteria for assessing progress. Furthermore it is commonly considered that science progresses if it prolongs human life in good health conditions for an undetermined amount of time, a goal that has been increasingly valued since Bacon first announced it as one of science's most important future conquests.

5.10   Antiscience, Politicised Science, Post-normal Science, and the Crisis of Science

Scientific progress is nonsensical for those that criticize or reject science and the scientific method in different degrees and for various reasons (Holton 1993). One common concern in anti-science movements is that science makes a systematic use of methodological reductionism in the physical and natural sciences, especially in physics, chemistry, and biology, coupled with the supposition that these sciences have a privileged role in our understanding of the world. The concern stems from the idea that the behaviour of complex systems is essentially and inherently irreducible, so that a holistic approach must be used to fully understand them. This debate is particularly apparent in psychology. Nevertheless, the successes of science in many domains, as for instance in medicine, which depends strongly on physics, chemistry, and biology, is unquestionable.

The greatest risks for and attacks on science in contemporary societies stem from its successes. The links between science, technology, business, and government are numerous and increasing in number and complexity, which reveals the growing importance that is attributed to the role played by science and technology in promoting economic growth and in the competitiveness of the major economies. Businesses and governments emphasize that this is a very positive development, essential to fulfilling every citizen's expectations of economic prosperity and well-being in the societies of all countries around the world. However, there is increasing evidence that the symbiotic relationship does not have only positive results, but brings in a range of challenges and harmful impacts on society. The main sectors where such impacts are more evident and disturbing are biomedical research, biotechnology, pharmaceuticals, tobacco products, military and defence, and fossil fuels. So what are these negative impacts?

Large scale commercial involvement in university-based research, engineering, and technological innovation can introduce a significant bias in research activities that can be detrimental and marginalise research pursuing environmental and social benefits (Langley and Parkinson 2009). Direct commercial funding, especially in the biomedical, biotechnology, and pharmaceutical sectors, increases the likelihood that the outcomes of the research will be favourable to the interests of the funders (Bekelman et al. 2003; Fava 2016). Sheldon Krimsky, who analysed corporate funding for biomedical research in the USA since the 1980s, finds that such funding has the capacity to corrupt the academic scientific endeavour if there is no clear separation between the roles of those that produce knowledge and the roles of stakeholders who have a financial interest in the applications of that knowledge (Krimsky 2003).

A second very important issue arises when scientific evidence emerges about harmful health and environmental impacts of the products of a given industry. In this situation some of the large companies will fund campaigns to spread doubt, confusion, and pseudoscience theories over scientific studies and conclusions that advise against the use of such products. Tactics include all sorts of open and covert actions, such as funding lobby groups to present the industry as being for "good science" and the opponents for "bad science". The best known examples are tobacco smoking, acid rain, stratospheric ozone depletion, and climate change. Naomi Oreskes and Erik Conway, both USA historians of science, have identified remarkable analogies between the campaigns promoted by the industrial corporations and the so-called contrarian scientists that challenge the scientific consensus on behalf of their sponsors (Oreskes and Conway 2010). David Michaels has shown (Michaels 2008) how the scientific literature has been shaped and skewed to the advantage of polluters and the manufacturers of dangerous products in order to avoid regulatory legislation and shows how the current regulatory system could be changed to ensure that it defends public safety, rather than private profits.

Some corporations fund research on their behalf and selectively publish results that suit their interests, often using highly imaginative and innovative ways and means designed to keep the funding outside public scrutiny. Corporations tend to view science, not as an activity to generate knowledge, but as one among many inputs into production. When the profitability risks are high, the scientists that produce unfavourable findings may be discredited and sometimes intimidated (UCS 2012). Science is particularly vulnerable to the process of obfuscation, spreading of doubt, and pseudoscience construction and dissemination because scientific knowledge is not based on human intuition but is the result of a long and demanding learning process that frequently leads to counter-intuitive conclusions. It is relatively simple to lead a person to believe in pseudoscience statements and theories if that person does not have a basic level of scientific training and knowledge, and most importantly some scientific curiosity that facilitates the search for and the evaluation of information sources.

One outstanding example of the lack of confidence or opposition to science's advice is the reluctance or refusal to vaccinate, despite the availability of vaccines, an attitude called vaccine hesitancy. Vaccination is one of the most cost-effective ways to avoid disease, and currently prevents two to three million deaths a year worldwide. However vaccine hesitancy is considered to be one of the top ten global health threats in 2019 (WHO 2019a). The "anti-vaxxer" movement believes that there is a connection between vaccination and brain disorders, especially autism, despite their being no scientific evidence supporting that theory. Anti-vaxxers tend to believe that vaccinations are a human rights violation. Some are full-time activists who oppose vaccinations through aggressive campaigns in the social media networks that collect tens of thousand of dollars in donations (Arciga 2019). The crucial point here is that some people are sufficiently credulous and misguided to donate to this anti-science cause which, as a matter of fact, endangers public health.

Discrediting and denying science is a highly selective process which, in some extreme contemporary situations, especially in the USA, means establishing a distinction between "good science" and "bad science", or beween "solid science" and "politicized science" (Milman 2016). Those who make this unwarranted distinction usually consider that bad science or politicized science is harmful to the interests of industrial or technological corporations, while the good science and the solid science is either neutral or favourable to those interests.

Examples of good science or solid science include: biomedical science; pharmaceutical sciences; space science, which enables space to be explored and used for civil and military objectives, and to prepare for the future settlement of human colonies throughout the Solar System; nuclear science, which can be applied to many useful purposes, including the generation of nuclear energy, and also to the fabrication of new nuclear weapons, including atomic bombs adapted to the changing needs dictated by the nature of current conflicts; science that supports ICT, robotics, automation, and artificial intelligence; quantum computation; biology for increasing agricultural productivity, including synthetic biology and, within certain limits, human genome editing, which may be able to extend human life in good health conditions, among many others.

The most emblematic example of bad science or politicized science is climate change science. There are others in the USA, however, such as when social sciences establish a cause and effect relationship between the permissiveness of firearms laws and the mortality resulting from their use.

A more discreet way of talking about the difference "between sciences" in the USA is to acknowledge that public policy has recently selected and used some of science's results and recommendations and ignored others. As Rush D. Holt, one of the executive directors of the American Association for the Advancement of Science (AAAS), has put it (Holt 2016): "Over recent decades, a disturbing trend in the US government has been for ideological assertions to crowd out evidence".

The trend of establishing a division between robust science and politicized science is strengthened by the opinion that there is no consensus among scientists. According to a Pew Research Center report (PRC 2016), only 27% of Americans say that there is extensive consensus among scientists about climate change and its anthropogenic origins. However, research published specifically on the topic concludes that 97% of climate scientists worldwide share the conviction that the global warming observed in the 20th and 21st centuries is anthropogenic (Cook et al. 2013). The level of political polarization on this issue is surprising. In fact 45% of Americans who identified as conservative Republicans said that they had little or no trust in climate scientists, while only 6% of Americans who identified as liberal Democrats held the same opinion (PRC 2016). Only 15% of the former group say that they trust climate scientists "a lot".

These results seem astonishing because science and scientists were generally respected and recognised as having made a fundamental contribution to the USA's power and wealth. But climate change conflicts with essential aspects of the American model and the country's identity. The coal, oil, and gas industries constitute one of the major pillars of US power, world hegemony, and national pride. How can it be possible that the exploration, use, and global trade of fossil fuels, mostly made in USA dollars, could have serious harmful effects for humanity? If it does, the prevalent solution to the problem is to deny that it exists.

Climate change generates responses that tend to be aligned with political opinions, especially in countries whose economies are highly dependent on the production and exportation of fossil fuels. Generally speaking, parties closer to neoliberalism believe that mitigating climate change would go against their political doctrines by demanding greater governmental regulatory intervention in the energy field. This intervention would eventually lead to an energy transition to renewable energies, harming the powerful fossil fuels industry, especially oil and coal, which have strong ties with neoliberalism. The fossil fuel industry is by far the world's largest industrial sector. Just the oil and gas industry had total revenues of two trillion dollars in 2017, which represents 2.5% of the world GDP in that year.

Based on the ideological stance of the fossil fuel industry, the concept of anthropogenic climate change is fought against and discredited in public opinion. On the economic side, emphasis is placed on mitigation harming the economy and creating unemployment. On the science side the lack of scientific consensus is promoted mainly by funding scientists willing to publicly deny that it exists or to amplify the uncertainties. In the USA, this process is generally carried out through think tanks directly or surreptitiously funded by many different types of organisations, such as politically conservative foundations (Brulle 2014). By promoting, commissioning, and paying for opinions that go against the scientific consensus (Cook et al. 2013), the opinions of the consensus scientists appear to be politicized and it becomes credible to public opinion to say that "politicized science" exists.

If there is an interest in politicizing science, that means that there are benefits arising from the process. Society does not create movements that deny the existing consensus on nuclear science, which made it possible to build atomic bombs and now enables them to be improved for specific aims. People have different opinions about the need to keep their national arsenals of atomic bombs and about continuing to manufacture them, but those differences have nothing to do with the fundamental nuclear science that had to be developed in order to be able to build the first atomic bombs in various countries. It is not nuclear science that is politicized but opinions on the military applications of that science. In the climate case it is climate change science that is politicized, and the differences are over scientific issues, namely the observation of climate change, the anthropogenic nature of that change, and the ability to project Earth's climate in the future.

By encouraging scepticism about the scientific justification of anthropogenic climate change, public opinion becomes open to post-truth politics. One extreme example is this tweet by businessman Donald Trump from 6 November 2012 in which he states: "The concept of global warming was created by and for the Chinese in order to make U.S. manufacturing non-competitive". As a candidate to become president of the United States, he was confronted with his surprising statement in a debate with Hillary Clinton on 26 September 2016. However, he repeatedly denied having said what he said in his 2012 tweet, which is typical of those who practise post-truth politics when they perceive that there is more harm than benefit.

It is impossible to conclude that negating science, distorting or manipulating specific scientific results, and trying to promote post-truth in science is an exercise in intellectual honesty. In fact, it is a systematic process to maintain ignorance and confuse and deceive people. In most cases such a process prevents people from being fully aware of issues that are harmful to their health and to the environment at the local, regional, or global levels. It means presenting truncated and biased narratives through omissions and distortions, which lead to a false interpretation of the facts. Science and its progress is not compatible with intellectual dishonesty.

There are cases of fraud in scientific research, but sooner or later they tend to be uncovered by other scientists due to the scrutiny of the scholarly peer review process and the intense competition among scientists. The refereeing system has shortcomings and sometimes it is unable to prevent the publication of invalid or dishonest research, but it provides a means to control the academic quality of published papers that is improvable. For instance, anthropogenic climate change is soundly established in the laws of physics, the discovery of which was the result of highly demanding intellectual honesty. Nietzsche referred to this when he wrote (Nietzsche 1882): "So, long live physics! And even more, long live what compels us to it – our honesty!"

At the beginning of the 1990s, various authors, mostly European, explored the consequences that the accelerated pace of change at the social, political, cultural, technological, and environmental domains would have for science (Funtowicz and Ravetz 1993). They argued that "science is now called to remedy the pathologies of the global industrial system of which it forms the basis". In a way they assumed that science had some responsibility in the current state of the world and that it was now necessary for it to reform itself in order to address the new challenges that it had helped to create. The present situation is partly the result of the use that humans have made of the scientific method and of the human goals that have been pursued with the help of scientific research, but not an inevitable consequence of science itself, which is essentially a methodology that can be used for whatever purpose. It is true that science is currently called upon more often to study complex systems, to deal with the uncertainties involved in predicting their behaviour, and to provide advice for public policies on issues of risk and environmental sustainability.

These new functions were assumed to justify the identification of a new type of science that was called post-normal, to emphasize the fact that it must adapt itself to the challenges of the "Post-Normal Age" or "Postnormal Times" (Funtowicz and Ravetz 1993; Sardar 2010). According to this view "the leading problems for science now derive from the challenges (and threats) presented by the hitherto blind and uncontrolled growth of our total scientific–technical–industrial system" (Ravetz 1997). It is argued that, when uncertainties in complex systems become great and when the stakes are high in decisions that must be taken, the "quality assurance of the whole process requires 'extended peer community' including all the relevant sort of concerned lay persons" (Ravetz 1997). Furthermore it is proposed that "the 'extended facts' provided by investigative journalism can be as crucial as the selected sample of the research results which is permitted to enter the public domain" (Ravetz 1997).

Adopting this parallelism in value between peer-reviewed research papers and journalist's accounts is not the best way to understand and monitor objective facts, and it can be negative for the political process. The pressing issues of the so-called post-normal world must be addressed primarily by governance through participatory and deliberative democracy, using public policies based on peer-reviewed scientific research. It is very doubtful that mixing the outcomes of scientific research with "extended facts" and contributions from the "extended peer community" into the inputs to the public policy process would have any positive effect on reaching sustainability, specifically as regards social and economic issues, the environment, natural resources, and climate change (Wesselink and Hoppe 2011). On the contrary, it may be interpreted as a contribution to the current tendency to devalue and negate science. Such a tendency arises mainly in a different context when science advises against practices that promote economic growth and profitability, but are harmful to health and the environment and more generally to sustainability.

Some researchers have claimed that there is a "crisis in science" over replicability and reproducibility, especially on pre-clinical and clinical medicine, psychology, organic chemistry, and applied economics (Ioannidis 2005; Saltelli 2018). It is undeniable that reproducibility issues pose serious challenges for scientific communities. Other researchers emphasize that measures are systematically taken by science to fight science's mistakes and misbehaviours, and that the claims that "science is in crisis" or "science is broken" are misplaced or unwarranted generalizations (Jamieson 2018). However, the fact is that the concept of a crisis in science is increasingly touted, in particular by those that have a vested interest in negating and distorting scientific findings, and by political activists. Industrial corporations, in their fight to deter regulation, are now invoking the reproducibility crisis.

The US Environmental Protection Agency (EPA) has recently proposed a new rule to "ensure that the regulatory science underlying Agency actions is fully transparent, and that underlying scientific information is publicly available in a manner sufficient for independent validation", with the justification that there is a reproducibility crisis. Oreskes interprets this apparently reasonable regulation as an attempt to prevent the use of "long-term epidemiological studies that linked air pollution to shorter lives and were used to justify air-quality regulations" (Oreskes 2018).

The assault on science promoted by some corporations when the scientific results are considered to be a harmful input to production and therefore to have a negative impact on the profitability of the industrial or technological enterprise is fast evolving to new forms. The techniques used to misinform, manipulate, and deceive people are increasingly sophisticated and effective. Eventually it is universally accepted that a certain product is harmful to human health, to the environment, or to both and its production stops, but the process may take a very long time depending on the amount of investments, assets, and potential revenues involved.

5.11 Technological Progress

In neoclassical economics, as already mentioned, technology is a key component of total factor productivity (TFP). This relation defines a criterion to evaluate progress in technology. Technological progress may be measured by how much more output can be produced from the same level of labour and capital. Interpreted in this way, technological progress is truly wonderful since it can increase the productivity and usually the profitability of economic activities without increasing the inputs. It has been one of the major drivers of economic growth in the world since the middle of the 19th century. Technological progress can be measured in terms of other objectives such as efficiency if it reduces resource consumption, reduces the risk and improves the security of activities, processes, and systems, and improves the quality and safety of products and services; it can also be measured in terms of environmental protection if indeed it helps to protect the environment.

In some cases technological progress generates side-effects that are harmful for society, public health, or the environment, such as various forms of pollution. Thus the implementation of a new technology usually brings a variety of benefits, but it may also potentially have known or unforeseen negative effects for human economic prosperity, well-being, and quality of life, especially in the medium and long term, but even in the short term. There is a natural tendency to emphasize the positive outcomes of using a new technology and disregard the negative side effects, largely because of the profitability expectations that it generates for those who have invested in its development.

The range of criteria that are used to define and evaluate progress tends to be much more diversified in technology than in science. In science the cumulative epistemic criterion continues to prevail in fundamental research and in some areas of applied research. For instance, progress in particle physics and astrophysics is primarily measured by advances in cumulative knowledge. The strong and diversified relationship that science has with technology (Brooks 1994), in particular its crucial contribution to the generation of technological progress, creates new criteria for progress in science. Some areas of scientific research have a greater potential to generate technological progress and economic growth than others. For instance, environmental science and sustainability science, by addressing the challenges that threaten the future of humanity, such as biodiversity loss, pollution, land and water degradation, and climate change, are more concerned about the impacts of certain industrial activities and products on the environment and on sustainability than in generating technological progress to enhance economic growth.

The "science in crisis" narrative is presented as having no analogue in technology, and technology is considered to be immune to crisis. It is one of the main drivers of economic growth, while the relation between science and economic growth is indirect, complex, and sometimes critical. At present, science is increasingly gaining evidence for and pointing toward dangerous unsustainabilities, such as the harmful effects of anthropogenic climate change. However, combating climate change is frequently viewed as being detrimental to short term economic growth and employment. The "science in crisis" narrative fosters cynicism and indifference in science, and risks discrediting the value of scientific evidence and feeding anti-science agendas (Fanelli 2018). According to Daniele Fanelli: "Scientific misconduct and questionable research practices occur at frequencies that, while nonnegligible, are relatively small and unlikely to have a major impact on the literature" (Fanelli 2018).

Technology also suffers from unethical practices and technology professionals who become whistleblowers risk being fired, blacklisted, and suffering severe financial, personal, and professional setbacks (Unger 1994). Whistleblowers are often torn between reporting inadequate testing, faulty or dangerous products, security violations on databases, conflicts of interest and many other forms of skulduggery and technological corporate misconduct for the benefit of the public interest, or simply keeping quiet for their own self interest, protection, and employment security. Contemporaneous social time does not see any signs of crisis here.

Technological progress can take the form of an invention, which is the discovery of a new idea regarding products, processes, or services, or an innovation, which primarily aims to improve inventions and make them useful and attractive for the market. Inventions usually have a higher social status than innovations, but the latter account for most of the economic growth generated by technological progress. In the history of technology, one finds macroinventions, which represent radical changes able to produce new technological and economic paradigms, such as the use of fossil fuels, nuclear energy technologies, and ICT, and microinventions, which generate new processes, products, and services linked to a technology that has already been established (Mokyr 1990).

Innovation in a given technology leads to progress in terms of production efficiency for long or short periods of time, depending on its capacity to remain competitive. It is therefore of great importance for engineers, policy-makers, and investors to attempt to predict innovation in a given technology, such as ICT, energy, transport, or agriculture. The earliest approach to model the evolution of production costs was developed for airplanes by T. P. Wright, concluding that costs decrease as a power law of cumulative production (Wright 1936). Recent studies that tested the ability of six different laws to predict production costs for a group of 62 different technologies, using annual data wth timespans of 10 to 39 years, show that Moore's and Wright's laws are those that get closest to describing the observed cost evolution (Nagy et al. 2013). The same study showed that production in each of the 62 technologies grew exponentially in the time periods under consideration.

Moore's law (Moore 1965), drawn up by Gordon Moore, one of Intel's co-founders, states that, in the process of improving the performance of an integrated circuit, the number of transistors will grow exponentially, doubling every two years. According to Intel, the transistors produced today are 90 000 times more efficient and 60 000 times cheaper than the first one made by the company in 1971. However, Moore's law describes a situation where consumer demand increases in tandem with production. If demand had failed, reducing the cost of a chip would have shrunk the market, and companies would not have invested in research to drive the costs down. Thus Moore's law applies to technologies that are successful in generating consumer demand.

Wright's law (Wright 1936) had its origin in the conclusion that, for every cumulative doubling in the number of airplanes produced, there is a cost decline in percentage terms. It is based on the principle that, the more we make, the more we earn. Cumulative production is a good proxy for the effort invested in innovation. Wright's law provides a better description of the historical data, in particular because it is better able to describe cases in which growth in production is super-exponential over long periods of time, a situation that occurs in information technologies (Nagy et al. 2011, 2013).

Innovation in a well-established technology has its limits, however. This limitation is becoming clear precisely in the technology for which Moore's law was originally formulated. It is still possible to increase the number of components in an integrated circuit by reducing the size of the transistors, but the cost of making them smaller is gradually going up. Transistors currently measure around 14 nm (1 nm $= 10^{-9}$ m) in length and it is possible to make them shorter, but for lengths of less than 5 nm, a length smaller than the thickness of a cell membrane, the manufacture of transistors becomes too expensive to keep production profitable. There will still be innovation in the manufacture of integrated circuits, but Moore's law, as initially formulated, will probably cease to be true within the next decade. Furthermore, the emergence of new technologies, such as cloud computing, mean that increasing the processing speed of personal computers is becoming less important. In the field of computing, quantum computing is expected to be one of the new macroinventions, able to generate new technological cycles.

Some new technologies cause deep economic transformations, producing a process that the German economist Werner Sombart (1863–1941) called "creative destruction". This concept was further explored by the Austrian economist Joseph Schumpeter (1883–1950) (Schumpeter 1942) in the context of a Marxist critical analysis of the process of accumulating and destroying wealth that is typical of capitalism. For Schumpeter, radical technological changes are the driving force of long-term economic growth in capitalism, even though they destroy many industries, companies, and jobs as they are overtaken by such changes. For him, "this process of creative destruction is the essential fact about capitalism" (Schumpeter 1942). Regarding the future, he wrote that "capitalism's very success undermines the social institutions which protect it, and 'inevitably' creates conditions in which it will not be able to live and which strongly point to socialism as the heir apparent" (Schumpeter 1942). Schumpeter also argued that creative destruction generates groups of unemployed people, many of whom become intellectualised by higher education due to the progress in education, and their discontent and resentment is rationalised through social criticism and "moral disapproval of the capitalist order" (Schumpeter 1942).

Nevertheless, the paradox of progress persists because society's economic and financial elites are convinced that they cannot harvest the rewards of creative destruction without some citizens being worse off in the short term and probably in the long term. Protecting industries that are declining and trying to preserve the jobs they provide obstructs the supreme goal of progress. The flourishing of the new industries requires the collapse of the old industries.

The expression "creative destruction" has fallen into disuse, but the concept re-emerged in the milder, self-assured forms of "disruptive technologies" and "disruptive innovations", introduced by Clayton Christensen (1997; 2003). Christensen distinguished between sustainable technologies that are already well established and that remain in place thanks to gradual improvements, and recent innovative, unstable, high-risk disruptive technologies that have not yet proved their profitability on the market. In his books, he tried to identify a way for large companies to develop, apply, and benefit from disruptive technologies, which could potentially become very profitable. Disruptive technologies cause significant reorganisation within an industry, and can even cause a new industry to emerge. In ICT, personal computers, smartphones, tablets, and cloud computing are examples of disruptive technologies. Robotics and artificial intelligence are examples of emerging disruptive technologies.

In the fossil fuel industry, fracking is a disruptive technology. It was discovered, developed, and first used in the USA in 1950, and made it possible to exploit oil and natural gas in several types of organic shale. It has significantly increased exploitable reserves, creating a new industry and disrupting the world fossil fuel market by making the USA practically self-sufficient in oil and natural gas. The use of fracking to access unconventional oil and natural gas resources is known to have adverse effects on public health (Kibble et al. 2014; Hill and Ma 2017) and environmental implications (Broomfield 2012; Tatomir et al. 2018). However, in the countries using fracking, such risks are considered to be manageable and the benefits to the economy to be far superior to the residual health and environmental losses that it may be impossible to avoid.

In the medical field, the gene editing CRISPR (clustered regularly interspaced short palindromic repeats) technologies (Lander 2016) are likely to become disruptive. A more futuristic set of disruptive technologies are those that would enable the development of outer space colonization on the Moon, Mars, and other celestial bodies, possibly with the aim of improving the survival prospects of *Homo sapiens* (Tierney 2007).

Those involved in the process of investing to promote technological progress firmly believe that progress has to move as fast as possible to avoid any form of stagnation, and disruption is one of the instruments that helps to fight stagnation. Through technological progress, they drive the acceleration of operative social time. Mark Zuckerberg, the founder of Facebook, wrote in a 2012 manifesto designed to attract researchers, that it was necessary to "Move fast and break things. The idea is that if you never break anything, you're probably not moving fast enough." The increase in speed and the associated disruption are considered to be indispensable conditions for economic and financial success.

5.12   The Environmental Crisis, Ethics, and the Expectations of Economic Growth and Technological Progress

Global anthropogenic environmental changes and the overexploitation of natural resources are two factors that are relevant when analysing and assessing the prospects for human progress. Both have direct and indirect influence on the future of humanity through their impacts on human health, social development, and the economy. The very positive expectations that most people have about technological progress in the coming decades are not matched by similar expectations of environmental sustainability, sustainable use of natural resources, and control of climate change. A few currents of opinion in society, arising especially among scientists and environmentalists, believe that humanity is facing what can only be called an environmental or ecological crisis (van Noort 1991; Berger and Kelly 1993; Hoff and McNutt 1995; Laybourn-Langton et al. 2019). Although the evidence for a global environmental crisis is abundant and indubitable from the scientific point of view, the concept of a crisis is far from being universally accepted. It has not penetrated the bulk of society, and is dismissed as over-pessimistic and damaging to the cherished expectations of increasing economic prosperity and well-being for billions of people. Societies have been acknowledging the need to protect the environment, to implement a circular economy and to combat climate change, but their actions remain insufficient and would be unable to stop or even slow down the environmental crisis.

There is a deep divide between those that consider the environmental crisis to be a crisis of modernity and modern civilization and those that consider that we are not really facing an environmental crisis, but just a set of problems that will be solved one by one with the appropriate new technologies, without endangering the future of humanity, and most importantly, without putting at risk the viability of continuing along the path of robust economic growth and economic prosperity. The first group of people think that the fundamental assumptions of our civilization are increasingly at odds with the sustainability of environmental and natural resources. The second look at our home, the Earth system, as a resource that can be exploited, transformed, and polluted without limit.

The two points of view regarding the environmental crisis are mirrored in neoclassical and ecological economics. Neoclassical economics adopts the weak sustainability concept, according to which the human use of the environment and its natural resources is viewed as a purely economic problem. According to this view, an economy is sustainable if the value of economic output does not decline with time and there are no foreseeable limitations regarding the substitution of natural capital by human made capital. Ecological economics adopts the strong sustainability concept, according to which sustainability is viewed not only as an economic problem, but also as a problem of maintaining essential and non-substitutable environmental assets. Natural capital cannot be completely replaced by human made capital, and it is essential for long-term economic stability. The two points of view are also correlated with political views on the left and right of the political spectrum. In general, left-wing parties and individuals are more concerned with inequalities, distributional issues in society, and market overvaluation, and they have more pro-environmental views and attitudes, so are more likely to support ecological economics than their right-wing counterparts (Neumayer 2004). This divide implies that governmental environmental policies are strongly correlated with the political ideology of the political parties that support the government and also with the prevailing economic conditions in the given country.

There is an enormous effort to transmit the idea that technological progress can solve what is presented by some as the environmental crisis. The conceptual groundwork of this movement, sometimes called environmental modernism, ecological modernism, or ecomodernism, has been presented by Jesse Ausubel as the only realistic and pragmatic way of acting to defend the environment in the medium and long term (Ausubel et al. 1989). Ecomodernism counters environmentalism, harshly criticised by Ted Nordhaus and Michael Shellenberger in their well-known essay *The Death of Environmentalism* (Nordhaus and Shellenberger 2004) and later in *Break Through* (Nordhaus and Shellenberger 2007). They denounce environmentalism for developing and encouraging "eco-tragic narratives", "the eco-apocalypse", and "limits to growth" without building an attractive view of the future based on a set of values that represent what they consider to be true, inalienable human expectations. According to these authors, environmentalism cannot resolve environmental problems because it is unable to win most people over, since they are much more interested in guaranteeing economic prosperity and the "good life", whatever the present and future environmental consequences of the current paradigm of development.

Ecomodernism considers that, in order to defend the environment, it has became necessary to attract people to the "politics of possibility" rather than confronting them with the "politics of limits". For them, the path of the possible is based on increasing public and private investment in technological innovations aimed at simultaneously improving the quality of the environment and economic growth, and improving economic prosperity and well-being using the Western model as the reference. Achieving this combination of goals is unlikely, but since it favours the current dominant political and economic narrative, it is unsurprising that Nordhaus and Shellenberger have been honoured as "heroes of the environment" by the American magazine Time in 2008 (Walsh 2008). Not long after, in 2009, Stewart Brand defended nuclear energy and geoengineering to fight climate change, the use of Genetically Modified Organisms (GMOs) to guarantee food security at world level, and the development of cities with higher population densities to free up land that could then be used to "restore nature" by rebuilding habitats and biodiversity (Brand 2009).

In April 2015, "An Ecomodernist Manifesto" was produced, signed by eighteen ecomodernists (Asafu-Adjaye et al. 2015). The manifesto defends the possibility of technological progress decoupling humanity from nature, allowing the current model for increasing economic growth, prosperity, and well-being to be applied at global level, but without significant environmental impact. Once this future transition has been guaranteed, the opportunity to "re-wild" and "re-green" the planet will emerge. The final goal is for humanity to have "a good, or even great, Anthropocene". The manifesto presents a techno-optimist view that is somewhat simplistic, bearing in mind the dimensions, intensity, and dynamics of the contemporary environmental crisis, besides containing some unfounded assumptions.

One of them is accepting that cities exemplify and encourage the decoupling from nature because they reduce environmental impacts. A recent study shows that people living in a set of 27 megacities (with populations greater than 10 million) use more electricity and oil and produce more waste material per capita than the per capita world average of the same indicators (Kennedy et al. 2015). Furthermore, most of today's large cities are not examples of social sustainability. In countries with advanced economies, cities created 60% of jobs in the last 15 years and family income is on average 18% higher than in rural areas (OECD 2016a). Nonetheless, the growth in wealth in large cities throughout the world has not been inclusive. According to the same OECD study, inequality, measured using the Gini coefficient, is worse in cities than the national average.

The horizons of expectation of environmentalism and ecomodernism differ profoundly. The former seeks to maintain the integrity of an already deeply transformed natural environment and its ancestral relationship with humans as far as possible, while the latter assumes a dissociation between humankind and the natural environment, replacing it with a new "nature" rebuilt and controlled by technology. Environmentalism is a well-established, robust movement that is attracting an increasing number of followers but with horizons of expectation that tend to be downgraded or frustrated as time goes by, because a reversal of the present global environmental crisis is beginning to look extremely difficult. There have been many successes, but only at the local level, scattered throughout the world. The progressive downgrading of the global goals of environmentalists contributes to the acceleration of their operative social time.

On the other hand, ecomodernism is a movement that opposes environmentalism, has much less capacity to attract followers, but enjoys the protection and support of the economic and financial establishment. Its horizons of expectation are continuously renovated and enlarged by the promise of technological progress. There is a permanently renewed expectation to re-wild and re-green the planet, although the evidence for the environmental crisis is accumulating. Once again, time is an essential factor in the debate about the environmental crisis and possible solutions to it. From a moral and ethical point of view, it is hard to explain the lack of action to control this crisis, considering that it disproportionately impacts poor people around the world. Maybe we could say that after a relatively short period of time, compared to the timescale of the impacts of the global environmental crisis, we will all be dead, meaning that we will avoid experiencing its most serious future consequences for human populations and the environment.

In a very different context, John Maynard Keynes wrote (Keynes 1923): "The long run is a misleading guide to current affairs. In the long run we are all dead. Economists set themselves too easy, too useless a task if in tempestuous seasons they can only tell us that when the storm is past the ocean is flat again." Keynes criticises economists for arguing that economic crises are transitory, rather than facing and dealing with them. The environmental crisis, unlike most economic crises, is not a short-term anomaly, but the manifestation of a growing political problem that will get worse in the medium and long term unless we act. If no consequent action takes place, the contemporary social generation loses relatively little but seriously harms the prospects for later generations. When the horizons of expectation of a generation are limited to its generational time, it becomes very difficult or even impossible to resolve multi-generational issues. If we could indeed reach the goal, desired by many, of indefinitely extending human life with quality of life, the justification not to act in the short term would no longer be valid. We would like to live much longer into the future, but we don't care much about the long term implications for future generations of our present individual and collective actions.

Time also intervenes crucially in the way we depend on and interact with the environment. The time taken to respond to anthropogenic interference in the biogeophysical subsystems of the Earth system is generally longer than the time taken to respond to disturbances generated within the human subsystem. This difference in timescale tends to underestimate the importance attributed to anthropogenic interference. However, the scientific knowledge that has been accumulated and the current scientific research on the Earth system allows us to make, not only short-term forecasts, but also medium and long-term projections about their behaviour with decreasing uncertainties. We are now able to identify which of the different components of the biogeophysical subsystems of the Earth system have a relatively fast response time—hours or days—and which only react slowly, over tens, hundreds, thousands, or millions of years.

The fact that some biogeophysical subsystems have relatively long response times has implications for the way in which we react to the risks that are being generated by the environmental crisis. From a human point of view, it is very different if we can foresee that a risk will manifest itself severely in the coming hours, days, or months, or indeed in the next few decades or centuries. In the case of the former, we are faced with a risk associated with an event which, if it takes place, is very likely to affect us and our family. In this situation, response to the risk is driven by an instinct for survival and to protect our lives and the lives of those who are close to us. We share our concern with all those who are subjected to the same risk, and we are willing to help them.

The situation changes when the risk only becomes severe in the long term, for example, in time intervals of 50 to 100 years, as is the case for anthropogenic climate change. In this case, the relatively long time interval that separates us from more serious situations in the future makes it unlikely that they will significantly affect our lives, our well-being, and our economic prosperity. However, future generations, including our descendants, will very probably be more severely affected. The justification for acting in the present to reduce future risk in the long term is no longer our survival, but solidarity between generations, which involves ethics and morals, since it is above all about others and not ourselves.

There is another factor related to time that makes it harder to find an answer to the question of medium and long-term environmental risks. Our consciousness builds an adaptive memory that enables us to react to a dangerous situation that we are already familiar with, because we have had to face it in the past. Past experience has taught us to recognise and assess the risk of similar situations that appear on the horizon and has prepared us to act to reduce or remove that risk. Nonetheless, if the dangerous situation is in the future and is entirely new, and bears no similarity to situations we have already been through and dealt with, we are forced to imagine such a situation and the consequences that it involves for us. Our memory file does not contain anything similar and all the work to prepare ourselves to face the new situation and act to try and avoid it is based solely on our imagination. We are poorly prepared from a biological standpoint for dealing with future situations that have no correspondence or analogy with those we know from the past (Boyer 2008; Schacter et al. 2012). Such situations are similar to many linked to medium and long-term environmental risks. We find it very difficult, for example, to imagine the state of coastal areas around the world when the global mean sea level is around one, two, three, or even four metres higher than it is now. Since past personal or historical experience does not involve anything like this, it will be hard to respond to the risk and take the decision to remove its causes.

The environmental crisis is essentially a political crisis that results from the current tendency to depreciate ethical and moral values. This is likely to be a long term trend, and we may therefore expect a long and damaging environmental crisis. There is, however, a possible mechanism for readjusting this within the current economic system that may help slow down the environmental crisis. It is becoming increasingly noticeable that the impacts of environmental degradation, overexploitation of natural resources, and climate change are beginning to have a negative impact on the economy. The insurance industry, which is particularly affected by extreme weather events, has recorded a tendency over the last few decades toward an increase in the number of global weather disasters and in the associated insurance losses. The number of events causing economic losses tripled between 1980 and 2014. The year of 2017 was the costliest year ever in terms of global weather disasters, with global overall losses amounting to about 320 billion US dollars and insurance losses of 133 billion US dollars (Munich 2019). The explanation for the extreme costs in 2017 were the strong tropical cyclones that hit the USA in that year. The observed increase in global weather disasters cannot be attributed solely to climate change. It is most likely also caused by population increases and the continuing use of vulnerable locations for homes and economic activities, especially in less developed countries.

Climate change risks for companies are rising, but company premiums and credit have not yet become significantly more expensive. A case that recently grabbed the headlines was the California utility PG & E (Pacific Gas and Electric Company), which was forced into bankruptcy protection in January 2019, because insurers and creditors concluded that it had probably accumulated billion of dollars in liabilities over its likely role in sparking California wildfires. The more frequent droughts in California, caused by climate change, have dried out much of the state, decimated forests, and increased the risk of wildfires. PG & E was considered to be the first climate change company bankruptcy, but it is likely that others will follow (Gold 2019).

Climate change creates two types of risk to the financial sector: physical risks that expose assets to the impacts of climate change and transition risks related to the mitigation of climate change (Lepousez et al. 2017). Preparing companies to participate in the energy transition involves short term costs, while the events that involve physical risks are to a large extent aleatory and may only require significant expenses in the medium and long term. In general, markets tend to punish companies that calculate previously unacknowledged risks, so most companies prefer not to calculate the physical risks of climate change, and avoid the energy transition as far as possible. However, this may become an increasingly dangerous policy as time goes by.

The most likely mechanism for decelerating the environmental crisis is the increasing cost of its harmful impacts on economic activities, and eventually its capacity to slow down economic growth. A strong reluctance to recognize and assume that capacity should be expected since it strikes at the core narrative of the current economic paradigm. Eventually, it will become extremely difficult to deny the effect of the crisis on economic growth, and the way it reduces the well-being and economic prosperity of billions of people. At this stage, assuming a gradualistic scenario, it becomes likely that a stronger tendency to promote sustainability may emerge. In a non-gradualist scenario, it is much more difficult to project the future course of events.

Invoking ethical and moral arguments to attenuate and possibly revert the environmental crisis is very important and should be pursued vigorously, but its contribution to changing the overall trends is likely to be tiny. The impacts of the recent strong tropical cyclones in different parts of the world reveal just how insensitive the world is to the ethical issues underlying climate change.

The 2017 tropical cyclone season was the costliest on record, amounting to losses of 215 billion US dollars (Munich 2019). In that year, the USA alone accounted for 50% of losses due to global natural disasters, as compared to a long-term average of 32%. When considering the continent of North America, including Central America and the Caribbean, the share rises to 83%. This disproportion results from two main factors: first, three major tropical cyclones, namely Harvey, Irma, and Maria, hit the USA in 2017; and second, living standards are relatively high in the USA, which increases the value of the assets exposed to tropical cyclones. Harvey reached category 4 on the Saffir–Simpson wind scale, while Irma and Maria reached category 5.

There is growing evidence that climate change is increasing the intensity of tropical cyclones, with the strongest storms increasing the most (Bhatia et al. 2019). This tendency is related to the increase in sea surface temperatures caused by global warming. There is evidence that tropical cyclones are tending to move more slowly, which implies that they are likely to release greater amounts of rainwater in the regions where they pass (Kossin 2018). Furthermore, these storms are slowly migrating poleward, and the area where they occur is expanding. Five out of the seven ocean basins where tropical storms occur have had the strongest tropical cyclone on record in the past five years. Tropical cyclone Irma sustained wind speeds of 300 km/h for 37 hours, longer than any storm on record. Tropical cyclone Harvey released more rain in the USA than any other storm on record.

When a tropical cyclone hits the least developed countries in the world, as happened recently in Mozambique, Zimbabwe, and Malawi, the situation is very different. Idai was a category 3 tropical cyclone that landed in the city of Beira in Mozambique on 15 March 2019 and left many hundreds of people dead, many more missing, and about 1.8 million displaced or affected by the storm. The storm surge coupled with torrential rains, including earlier rainfall, led to disastrous flooding throughout the region. The destruction of poorly built houses and the devastation of about 800 000 ha of crops that were near harvest left people living in squalid conditions, some with very difficult access to drinkable water and food, and battling cholera, malaria, and other diseases. The losses caused by Idai have not yet been accurately estimated, but they have been estimated to be of the order of 2 billion US dollars, much less than the losses of 125, 53, and 91 billion US dollars resulting from the tropical cyclones Harvey, Irma, and Maria. However, the number of victims, the number of people with destroyed homes, and the human suffering are considerably greater. Idai was one of the most damaging and strongest cyclones ever to hit the Mozambique coast, and climate change is increasing the probability that similar tropical cyclones will follow. Mozambique is much more vulnerable to climate change than the USA, but the latter contributes much more to climate change at the individual level than the former. In fact, greenhouse gas emissions per capita in Mozambique and the USA in 2013, were 1.04 tCO2e and 19.9 tCO2e, respectively (WRI 2014).

Is technological progress going to resolve the environmental crisis? It will surely contribute to mitigating some of the more harmful impacts of the crisis, but it is doubtful that it will be able to slow it down in any significant way. The main reason is that the main driver of technological progress is the free market, which is to a large extent disconnected from environmental concerns. Only those technologies that are likely to maximize profits are able to attract the investment that will enable them to be developed and reach the market. If a technology sells well and turns out to be profitable, society feels obliged to accept it. Invention and innovation have become ends in themselves and tend to drive the economy. Today, human hubris is expected to transcend all limits with brain power, artificial intelligence, and money, and it is difficult to fit reverting the environmental crisis into such a program.

5.13   Indigenous Environmentalism and the Progress of the Modern Economy

Environmental degradation at local, regional, and global scales, overexploitation of some natural resources, pollution, and anthropogenic climate change are essentially ethical and moral problems. They are not only moral but also ethical because as well as involving individual behaviour, they entail ethical principles built up and agreed upon collectively. It will only be possible to stop these trends if humanity changes its ethical behaviour, attributing greater value to solidarity with the most vulnerable contemporary generations and future generations. In fact, it must not only be valued, but also guaranteed through action.

This moral progress is highly desirable and it could be achieved, but that remains unlikely, although it may become more likely as the effects of a growing environmental crisis become more serious. The impact that environmental issues have will be gradually amplified by their negative effect on the social and economic sectors. For instance, they will hinder national and global economic growth by increasing inequalities between countries with advanced economies and the other more vulnerable and fragile states, thereby encouraging migration.

Alongside the ideological debate about the environmental crisis, one finds many examples of resistance against environmental degradation and destruction in tropical and subtropical regions, where there is still some harmony between the local communities and the ecosystems that directly support their lives. The extraordinary expansion of agriculture, particularly cattle farming, soybean and palm oil production, the construction of dams to generate electricity and irrigated farming, and the growing exploitation of natural resources, particularly forest products and mineral resources, is frequently done with serious damage to the living conditions of local communities. Some of the leaders of those communities end up reacting.

According to the Global Witness annual report for 2016 (Global Witness 2016), 185 environmental activists and defenders of their land, forests, and rivers, spread over 16 countries, died in 2015 while defending their territories, representing a 59%

increase compared to 2014 and the largest total since 2002. Around 40% of these activists belong to indigenous peoples that live in remote regions whose natural resources are coveted by governments and companies. These murders, performed by military or police officers, the security forces of state or private companies or contract killers, are hard to investigate because of the protection and assistance given by public and private economic interests. The victims' actions to defend the environment where their communities live are presented to public opinion as attempts to stand in the way of the country's development and progress, and this makes the killings politically acceptable, particularly in some regions of Central and South America, Africa, and South-East Asia. The countries with the highest number of deaths related to environmental activism in 2015 were Brazil, with 50, the Philippines, with 33, and Colombia, with 26 (Global Witness 2016).

One of the most dangerous regions in the world is the island of Mindanao in the Philippines because of its great wealth in minerals, especially copper, gold, aluminium, iron, nickel, oil, and natural gas, and the resistance that indigenous peoples have led against occupation of their land and environmental degradation. The mines are exploited by some of the biggest multinationals, including BHP Billiton, Anglo-American, Russell Mining, Xstrata, Sumitomo Metal Mining, and Toronto Ventures.

A good example that reveals the complexity of the current situation in some developing countries that possess profitable natural resources to be exploited, is the case of Berta Cáceres (1971–2016), a human rights and environmental activist from Honduras. After receiving death threats, she was murdered just before midnight on 2 March 2016 at her home in La Esperanza, in south-west Honduras, for opposing the construction of the 21.7 MW Agua Zarca hydroelectric dam on the Rio Gualcarque, located in the region of her indigenous people, the Lenca. The Mexican environmental activist Gustavo Castro Soto witnessed the crime and was injured, but survived by pretending to be dead. He was then stopped from leaving Honduras. The private firm behind the dam project was Desarrollo Energético Sociedad Anónima (DESA) whose president Roberto David Castillo Mejia studied in the USA West Point Military Academy, served as assistant to the director of the Honduran Army Intelligence and was prosecuted for corruption charges while he worked for the Honduran goverment.

Around 30% of the territory of Honduras is granted to mining companies, creating a large energy demand which, according to government plans, will be met by a range of dams. Mining and dam construction has caused local communities to be displaced and has transformed and degraded the environment, profoundly changing the living conditions of local populations. In June 2009, President Manual Zelaya Rosales was ousted in a *coup d'état* and the new regime unleashed a wave of violence against community leaders, lawyers, journalists, and human rights defenders. In August of the following year, 47 hydrolectric dam concessions were granted by the new regime. The concessions were given without any previous consultations with the people who would lose their lands, their forests, their culture, and some of them even their lives. It is estimated that in Honduras, since 2010, at least 124 environmental and land activists opposing mines, dams, logging, and tourist resorts have been murdered, making it one of the most deadly countries in the world for such campaigners (Global Witness 2016).

However, the supply of mineral resources at reasonable prices is really essential to sustain global economic growth and to drive technological progress. This recognition is shared throughout the world by governments and companies alike. The mining projects in Honduras are just an example among many others around the globe. In Honduras, as elsewhere, they have been backed by leading politicians, business figures, and army officers, but many are bankrolled by international funders.

The Agua Zarca hydroelectric power station was to be built by the Chinese state-owned Sinohydro Group, the world's largest dam developer, and financed by the Dutch Entrepreneurial Development Bank (FMO), the Finnish Fund for Industrial Cooperation Ltd (FinnFund), the Germany company Siemens, the US government's USAID, the Central American Bank for Economic Integration (CABEI), which brings together several countries in Central and South America, and several other institutions. In 2013 Berta Cáceres wrote to FMO, asking the bank not to finance Agua Zarca, but the bank did not respond and granted the loan for the construction of the dam.

Berta Cáceres' mother was a social activist who helped many refugees from El Salvador. She served as a two-term mayor of La Esperanza, as a congresswoman, and as governor of the Department of Intibucá. Berta Cáceres was very close to her mother and grew up to become a university student activist, graduating as a teacher. In 1993, she cofounded the National Council of Popular and Indigenous Organizations of Honduras (COPINH) and was listed in 2009 by the Inter-American Commission on Human Rights as being under threat during the 2009 Honduran *coup d'état*.

The Lenca people witnessed the arrival of machinery and construction equipment to the Rio Blanco communities but had no idea what the construction was for and who was behind it. They approached Berta Cáceres at the COPINH who identified the Aqua Zarca dam project and its promoters and worked together with the communities to mount a protest campaign. Construction began in 2011 but the developers had violated international law by failing to first consult with the local people about their project in Rio Gualcarque. The communities suspected the mayor of Intibucá Martiniano Dominguez of making illegal sales of areas of land, thus allowing DESA to claim that it had the title to the area for the Agua Zarca dam. DESA has systematically refused to share the documentation for the land rights they claim to possess.

The Lenca people were afraid that the dam would cause the Gualcarque, which is a sacred river to them, to dry up, leaving them without access to water and unable to irrigate their common farmland. Berta Cáceres organized legal actions and community meetings and led a protest where people peacefully demanded their rightful say in the project. In a municipal assembly convened by Dominguez, the project was rejected, but he went ahead and signed the contract with DESA.

From 1 April of 2013 onwards, resistance to the project increased. The military and police forces responded by deploying a massive repression and criminalization of the protesters and COPINH members. On 15 July 2013, the indigenous leader Tomás Gracia was gunned down by an army officer when the Lenca people blocked the access roads to the dam's construction facilities. Others were attacked with machetes, detained, and tortured. The soldier who killed Gracia and wounded his son was initially acquitted but later, following COPINH intervention, he was given a prison sentence of less that 90 days. Nevertheless, the road blockade to the DESA construction site persisted, and in October 2015 the dam project was relocated to the west side of the Gualcarque to avoid affecting the lands of the La Tejera community, one of the strongest opponents to the dam. It was also the community that was most affected socially by the project and by the reaction to its opposition. Violence and crime had risen and several villagers had been attacked and murdered under mysterious circumstances. Conflict between communities increased in part because members of other communities were reported to be paid by DESA to guard the construction site.

The persistence of the protests, the repression, and the killings obtained visibility in the international media, in the social media networks, and in NGOs. In 2015 Berta Cáceres won the Goldman Environmental Prize for her campaign against the building of the Agua Zarca Dam. Only 12 days after Berta Cáceres' murder, on 15 March 2016, another Honduran environmental activist, Nelson García, was killed after his protest and failed attempt to avoid the violent expulsion of a group of Lenca families from their traditional indigenous houses and the destruction of their crops at the Río Lindo settlement, where the plan was to install a cattle farm promoted by the Honduran government.

The silence of the foreign banks and companies that were financing and building the hydroelectric Agua Zarca project became untenable. Most of the institutions suspended their activities and terminated their contracts including Sinohydro. The case was slowly and reluctantly brought to justice and on November 2018 a court in Tegucigalpa ruled that Berta Cáceres' homicide was carried out by a group of seven hitmen on the orders of DESA's executive director Roberto David Castillo Mejia, because there was frustration over the costly delays in the construction of the Agua Zarca dam that resulted from the protests. Roberto Castillo Mejia was the mastermind behind the homicide, providing the necessary logistics and other resources. Most of those indicted had strong connections with DESA and the USA military. Douglas Bustillo was a US-trained former army lieutenant and Mariano Díaz Chávez an active US-trained special forces major who had previously served in the Iraq war.

The case of Berta Cáceres is paradigmatic of the pressures exerted on developing countries to force their transition to the contemporaneous model of progress as promoted by the advanced and emerging economies and the reactions it may generate among their citizens. There are many projects throughout the world similar to this one in Honduras, but we hear nothing about them. The case of the Agua Zarca dam didn't evolve according to plan, and it is because it went wrong that it ended up revealing the inner workings of the wave of progress that overwhelms the world and ensures the persistence of the environmental crisis. As often happens, what made this case exceptional was the clumsiness and lack of experience of the promoters and the strong personality, vision, determination, and endurance of Berta Cáceres.

In retrospect it may look surprising that the Rio Blanco communities rejected the progress that was heading their way and did not finally agree to the construction of the modified version of the dam, which involved run-of-river hydroelectricity. As the case evolved and gained visibility, some investors got much more careful and smarter. They redesigned the project to minimize the environmental impacts to standards much closer to those required in their more developed countries. They also pointed out their full respect for human rights in all their activities and repeatedly emphasized the economic and environmental advantages that would result from building the dam (FMO 2016), for instance, providing renewable electricity to various communities, access routes, drinking water systems and street sewers in one of the communities, Santa Fé (FMO 2016). Nevertheless, it was too late because there had been various high profile murders and clear evidence of corruption, and trust had been destroyed. FMO and FinnFund left the project in 2017, and only CABEI remains invested in it (Dumas 2017).

It is highly probable that DESA or some other company will eventually build a dam on the rivers of the Rio Blanco communities, and that they will all have electricity, sewage systems, and a new road system. Companies will arrive to exploit forest and mineral resources, agriculture will expand, cars and trucks will arrive on the roads and streets of the villages, the number of shops in the villages will increase, and they will sell an increasing variety of goods and services, whereupon most traditional ways of life will fade away in just a few generations.

However, this type of model may take quite different forms and there are many different ways to reach them. In Honduras, as in other coutries with similar levels of social and economic development, there is a strong tendency to avoid consultations with indigenous or local communities before taking decisions on how to exploit their land, forests, and rivers. Once again, we find a problem intimately related with time. The intention is, as fast as possible, to bridge the gap between two cultures and two visions of the world belonging to operative social times separated by centuries. There is no interest, opportunity, or time to find out whether people want to change their operative social time so fast, and no consideration of how best to do it. Furthermore, people know intituively that the main casualty of the transition will be their beloved nature. They agree with Berta Cáceres when she says that: "Mother Nature – militarized, fenced in, poisoned – demands that we take action" (Cáceres 2015). The dominant vision in the world is insensitive to this attitude and firmly believes that disruption is one of the main secrets of progress.

Honduras has one of the highest levels of inequality in the American continent with about 60% of households in rural areas living in extreme poverty, on less than 2.50 USA dollars a day (Global Witness 2017). The exploitation of its valuable forests and minerals mainly benefits a small section of society that can obtain lucrative contracts and official licenses to start projects through political influence and corruption. Although international and municipal laws guarantee that communities should be consulted on the use of their land, these rights are not being protected, because governments, companies, and foreign investors take advantage of the powerlessness that arises from extreme inequality to avoid engaging meaningfully with the affected communities. Rural communities that stand against the imposition of new dams, the exploitation of mines, and intensive logging or agriculture suffer shocking levels of violence and intimidation. Honduras has been considered one of the most dangerous places in the world to be an environmental activist (Global Witness 2017).

It is very unlikely that the future will leave the Rio Gualcarque and its surrounding areas in the pristine state that the community of La Tejera remembered and invoked during the protests. However, that may have nothing to do with local development but with climate change, which is a global issue. It is now well established that climate change is increasing extreme-weather phenomena, some related to ENSO events, especially droughts, flash floods, landslides, and more intense tropical cyclones in the region that extends from the Pacific Coast of Chiapas in Mexico down to Guatemala, Honduras, El Salvador, Nicaragua, and the western provinces of Costa Rica and Panama (IPCC 2013; Imbach et al. 2017, 2018).

The region has been named the "Central American Dry Corridor", because one of the most significant features of climate change to those that live in rural areas are the dreaded droughts that bring famine or force emigration. Honduras, Nicaragua, and Guatemala are among the ten countries with highest long term climate risk (Kreft et al. 2016). Furthermore, the vulnerability of these countries to climate change is even greater because about 75% of the population depends on agriculture, which greatly reduces their capacity to adapt. In the arid, highland regions of Guatemala the malnutrition rate is about 65%, among the highest in the western hemisphere. According to the World Food Programme (WFP 2017), nearly 50 percent of children under five years old are considered chronically malnourished in Guatemala, and this percentage increases to 90% in some rural areas. About 60% of those migrating from the most affected regions of Guatemala mention climate change and food security as their reason for leaving (Brigida 2018). A large proportion of the immigrants are women with their children, probably because they have lost hope of surviving in their ancestral lands. From 2015 to 2017, more than 158 000 unaccompanied minors and more than 175 000 women and children from Central America's so-called Northern Triangle—Guatemala, El Salvador, and Honduras—crossed the US border (Brigida 2018).

These migrations have generated one of the most highly developed networks of illegal immigration, with "coyote" smuggling guides who deliver families to the US border. However, significant proportion of those that migrate are forced to return. Two to three planes a day land in Guatemala City, each carrying around 150 Guatemalan citizens who have been deported or intercepted as they attempted to cross illegally into the USA (Steffen 2018). Guatemala, Honduras, and El Salvador are now the countries of origin of the majority of migrants crossing the Mexico–US border.

To stem this flow, the present US Government is planning to build a wall along that border with Mexico, similar to the one that Israel built to separate the country from Palestinian Territories. It is currently a hot topic of political debate, the question of how to stop this flow of illegal emigrants into the USA, which some consider to be "stone cold criminals" while others have different opinions (Levitz 2018). Anti-immigrant policies are being ruthlessly implemented at the Mexico–US border. One recent policy involved separating young children from their parents by force and housing them in wire cages (Mason 2018).

Rising poverty rates, high levels of violence associated mainly with the traffic of drugs between South American producers and the US consumer market, and increasing food insecurity produce a bleak outlook for the future of people living in the Dry Corridor. An important part of the solution to these problems would be to mitigate climate change at the global scale, increase the resilience of the populations to the effects of the changing climate, and more generally invest in development aid and provide direct foreign investment at a more adequate level. Some countries and international organizations are funding programs to support food security and strengthen community resilience to climate change (WFP-EU 2017; WFP 2017), but these highly commendable efforts remain insufficient.

The situation in the Dry Corridor of Central America provides a further example of the dysfunctionality of the current global social and economic system. The economic narratives provided by the large corporations for the countries in this region tend not to include the increasing food insecurity and poverty, nor the rising inequality, and they don't usually acknowledge the impacts of climate change. A characteristic example of this standpoint is provided by a recent report by McKinsey (Cadena et al. 2019), the iconic American management consultancy firm which works for about 80% of the world's largest corporations and for many governments, and has significantly contributed to shaping the evolution of the US economic model and its influence around the world (McDonald 2015). The "McKinsey Way" of management has developed many of the concepts and practices applied by US business.

The report, entitled "Unlocking the economic potential of Central America and the Caribbean (CAC)" is focused on economic growth and compares the potential for high GDP growth across the various countries in the region. They are divided into four groups—stars, falling behind, emerging, and laggards—according to the GDP per capita and the GDP growth in the period 1987–2017. The laggards include Central America's Northern Triangle, countries where the USA has been involved for many decades in regime change and the support of military and authoritarian regimes that have waged a brutal war of repression on the indigenous way of life (Granjon 1982; Kinzer 2007; Grow 2008).

According to this report, the reason why some countries, such as Guatemala, Honduras, El Salvador, and Nicaragua, are laggards is their high dependence on commodities and low levels of investment. In the whole report there are no references to the words poverty, food security, or climate change. Migration is referred to and it is emphasized that "flows into CAC have been mostly from within the region, from South America, and from the USA to the three-star economies (Panama, Dominican Republic, and Costa Rica) in CAC (which accounted for 45 percent of all immigrants to CAC in 2017)" (Cadena et al. 2019). As regards the "outward flows, in 2017, 92 percent of all emigrants from Nicaragua and the countries of the Northern Triangle had Costa Rica, Mexico, or the United States as a final destination" (Cadena et al. 2019). However, no analysis of the reasons for these flows is forthcoming.

The report is strictly focused on application of the mainstream modern economic model to the identification and assessment of the opportunities for high return on investment in CAC. The political and military history of the region, the growing inequalities, the precarious lives of the indigenous peoples, the environmental degradation and the harmful impacts of climate change are considered to be utterly irrelevant to the purpose of the report. The objective is totally purged of such considerations, focussing purely on ways to increase economic growth and make more money in CAC. No distractions, no nuances, just straight to the point. No need to mention that all the ignored socioeconomic problems in CAC will be swiftly solved once the objective of very strong economic growth is achieved throughout the whole CAC region. The report is optimistic about the possibility of achieving this goal and of diminishing the disparities in economic growth and GDP per capita in the region. It estimates that what has been learnt "from regional success stories could boost exports of goods and services (excluding commodities) across CAC by up to US$ 140 billion by 2030" (Cadena et al. 2019). The report is just a reflection of McKinsey core ideology, which defends the idea that the same management theory that makes business more profitable can and should be applied to further the public interest. It assumes that progress in the modern economy will necessarily further the common good.

There are many examples showing that the large multinational corporations which currently play a dominant role in the global economic system consider that the remaining indigenous peoples throughout the world practice an anachronistic way of life that has no future and frequently constitutes an impediment to economic progress. Furthermore, they don't acknowledge any form of environmental crisis that would disproportionately affect those indigenous nations.

One region of the world where these beliefs are particularly evident is the Amazonian forest in Ecuador. Over the last 50 years, multinational oil companies and the Ecuadorian state oil company have opened roads through the rainforest to construct oil platforms and pipelines, causing deforestation and contamination of the environment, and triggering an influx of illegal loggers and gold miners. In 2019, a lawsuit was filed by the Waorani peoples against the Ecuadorian government. They argued that the Ecuadorian constitution had been violated because the proper consultation process had not been observed prior to an oil auction that would offer up the Waorani's lands in the Pastaza region, one of the world's richest in biodiversity, to the highest bidding oil company. The government auction included 16 new oil concessions covering about 28 000 km^2 of primary Amazonian forest in southeast Ecuador (Loki 2019). Where the oil companies see oil deposits and huge opportunities for profits, the Waorani peoples see forests, medicinal plants, jaguar trails, fishing holes, sacred waterfalls, ancient cave carvings, and historical battle sites (AF 2019).

From the sustainability viewpoint, oil exploitation in the Amazonian rainforest is surprising, to say the least, because it has the doubly negative effect of contributing to climate change and to the destruction of the forest. However, oil companies and most governments are indifferent and unresponsive to such arguments. In the short term investing in oil exploitation increases economic growth and employment opportunities, which are essential advantages when seeking to win elections. No wonder it has become much easier to deny the environmental crisis than to acknowledge it and engage in fighting against it.

The Ecuadorian government is more sensitive to the predicament of the Waorani peoples than the multinational oil companies, because it has the constitutional obligation to represent and protect them, but it slowly and inexorably succumbs to their demands. The mechanism driving this process is rather simple. Ecuador defaulted on around US$3 billion in 2009 and the then President Rafael Correa made an oil-for-cash agreement with China, selling his crude oil to Petrochina in exchange for a US$1 billion loan. The current President Lenín Moreno wants to increase the country's economic growth and the straightforward way to do this is to attract foreign investment to promote public–private partnerships for mining, energy, infrastructure, and telecoms. He plans to obtain an investment of US$800 million to increase oil production and a total investment of US$7 billion by 2021 (Loki 2019). Thus it is likely that oil exploitation in the Ecuadorian Amazon forest will continue to expand in the foreseeable future. The Waorani nations will slowly lose their identity, their pride, and their ways of life, while their forest will be further destroyed and degraded, and the risk of biodiversity loss will increase.

Countries that default can generate many opportunities for huge returns on investment because they are in a weak negotiating position. If they accumulate an increasing debt, that is usually good news for the modern global economic system since it disrupts and renovates the economy of the country and creates the conditions for multinational corporations and financial, hedge fund, consultancy, and law firms to intervene, making large profits and aquiring greater power. An interesting example of this situation is given by the island of Puerto Rico, which began to amass a huge debt as of 2006, while being devastated by the tropical cyclone Maria in 2017, one of the deadliest storms to hit the USA, and subsequently going bankrupt, with US$123 billion in debts. Following a Congress law of 2016, the island's finances are now managed by a federally appointed oversight board that hired McKinsey as their strategic consultant.

The firm produced a fiscal plan to reset the economy, imposing very strong austerity measures of "right-sizing", service reductions, and agency consolidations. The fiscal plan sets aside US$1.5 billion to pay for the restructuring process, with more than US$1 billion to pay the lawyers, bankers, and consultants, including McKinsey (Rice and Ortiz 2019). These fees will be paid by Puerto Rico taxpayers. One of the main questions for the oversight board is to determine how much money the creditors recoup on their investments, especially the hedge funds that bought bonds at a discount after Puerto Rico defaulted on its debts.

The US district court in San Juan, the capital of Puerto Rico, showed that affiliates of McKinsey held bonds issued by Puerto Rico totalling at least US$20 million. Since McKinsey wants to make as much money as possible on the bonds, it has a conflict of interest, which was not previously disclosed (Walsh 2018). Strangely, Congress excluded advisor to the oversight board from the disclosure requirements that are normally included in this type of contract (Walsh 2018). While Bertil Chappuis, who directs McKinsey's advisory role to the board, believes that he can help the island pull off an economic miracle (Rice and Ortiz 2019), Puerto Ricans are exhausted and demoralised by the austerity measures. But they must endure them, since they have no way to avoid or combat them. To what extent are they resposible for their own precariousness? There is no convincing answer available.

The sustainability of the modern economic system would require the establishment of a new ethical foundation based on the "Ethical State" (Collier 2018) that characterized social democracy when it reached its apogee. However, this is an unlikely path because mainstream neoclassical economics is increasingly supported by rational egoism, which augments its resilience. It seems likely that the system's dysfunctionalities will develop much further, and decisive responses will only spring up in a reactive way when critical situations have to be confronted.

5.14   The Overriding Importance of Military Technologies

When analysing technological progress, the usefulness, advantages and negative side-effects of new technologies should be considered, paying particular attention to their impact on human well-being, quality of life, security, and the solution of human problems from the perspective of historical time. Technologies developed for medicine, pharmacology, and public health, aiming to improve human health and well-being and achieve longer lifespans in good health conditions, have seen notable advances in recent centuries. Their use is still deeply unequal around the world, with great benefits for countries with advanced economies and relative losses for the poorest and most needy countries. One of the most extraordinary successes of our time is the extent to which infant mortality has fallen and life expectancy has risen around the world. Most diseases that were considered deadly in the past have now been practically eradicated or are curable, including tetanus, rabies, polio, yellow fever, diphtheria, malaria, meningitis, typhoid, whooping cough, smallpox, scarlet fever, and many others.

Applications of technology to domestic appliances, such as the refrigerator, washing machines, dishwashers, gas hobs, microwave ovens, sewage systems, and heating and cooling systems, have greatly improved comfort and safety at home. There have also been remarkable advances in technologies for education, training, leisure, entertainment, and all aspects of digital ICT, not to mention mobility with the invention and development of fast road, rail, and air transport.

Technologies for the less noble, darker sides of human nature, such as conflict, violence, and vice, have also evolved immensely. The main goal of military technologies is simply to produce weapons as deadly and destructive as possible. The evolution of the destructive power of weapons has been remarkable since the discovery of gunpowder in China in the 9th century. The first firearms were invented by the Chinese at the beginning of the 13th century (Chase 2003). Shortly afterwards, the Mongols, in their battles with the Chinese, captured the experts who were able to make them. They also began to produce firearms and used them against the Arabs during their advance towards the west. From then on, these ideas spread rapidly around the world and were further developed. There are records of portable cannons used against the Mongols by the Egyptian Mamluks at the battle of Ain Jalut in Galilee in 1260, in order to scare enemy horses and cause them to flee.

Heavy artillery was used frequently from 1750 onwards, and from 1804 the English army began to use Shrapnel shells in combat. Invented by Major-General Henry Shrapnel (1761–1842), this was an anti-personnel weapon that was very effective because of the multiple bullets it carried. In 1863, the American Samuel Colt (1814–1862) patented the revolver, which offered the advantage of being quick to reload. In 1851, the first machine guns appeared. In 1884, Hiram Stevens Maxim (1840–1916), also American, invented a fully automatic machine gun and, in 1909, his son Hiram Percy Maxim patented the silencer for pistols, which meant their shots could be more discreet.

The British used tanks for the first time in the First World War. Both sides involved in the conflict undertook aerial bombing, a technology that became highly effective and continues to be used intensively, extensively, and constantly. The term weapons of mass destruction was used for the first time by the Archbishop of Canterbury Cosmo Gordon Lang (1864–1945) in a letter published in the Times on 28 December 1937, when discussing the terrible aerial bombings in the Spanish Civil War, especially at Guernica, and in the Second Sino-Japanese War, warning of the horror that would be inflicted if they were used in a new world war.

Trinity, the first nuclear bomb test, was successfully deflagrated on 16 July 1945 in the Jornada del Muerto desert in New Mexico, USA. Immediately afterwards, on 6 August, a nuclear bomb named "Little Boy" was dropped on Hiroshima, and on 9 August another nuclear bomb named "Fat Man" was dropped on Nagasaki. Together these killed roughly 120 000 civilians and forced Japan to surrender. The horror of war between humans increased to previously unimaginable levels.

Intercontinental ballistic missile (ICBM) technology took its first steps in Germany during the Second World War with the aim of reaching US cities, but it was only developed later during the Cold War. ICBMs have now evolved into multiple independently targetable reentry vehicles (MIRVs). These can transport several nuclear warheads that can be directed at different targets and launched from underground silos, submarines, or mobile land-based launchers. Further to nuclear, biological, and chemical weapons of mass destruction, laser weapon systems and satellites have been developed for several specific military purposes. One emerging technology that is being used increasingly, especially in the war on terror, is drone technology, known as unmanned combat aerial vehicles (UCAVs).

In terms of social factors, in some countries, particularly the USA, easy access to a wide range of firearms—including semi-automatic and automatic weapons, which have benefited from constant technological progress—tends to increase violent criminal acts and the number of fatal victims. Military technologies, from missiles to bombers, from aircraft carriers to submarines, from atomic bombs to semi-automatic weapons, continue to progress through disruptive technologies and cycles of invention and innovation, just as happens with other forms of technology.

The launch of the Sputnik 1 satellite by the Soviet Union on 4 October 1957 created a new field of science and technology and marked the beginning of an intense rivalry between the USA and the Soviet Union for strategic control over the space around Earth. From that year on, the technologies for using and exploring outer space have progressed astonishingly. They are now used by dozens of countries for peaceful means and bring great benefits to humanity.

Faced with the danger of the militarisation of outer space by placing weapons of mass destruction or military facilities there, the General Assembly of the United Nations reacted quickly and on 14 November 1957, through Resolution 1148 (XII), urged its Member States to establish an inspection system designed to ensure that the sending of objects through outer space would be exclusively for peaceful and scientific purposes. Two years later, the General Assembly created the United Nations Committee on the Peaceful Uses of Outer Space (COPUOS) through Resolution 1472

(XIV), the main aim of which was to avoid the militarisation of outer space and to encourage its use for peaceful purposes and the benefit of all mankind.

Satellites have become essential for ensuring the continuity of our global development model. Global Navigation Satellite Systems (GNSS) is permanently used to access UTC and to find one's position in space, as well as the geographical position of all sorts of sites and places and ways to navigate toward them. Communication satellites provide all kinds of digital communications, such as telephone, television, radio, internet, telemedicine, and tele-education. Remote Earth observation satellites have become indispensable to monitor farming and forestry activities, water resources, oceans, fisheries, and the climate system, but also to manage natural disasters related to tropical cyclones and extratropical storms, earthquakes, tsunamis, volcanic eruptions, and many other types of disaster. There are also weather satellites specialized in weather forecasting and satellites for studying astrophysics and cosmology, which carry different kinds of telescopes designed to observe the Solar System, galaxies, clusters and superclusters of galaxies, the large scale structure of the Universe, and the immense variety of phenomena that happen in deep space.

On 10 October 1967, ten years after Sputnik 1 was launched, the Outer Space Treaty came into force. One of its main objectives is to prevent states partie to the treaty from "placing in orbit around the Earth any objects carrying nuclear weapons or any other kinds of weapons of mass destruction, installing such weapons on celestial bodies, or stationing such weapons in outer space in any other manner" (UN 2002). Up to now it has been possible to forestall weapons and weapon systems from being deployed in outer space, but a large fraction of satellites in orbit do nevertheless have military purposes. They are used for military command and control of operations, for navigation, particularly to locate enemy targets and guide bomb and missile launches, for espionage, surveillance, and reconnaissance by obtaining high-resolution images, and for early warning systems for missiles launched by opposing forces.

According to the records of the United Nations Office for Outer Space Affairs (Andy 2019), there were 4994 satellites in space in January 2019, of which only 1957 were active. They wear all orbiting the Earth except for 7 which orbit other celestial bodies. A total of 8378 objects have been launched into space since 1957. It is estimated that, at the end of 2018, there were are around 320 military satellites orbiting in space, very unevenly distributed among countries, which reveals the strategic importance of outer space for achieving military competitiveness and supremacy in the world. The number of military satellites owned by a country is a good proxy for evaluating their military ambitions and power. The USA, with 123, is the one with the most military satellites, which confirms its current military hegemony. It is followed by Russia, China, France, Israel, India, the UK, Germany, and Italy with 74, 68, 8, 8, 7, 7, 7, and 6, respectively (WA 2019).

The intensification in the use of outer space for very different purposes by a growing number of countries is creating sustainability problems for those activities in the medium and long term. One of the main problems is the accumulation of debris in space, formed from non-operational satellites that remain in space, objects from launches, or fragments produced when satellites are accidentally or intentionally destroyed. It has been calculated that in January 2019 there were 128 million objects measuring between 1 mm and 1 cm, 900 000 measuring between 1 cm and 10 cm, and 34 000 measuring more than 10 cm (ESA 2019). These numbers are continually growing with the increasing numbers of satellites that are launched into outer space. Collisions between the larger debris and space vehicles, the ISS, or operational satellites may disable or destroy them. The total mass of the debris in orbit around the Earth is currently estimated at 8400 tonnes.

Human activity in outer space is an interesting example of an area where scientific and technological progress has been swift over the past 60 years, but where the medium and long-term sustainability of that activity is not guaranteed and requires globally coordinated action. Within the United Nations, the COPUOS decided in 2010 to build a set of guidelines on the long-term sustainability of outer space activities (UNGA 2010). However, it has been an arduous and protacted process, because countries prioritise their geostrategic national interests in outer space, particularly the great space powers, over those of the community of nations, one further example of the tragedy of the commons. In spite of the fact that the accumulation of space debris constitutes an increasing risk for outer space activities, and the fact that there are various known technologies to solve the problem (Froehlich 2019), the principal space powers are unable to reach agreement on what strategy should be used and on how to distribute the costs of implementing it.

After nine years working on this project, COPUOS was unable to reach a consensus on all guidelines for the long-term sustainability of outer space activities that had been identified, nor even on how to write its final report or how to refer the preamble and guidelines to the General Assembly. The effort will surely continue, but the present situation reveals the incapacity of the most active space-faring nations, which are also the most powerful from the military point of view, to negotiate for the common good of all nations. In the medium and long term, a lack of agreement may put the use of outer space at risk for several activities that provide crucial support for contemporary civilisation, particularly those that contribute to sustainability on Earth (Santos 2016).

The development and implementation of military technologies and of all other technologies that may have military applications is the foundation of military power, and this is essential to the most powerful countries if they are to consolidate and augment their dominant position. Thus, for them, it is essential to keep on developing new military technologies adapted to the new tyes of warfare and enemies. The countries that have the most mortiferous and powerful military technologies strive to keep them out of reach of other countries through the implementation of international agreements preventing their proliferation. The Treaty on the Non-Proliferation of Nuclear Weapons and the United Nations Security Council resolution 1540, regarding the non-proliferation of weapons of mass destruction, constitute examples of such agreements.

5.15   The History of Tobacco Consumption and Technology

Psychoactive substances have been used for more than 3000 years for spiritual experiences, aimed at producing visions or mystical insights in shamanic or religious rituals, in which case they are usually referred to as entheogenic. Entheogens are also used for traditional medicines and for recreational purposes, and to provoke well-being and pleasure through altered states of consciousness. Most of these drugs are extracted from plants and fungi, and also from the secretions of animals, such as toads. Important plants of this type include the opium poppy, *Papaver somniferum*, whose active substance is morphine; coca, *Erythroxylum coca*, from which cocaine is extracted; hemp, *Cannabis sativa*, used to prepare hashish; the tobacco plant, *Nicotiana tabacum*; salvia, *Salvia divinorum*, used by shamans in Mexico; peyote, *Lophophora williamsii*, a small cactus found in the deserts of south-west USA and northern Mexico, which contains several psychoactive alkaloids; the areca nut palm, *Areca catechu*, a psychoactive drug used in the countries of south-east Asia, the fourth most widely used in the world after caffeine, nicotine, and alcohol; and ayahuasca, *Banisteriopsis caapi*, from South America, used by some indigenous peoples in the Amazon and recently by Westerners, which enhances well-being and mindfulness-related capacities.

The history of tobacco use is an extraordinary example of how technological progress has influenced the human relationship with drugs. It also shows how technology can assist the development of some of the most essential human psychological and behavioural characteristics. Genetic studies show that the tobacco plant originated in the Andes, in the region that is now Peru and Ecuador. It was probably first farmed between 6000 and 5000 BC (Gately 2001) and was used by the indigenous Aguaruna or Awuajún people of Amazonia around 2000 years ago. Its use spread rapidly through South and Central America, where the art of snorting snuff and smoking tobacco and other plants was invented. We have records showing how the Maya smoked in the Madrid Codex, on a stela at the temple of Palenque, in Mexico, on ceramic vases, and in wall paintings. The images show gods and animals smoking different forms of pipes. A large number of Mayan languages, living and dead, use words close to *siik* for tobacco and *siikah* for "to smoke", which is likely to be the origin of the Spanish word *cigarro*, source of the English word "cigar" and the French *cigare*.

According to the Dominican friar Bartolomé de las Casas, a defender of the indigenous people's human rights (Las Casas 1552), *tabaco* in Spanish derives from the word in the dialect used by the Taíno in the Caribbean for a roll of dry tobacco leaves. As well as being used in snuff and smoked in the form of dry leaves, tobacco was also chewed, drunk in infusions, and used as a tobacco smoke enema. As a powerful insecticide, it was used in washes against parasites and in agriculture to avoid infestations. Recently, nicotine was found in a porcelain vase from the Maya late Classic period, dating from between the 7th and 10th centuries, with the inscription "the home of his/her tobacco" (Zagorevski and Loughmiller-Newman 2012). Tobacco was considered by the Maya and other Central American peoples as a sacred plant with curative powers, and it was used by priests and shamans. It served to relieve pain and to communicate with the spirits.

Westerners' first contact with tobacco occurred immediately after Christopher Columbus' arrival in the Americas on 12 October 1492, on the beach of an island of what is now the Bahamas, to which he gave the name San Salvador and which the indigenous locals called Guanahani. In his diary, Columbus recorded that the Indians offered him fruit, wooden spears, and dried leaves, probably from the tobacco plant. The latter had a very pungent smell and they threw it away. A few days later, Columbus was on the north coast of the island of Cuba and was convinced that he had arrived in China, so he intended to deliver the letter that the Catholic Monarchs, Ferdinand II of Aragon and Isabella I of Castile, had written to the Great Khan. In his diary (Columbus 1492), he tells us that he has decided to send Rodrigo de Jerez and Luis de Torres to find the city of Guisay, Kinsai, or Quinsai, described by Marco Polo as the "City of Heaven", "greater than any in the world" (Polo and Pisa 1350), and nowadays called Hangzhou. However, the men sent by Columbus found only a settlement of about fifty huts where they saw for the first time indigenous people almost nude smoking tobacco. Literally speaking, rather than opening up a maritime route to China and India, Columbus had landed on an island close to an as yet unknown continent and procured for the West a substance that produced a feeling of pleasure, increasing the joys of living, but also increasing mortality and morbidity for thousands of millions of people. And this was the starting point of the powerful tobacco industry.

Rodrigo de Jerez became a smoker and maintained the habit in Ayamonte, his birthplace and place of residence, when he returned to Spain in 1493. The cigarette smoke emerging from his nose and mouth scared his neighbours and the Inquisition imprisoned him for seven years because of his sinful and infernal habit, which could only be explained as possession by the devil. When he left prison, the habit of smoking had already won over some members of the aristocracy and high clergy in Spain and was spreading throughout Europe. By 1559, the tobacco plant was being grown in a farm called Los Cigarrales close to Toledo by order of King Philip II.

When they arrived in South America in 1500, in what is nowadays Brazil, the Portuguese were introduced to the tobacco plant by the Tupi people, who called it *petum*. Luís de Góis, one of the first colonisers of Brazil, saw how the Tupi rolled up a leaf of *petum* into a kind of cigar and smoked it at religious ceremonies. Fascinated by the medicinal properties of *petum*, which he called *erva santa* ("holy grass"), he brought it to Lisbon in 1542 (Góis 1566), and from then on it was grown at the gardens of the royal palace in Santos, Lisbon, built by King Manuel I. In the meantime, the holy grass was already being grown in Brazil for export by 1548.

Jean Nicot, the French ambassador in Lisbon, came across the plant through his friend the humanist philosopher Damião de Góis (1502–1574). He understood its value and sent it to the French court by way of Cardinal Charles de Lorraine. It was greatly appreciated by Catherine de Médicis to treat her migraines and became known as *herbe à la Reine* or *herbe à Nicot*. This is the origin of the name *Nicotiana* for the genus of the tobacco plant, given by Carl Linnaeus (1707–1778). Nicotine was then the name given to the alkaloid that can be found in varying amounts in plants of the

Solanaceae family, although occurring in higher percentages in the *Nicotiana* genus. In the 16th and 17th centuries, the Portuguese took tobacco to Africa and probably also to India, but kept their lucrative farming mostly in Brazil.

In the 18th century, the tobacco industry flourished in Europe and the Americas, boosted by the growing number of people who consumed tobacco in different forms, from snuff to cigars made in Cuba to tobacco for pipes and for chewing. Tobacco, as well as being considered a miraculous drug for curing every illness from headaches to bubonic plague, was also a form of recreation and source of pleasure. The fascination with nicotine is partly the result of its biphasic nature: in small doses it stimulates the central nervous system, and this, by increasing dopamine levels in the brain, may lead to addiction; in high doses it slows down transmissions between neurons and produces hallucinations and trances, and may even cause death.

Tobacco use also saw opposition in Europe and Asia. King James I of England (who was also James IV of Scotland) wrote a book in 1604 pointing out the evils of tobacco, called *A Counter-Blaste to Tobacco* (James VI and I, 1604). In Turkey, the Ottoman Sultan Murad IV condemned tobacco smokers to death because he believed that it caused sterility, while in Russia the sentence for smokers was to have their nose cut off. The Church also campaigned against tobacco. Pope Urban VII banned the use of tobacco in churches and surrounding areas in 1590 and threatened transgressors with excommunication. The Calvinists believed that it violated the Ten Commandments. Nevertheless, opposition waned throughout the world as time went on, since it was such a pleasant, addictive, and profitable drug.

In the Americas and in Europe, cigars were rolled in corn leaves or leaves from other plants, but by the end of the 17th century, they were being made in smaller sizes and rolled in thin paper to create the cigarette, the French diminutive word for cigar, known as *papelate* by the Spanish. Cigarettes were hugely successful, partly because they were a quick way of getting nicotine to the brain, an advance that was fully adjusted to the ongoing acceleration of operative social time. Only seven seconds after inhaling cigarette smoke, nicotine is already acting on the brain, having passed through the lungs and been absorbed into the bloodstream.

Cigarettes were made by hand, but each worker could only roll three or four cigarettes a minute. In 1875, the Allen and Ginter Company, headquartered in Richmond, Virginia, USA, launched a contest with a $75 000 reward for anyone who could invent a cigarette rolling machine. James Bonsack (1859–1924), an ingenious mechanic aged only 16, abandoned his secondary education and devoted himself to the matter. He built a cigarette rolling machine which he patented in 1880 and which subsequently revolutionised the tobacco industry. Bonsack's machine was able to produce around 200 cigarettes per minute and therefore hugely increased productivity in cigarette factories. It broke down regularly, however, and Allen and Ginter ended up not using it. In the meantime, James Buchanan Duke (1856–1925) inherited the W. Duke, Sons & Company cigarette factory in Durham, North Carolina, from his father. In 1881, he hired around 125 Jewish cigarmakers who had come from Eastern Europe to New York and already had experience rolling cigarettes. In 1884, Buck Duke, as he was known, heard of Bonsack's machine. He acquired one and, with the help of the mechanic Tim O'Brien, managed to improve it.

Two problems emerged, however. The first was that the recently hired cigarmakers saw their jobs seriously threatened and tried to defend themselves. They threatened Tim O'Brien and formed the Cigarmakers' Progressive Union. The problem was quickly solved by compensating and replacing the most active protesters among the workers, so that by 1888 all the cigarmakers at the Duke Company had been replaced by machines. The second problem, which was harder to solve, arose because Bonsack's machine produced around 120 000 cigarettes per day, an output that was higher than the amount Duke was able to sell. The solution was to create a wider market for the product by increasing demand for cigarettes through advertising and marketing.

At the time, women smoked much less than men and it was not socially acceptable for them to smoke in the street. A cultural change was needed to win over this market. To do this, Buck Duke gave away cigarettes at beauty pageants and placed adverts in the most popular women's magazines. A woman smoking cigarettes was represented as emancipation and freedom from men, a hot topic in the first few decades of the 20th century. The American Tobacco Company, founded by Duke in 1890, made Lucky Strike the dominant brand in the USA in the 1920s through an aggressive advertising campaign. The advertisements said that tobacco underwent a toasting process that helped reduce the throat irritation and cough caused by smoking cigarettes, although the scientific justification for this statement was not firmly established. Doctors were encouraged to defend smoking certain cigarette brands in exchange for undisclosed forms of retribution, possibly complementary cartons of cigarettes (Gardner and Brandt 2006). A 1930 advert shows a smiling doctor with a pack of Lucky Strike in his hand and the tagline reads "20 679 physicians say 'Luckies are less irritating' ".

Although Duke was a cigar smoker, his strategy to increase cigarette consumption was very successful. He solved the excess supply problem created by Bonsack's machine and managed to make cigarettes the main form of tobacco consumption in the USA from the 1920s onwards, overtaking all other forms: chewing tobacco, cigars, pipes, and roll-your-own cigarettes. Annual cigarette consumption per capita in the USA rose from 54 in 1900 to a high of 4345 in 1963.

The history of the astonishing rise in cigarette consumption in the USA between 1885 and 1960 is an excellent illustration of the forces that drive the economic model of our current civilisation. The invention of one technology, in this case the automation of cigarette manufacturing, reduced employment but created the chance to generate a more lucrative activity that required the creation of a market through advertising and marketing. Consumption of the product did not meet any vital human need and turned out to be harmful to human health but since it almost immediately gratifies the consumer it became fashionable and therefore highly profitable for the manufacturing companies. Success creates positive feedback in which the increase in demand ensures profit, which further increases consumption through advertising and marketing. The technology itself does not guarantee growth for the economy but when growth is reached it becomes essential to safeguard it.

Of course, like everything human, products fall into disuse. Habits, social behaviours, and expectations are transformed in unpredictable ways. To cope with this transformation, technology makes an effort to create innovative processes and products for niches in the market that can generate profit under the new conditions and keep the existing economic system healthy. However, the system itself is also transformed in ways that are impossible to foresee. The most recent evolution of the cigarette industry around the world allows us to further explore other aspects of the relationship between human nature and technologies.

For four centuries, from Europeans' first contact with the tobacco plants on the Caribbean islands until the beginning of the 20th century, tobacco was generally considered to be beneficial to health. One of the first signs that this was not the case was that lung cancer rapidly became a global epidemic, whereas it had been a rare disease up to the end of the 19th century (Proctor 2012b). In the 1940s and 1950s, epidemiology and cellular pathology studies and experiments on animals performed in Europe and the USA led to the conclusion that smoking cigarettes was the reason behind the increase in cases of lung cancer. It is estimated that cigarettes cause about 1 lung cancer death per 3 or 4 million smoked (Proctor 2012b).

Initially, this cause and effect relationship was denied by the tobacco industry, just as similar relationships have been denied by practically all successful and highly profitable industries which depend on products that scientific evidence proves to be damaging to health or the environment (Oreskes and Conway 2010). Cigarette manufacturers disputed the scientific evidence and continued to develop their aggressive marketing because they had factual inside information that their denialist propaganda continued to be successful.

Cigarette consumption in the USA only began to fall when a report by the Surgeon General, Luther Terry, on the consequences of tobacco consumption for health was released on 11 January 1964 (USDHEW 1964). The extensive 387-page report, based on dozens of scientific articles, stated that the likelihood of a smoker developing lung cancer was roughly 1000 times greater than for a non-smoker. The report also indicated that the incidence of bronchitis, pulmonary emphysema, and coronary atherosclerosis was much higher in smokers.

At the beginning of the 20th century, Duke decided to expand his tobacco market with investments abroad in Australia, Canada, and Japan. When his involvement in the industry in Japan ended in 1904, forced out by the country's political leaders who accused him of being "the capitalist (who) is intending to monopolise the whole world", Duke turned his attention to China. In 1905, as president of the new British American Tobacco Company, he appointed James Thomas to launch his new tobacco empire with its headquarters in Shanghai. At first, the cigarettes were produced in the USA and exported, but shortly after that, the British American Tobacco company moved production to factories located in China. this is one of the first examples of market globalisation in which a multinational company transferred production technology from the country that developed it to another one. Cigarette production at Duke's factories in China grew quickly, and by 1919, it had already reached 243 million per week in Shanghai alone (Tinari 2005).

The number of cigarettes consumed in China never stopped growing after the beginning of the 21st century, although this was not the case in more developed countries where consumption had begun to fall at the end of the previous century due to concerns about the consequences for public health. At the beginning of the current century, demand for cigarettes in less developed countries was still growing at a rate of 3.4% per year (WHO 2002). Presently, around 80% of the more than 1 billion cigarette smokers in the world live in lower and middle-income countries, which are therefore less ready to deal with the social and economic consequences of the morbidity and mortality rates caused by cigarette consumption (WHO 2016a).

The history of cigarette production and its expansion throughout the world reveals several essential aspects of human nature and the humans relationship with the economic model. To begin with, the astonishing increase in cigarette production was due to technological advances initially made by Bonsack's machine. That progress continued, and machines now produce around 20000 cigarettes every minute, 100 times more than Bonsack's machine, and an enormous variety of types of cigarette for many different tastes. Large-scale cigarette production was first developed in the USA and then spread around the world largely thanks to James B. Duke's entrepreneurship and understanding of human psychology. Duke is also remembered for creating a $ 40m endowment fund in 1924, part of which was attributed to Trinity College, Durham. It subsequently changed its name to Duke University, and it is now the seventh-richest private university in the USA. As already mentioned, the spread of cigarette production around the world at the beginning of the 20th century was one of the first examples of the globalisation process encouraged by the West. This process has accentuated the disparities in the levels of development between countries, because the spreading addiction to cigarette smoking around the whole world is having much more serious consequences in poor countries.

The most noteworthy aspect of the history of tobacco consumption, however, is that cigarettes are probably the deadliest artefacts of modern civilisation (Proctor 2012a), while at the same time average daily sales total around 15000 million cigarettes (WHO 2016a). According to the World Health Organisation, 100 million people in the world died prematurely in the 20th century due to tobacco consumption (WHO 2012), more than the estimated 20 million deaths of the First World War and the estimated 90 million deaths of the Second World War combined. It is currently believed that tobacco smoking causes around 5 million premature deaths every year throughout the world, and that there are more than 600000 premature deaths in non-smokers who breathe in smoke from others, known as passive smokers (WHO 2016a). Roughly half of those who smoke regularly die prematurely (WHO 2016a).

The current situation is partly the result of smokers' lack of knowledge about the effects of smoking on health, but there are many who keep on smoking despite knowing the consequences. In the latter case, the role of time is crucial. It is not pleasant to imagine that one is going to die of lung cancer or another serious illness caused by smoking, but it may only happen some time away and it is not at all certain that it will happen at all. That uncertainty about a future event, which has a probability that is impossible to calculate with any precision, is decisive in the choice to carry on an activity that provides pleasure and is addictive and therefore extremely difficult to drop.

Smoking tobacco, and smoking cigarettes in particular, is a social habit that has deep social roots throughout the world. People generally start smoking in their teens. In the USA, nearly 90% of smokers started smoking before they were 18

(USDHHS 2019). Once the habit has been picked up, many get addicted to tobacco and are unable to give it up. The main motives that lead teenagers to smoke for the first time are copying friends and family members who smoke, a form of initiation that provides access to older groups, proving maturity, and satisfying the desire to experience something new, sometimes with the added thrill that it may be forbidden. These are all natural impulses that form part of our genetic heritage and are therefore very hard to resist, and impossible to overcome completely.

But there is another factor that makes the fight against tobacco consumption harder and it is linked to our cultural software rather than our biological hardware. The advanced forms of advertising and marketing used very successfully by Duke to sell more cigarettes have not only been kept up but have progressed immensely, and are now much more sophisticated and efficient. Their cost continues to rise rapidly the world over. In 2013, American cigarette companies spent $8.95bn on advertising and promoting their products in the USA alone (FTC 2016). The cigarette industry remains highly profitable and spreads around the world in search of new markets in less developed countries, using the most modern advertising and marketing techniques. The techniques are adapted to each region and country in which they are used, regardless of what we now know from a scientific point of view about the consequences of becoming addicted to tobacco.

In less developed countries, the main targets for advertising are often, as in the past, women and young people, although for the latter care is generally taken to ensure that the advertising message is subliminal. In the end, the tobacco industry is only applying the unquestionable principle of today's economic model: if there is a demand for a product, and if its production is profitable, nothing can stop people from taking the opportunity to produce, sell, and profit from it, unless there are very strong and widely agreed-upon reasons for banning consumption, which is clearly not the case. The powerful economic interests connected to the tobacco industry all over the world now make it unfeasible to ban people from consuming it. Prohibition would be a form of restricting freedom and would also tend to bring the long term end to the pleasure that many get from smoking cigarettes or a good Cuban cigar, one of the symbols used most frequently in the West to represent capitalists.

The alternative adopted by the United Nations with the World Health Organisation Framework Convention on Tobacco Control (WHO FCTC), in force since 27 February 2005, is a way to mitigate the problem that tobacco consumption represents to human health all over the world. Its main goal is to put tobacco consumption control into practice, following a rationalist vision but without directly threatening the tobacco industry's economic interests or people's freedom to smoke. The WHO FCTC is another example of an application of the model of international law treaties, conventions, and agreements, under the aegis of the United Nations, relating to a huge range of global issues, including human rights, refugees, terrorism, organised crime and corruption, disarmament, the oceans, and the environment. The creation and entry into force of all these instruments of international law, which almost all countries in the world have ratified, is a significant step forward in civilisation compared with past eras. One of the most remarkable factors of this step forward is the fact that all the information on the problem—the risks, goals, actions, and challenges relating to each of the instruments—is available on the internet, on the United Nations website, together with an extensive bibliography. With a little effort, anyone with internet access can get informed, build their own opinion, and remain up to date.

However, it is fair to ask to what extent such treaties, conventions, and agreements overseen by the United Nations actually achieve the goals they aim for and, more specifically, how compatible those goals are with the current global economic model. The implementation of several sectoral goals frequently produces contradictions between them and, furthermore, some incompatibilities with the model itself. It may be that the best possible solution is not fully consistent and compatible with the inner workings of the economic model that is currently capable of delivering the satisfaction of the contemporaneous generation's horizons of expectation.

The WHO FCTC was the first treaty established by the World Health Organisation based on Article 9 of its Constitution. It aimed to establish control of production and measures for reducing the demand for tobacco, such as increasing taxes on tobacco, informing the public about the damage to health resulting from consumption, regulating images on cigarette packets, restricting advertising and sponsorship, and tobacco control and prevention programmes. Many of these measures have already been implemented in countries with advanced economies, and this has reduced cigarette consumption per capita in those countries. The World Health Organisation believes that if they are not gradually taken up by less industrialised countries, 1 billion people may die in the 21st century due to tobacco consumption (WHO 2016a), which is ten times as many as in the last century. The problem is not easy to solve, however, and this for several reasons.

Let us look at one particularly significant case. From the 1960s on, the largest tobacco-growing areas shifted from the Americas to Asia and Africa, largely to reduce labour costs for growing and producing tobacco. Malawi, which according to the World Bank had a Gross National Income per capita of $250 in 2015 (WB 2015), the lowest in the world, is also one of the countries whose economy most depends on tobacco growing. In 2009, it provided 70% of income in foreign currencies (Otañez et al. 2009). According to the Tobacco Association of Malawi, the tobacco industry is the second largest employer in the country with 350 000 farmers, 70 000 temporary workers, and 10 000 people employed in tobacco leaf processing plants (Vidal 2015). The companies that buy the tobacco produced in Malawi are large tobacco multinationals, including the British American Tobacco Company, founded by Buck Duke, and Philip Morris International Inc.

The reason behind the relocation of tobacco growing to Africa was to lower production costs and therefore guarantee high profits. In the 1980s, the two companies were concerned about negotiations within the United Nations to create the WHO FCTC, since if it was approved it would clearly threaten to reduce their market (Hirschhorn and Initiative 2005). The greatest ally of those two leading tobacco companies, and of a consortium of companies established at the time to defend the interests of the major tobacco companies against the WHO, was the Malawian government, despite tensions between the two sides over prices, tobacco quality, controls on processing, and distribution. On the one hand, the Malawian government was convinced that reducing the tobacco sector would worsen the country's economic situation in the short term, despite the serious social, economic, and environmental consequences caused by the sector in the long term. On the other hand, the multinationals have to stay in Malawi no matter what, because there are probably not many countries in the world where the tobacco industry can make higher profits than they make in Malawi. Pressure on the WHO was unable to stop the WHO FCTC from being passed, but both the multinationals and the Malawian government continue to make complaints.

The Malawian Minister of Agriculture, Allan Chiyembekeza, acknowledged the harmful effects of tobacco in 2015 (Vidal 2015), but he also asked why the WHO, rather than trying to reduce the consumption of other products that have negative effects on health, such as sugar and alcohol, concentrated on tobacco, damaging the economy of a country as poor as Malawi. The Tobacco Association of Malawi complains that campaigns in some countries to use brand-free cigarette packets, i.e., a white-label product, would inevitably lower the price of tobacco and harm Malawian farmers. The big tobacco multinationals have joined these protests and warn of the danger that WHO measures may contribute to increasing poverty in the African countries that are most dependent on tobacco growing, such as Malawi, Zambia, and Zimbabwe.

Large-scale tobacco production in Malawi also has serious consequences for the environment. Tobacco leaves are cured in smokehouses heated by fires that use roughly 10 kg of wood to produce around 1 kg of cured tobacco leaves. The wood is obtained by clearing forests around the farmers' small plantations. The forested area of Malawi fell by 749 000 hectares, corresponding to 19.2% of its total area, between 1990 and 2015, one of the highest deforestation rates in the world (FAO 2015). Deforestation increases soil erosion, lowers soil quality, causes biodiversity loss, and makes the country poorer and less sustainable. The example of tobacco offers a clear demonstration that the various problems of a globalised world with highly pronounced and increasing inequalities are intimately linked together and that sectoral solutions are often insufficient to solve the problems, and often inconsistent with one another.

The unbounded greed that drives the global economic model becomes specially perverse and damaging in the very lucrative markets that exploit, with the precious help of technology, human indulgence and unquenchable desire to consume all sorts of drugs that bring new emotions, pleasures, experiences, insights, perceptual anomalies, hallucinations, and many other types of perturbed states of consciousness. We are facing a perennial problem because the unbounded greed, the indulgence, and the unquenchable desires will all accompany us until the end of *Homo sapiens* time. The same greediness also becomes specially perverse and damaging in the very profitable markets that exploit the natural resources of the less developed countries.

5.16   Technological and Marketing Progress in the Synthetic Addictive Drug Market

Chemical technologies now make it possible to synthesise natural addictive drugs, such as morphine, cocaine, and hashish, and synthesise new drugs that are sometimes very powerful and cause addiction. These substances, when they are launched on the underground market, have the attraction of being new, and are eagerly consumed without full knowledge of their effects. They often cause deaths by overdose and big financial gains for producers and sellers.

One striking example is methamphetamine or crystal meth, a powerful, addictive drug that causes euphoria, intense pleasure, insensitivity to risk, hunger, thirst, tiredness, greater self-confidence, self-esteem, and sexual desire by increasing dopamine levels in the brain. Prolonged use by those addicted to it causes changes in brain structure and functions (Krasnova and Cadet 2009). Methamphetamine was synthesised for the first time in 1893 by the Japanese organic chemist Nagai Nagayoshi (1844–1929) from ephedrine, a psychotropic substance taken from the plant *Ephedra sinica*, used in traditional Chinese medicine. In 1919, one of Nagayoshi's disciples managed to synthesise a crystalline form of methamphetamine and gave it the name of crystal meth.

This drug, produced under the name Pervitin by the pharmaceutical company Temmler Werke since 1938, was widely used by the Nazi regime during the Second World War. Pervitin could be bought at pharmacies without a medical prescription and was often used by the civilian population and especially by German troops. The military command forced front-line soldiers and air force pilots to take Pervitin to fight more forcefully and for longer, without worrying much about their personal safety. Considerable amounts of Pervitin were consumed during the Blitzkriegs against Poland and France (Ohler 2015). Many surviving soldiers became drug addicts and found it very difficult to recover.

Other examples of synthetic drugs include MDMA, commonly known as ecstasy, molly, or mandy, NBOMe, also called N-Bomb or Smile, and 4-methylaminorex, known as euphoria or ice. 2C-E is very successful because it is one of the drugs that has the greatest hallucinogenic power. It was synthesised by the American biochemist Alexander Shulgin (1925–2014), who spent most of his life synthesising psychoactive substances in the lab at his house near San Francisco, California. He described their properties and how they are manufactured in two books he published himself in the 1990s (Shulgin and Shulgin 1991, 1997) with the aim of making them accessible to anyone who wanted to try them. Among the hundreds of drugs described, there are paths to a huge range of new experiences: there are stimulants, depressants, aphrodisiacs, drugs that cause convulsions, vision or hearing alterations, or changes to the perception of time, drugs that produce euphoria, feelings of increased energy, pleasure, and emotional warmth, or dampen emotions, drugs that cause violent outbursts, and many other effects. These are recipe books for exploring the limits of one's mind, senses, and emotions. Two years later, in 1993, agents from the Drug Enforcement Administration (DEA), with whom Shulgin had always had a good relationship, entered his laboratory, took away what they considered to be illegal drugs, and cancelled the drug production licence that the DEA had previously granted him.

The production and consumption of new synthetic drugs continues to be a flourishing business. The number of new synthetic cannabinoids and opioids has grown significantly in recent years. This diversity is attractive for consumers who are always keen to try the latest novelty. One of the main ways to explore and enjoy this diversity is the surface web, often used for illicit medicines and new psychoactive substances.

However, since the 2010s, "psychonauts" tend to use the deep web, which provides access to the dark net markets or cryptomarkets, supported by innovative technologies to protect privacy, such as the encrypted browser named Tor (The Onion Router) (Mounteney et al. 2016). Payments for the drugs, most of then banned by the authorities, can be made by decentralised and relatively untraceable cryptocurrencies, such as bitcoin and litecoin. Consumers can afterwards share their experiences through encrypted communication.

The increase in opioid consumption is a global problem, but in the USA it has reached the level of what is considered to be a very serious crisis (CDC 2019; SAMHSA 2018). On 26 October 2017, the US government was driven to declare a public health emergency. The opioid epidemic is now considered the deadliest drug crisis in American history. The number of fatal victims of overdoses involving opioids obtained through medical prescriptions or other means has more than quadrupled since 1999, climbing to 218 000 people between then and 2017. Excessive drug use and alcoholism were responsible for an 8% increase in mortality for the 25–44 age group between 2010 and 2015 (CDC 2019). In 2016, 63 632 people died in the USA as victims of drug overdoses, more than the total number of victims in the Vietnam War. Of these cases, roughly two thirds were caused by opioids (CDC 2019). Life expectancy in the USA has decreased slightly over two consecutive years in spite of all the advances that have been made against heart disease and cancer. At birth, life expectancy was 78.9, 78.7, and 78.6 years in the years 2014, 2015, and 2016, respectively (NCHS 2018). In 2017, life expectancy at birth fell again. This new trend is attributed mainly to the increasing death rates from opioid drug overdoses, suicides, and chronic liver disease.

One of the new, most sought-after drugs is fentanyl or fentanil, of which there are more than 30 varieties, some 50 to 100 times more powerful than morphine. Its enormous power means that, using very small amounts, even just 1 g, hundreds of doses can be produced. This miniaturisation makes illegal trade easier and greatly increases dealers' profits. Recently, there have been reports that carfentanil, an extremely strong variant of fentanil that is sometimes used to sedate elephants, is being used on humans and provoking overdose deaths in the State of Ohio, USA (Associated Press 2019). There is likely to be no limit to the ambition of creating ever more powerful and lucrative drugs.

In 2017, more than 47 000 Americans died as a result of an opioid overdose, including prescription opioids, heroin, and illicitly manufactured fentanil. That same year, an estimated 1.7 million people in the United States suffered from substance use disorders related to prescription opioid pain relievers, and 652 000 suffered from a heroin use disorder (NIH 2019). It is estimated that 72 287 people died from overdoses in 2017, an increase of about 10% from the year before (CDC 2019). About 49 000 of the deaths are attributed to opioids. In spite of all the efforts being made by the US public health authorities and institutions, this crisis has still not been quelled.

One major factor that contributed to the crisis was the authorisation granted in 1995 by the Food and Drug Administration (FDA), the body that regulates the use of medicines in the USA, for the use, with a medical prescription, of OxyContin, a drug designed to relieve severe or moderate pain. OxyContin's active ingredient is oxycodone, a semi-synthetic drug derived from heroin that is more potent than morphine. OxyContin's main aim is to alleviate extremely violent pain caused by serious illnesses such as cancer. However, thanks to a very expensive marketing campaign supported by physicians, the FDA agreed to liberalise its use for all types of pain, including toothache. The decision was taken although it was well known that the drug had highly addictive effects. It is estimated that the "economic burden" of prescription opioid misuse in the USA is $78.5 billion a year, including the costs of healthcare, lost productivity, the treatment of addiction, and the involvement of criminal justice (Florence et al. 2016).

The marketing campaign was paid for by Purdue Pharma, the private pharmaceutical company that produced and sold the drug. It argued that there was no risk of addiction because of a mechanism that had been introduced into the drug to delay the absorption of the oxycodone, ensuring effectiveness for 12 hours. Despite its relatively high price, there was enormous demand for the drug and it was an extraordinary financial success, with estimates that it has generated profits of around $35 billion. The other side of this story is that a significant number of people have become addicted to OxyContin and many of them have died from overdoses, especially in the most socio-economically depressed regions of the USA. OxyContin pills can be inhaled once they have been crushed into dust and they can also be injected if they are dissolved in liquid. In its marketing campaign, Purdue, with the help of the IMS Health database, sought out and targeted population centers, towns, and cities with high poverty and low education and job opportunity levels (Keefe 2017). The economically distressed region of Appalachia was one of the targeted regions and is one of the most severely affected by the opioid crisis. The company initially created a programme for doctors to prescribe the drug free of charge using a coupon system to make it more affordable.

These aggressive marketing methods are comparable to those used by the Xalisco Boys, Mexican heroin traffickers who roam the USA in search of places with the highest number of potential consumers and then drive them to take the drug (Quinones 2015). In August 2010, when the reaction to the OxyContin epidemic became more acute, joined by numerous lawsuits against Purdue, the company replaced it with another, slightly reformulated drug with the aim of making it harder to become addicted to oxycodone. However, the result was that many users, when denied Oxy-Contin, went on to take heroin as a cheap and effective alternative. A study carried out by Evans et al. (2018a) showed that the reformulation did not lower the mortality arising from joint consumption of heroin and the synthetic opioid. Deaths from OxyContin overdoses were replaced by deaths from heroin overdoses.

Thousands of lawsuits have been filed since then against Purdue and other pharmaceutical companies to pay the costs of the opioid crisis, including hospital visits, foster-care systems, and coroner's offices. In 2007, Purdue Pharma and three of its top executives pleaded guilty to federal criminal charges that it had misrepresented the dangers of OxyContin, and they paid $634.5 million in fines. In March of 2019, Purdue Pharma and its owners, agreed to a $270 million settlement in just one lawsuit, to avoid going to trial in the state court in Oklahoma. In business terms, the company has responded to these challenges by globalising its market and seeking out consumers for its drugs throughout the world, following the same model adopted by the tobacco industry.

A significant feature of this story is that the Purdue Pharma pharmaceutical company is owned by a family with the surname Sackler, one of the richest in the USA and widely known and respected for the generous philanthropic actions carried out by several generations of the Sackler dynasty, particularly as benefactors of art and science museums, universities, and medical research institutions. Isaac Sackler and his wife Sophie Greenberg, a Jewish couple originally from Ukraine and Poland respectively, migrated to the USA and acquired a grocery in Brooklyn, New York, at the beginning of the 20th century, before the First World War.

Their three children, Arthur, Mortimer, and Raymond, were very successful psychiatrists. In 1952, they acquired a small pharmaceutical company called Purdue Frederick in Greenwich Village. It produced laxatives and cotton buds for removing earwax and they turned it into the Purdue Pharma Company. In the 1960s, Arthur Sackler, who died in 1987, applied a pioneering and aggressive direct-to-physician marketing approach to the tranquilizer Valium, and it became the first pill ever to generate $100 million in annual sales. In the 1990s, Purdue Pharma introduced onto the market OxyContin and other painkillers, containing hydromorphone, oxycodone, fentanil, codeine, and hydrocodone, and generated very large financial returns using the same marketing strategies.

The Sacklers have always been very secretive about their connection to Purdue and have avoided talking about the topic as much as possible. Currently, the members of the Sackler dynasty, with a fortune valued at between $13 bn and $14 bn, are no longer directly involved in the everyday running of Purdue Pharma. Nevertheless, their involvement in the OxyContin business is very clear and abundantly documented.

In a 1999 email to a Purdue executive, Richard Sackler, son of Raymond Sackler, wrote: "You won't believe how committed I am to make OxyContin a huge success. It is almost that I dedicated my life to it." In spite of all the lawsuits directed against Purdue Pharma, the Sacklers were not accused of any wrongdoing and have not faced personal legal consequences because of the drug. However, in June 2018, Maura Healey, the Massachusetts attorney general, sued eight members of the Sackler family, alleging that they had misled doctors and patients about the risks of using OxyContin and that the company had been involved in dishonest marketing schemes to promote sales of the drug. Mary Jo White of the New York law firm Debevoise & Plimpton, who represents four members of the Sackler family in the lawsuit stated that the family believes the litigation against them and their company is legally dubious, factually misleading, and politically motivated (Walters 2019).

Some of Purdue's profits have been generously donated to several cultural and scientific institutions. The list is very long, covering mainly the USA, Israel, and Europe and includes: the Sackler Library at Oxford University, the Sackler Faculty of Medicine in Tel Aviv, Israel, the Mortimer and Raymond Sackler Institute of Advanced Studies, Tel Aviv University, the Sackler Institute of Biomedical Science at New York University, the Sackler School of Graduate Biomedical Sciences at Tufts University, the Smithsonian Arthur M. Sackler Gallery, the Sackler Wing of the New York Metropolitan Museum of Art, the Sackler Center for Arts Education at the Guggenheim, the Sackler Wing at the Louvre in Paris, and the Sackler Gallery in Washington. The Sackler family has also made many other donations including to Kew Botanic Gardens, the Natural History Museum, the Royal College of Art, the Old Vic Theatre and the Royal Opera House, all in London. The cultural philanthropy of the Sackler family is presented as an outstanding example of the very good cultural, scientific, and educational benefits provided to society by the major actors of the present political, financial, and economic system.

For more than 20 years now, the pharmaceutical empire's discreet philanthropic champions who have made a decisive contribution to creating the current opioid crisis in the USA, presently spreading to other parts of the world, only needed to hide the origins of their huge wealth. Now that their involvement in the crisis has become more visible in the news and to society (Keefe 2017), it is less attractive for the beneficiary institutions to continue receiving their donations. In March 2019, London's National Portrait Gallery, the British gallery group Tate, and the Solomon R. Guggenheim Museum in New York said that they will start rejecting financial donations from the Sackler family due to their direct involvement in the opioid crisis. The future will tell whether other beneficiary institutions will follow suit or not.

It is important to try to understand the logic of the Sackler family's thinking and behaviour, although there are no public statements about their contribution to the opioid crisis. Mark Sullivan, a psychiatrist at the University of Washington, expressed the Sackler family's argument when he said "Our product isn't dangerous – it's people who are dangerous" (Keefe 2017). In one email written by Richard Sackler, which was uncovered by investigators working on the Massachusetts lawsuit, he says that "we have to hammer on abusers in every way possible. They are the culprits and the problem. They are reckless criminals" (Brico 2019). We may infer that the creators of OxyContin were fully convinced that they were fulfilling a legitimate consumer need. The fact that the drug is addictive is a problem that has to do with the immutable human nature, but that does not affect everyone. Only some consumers get addicted, and it is entirely their own fault. They are dangerous. They are "the culprits of the problem" and in fact "they are reckless criminals".

The Sacklers' reasoning is likely therefore to be the following. Our political, financial, and economic system allows us to tempt consumers into consuming a drug that alleviates their pains even though it is likely to generate addiction in some cases. We have discovered a niche to make huge amounts of money, we were clever enough to be able to implement remarkable and aggressive marketing strategies to almost completely dominate the competition in the pain reliever market, and we have been extremely generous philanthropists for the arts, education, and science research in our own country, in Israel, and in some other countries. Our name is forever associated throughout the world with magnificent works of art, with the performing arts, and with higher education and research institutions. For more than 20 years, US institutions and society have complacently and silently accepted our pharmaceutical enterprise, and the social status of our family has increased immeasurably during those years. We are now a very large, powerful, and respected family, with very talented young people that have benefited from some of the best opportunities in the world, and we are also one of the richest families in the country due to our intelligent and persevering work, especially on tranquilizers and prescription opioids. What have we done wrong?

It is estimated that around 250 million people, roughly 5% of the world's adult population, took drugs in 2015 (UNODC 2017). Of these, 29.5 million have symptoms of addiction and need help. Opioids are responsible for around 70% of the negative impacts of drugs on human health. The number of psychoactive drugs in use has increased significantly in recent times, from 260 in 2012 to 483 in 2015 (UNODC 2017). Many of them are produced illegally on a large scale or are diverted from the legal pharmaceutical market, while some are adulterated to boost their strength.

Dependence on addictive drugs has ruined millions of human lives and there is an enormous effort being made around the world by governments, national public and private institutions, the United Nations, and other international organisations to control and reduce drug addiction. Nevertheless, the progress of biochemistry technologies will continue to ensure the production of new addictive substances with differing levels of potency and addictive capacity and with renewed, surprising, and enticing properties. The major driving force behind this progress is the unquenchable human desire to consume all sorts of drugs, some of which become fashionable and immensely profitable for the pharmaceutical industry and trafficking barons.

5.17 Inevitability of the Fourth Industrial Revolution

The contemporary social time is probably the one that is most turned towards the future because average expectations in the world for a better or even magnificent future have never been greater. Humanity is now very widely removed from the religious view that used to dominate the West in the past, in which a better world could only be found after the last judgment. The current focus on the promise of a better future includes all areas of life: health, employment, the economy, pleasant and healthy urban areas, quality homes, well-designed and comfortable infrastructures, good and fast transportation, low-priced, diverse, abundant and attractive consumer products, services, leisure and entertainment, and many other modern facilities. Sometimes events contradict these expectations but they are considered to be circumstantial, temporary setbacks.

This is a reassuring vision of the future. However, it is not a universal vision because some regions in the world, particularly in more fragile states, suffer from water, food, and health insecurity, armed violence, civil war, terrorism, environmental degradation, and droughts and other extreme weather related events. On the other hand, the vision of inevitable and fast progress does prevail in many regions of the world, mainly in the developing and emerging economies.

In the advanced economies, especially in Europe and the USA, a significant part of the population is now suffering from an epidemic of nostalgia. Many people who don't belong to the younger generations think that the world in the past had many things that were better than now. But that past is not well localised in time and the references associated with it are very far from being coincident or coherent between them. Although everyone has different memories, they nevertheless tend to generate a feeling of nostalgia. They are afraid that their sons and grandsons will have less quality of life, well-being, and economic prosperity than they had, but they are confused about the causes. They follow very different or even opposite paths to identify the causes and they usually reach different and incoherent conclusions in terms of political interpretation and action. As regards governance, there are democratic regimes with many diverse political parties defending different ideologies, ranging from the extreme left to the extreme right, from various forms of socialism to nativist, nationalistic, and neoliberal policies, as well as one-party regime states, such as China, all with the common objective of improving the well-being and prosperity of their citizens. Thus, there are many different political ways to achieve rather similar or comparable objectives at the national level.

To support their vision of a deterministic advancement towards more well-being and a better economy, all governments invoke technological progress. In a world with a growing population, which is overexploiting natural resources, producing increasing amounts of waste and pollution, gradually degrading the environment, and changing the global climate, technology has become indispensable to maintaining some equilibrium between human and natural systems. But the current relationship with technology goes far beyond the dependence on natural capital and reveals a deep belief in its power to transform, substitute, save, and hopefully create a continuously better future.

The topic of the 46th annual meeting of the World Economic Forum, which took place in Davos, Switzerland, in 2016, was "Mastering the Fourth Industrial Revolution". The announcement of an imminent Fourth Industrial Revolution, or, preferably, the fourth stage of the Industrial Revolution, attracted more attention than usual from the media and governments all over the world, as if it were the revelation of a prophecy by an unerring oracle. The Fourth Revolution was presented as unstoppable and unavoidable. According to Klaus Schwab, founder and executive chairman of the World Economic Forum, which meets every year in Davos, it will profoundly change the way we live, work, and relate to one another (Schwab 2016). Also according to Schwab, "the changes [that the Fourth Revolution will bring] are so profound that, from the perspective of human history, there has never been a time of greater promise or potential peril" (Schwab 2016). Why this kind of dramatization?

The main message is that there is no alternative to the Fourth Industrial Revolution. Whether it will bring promise or peril depends on whether or not rulers, electors, and people in general are able to accept, adapt, and benefit from the forces of technological disruption that unrelentingly spread around the world. It is as if the Fourth Industrial Revolution were foretold by superhuman beings, who play the role of the demigods in the ancient Greek and Roman cultures and in Hinduism. They better understand the destiny of mankind, and mortal human beings can either accept it unquestionably or run the very dangerous risk of trying to go against a predetermined fate.

But what is the Fourth Industrial Revolution and what are the forces behind it? The First Industrial Revolution, which was probably the most important transition in the history of humanity, was only identified in the mid-20th century and is now believed to have started in the mid-18th century. The term "Industrial Revolution" was first used in France (Bezanson 1922) around 1820 to refer rather mockingly to the mechanisation of the textile industry in Normandy, French Flanders, and Picardy, in comparison with what was considered the true political revolution of 1789 (Fohlen 1971). It was clearly acknowledged that the French Revolution, by valuing freedom of thought and the use of reason, had opened the way for science, technology, and industry to develop. But it was only later, in 1837, that the economist and historian Adolphe Blanqui (1798–1854) assigned a broader meaning to the Industrial Revolution, closer to the meaning associated with it today. Engels (1845) and Marx (1867) also used the term, saying that it represented the emergence of the machine to replace manual work and that the accumulation of capital had been accelerated by the mechanisation, growth, and concentration of industrial business in the second half of the 18th century.

In Great Britain, unlike France and Germany, the term "Industrial Revolution" was rarely used. However, when the economic historian Arnold Toynbee (1852–1883), uncle of the well-known historian of civilisations Arnold Joseph Toynbee (1889–1975), analysed economic development in Great Britain from 1760 to 1840, identifying and describing a social, economic, and technological transformation that deeply marked the country and later served as a development model around the globe, he gave it the name "Industrial Revolution". The expression began to have a specific, well-defined meaning and became well known worldwide, partly with the posthumous publication of *Lectures on the Industrial Revolution in England* (Toynbee 1884). According to Toynbee, the essence of the Industrial Revolution was "the substitution of competition for the medieval regulations which had previously controlled the production and distribution of wealth" (Toynbee 1884). There were three fundamental components that came into play at the same time: the invention of several new technologies that increased production, sometimes to an extraordinary extent, the concentration of capital due to the growing mechanisation of production, and the industrialisation of the economy together with a new social organisation of labour (Mantoux 1906).

The First Industrial Revolution was only identified and characterised around a century after it began, but the Fourth Industrial Revolution, although it has not yet fully arrived, has already been widely announced and described in great detail. This is a revealing example of the acceleration of operative social time and the increasing importance of the future in our lives and civilisation. The habit has been formed, leading to the need to expect that the future must hold "the greatest promise in the history of humanity". From another perspective, the First Industrial Revolution was so remarkably successful that it became tempting to believe in the possibility of summing up the future as a series of industrial revolutions. After the Fourth there will the Fifth Industrial Revolution that will bring a renewed grand promise for the well-being and economic prosperity of humanity. It may happen, however, that the future will turn out to be more complex than a regular succession of industrial revolutions, and for many reasons, one being the problematic interactions between natural and human systems.

The First Industrial Revolution was mainly characterized by the use of new energy sources, including new fuels and motive power, such as coal and the steam engine, and the invention of new machines, such as the spinning jenny that started the factory system. The Second Industrial Revolution began in the latter half of the 19th century and was based on the use of electricity, which made it possible to develop production lines in factories. At that time remarkable technological inventions were made using the discoveries of modern science. These included the internal combustion engine, the telephone, photography, cinema, the electric light bulb, the aeroplane, television, antibiotics, and many others.

The third began in the 1960s, based on the Digital Revolution and the rapid development of ICT, and made it possible to further the automation of production in many different areas of economic activity. Finally, the Fourth Industrial Revolution is an extension of the third but differs from it in terms of the new opportunities and plans for automation, interoperability, the use of intelligent technical support systems, and decentralised decision-making and information exchange in production technologies for goods and services. It is characterized by the development of emerging technologies and the production of new products and services based on those technologies. In the Fourth Revolution, information technology, which automates business and office processes, will be coupled with operational technology in the internet of things so that industrial and operating processes in factories, infrastructures, and homes can be fully be automated. One of the paradigms of this revolution is machine-to-machine communication, abbreviated as M2M, which takes place not just among machines in one factory, but between all kinds of devices and systems with a wide variety of functions and in different locations. The main areas of activity and emerging technologies that will support the Fourth Revolution are ICT, artificial intelligence, robotics, the internet of things, big data, and additive manufacturing, including 3D printing, rapid prototyping and direct digital manufacturing, blockchain, digital currency, autonomous vehicles, drone technologies, precision agriculture, nanotechnology, genetic engineering, synthetic biology, and geoengineering.

Some of the founding ideas of the Fourth Industrial Revolution emerged in 2011 with a German initiative to build a strategic programme running up to 2020 led by businessman, politicians, and academics, known as Industry 4.0. This aimed to increase the competitiveness of German industries through the increased integration of "cyber-physical internet-based systems" in manufacturing processes (Kagermann et al. 2013). In the USA, there were several similar projects, such as the Industrial Internet Consortium, founded in 2014, which brings together General Electric, AT&T, IBM, and Intel.

The wide dissemination of the Fourth Industrial Revolution in all types of media communication, business and government activities, and business programs and plans around the globe is largely promoted by the countries with the most advanced and competitive economies in the world. Their main driver is the need to establish, maintain, or strengthen their technological and economic leadership. The main fight for leadership is between the USA and China, with the former desperately trying not to be overtaken by the latter. But all countries strive in their own way to maintain an unstoppable momentum of technological inventions and innovations by creating and promoting demand for new products, services, and markets. The fourth Industrial Revolution and in particular artificial intelligence are being hailed as the engines of future economic growth.

Under the national policies and regulatory frameworks that prevail all over the world, especially in the more advanced economies, the Fourth Industrial Revolution runs the risk of making a number of current jobs useless or redundant, mostly as a result of the rapid pace of technological change, the disruptive nature of new technologies, and the progress of automation and robotisation. It is likely that there will be more unemployed people and the social, political, and economic concepts of employment and income will tend to change. The role of employment in the distribution of wealth, which was strong in the first half of last century, has decreased and is now called into question. However, those that are directly involved in the development of robotics and artificial intelligence have a different view and express their conviction that in the medium and long term, technological progress never implies a decrease in the number of available jobs or an increase in unemployment (Lee 2018).

Work that is repetitive or involves manual activity runs a greater risk of eventually being automated. On the other hand, work that depends largely on thought, abstraction, curiosity, creativity, ability for reflection and analysis, or empathy, fellow feeling, and compassion for others and social interactions runs a lower risk of becoming automated. For instance, there will be less demand for drivers, divers, translators, accountants, paralegal workers, back-office workers, haematologists, nurses, factory workers, and farmers. Even so it will remain difficult or may be impossible to replace psychiatrists, criminal lawyers, physical therapists, mathematicians, scientists, philosophers, writers, and other types of artists and freethinkers by machines.

In medicine, the new model will tend to focus on monitoring, prevention, and proactive health instead of reactive health based on intervention by doctors and nurses. Using mobile applications like Apple's Health App, it will soon be possible to constantly monitor one's state of health and wellness using sensors for a huge range of indicators, including body temperature, heart rate, and blood pressure, and to analyse blood, including sugar and oxygen levels, saliva, and urine, and obtain electriocardiograms. The sensors and measuring devices are all easily connected to a smartphone which will then send the data automatically to a clinic for diagnosis. The "medicalised" smartphone will become an essential part of individual and public health care. The idea is to transform healthcare into a patient-centered activity based on ICT, and especially automation and robotisation.

One of the main goals of artificial intelligence is the creation of entities or beings (sometimes called androids) that are not alive in the biological sense but are nonetheless sentient, socially active, and intellectually superior to humans. To achieve this objective, computers must assure that such entities are able to communicate with humans and understand communication between them, virtually becoming another active "human" in their midst in cities, in the workplace, and in homes. Thus they must be able to recognize people, using face and body recognition technologies, understand what they say, and know how to interpret the meaning of facial movements, gestures, and attitudes. The creation of such new beings may be far away in the future or may never happen. But many people firmly believe that their existence is within reach and very likely to occur, which makes it important to get this matter on the agenda.

A promising area of artificial intelligence that will be critical if such goals are to be reached is "deep learning", which enables computers to acquire new skills through self-learning based on the functioning of neural networks and with the help of big data technologies. These types of learning algorithms seek to provide the new entities with social skills that will significantly shorten the distance that separates them from humans. Besides behaving like humans they will have the mathematical calculation and data processing powers of computers, which are incomparably superior to those of humans. According to Kai-Fu Lee (2018), a well known Chinese ICT investor, previously employed by Google, Apple, and Microsoft, artificial intelligence (AI) will come in four waves that he has called Internet AI, Business AI, Perception AI, and Autonomous AI. With the fourth wave, machines will finally become entirely autonomous, potentially capable of being transformed into the new human-like entities with some powers superior to those of humans.

Deep learning can also be used in a wide variety of activities, such as instant translation from one language to another, dealing with the due diligence process, forecasting litigation outcomes, and providing legal analytics in legal practices, but also digital communication and the analysis and interpretation of images of patients' organs obtained using X-rays, magnetic resonance, and other types of imaging. At a business level, it can be used in company management to optimise production and to develop new marketing strategies and procedures.

Bringing together robotics and artificial intelligence, which will allow robots to carry out all kinds of tasks (see Fig. 5.7), adapting and reacting to what goes on around them, has the potential to replace a large number of jobs and eventually transform society. Recent estimates indicate that 47% of the total number of jobs in the USA are at a high risk of disappearing in the next two decades (Frey and Osborne 2013). A fall in the number of jobs in areas where robotics and AI-based technologies have penetrated can already be observed. According to Carl Frey (Frey and Osborne 2013), in the 1980s, 8.2% of the US workforce was employed in the new technologies introduced during that decade. However, in the 1990s, that percentage dropped to 4.4%, and in the 2000s it was less than 0.5%, including new industries, such as online auctions, video and audio streaming, and web design. In the USA, Germany, and Japan it is estimated that more than one third, or even half in Japan, of the workforce will need to learn new skills to find employment in other occupations (Manyika et al. 2017).

Technological unemployment is a concept that already has a pedigree of almost a hundred years. Keynes defined it in 1930 when he wrote about "technological unemployment [...] due to our discovery of means of economising the use of labour outrunning the pace at which we can find new uses for labour" (Keynes 1933).

Fig. 5.7 Robots dance for an audience on 8 July 2017 at the opening of the Beijing International Consumer Electronics Expo held in Beijing China National Convention Center. In 2014, in a speech to the Chinese Academy of Sciences, President Xi Jinping called for a "robot revolution" that would transform China, and then the world, saying: "Our country will be the biggest market for robots." According to the International Federation of Robotics, China added 87 000 industrial robots in 2016, slightly below Europe and the United States combined. Robots have the advantage of working 24-7, without requiring holidays and doing exactly as programmed, which improves efficiency and quality and reduces risks. The Chinese are aware that automation will threaten a high percentage of jobs, estimated at more than 70%, but they are the world's most optimistic about the impact of artificial intelligence on their lives (Bateman 2018). Humans, besides being more adaptive and responsive, are still preferred, and indeed are essential to code, monitor, regulate, manage, and repair machines. Photo credit: Zhang Peng, LightRocket, Getty Images

Although the current problem is much more severe than in Keynes' day, most contemporary economists rule out concerns about technological unemployment. They argue that automation and robotisation, by increasing productivity in industries, increase salaries and demand for new goods and services, which creates new jobs for unemployed workers. Other authors, such as Sachs and Kotlikoff (Sachs and Kotlikoff 2012), believe that such an increase in productivity is likely to be harmful for future generations in average economic terms. According to estimates in a 2015 Bank of America Merrill Lynch report (BAML 2015), the global market for robots and artificial intelligence will be worth $152.7 billion in 2020 and will increase productivity in some industries by around 30%. In 2014, robot sales around the world hit 229 000, a 29% increase compared with 2013. It is also estimated that between 400 and 800 million workers will lose their jobs due to automation around the world by 2030 (Manyika et al. 2017).

Even without the potential problems of robotics and artificial intelligence, unemployment already threatens social cohesion and causes a great deal of harm for unemployed individuals and their families, as well as being politically problematic and negative for the economy in the medium and long term by reducing the potential for output. There have been high unemployment rates, especially among young people, reaching more than 30% of the active population (ILO 2016) in developing countries. According to a 2015 International Labour Organisation report (ILO 2015), 61 million jobs have been lost around the world since the West's economic and financial crisis. In recent years unemployment has been reduced in most countries with advanced economies, especially in the USA, Canada, Japan, and some European countries (OECD 2016b; ILO 2018). However, the process of reducing vulnerable employment, meaning jobs with low wages and no security or guarantees, has stalled in recent years, with 1.4 billion people with vulnerable jobs in 2017 and an additional 35 million expected by 2019 (ILO 2018).

Specialists can be divided into techno-optimists and techno-pessimists regarding the future of employment in the Fourth Industrial Revolution. A recent Pew Research Center survey (Smith and Anderson 2014) asked 1896 specialists if networked, automated, artificial intelligence applications and robotic devices will have displaced more jobs than they have created by 2025. The answers were split practically down the middle: 48% believed that a significant number of blue-collar and white-collar workers will be out of a job and expressed their concern about the increase in inequality in society, the large number of people who will not be able to find a new job, and breakdowns in social order.

The other 52% believe that the overall balance for employment will be positive rather than negative in 2025. The justification they give is based on a conviction that human ingenuity will be able to create new jobs, just as it has done since the start of the Industrial Revolution, even though the characteristics of the Fourth Industrial Revolution are very different from the First. While during the first, a large proportion of output was mechanised, in the Fourth Revolution one of the goals is to replace wherever possible the activities performed by humans by functionally equivalent activities performed by intelligent systems, with the advantage of ensuring more regularity and better reliability and efficiency.

Even if the overall employment balance is neutral, the new jobs that emerge, as well as those directly connected to technological advances, are mostly low-wage, service-sector jobs, precisely the positions that are most vulnerable in the medium and long term. In fact, this trend is already being observed in the USA (BAML 2015). A current example of this is the future of the roughly 3.5 million people who work as professional truck drivers in the USA in the face of the potential introduction of self-driving trucks. Automation lowers costs and increases profits because the cost of installing a self-driving system in a truck is comparable to the annual wage for a professional driver, currently around $40 000.

In 2015, in the Pilbara iron ore mines in Australia, Rio Tinto was already using a fleet of 69 self-driving trucks to transport ore 24 hours a day and 365 days a year in order to lower costs (Smyth 2015). However, the company primarily points out the additional advantage of avoiding potential accidents involving people. The automation and robotisation of mines is an ongoing process and one that is moving forward quickly throughout the world.

The Fourth Industrial Revolution will be eagerly used in countries all over the world to try to improve their relative positions in the world hierarchy of military power. Currently, one of the Pentagon's main strategies to keep the USA as the world's greatest military power is to develop the military applications of artificial intelligence and in particular to produce autonomous or semi-autonomous weapons and killer robots. The latest Pentagon budget has allocated $18 billion to developing this military technology over the next three years (Rosenberg and Markoff 2016). The dilemma over whether or not to develop autonomous weapons and thereby put science fiction into practice, as in stories of the Terminator kind, is known at the Pentagon as the "Terminator Conundrum". There are already semi-autonomous and autonomous robots, drones, aeroplanes, and tanks that are able to recognise, attack, and kill soldiers and other humans and infrastructure targets.

Swarming autonomous drones are likely to become a particularly powerful weapon very soon. China, Russia, the United Kingdom, Israel, and South Korea, among others, are developing these technologies. Besides the military institutions of these countries, some of the "Frightful Five" technological giants—Amazon, Apple, Facebook, Microsoft, and Alphabet, the parent company of Google—, universities and science and technology research institutes are also working on them. Autonomous weapons systems and killer robots are also being developed at a great rate. All these weapons will have a much greater capacity and precision to kill and destroy than those handled and controlled by humans. They also have the further attraction of potentially reducing the number of soldiers killed in a conflict or at war. The development of such weapons will very likely lead to a new arms race between the main military powers of the world, similar to the arms race over nuclear weapons.

The repeated calls for the prohibition of autonomous weapons systems, for a range of different reasons, have been unable to prevent their development (Sharkey 2019). Among the main arguments against them are the increased risks that they create for civilians in a conflict, their inability to comply with international humanitarian laws on armed conflict, known as the law of war, and the threat that they pose for national and global security (HRW 2018). How can fully autonomous weapons systems and killer robots have compassion and how can they use it to avoid killing civilians that are not directly involved in a conflict or in war? On 28 July 2015, an open letter was published, signed by artificial intelligence and robotics researchers worried about the negative image that the development of autonomous weapons could have on their area of research. The letter called for "a ban on offensive autonomous weapons beyond meaningful human control" (FLI 2016). The combination of nuclear, autonomous weapons systems, and killer robot technologies significantly raises the potential destructiveness of armed conflicts in which they are used.

Those that are likely to benefit the most from the Fourth Industrial Revolution are the investors. Due to its intrinsic characteristics, this kind of investment encourages work to be replaced by capital, which further increases the return on capital (Piketty 2013) and worsens the trend of increasing inequality that has been seen in recent decades. Businessmen who discover a successful product or service are likely to get rich rapidly if they use the new technologies, because with them the marginal costs per unit of output tend to become very low or almost zero and the economies of scale are high. Companies are now avidly searching for such new technologies since they help them increase productivity and lower the marginal costs of producing and distributing goods and services, so they can bring down prices, attract more consumers, and increase profits. Obviously, the principal driver of the Fourth Revolution is money.

Technological multinationals are in a privileged position to benefit from moving the new frontiers of technological inventions and innovations and to tap into the changing markets, while governments and the legislation and regulations that they produce have been incapable of anticipating and addressing the ethical, social, and political consequences of such advances. The giant technological multinationals, which have campaigned successfully for lowering taxes, are also excelling in their efforts to deprive governments of the ability to acknowledge and respond to the social anxieties created by robotics and artificial intelligence because they fear it may slow down their development.

On the other hand the idea that artificial intelligence can solve many social problems is promoted, in particular by the technological giants. It is emphasized that social artificial intelligence can be used to develop social acuity, to resolve social conflicts, and to produce social robots that help vulnerable and lonely people. Artificial intelligence has an enormous potential to help human development and well-being. Furthermore it can be a powerful driver of sustainability. Its successful application depends on the willingness of society and governments to invest adequately to assure that artificial intelligence does not endanger human values, and in particular the values of security, solidarity, compassion, social cohesion, and wisdom.

From a philosophical standpoint sharing the social world with new non-living types of beings that in some respects are superior to humans calls into question the concept of human rights (Risse 2018). The underlying assumption in this concept is the hierarchical superiority of humans with respect to other forms of life meriting lesser protection. Will the new non-living beings respect this hierarchy and respect what are now considered to be human rights? Will human rights remain central or will they change with the application of artificial intelligence?

Present and future progress in many areas of the Fourth Industrial Revolution, especially ICT and artificial intelligence, depends crucially on the contributions of highly creative mathematicians. Some of those engaged in quantitative analysis and various hedge fund activities that use that technique, usually called "quants", have made billions in Wall Street using sophisticated algorithms for high-speed trading (Patterson 2010). Mathematics has become increasingly important because it is the language of science and technology, and also because, with increasing access to supercomputers, much more complex problems can be modelled mathematically and solved using appropriate alghorithms. Furthermore, the ingenuity of mathematicians is ever more crucial for creating the methodologies and algorithms that allow one to extract the useful information contained in big data sets. The dependence on highly specialized, creative, and well paid people working in various fields of mathematics, statistics, and computer sciences is increasing in all domains of the Fourth Industrial

Revolution. In particular, the likely creation of beings that are not alive but are sentient, socially active, and intellectually superior to humans will require long and devoted work by many outstanding human minds who will be immeasurably proud if ever they achieve success.

5.18   The Legacy of Thomas Paine and the Quest for a Universal Basic Income

Those who believe that the Fourth Industrial Revolution will profoundly transform the labour market and increase unemployment have revisited the relatively old concept of a universal basic income, or simply basic income, as one possible way to address the problem. A basic income or citizen's basic income is an individual, unconditional, automatic and non-withdrawable payment delivered regularly by the government or a public institution to every citizen in a specific region or country. Thomas More was probably one of the first to discuss this topic from a moral point of view when he put the following words in the mouth of Raphael Hythloday: "[…] no penalty on earth will stop people from stealing, if it is their only way of getting food. […] Instead of inflicting these horrible punishments, it would be far more to the point to provide everyone with some means of livelihood so that nobody's under the frightful necessity of becoming first a thief and then a corpse" (More 1516).

     The English-born American political theorist and activist and philosopher, Thomas Paine (1737–1809), defended a structured and sustainable form of egalitarianism to respond to his concerns that the majority of people would lose their natural inheritance and means of independent survival because of the agrarian laws and the agrarian monopoly. He was the first to advocate a form of basic income in the *Agrarian Justice* pamphlet (Paine 1797), published while he was in France during the Revolution. He proposed that the Government should create "a National Fund, out of which there shall be paid to every person, when they arrived at the age of twenty-one years, the sum of fifteen pounds sterling" (worth about 1390 pounds sterling in 2017) and also "the sum of ten pounds per annum, during life, to every person now living, of the age of fifty years, and to all others as they shall arrive at that age" (Paine 1797). According to him every landowner "owes to the community a ground-rent […] for the land which he holds; and it is from this ground-rent that the fund proposed in this plan is to issue" (Paine 1797).

     Paine was a remarkable man who also lived at an extraordinary social time. He argued for an egalitarian, republican, free society and defended the free market and the inevitability and benefits of scientific and technological progress. Furthermore, he was the first to promote the Trans-Atlanticism and cosmopolitanism of the Western revolutionary process of the latter half of the 18th century. In 1774, he went to British America, the British Empire's colonial territories in North America, following a suggestion by Benjamin Franklin, who had a chance meeting with him in London and was very impressed with his talent and publications. Shortly after arriving, he published the famous *Common Sense* manifesto (Paine 1776), the first public call for the independence of British America, in which he attacked the British monarchy and argued for a peaceful, democratic path to achieving independence. The book was very successful, helped inform the public debate on independence, and became the most influential essay of the American Revolution.

In 1790, he visited France, became friends with Condorcet and other French politicians and philosophers, and came to be an enthusiastic supporter of the French Revolution, which had just begun. When, in November 1790, the Anglo-Irish statesman, political theorist, and philosopher Edmund Burke (1729–1797), regarded nowadays as the philosophical founder of modern conservatism, wrote the political pamphlet *Reflections on the Revolution in France*, one of the best known attacks on the French Revolution, Thomas Paine was immediately and passionately inspired to write the book *Rights of Man* (Paine 1791) as a scathing response. The book, published in two parts in Great Britain, argued for the right to revolt against leaders who do not protect the natural rights of the people. It was hugely successful and sold more than a million copies. However, it was harshly criticised by the British government led by William Pitt (1759–1806) for suggesting a representative democracy and establishing a republic in Great Britain.

Paine was accused of sedition and fled to France in 1792; he was tried in absentia and convicted. His effigy was burned at various public events all over Great Britain. He defended himself from the accusation, writing; "If, to expose the fraud and imposition of monarchy [...] to promote universal peace, civilisation, and commerce, and to break the chains of political superstition, and raise degraded man to his proper rank; if these things be libellous [...] let the name of libeller be engraved on my tomb" (Paine 1792).

In the same year of 1792, he became an honorary French citizen and was elected to the National Convention. Because he sided with the Jacobins, he was arrested, sentenced to death, and imprisoned on 28 December 1793, at the Prison of Luxembourg, in Paris, on the orders of Robespierre, despite his protests that he was an American citizen. He was saved from the guillotine because the gaoler marked the inside of his cell door with a chalk mark instead of the outside when the door was open for Paine to receive official visitors. While in prison he began to write *The Age of Reason; Being an investigation of true and fabulous theology*, published in three parts in 1794, 1795, and 1807 (Paine 1794).

It is a deist text that challenges institutionalized religion and the inerrancy and validity of the Bible, and argues for a natural religion based on reflection, exploration, and the individual search for spirituality. The *Age of Reason*, probably as a reaction to its great success, was viciously attacked in the US, and the British government prohibited the book and prosecuted anyone who tried to publish or distribute it, from 1795 to 1822. In 1795 Paine was freed by the US ambassador in France, James Monroe (1758–1831), and he blamed President George Washington (1732–1799) for the time he spent in prison.

He returned to the USA in 1802, on President Thomas Jefferson's invitation, disappointed and highly critical of the government of Napoleon Bonaparte (1769–1821). However, he was not made particularly welcome because of his attacks on the institutionalized Christian Churches published in *The Age of Reason*, his close connection with the French Revolution, his friendship with Thomas Jefferson, and his fight for the abolition of slavery. Nevertheless, in 1805, the then famous Adam Smith said that "I know not whether any Man in the World has had any more influence on its inhabitants or affairs for the last thirty years than Tom Paine" (Peabody 1973). His previous friends left him, he suffered from alcoholism, and died in poverty in Greenwich Village, New York, on 8 June 1809. Paine wanted to be buried in a Quaker cemetery but this was not allowed because he had refused to renounce his deist convictions. Only six people attended his funeral, including two grateful black freedmen. A very brief obituary published on 10 June 1809 in the New York Evening Post stated: "He had lived long, done some good, and much harm". Thomas Paine's mortal remains were placed in a grave at his farm in La Rochelle, New York, but in 1819, the British journalist William Cobbett, who was an admirer, covertly exhumed Paine's body and shipped it to Great Britain in the hope of building a proper memorial. However, he was unable to raise enough money. Paine's remains were left in a trunk in the attic of his home and eventually disappeared.

For more than a century following his death, Thomas Paine continued to be hated or simply forgotten. In 1888, Theodore Roosevelt (1858–1919) said that Thomas Paine was a "filthy, little atheist" and never retracted this statement, in spite of the protests that reached him. It seemed that oblivion would have to be the final historical verdict. But surprisingly, on 30 January 1937, an article in the Times of London praised his remarkable achievements, referring to him as the "English Voltaire". On 18 May 1952, seven years after Paine had been elected to the New York University Hall of Fame, his bust was unveiled in the allotted niche in the Colonnades. Now he is regarded as a towering figure of American history, one of the most important protagonists to the American Revolution and Independence (Nelson 2006).

Nicholas de Condorcet and Thomas Paine were two exceptional thinkers who generously seized the opportunity to engage in the accelerated social, political, and economic transformations of the social times they lived in. They reflected deeply on the human condition and were visionaries, full of optimism and confidence about the future of mankind. We should be grateful for their efforts, which cost them dearly due to internal contradictions and the relatively slow rate of progress of social and political transformation processes. The main difference from the present social time is that the transformation of society is now almost exclusively determined by a particular type of strongly supported scientific and technological progress that is supposed to satisfy increasingly monetary and frequently egoistic human expectations. This process tends to transform citizens into mere consumers, whose lives, behaviour, and opinions are controlled by the social media networks.

What we should now be thinking about is the role of science and technology in the future of human societies. Much more knowledge and experience have been accumulated in the social sciences since the 18th century, but social and political ideas no longer play such an active role, and they no long have the influence they once did. The implementation of human rights and ethics has slowed down in many parts of the world. Social cohesion, cooperation, and solidarity have been institutionalized in the various forms of the welfare state and tend to move away from the personal responsibility and engagement of citizens. Compassion is likely to be increasingly considered superfluous for those that believe that the world has never been a better place to live in than now.

On the other hand social sciences have been promoting a criticism of the methodologies of the natural and physical sciences, insisting that the time has come to develop a post-normal science and that science is in crisis. Furthermore, an alliance between some recent technologies, such as ICT, big data, and artificial intelligence, and some social sciences, in particular psychology, political science, data science, and marketing, a domain especially close to the social sciences, have been improving methodologies and procedures to successfully misinform, manipulate, and deceive citizens. These techniques are now applied consistently to influence citizen's political opinions, actions, and voting intentions and to make an assault on specific areas of science that produce results considered to be negative for the progress and profitability of certain industrial and business activities.

The idea of a basic income proposed by Thomas Paine lay practically dormant in the 19th century, but since the mid-20th century interest has increased with arguments from defenders on various sides of the political spectrum. Friedrich Hayek (1899–1992), one of the main ideologues behind the liberal economic theories of the Chicago School, believed that it was necessary to give people the security of a minimum wage. In his book *The Road to Serfdom* (Hayek 1944), he wrote: "There can be no doubt that some minimum of food, shelter, and clothing, sufficient to preserve health and the capacity to work, can be assured to everybody. [...This is] no privilege but a legitimate object of desire [...that] can be provided for all outside of and supplementary to the market system". Milton Friedman (1912–2006), who taught economics at Chicago University for 30 years, had the same ideology. He also argued for a form of guaranteed income that would provide a minimum to everyone, regardless of the reasons for their need. This income, which he called negative income tax, would cause as little harm as possible to self-esteem, independence, and incentives to improve the economic conditions of those who would receive it (Friedman 1962).

More recently, the idea of a basic income has been defended by André Gorz (1923–2007), Guy Standing (2011, 2014), Charles Murray (2006), Jean-Marc Ferry (1995), and Philippe Van Parijs (1995). Standing uses the term "precariat" (a neologism which combines 'precarious' with the suffix -iat from "proletariat") to describe a new social class of people who have no job security and are underemployed or have intermittent or unstable jobs. This class contrasts with the salariat, formed of people with medium and long-term job security, pensions, and holiday and health systems paid for by employers. The situation of the precariat naturally has an impact on their well-being and lifestyle and on their political, social, and economic values and convictions. Although it is hard to calculate, due to a lack of reliable data, Standing estimates that the precariat represents 30–50% of the workforce available in countries with advanced economies, with the highest numbers found in Southern Europe, Japan, and South Korea. The solution that he suggests for this problem is to implement a basic income instead of continuing to develop an increasingly complex social security system that divides society in often unfair ways and that has been unable to end deprivation and poverty.

There have already been some attempts to put it into practice. In Switzerland, the introduction of the concept of an unconditional basic income in the Federal Constitution at an amount of 2500 Swiss francs for adults and 625 for children was the subject of a referendum on 5 June 2016. However, it was rejected by a clear majority of 77%. The proposal was opposed by the government and most political parties. It was argued that it was a "Marxist dream", which forgets that the idea of a basic income has also been supported by the ideologues of the liberal Chicago School, although they never seriously thought about how to put it into practice. One of the main arguments against the proposal was that, if people were paid to do nothing, then they would end up really doing nothing.

One of the crucial points in the ongoing discussion about the universal basic income is whether or not there are substantial market failures in labour markets that the welfare system is unable to fully address (Murray 2006). Some authors point out that the welfare system in most advanced economies leaves millions of people without a confortable retirement, without adequate healthcare, and living in poverty (Murray 2006). Furthermore, they note that the fact that many people work from need and not from motivation is a market failure that can be addressed by basic income, which maximizes welfare by allowing job uptake to be made according to personal preferences (Van Parijs 1995). Others are convinced that the most important objective is financial security and that holding a job is the only reliable way to reach that end. When the financial incentive to work is removed the state encourages idleness, which is contrary to the entrepreneurial drive that assures the progress of the economy. Furthermore, they argue that the implementation of large-scale programs of basic income would be much more expensive than the current welfare systems and would require substantial new state revenues (Hoynes and Rothstein 2019).

Support for basic income has led to the implementation or planning of several experiments in four continents, in countries in different stages of development and with very diverse welfare systems. Experiments have been performed in Finland, Italy, The Netherlands, Kenya, Uganda, Brazil, and the USA and more are planned for Iceland, Germany, Spain, UK, Ukraine, India and Canada (BIEN 2019). In the experiment performed in Finland, it was concluded that there was no statistically significant difference between the test and control groups as regards employment, but the test group reported improvements in several aspects of well-being. We are in the very early stages of the implementation process of reliable, beneficial, and cost-efficient basic income systems, but in view of the current dynamics of labour markets and the pressures exerted by technological progress, the movement is likely to expand.

5.19 Technology, Human Rights, Economic Growth, Democracy, and the Future of Progress

Some of the main actors of the Fourth Industrial Revolution who recognise the problem of its potential effects of increasing inequality, unemployment, and vulnerable employment, suggest that the solution is to reinforce human rights around the world (Schwab 2016), which is a sensible, indisputable, and politically correct suggestion. Nonetheless, it is unlikely to be an effective solution on its own, because of the enormous difference in the pace of technological innovations compared with the pace of human rights improvements throughout the world. Human rights are essential if we are to build a more sustainable world with more justice, peace, and well-being, and with less economic inequality, but it is unlikely in the near future that the implementation of human rights will be fast enough to redistribute the benefits of the Fourth Industrial Revolution equitably.

On the contrary, there are signs in the opposite direction. For instance, in the USA, which is fast promoting the next technological revolution, the administration has ceased to respond, since 7 May 2018, to official requests from United Nations special rapporteurs on issues such as poverty, migration, freedom of expression, and justice, a development that represents a clear departure from their usual practice during many decades (UN Water 2019). Being the world's first military and economic power, this attitude is an unfortunate precedent for the authoritarian regimes that do not have a good track record as regards respecting human rights. These governments will also feel free to opt out of routine processes for monitoring human rights.

The increasing number of migrants that are forced to leave their countries for various reasons, including persecution and conflict, poverty, lack of water and food, environmental degradation, and climate change are particularly vulnerable to human rights violations. Again, the USA is a case for concern. According to official statistics more than 300 000 migrants were apprehended along the Mexico–USA border in the first four months of 2019 and the monthly numbers are rising. The Central American migrants at the border are kept in cramped pens and poor quality accommodation, where children have been separated from their parents and six died in US custody between September 2018 and May 2019.

In Europe there was a wave of public sympathy for the refugees coming from Africa and the Middle East. However, when the number of people applying for asylum reached almost three million in the years 2015 and 2016, this generated a backlash, raising nativist fears of terrorism and Islamization and promoting anti-immigrant political parties. The human development problems in Africa and the Middle East are immense and far from being solved. Non-governmental organizations, local experts, journalists, and researchers in the field of migration have recorded 34 361 deaths of migrants attempting to reach Europe from those regions (UNITED 2018). Most deaths occurred at sea, but some were in detention blocks, asylum units, factories, and town centres.

The overriding characteristic of the Fourth Industrial Revolution is the increased speed of technological progress, especially in the field of ICT. It began in the mid-

20th century and led to an acceleration of operative social time and an increased pace of transformation in social and political attitudes and structures. Technological and scientific advances have never been made faster and they are likely to be made faster still in the future. One of the main drivers behind this acceleration is the military, economic, and technological competition between the major world powers, especially between the USA and China.

Companies, particularly in countries with advanced economies where the Fourth Revolution is gaining ground, are forced to continuously renew themselves just in order to survive and retain their competitiveness in a market that is increasingly dynamic, diversified, and globalised. A study that analysed the performance of 35 000 USA corporations concluded that their average life expectancy had decreased from around 55 years at the beginning of the 1970s to around 30 years in 2010 (Reeves and Püschel 2015). Currently, the risk of a corporation closing within five years is 32%, compared with 5% 50 years ago (Reeves and Püschel 2015). On the other hand, some companies have achieved enormous success and globalised in very short periods of time. This is the case for Uber, which has become the largest transport network in the world in only five years.

Since the begining of the 21st century, US industry has become more concentrated because of a decline in the number of public companies. The total number of companies is now lower than at the beginning of the 1970s, when the real GDP was roughly a third of what it is now (Grullon et al. 2015). In some industries there is a worldwide trend towards a wave of mergers and acquisitions leading to the creation of large multinational corporations, the aim being to increase competitive power and profitability. Firms with more than $1 billion in annual revenue account for nearly 60% of global revenue and 65% of the market (Dobbs et al. 2015). The global creation of value is driven by a relatively small number of firms: just 10% of the world's public companies account for 80% of the profits (Dobbs et al. 2015).

Market concentration is particularly noteworthy in ICT. Alphabet (which now controls Google), Amazon, Apple, Facebook, and Microsoft dominate the global consumer ICT industry. The supremacy of the "Frightful Five" comes from their capacity for competition, from increasing consumption of ICT products, growing connectivity of networks, rising industrial investment, and availability of low-cost labor associated with more globalized supply chains. In the past, corporations reinvested most of their earnings but now they are holding on to more of their profits. Since 1980, corporate cash holdings have increased to 10% of GDP in the USA, 22% in Western Europe, 34% in South Korea, and 47% in Japan (Dobbs et al. 2015).

On the consumer side, the acceleration is caused by an increasing pressure to consume and a growing range of ever more diverse products and services. Using digital systems, consumers are almost continuously connected to a market that is constantly changing and providing new and enticing products and services. People must therefore adapt to this irresistible flow in order to make their choices and get the most out of them, as well as keeping up with their friends. Should anyone disregard that pressure, they will fall behind in a race which now involves the vast majority of consumers. If they do drop out, they will tend to become isolated from friends and colleagues and lose self-esteem.

Neoclassical supply and demand economics is morphing into new forms of unconventional economies known as collaborative consumption, which includes the sharing economy, the peer-to-peer economy, and the on-demand economy. Collaborative consumption is now changing a range of industries from Uber to Airbnb to the power industry. Airbnb, based in San Francisco, is the largest peer-to-peer hospitality service and processes more accommodation for travellers per day than the biggest hotel groups in the world.

In the financial sector, fintech helps businesses and consumers in general to better manage their financial operations and everyday lives by using specialized software and algorithms in their computers and smartphones. There is a strong growth of fintech, crowdfunding, and virtual currencies, of which bitcoin is a well-known example. According to a PricewaterhouseCoopers report (PwC 2015), the world revenue of the current five largest sectors of the sharing economy—P2P funding, online human resources, P2P accommodation, car-sharing, and music and video streaming—will grow from today's $15 bn to $335 bn in 2025. We are dealing with an accelerated economic, social, and cultural transformation in which consumers prefer to deal directly with and trust suppliers rather than turning to traditional retail companies.

The number of technology-focused venture capital funds is also increasing. Vision Fund, directed by the Japanese investor Masayoshi Son, has over $100 billion in capital. One of its aims is to create virtual digital worlds that help people find entertainment, products, services, jobs, and business opportunities. Vision Fund is active in most of the Fourth Industrial Revolution technologies, such as ICT, fintech, artificial intelligence, robotics, computational biology, and other data-driven business models.

It is still too soon to say whether or not the Fourth Industrial Revolution will cause a period of greater economic growth at the global level. Considering past experience, the outlook is not entirely favourable because the Third Revolution suffered from Solow's Paradox. While on the one hand we are seeing a growing pace in technological innovations, which some believe will lead to the technological singularity, on the other hand, economic growth has been weak in countries with advanced economies. A 2016 McKinsey study (McKinsey 2016) shows that the wage and capital income of around two-thirds of families in 25 countries with advanced economies fell or remained roughly the same between 2005 and 2014. However, in the previous decade, there was an increase in income for 98% of families. What was an exception from the turn of the century until 2004 has become the norm in the ten years since. The slowdown in income has been more pronounced among younger workers and those with fewer or weaker professional qualifications.

This trend alerts us to the risk of the new generation being poorer than their parents' generation. Surveys carried out for the same study reveal that around 40% of those who took part believed that their economic situation had got worse during the same period of time. It would be deeply traumatic if the mythic relationship between increased scientific research and technological innovation and robust economic growth were to fail.

There are three main reasons commonly mentioned to explain the current situation. Firstly, the fact that the recession caused by the 2008–2009 financial and economic crisis was a deep one and recovery around the world has been slow. Secondly, there are demographic factors caused by a decrease in the number of people of working age per household in countries with advanced economies. This decrease is due to the ageing population, lower fertility, and also social changes that affect family structure. The third reason arises from labour and market factors, namely the variation in the distribution of salaries and the relative importance of labour and capital in GDP.

During the period in question, highly qualified and specialised workers' incomes increased, while workers with middle or low levels of qualification had negligible increases in income or saw their income fall. On the other hand, the average income was brought down by an increasing number of part-time or precarious workers, whose wages are lower than those in full-time work. In other words, the growing precariat and shrinking salariat produces a reduction in the contribution made by labour to GDP. These two trends partly find their roots in the globalisation of the economy, the shift in production to countries that have lower wages, emigration to countries with advanced economies, and automation and robotisation. To better understand the current situation, one must also analyse the issue of inequality.

From the end of the Second World War until the 1970s, for the first three decades of the Great Acceleration known as *Les Trente Glorieuses*, middle class incomes in countries with advanced economies grew strongly, reducing inequality in society. This growth was caused mainly by extensive industrial development, which created a high number of jobs in the manufacturing, construction, and service sectors for workers with middle levels of qualification. Meanwhile, most of the less developed countries continued to have a relatively low levels of economic growth and the distance between them and the more advanced economies grew.

From the 1980s onwards, a profound transformation took place. China, followed by India and other emerging countries, joined the global economy and began to provide goods and services at low prices to consumers in advanced economies, thanks to their cheap workforce. This process, together with the increase in the price of oil and other essential commodities for economic growth, exported by developing countries, increased the rate of economic convergence between the two groups of countries and freed hundreds of millions of people from poverty in Asia, Latin America, and some regions of the Middle East and Africa.

China is the most noteworthy case. According to International Monetary Fund statistics (IMF 2016), its GDP in purchasing power parity (PPP) in 1980 was only 8% of the US GDP in PPP, while it had slightly overtaken it in 2014. In terms of per capita averages, the trend has been similar. The GDP per capita in China as a percentage of the USA's GDP grew from 3.5% in 1980 to 35.9% in 2015. In 2016, of the twenty largest economies in the world, nine did not belong to the OECD: China, India, the Russian Federation, Brazil, Indonesia, Saudi Arabia, the Islamic Republic of Iran, Thailand, and Nigeria (WB 2016). Countries with advanced economies continue to have average per capita incomes in PPP that are higher than those of countries with emerging economies, but the differences have become drastically smaller, especially this century, and will eventually disappear in several cases. In 1980, the average income of the middle class in advanced economies began to grow more slowly or stagnate. A "global middle class" began to emerge, and the difference in its wealth between the rich and some of the non-rich countries began to get smaller.

At the same time, globalisation, by serving as a driver of growth in the world economy, dramatically increased the income of the financial and business elite in countries with advanced economies, and in particular those who were involved in funding and implementing global development. This process, together with the implementation of neoliberal policies in the 1980s and 1990s, policies that dismantled US financial and banking regulations in place since the Great Depression, increased corporate market power, produced a decline in workers' bargaining power, reduced public expenditure for social services, made labour laws more flexible, and privatised public services. These processes explain to a large extent why inequalities have become so great and growth so lukewarm, since economies with less inequality tend to perform better. In the USA, the percentage of income of the top 1% formed a "great economic arc" (Krugman 2007), decreasing from high values on the order of 20% before the Great Depression to reach a low point in the 1970s, and deviating again until the present day. In the USA, the top 0.1% wealth share increased from 7% in 1979 to 22% in 2012 (Saez and Zucman 2014).

It is likely that the economic convergence between emerging countries and advanced economy countries will continue. Nonetheless, the inequality within advanced economies will also continue to grow, in part due to the future effects of the Fourth Industrial Revolution. At the same time, the relative stagnation of the middle and lower classes, is producing increasing social and political instability in countries with advanced economies, particularly in Europe and the USA. Large sections of the population who have low and middle incomes are beginning to become convinced that the future will not necessarily bring them or their children and descendants better well-being or greater economic prosperity and job security. This feeling goes squarely against the expectations of assured and perpetual social and economic progress inherent to the culture of optimism with which most politicians seek to convince the electorate. Many citizens have lost confidence in democracy's ability to solve their social, economic, and environmental problems.

The case of the USA is paradigmatic of the challenges facing the democracies of the advanced economies. Governance in the USA is increasingly dominated by the power of the wealthy since there are no effective checks on the mechanisms by which the large corporations capture the institutions, including the courts, Congress, regulatory agencies, and a large part of the media. Universal healthcare and more progressive taxation are systematically blocked by the very rich members of Congress. What really matters is the opinion of the financial and economic elites who did not win the popular political vote. This malfunction of democracy is aggravated by the fact that the procedural rules of the Constitution provide white men with considerably more power than women and people of color. Although white men constitute only 31% of the population, 97% of all Republican elected officials are white and 76% are male (RDC 2018). Concerning Democratic elected officials, 79% are white and 65% are male.

The tensions that weaken democracy are exacerbated by a decreasing white majority that is predicted to disappear by about 2045 and by the increasing pressure of illegal immigrants. A person is defined as white in the USA if he or she identifies as being only white and non-Hispanic. It is estimated that when Christopher Columbus arrived in America in 1492, there were about 10 million American Indians or Native Americans living north of what is today Mexico. The Native Americans are now the smallest race group in the USA, representing about 1% of the total population. The white share of the US population increased from roughly 80% in 1776, when the country became independent, to a little under 90% in 1950, then decreased to 60% in 2018 and will drop below 50% in another 25 years. The belief that white people are superior to other races, and should therefore be dominant, may lead a white supremacy minority to override democracy, taking extreme steps to cling to power (Robinson 2019).

The situation in Europe as regards democracy has some similarities with the USA, but is more complex and diversified. The main instability drivers are weak economic growth, rural abandonment, increasing inequalities, religious tensions, and the pressures of illegal immigration.

The decline of democracy is not irreversible and there are many ways to adapt it to the current social time and improve the way it functions. However, it is unlikely that this program can be achieved without a profound reform of the prevailing economic system, which has been strongly linked with democracy and is now under stress. This stress is recognized by some billionaires as a problem for the future of capitalism. Raymond Dalio, a billionnaire US investor and hedge fund manager wrote recently that the current system is "producing a self-reinforcing feedback loop that widens the income/wealth/opportunity gap to the point that capitalism and the American Dream are in jeopardy" (Dalio 2019).

Capitalism's essential and greatest promise is intergenerational progress, or in other words, the promise that each social generation will improve on the one before as a result of the performance mechanisms of the market economy. When this promise begins to be put in question, capitalism is in trouble and defends itself by developing extreme and anomalous forms. It is significant that Joseph Stiglitz advocates a reform of capitalism that he calls progressive capitalism to preserve its unassailable promise of progress inherent in capitalism (Stiglitz 2019).

Social divides are becoming deeper and this feeds into a trend towards greater polarisation of political beliefs. In the meantime, anti-globalisation movements have begun to emerge that argue for various forms of protectionism, particularly in the USA. All these signs are part of a political response to the current social and economic problems within the societies of countries with advanced economies. Populist forms of neoliberalism oppose all kinds of socialism using every means available to them, including disrupting the establishment and attacking intellectual elites and the media. At its most extreme, contemporary neoliberalism runs the risk of converting representative democracy into a plutocracy. Although neoliberalism aggravates social and economic inequalities, its economic policies are sometimes presented and defended as having precisely the opposite effect through what its critics refer to as "trickle-down economics" or "voodoo economics". This is a form of supply-side economics that argues for a reduction in the progressiveness of taxes to promote economic growth, the idea being that this will redistribute wealth in society by trickling the growth down through the economy. However, several studies by international organisations have shown that the application of the neoliberal economic theories reduce the percentage of total wealth shared by the lowest income brackets (Dabla-Norris et al. 2015; Keeley 2015), instead of generating a trickle-down effect.

After this brief digression into the history of the idea of progress, born in the West in the 18th century, it is time to sum up. Among the different sectoral forms of progress, science and technology are the areas where the concept is clearest and where the signs of progress are most obvious and unmistakeable. Social and economic progress has also been remarkable in the last 300 years for a large part of mankind, despite some dramatic interruptions, but its propagation throughout the world and its global success has been clouded in the last 40 years by increasing inequalities and fragmentation propelled by a culture of greed and egoism. Billions of people continue to live in a situation of poverty and deprivation of human rights, frequent conflicts, natural resource depletion, environmental degradation and pollution, worsening climate change, and growing human migration movements. The West's trust in the continuity of its economic progress is beginning to falter.

In contrast, the progress of the economic and financial elites has been meteoric, enabling lifestyles and luxuries that are unimaginable for the overwhelming majority of the human population. These elites are advised by the best experts with a realistic and matter-of-fact view of the present and future challenges facing humanity, and their job is to guarantee that their employers are protected and cloistered from the increasing uncertainty, instability, insecurity, and threats that proliferate around the world. They live in a progress bubble believing that they are the quintessence of modern civilization and the guardians of the last refuge of progress. However, it is unlikely that the bubble can be forever detached from the world. They depend on the survival and the consumer capabilities of the rest of the world to assure a healthy progress bubble.

The progress of science and technology that helped the Industrial Revolution to flourish held the promise of inclusive social and economic progress for some time. It was thought that it would be possible to achieve human progress for all of humanity, and it was argued that this could be achieved by way of the Enlightenment. However, technological progress presented an irrefutable opportunity to develop an immense and insatiable ambition and greed. Now, without a shift to a more sustainable economic and financial system, the evolution of the current situation is likely to harm or reverse social, political, and economic progress for a large part of humankind.

Nevertheless, scientific and technological progress will remain intense. That intensity is the result of two factors. The first is the military competition between nations and the second results from the fact that human population growth and the continuous increase in exploitation of natural resources needed to fulfil the expectations of economic prosperity and well-being can only be achieved by extending technological progress to every region of the world. To guarantee technological progress, it is essential to maintain scientific progress. Furthermore, in the current system, scientific and technological progress are only possible with economic progress, understood as robust GDP growth. Without technological progress, the likelihood of collapses in food security and people security rises perilously.

In the triad of economic, scientific, and technological progress, the weakest link is science. The supremacy of economic and technological progress tends to depreciate the other values, particularly the values of truth, ethics, human rights, political freedom, and the environment. Increasingly free from social, political, and ethical values fallen into disuse, the supremacy of technological progress also has the extraordinary advantage of creating an opportunity to expand horizons of expectation to enhance the human race. Studies on morals and ethics regarding new disruptive technologies and its consequences will continue and even be expanded, many supported by public institutions. They will above all represent a ritual kept proudly and with great respect, but inconsequential and without practical effects regarding the transfer of their conclusions into rules and behaviour so as not to disturb or impair the economic advantages of technological progress. A ritual that many practise genuinely and purposefully, but that the vast majority disregard.

In the current view of the future, stasis or regression often tends to replace progress. There is an increasing gulf between the sources of power in society and the politicians that conventionally hold the power. The perception that the new forces appear to be malign and uncontrollable redirects the arrow of time to the past since the best option is to think that something went wrong and that by turning to the past we can find a solution for the future. Following this way of thinking implies that we no longer believe in imagining an ideally perfect society somewhere, or a utopia, but content ourselves with reviving "visions located in the lost/stolen/abandoned but undead past", or a retrotopia (Bauman 2017). This would be a bad sign for the future of *Homo sapiens*, since it would signal that the whole field of imagined societies would have been exhaustively and unsuccessfully exploited and that we should content ourselves with reviving the more memorable solutions of the past.

However, the "Age of Nostalgia" is not the way forward but a mere human reaction. The future will be determined by where progress takes refuge. It has lost its past plasticity and flexibility. Our grandiose past successes have made our path into the future progressively narrower. Our narrative of the future is losing its imagined diversity and becoming increasingly triter, but full of changeability, uncertainty, and threats.

To try to understand its direction it may help to revisit Hegel, who believed that "reason in history" operates cunningly (Hegel 1830). It is therefore likely that that reason will lead progress along the path of transhumanism and the construction of transhuman beings, thereby freeing us from the human condition of dependence on temporal, physiological, psychological, social, environmental, and climatic constraints. However, as it will only be possible to guarantee this transition for a small part of humanity, *Homo sapiens* will remain a being denied progress, in decline, a kind of zombie. Our potential extinction would be interpreted as progress, representing the triumph of technological progress over all other forms of progress. It would ultimately be yet another manifestation of the unescapable reason of history, which, in short, is the history of hubris and consists of humans generating gods or a god to follow, comfort, and guide them and for humans to later imitate, incarnate, and forget them.

5.20   Modern Utopias and Dystopias and the First Steps in Science Fiction

Utopias give us the freedom to imagine the future and exaggerate some features of the world that trouble us without worrying overly about consistency. They fall into the area of transition between subjective understanding of the world and fiction and they reveal a great deal about the horizons of expectation of successive social generations. While religion was the topic of utopias in the 16th century and science was the theme in the 17th century, the focus in the two following centuries was on political, social, and economic themes. The history of totalitarianism in the 20th century can be considered to have been dominated by a series of "utopists"— Vladimir Lenin (1870–1924), Joseph Stalin (1878–1953), Benito Mussolini (1883– 1945), Adolf Hitler (1889–1945), and Mao Zedong (1893–1976)—who forced their countries to conform to their ideological views. They did not see themselves as the architects of utopias. They believed that the State could rework human nature and social behaviour to achieve what they thought was an ideal society, putting into practice what others thought were mere utopias. In the words of Benito Mussolini the plan was essentially described as "Everything within the state, nothing outside the state, nothing against the state".

The failure of these attempts helped create a new literary genre, dystopian literature, sometimes called counter-utopia or cacotopia. These imagine and describe societies in which there is great suffering or injustice. The other important reason for creating dystopias was to denounce the dehumanisation of societies dominated by the perverse use of science and technology or by environmental crisis and disasters. The difference between utopias and dystopias depends to a large extent on the intention of the author and the reader's subjective interpretation. If we think about what they would be like in practice, utopian societies would probably be as unpleasant, inhuman, and harmful to human values as dystopian societies. Building a society with all the details that a utopia or dystopia assigns is of course impossible in practice and, if it were possible, it would be a nightmare.

However, one of the important contributions made by utopias and dystopias is that they show the ways that human imagination reacts to its operative social time and explores future answers to its expectations and preoccupations often by the method of *reductio ad absurdum*. Some scattered, incoherent, and unpredictable elements of the responses presented in the form of utopias or dystopias end up later being partially realised in historical time.

When we try to imagine the social and political consequences of current and future scientific and technological advances, we enter into the realm of science fiction, which has close connections with utopia and dystopia. In science fiction, the narrative often takes place far from Earth, somewhere else in the universe, rather than a remote island in the ocean, like More's island of Utopia in the Atlantic or Bacon's Bensalem Island in the Pacific.

The first utopia involving science fiction and located outside the Earth was an alien society described by Francis Godwin (1563–1633), the Anglican Bishop of

Hereford, in his book *The Man in the Moone* (Godwin 1638). Godwin used the pseudonym Domingo Gonsales, a Spaniard, and the book, although written in 1620, was published posthumously in 1638, probably because the theme and style were so far removed from the orthodoxy and formal posture that was expected of a bishop at the time. Nevertheless, the book is noteworthy because it uses science fiction to explore the "new astronomy" of the revolutionary theories developed by Nicolas Copernicus (1473–1543). Gonsales, fleeing Spain, and following several adventures, arrives at the island of Saint Helena, where he builds a machine pulled by wild geese, the gansas, which enables him to fly. Later, after more adventures, he takes off in his flying machine and twelve days later arrives on the Moon. Observing the Earth from space, Gonsales observes its rotation and uses it to measure time by watching the position of Europe, Africa, and the Atlantic Ocean relative to him. On the Moon, he encounters a people of giants, the Lunars, living in a paradisiacal society inspired by the myth of the golden age. After six months, missing Earth and worrying about his gansas' health, he decides to return and lands in China, where he is able to win over the mandarin's trust and tell his extraordinary story to Jesuit missionaries.

Some years earlier, in 1608, the astronomer Johannes Kepler (1571–1630) wrote a posthumous book called *Somnium* (Kepler 1634) about a journey to the Moon. It had some similarities with Godwin's book but was more focused on the science of lunar astronomy and less on utopia. Both books try to make the new facts discovered by astronomy plausible and comprehensible, particularly the Earth's daily rotation. The journey to the Moon created a new reference frame that made it possible to observe the Earth and view its movements. Curiously, in Kepler's book the narrator Duracotus makes his journey to the Moon with the help of demons, following the cone of shadow projected by the Earth into space, and it takes only four hours. Both Kepler and Godwin accepted that the Moon was populated and believed that the light and dark areas of the surface were oceans and continents, contradicting the observations made by Galileo Galilei using a telescope, and reported in *Sidereus Nuncius* (Galilei 1610), which led him to conclude that they were mountainous regions and low-lying plains, respectively. Both authors mention that, at the start of the journey, a huge effort was required to overcome what they believed was the Earth's magnetic attraction, only later identified as gravity by Newton in 1687. Godwin and Kepler's books began to build the road to interplanetary travel and particularly to the Moon, where Neil Armstrong (1930–2012) and Buzz Aldrin arrived on 20 July 1969, almost 330 years later, on the Apollo 11 mission. Since the 17th century, the universe has gradually become more open to observation and human exploration, as well as a realm for science fiction.

5.21   Wells' Experiment in Prophecy and the Contemporary New Republic

Jules Verne (1828–1905) and Herbert George Wells (1866–1946) were two remarkable pioneers of science fiction. Wells devoted himself to writing books about history, politics, social issues, novels, utopias, and dystopias and was a visionary of the future, especially the way the future might be influenced by science. He was highly successful in his time and a friend and confidant of the intellectual and political elites in Great Britain and around the world. Winston Churchill (1874–1965) was a Wells admirer and appeared to use several of his ideas, saying "I owe you a great debt" (Toye 2010). He benefited from a scientific education in biology, partly dispensed by Thomas Henry Huxley (1825–1895), an ardent Darwinist, grandfather of the brothers Aldous (1894–1963) and Julian Huxley (1887–1975).

In his first non-fiction bestseller *Anticipations of the Reaction of Mechanical and Scientific Progress upon Human Life and Thought* (Wells 1901), first published as a series of articles in *A Fortnightly Review* magazine under the subtitle "An Experiment in Prophecy", Wells was one of the first writers who directly tried to foresee the future in the context of the scientific and technological progress that would dominate the 20th century and beyond, insofar as it could be predicted at the very beginning of the century. It is a remarkably unreserved sibylline text that Wells considered to be "The keystone to the main arch of my work".

Wells forecasted the end of nation states and the emergence of a World State from "a great federation of white English-speaking peoples", with a single system of laws, English as a universal language and led by the New Republic. The New Republic is introduced in the text as follows: "I have sought to show that in peace and war alike a process has been and is at work, a process with all the inevitableness and all the patience of a natural force, whereby the great swollen, shapeless, hypertrophied social mass of to-day must give birth at last to a naturally and informally organized, educated class, an unprecedented sort of people, a New Republic dominating the world. It will be none of our ostensible governments that will effect this great clearing up; it will be the mass of power and intelligence altogether outside the official state systems of to-day that will make this great clearance [...]. The men of the New Republic will be intelligently critical men, and they will have the courage of their critical conclusions. [...] In all sorts of ways they will be influencing and controlling the apparatus of the ostensible governments" (Wells 1901). In Wells' view, the New Republic would be put into practice around the year 2000, "by which the final peace of the world may be assured for ever". Wells acknowledged that he could not say much about the details, but stated that "In its more developed phases I seem to see the New Republic as a sort of outspoken Secret Society, with which even the prominent men of the ostensible state may be openly affiliated" (Wells 1901).

On the darker side of the future, Wells foresees that the working classes are "more properly speaking [...] the People of the Abyss". Further along he states that "the peasant of to-day will be represented to-morrow by the people of no account whatever, the classes of extinction, the People of the Abyss". The New Republic adopts eugenic social policies that deal with the "People of the Abyss" and lead to a "reconstructed ethical system" and finally a "new ethics", which can be interpreted as Wells' way to address his Malthusian preoccupations. In his own words: "The new ethics will hold life to be a privilege and a responsibility, not a sort of night refuge for base spirits out of the void; and the alternative in right conduct between living fully, beautifully, and efficiently will be to die. For a multitude of contemptible and silly creatures, fear-driven and helpless and useless, unhappy or hatefully happy in the midst of squalid dishonour, feeble, ugly, inefficient, born of unrestrained lusts, and increasing and multiplying through sheer incontinence and stupidity, the men of the New Republic will have little pity and less benevolence."

Wells also devoted himself to the utopia genre in his book *A Modern Utopia* (Wells 1905). After studying the previous utopian literature, he presented a society located on a twin planet of Earth, exactly like our own, which replicates every man and woman alive on Earth but with different habits, traditions, knowledge, and ideas. It is an openly unequal society, led by the Samurai, intellectually and spiritually superior beings. The social theories of the utopia built by Wells are not based on distinctions regarding work or capital, considered to be accidental, but on classes of mind which, according to Wells, reflect differences in the reach, quality, and character of individual imagination. There are four classes of mind: the Poetic, the Kinetic, the Dull, and the Base. Education is the same until the differences become clear and each man and woman must then decide to which class of the mind he or she belongs.

In *A Modern Utopia*, Wells describes his design for the World State in great detail. The origins of the concept go back to the first great civilisations as it realizes the aim of governing and dominating "everything illuminated by the Sun", as the Egyptians said, or "under the sky", as the Chinese said, thereby creating a universal state of peace. Wells believed that the World State was inevitable for creating the future ideal society and concluded that it would likely arise as the natural outcome of Anglo-Saxon hegemony. The citizens of the World State would share the same language, currency, customs, and laws and have complete freedom of movement.

As for the economy, the World State, local governments, and municipalities would together own all natural resources, including energy and water, in order to assure sustainability through their most efficient management. Although some of Wells ideas are today considered as liberal-leaning, most are socialist-leaning, such as his apology of State support for ordinary citizens, saying that: "The State will stand at the back of the economic struggle as the reserve employer of Labour". Regarding the domain of science and technology, Wells stated that machines would entirely free humanity and did not encounter limits to the incursion that they would make into human life. On the contrary he believed that with automation, arduous routine jobs will eventually disappear and "a labouring class—that is to say, a class of workers without personal initiative—will become unnecessary to the world of men" (Wells 1905). A blunt view of the consequences that the Fourth Industrial Revolution and its successors may have for employment, written more than a hundred years earlier.

Wells had many foresights and utopian ideas that are clearly relevant in the present day. There are similarities between Wells' vision of the New Republic and the vision of the world pursued by contemporary economic and financial elites, including the transnational business elites who control the large multinational corporations (LMNCs). These elites share a disproportionate part of the economic, financial, and political power in the world. It is curious that, following very different paths, more complex and subtle than those that Wells imagined, humanity has created an entity resembling the New Republic. The fluid contemporary New Republic controls, sometimes subliminally, social, economic, and political development for a significant part of humanity, as well as aspects that are crucial to the future of our common environment and well-being. It is a very robust system that derives its strength from the ambition, ability, devotion, and intelligence of its leaders and the support they get from the shareholders of the companies and banks they succeed in managing so profitably.

Over the last 40 years, multinational corporations have proliferated. The number has gone from around 7000 at the beginning of the 1970s to 60 000 in 1999 (Held and McGrew 2001). Most have their parent company in countries with advanced economies and control more than 500 000 subsidiaries abroad. There is a current trend towards the formation of "metanational" corporations, which have interests, operations, and assets spread over many countries, so that they no longer have a privileged relationship with any particular one. Currently, the large multinational companies have interests that are well above national interests. The CEOs of the New Republic are highly active in seeking to influence to their advantage the governments of each country where the company operates, by all available means, but as discreetly as possible. The crucial features of this enterprise are that the market is global, competition between companies is extreme, and survival is always under threat, whence the challenge is to innovate continuously, in particular by learning from each country where the company operates in order to increase market opportunities and profitability.

The main criteria for metanational corporations when choosing a country in which to invest are the presence of favourable tax laws, regulations, and incentives, cheap and abundant resources, cheap labour, some political stability, and reasonable infrastructures and communications. The corporation headquarters, managers and directors' offices, administrative staff, factories, banks, and other subsidiary companies with financial assets are located in countries that offer the most advantages. It is estimated that in 2012, multinational corporations administered 80% of the $20 tn in global value chains for world trade (UNCTAD 2013). There is a well-known tendency in most industries for the market to merge into a small number of large corporations that manage to concentrate and harness profits.

The process of concentrating power is fractal and is also found among all multinational companies from smaller to larger. An analysis published in 2011 (Vitali et al. 2011) covering 43 060 multinational corporations and banks identified a group of 1318 that controlled a substantial part of the global economy. The study was carried out by three experts in complex systems that mapped the interdependence networks among the 43 060 corporations and financial institutions to determine the structure of economic and financial power at global level. A more detailed analysis of ownership and control relations revealed a "super-entity" of 147 corporations and banks with close connections between each other that controls 40% of the total network. The many financial institutions in this super-entity include Barclays, Goldman Sachs, JPMorgan Chase, and UBS AG. This concentration in a group whose members are heavily connected and dependent on one another is potentially unstable if one of the members is affected or collapses, as was the case of the Lehman Brothers on 15 September 2008. Concentration of economic and financial power does not result from an intentional global command or policy, but is the outcome of an organic and competitive process to optimise the profit–power binomial that benefits the group.

In 2015, the annual revenues of the 25 multinational corporations with highest revenue values—Walmart, ExxonMobil, Royal Dutch Shell, Apple, Glencore, Samsung, Amazon, Microsoft, Nestlé, Alphabet (Google), Uber, Huawei, Vodafone, Anheuser-Busch, Maersk, Goldman Sachs, Halliburton, Accenture, McDonald's, Emirates, Facebook, Alibaba, Blackrock, McKinsey, and Twitter—ranged from $485 billion for Walmart to $2.2 billion for Twitter (Khanna and Francis 2016). Most of these revenues were greater than the GDPs of several countries in the world. Furthermore, a large percentage of the world's population directly or indirectly use the goods and services provided by the LMNCs and this dependence is likely to continue growing.

According to Forbes' 17th annual ranking of the world's largest public companies for 2019, the top ten are: ICBC (Industrial and Commercial Bank of China), JPMorgan Chase, China Construction Bank, Agricultural Bank of China, Bank of America, Apple, Ping An Insurance Group, Bank of China, Royal Dutch Shell, and Wells Fargo. The four largest state-owned banks of China are in the top ten positions and ICBC has been in the first position for the last 7 years. This bank has more than $US 4 trillion in assets and employs nearly half a million people. The increase in the economic power of the Chinese companies has been very clear over the last two decades. Of the 61 countries that are present in the 2019 ranking, 576 companies are from the USA, 309 from China, and 223 from Japan. However in 2000, the USA had 776 companies while China had only 43.

The likely possibility that China's economy will surpass that of the USA around 2030 or before is a motive of concern in both the USA and more generally for most people in the West. This development is a direct consequence of the fact that China is catching up with the productivity of labour and capital in the USA. According to the United Nations, the total populations of China and the USA in 2017 are estimated at 1409.5 and 324.5 million, respectively (UNDESAPD 2017a). One way to prevent China's economy from overtaking that of the USA would be a stagnation or decrease in its economic productivity, maintaining it at a level about four to five times lower than that of the USA, or a major decrease in China's working-age population. The currently favoured way to slow down the increase in China's economic power is to wage a trade and technological war. In this confrontation, the USA has the extraordinary advantage that the US dollar and the Federal Reserve Bank play a dominant role and create a unipolar financial world.

It may happen that this trade and technological war will force the annual economic growth of China to fall rapidly from its present value of around 6% to less than 4%, while the annual economic growth of the USA may increase to values over 3% (Fickling 2019). This highly favourable scenario for the USA is implausible without triggering a global crisis that would probably tip the world into a recession, and given the global scale of public and private debt, lead to another global financial crisis. Since central banks are increasingly unable to serve as last resort lenders, illiquid financial markets tend to become more vulnerable to all sorts of economic, financial, technological, and military disruptions.

Another possibility is that China could suffer a sudden demographic collapse that could significantly reduce the imbalance in the size of the working-age population relative to the USA, from about five to one to less than three to one (Guillemette and Turner 2018). However, the Chinese will no doubt try very hard to solve their future demographic challenge and there is every reason to think they will succeed as well as the West. After all they have succeeded in maintaining a high rate of economic growth for more than three decades, in spite of all the Western forecasts that it would be impossible.

In terms of GDP in PPP, China already leads the world and is likely to maintain that position in the foreseeable future. The countries with emerging economies will probably dominate the top 10 positions in terms of GDP in PPP. The 10 first places are occupied by China, India, the USA, Indonesia, Brazil, Russia, Mexico, Japan, Germany, and the UK (PwC 2017).

Adjusting peacefully to the likely change in global power shift between the USA and China will require great skill on both sides. However, if the USA persists with its goal of impeding China's economic progress at any cost, it may use its military hegemony and engage in war. This possibility and the way to make it happen is openly discussed and analysed and often referred to as the "Thucydides Trap" where the USA provokes a war with China (Allison 2017; French 2017), just as the Peloponnesian War became inevitable because of the growth of Athenian power and the fear this caused in Sparta, which was the dominant power at the time. It would be more rational to find a peaceful *modus vivendi* between the two blocs carrying the heritage of different cultures, which would be beneficial to both countries in the medium and long term, but very difficult to put it into practice because it goes against other essential traits of human nature.

All the medium term scenarios of economic growth and global economic order mentioned here assume that the current global economic and financial system perdures with the adaptations required by an ever-changing environment, but without any profound structural changes regarding the crucial role played by LMNCs.

One of the main features of LMNCs is transnational mobility in the search for opportunities to produce goods and services at low cost and thereby optimise profit and accumulate capital. LMNCs often compete for economic and financial power with the governments of the countries where they operate and, as a result, they challenge and transform the concepts of state sovereignty, governance, and democracy. They also hold part of the political power that previously belonged to national governments, or ostensible governments, in Wells' terms, although it is difficult to accurately quantify the distribution. The causes behind this situation are primarily the result of the power that large corporations have to lobby governments, parliaments, politicians, and members of the central administration, to influence international trade agreements, to use biased criteria to choose the countries where they make investments, and to control and often prevent technological transfer and the sharing of their technological knowledge and know-how.

Power-sharing between governments and large corporations has been particularly ominous in the USA as a result of the lobbying mechanism, as described extensively by Lee Drutman (2015). However, there are some fields of activity in which the direct responsibilities of the contemporary New Republic and governments are quite distinct. This separation happens for instance in war engagements, which are led by governments. Because of their enormous cost in terms of people, resources, and wealth, the contemporary New Republic tries to influence decisions about major military interventions, especially when they perceive that they are likely to benefit from them, but that field of action is mostly in the hands of the ostensible governments.

The contemporary New Republic is not a secret society, as Wells imagined at the start of the 20th century, but the transparency of their actions is limited, for example in terms of their transactions and tax payments in the countries where they operate. LMNCs systematically use direct investment abroad to reduce tax and increase profits by way of complex networks of holding companies created in countries where the tax system is more favourable, or in tax havens. The amount of money that passes through those havens on the way to its final destination where it comes exclusively under the management control of LMNCs has increased considerably this century and was estimated to have reached 30% of the total in 2012 (UNCTAD 2012). The same year, the British Virgin Islands, a Parliamentary dependency under the constitutional monarchy of the UK, received the fifth largest amount of direct foreign investment in the world, totalling US$72 bn, while the UK itself, whose GDP is almost 3000 times larger, received US$46 bn (UNCTAD 2012).

The issue of tax evasion by multinational corporations and financial institutions has been discussed repeatedly at G20 meetings, although the practical results have been limited. In 2014, the OECD (2016c) launched a framework called "Base erosion and profit shifting" (BEPS), designed to avoid artificially shifting profit to places with lower or non-existent tax. In 2016, BEPS had been joined by 100 countries and was supported by the G20 summit held in Hangzhou, China, on 4–5 September 2016.

Tax havens are used not only institutionally by most LMNCs, but also by the leaders of the contemporary New Republic, with interests or dominant positions in the corporations. A study carried out by James Henry for the Tax Justice Network (Henry 2012), using data from the World Bank, the IMF, the United Nations, and central banks, concluded that, since the 1970s, the private elites of 139 countries had held between $ 7.3 tn and $ 9.3 tn of financial assets in tax havens to avoid paying tax in their countries. These assets do not include property assets, gold, yachts, racehorses, works of art, etc. The same study estimates that there are between $ 21 tn and $ 32 tn in tax havens, which means a loss of tax in their countries of residence somewhere in the order of $280 bn. To get a more objective idea of what those amounts mean, the US national debt in June 2019 was about $ 22 tn and its GDP in 2018 was near $ 20 tn.

The "prominent men of the ostensible state" are not members of the New Republic, as Wells shrewdly anticipated, but easily jump from one group to the other, sometimes through skilfully disguised movements. Every year, some members of the two groups meet in Davos at the World Economic Forum, whose slogan is "Dedicated to improving the state of the world". The key players and many hundreds of millions of supporters of the contemporary New Republic believe it to be a close-to-ideal system and conclude that the alternatives are much worse and more dangerous because they may hamper or impede economic, scientific, and technological progress and lead to chaos. The current system is marked by the priority given to economic and financial values as the only way to construct a world that is supposed to provide more well-being and economic prosperity through a rising dependence on science and technology.

Although money has been central in civilizations from early Imperial China to ancient Greece and to the Modern Era in Europe up to the French Revolution, it was not until the 19th century that monetary earnings were first used systematically to measure a citizen's well-being. Up to that time the dominant form of social measurements were the so-called "moral statistics", which were social indicators of literacy, education, insanity, disease, life expectancy, poverty, prostitution, and incarceration. These social, health and mental indicators were replaced by money-based economic indicators both in Europe and the USA during the 19th century. The main reason for this change was the full institutionalization of capitalism in Western societies. This process meant that natural resources, technological inventions, innovations and devices, urban areas, real estate and all other tangible assets, institutions, organizations and associations, human beings, human populations, and nations were capitalized into income-generating assets that were valued according to their capacity to make money and yield future returns.

The ascent of money has penetrated all domains of human life. One can find many manifestations of its increasing ascendancy in the historical record. The essayist and journalist of Romanticism, Heinrich Heine, born of Jewish parents, acknowledged it in a way that became very public in an article sent to the Allgemeine Zeitung newspaper on 31 March 1841, during the long period of time that he lived in Paris (Heine 1854a, 1855). Heine had some acquaintance with James de Rothschild (1792–1868), son of the founder of the Rothschild banking dynasty, the German Jewish banker Mayer Amschel Rothschild (1747–1812), founders of the international financial system. James de Rothschild ran the French branch of the banking empire and had enormous financial and political power in France.

Heine's article described the reverence, admiration, and deep veneration that people held for James de Rothschild when he deigned to receive them in the halls of his lavish Hotel d'Otrante, 19 Rue Laffitte, in Paris. Interestingly, this palace had been home to Joseph Fouché (1759–1820), the first Duke of Otranto, Napoleon's feared Minister of Police and a former member of the Robespierre's Montagnards during the French Revolution, which clearly illustrates the human capacity for adaptation. In the letter, Heine wrote the following about conduct at the Hotel d'Otrante: "I like best to visit him in his office at the counting house, where I can make philosophical observation of the way people—not alone the chosen people of the Lord, but all other people—bow and cringe before him. There is a bending and twisting of spines that a contortionist could hardly beat. I see some who jump as they approach the great baron as if they had touched a voltaic pile. At the door of his room many are taken with a shudder of awe, such as seized Moses on Horab when he stood on holy ground.

And as Moses forthwith took off his shoes, so many a broker and agent de change would certainly pull off his boots before daring to step into Herr von Rothschild's office but for the fear that his feet might smell worse than his boots and the odor might be unpleasant to the Herr Baron. This private office is truly a wonderful place, exciting noble thoughts and emotions, like the sight of the sea or the starry heavens. There we see how small man is and how great is God, for money is the god of our era, and Rothschild is his prophet" (Heine 1854b).

Comparisons between the contemporary power of the New Republic and divine power are very rare but revealing when they occur. One example was Lloyd Blankfein, when he stated in an interview with the UK's Sunday Times, on 9 November 2009, during the financial crisis in the USA, that he was "doing God's work", referring to his work as CEO of Goldman Sachs (Arlidge 2009). As expected, a short time later, on 17 November, Blankfein stated at a meeting in New York that he regretted making the statement and said that he only meant it as a joke (Guerrera et al. 2009). At the same meeting, Blankfein apologised for his bank's role in the crisis that was being experienced at the time and a few hours later Goldman Sachs revealed that it was assigning a $500 million fund to help 10 000 small companies in the USA to recover from the recession (Guerrera et al. 2009). The amount represented around 2.3% of Goldman Sachs' payroll and year-end bonuses in 2009 (Guerrera et al. 2009).

Currently, ideas relating to the cultural heritage of contemporary civilisation but having little or no economic or financial value are conserved as precious relics of the past. New ideas without that value are viewed with suspicion or lack of interest; only a fringe of intellectuals, well tolerated but practically irrelevant, take an interest. Morals and ethics will continue to draw the attention of many people all over the world, including academics, but their power to influence the transition into a future society dominated by the dependence on technology and the desire to enhance the biological and mental functioning of human beings will likely be evanescent.

Wells was a committed defender of the World State, but throughout his life he recognised the difficulties involved in building it. After the First World War, in his book *In the Fourth Year* (Wells 1918), he argued for a League of Nations to be founded and actively helped draw up the Sankey Declaration of the Rights of Man, which served as the basis for the Universal Declaration of Human Rights adopted by the United Nations in 1948, shortly after his death.

The world is moving into the opposing direction of a World State. Nationalistic, nativist, and protectionist policies are increasingly popular, probably as a defensive response to the perception of a more adverse social, economic, financial, and environmental outlook on the global level. The global cooperation that is essential for solving global problems, especially those regarding the environment, is becoming more difficult. Implementing a World State, or simply strengthening the United Nations to deal with growing global problems, has become highly unlikely. Contrary to Well's beliefs, the contemporary New Republic has been unable to guarantee that "the final peace of the world may be assured for ever".

As regards the people of the abyss, Wells was much closer to contemporary reality. Nevertheless, "the alternative in right conduct between living fully, beautifully, and efficiently will" not be to die, as he wrote, but to flee and migrate. All human beings, no matter how dire the situation in which they find themselves, continue to be inherently human and therefore full of hope. They know that they share the same essential humanity as all other human beings and will not be willing to die due to the predicaments they suffer. They do unimaginable things to survive, as they have done in the past and will continue to do until the end of *Homo sapiens* time. One finds contemporary examples in those that migrate from the Dry Corridor of Central America to the USA or from the hunger and violence of the Sahel countries to Europe or from the corruption and poverty of some countries in the Equatorial Region of Africa or in the more than "70 million women, children & men who have been forced to flee war, conflict and persecution" (Guterres 2019). Many migrants go through unthinkable maltreatment in their journeys fleeing homeland. They are tortured, raped, and held for ransom, and some see others dying by their side or killed, some by the smugglers that transport them for a very high price relative to their possessions. All these events happen with almost total impunity. Most of the factual description of these modern migration movements is obtained from journalists that accompany them in their difficult journeys (Sengupta 2016).

The problems of the fragile and failed states in the world are increasing and so is the number of people in the abyss. Where are the signs that the leaders of the most powerful LMNCs or the governments of the most powerful economies would like to, and are prepared to, do what is necessary to decrease that number in the coming decades? Are the people of the abyss good and profitable consumers? Are they likely to join customer loyalty programs? Here, Wells' exercise in prophecy was right when he wrote that "the men of the New Republic will have little pity and less benevolence" for the people of the abyss (Wells 1901). However, one may ask what would happen if the people of the abyss, instead of being a collection of localized minorities, a small percentage of the world's population, were to grow in number and become comparable in size with the totality of world consumers? What would be the consequences for economic growth? The good news is that the people of the abyss can count on help and protection from many individuals, associations, and organizations at the national, regional, and global levels, scattered all over the world.

There is another kind of people in the abyss, emerging first in countries with advanced economies and eventually throughout the whole world. The new algorithms and the increasing automation and robotization of economic activities are replacing an increasing number of humans in their jobs. The world may evolve along a path where the only jobs left for humans will be those where they are definitely economically advantageous relative to the new algorithms and hardware of the Fourth Industrial Revolution. What to do with all the excess of human beings, especially in a world where the total population is likely to continue increasing at least until the end of the century? This is the new class of people that Yuval Harari calls the "useless class" (Harari 2015), since they will be not merely unemployed but unemployable. Being unemployable, most of them will lose the status of good and profitable consumers.

The future vision of the world in utopias is fragmented, partial, disconnected, and improbable, but with time, certain features eventually occur in forms that bear some similarity to their utopian counterparts. The examples above show how we are incapable of imagining and building a vision of humanity's future with the complexity and consistency that it will actually have when it happens. We always fail to have a clear view of the future.

5.22 Contemporary Operative Social Time and the Twentieth Century Dystopian Novels

Wells' prophecies and utopias endeavour to highlight and project a more developed and peaceful modern society that becames highly dependent on science and technology, but many would in practice create a dystopian world. It is unsurprising, then, that some authors, particularly Yevgeny Zamyatin (1884–1937) and Aldous Huxley (1894–1963), were inspired by Wells and decided to nurture the more perverse seeds of modernity to build their dystopias. Zamyatin, in the book *We* (Zamyatin 1924), describes a totalitarian World State in the 30th century, where perfect equality is achieved at the price of repressing all human feelings and individual freedom. The use of time and all human activities, including private ones such as sex, are precisely and unquestionably regulated by the Table of Hours, inspired by the influential methods invented by the American mechanical engineer Frederick Winslow Taylor (1856–1915) to improve industrial efficiency. Taylor believed that "With the triumph of scientific management, unions would have nothing left to do, and they would have been cleansed of their most evil feature: the restriction of output" (Montgomery 1989), which would mean the final reconciliation between labour and capital.

The narrator in Zamyatin's book is the engineer D-503, whose work involves building the Integral, a spaceship designed to convert alien civilisations to the new happiness. However, D-503 is attracted by the freedom and unpredictability of happiness in the old world that is still our present time. The book, written in 1920–1921, was immediately banned by the Goskomizdat, the Soviet organisation in charge of controlling publications in the Soviet Union. It was later translated into English and published in the USA in 1924.

Huxley, in his famous book *Brave New World* (Huxley 1932), sought to imagine the limits that science could reach to forge the man of the future, a topic that remains just as relevant today. The World State imagined by Huxley for London in 2540 was inspired in part by the legacy of innovation, automation, and progress of the American industrialist Henry Ford (1863–1947). Ford was able to make the automobile into a consumer product accessible to the American middle classes by using production lines and mass production. Nowadays, humans continue to be utterly enchanted by the automobile.

Huxley recognised the revolutionary nature of Ford's feat and imagined its possible future applications, creating a production line where human embryos are artificially developed in hatcheries and conditioning centres. Five castes are produced: alphas and betas, who form the elite, and gammas, deltas, and epsilons, who are conditioned to carry out lower tasks. Huxley's World State is a demographically stable society, since natural reproduction has been abolished. Emotional and romantic relationships are obsolete and spiritual needs are satisfied in meetings to take soma, a hallucinogenic drug to crate euphoria that is provided by the World State and generally ends with a sexual orgy. The secret to social stability is being able to mould society so it believes it is happy. There is no rebellion because it is not needed. The main values of Huxley's World State are efficiency, production, and consumption.

Huxley's dystopia is based on the exploration of our weakness to crave for affluence and pleasure at any price. Choosing dictatorship and losing freedom is not too high a price for the immense pleasure derived from unlimited consumption of goods and services. The words religion, family, mercy, and honour lose their meaning and usage. The concepts of greed and egoism are viewed much more positively. Non-economic values, such as morals, ethics, compassion, the various forms of artistic expression, and scientific research are practically non-existent or are considered subversive. However, channels for entertainment—cinema, radio, and others–abound. Everything in the Brave New World is designed to encourage consumption. In Huxley's words: "Primroses and landscapes, he pointed out, have one grave defect: they are gratuitous. A love of nature keeps no factories busy. It was decided to abolish the love of nature, at any rate among the lower classes; to abolish the love of nature but not the tendency to consume transport" (Huxley 1932).

The book 1984 (Orwell 1949), written by Eric Arthur Blair (1903–1950) under the pseudonym George Orwell, is another influential dystopian novel critical of the totalitarian regimes of the 20th century and inspired by Zamyatin's We. The central idea is to imagine how a society can be controlled to give it absolute political stability. Orwell confronts the stability problem before arriving at its origin in operative social time and describes how it is possible to gradually solve it and finally make it vanish. In Chap. 3, he writes: "'Who controls the past', ran the Party slogan, 'controls the future: who controls the present controls the past'. And yet the past, though of its nature alterable, never had been altered. Whatever was true now was true from everlasting to everlasting. It was quite simple. All that was needed was an unending series of victories over your own memory. 'Reality control', they called it: in Newspeak, 'doublethink'."

Orwell highlights one of the paradoxes of our relationship with time that became particularly visible at the beginning of the 20th century. The main goal of political power is to keep it, to ensure the stability and continuity of the exercise of power by those who hold it. Power always attempts to be timeless, but modernity is inextricably linked to the acceleration of operative social time. The totalitarianisms of the 20th century, which Orwell sheds light on and criticises by reducing them to the absurd through a dystopian view, were an attempt to try to slow down operative social time within the sphere of power, without, however, losing the benefits of the acceleration of that same time caused by scientific and technological progress. They sought to be definitive "political solutions" able to remain stable indefinitely, but fully compatible with the permanent changes brought by time's acceleration. However, the mistake here is that operative social time cannot be divided. Politics is an integral part of modernity and inevitably shares the acceleration of operative social time. By trying to achieve political timelessness, power takes on an unsustainable absolute value that

Orwell acknowledged when he wrote (Orwell 1949): "The Party seeks power entirely for its own sake. We are not interested in the good of others; we are interested solely in power. Not wealth or luxury or long life or happiness: only power, pure power."

This tension between power and time also manifests itself in international relations. Since the Treaty of Chaumont, signed in 1814 between Austria, Great Britain, Prussia, and Russia to contain Napoleonic France, the distinction between great and small powers has frequently been used to distinguish between sovereign states with greater or lesser military and economic power. At the present time, the USA, as the world's major superpower, and indeed all the other great powers, attempt by every means available to increase or at least maintain their status. The international system seeks a stable equilibrium as a way of fostering world peace in spite of the various factors that accelerate time and bring instabilities into the power structure. Before the 19th century the substitution of a world superpower by a new power on the rise was not seen as a dangerous threat to geopolitical equilibrium. The situation nowadays, in a globalized world, is different.

The current case of the USA and China is particularly interesting because their elements of operative social time are quite different, while their contemporaneous operative social times are highly convergent because of the globalisation of the economy, finance, science, and technology. China's element of operative social time began in the 16th century BC with the Shang dynasty, which left the first known records of Chinese writing, while the USA's began formally on 4 July 1776. If these beginnings are accepted, although both countries have a long history behind them, China's historical time is at least 15 times longer than the USA's. In the Sung dynasty (960–1280), China experienced intensive and extensive growth and it seems likely that from 1500 on its constant-price GDP became the world's largest, at least until about 1840 (Maddison 2007a) when the bloc of divided and often competing great and small powers making up Western Europe became the next economic superpower, essentially as a consequence of the development of modern science and the Industrial Revolution. From 1840 onwards China suffered from weakness of governance, internal conflicts, foreign political and military interventions, and technological backwardness. Its GDP decreased, was overtaken by the USA's before the end of the 19th century, and only started to grow again after 1950. Now China excels in science and technology, having assimilated all the inner workings of capitalism and using them selectively to vindicate and build upon its past world status.

The USA takes advantage of the relative brevity of its element of operative social time and privileges the short term when promoting the acceleration of operative social time through the tribalization of politics and disruptive technological progress. The longevity of China's element of operative social time privileges the long term because of its richer experience with longer time frames and long-term strategies, and favours the stability and resilience of its operative social time. In their struggle for hegemonic power, the USA has less time, because its element of operative social time is shorter, and it counts on the continuous acceleration of social time, while China has more time, because China's element of operative social time is longer, and counts on the long-term resilience of social time.

The world in Orwell's novel *1984* has three superstates Oceania, Eurasia, and East-asia. However, Oceania, which comprises most English-speaking countries including the UK, is the superpower that controls in various ways the other superstates. The unofficial language of Oceania is English but the official language, which is supposed to eliminate the capacity to express incorrect political and social thoughts, is Newspeak. One of the most distinguishing characteristics of Oceania is the preoccupation with surveillance. The Party is constantly watching all citizens, looking for any sign of thought-crime or revolt, but tries to appear kind and concerned rather than ruthless and invasive. Oceania's residents do not have privacy. Apartments contain telescreens that make it possible not only to see and hear people all the time but also to broadcast state propaganda, controlled news, and disinformation. Any sign of rebellion, however slight or masked, may land someone in prison.

The world of *1984* is in various ways a metaphor for today, in particular as regards surveillance. Governments and companies spy on the citizens of the world as never before and this activity is accepted as a normal part of our social, economic, and political lives. For instance, most people are deeply dependent on their smartphones, and these carry sensors tracking them wherever they go. Besides their location they contain accessible information about consumer habits and preferences, personal and social relations, and the political leanings of their owners. Edward Snowden's revelations showed that the NSA, and similar organisations in various countries, were engaged in mass surveillance and in particular on specific targeted surveillance of world politicians.

Current surveillance technology is significantly more powerful and sophisticated than Orwell could ever have imagined, but the main ideas and central objectives have obvious similarities. Another crucial area of activity for the survival of Oceania's superstate, in which current technology is fast progressing, is political propaganda and disinformation. Today, in our world, people can become first-hand witnesses of events without the need to decide whether or not to trust accounts given by others, through audio and video recordings that appear in the virtual digital world provided by smartphones and computers. But there is no easy way to ascertain whether those events correspond to *de facto* events in the physical world of the Earth system.

In fact they can be deep fakes produced by altering the video or image content using artificial intelligence technologies (Chesney and Citron 2019). Deep fakes are produced by a specific type of deep learning in which pairs of algorithms are set against each other in generative adversarial networks or GANs. A deep fake video is created by using two competing AI systems, one called the generator, which creates content modeled on source data, and a second called the discriminator, which tries to detect the artificial content. When the discriminator discovers that the video clip is a fake it gives the generator who produced it a clue about how to correct it in the next clip. The algorithms learn together how to improve their performance. The self-learning process can then proceed to the desired level of accuracy.

Deep fakes have been used to create opportunities for blackmail, intimidation, and sabotage, but their more likely and promising future use will be in politics, in particular to influence election results, and in international affairs. The success of using disinformation, biased and fake news, and deep fakes depends on people's propensity for confirmation bias, the tendency to interpret, favour, recall, and believe information that confirms one's pre-existing beliefs and narratives. Confirmation bias is particularly strong for emotionally charged issues and deeply entrenched beliefs. It is also stronger if people have limited information, are unwilling to improve their knowledge, and react emotionally rather than investigating in a neutral way. Thus ignorance increases confirmation bias. In Oceania, the ignorance of the people is the Party's greatest strength and the Party slogan is simply "Ignorance is strength".

The deep fake technology is available, but up to now it hasn't been widely used around the world. Deep fake detection will improve and prosecutors will became more active against the creators of harmful fakes, while social media platforms will improve ways to remove fraudulent content. However, artificial intelligence technologies are likely to continue improving their products and making deep fake videos increasingly difficult to identify. Societies will have to learn to live with and become resilient to the future higher levels of misinformation, disinformation, and fake news.

Orwell understood human nature well, especially in extreme situations, because he lived at the time when the totalitarian regimes were formed in Europe, experienced the two World Wars, and took part in the Spanish Civil War, fighting for the Republican government in Catalonia. Towards the end of this life, Orwell wrote in his book *1984* "that the choice for mankind lay between freedom and happiness, and that, for the great bulk of mankind, happiness was better" (Orwell 1949). Although there are many changing interpretations of what constitute freedom and happiness the meaning of Orwell's statement reveals an existential challenge at the heart of human nature and the most common answer to it.

Meanwhile the world has evolved and it is possible to identify the analogous challenge today. In the broad sense, happiness is increasingly an outcome of economic progress, and indirectly of technological progress. There are other drivers of happiness and many people throughout the world would not agree with the exclusiveness of the economic characterization, but at a globally averaged level there is a strong belief that economic prosperity is a necessary condition to reach that vague and ill-defined realm of happiness. Redefining prosperity by decoupling it from the economic component is not a priority in the operative social time of the contemporaneous generation. The problem now is that economic progress is increasingly demanding in a world with a growing population that overexploits natural resources, pollutes and degrades the environment, and is unable to control anthropogenic climate change. On the other hand the economy and finance at the global and national levels are dominated by the thriving contemporaneous New Republic, a dependence that contributes decisively to increasing inequalities.

Most citizens feel powerless towards the newly established powers and equate freedom with unrestrained consumerism coupled with ceaseless navigation in the virtual digital world. Democracy tends to become superfluous in a plutocracy that provides increasing economic prosperity to consumers and a progressive view of the future, full of optimism. Most of the remaining political opposition to the fall of democracy can be disabled by disinformation and fake news in the virtual digital world, the current form of Newspeak. All the necessary ingredients to build totalitarian dystopias are in place and can be readily used. Current economic and technological progress is not necessarily an ally of liberalism and political freedom. There is no danger of an immediate transition to totalitarian regimes in the Western world but there are signs of the demise of liberalism and a degradation of democracy in some countries, such as the USA (Tomasky 2019). On the other hand, there signs of democracy's resilience in other parts of the world, such as Hong Kong, a special administrative region of the People's Republic of China.

After the inauguration of the US President, Donald Trump, on 20 January 2017, Orwell's novel *1984* witnessed a 9500% increase in sales in the USA. The book rapidly ascended to the top of Amazon's bestseller list and sales also increased in other English-speaking countries. The same phenomenon happened with *Brave New World* and most of the other well-known dystopian novels. There are probably many different motivations for this revival of interest. One of the most likely is that people are trying to interpret and make sense of a world that is exhibiting unrecognizable events and problems through the coherence of dystopian narratives. It is not a sign of any willingness to address and solve the new problems, but mainly the hope of satisfying our curiosity about the way the future is taking shape.

5.23   The Unlikely Ecotopia

Time is an essential ingredient of utopias and dystopias. They always address the future in some form or another, and select a particular domain of human activity to focus on, such as religion, politics, science, or technology. In *Ecotopia*, published in 1975 (Callenbach 1975) by the American environmentalist writer Ernest Callenbach (1929–2012), the main goal is to achieve a stable harmony between society and the natural environment. Ecotopia is founded when the states of California, Oregon, and Washington secede from the USA to form a social-ecological system in a "stable state". The state becomes isolated and little is known about what is going on there until the American journalist William Weston from the Times-Post newspaper is allowed to visit it. He subsequently tries to make an objective report what he has seen and heard.

He finds a country where the large cities have been replaced by groups of minicities that are full of trees and energy efficient, where solar power is used extensively, waste is systematically recycled, and pollution has been reduced to a minimum. The Ecotopians have a devotion to trees which, in Callenbach's words, "can almost be called tree worship" (Callenbach 1975). People get around on foot, by bicycle, or by public transport and there is practically no traffic except for a few electric vehicles. The level of comfort inside houses is much lower than in the USA. The head of state is a woman, Vera Allwen, and the Survivalist Party which supports the government is dominated by women and played a crucial role in the fight for independence. The Ecotopian government has managed in a peaceful way to pass laws that make employees the owners of the companies or farms in which they work, and it has also introduced a 20-hour working week. All Ecotopians earn the same amount, except for artists, scientists, and doctors, who make a little more, implying that economic inequality is no longer an issue. Weston is surprised by the Ecotopians' sexual disinhibition, which contrasts with the puritanism that he knows and describes as the "American psychodrama of mutual suspicion between sexes" (Callenbach 1975). He also mentions that they perform violent and bloody "ritual war games", probably to release tensions, he thinks.

In the book, Callenbach reveals the crucial aspect that makes Ecotopia stand out during a conversation about its established stability when Weston asks Bert Luckman, a "bright Jewish kid from New York" who was "studying at Berkeley at the time of Independence": "But it means giving up any notions of progress. You just want to get to that point and stay there, like a lump". Bert's reply does not contradict the idea that progress is weakened but emphasises that they have yet to reach the point of stability and that the search for it continues within Ecotopian society: "We're always striving to approximate it, but we never get there. And you know how much we disagree on what exactly is to be done—we only agree on the root essentials, everything else is in dispute" (Callenbach 1975).

In Ecotopia, economic, scientific and technological progress is no longer the main driving force for society's dynamics. The Ecotopians use science and technology but do so selectively, for example to generate energy in a way compatible with the environment. Consumption of goods is kept to a minimum and "'consumers' is not a term used in polite conversation here" (Callenbach 1975). Weston realises that the devices manufactured in Ecotopia look primitive compared to those in the USA, probably because, by law, they are designed to last a long time and some of them can be easily repaired by their owners.

In this utopian world, the main driver accelerating operative social time was relegated in favour of other values and eventually extinguished. A world that prefers stability, in which social time has decelerated and people have begun to have time to cultivate their relationships with fellow humans and with nature to reach a balanced social–ecological sustainability.

Callenbach does not enlighten us as to how it is possible to maintain Ecotopia's economy, security, defence, and international relations with the rest of the world as he understands it, but goes into some detail about the transition process that led to its independence, in other words, how it managed to slow down operative social time. He tells us that the formation of Ecotopia represents the revival of the protestant work ethic that formed the basis for building the USA. With a 20-hour working week, Ecotopia was forced to isolate its economy from the competition exercised by other, more hard-working countries. Even so (Callenbach 1975): "There was a drop in Gross National Product by more than a third", and furthermore, "humans were meant to take their modest place in a seamless, stable-state web of living organisms, disturbing that web as little as possible. This would mean sacrifice of present consumption, but it would ensure future survival—which became an almost religious objective, perhaps akin to earlier doctrines of 'salvation'." Callenbach also writes that the economic disaster required to make the transition can be survived and that "a financial panic could be turned to advantage if the new nation could be organized to devote its real resources of energy, knowledge, skills, and materials to the basic necessities of survival" (Callenbach 1975).

Callenbach's Ecotopia, despite being a utopia, subliminally reveals a political and economic naiveté regarding the possibility of making the transition to an economic system that achieves regional sustainability in part of a nation like the USA. Later, Callenbach wrote a prequel to *Ecotopia* called *Ecotopia Emerging* (Callenbach 1981), classified as suitable for young adults by the influential Publishers Weekly, describing how the three states on the west coast of the USA went about achieving secession. A transition to Ecotopia by a part of the USA would generate a strong reaction by large companies and the powerful economic and financial system which, using their influence, would support the measures necessary for the government and armed forces to avoid it. This likely reaction is practically absent from both of Callenbach's ecotopist books. It is as if most Americans suddenly forgot the success of their economic and financial system in creating economic prosperity and the American way of life and focused instead on the problems of pollution, environmental degradation, and lack of sustainability, not to mention the negative aspects of American capitalism, so that they could give democratic support to such a transition to a kind of ecotopia.

Despite these features, the fact is that Ecotopia had great success, and nearly a million copies were sold. It has influenced the development of the sustainability movement in the USA and the world. None of this was anticipated by American publishers in the 1970s. In fact the Ecotopia manuscript was rejected by 25 publishers and ended up being self-published. Some of the ideas that Callenbach presents on energy and the environment have been used, mostly in countries with advanced economies, and put into practice locally to some extent. Electric cars have been developed but are still a long way from replacing internal combustion engines; modern renewable energies, are also being developed fast in many countries and represented an estimated 10.4% of global total final energy consumption in 2017 (REN21 2018). Furthermore, recycling has become systematic in several countries and there are various local food production initiatives, although their importance in the world is still limited. On the other hand most of Callenbach's social and economic suggestions have shifted further away from today's reality to become even more utopian, and this includes the economic equality of Ecotopians.

In short, the dominant economic and financial system has become more powerful and its influence upon society at the national and global levels is becoming ever stronger, although it now lives alongside sustainability practices and experiments, spread throughout the world. What remains practically unalterable and has likely got more pronounced since Ecotopia was published is the conviction, widely held around the world, that progress, particularly economic progress based on scientific and technological progress, is not only possible but strong, and will remain strong in the foreseeable future. Callenbach's Ecotopia represents the denial of progress as it is seen by the vast majority of the human population. Only a few proposals in the field of ecology can be implemented locally or regionally without revolutionising the existing system.

Ecotopia was written at the end of a period of strong economic growth and increasing quality of life in the USA, which began in the 1940s and during which the TFP saw very high growth (Gordon 2000). Although acknowledging this economic prosperity and the high standards of civilization that had been achieved in his country, Callenbach became convinced that the system contained unsustainable aspects, so it was necessary to create new horizons of expectation to achieve sustainability. These horizons are less attractive because they diverge from both the security of salvation promised by religious faith and the comfort of believing in unlimited economic progress which has prevailed since the 18th century. The new horizon of expectation for life in a stable state, in harmony with the environment, is unable to compete with the current horizon of expectation of a never-ending increase in economic prosperity.

Many people may be aware of the likely unsustainability of the current economic and financial system as regards environmental destruction and degradation, overexploitation of natural resources, and climate change, but they are often also convinced that the serious crisis that might follow will eventually be resolved with appropriate technologies, or otherwise that this crisis will occur at a time when they will not be personally affected. The vast majority of people are far from being convinced that the main threat to the global economy and to their future economic prosperity is the unsustainability of the economic and financial system. Their operative social time is focused in the short term and not on the medium and long term. This essential human tendency can be and is in fact opposed through a rationalistic voluntarism by many individuals, collective initiatives, and institutions, but their aggregate success is statistically limited and has no expression at the macroscale. Furthermore, the prevailing emphasis on the short term tends to generate a perverse positive feedback. If people become convinced of the future dangers of unsustainability and of the impracticability of the necessary actions to avoid it, they will be inclined to further enjoy the present windfall and ignore the plight of future generations. Meanwhile the destruction of the environment, the overexploitation of natural resources, and climate change go on unabated. There are no winners or losers in this process, just *Homo sapiens'* monologue. The main consequence is the quality and duration of *Homo sapiens* time, nature being truly indifferent.

Chapter 6
Prosperity, Egoism, Voluntary Simplicity, Sustainability, and Expectations of Immortality

6.1 Happiness, Well-being, Prosperity, and the "Good Life"

The presence or absence of states described by the concepts of happiness, well-being, and prosperity play a crucial role in determining behaviour and valuing the space of experience and horizon of expectation at the individual and social group levels. On the other hand, the characterisation and value assigned to these four concepts change over time and vary throughout the world depending on social and economic development levels and on cultural and religious factors. Nonetheless, before attempting to perform a specific analysis of each concept, it is worth pointing out that the desire to enjoy happiness, well-being, and prosperity is universal, even if the relative value assigned to each varies over time and around the globe. Let us begin with a brief description of each concept.

In the West, the meaning of happiness finds its roots in the Greek notion of *eudaimonia*, a composition of *eu*, meaning "good", and *daimon*, meaning "spirit". This concept, together with others corresponding to *arete*, for excellence and goodness, and *phronesis*, for practical wisdom and morality, were extensively discussed by Greek philosophers. Living with *eudaimonia* meant being possessed by a good spirit, in other words having the ability to flourish, be happy, and have a "good life". In Ancient Greece, having *eudaimonia* was interpreted as having the favour of the gods and hence having access to good luck. Modernity has developed the description of psychological states and has put more emphasis on emotional happiness characterised by pleasant but impermanent mental or emotional states that can range from simple contentment and a good mood to intense joy.

Well-being is a more general and inclusive concept that can be defined as a state of being healthy, happy, and prosperous. Since well-being is a changeable state it can also be characterized in a dynamical theory as the equilibrium point between an individual's resource pool and the challenges he or she faces in the social, psychological, and physical domains (Dodge et al. 2012). Wellness is the condition of good physical and mental health while living a good social and spiritual life. According to an estimate by the Global Wellness Institute, promoting wellness has become a

© Springer Nature Switzerland AG 2021
F. Duarte Santos, *Time, Progress, Growth and Technology*, The Frontiers Collection, https://doi.org/10.1007/978-3-030-55334-0_6

global business that was worth US \$4.2 trillion in 2017, which is more than the pharmaceutical and diet industries (GWI 2018). Due to its broad and holistic nature, out of the four concepts, well-being is the one that has been the object of the greatest attention, analysis, and study.

Regarding prosperity, the etymology of the verb "prosper" goes back to the Latin word *prosperare*, which means to be successful and happy and to make something favourable. With the economic growth that followed the Industrial Revolution, the meaning of prosperity shifted to become more frequently connected with wealth, interpreted from a material standpoint or in terms of financial capital. More recently, and probably as a reaction to this trend, the word "prosperous" has also been used to qualify a rich life but in the sense of completeness, fulfilment, and happiness, or indeed a life in which a person has flourished.

Finally, the expression "a good life" essentially refers to well-being, but it goes further, and this makes it relatively difficult to define and characterise. A good life may be considered a practically consensual objective for humanity, but it is constantly transforming in different ways throughout the world. It is a way of living that incorporates glamour, temptations, and opportunities to enjoy the satisfaction, pleasure, and luxury that social and economic development, technological progress, and fashions provide in the most varied of ways in a highly diverse world with profound inequalities.

The concept of well-being has been extensively analysed and studied by a large number of authors from the psychological, social, economic, and philosophical viewpoints (Fletcher 2015). Research into well-being has been divided into two main approaches. One is focused on aspects of well-being that arise from satisfying desires and different forms of hedonism, while the other, called the eudaimonia approach, is focused on personal fulfilment, on sharing trust and identity, and on social and environmental integration (Ryan and Deci 2001). These two features of well-being probably coexist in all of us but elicit different levels of attention and interest in different fields of work. Economists pay special attention to satisfying preferences and desires as a way of achieving people's well-being. In English-speaking countries, economists often use the synonym "welfare" instead of "well-being", which is also used more specifically to refer to material prosperity and social, economic, and financial assistance. The preference for satisfying the desire to have and consume is behind the origins of utility theory, which plays a central role in microeconomics; methods were developed within it to measure consumer satisfaction levels in relation to different products and thereby establish utility functions. From an economic standpoint, human beings are by their nature agents that maximise utility. The eudaimonic aspect attracts above all those who value the inclusive, holistic, and transformative concept of well-being.

From the point of view of psychology, there was an important development at the end of the 1990s, promoted by the North American psychologist Martin Seligman, who launched a new area of research called positive psychology. This uses the scientific method to analyse what "authentic happiness" might be and how it might be achieved, understanding it as a form of eudaimonia (Seligman 2002). This movement was partly a reaction against 20th century psychology, which was highly focused on psychoanalysis and mental illnesses. More recently, a model was developed, referred to by the acronym PERMA, in which the psychology of well-being is considered to comprise five components (Seligman 2011): "positive emotions", which include the hedonic aspects of happiness, such as joy, excitement, contentment, satisfaction, wonder, and good mood; "engagement", which reflects the flow of experiences, capacity for concentration and persistence with professional, institutional, leisure, and other activities; "positive relationships", which include the feeling of being socially integrated, acknowledged, respected, and safe; "meaning", which corresponds to adherence to purpose, valuing goals, and acknowledging the transcendent aspects of life; and "accomplishment", which refers to the ability to persevere and struggle to achieve one's goals, sometimes successfully.

Establishing the well-being of a community or a nation has become a highly relevant social and economic goal, but it remains above all a political one. Methods for assessing subjective well-being have been developed and implemented for more than fifteen years in over 150 countries (OECD 2013; GHWBI 2014; Helliwell et al. 2017). Research on this topic distinguishes between two different features of subjective well-being. The first is emotional well-being, which reflects the quality and intensity of transient experiences of joy, satisfaction, contentment, stress, concern, sadness, rage, and revolt. The second is the overall evaluation of life satisfaction. To assess this component, the question normally asked in questionnaires is: "How satisfied are you with your life as a whole these days?"

Kahneman and Deaton (Kahneman and Deaton 2010) assessed the answers to these two components of subjective well-being according to growth in income in a group of 450 000 Americans living in the USA between 2008 and 2009. The answers showed that the life satisfaction indicator rises consistently with increasing annual income. Emotional well-being, however, stops increasing above an annual income of around $75 000. The authors conclude that the stagnation of emotional well-being arises because, from a certain annual income threshold upward, constraints emerge connected to temperament and life circumstances which generate an increase in negative emotions over positive ones. Greater wealth does not guarantee greater emotional well-being, although lower income leads to lower emotional well-being. Similar conclusions were obtained on a group of working adults in another study which intended to establish the ability to enhance and prolong or, as the authors define it, "savour" positive emotional experiences (Quoidbach et al. 2010). The richer participants manifested less ability to enjoy the simple pleasures of everyday life. It was also deduced that this weakness in comparison with less wealthy participants reduces the positive effect of wealth on the emotional well-being of the richest. The researchers thus concluded that, while money provides emotional benefits, it may harm our ability to enjoy the simpler pleasures of life.

An important issue is to identify the reasons why, according to Kahneman and Deaton's study, life satisfaction does not stop growing as annual income goes up. There is a common intuitive belief that this correlation is natural and stems from the most essential characteristics of human nature. However, the correlation is weaker than intuition would lead us to believe; it is complex and involves different factors and temporal, spatial, social, and economic processes (Lucas and Schimmack 2009).

For example, the correlation tends to be stronger in less developed countries (Howell and Howell 2008) and gets weaker as people get older. Within this complex context, it is important to identify and understand the process that keeps the correlation strong even when income continues to grow to higher and higher levels.

The answer can probably be found in an article that seeks to assess the correlation between life satisfaction and income in a group of people in absolute and relative terms by way of their relative position in a ranking of the group's income (Boyce et al. 2010). The study was carried out with a group of 12 000 British adults. As a predictor of life satisfaction, a ranking of each person's income in the group for a specific year was compared with the absolute value of that income. The study concluded that ranking income can predict life satisfaction, while absolute income or income in relation to a reference group cannot. The higher a person's position in the ranking, the greater his or her life satisfaction. The study also showed that, in accordance with psychological intuition, the individuals in the group placed more value on comparisons with higher positions in the ranking than with lower positions.

This social effect can be visualised by picturing a stairway. We don't know where it leads but it has people standing on every step, although there are many more people on the lower steps. Satisfaction with life increases as we go up the stairs and it does not stop growing, however high up the steps one is. If there are people on steps further up, most people will try to climb higher than those steps, although a few will be satisfied and not wish to climb any more. The relativist reading of non-satiation of life satisfaction with increases in income has only a statistical value, so it does not apply to everyone.

From the study by Boyce et al. (2010), we could conclude that life satisfaction, because it has a relative rather than absolute dependence on income, is not responsive to income distribution, but this is not the case. Studies carried out by several authors have shown that inequality in income distribution has a negative effect on life satisfaction (Oshio and Kobayashi 2010; Schneider 2012; Graafland and Lous 2018). Schneider showed that the degree of influence that inequality has on life satisfaction depends on the cultural perception of socio-economic issues. If inequality is perceived as a demonstration of a high potential for social mobility, it has a positive effect on life satisfaction (2012). According to some authors, this perception, right or otherwise, which corresponds to the view that society is a meritocracy in which social mobility is the result of merit, can be found in the USA more often than in Europe (Alesina et al. 2004). This is one of the arguments most frequently used to downplay the negative effects of growing income and wealth inequality in today's societies. Schneider (Schneider 2012) also shows that if inequality is perceived as being connected to considerable social immobility then it has a negative effect on life satisfaction.

A more recent study based on an empirical analysis of databases from 21 OECD countries concludes that income inequality reduces average life satisfaction (Graafland and Lous 2018). The same article also shows how net income inequality is positively related to fiscal freedom, free trade, and the deregulation of labour, production, and financial markets. It is further possible to conclude that income inequality indirectly affects life satisfaction due to its effects on mental and physical health (Wilkinson and Pickett 2010). Hajdu and Hajdu (2014) showed that the relatively greater redistribution of income by way of tax in Europe, compared with the USA, tends to increase life satisfaction because it reduces inequality. It is important to stress that there are several dimensions of economic freedom that produce positive effects on life satisfaction, especially through their contribution to encouraging economic growth (Graafland and Compen 2015). However, a high level of economic freedom also implies low marginal tax rates that tend to limit redistributive policies, thereby increasing inequality. Some authors have carefully tried to translate the results obtained into recommendations for public policy by stating that a radical free market based on neoliberal economic policy tends to decrease life satisfaction by increasing inequality (Graafland and Lous 2018).

6.2 The Progress of Ultra High-Net-Worth Individuals

At the highest levels of wealth, the stairway effect manifests itself openly in the formation of social groups defined by the progressively higher wealth of their members and their social intergroup behaviour. These groups correspond to segments with an increasing amount of personal assets net of liabilities. Ultra high-net-worth (UHNW) individuals are defined as those with net assets of more than $30 million (Wealth-X 2017a). The lowest segment of UHNW individuals ranges from $30 million to $49 million. This is followed by those within the $50–99 million bracket, and then segments of increasing wealth up to the top group of billionaires who have a net worth of more than $1 billion.

For UHNW individuals, rising to the billionaire group and remaining in it is very likely seen as a glorious achievement. In an operative social time when the value of money has become supreme and irreplaceable to the largest part of humanity, belonging to the group of a few hundred people who have the greatest accumulated wealth among the current 7.7 billion individuals, must be a motive for feelings of superiority, pride, and power.

But the ambition that made it possible to reach the status of a billionaire continues to grow with time and stimulates a desire to climb further up the stairway to the even more exclusive group of those who have a net worth of more than $10 billion. Human ambition is and will always be insatiable. It drives some people to make their way up to the group of people with net assets worth more than $50 billion, a group that had only six members in 2016 (Wealth-X 2017b). To have an idea of what $50 billion means in today's world, notice that 109 of the 191 countries in the IMF database had a GDP lower than that amount in 2016 (IMF 2017a).

Wealth-X publishes annual reports that provide an up-to-date database and an analysis of the world's UHNW population, the global trends that impact them, and their future outlook, especially their capacity to grow and become stronger. Furthermore, the company tries to attract the flock of the faithful to the stairway of the UHNW elite by "providing intelligent solutions to help uncover, understand, and engage with the world's wealthiest individuals" (Wealth-X 2019a). It describes arrivals and departures into and out of the wealth tiers, statistics relating to sectors of activity, and the geographical distribution of members and group dynamics at the world level. They also include financial, economic, technological, and political analysis of the state of the world and also recommendations on how billionaires should proceed and act to increase their wealth.

This is a global social group that intentionally displays its grandeur and power, describes its desire to keep growing and getting stronger, and analyses the opportunities and obstacles that today's world puts in its path. From 2015 to 2016, the total number of billionaires fell by 3.1% to 2397; this was the first year with a fall in numbers since the West's financial and economic crisis in 2008–2009. Billionaires' total wealth also fell by 3.7% to $7.4 trillion, but both the number of billionaires and their total wealth are higher than in 2014 (Wealth-X 2017b). In other words, the report explains that the 2016 decrease should be seen as a circumstantial fluctuation and not as a mid or long-term trend. In fact the billionaire population reached a maximum in 2017, but in 2018 declined again by 5.4% to 2604. The decrease is attributed to market volatility, global trade tensions, and a slowdown in economic growth (Wealth-X 2019b). Despite such fluctuations the progress of billionaires has been unstoppable over the last three decades. According to Forbes magazine list of billionaires, an independent source from Wealth-X, there were 140 billionaires in the world in 1987, with fortunes totalling $295 billion, and in 2017, there were 2043, with a total fortune of $7.7 trillion.

In 2016 the USA was the only one of the nine countries that had the highest number of billionaires that saw an increase in members of the group, and their total wealth reached $2.6 trillion, the same as the total worth of all the billionaires in Europe and the Middle East put together. The report states that this significant behaviour is due to the "vigorous 'Trump reflation trade' in the financial markets" (Wealth-X 2017b), an expression that means a return to investment with the hope that economic growth and inflation will rise and the group's interests will be better defended. According to US official statistics this expectation has been accomplished and is reflected in both economic and employment growth indicators and in Wall Street stock market indicators. The profits of multinational corporations, major businesspeople, and investors are increasing with the new less distributive tax laws that favour the wealthiest social groups, particularly those with inherited wealth. A step forward in American democracy's shift towards a plutocracy. It should be noted, however, that this bias towards a plutocracy has much deeper roots and a strength and consistency that goes well beyond the current political context in the USA. The present situation is just an epiphenomenon of the decline of democracy in the USA.

In 2018, North America was the only region to have an increase in billionaire population. The Asia-Pacific region, which accounted for the billionaire growth in 2017, saw a decline in 2018 (Wealth-X 2019b). New York is the city in the world with by far the most billionaires. It has more billionaires than any country except China and Germany (Wealth-X 2017a). It would be interesting to investigate in a systematic way the extent to which this exceptional situation contributes to cosmopolitanism, the energy of New York's cultural, artistic, and scientific life, its museums, arts, and shows. The university that has produced the highest number of billionaires is Harvard, followed by Stanford. Curiously, almost a quarter of all the Harvard billionaires' wealth belongs to Bill Gates and Mark Zuckerberg, two Harvard students who did not complete their university courses.

Nevertheless, San Francisco is now the top city with the largest population density of billionaires, with about one billionaire for about 11 600 residents, ahead of New York, Dubai, and Hong Kong which occupy the second, third, and fourth places, respectively. Billionaires tend to congregate in a relatively small number of cities around the world, rather like rare biological species, which tend to occupy just a few specialized habitats.

One interesting aspect revealed by the statistics presented in the Wealth-X billionaire census is the impact of ICT on the age distribution of billionaires. In 2016 around 50% of the 115 billionaires in the technology sector, mostly related to ICT, were aged under 50, while that percentage falls to 14% for the group as a whole (Wealth-X 2017b). The ICT sector is very efficient at creating relatively young millionaires, and in this case Solow's Paradox does not apply to them. The technology sector, despite its importance, is far from having a dominant position. Globally, 20% of billionaires are in the financial sector, 14% in the industrial sector and only 4.8% in the technology sector (Wealth-X 2017b).

Youth is one of the characteristics of Chinese billionaires, a third of whom are aged under 50. "Self-made men" also predominate. Unlike in the West, there are few Chinese billionaires who have inherited fortunes, and their average net worth of $2.7bn is relatively low, indicating difficulty in "cashing out of their primary business" in comparison with their Western counterparts (Wealth-X 2017b).

According to Wealth-X, "philanthropy is the number two passion of the world's ultra rich". Philanthropists can be sure that, when they announce massive donations, society will hail them as heroes. The situation is quite different from what happened in 1909 when John D. Rockefeller proposed the Rockefeller Foundation and found considerable opposition from Congress. The Unitarian pastor John Haynes Holmes testifying before Congress said "it seems to me that this foundation (Rockefeller Foundation), the very character, must be repugnant to the whole idea of a democratic society" (Reich 2018). President Theodore Roosevelt was also critical, saying about the "robber barons" that "no amount of charity in spending such fortunes can compensate in any way for the misconduct in acquiring them" (Reich 2018). Finally, since Congress did not approve the Rockefeller Foundation on Rockefeller's terms, he turned to the State of New York, which accepted them.

Philanthropy is a social and political activity that involves questions of personal morality, which implies that all the motives for a donation matter, not only its good consequences (Reich 2018). It is an exercise of power in the realm of society. In a democratic regime the exercise of that power must be scrutinized in order to know to what extent it serves or undermines democracy. Philanthropists choose the social or public benefits that they want to create, enjoy various tax breaks, for instance when making charitable donations, and expect to be revered by society for their genius and generosity.

Health, social, and economic organizations, high schools, universities, and other educational and research institutions, museums, and cultural institutions, usually accept donations without further scrutiny regarding the nature and lawfulness of the activities that allowed the fortune to be amassed. Only when the actions of the donors became visibly controversial and suspicious in the media do the beneficiaries look more carefully at the matter and in some cases turn down the donations. An example of this situation was provided in March 2019 when Britain's National Portrait Gallery turned down a $1.3 million donation from a charitable arm of the Sackler family linked to the opioid crisis in the USA. Various cultural institutions in Europe and the USA have adopted the same attitude, including the Tate group of museums in Britain and the Guggenheim Museum in New York. More recently, in July 2019, the Louvre Museum in Paris removed the Sackler name from the so-called Sackler Wing of Oriental Antiquities, following protests outside the museum's glass pyramid by the activist group Prescription Addiction Intervention Now (PAIN) (Marshall 2019).

The increase in the number of billionaires and super billionaires in recent decades is one of the signs of worsening wealth inequality at the global scale. According to reports by the Credit Suisse Research Institute, economic inequalities measured by the percentage of global wealth belonging to the richest 1 and 10% of adults has become more pronounced in recent years (Shorrocks et al. 2018). The 50% poorest had less wealth in 2016 than the richest 1% and the richest 10% possessed 89% of global wealth. This very significant concentration of wealth provides those who have it with a financial, economic, and political power that are incommensurate with the rest of the world. A large part of the world economy and finance is run and controlled by about that 10% of the human population.

During the first years of this century, annual growth in global wealth was strong, often exceeding 10%. This growth was to some extent socially inclusive since all levels of society benefitted from it. The 2008–2009 financial and economic crisis changed that situation. In 2008 there was a decrease in global wealth of around 12%, but it picked up again and was on the order of 5% until 2013. In 2014 and 2015 there was a new reversal and the growth in wealth dropped. In spite of this situation there was a rise in financial wealth that mainly benefitted the wealthiest.

Wealth inequality increased all over the world and median wealth stopped rising in most countries except in China. In 2016, median wealth in China exceeded the median value in Europe and remained close to the European level in 2018. In the same year global wealth grew by only 1.4%, reaching $256 trillion (Shorrocks et al. 2018). This percentage growth is close to the annual and global growth in the number of adults, which implies a stagnation in the wealth per adult, which was $52 800 in 2016. This stagnation, however, was short-lived, since during the twelve months to mid-2018 global wealth rose 4.6% to a total of $317 trillion, outpacing population growth (Shorrocks et al. 2018). Wealth per adult grew to a record high of $63 100. Some analysts believe that the growth pattern has recently shifted back toward the pre-crisis pattern and that inequality will start decreasing (Shorrocks et al. 2018).

Not all countries in the world have seen stagnation in their wealth since the financial crisis. In 2016, the countries that had the greatest growth in wealth were Japan, followed by the USA, while the greatest falls were in Ukraine and Argentina (Shorrocks et al. 2018). The USA has clearly been the world champion in the rise of household wealth growth. The total wealth and wealth per adult have grown every year since 2008, even in the 2014 and 2015 years, when global wealth growth decreased. The financial sector has been the major driver of growth, but the wealth thus generated has disproportionately benefited the richest.

The USA commands the generation of UHNW individuals at the global level. Advanced economies and emerging markets, especially China, follow the trend enthusiastically. In 2018, UHNW accounted for just 1.2% of the world's millionaire population but held 34% of their collective wealth. This polarization tends to become more and more extreme since the ultra wealthy continue to experience faster growth than the lower wealth tiers. In 2018, those with more than $30 million in net worth had an increase in their total wealth of 16.3%, which is higher than the 12.7% average rise of the three lower tiers (Wealth-X 2019a).

Inequality is not considered to be a problem at all for the majority of the UHNW individuals or for their businesses, companies, and governments. There is no organized opposition to the progress of inequality. In most countries with developed economies and in many other countries some people are acutely aware of the negative social, economic, and political implications of advancing inequality, but they are politically divided and feel powerless to bring about a solution. In the USA the UHNW individuals are emblematic of what is considered the unparalleled success of American capitalism. The symbolism is incomparably more relevant than any concerns about inequalities. Any alternative to capitalism is viewed by the majority of Americans as a death sentence to the exceptionalism and unrivalled economic and military power of their country. A far greater emerging risk is that China now firmly occupies the second place in the world's wealth hierarchy.

6.3   Acceptance of Tax Havens

One mechanism that contributes to the rise in wealth inequality is the use of tax havens or offshore financial centres, or simply offshores, that offer very low "effective" tax rates to foreign individuals and companies without requiring residence or business presence, and in a politically and economically stable environment. The use of offshores allows tax avoidance and tax evasion, which promotes tax injustice, increases wealth inequality, and distorts the economy, especially in low-income countries. Furthermore, offshores offer financial secrecy, allowing for opaque or secret financial transactions. This feature makes them highly attractive places for laundering money from corruption, drugs, and organised crime or for depositing funds connected to all sort of illegal activities including terrorism.

It is estimated that only multinational corporations, businesses, and investors with a net worth of more than $50 million have the resources needed to make full use of the services offered by tax havens. However, the use of offshores is internationally accepted, being viewed as a perfect legal thing to do to lower tax liabilities, and is supported by an army of generously remunerated lawyers and consultants who specialise in such placements. Large multinational corporations, such as Apple, Microsoft, Alphabet, Cisco, and Oracle keep billions of dollars in offshore accounts with tax rates in the low single digits.

The French economist Gabriel Zucman has estimated that assets equivalent to 10% of world GDP are deposited in offshores (Zucman 2013). In 2016, this corresponded to $7.6 trillion. Switzerland manages around 40% of that total, while the rest sits in places like Luxembourg, Hong Kong, Singapore, Panama, Mauritius, the Cayman Islands, British Virgin Islands, the Bahamas, Bermuda, Jersey, and the Isle of Man. The value of offshore deposits from the USA and Europe is so high that, when properly estimated, it reduces their debt and helps solve the mystery of the West's missing wealth in comparison with China (Zucman 2013). Recent studies based on audits and databases of two offshore financial institutions—HSBC Switzerland ("Swiss leaks") and Mossack Fonseca ("Panama Papers")—reveal that tax evasion increases rapidly with wealth in countries with advanced economies (Alstadsæter et al. 2017).

In Scandinavia, about 3% of personal taxes are evaded on average, but the evasion rate rises to near 30% in the top 0.01% of the wealth distribution. The wealth inequality seen in the official tax data of rich countries since 1970 increases markedly when the financial assets that evade taxes using offshores are included in the calculation (Alstadsæter et al. 2017).

In countries with advanced economies, and especially in the USA, the attitude towards compulsory payment of taxes to the state has not changed much since the beginning of the 1980s. The period of highest tax evasion began during Ronald Reagan's presidency, founded on the neoliberal ideas that the greedy and insatiable state should stop being fed and private initiatives and privatisations should be encouraged. Several of America's Republican politicians have recently argued for the abolition of income tax and the Internal Revenue Service in charge of collecting it. The proposal would mean rejecting the Sixteenth Amendment of the US Constitution ratified in 1913. The American politician Ron Paul, a congressman from 1997 to 2013 and one of the driving forces behind the Tea Party movement, said on 15 April 2009 that "an income tax is the most degrading and totalitarian of all possible taxes" and that its progressive nature was directly based on the Communist Manifesto.

According to Zucman, only around 20% of the $7.6 trillion in financial assets deposited in offshores pay tax. Taxes are not levied on the remaining 80%, representing an estimated annual loss of $200 billion for governments (Zucman 2015), which some consider a violation of the law. The percentage of financial assets deposited in offshores varies according to country, and is generally greater in developing countries. The percentage is estimated at around 4% for the USA, 20% for Latin America, 30% for Africa and 50% for Russia. As for multinational corporations, Crivelli et al. (2016) calculate that the overall loss of taxes is around $650 billion, of which roughly $450 billion affect OECD countries and $200 billion affect other countries. More recent estimates suggest a global amount of less than $500 billion but a greater percentage of losses in countries outside the OECD (Cobham and Janský 2017). Estimates of lost taxes involve a great deal of uncertainty and are very difficult to make because, clearly, the goal of those who make deposits in offshores to lower tax liabilities is also to cover up financial movements and take advantage of all the gaps and ambiguities found in the law and in national and international regulations.

Clearly, the individuals and companies which have the greatest financial power also have the most resources to engage in and indeed cover up tax evasion, as well as the means to defend themselves in the lawsuits that may arise from it. It is not surprising, then, that the tax losses associated with UHNW individuals and large companies are sometimes defended in the media as being irrelevant (Worstall 2016). The main argument is that those tax losses are of relatively small value to governments in relation to their total tax revenue and that the illegal nature of evasion is controversial and increasingly hard to prove because of the system's growing complexity.

Solutions for reducing tax evasion are known but depend on agreements and active cooperation between governments of all countries involved, and this is difficult to achieve. One of the difficulties is that the money flows through offshores have likely become so large that eliminating them or supervising them effectively would be dangerous to the current global financial and economic system, the construction of which is largely down to the USA and the UK. This is one of the plausible explanations for interpreting the USA and the UK's inability to eliminate or effectively supervise the large and powerful group of offshores in English-speaking territories, located within their spheres of influence.

6.4 Acceptance of Increasing Inequalities

Can one find out what the 1% wealthiest members of the world's population believe about the other 99% and about the increase in inequalities that have separated them ever more over the last few decades? The task is difficult because there is little research on the topic and public declarations are rare. In the 19th century and most of the 20th century, many rich people's overriding attitude was to ostentatiously show off their wealth and luxurious lifestyle, something which Thorstein Veblen (1857–1929) called "conspicuous consumption" (Veblen 1899). There is now a huge variety of behaviour. A recent book that included interviews with some UNHWs from New York shows that American society and culture readily accepts them, provided that they have an intense working life, live relatively discreetly, and make philanthropic donations to society for social, cultural, artistic, or scientific activities (Sherman 2017). However, some of the UHNWs interviewed appeared to be ill at ease, embarrassed, or even ashamed about their extreme economic advantage.

Others passionately defend their economic and social status and do not appear to be concerned by inequalities. Nevertheless, they generally take great care to ensure that their public statements are politically correct. There are, however, a few exceptions. At the end of 2013, the wealthy technological workers of the San Francisco area were driving up the value of housing, pricing out the lower and middle classes. Some of those affected reacted by stopping a bus filled with Apple employees in San Francisco on the way to work and another with Google employees in Oakland.

Furthermore, they demanded that the tech companies kick in $1 billion for affordable housing. These incidents were sufficient reason for the Silicon Valley billionaire Tom Perkins, with an estimated net worth of $9 billion, to write a letter to the Wall Street Journal on 24 January 2014, where he states that "Regarding your editorial 'Censors on Campus' (Jan. 18): Writing from the epicentre of progressive thought, San Francisco, I would call attention to the parallels of fascist Nazi Germany to its war on its 'one percent,' namely its Jews, to the progressive war on the American one percent, namely the rich". He also writes in the letter that "I perceive a rising tide of hatred of the successful one percent [...] This is a very dangerous drift in our American thinking. Kristallnacht was unthinkable in 1930; is its descendent 'progressive' radicalism unthinkable now?" Many commentators and analysts from Silicon Valley and elsewhere manifested their displeasure, rebuttal, and hostility to the article. Tom Perkins later regretted the use of the word Kristallnacht but stood by his letter's message in an interview on Bloomberg TV on 27 January. The irrelevance or relevance of the letter published by the Wall Street Journal will doubtless linger on in the internet for many decades.

Politicians that openly defend the current most extreme trends of neoliberal capitalism are also very careful with their public interventions. Few have publicly declared their opinions on the growing socio-economic inequalities. One example was the Canadian entrepreneur and television personality Terence Thomas Kevin O'Leary, better known as Kevin O'Leary, who on 18 January 2017 announced his intention to run for leader of the Conservative Party of Canada. His electoral platform was to lower corporate tax and remove the tax on CO_2 emissions that the Canadian government proposed to implement in 2018. In an interview with Amanda Lang on the CBC News Network television channel on 22 January 2017, when confronted with an Oxfam report (Fuentes-Nieva and Galasso 2014) according to which the world's 85 richest people hold the same amount of wealth as the 3.5 billion poorest people, many of whom live on less than 2.5 dollars a day, he said that it was "fantastic news!" He then explained: "This is a great thing because it inspires everybody, gets them motivation to look up to the one per cent and say 'I want to become one of those people, I'm going to fight hard to get up to the top.' This is fantastic news, and of course I applaud it. What can be wrong with this?" Amanda Lang looked surprised and exclaimed "Really?" "Yes, really. I celebrate capitalism," replied Kevin O'Leary. Most commentators agreed that the remarks were ridiculous. However the opinions expressed by O'Leary are well received and supported by the most conservative and white supremacist sections of US and Canadian societies. Why do they implicitly reveal such a deep indifference and disdain about the significant portion of humanity that lives in the worst conditions of poverty, malnutrition, and lack of health care, education, employment, and quality of life? In fact the situation is getting increasingly grotesque. A more recent report from Oxfam (Hardoon 2017) indicates that "Eight men now own the same amount of wealth as the poorest half of the world".

The redemption fallacy puts out the idea that these people, or at least most of them, can—if they have the necessary motivation and strength of will—rise up to the highest levels of wealth and power. According to this theory the onus in the situation of growing inequalities is on the passiveness and incapacity of the poorest and not on the current economic and financial system built up by the countries with the most advanced economies. Theoretically, the system allows people to rise up economically, so it does not need to be reformed to achieve a more equitable world. If that does not happen, it is just because most of those 3.5 billion people do not individually and collectively have the motivation, ability, or qualities necessary to lift themselves out of the situation they are in through the use of all the opportunities for development that the economic and financial system generously provides in conjugation with science and technology. According to this political argument, these are the "People of the Abyss" who have always existed and will always, unavoidably, continue to exist. It is very commendable and recommendable to practise charity and philanthropy, but there is no need to try to stop the increasing inequalities. These arguments apparently indicate that the present economic and financial system is by far the best that *Homo sapiens* can construct for himself. But can humankind be satisfied with an economy that is biased towards benefiting the richest 1% and not everyone?

Billions of people feel powerless to change their condition of poverty, despite the immense force of will that many have, and that they would be able to develop further to improve their living conditions, if they were given the most basic opportunities to do so. Nevertheless, the 3.5 billion poorest people do not, individually or as a whole, pose any danger, risk, or threat to the 1% richest people or to the economic and financial system that supports them. This may seem surprising, but it is essentially the result of two reasons.

The first is that the current development model, based on acceleration in the production and consumption of goods and services, has been an undeniable success, has become hegemonic, and is sought by almost all. The system is extremely able to defend itself and to convince all concerned that there is no attractive and credible alternative. In any case the transition to another system would likely be chaotic and the new one, whatever it may be, is unknown and therefore risky. The only alternative seems to be to make adjustments to the current system, but these have not been very successful over the last few years. Furthermore, the human and environmental context is changing and the system has so far been unable to adapt.

The second reason concerns the characteristics of the structural organisation of human societies. The social, political, religious, financial, economic, and trading systems, in terms of fighting for, gaining, and sharing power, proselytism, marketing, cooperation, competition, conflicts, and corruption, has a fractal structure based on inter-human group and intra-human group relations. These human groups, or simply groups, have different natures, sizes, identities, goals, and functions. Nonetheless, inter-human group behaviours and intra-human group relationships between homologous groups, which means that they have similar natures and goals but exhibit different identities and may compete for different but analogous objectives, are reproduced like fractals. In other words, the homologous human groups are reproduced in an almost self-similar way, at all scales of human activity from sovereign states with different populations and economic power, to cities or urban areas, to villages or rural areas, large worldwide religions and religious denominations, banks and financial institutions, multinational corporations, companies, corporate groups, academic institutions and organizations, professional groups and associations, political parties, sports clubs, social groups, including local communities, ethnic groups, social clubs, and family clans. Connectivity and interaction between non-homologous groups vary significantly but they are rarely strong, and are in fact generally weak compared with the relations between homologous groups.

The system as a whole is very stable because conflicts and tensions are manifested primarily within and between homologous groups, so they are organised and confined to those group subsets. The social energy required to start a meaningful reaction to a controversial issue involving various non-homologous groups is greater than what is required to start a comparable reaction involving only homologous groups.

The 3.5 billion people poorest people are dispersed among thousands of different homologous and non-homologous groups. But the capacity to react, protest, and make demands is exercised mostly within a small subset of the most important of the homologous groups to which people belong. It is in this very limited subset that most people feel capable of defending their interests. Conflicts, risks, and threats then acquire the same fractal structure. The structure of reactions, protest, and conflict also become fractal and therefore disconnected. The threat to the installed system does not come from fractal protest but from a generalized protest involving a large scale set of non-homologous groups. However, the mobilization to achieve this synergy is unlikely to happen because the fractal protest is closer to most people's interests and therefore dominant. A more significant threat to the system is the possibility of chaos, in other words the risk of destroying the stability, consistency, and cohesion of the fractal structure itself.

There are currently five main issues capable of generating the last kind of threat: globalization, migrations, all kinds of terrorism and vandalism, geopolitical and geo-economic tensions that threaten the established world order, and environmental degradation and pollution, over-exploitation of natural resources, and climate change that threaten the sustainability of the current civilization. Far from being isolated and autonomous, these issues are strongly interrelated and generate powerful feedbacks.

The increasing inequality at global level that the present financial and economic system is bringing about is a rather abstract global issue that involves all human social groups, including all countries, although in different degrees. Human social groups are concerned with the issue of inequality but at the level of the most important human social groups to which they belong, such as the company or institution where they work, their community, their region and their country. For them the relation between the global financial and economic system and inequality is remote and out of reach. The defenders of the current financial and economic system know about and explore these shortcomings. In conclusion it is likely that an increasing number of people, especially in the fragile states, will be structurally condemned to be outcasts to the system, or People of the Abyss. Some UHNW individuals feel erroneously that the 1% is under threat and have nightmares about an unbelievable 21st century Kristallnacht.

6.5 The Emergence of Individualism and Egoism

Individualism–collectivism is an important psychological, philosophical, sociological, political, ethical, and cultural dimension that affects values, norms, and practices in all domains of human activity. It is therefore important to identify the current tendencies in the movements along that dimension and their origins. Individualism is a broad concept cutting across the above-mentioned areas. It describes the tendency to privilege the rights, interests, practices, and values of the individual relative to those of the group. It promotes a view of the self as self-directed and autonomous, while collectivism promotes a view of the self as overlapping with others so that one's thoughts, feelings, and behaviours are embedded in social contexts (Santos et al. 2017).

Collectivism also refers to the principle of centralization of social and economic power in the people collectively, and more specifically to the political or socio-economic theories or practices that encourage communal or state ownership of the means of production and distribution. A wide variety of models have been proposed to construct collectivist organizations, and political systems of government have included collectivism with varying degrees of expression since the 19th century, in such movements as socialism, communism, and fascism.

A recent study that examined 51 years of data on individualistic practices and values across 78 countries concluded that individualism is rising in most of the society's analysed (Santos et al. 2017). It also found that socio-economic development emerged as the most important factor in the move towards individualism, explaining 35% of the 58% variance in change in individualism over time. Furthermore, they found that socio-economic development preceded increases in individualism, thereby suggesting a causal relationship between them. Ecological factors, such as the frequency of natural disasters and climatic stress had a less clear effect on individualism.

However, the correlation between socio-economic development and increasing individualism was not found for all countries. Armenia, China, Croatia, Ukraine, and Uruguay showed a non-negligible decline in individualistic values, which indicates that the individualism–collectivism dimension is multifaceted, involving many diverse factors, such as history, politics, and culture. China is a particularly interesting case since it showed no increase in individualistic values, in spite of the strong economic growth of recent decades. Although further research is required, the result may mean that individualism is not a strong cultural value in China and is probably not viewed and felt as a better stance to maintain growth and improve well-being.

There is a close connection between the modern concepts of individualism and egoism. Both refer to the same goal of acting in self-interest and devaluing the interests of social groups, whatever they may be, but while individualism is a stance, egoism may be a moral and ethical belief. Besides individualism has branched out into applications in many domains, while egoism has been more centered on psychology and philosophy. Curiously, both words are fairly recent inventions. They were first used after the beginning of the Industrial Revolution and initially had a pejorative meaning. A long transformative process has occurred up to the present operative social time, where individualism lies at the foundation of the financial and economic system and egoism is defended and hailed by some as the driving force of progress and growth.

The word "individualism" was probably used systematically for the first time in France in 1826 by the defenders of the sociological theories of Saint-Simon in the review *Le Producteur, Journal de l'Industrie* to characterize the undesirable excesses of individual freedom and of thinking and acting merely out of self-interest (Piguet 2008). In England, the word "individualism" was also used pejoratively in the late 1830s by Robert Owen, who defended cooperative social ideals and "a new organization of society, on the principle of attractive union, instead of repulsive individualism" (Owen 1836).

"Egoism" has been used since the 17th century, rather surprisingly in connection to the Counter-Reformation. The meaning and use of the word broadened and entered the realms of philosophy and psychology, and more recently those of economics and politics. It describes behaviours that have likely been seen since the origins of *Homo sapiens* and are part of our very nature, but have developed and become more pronounced in the modern world. The modern proliferation of egoistic behaviour probably comes from the increasing abundance and variety of opportunities, incentives, and rewards.

The word "egoism" comes from the Latin word *ego* for "I" and is found for the first time in an entry in the French Encyclopaedia of the Enlightenment published in 1751 (Diderot and D'Alembert 1751). In its initial meaning, "egoism" described the behaviour of people who cite and talk about themselves a great deal, exalting their virtues, their lives, and their achievements. This type of behaviour, previously known as *amour-propre*, was a form of "vanity, haughtiness, pettiness of spirit and, at times, rudeness" (Diderot and D'Alembert 1751). The Encyclopaedia also states that the "Gentlemen of Port-Royal", the famous Parisian abbey that played a significant role in the Counter-Reformation and the development of Jansenism, stopped writing their works in the first person because they believed it "came from a principle of vainglory and an overly good opinion of oneself" (Diderot and D'Alembert 1751). To mark their distance, they mocked authors who wrote in the first person, claiming that they were performing *egoisme*, an expression that had not been used until then. In 17th-century England, the Presbyterians began to use the word "selfish" with the meaning of looking to satisfy one's own interests, even if that was to the detriment of others' well-being, which is closer to the meaning that the term *egoiste* eventually acquired. Words with the same formation and meaning are used in Germany—*selbstisch*—and Denmark—*selvisk*.

Egoism can be considered from a descriptive or a normative standpoint. The former, usually called psychological egoism, claims that the motivation for any human action is always to satisfy one's own interest, even when some may see that as an example of altruism. Ethical egoism is a normative theory according to which the condition that is necessary and sufficient for human action to be morally right is to maximise self-interest (Shaver 2015). The definition of rational egoism can be obtained by replacing the expression "morally right" by "rational", in the previous definition (Shaver 2015). Putting this another way, ethical egoism claims that I ought morally to perform some action if and only if, and because, performing that action maximizes my self-interest and rational egoism claims that I ought to perform some action if and only if, and because, performing that action maximizes my self-interest (Shaver 2019).

In Christianity and in the moral principles that it has bequeathed to the West, helping others, and particularly the poor, the starving, the sick, and the abandoned, doing what Auguste Comte later called altruism (Comte 1852), is seen as a redeeming virtue. The Bible often insists on the duty to help the poor and share one's wealth with them. Ephesians 4:28 reads: "He who steals must steal no longer; but rather he must labour, performing with his own hands what is good, so that he will have something to share with one who has need".

On the other hand, throughout the history of the West, several philosophers and theologians, including Aristotle and Thomas Aquinas (1225–1274), stressed that it was rational for man to seek the greatest happiness possible for himself, in the sense of eudaimonia. Baruch Spinoza (1632–1677) went further in his posthumous book *Ethics*, written between 1664 and 1665, by arguing that every person should seek what is truly useful for him or herself and desire what will take him or her to a state of greater perfection (Spinoza 1677).

At the same time, on the other side of the English Channel, the philosopher Thomas Hobbes (1588–1679) developed a theory about human nature and morality which allowed egoism to become further rooted in Western culture. According to him, mankind is an essentially selfish, quarrelsome, and aggressive creature, so only a strong state could stop a "war of all against all" (Hobbes 1642). Hobbes was the first to believe that all human behaviour was ultimately selfish, and therefore to defend the theory of psychological egoism, although he did not use the term "egoism" itself. For Hobbes, everyone is not only allowed to do what best satisfies their personal interest, but they actually have a duty to do so.

Joseph Butler (1692–1752), an English bishop theologian and philosopher, opposed Hobbes' view of an essentially selfish human nature, believing human motivations to be more complex. He was the first to analyse the theory of psychological egoism in detail, demonstrating that it cannot be proven from a philosophical standpoint (Henson 1988). Nonetheless, Butler believed that it was only possible to provide harmony between virtue and personal interests with a fair God that rewards or punishes people in the afterlife according to their just deserts (Butler 1726).

In the 19th century, Henry Sidgwick (1838–1900), an English utilitarian economist and philosopher, published the famous book *The Methods of Ethics* (Sidgwick 1874), which contains one of the fullest analyses of the ideas surrounding ethical egoism. In it, he argues that both morality, in the common and intuitive sense defended in the West at his time, and egoism are rational normative systems. If this is the case, he asks, how is it possible and how can we make these compatible with one another? The question remains relevant today, and finding out whether ethical egoism is a genuine and acceptable ethical theory is still a topic of great philosophical interest (Österberg 1988).

Setting aside the theoretical reasons behind ethical egoism and following practical reason, one finds that egoism is intimately related to the individualistic values and practices that have grown steadily throughout the recent history of the West. This growth is largely the result of a transition to a consumer oriented society that provides easy access to increasing consumption of goods and services. This ease of access can be seen at the global level, although with varying intensity, and is based on the extensive development of industry, international trade, globalization, science, and technology. With endless opportunities to acquire and use increasingly attractive and diverse goods and services that would have been unimaginable in the past, people are constantly invited to maximise utility, in other words, to develop and satisfy their preferences, interests, and desires. Currently, the belief that a person only acts well when seeking to maximise utility has become dominant and is producing a society of egoism. We are dealing with an extreme form of the theory of Utilitarianism initially proposed by Jeremy Bentham (1748–1832). This new type of utilitarianism is a form of ethical egoism that Osterberg (1988) calls non-ideal ethical egoism. It is non-ideal because it is systematically used in any situation, disregarding the generally limited information and knowledge held about each specific situation where it is used and therefore about its positive and negative consequences for the person using it, and especially the longer term consequences. By disregarding that information and knowledge, the use of non-ideal ethical egoism cannot meet one's own interests, nor maximise utility, especially in the medium to long term.

6.6   Adopting Ethical and Rational Egoism

The 20th century has been especially fertile for the ideological defence of ethical and rational egoism. One of the most important contributions were the principles of Objectivism developed by Ayn Rand (1905–1982), and framed erroneously as a contribution to the philosophy of egoism, which had an enduring influence on the ideological and moral grounding of neoliberal policies. Her life, writing, and impact on society are one of the most remarkable examples of the acceleration of social time in the West and the consequences of its extreme dysfunctions in the first half of the 20th century.

Alissa Rosenbaum, who later adopted the name Ayn Rand, was born in St. Petersburg to a bourgeois Jewish family in 1905. Her father was a pharmacist and owned a pharmacy and the building where they lived. Very early on, she revealed an extraordinary ability for writing plays and novels, which she began to produce at the tender age of eight. She witnessed the Russian Revolution at the age of twelve, and a few years later the Bolshevik rulers confiscated the Rosenbaum family business and assets for the good of the Russian people, and the family fled to Crimea, where they endured great difficulties, poverty, and malnutrition. It is very likely that this deprivation during her adolescence played an important role in forming the extreme libertarian ideas she later defended (Heller 2009). When she was 20, Alissa managed to obtain a visa to visit relatives in Chicago, and she cried when she saw Manhattan from the boat, later stating that she had cried "tears of splendour". Her great ambition to be a successful writer and screenwriter took her to Hollywood, where she met Cecil B.

DeMille, who gave her work. She finally gained American citizenship in 1931 after marrying an American actor.

Nietzsche was one of Ayn Rand's main philosophical influences, and she appreciated the praise for the "superman" (*Ubermensch*) in his work. Rand's political worldview took hold after the 1929 Great Depression, when she considered the financial reform policies and development programmes of the New Deal as a form of "collectivism", and therefore a dangerous imitation of the Bolshevism she viscerally opposed.

The theories she developed can be interpreted as an effort to morally justify a policy that literally inverts Marxism. In this case, the workers produce all the value while capitalists take out all the fruits of their labour. In Ayn Rand's analysis, explained in her highly successful books *The Fountainhead* (Rand 1943) and *Atlas Shrugged* (Rand 1957), capitalists are businesspeople with remarkable leadership qualities, initiative, and intelligence who create wealth for the whole of society, while workers in their companies benefit from that wealth and would have a much worse life if they were left to fend for themselves.

The essence of Ayn Rand's message is extreme radicalism, and in that regard she was influenced by the book *What is to be done? Some Stories about the New People* (Chernyshevsky 1863), written by the Russian journalist and philosopher Nikolai Chernyshevsky (1829–1889) while he was imprisoned in the Peter and Paul Fortress in St. Petersburg. This book, of no literary value and highly criticised by Fyodor Dostoyevsky (1821–1881), defends rational egoism and is full of revolutionary ideas and characters that sacrifice everything in life for the cause of the revolution. It had an enormous influence on the Russian revolutionary intelligentsia at the time of the transition from the 19th to the 20th century and the radical emotional dynamics of the Revolution. Vladimir Ulyanov (1870–1924), who adopted the name Lenin, was an admirer of Chernyshevsky's book, and said he had read it many times. He adopted its title when he wrote the manifesto *What is to be done? Burning questions of our movement* (Lenin 1902) between 1901 and 1902. In it, he argues for the formation of a front line of revolutionaries to spread Marxist ideas among workers. Although she loathed the Russian Revolution, its ideals, and the Red Terror, Alisa Rosenbaum was inspired by the rational egoism, the extreme radicalism of Chernyshevsky's book (Weiner 2016), and the revolutionary superheroism of one of its best known characters, Rakhmetov, who also inspired Lenin.

Ayn Rand organised her ideas around a set of principles that she aimed to provide with the respectability of a philosophical theory, and she gave it the name of Objectivism. Its central idea is to defend a form of ethical–rational egoism, that is, a combination of ethical egoism and rational egoism, since in Objectivism, egoism cannot be fully justified without an epistemological analysis based on reason. In her book *The Virtue of Selfishness* (Rand 1964), she explains in detail the type of normative egoism that she defends and encourages society to follow. Ayn Rand believed altruism to be immoral and irrational because she erroneously and simplistically interpreted it as a way of acting exclusively against each person's own interests. In her opinion: "If a man accepts the ethics of altruism, his first concern is not how to live his life, but how to sacrifice it; altruism erodes men's capacity to grasp the value of an individual life; it reveals a mind from which the reality of a human being has been wiped out" (Rand 1957). Rand defended an ethical–rational egoism that rejects voluntary sacrifice in all its possible forms and argues that "man exists for his own sake, that the pursuit of his own happiness is his highest moral purpose, that he must not sacrifice himself to others, nor sacrifice others to himself".

Most philosophers in the academic world reject or ignore Objectivism because of its incoherent and confused ideas (Österberg 1988). Nonetheless, this denigration has not been the slightest obstacle to its spread and impact on American society. On the contrary, Objectivism has made it possible to defend radical capitalism as an ethical system in which successful businesspeople have led society's progress. In the ethical–rational egoism framework defended by Ayn Rand, collecting progressive taxes from that enlightened, entrepreneurial, rich elite to be able to redistribute them in society, thereby providing more help to the people in need that belong to the lowest-income brackets, is considered a moral offence.

Although Objectivism has remained limited to a small group of followers of the Ayn Rand cult, ironically known as the "Collective", the most relevant point is its influence on right-wing political thought in the USA, the UK, and other countries and the fact that it remains significant to this day. This significance is primarily the result of having established a theory in which the radical capitalism of neoliberalism is presented as being morally justified.

The vigour and attractiveness of her ideas are particularly impressive because Ayn Rand was a Jewish atheist who rejected Judeo-Christian morals. These morals, despite all the social transformations that have taken place in the USA, still have influence on society. Ayn Rand's ideas about egoism have managed to convince many people and have come to form an alternative to mainstream morals. One example of this tendency can be found in the results of a survey done in 1991 by the American Library of Congress to identify the most influential book in the USA. *Atlas Shrugged* came in second place in the study, after the Bible (Heller 2009). Since many believers in Christian denominations, particularly the Evangelicals in the USA, support neoliberal policies, it became necessary to find a way to make the society of egoism compatible with the Bible's teachings. Countless books and articles have therefore emerged to defend the existence of several types of incitements to egoism in the Christian religion and the Bible. For example, the libertarian philosopher John Hospers (1918–2011), a great admirer and friend of Ayn Rand, although later rejected by her, believed that Christians, by establishing their behaviour with a view to salvation, through adhering to a doctrine of eternal rewards and punishments, are defending their own interests and are therefore using ethical egoism. At other levels, particularly on the internet, there is a growing number of texts that attempt to reconcile following ethical–rational egoism with the practice of Christianity (TCE 2017).

Currently, Ayn Rand's books are studied alongside those of Thomas Hobbes (1588–1679) and Edmund Burke (1730–1797) on political science and economics courses at several universities in the USA and around the world. In the United Kingdom, the pre-university politics syllabus (A-level Politics) included Ayn Rand's books for the first time in 2016. The bank BB&T made donations to over 60 colleges and universities in the USA to teach what it calls the "moral foundations of capitalism" (Beets 2015). These donations, typically of a million dollars, are made on the condition that courses will include the reading and analysis of Ayn Rand's books. In large companies, many businesspeople recommend that their employees read *The Fountainhead* and *Atlas Shrugged*.

There are many politicians in the USA, especially Republicans, who have declared their great admiration for Ayn Rand's books and say they have been influenced by her ideas. One example is Alan Greenspan, an economist who was Chairman of the Federal Reserve from 1987 to 2006 and belonged to the "Collective". Many believe him to be one of the main people responsible for the 2008–2009 financial crisis due to his policies of deregulation of financial markets, or laissez-faire policies, to use Ayn Rand's words (Canterbery 2006). On 4 September 1974, Alan Greenspan took office as the Chairman of the Council of Economic Advisers, having been appointed by President Gerald Ford (see Fig. 6.1). He would later recall the moment he took office in his autobiography (Greenspan 2007): "It did not go without notice that Ayn Rand stood beside me as I took the oath of office in the presence of President Ford in the Oval Office. Ayn Rand and I remained close until she died in 1982, and I'm grateful for the influence she had on my life. I was intellectually limited until I met her."

There are many politicians in the USA who claim to have been deeply influenced by Ayn Rand's ideas. Paul Ryan, who was Speaker of the US House of Representatives from 29 October 2015 to 3 January 2019, and a great admirer, said in an interview with *The Weekly Standard* on 17 March 2003 that he usually gives the book *Atlas Shrugged* as a Christmas present and tries to get all his staff to read it. In his opinion, Ayn Rand excellently set out, defended, and was able to communicate what she calls the "morality of capitalism", opposing the moral of altruism. President Donald Trump said in an interview on 7 April 2016 with the journalist Kirsten Powers that he is an admirer of Ayn Rand, mentioning that he had read *The Fountainhead* and complimented it, saying: "It relates to business, beauty, life and inner emotions. That book relates to …everything."

In essence, the worldview defended by Ayn Rand is an attempt to promote individualism and build an elite of "supermen" who will lead the advancement of civilisation. An eloquent example is provided by Howard Roark, the protagonist of *The Fountainhead*, a young and idealistic architect who constantly struggles to practise modern architecture against the reigning traditionalism and conventionalism, without even slightly indulging the majority who defends them. In his final statement to the court, he states that: "Civilization is the progress toward a society of privacy. The savage's whole existence is public, ruled by the laws of his tribe. Civilization is the process of setting man free from men."

This extreme form of individualism represents a dangerous attempt to counteract humankind's social nature. Ayn Rand is able to captivate her readers by grooming their egos and convincing each and every one of them that, whatever their talents and qualities, they can be supermen if only they would defend their own interests. Helping others, especially the weak and the sick, the poor, the displaced, and the persecuted is not recommended since it does not promote one's self-interest. The same applies to the contributions, commitments, and, at times, sacrifices made in building

Fig. 6.1 Photograph taken on 4 September 1974, after Alan Greenspan's swearing in as Chairman of the Council of Economic advisors in the Oval Office. *From left to right*, Rose Goldsmith, Alan Greenspan's mother, President Gerald R. Ford, Greenspan himself, and Ayn Rand with her husband Charles Francis "Frank" O'Connor. Greenspan was a devoted contributor to the theory of Objectivism and the most famous protégé of Ayn Rand, who dubbed him the Undertaker and the Sleeping Giant. *Photo credit* David Hume Kennerly/The Gerald R. Ford Library/Getty Images

a fairer and more equal society. Others should solve their own problems by promoting their own self-interest. This attitude justifies and encourages tax avoidance, the use of offshores, financial deregulation, the fight against progressive taxes, and in particular the increasingly varied, sophisticated, and ubiquitous forms of corruption that proliferate around the world.

Before the 17th century, there certainly were behaviours that we would today classify as selfish or egoistic, but since they were probably less clearly rewarded by society there was no reason to give them a name. Members of society were much more strongly dependent on one another due to the generally harsher living conditions. The cultural transformations that have arisen from modernity since the Industrial Revolution and the development of a society of consumers have significantly developed and diversified the practice of egoism. Generally speaking, a perception prevails nowadays that those who follow egoism are more successful in achieving better well-being and greater economic prosperity and accumulating wealth. It is therefore likely that the development and ennoblement of egoism have been a form of adaptation in the current phase of cultural evolution dominated by consumerism and technology.

Egoism promotes and is frequently linked to greed, a behaviour that has been clearly identified since times much further back than egoism. For example, Luke 12:15 in the Bible reads: "Watch out! Be on your guard against all kinds of greed; a man's life does not consist in the abundance of his possessions." Egoism and greed cannot be separated from the paradigm of our current civilisation and in practice they have contributed to its success and to its lack of sustainability in the medium and long term. How is it possible to have a sustainable world from a social, economic, and environmental viewpoint if selfishness and greed are the emerging values? Dazzled by success, we underestimate the importance of sustainability and we run as if we wished to reach the limits of egoism and greed without realising that such an illusion brings the risk of progressively weakening compassion, solidarity, social inclusion, and perhaps even destroying democracy and peace.

Emerging egoism does manifests itself not just at the individual level but also at the collective level, in a variety of social groups whose own interests conflict with those of others. It also manifests itself within the set of nations when one or more states defend their own interests to the detriment of the living conditions and interests of the group as a whole or of some nations in particular. A clear contemporaneous example is provided by countries which are large consumers of fossil fuels but have no plans to reduce their greenhouse gas emissions in accordance with the United Nations climate agreements through an energy transition to renewable energies and improved energy efficiency. A further example is provided by the continuous funding of fossil fuel subsidies by many countries (Coady et al. 2017). These policies satisfy the self-interests of the fossil fuel producing countries but harm living conditions everywhere in the world, especially in countries that are more vulnerable to climate change, which are generally the poorest and those that have fragile states.

Ayn Rand helped build a contemporary utopia of progress and growth driven by the combination of technology, ethical–rational egoism, and greed. Human enchantment with technology has evolved considerably and extensively since the Age of Enlightenment and the science demonstrations in the streets of Paris and London, having become less naïve, much more demanding, and stronger. There is, however, a fundamental difference. Fun and wonder with technology have been joined by an enormous dependence that has become increasingly necessary and irreplaceable. Scientific and technological advances are now crucial to ensuring the survival of a global human population which will probably grow to more than 11 billion by 2100 (UNDESAPD 2017a) and which above all will continue to have enormous expectations for economic prosperity and well-being. The road is hard and may lead to disaster if we use the uncontrolled force of ethical–rational egoism and greed to travel along it, supported by science and technology, and disregard the issues of sustainability, or systematically relegate them to the future. Attempting to deny our imminently social nature by ignoring or rejecting cooperation, mutual help, and solidarity is a form of perversity that leads to dystopia. Attempting to reject the human subsystem's dependence on biogeophysical subsystems will lead to deterioration in human well-being for most of the population, and to migrations and conflicts.

6.7 Concepts of Prosperity

As inequalities in income and wealth in the world become more extreme, we see new concepts of prosperity being developed and adopted, sometimes inspired by religious or philosophical sources. In 1984, the Indian writer Amartya Sen analysed different lifestyles using three concepts characterised by the words "opulence", "utility", and "human flourishing". These concepts play a central role in his work on social and economic development (Sen 1984) and can be used to define three notions of prosperity (Jackson 2009).

Opulence is the key to the mainstream interpretation, wherein prosperity means being successful and thriving financially. In particular, this condition concerns those who are rich enough to access and enjoy the rampant consumerism made possible by the immense abundance and diversity of goods and services available today. Regarding the second concept, linked to the word "utility", prosperity consists in being successful but frugal when it comes to the consumption of goods and services, which are accessed and enjoyed because of their utility, rather than seeking happiness and fulfilment through consumerism. These two forms of economic prosperity are related to the GDP via the economic concept of utility. As already mentioned, this describes the degree of preference or satisfaction derived by a person from the consumption of a good or service, and its maximization motivates all economic actions. If one assumes that economic prosperity is an indicator of subjective well-being, then there is likely to be a relationship between average life satisfaction and GDP per capita in any given country.

Generally speaking, life satisfaction grows at the personal level when income and wealth increase. However, when discussing national averages over time in countries with different cultural and historical backgrounds and varying degrees of socio-economic development, the relationship becomes less simple and more difficult to interpret. For instance, real income per capita in the USA has tripled since 1950 but the percentage of people who reported having a high level of life satisfaction barely increased during that period and began to decrease in the mid-1970s (Jackson 2009). Life satisfaction in Japan has seen only small variations over several decades. In the UK, the percentage of people who said that they were very satisfied with life fell from 52% in 1957 to 36% in 2008, while real incomes have more than doubled during that period (Jackson 2009). These collective trends reveal themselves predominantly in countries with advanced economies and are rather paradoxical since they apparently contradict the stereotyped correlation at individual level between life satisfaction and income growth.

Comparing countries with different levels of economic development shows that the general trend for the low per capita income countries is a very steep increase in the "mean of per cent happy and per cent satisfied with life as a whole" (Jackson 2009; The Worldwatch Institute 2008), which we shall abbreviate to "happy and satisfied with life". In other words, while for countries with per capita incomes between $1000 and $5000 (in PPP, 1995 USA dollars), the percentage of people that report being happy and satisfied with life varies between 30 and 80%, while above $15 000 that same percentage is consistently above 75% for all such countries. This result also means that above $15 000, the gains in happiness and life satisfaction produced by increases in per capita income become much smaller. Note, however, that these are average trends for a large group of countries and that there are many exceptions regarding the correlation between per capita income and happiness and life satisfaction. For incomes per capita close to $5000, one finds countries ranging from Russia with about a 40% level of happiness and life satisfaction to Peru with 60%, while Mexico and Colombia are closer to 80% (Jackson 2009; The Worldwatch Institute 2008). Above $15 000, there are also anomalies in the correlation. For instance, Finland, Ireland, the Netherlands, and New Zealand all have higher levels of happiness and life satisfaction than either Canada or the USA, although the latter have higher levels of per capita income.

In conclusion, marginal utility, measured as subjective well-being, diminishes rapidly at country level with increases in per capita income. Economic growth is really important to increase well-being in low-income countries but the returns on further growth in high-income countries are more limited (Jackson 2009). These results indicate that the sensible path towards global sustainability would be for the advanced economies to make room for the growth of the poorer countries, in particular as regards the exploitation of natural resources, since they are not unlimited.

More recent studies carried out by the Pew Research Center in 43 countries and involving 47 543 participants have reached similar conclusions (Simmons et al. 2014) to those highlighted by Tim Jackson. The study used the Cantril Self-Anchoring Striving Scale (Cantril 1965), where the question asked about well-being was: "On a ladder of life from 0 to 10, on which step do you stand at the present time?" The indicator is the percentage of people who reply saying they are on steps 7, 8, 9, or 10. The group of countries with advanced economies had a higher average value of the well-being indicator than the group of emerging economies, and the latter had a higher average value than the group of least developed countries. Many exceptions can be found, however, including saturation with increased income, which reveals the complexity of the processes and conditions that foster life satisfaction.

In the USA, the indicator used yielded a result of 65%, less than El Salvador with 66% and close to Vietnam and Colombia, both of which stood at 64% (Simmons et al. 2014). In general, Latin American, African, and Asian countries have higher satisfaction indicators, while countries in the Middle East have abnormally low ones. One relevant time element is that the average well-being indicator in countries with advanced economies fell between 2007 and 2014, but rose in emerging economies and the least developed countries.

These studies show that life satisfaction does not depend exclusively on economic prosperity, interpreted as opulence or utility, but also involves what is known as "social capital", which describes the quality of the social environment and the community. More specifically, it involves access to education and health, social, cultural, and psychological factors related to shared identity, inclusion, trust, solidarity, integration, and participation in the community. It also involves personal fulfilment and the perception of security, and generally speaking it depends on a positive assessment of spaces of experience and horizons of expectation. Finally, it depends on aspects of well-being and harmony with the environment, which implies respecting the finite ecological limits of the planet. These environmental and social capital factors form and characterise a concept of prosperity grounded in the possibilities for human flourishing, which we shall call here sustainable prosperity. It involves the ideas of shared prosperity, lasting prosperity, and prosperity without growth discussed by Jackson (2009).

Economic prosperity is a modern construction, which tends to separate and rank people on a one-dimensional scale. This is socially damaging because it is based on comparing what a few have and others don't. The one-dimensionality of the scale promotes the stairway effect and functions as a social trap where there are no limits to poverty and wealth. It encourages consumerism and a society dominated by egoism. On the other hand, the prosperity that emerges from developing the possibilities for human flourishing is part of a multi-dimensional social and environmental space, encouraging human diversity and enriching the social values of solidarity, cohesion, generosity, compassion, and mutual help, as well as encouraging well-being and environmental harmony. Sustainable prosperity does not exclude economic values but it does not make them a priority. Sustainable prosperity includes utility but not opulence. However, some circumstances require that utility become secondary to the priority goal of contributing to sustainability at the local, national, and global levels. This is the best option we have to guarantee the social, economic, and environmental sustainability of a civilisation that is able to respect freedom, human rights, tolerance, solidarity, justice, and social inclusion in the medium and long term.

6.8   Will Sustainable Prosperity Lead to Sustainability?

The relationship between prosperity and environment is becoming increasingly important and will very likely become decisive in the future. If it is not possible to significantly reduce the dangerousness of our interference in the Earth system, our common future will have increasingly frequent, intense, and prolonged social, economic, and environmental crises, making it more uncertain, more stressful, and more violent. If humankind is to leave this dangerous course, it will be necessary to put in place new visions of sustainable prosperity, free from an exclusive dependence on the motivations of economic utility and opulence.

A lot of work has been done and much knowledge has been accumulated about mechanisms and incentives that encourage a transition in behaviour towards sustainable prosperity, but the cultural, political, economic, and financial obstacles are considerable. On the other hand, there have been many successful examples of communities that have adopted concepts of sustainable prosperity, although they are mainly local in scope and have had little impact at the global scale up to now.

One of the major obstacles is that the lack of sustainability manifests itself at global level, where it is difficult for countries and their communities to provide a coherent, coordinated, and fair response because of the enormous diversity in their socio-economic development, culture, religions, political systems, demographics, environmental characteristics, and other concerns. Generally speaking, awareness of this diversity and of the related problems is rather limited in countries with advanced economies. Most of the world population belonging to the near 20% that enjoys substantial economic prosperity does not realise what an extraordinary asset this is, placing it in a privileged position in relation to the rest of the human population, who are still striving to survive. On the contrary, most believe their privileged situation to be perfectly normal and demand even more economic prosperity. If there were a world plan for development to move in the direction of increasing sustainability, the citizens of countries with advanced economies would prefer to adopt the concept of sustainable prosperity, which would reduce those countries' impact on natural resources and on the environment to create the space needed for poorer countries to achieve convergence of their economies with the others. It is perfectly natural for citizens in poorer countries to prefer economic prosperity until they converge with the others. In practice, we are already seeing economic convergence between the countries outside the OECD and those in the OECD because the latter have been experiencing weaker growth. However, the pace at which this convergence is happening is too slow to significantly reduce either the deep income and wealth inequalities between the two groups of countries or the risks of unsustainability.

The enormous breadth of this dual problem becomes clearer if we quantify it and apply a *reductio ad absurdum* argument. In 2017, global GDP at current prices was $79.3 trillion, which corresponds to a world GDP per capita at current prices of $10 700 (IMF 2017b). However, GDP per capita varies greatly throughout the world. In 2017, GDP per capita at current prices in countries with advanced economies was $45 070, roughly 4.2 times higher than the world average and 9.1 times higher than the average in countries with emerging or developing economies (IMF 2017b).

To simplify the argument, let us assume that economic convergence is reached when global GDP per capita reaches $45 070. If the world's population did not grow, this would represent an increase in global GDP by a factor of 4.2. Will technology be able to assure a global GDP of $333 trillion that is sustainable in terms of natural resources and environmental impact, including climate change? What are the environmental planetary boundaries (Rockström et al. 2009a) that will resist being surpassed in this scenario?

Consider now the demographic factor. If the timescale for convergence is 2050, the world's population will be around 9.7 billion (UNDESAPD 2017a), meaning that global GDP at that time will be around $437 tn. If the timescale for convergence is 2100, the world's population will be roughly 11.2 billion (UNDESAPD 2017a) and global GDP will be on the order of $504 tn, roughly six times what it is today. These scenarios are highly likely to lead to natural resource and environmental breakdown situations before they come to pass. If we intend to make the future journey less dangerous and uncertain, countries with advanced economies will need to make a transition from economic prosperity to sustainable prosperity.

It should be emphasized that both scenarios of convergence and global demography, serve only as an illustration of the present conundrum. They are extremely unlikely to happen since medium and long term GDP growth requires the availability of natural resources and ecosystem services, as well as control over climate change. The most likely scenario is the continuation of the recent tendency for slowdown in the world's real GDP growth, but the global population is very likely to continue increasing as projected.

Let us look at the proposed transition in more detail. The practice of sustainable prosperity should be encouraged and coupled with public policies aiming at sustainable development (Sachs 2015). However, it would be a mistake to think that sustainability depends only on implementing a well-designed and efficient political agenda to promote social, economic, and environmental sustainability. Because of the strong attraction of economic prosperity, the implementation of that agenda will be very unlikely to succeed in democracies with advanced economies before a transition to the collective practice of sustainable prosperity. Any indication or news story that such public sustainability policies would eventually lower GDP per capita and thereby reduce voters' ability to consume goods and services would be fatal for the government. Voters will only allow sustainability goals to come to the foreground if a majority have already made the transition to concepts of sustainable prosperity. It is not governments that can guarantee sustainability, but each and every one of us taken together, if we are able to adopt and defend values other than those that are strictly economic and dependent on consumerism and the accumulation of wealth. In critical situations, where survival becomes the primary objective, it is conceivable that a government could initiate a transition to sustainability against the will of its citizens, forcing them to adopt the practice of sustainable prosperity.

The transition to sustainable prosperity will be easier if it becomes fashionable or a trend. However, trends are difficult to establish. The newly emerging behaviour should be as unique, innovative, and interesting as possible. One of the most important aspects would be the understanding that such behaviour will bring advantages for those who follow it. Of course, favouring sustainable prosperity instead of economic prosperity is initially difficult and relies on a high level of knowledge and awareness of the social, economic, and environmental problems that we are faced with at the local, national, and global levels. This in turn depends primarily on having a realistic and balanced view of the major world problems and having the openness and flexibility of mind to be critical, and not rely on identity-based, short-term visions. It depends on accepting science, on recognising the advantages of public policies based on science, and on having a sufficient level of scientific and technological literacy to understand those problems. Finally, it depends on the ability to value medium and long term issues and goals.

An effective way of implementing the transition to sustainable prosperity is to develop sustainable communities that aim at sustainability as specified by several crucial indicators, such as social cohesion, employment, energy, water, and food. When they achieve their goals, the members of sustainable communities have reason to be proud of such an achievement, and the proliferation of these groups tends to create a movement, or even a trend, which in turn attracts more communities who wish to innovate and compete for such notoriety.

In countries with emerging or developing economies, sustainability is far from being a priority for most of the population. This does not mean that those countries do not have many people who are fully aware of the problems of unsustainability, such as overexploitation of natural resources, environmental degradation, pollution, climate change, and the consequences that it will cause for their lives, families, and communities. Although there are exceptions, the population's first priority is naturally economic convergence with countries with advanced economies. It will therefore need to be the latter that lead the movement towards sustainable prosperity, based on wisdom, responsibility, and solidarity with their fellow men and women around the world.

It is likely that many people living in countries with advanced economies are indifferent to the questions related to the global human society or are not aware that their concept of prosperity leads to behaviour that involves greater or lesser interference in the Earth system, with consequences for our common future. Each individual tends to focus on his or her personal interests and seeks to increase his or her own well-being and economic prosperity as far as possible. This behaviour is seen as natural and is welcomed and encouraged by society, especially when egoism becomes prevalent.

An example of the new tendencies toward individualism can be seen in the eager search for self-improvement, an emerging behavioural trend and industry in several countries, particularly in the USA. Self-improvement aims at improving one's life to near perfection, especially as regards relationships, health, body, sex, pleasure, brain, creativity, productivity, money, happiness, spirituality, and attention in a focused, efficient, and effective way, with planned, quantifiable targets (Seligman 2007; Duhigg 2016; Ferriss 2016; Cederström and Spicer 2017; Storr 2017). Dedicated practitioners of self-improvement should be fitter, healthier, wealthier, more productive, and happier than others. They must be the leaders, showing how to lead the perfect life (Cederström and Spicer 2017), not just a good life. The development of this new form of behaviour is supported by academic studies, experts, and gurus and is generating a powerful industry of products and services specially designed for self-improvement.

Periods of intensified individualism and narcissism have already occurred in the past, as for instance the trend toward self-improvement in the 1930s, in the wake of the Great Depression in the USA. It can be argued that in a time of social and economic depression people tend to concentrate on self-improvement since they have to rely more on themselves. Nevertheless, the current era of hyper-individualism and the "selfie" generation (Storr 2017) is likely to have more complex and deeper roots and to last longer. The most significant message from the present trend is indifference toward others. Although there are many exceptions, most people in advanced economies tend to concentrate more on improving themselves and less on trying to strengthen and improving the community and society where they live. They are inclined to behave as though all social problems in their environment can and will be solved by the public and private institutions dedicated to welfare and social security, so that their personal contribution is superfluous. Now that human successes have been so outstanding and far-reaching, humans can finally relax their social nature and explore the limits of solipsistic pleasures.

The time factor plays a crucial role in the pursuit of self-improvement. The acceleration of operative social time requires each practitioner to focus on making use of every opportunity to satisfy his or her interests and short term desires as fast as possible. There are various reasons for that haste: speed makes it possible to accumulate more experiences, achievements, pleasures, and successes, to enjoy being more competitive than others, and also to avoid the contingencies and frustrations that an unpredictable future may bring. In other words, the acceleration of time favours immediacy and produces high time discount rates.

Merely alluding to the idea that the medium and long-term sustainability of the remarkable civilisation people are currently enjoying, especially those people who are benefitting from robust economic prosperity, could be somehow dependent on their everyday behaviour is considered by many as absurd and an invasion of personal freedom and privacy. Those who share the conviction that it is possible to maintain economic growth indefinitely believe that future generations will have progressively much better conditions than the contemporary generation for facing up to the sustainability problems that may eventually appear. No need to imperil our present economic prosperity because of an uncertain future. On the contrary, if that uncertainty is really felt, then it should be taken as an imperative to improve economic prosperity as much as possible now, while it is possible. This positive feedback devalues the future, making the transition to new forms of sustainable prosperity unnecessary.

In spite of these arguments there are many people around the world who are conscious of the present risks of unsustainability and prefer to take a different approach and to contribute however modestly to mitigating the problem by practising sustainable prosperity. Those that adopt and those that do not adopt sustainable prosperity probably both consider that their choice of behaviour is rational. Thus, in conclusion, sustainable prosperity is an important step forward to achieve sustainability at different spatial scales, but the likelihood that a large part of the world population will make the transition to sustainable prosperity is still very slim. Perhaps the harshness of some future crisis will convince people of the benefits of such a transition.

6.9 Voluntary Simplicity

One of humanity's most notable characteristics is the uncontrollable need to explore every path that appears before it or that it discovers. Nothing is left unexplored, but only a very few such paths are adopted and eventually used by the masses. Those paths leading to increased economic prosperity are the most popular and are taken up by most of humanity. In the realm of sustainable prosperity, there are many different paths that can be discovered and followed but the number of fellow travellers is generally rather limited. Another interesting human characteristic is that, throughout history, certain very similar paths may arise in a recurrent manner, although they will differ in some ways because they belong to different operative social times.

In 1936, the American social philosopher Richard Gregg (1885–1974), a scholar and great admirer of Mahatma Gandhi (1869–1948), introduced the concept of "voluntary simplicity" in an article published in the Indian journal *Visva-Bharati Quarterly* (Gregg 1936). For the author, voluntary simplicity was a way of reacting to the consumerism that was emerging at the time in the USA, working against the idea that he attributed to Henry Ford that "civilization progresses by the increase in the number of people's desires and their satisfaction" (Buell 2015). In his own words, voluntary simplicity means "singleness of purpose, sincerity and honesty within, as well as avoidance of exterior clutter, of many possessions irrelevant to the chief purpose of life" (Gregg 1936). Gregg does not defend asceticism as an end in itself, but advocates a simple life, resisting excessive consumption. His idea was not to completely reject the amenities of the modern world but to seek a higher quality of life.

The essence of these goals is ancient, with examples in both the secular and the religious worlds, but the reasons that drive them are new. Socrates (470–399 BC), condemned by Athens to drink hemlock, gave us what is probably the first recorded example of a simple lifestyle, without material goods, based on self-control and dedication to knowledge and wisdom. Diogenes of Sinope (404–323 BC), the central figure of the Cynic philosophers, took the principle to its most extreme, living in Athens as a beggar in the utmost poverty. This led Plato to say that he was a "Socrates gone mad", according to Diogenes Laërtius, a biographer of the Greek philosophers who was active during the 3rd century BC.

In the sphere of religion, the simple life and asceticism were advocated and practised by the founders of the main religions of the Axial Age between 800 and 200 BC, including Siddhārtha Gautama (Buddha), Zoroaster or Zarathustra, the founder of Zoroastrianism, who lived in Persia somewhere between 1000 and 600 BC, Laozi, also known as Lao Tzu, who is believed to be the author of *Tao Te Ching*, the most sacred book of Taoism, written in about the 5th century BC, and Confucius. The same principle applies to the following period up to Jesus Christ at the beginning of the Christian era. In Christianity, simplicity, poverty, detachment from material goods, asceticism, and above all serving your fellow men have been recurrent concerns in a dynamical context, although always recognising that asceticism should not obliterate all desire, because desire is the driving force of life.

Notable early examples include the movement of the Desert Fathers in Egypt from the 3rd century on, led by Saint Paul of Thebes (c. 227–c. 341) and Saint Anthony of Egypt (c. 251–356), which served as a model for Christian monasticism and the Franciscan Order, founded by Saint Francis of Assisi (1181/1182–1226) in the 13th century.

Following the Industrial Revolution, the American essayist and philosopher Henry David Thoreau (1817–1862) argued, even before Richard Gregg, for a simple and sustainable life in his famous book *Walden; or, Life in the Woods* (Thoreau 1854). In this, he describes his experience of living for almost two years in a cabin he built himself at the edge of Walden Pond in the forests of Massachusetts. The pond belonged to his friend and mentor Ralph Waldo Emerson (1803–1882), also an American essayist and philosopher, one of the founders of transcendentalism, a form of romanticism and individualism. Thoreau advocated slowing down the pace of life until we are totally available to enjoy the existential value of full awareness of every moment and to let thought be utterly free, without any time limits. The objective is not asceticism in itself, but rather to optimise the capability for human flourishing by way of a simple life, meeting basic needs, reducing stress, and always focusing on what is truly essential. The aim is to be much more concerned with what one has and does than with what one wants and owns.

Voluntary simplicity and similar previous movements, focusing as they do on religious kinds of non-indulgence and austerity, have the common characteristic of slowing down operative social time, letting it flow freely, uncluttered by events, projects, and superfluous obligations. Letting time flow in a leisurely manner, without set times, to free the body, the senses, and thought to concentrate on what has been elected as the main goal of life. To assume that the main goal of life can be reached by diluting or rarefying operative social time goes frontally against the prevailing practice of accelerating operative social time to gain success, interpreted as increasing economic prosperity and well-being. For the majority of those that live above the poverty line, and especially for those that live well in the advanced economies, operative social time is constantly densified by work, projects, deadlines, and all sorts of commitments, except during a few blissful periods of quality time. For those that live below the poverty line, especially the unemployed and those that live in severe and extreme poverty, operative social time is rarefied, but also inhospitable and toxic. To reach a situation where life's goal does not depend necessarily on the limitless densification of operative social time requires a profound cultural change. Most will likely say that this cultural transition would lead to generalized poverty.

Voluntary simplicity gained more public visibility in the 1980s with the contributions of Elgin (1981) and the popular simple living guide by Luhrs (1997). This led to its being recognized as a movement. A Simplicity Institute was established by Ted Trainer in Australia and the theoretical foundations of the movement were more firmly established (Etzioni 1999; Alexander and Ussher 2012). The main objective is to address the current global problems by making a transition from the consumer society to a simpler, more cooperative, just, and ecologically sustainable society (Trainer 2015). The movement is based on the recognition that consumerism in advanced economies and its rapid globalization are leading to an unsustainable overexploitation of natural resources and to the pollution and degradation of the environment, but also the fact that high consumption is unethical in a world where hundreds of millions of people are living in great human need. Furthermore, it considers it a mistake to look for meaning in life through the consumption and accumulation of material things. A sustainable world would require a simpler way of life, with a voluntary downscaling, learning to live happily with less, although without necessarily settling for deprivation or hardship. It would depend on a generalized willingness to establish a new approach to production and consumption, living closer to nature, and indeed embracing nature once more, but now, after so many centuries, with much greater knowledge, respect, and curiosity.

But is such a program feasible, and if so how can it be implemented? The leaders of the simplicity movement concede that they still have very few adherents, but they believe that their numbers will increase as the negative consequences of consumerism grow more conspicuous. By that time, people should at least be acquainted with the possibility that simpler ways of life would make more sense, and it would be easier to adopt them (Trainer 2015).

Voluntary simplicity carries at its core the stigma of rejecting consumerism, the main driver of the global economy at the present time. It was in the USA after the Second World War that mass consumerism began to be developed, as so eloquently described by Cohen (2004). Now, although unsustainable in the middle and long term, and even more so with a fast-growing human population, it is considered indispensable, and many consider it beyond question if we are to ensure the economic prosperity of humankind. Furthermore, voluntary simplicity does not depend intrinsically on the development of technology, so could be interpreted as a return to the past, a form of denial of what is frequently considered to be humanity's irreversible and triumphant ascent in modern times, thanks to the impetus given by science and technology. But of course technological development can still help to enrich and diversify life in a regime of voluntary simplicity.

Voluntary simplicity can be exercised in many different ways and with varying emphasis on its different components, from a more radical behaviour of extreme frugality to a simple life free from consumerism and superfluous accumulation of possessions, enjoyed as far as possible outdoors in contact with nature. In all these ways of life, technologies of certain kinds, for instance, all technologies associated with health and ICT, are still an option. The relation between voluntary simplicity and science is normally a privileged one, since science is all about curiosity, understanding, and knowing what is around us, including ourselves, fellow humans, humankind, the Earth system, and the Universe, both now and in the past, and it is also about predicting future behaviour. All these goals are compatible with voluntary simplicity, and moreover they are recommended by it.

Fostering contact with the natural environment is one of the most valuable components of voluntary simplicity. There is a sound epidemiological evidence that this kind of behaviour is associated with better health and well-being, especially in the largely urbanised populations of the advanced economies (White et al. 2019). Various studies have shown that living in neighbourhoods that have areas with parks, open fields, woodlands, or beaches induces better self-reported health and better well-being, and tends to reduce mortality, particularly that associated with cardiovascular disease (Gascon et al. 2016).

There are noteworthy individual examples of voluntary simplicity that have had a large-scale influence on humankind. One of the most remarkable examples is Mahatma Gandhi, who had a revolutionary political, social, and environmental impact in India and beyond. More recently, one finds written testimonies by people who have chosen lifestyles of extreme restraint, deprivation, and simplicity, or just made the transition to a simpler lifestyle. In 1977, Charles Gray, then a 52-year-old university lecturer, aware of the "crisis of late twentieth century humanity", decided to make his own personal contribution (Gray 1995). He lived exclusively on an amount equivalent to the world GDP per capita at the time, that is, US$142 per month. The resulting state of severe poverty led to his divorce, isolation, and depression, but he eventually adapted to his new lifestyle.

Less radically, Colin Beavan wrote his book *No Impact Man* (Beavan 2009) to describe the incremental downscaling that he, his wife, and his daughter imposed upon themselves in a New York apartment, practically reducing their waste to zero, travelling only on foot or by bicycle, reducing their electricity consumption to a bare minimum, and only eating food produced locally, that is, within a radius of approximately 400 km.

Voluntary simplicity makes it possible to explore and promote the values of material frugality, willpower and self-control, and ecological awareness. It also provides a way to explore new dimensions of human experience and an opportunity to develop further empathy, cooperate more with others, and flourish physically, intellectually, artistically, and spiritually. But will these individual initiatives be able to help solve the large-scale problems of consumerism and promote greater sympathy and compassion in humankind? Will they thus provide a way to mitigate certain social problems?

For that to happen, there need to be examples of voluntary simplicity that are attractive enough to encourage new practitioners and therefore increase their numbers. There are two main arguments in favour of adopting voluntary simplicity. The first is the personal advantage that comes from material frugality, since it promotes healthier lifestyles, greater cooperation and empathy with others, physical well-being, and intellectual flourishing. The second involves the satisfaction of actually contributing to sustainability, including its social, economic, and environmental components, while remaining aware that it is a limited contribution because of the scale of the global human population. One must be willing to contribute while accepting that the contribution is necessarily small from the material point of view, even though it may be very valuable psychologically and symbolically. The main risks of voluntary simplicity are identification with poverty, the relative lack of material comfort, a feeling of disconnection from mainstream society, and the stigma of going back to bygone ages. To be attractive, voluntary simplicity must carefully accept the comfort provided by a sensible but frugal access to contemporary goods and services.

In the mid-2000s, the development of initiatives to transfer wealth to the poor led to the "effective altruism" movement, defended by the British philosopher William MacAskill (2015) and later by the American philosopher Singer (2009, 2015). It stands out from charitable initiatives by seeking to optimise that transfer through a careful analysis of the resources available and the current priority needs at global, national, regional, and local levels. Although taken far enough these voluntary donations could force more generous donors into a life of voluntary simplicity, the practice of effective altruism does not necessarily imply voluntary simplicity.

One of the main aims of the movement is to change the current culture of giving in affluent countries, to raising donations from those countries for the people living in extreme poverty in developing countries, using selected non-profit organizations that maximise the positive impact. Effective altruism is inspired by the utilitarianism of Jeremy Bentham and John Stuart Mill (1806–1832), which was in turn founded on the principle that "it is the greatest happiness of the greatest number that is the measure of right and wrong" (Bentham and Harrison 1776), often quoted as "the greatest amount of good for the greatest number".

The movement is mainly active in Anglo-Saxon countries and does not challenge the present financial and economic system based on consumerism, but instead offers optimized technical and economic solutions to the problems of extreme poverty. It is a valuable initiative to alleviate the suffering of many people in need, but it does not address the political problems that contribute to such situations (Rubenstein 2016).

Naturally, given that effective altruism belongs to the established paradigm, albeit somewhat closer to its fringes, the establishment is eager to give it visibility and support. Since 2013, the Centre for Effective Altruism at Oxford University has been running a series of Effective Altruism Global (EA Global) Conferences. The one in 2015 took place at Google's Quad Campus in Mountain View, California, and focused on "global poverty, animal advocacy, cause prioritization research, and policy change". It was attended by tech billionaire Elon Musk, considered by Tyler Alterman, president of EA Global, to be "almost the perfect effective altruist". It also included artificial intelligence expert Stuart Russell and philosopher Nick Bostrom, who works on existential risks. While effective altruism is mostly about combating poverty, EA Global has broadened its interests to address other perceived global risks, especially to fund computer science research to avoid the kind of apocalypse that might be provoked by artificial intelligence. According to Dylan Matthews (Matthews 2015), who took part in the 2015 Conference, many attendees manifested the opinion that, compared to such a risk, global poverty is a "rounding error". This opinion is natural and to be expected from "nerd altruists", as Nico Pitney has called the participants (Pitney 2016), or indeed to be expected of a culture based within the quantitative fields of technology and finance.

The impact and attraction of voluntary simplicity becomes stronger and takes on a new dimension if it is followed coherently and effectively by a community. There are currently many examples of intentionally sustainable communities that have the objective of becoming more socially, economically, and ecologically sustainable in countries with advanced economies and in other countries. The number of these initiatives is increasing and they represent an effective way to integrate sustainable prosperity into social behaviour. Their members tend to share the same ecological, socio-economic, and cultural–spiritual values. Although the visibility of such communities in society at large is still rather limited, they are accumulating evidence that the health and well-being of their members improves. The most important asset of the sustainable communities is that they are likely to be better prepared for and more resilient to the impacts of the various types of crisis that are likely to result from the way the global community continues to follow an unsustainable pathway.

6.10   Post-Cooperation Utopia and the Future of Capitalism

Sustainability necessarily involves future generations, so the crucial issue is finding out if there is any basis or support for cultural evolution towards post-cooperation, in other words, towards a form of cooperation, or altruism, between members of one social generation and later social generations, or cooperation by the living with those who have yet to be born. Sustainability is interpreted here as strong sustainability and not as weak sustainability. Post-cooperation is an active form of intergenerational solidarity, usually practised with a specific focus and objective. It is motivated by the desire to achieve intergenerational justice, or intergenerational equity.

There is a profound difference between cooperation and post-cooperation, or if one prefers, between altruism and post-altruism, because the latter represents a type of behaviour with an effective cost for those who practise it but only an abstract benefit for undefined beneficiaries situated in the future. In fact, although the benefits and beneficiaries can be characterised, to a greater or lesser extent, their definition will always be somewhat abstract, incomplete, and unforeseeable. It is natural for there to be limited motivation to perform an act that involves a cost in exchange for a benefit that is uncertain, to help beneficiaries who have yet to be born.

In spite of these difficulties, it is possible to identify types of behaviour that include forms of cooperation and post-cooperation in the actions of some people who sacrifice their potential enjoyment and well-being to invest their money and possessions to benefit their descendants, both those who are already alive and those who have yet to arrive. Post-cooperation may also occur outside family relationships when the donor invests with the objective of benefiting future generations, but without expecting any return for himself. These situations are probably uncommon and hard to identify.

Since the 1980s, there has been an increasingly precise and wide-reaching belief that the behaviour of the contemporary social generation will affect the opportunities and development conditions of future generations. One of the main consequences of this recognition was the introduction of the concept of sustainable development. The World Commission on Environment and Development, created in 1983 by the United Nations General Assembly and presided by Gro Harlem Brundtland, defined it in its *Our Common Future* report as "development that meets the needs of the present without compromising the ability of future generations to meet their own needs". It is implicitly acknowledged that the current social generation may follow development models that are not sustainable and can therefore harm the capacity for development of future generations. Sustainable development stems from the imperative of intergenerational justice.

There is an evolutionary coherence in social generations that results from the influence that each generation exercises over those that follow, especially the one immediately after it, and the connections through family, social, economic, political, and cultural ties that are established between successive generations. However, throughout the history of human societies and up until the end of the 20th century, there was no systematic concern with the possibility that the behaviour of people in one social generation or set of successive generations might have the potential to hinder the development of following generations, considering and integrating its social, economic, and environmental components. On the contrary, the Enlightenment established the idea that generations can be supported by previous generations in the creation of an upward momentum in the progress of humanity that can procced indefinitely, without any structural constraints or limitations.

The United Nations General Assembly, when it approved the Brundtland Report on 11 December 1987, gave international expression to concerns about the path of social, economic, and environmental development. The report from that session begins by pointing out the rapid deterioration in the environment and natural resources, and the consequences that this may have for social and economic development. It also considers that sustainable development, with the objective of reaching strong sustainability, should be a central guiding principle for the United Nations, governments, and private institutions, organisations, and enterprises. The General Assembly report also states that the key to achieving sustainable development is equitable sharing of the environmental costs and benefits of economic development in each country and between countries, and between present and future generations. After more than three decades, a period of time close to the length of a generation, humankind is still far from achieving the goal of sustainable development at the national and global levels, although some countries are considerably more advanced than others. The concept of sustainable development is now likely to be more generally recognised as relevant, but some sectors of society in some countries ignore it in their activities.

For sustainable development to achieve strong sustainability, which ideally describes a stable sustainable state in the short, middle, and long term and at all spatial scales within the Earth system, including its social, economic, and environmental components, the individual and collective behaviour of the current social generation must be changed into behaviour that shares the property of contributing to sustainability and that will be called sustainable. Such behaviour will be partly determined by the needs of the future generations and will therefore require some degree of post-cooperation with the members of such generations. The question is how to define and induce such a behavioural transition.

Sustainable development demands costs from the current social generation in exchange for a benefit that, in most cases, will only be tangible in future generations. Not all acts that contribute to sustainable development necessarily have a cost for those that perform them. Among the acts that work towards sustainability, some have a benefit, some have a cost, and some are neutral for those that perform them. The differences between these three types are related to social, political, economic, cultural, and environmental issues and involve individual and collective behaviour, governance at all levels from global to local, and science and technology.

It is clear that the costs and benefits of post-cooperation acts cannot be measured in terms of Darwinian fitness because the move towards sustainable development is not based on a process of natural selection. The cost of a consciously performed act of post-cooperation is measured relative to the highest standard of well-being and consumption in the lifestyle that the person who performs it could achieve in the generation to which he or she belongs and with the means that are available to him or her. The given act must have a high probability of generating a benefit for future generations and will necessarily involve a renunciation of the satisfaction and pleasure provided by that highest standard. It may consist of a continued effort to change individual or collective behaviour, involving, for example, material or social costs, the transition to the practice of sustainable prosperity, or a rejection of the consumer standards that are currently accessible and encouraged by family, friends, and society.

The act of post-cooperation may also consist of engaging in activism to promote intergenerational justice, to change social, economic, financial, and environmental policies, and to move closer to local, national, or global standards of sustainability. For activists, these acts may be interpreted as providing satisfaction and personal fulfilment but they may also imply some material or social costs. In any case, the measure of cost is to be found at a collective level and reflects societies' resistance to adopting the measures necessary to reach sustainable development within the available time. Experience has taught humankind that sustainable development involves a high cost, or in other words, is difficult to achieve.

The benefit is increased sustainability for future generations and is measured by comparison with the situation that would result from not performing post-cooperation acts. Given that the beneficiaries are not fully defined and the benefits are unforeseeable, since they will only exist in the future, acts of post-cooperation are affected by uncertainty and their effectiveness can only be measured with hindsight, in the operative social time of a set of successive future generations. Acts of post-cooperation can also favour the contemporary social generation of those who perform them. However, the essential nature of post-cooperation is to benefit future social generations in the medium and long term. As a result, motivations for post-cooperation, although they may include the same motivations as cooperation, go deeper in time.

In a community where the highest standard of well-being is extremely low, at the level of mere survival, as is the case for people who live in situations of severe or extreme poverty, acts of post-cooperation are unlikely or impossible. Nonetheless, people who live in such situations often act altruistically to ensure the survival of their families and social groups, and they are likely to be very concerned about their future survivability. Increasing standards of living and well-being facilitates post-cooperation, but there is no reason to believe that there is any law of proportionality regarding this relation. While altruism usually involves a close relationship between those who perform it and those who benefit from it, that relationship is of a different nature in the case of post-cooperation since the beneficiaries are yet to be born.

There is a further difficulty regarding post-cooperation contributions to sustainability. The concepts introduced above assume that it is possible to identify and describe what sustainable development is and the actions that work towards it. But sustainable development is not a scientific concept whose meaning can be established in a definition free from ambiguity. It is, instead, a social, economic, and environmental discourse that can be applied locally, regionally, and globally, and is distinct from both the limits discourse and the Promethean discourse (Santos 2012). Sustainable development is today a conceptual framework in which different and sometimes conflicting interests search for dialogue while attempting to defend their territory. Those who are most interested in the environment stress the need to conserve nature and stop biodiversity loss, to ensure the sustainability of services provided by ecosystems, to support sustainable exploitation of natural resources, and to fight climate change. Those who are most concerned with the deep and increasing social and economic inequalities focus on the need to respect human rights, to combat hunger, malnutrition, poverty, disease, lack of water, poor sanitation, and inadequate healthcare, to make electricity freely accessible, and to create institutional capacity for education and training. Finally, those who are more closely linked to the economy tend to believe that the key feature of sustainable development is to ensure an economic growth that is preferably strong and stable, and that this will end up solving all other social and environmental issues.

The difficulty in reconciling these differing points of view has led to the belief that sustainable development requires a transition to new paradigms of social, economic, and financial development. In spite of the complexity of this issue, let us assume for the sake of argument, and to put it in terms of the concepts used in the Brundtland Report, that if sustainable development is not achieved, future social generations will see their opportunities for development, well-being, and the quality of the environment where they live steadily reduced. Let us also assume that it is possible to define the overall average change in socio-economic development, well-being, and quality of environment opportunities for each generation in relation to the previous generation, which we shall call the variation in strong sustainability between two generations. This variation expresses the average difference in sustainability between the two generations. When we consider that we have not yet achieved sustainable development since the 1980s, we are implicitly admitting that there has been a negative variation in strong sustainability between the Millennial generation and the Post-Millennial generation. If the trend remains the same, the deficit in sustainability for future generations will continue to increase, compared with the sustainability deficit of the Millennial generation.

Had there not been a continuous flow of sustainable development initiatives and actions throughout the world since the 1980s, the sustainability deficit between the Post-Millennial and the Millennial generations would have been even greater. Climate change is a conspicuous example: the atmospheric concentration of greenhouse gases emitted by anthropogenic activities has been increasing continuously since well before the 1980s and right up to the present time, in spite of the mitigating actions that some countries have been implementing. Stabilizing the concentration of such gases in the atmosphere is a necessary condition to reach strong sustainability. Finally, it is noted that the situation would probably be different for the purely economic concept of weak sustainability, which disregards natural capital, considering that it can very likely always be replaced by human-made capital. If there was no decline in the value of economic output between the two generations, there would be no weak sustainability deficit between them.

The Millennial generation is chosen as the reference to measure the variation in sustainability of the following generations, since it was in the 1980s that the concept of sustainable development was first defined. However, this is just a convention. The causes of unsustainable development began long before the 20th century and it is likely that the variation in sustainability was negative at least throughout the century.

The sustainable development of a continuous set of successive future generations in relation to a given social generation is achieved when the variation in sustainability of each generation in the set is positive or zero. Let us consider the specific case of the Millennial generation as a reference. To achieve sustainability in relation to that generation, the current trend towards negative variations in sustainability must be stopped and inverted to overcome the accumulated sustainability deficit and later the sustainability variations must be kept positive or zero.

Let c_i designate the global sum of the costs of post-cooperation acts performed by generation i, attributing the value $i = 0$ to the Millennial generation. The cost c_i is the contribution of generation i to the sustainability of coming generations, so a future generation k can receive a sustainability benefit of b_{ik} from generation i :

$$c_i = \sum_{\substack{k = 1, 2, 3, \dots \\ k > i}} b_{ik} , \quad i = 0, 1, 2, 3, \dots .$$

The benefit b_{ik} is linked to the deficit in sustainability acts d_k between generation k and the Millennial generation through the equation

$$b_{ik} = r_{ik} d_k ,$$

where r_{ik} is the post-cooperation relationship coefficient between generation i and the later generation k. The value of this coefficient is a positive number less than 1, and equal to 0 when there is no post-cooperation.

A generation k reaches the limit of sustainability when the sum of the benefits from post-cooperation acts received from past generations is greater than or equal to the deficit in benefits from sustainability acts between that generation and the reference generation, in other words, when

$$\sum_{i=0,1,\dots,k-1} r_{ik} \geq 1 .$$

To maintain sustainability it is necessary to keep performing post-cooperation acts, which means that there is an unavoidable intrinsic sustainability deficit of a given generation relative to the previous one that has to be addressed. This situation results mainly from the current high levels of global consumption of natural resources, pollution, waste, and emission of greenhouse gases. Without post-cooperation, the sustainability deficit d_k increases very quickly with k. However, the deficit of a future generation relative to the previous one can be gradually reduced to a minimum, which implies that sustainability could be reached with a minimum amount of post-cooperation in the preceding generation.

If we assume for the purposes of simplicity that the post-cooperation coefficients of the relationship are not time-dependent, this implies that sustainability is necessarily reached in generation k when

$$\sum_{i=1,\dots,k} r_i \geq 1 \,,$$

where the post-cooperation coefficient r_{jk} now only needs to have an index i equal to the difference between the generations k and j :

$$r_i = r_{k-j} = r_{jk} \,.$$

Naturally, the time it takes to reach sustainability depends on the constant value of the post-cooperation coefficient between two successive generations. As previously mentioned, once sustainability has been achieved in generation k, post-cooperation must be pursued by that generation to compensate for the intrinsic sustainability deficit of generation $k + 1$ relative to generation k. This result shows that sustainability can be reached if it is pursued steadfastly throughout a continuous finite set of generations. Furthermore, it shows that, once sustainability has been reached, the post-cooperation required to maintain it can be reduced to a minimum. If post-cooperation is not implemented, the sustainability deficit will increase for each future generation as time goes by.

Why is the briefly described post-cooperation framework utopian and unlikely to be implemented? The main reason is that sustainability is to a large extent a global challenge that must be addressed by all countries "on the basis of equity and in accordance with their common but differentiated responsibilities (CBDR) and respective capacities" as stated in article 3, paragraph 1 of the United Nations Framework Convention on Climate Change (UNFCCC). The implementation of the CBDR principle is difficult because an essential human trait is to feel and behave almost exclusively as a citizen of a given country and as a member of various social groups within that country, but not as a member of the community of human beings. Such a community is not really a social group since it does not have rival groups.

The CBDR principle recognizes the common responsibility but does not provide a roadmap on how to define, reach agreement, and implement the differentiated responsibilities of each country. Each country will tend to defend as much as possible what it considers to be its own interests, and therefore it becomes difficult to agree on the differentiated responsibilities to support the costs of post-cooperation. One finds once again the problem of free-rider countries that benefit from the higher level of post-cooperation adopted by some of other countries. Thus the advancement of post-cooperation is more likely to succeed at intracountry level through the establishment of sustainable communities.

The second main reason why the global framework for post-cooperation is utopian is that the level of consumerism, which is the main driver of the present globalised financial and economic system, is unsustainable in the medium and long term. The majority of citizens in most countries want more economic prosperity and are not interested in embarking on a transition to sustainable exploitation of natural resources through the effective implementation of a circular economy, nor in taking decisive actions that would strongly reduce environmental degradation and pollution, nor in making a conclusive transition to a carbon-neutral global economy. They are mostly convinced that such changes would considerably reduce their personal chances of increasing economic prosperity, no matter how healthy they already are. They identify the solutions to most of these global problems as dangerous initiatives defended by unrealistic environmentalists. They are not convinced that they should practise post-cooperation.

Despite all these tendencies, the current financial and economic system is under mounting pressure and there are signs that things could start to change. An increasing number of people recognize that capitalism, or more specifically, the neoliberal form of capitalism of the past few decades, has failed to build an equitable society, bringing instead extreme inequality, underemployment, and precariousness. Furthermore, it has brought financialization and economic stagnation to many countries, especially in the advanced economies. The worst effect of the current form of capitalism, however, is that it is failing not just the current social generation but future generations through an accelerated environmental crisis that could only be stopped by radical action. However, those belonging to the 1% that have amassed an unprecedented amount of wealth at the very heart of this system are well protected against the effects of unsustainability and deficits and they are well prepared to resist attempts to change the present financial and economic system using the power and means at their disposal.

It is impossible to predict the future evolution of capitalism, but one may say that being incompatible with sustainability will force it to change. Some go as far as to consider that climate change, one of the most potentially dangerous aspects of unsustainability, "is a characteristic and intrinsic symptom of neoliberalism" (Elliott 2016). We are compelled to ask how capitalism can renew its promise to provide a better future for humankind, as it has done in the past. In the last 200 years, capitalism has been extremely successful in creating the modern lifestyle of well-being and economic prosperity that is enjoyed and shared by about 20% of humankind. Furthermore, it has been the main driving force behind the collapse of extreme poverty from values close to 90% of the world population to current values of the order of 10%. Nevertheless, an unchanged continuation of the present version of capitalism over the next 200 years, although impossible to imagine in detail, is likely to be very damaging for humankind.

One answer that has been proposed is that neoliberal capitalism can be replaced by a return to a more regulated Keynesian type of capitalism. Jorgen Randers (Randers 2012) projects that, in 2052, the world is likely to have a modified capitalism with "less democracy and less market freedom", but it is likely to be more efficient and sustainable. In this scenario, capitalism would survive but at the expense of democracy.

In 1993, Drucker (1993) designed a likely evolution into a post-capitalist society where knowledge would be the new basis for wealth, instead of capital, land, and labour. Later, Paul Mason (Mason 2015) argued along the same lines that capitalism is in the process of transforming into a new type of economic model which he calls post-capitalism, based on free access to information through ICT. According to him, the market system will be replaced by a participatory economy sustained through the collaborative production of machines, goods, and services with zero marginal costs and the reduction of work time to near zero. He strongly believes that networking, by providing an easy and almost free access to information, goods, and services, will be able to replace the system of monopolies, giant technological multinational corporations, banks and financial institutions, and "governments trying to keep things private, scarce, and commercial" (Mason 2015).

The emphasis on collaborative production and networking establishes a possible link with sustainable communities in which the development of sustainable prosperity practices may also lead to forms of post-cooperation. However, there are no signs that post-capitalism is gaining ground on a significant social and economic scale throughout the world, and the power of the neoliberal centralised hierarchical structure of giant multinational businesses and financial institutions is getting stronger, not weakening.

From the political point of view, the situation as regards capitalism is also becoming more complex. Capitalism is criticised and often strongly rejected by those on the left who defend various forms of socialism, but also by some conservatives on the right. The main reason for this shift is that, for conservatives, the neoliberal devotion to creative destruction is destroying the social fabric of societies—families, stable communities, church communities, and labour unions—treating human labour as a commodity that moves around the globe solely at the beck and call of supply and demand.

If the present trends toward unsustainability are maintained during the next five to ten decades, the consequences are likely to be complex, upsetting, and traumatic for a significant part of humanity. The best way to address these challenges is probably to start right now developing a humble approach based on experiments for the construction of sustainable communities that explore and promote those essential features of human nature that can help to develop post-cooperation for sustainability. Inevitably, such experiments are politically, socially, and economically rudimentary and need to be scaled up to the regional and national levels (Bosch and Rotmans 2008) and eventually to address the challenges facing billions of people living in the most diverse levels of socio-economic development and cultural and political environments. The process of scaling up is likely to be difficult and perilous, and requires a suitably enabling environment. If such communities become successful and attractive, they will gradually become part of historical time and they will probably need to defend themselves against their enemies.

6.11 Expectations for the Unlimited Extension of Healthy Human Life

Historical estimates show that human average life expectancy at birth (Riley 2005) stayed approximately constant in all continents until after the more lasting social and economic effects of the Industrial Revolution began to be felt in about 1820. Life expectancy in Europe only started to increase significantly around 1870, when it was 36.2 years, while the global average was 29.7 years. The latter value is not much higher than the minimum below 25 years reached during the Neolithic and is probably below the life expectancy in the Bronze Age (Galor and Moav 2007). The increase that began in the 19th century was the result of a health transition that has occurred in all continents but at different times, Africa being the last, in the 1920s.

The extraordinary achievement is that global life expectancy at birth is now above 70 years, having reached 72.2 years in 2017 (WB 2018), and it continues to increase. Global life expectancy increased 5.5 years in the period between 2000 and 2016 (WHO 2018). At the country level there is still a wide variation in average life expectancy from a maximum value above 83 years in Japan to minimum values close to 50 years in Sub-Saharan Africa. The global gap in life expectancy of men relative to women has been stable with values slightly above 4 years.

The increase in life expectancy is primarily driven by the reduction in infant and child mortality. In the past, death was largely caused by infectious diseases, so the time when it occurred had little correlation with age. Currently, most deaths take place at older ages after a period of gradual, overall deterioration in the state of health. Chronic illnesses, particularly cancer and cardiovascular diseases, are the main causes of premature deaths (WHO 2009). The longest life recorded belongs to Jeanne Calment, a French woman who died in 1997 aged 122.

Although life expectancy at birth has greatly increased, there has been less success in mitigating, delaying, or avoiding morbidity during ageing. It is now usual to distinguish between healthy life expectancy and the years lived with a disability, which in 2016 represented about 13% of the average global life expectancy (IHME 2018). Better medicine and healthcare have increased the number of years that people live with a given disease burden or disability, but this increase is marginally less than the increase in healthy life expectancy.

To achieve longer lifespans in good health conditions, it is necessary to understand the biology of ageing. The subject has become an emergent and fast developing area of research, called biogerontology or geroscience. The idea that old age is a process programmed specifically and intentionally by the direct action of natural selection to lead to death, has been discredited (Vijg and Kennedy 2016). What appears to be a limit for human age is a collateral consequence of genetic programmes designed to boost the value and increase the opportunities in the initial and mature stage of life, to promote growth and primarily reproduction. The central aim of our existence as a biological organism is to live long enough to pass on our genes to our descendants. What happens once that goal has been achieved is almost irrelevant to biological design. In other words, the main "goal" of natural selection that led to multicellular organisms, including *Homo sapiens*, was to guarantee the strength and success of processes that ensure reproduction under the best conditions. It was not to indefinitely prolong lifespans.

The latter goal is a specific human ambition that emerged from our cultural evolution and is particularly natural and appealing to all those that enjoy a fruitful and pleasant life. Dreaming about a limitless lifespan is probably an inevitable consequence of the self-awareness and self-knowledge that emerged with symbolic thought, although this conjecture would be impossible to prove. Modernity, with the weakening of religious practices in the West and the emergence of a techno-optimist vision created by scientific and technological progress, brought the belief that the ambition of immortality could be achieved. Transhumanists argue that conditional immortality is within the reach of science and technology. Zoltan Istvan, a well-known American transhumanist, states with extraordinary naiveness in his popular science fiction book *The Transhumanist Wager* (Istvan 2013) that: "If a reasoning human being loves and values life, they will want to live as long as possible—the desire to be immortal. Nevertheless, it's impossible to know if they're going to be immortal once they die. To do nothing doesn't help the odds of attaining immortality—since it seems evident that everyone will die someday and possibly cease to exist. To try to do something scientifically constructive towards ensuring immortality beforehand is the most logical conclusion."

Some studies suggest that, although average life expectancy has increased considerably since the 19th century and continues to rise, the maximum limit of lifespans and the frequency with which it is reached have not increased in a similarly significant way (Dong et al. 2016). In other words, the number of people who reach ages very close to or above 122 years has not increased in the last few decades. There is a higher percentage of people aged over 90, but there are still no people aged 130 or more. Some scientists believe that there is a maximum lifespan for *Homo sapiens* that is established by and subject to natural limitations (Dong et al. 2016). Others, however, believe that the hypothetical natural limit may be largely overcome using appropriate diets and physical exercise, but mostly by way of drugs and other physical interventions specifically designed to prolong lifespans. They are convinced that in the future it will be possible for an increasing number of people to live far beyond 122 years.

It is not an exaggeration to say that humanity's most desired ambition for science and technology is to be able to use them to prolong human life in good health conditions for an undetermined amount of time. Many people's dream is to achieve conditional immortality. Is this possible? What kind of priority should be given to achieving such a goal? To answer these questions, one must start by understanding the cellular, genetic, and molecular mechanisms of ageing. We have already seen that ageing is related to a shortening of the telomeres in our chromosomes (Blackburn et al. 2015) and to the gradual inability of stem cells to repair the cellular tissues of the human body. Many cells in the human body regenerate regularly during our lives. Others, however, remain throughout, and are thus of the same age as the individual. This is the case for certain cells in the brain and heart. Furthermore, mitochondria, which play an important role in generating energy in cells, become increasingly dysfunctional with age (Wang and Hekimi 2015). Another important aspect that has been explored is the relationship between the fragility specific to old age and alterations in micro-organisms in the digestive system (Kumar et al. 2016).

A great deal of knowledge about the biology of ageing has been obtained by studying animals. The lifespans of several animals have been increased through lab tests. Some researchers use the fruit fly, *Drosophila melanogaster*, whose life expectancy is very sensitive to diet, access to reproduction, temperature and other environmental factors. Just like other animals, a restrictive diet, where feeding is limited to a minimum, increases lifespan (Tatar et al. 2014). By causing mutations in genes it is also possible to prolong the flies' lives from an average of 40 to 50 days to more than four years.

Mutations in the age-1 gene of the nematode *Caenorhabditis elegans* made it possible to increase its life expectancy to 145–190 days, which is more than nine times its normal life expectancy in a similar environment (Ayyadevara et al. 2009). More recent experiments with the same species showed that, by increasing or reducing the levels of a group of proteins that regulate gene expression, called Kruppel-like transcription factors (KLF), it is possible to increase or reduce lifespan in good health conditions (Hsieh et al. 2017). This result suggests that the same effect can be observed in mammals, since they also have proteins in the KLF family.

Studies of the biology of ageing in animals give us useful knowledge for the development of human geroscience. One crucial issue in geroscience is knowing whether delaying the process of biological ageing in humans also delays the start of the period of high risk of morbidity that is characteristic of old age (Kaeberlein 2013). Achieving this dual goal is essential to increasing healthy life expectancy, which means life in good health conditions, physical well-being, and with full mental capacities. If it is not possible to reduce the relative time difference between life expectancy and healthy life expectancy, the future artificial centenarians will suffer a disproportionately long period of life with chronic diseases or disabilities.

Several intervention strategies have been identified that have the potential ability to slow down or reverse the biological ageing process. The current challenge is to develop research that makes it possible to transfer knowledge acquired about animal ageing to humans. Before starting to use a certain type of intervention, it is useful to know the most suitable time of life for doing so, and also whether the benefits outweigh the risks. To answer these questions we must be able to determine with some certainty the state of human biological ageing so that the impact of interventions can be assessed. In order to do this, epigenetic biomarkers of the biological rate of ageing or biomarkers connected to metabolomic signatures can be used (Kaeberlein et al. 2015).

Interventions designed to increase lifespan also face a regulatory challenge by public certification agencies. There will need to be certification for drugs specifically designed to slow down ageing, since it is something that happens naturally to everyone and is not a disease. Government certification agencies require drugs prolonging healthy life expectancy to exhibit a high degree of safety, especially because people are likely to take them for the rest of their lives. If successful, everyone aged seventy and above would claim access to such drugs, and this would create severe financial problems for national health services or future national anti-ageing services. Meanwhile, the unrestrainable human desire to drive science to increase lifespan is creating a vibrant research activity for scientists and a powerful market for companies dedicated to geroscience.

There are essentially three types of intervention that can increase human lifespan: calorie-restricted diets, pharmacological drugs, and genetic interventions. In animals, restrictive diets cause a transition in cells from a regime in which they primarily concentrate on growth to another in which they focus more on cell repair. In the latter regime, the process of accumulating damage in cells becomes slower, and this leads to a longer lifespan for the organism. At the metabolic level, restrictive diets deactivate a protein kinase called mammalian target of rapamycin (mTOR) that sends growth hormone signals to the parts of the cell involved in protein synthesis. Through this mechanism, mTOR regulates growth in response to growth factors, such as nutrients, energy levels, and stress. Some drugs cause similar effects to those of a restrictive diet by reducing the action of mTOR.

One example is rapamycin, also called sirolimus, an immunosuppressant used to avoid the rejection of transplanted organs, which has the property of increasing the maximum lifespan in mammals (Ehninger et al. 2014). Rapamycin is a substance produced by the bacterium *Streptomyces hygroscopicus*, found in soil samples from Easter Island collected in 1965. Its name comes from Rapa Nui, the name of Easter Island in the local language. Rapamycin has a high potential to increase human lifespan, but its side-effects will need to be neutralised.

Another promising drug is metformin, extracted from the plant *Galega officinalis*, originally from the Middle East and naturalised in Europe and Western Asia, which has been used in Europe for centuries to treat diabetes, and is now used in the treatment of insulin-resistant type 2 diabetes. Metformin has multiple actions at the cellular level that can help slow down ageing, including decreasing insulin levels, inhibiting mTOR, inhibiting some mitochondrial disorders, and reducing reactive oxygen species (ROS). The TAME trial—Target Aging with Metformin—was recently launched in the USA. It involves a group of 3000 patients aged between 65 and 79 years and aims to test the effects of using metformin on these people's ageing. It is hoped that TAME will serve as a model for future evaluations of the effects of drugs on extending human lifespan in good health conditions or, as also mentioned, extending productive lifespan (Barzilai et al. 2016).

Despite all these efforts, discovering an anti-ageing pill that would be effective and available for all is at best far away in time. The big pharmaceutical companies are not especially interested in this field because the search for such a drug would require a long-term commitment to research and evaluation of side-effects, and the final result involves many uncertainties.

Another way of advancing geroscience is by combining the new digital technologies of big data and artificial intelligence with genetics. At the InSilico Medicine Institute, initially set up by Johns Hopkins University in the USA, deep learning is used to discover how patterns of genetic expression are transformed with age, comparing extensive databases of genome sequences belonging to young people and older people. The information obtained is then used to search in drug databases for molecules that may be able to block the transformations identified. Another company devoted to analysing large human genetic databases to understand the genetics of ageing is Human Longevity Inc, based in San Diego, USA, the founders of which include Craig Venter, the American biotechnologist who distinguished himself in the mapping of the human genome in 2001. These analyses make it possible to conclude that certain genetic variations missing in older people may characterise a shorter lifespan.

An intrusive way of increasing the lifespan in good health conditions, or health span, appeals to regenerative medicine, which uses processes for replacing, engineering, or regenerating human cells, tissues, and organs. One of the most promising techniques is to restore and rejuvenate the stem cell population of an ageing person, since these cells maintain and repair the tissue in which they are found. They can be used to develop therapeutic options for chronic diseases and regenerative treatment for various types of diseases, such as heart stroke and neurodegenerative diseases. The stem cells can be obtained with the least risk from one's own body, through autologous harvesting from bone marrow, adipose tissue, and blood, or else from the umbilical cord and placenta after birth.

Blood transfusions for rejuvenation were first tried in 1924 by Alexander Bogdanov (1873–1928), a Russian physician, philosopher, writer, and revolutionary, who co-founded the Bolshevik party but was expelled from it in 1909 as one of Lenin's rivals. Bogdanov experimented the blood transfusions on himself with relative success. However, he died a short time later, apparently due to a blood transfusion from a young man with malaria and tuberculosis that survived. One of his more interesting and far reaching proposals was the development of a discipline that he called "tektology" (Bogdanov 1922), where the social, biological, and physical sciences become unified by viewing them as systems of relationships that follow organizational principles common to all systems. Tektology was the foundation for what is nowadays called systems theory.

The insatiable desire to live a healthy life for longer and longer is particularly strong in the technological elites who are willing to pay whatever they are asked by burgeoning start-ups to achieve their goal. A recent example is provided by a company headquartered in Monterey, USA, called Ambrosia, run and founded by Jesse Karmazin a Stanford Medical School graduate that has been offering a "young blood treatment" since 2017 for $ 8000 for a litre of blood. The idea is essentially to use blood transfusions from young donors for rejuvenation just as Bogdanov had proposed about a century earlier, but also taking into account more recent research on parabiosis studies in mice, which showed that the exchange rejuvenated the older mouse who received the blood from the younger (Conboy et al. 2005). However, the animals were linked in other ways and were not only swapping blood, so the positive effects observed in the older mouse may also have been caused by the contact with the younger internal organs and circulatory system of the younger mouse. Researchers who have studied the procedures employed by Ambrosia say it poses risks for patients, including the development of several serious conditions later in life, such as graft-versus-host disease, which can occur when transfused blood cells attack the patient's own cells. In February 2019, the FDA warned against the treatments provided by Ambrosia and the company stopped operations. However, it was later revived by the founder of Ambrosia in the form of another company named Ivy Plasma (Brodwin 2019).

Forever Labs is another start-up in the business of extending human life. It extracts and freezes the stem cells of its young clients and stores them until they are later injected into their much older bodies to combat age-related illnesses. Forever Labs charges $ 7000 dollars for this plan.

Currently, the search for conditional immortality or, more modestly, regenerative medicine and translational research in geroscience, are attractive business opportunities for a growing number of biotechnology companies (de Magalhães et al. 2017). The global anti-ageing drug, product, and therapy market reached 140.3 billion dollars in 2015 and it is calculated that it will reach 216.52 billion in 2021 (Aspa 2019). One of the best known companies with the aim of reversing ageing is Calico—California Life Company. This was set up in 2013 by Larry Page, a computer scientist who was CEO of Google at the time, with a start-up financing of a billion dollars. There are many others specialising in different technologies to extend human life, including Alcor, AlloCure, Biomarker Pharmaceuticals, Centagen, Elysium Health, Gero, Geron, Genescient, Human Longevity, InSilico Medicine, Lifegen Technologies, Lineage Cell Therapeutics, Mesoblast, Nu Skin Enterprises, Legendary Pharmaceuticals, Prara Biotechnology, Retrotope, Sibelius, and Sierra Science. A significant number of these companies had their origin in, or are presently based in Silicon Valley, which became the main centre of the movement for conditional human immortality. One of the movement's leading figures is Aubrey de Grey, a British computer scientist and co-founder of the SENS (Strategies for Engineered Negligible Senescence) Research Foundation. He defends the idea that genetic engineering will drive humans to achieve "life extension escape velocity", allowing people to live more than 1000 years (De Grey and Rae 2007). De Grey identifies seven types of physical damage to cells caused by ageing that must be avoided, but each of them is difficult to repair and control.

6.12 Genetic Engineering for Conditional Human Immortality, Human Enhancement, Superintelligence, and Transhumanism

It seems likely that the greatest potential progress in geroscience lies at the genetic level in the identification of genes and epigenetics related to lifespans. The objective of this research area is to identify the gerontogenes. These are the ones that have a genetic expression or mutations with the property of increasing lifespan in animals or humans in good health conditions (Moskalev et al. 2014). With the knowledge obtained about gerontogenes, the next step will be to manipulate the human genome using genetic engineering.

Genetic engineering began with experiments on bacteria, then on multicellular organisms and is now used in medicine, where it is called gene therapy. One of the genetic technologies that is more likely to be used in the future is genome editing with site-directed nucleases, called gene editing with engineered nucleases (GEEN). GEEN is a type of synthetic biology in which DNA is inserted, removed, or replaced at precise points of the genome using nucleases programmed for that purpose. These interventions, which edit the genome and in the future will be able to rewrite much of it, induce artificial mutations in the organism. In 2017, 9 types of genome "editors" were used, one of which, viz., CRISPR-Cas9 (Lander 2016), is very promising in gene therapy for humans. The CRISPR-Cas9 system has the advantage of being able to intervene in chosen places of the genome with great accuracy and precision. These properties are important because a small mistake in an intervention or repair on a genome can completely change the sequence of proteins that codifies or even stops their production. GEEN technologies are also a valuable instrument for better understanding the functioning of the human genome. On the other hand, it is foreseeable that GEEN will come to be used extensively in agriculture to improve plants and animals, increase agricultural productivity, and fight diseases (Sovová et al. 2017). As previously mentioned, GEEN is an example of a disruptive technology.

Technologies for editing the human genome have practically unlimited applications for improving human health, increasing active lifespan, and human enhancement, defined as "any kind of genetic, biomedical, or pharmaceutical intervention aimed at improving human dispositions, capacities, or well-being, even if there is no pathology to be treated" (Giubilini and Sanyal 2016). When applied in somatic cells, the modified gene affects only the cells of the individual that received it. When germline cells are edited, the genetic change becomes hereditary and is passed on to later generations. These types of intervention can be used for therapeutic purposes and to prevent hereditary diseases. They can also be used for human genetic enhancement by generating human beings with physical, mental, or other faculties that are planned intentionally.

This dual opportunity and the establishment of a boundary between the two goals raises ethical questions that have been analysed extensively (Savulescu and Bostrom 2009; Agar 2010; Roache and Savulescu 2016; Gyngell et al. 2017). The US Academy of Sciences produced a report in 2017 (NASEM 2017b) with conclusions that are very cautious and do not recommend stopping the development of genome editing technologies on germline cells, or human germline engineering. It does not recommend stopping the future use of CRISPR to create babies with modified genomes provided that the goal is eliminating serious hereditary illnesses that could be passed on, and also that the technology is reliably safe. The threshold for producing super babies is hard to define precisely and make legally identifiable and binding. Legislation on genome editing policy varies greatly depending on the country, ranging from permissive regulations to explicit prohibition (Isasi et al. 2016).

The first genome editing experiments on human embryos with the aim of introducing mutations to produce immunity to HIV infection, were carried out by Chinese researchers in 2015 and 2016 using CRISPR technology, but the success rate was low, around 10% (Liang et al. 2015; Kang et al. 2016). The researchers involved underlined that there is still a long way to go until it is safe to edit an embryo and implant it to create a human being, and they argued that this technology should be banned until the ethical and scientific issues have been resolved. New experiments carried out in 2017 reported advances but suggested the need for more research to be carried out before the technology can become reliable (Tang et al. 2017). In 2017, researchers from the USA reported encouraging results in new experiments of the same type using CRISPR (Ma et al. 2017). In none of the cases published was the genetically modified embryo allowed to develop.

Genome editing technologies, particularly CRISPR, are being developed, driven by huge investments, but the genetic and epigenetic processes of the human genome are still not completely understood, so mistakes when editing the germline may lead to dramatic and entirely undesirable human situations. Several groups of researchers have identified the medical and social problems and risks of editing the genetics of the human germline, and they have recommended not using that technology or limiting its use (Lanphier et al. 2015; Baltimore et al. 2015).

However, in November 2018 the Chinese biophysicist He Jiankui of the Southern University of Science and Technology in Shenzhen, announced the birth of the world's first two gene-edited babies using the CRISP technology. The objective was to protect the babies from the HIV infection by editing the gene CCR5 that codes for a protein used by the HIV-1 virus to enter cells. On December 2018, China's National Health Commission ordered an investigation into He's research and in January of 2019 he was fired from his University. His achievement was considered unsafe and met with strong public disapproval in various institutions, generating a call for a global moratorium on creating gene-edited babies (Cyranoski 2019). Meanwhile, there is no information on whether or not He's intervention succeeded in modifying the girls' genes.

In spite of the criticisms levelled at the first experiment, it is likely that others are being actively planned. In fact in 2019, the Russian molecular biologist Denis Rebrikov at the Kulakov National Medical Research Center for Obstetrics, Gynecology and Perinatology acknowledge that he is also planning to edit the CCR5 gene so that the babies are HIV-negative, and claims that his method is entirely safe. A WHO international committee is presently working on establishing guidelines for human genome editing.

In spite of all the regulations and prohibitions that may be established, it is possible for eugenics to become an emerging industry providing considerable power to those that would use and control it. On the one hand, there is enormous public appetite for the use of germline engineering to avoid hereditary transmission of serious diseases, and on the other, there is heavy competition between countries to develop technologies that have the ideal conditions to create new areas of scientific and technological progress, likely to become very powerful and profitable. The ethical problems will be extensively and deeply analysed, but the emerging interests will probably circumvent most guidelines. The gap between what scientists proclaim are their values and what they are likely to do in favourable circumstances, although contradicting their professed values, is and will continue to remain wide open.

If and when eugenics becomes a current practice, the value of human life will be different from now. Some of the characteristics that make each of us unique will become reproducible. In the limit, humans could be produced in order to have specific characteristics and capabilities, just like any market product.

In the unlikely event that human enhancement progresses to the point where most humans become genetically optimised and therefore similar, one can start thinking of biological arguments about why that would not be advisable. The use of genetic engineering for human enhancement is a form of genetic evolution directed towards reducing the genetic diversity of the population, and this will in the long term tend to reduce its capacity for survival (Brosius 2003). This risk is well known in agriculture. Saving the genetic diversity of plants of agricultural value by way of gene banks, for example in the form of seed banks (Peres 2016), is essential to ensure food security for the human population, particularly in a crisis situation. Preserving the genetic diversity of cultivars is a necessary condition for adapting agriculture to climate change. Diversity is crucial to the development of varieties that are more resistant to a future climate that will be warmer, with droughts, floods, and other increasingly severe and frequent extreme events. Optimising the human species through genetic engineering would create the risk of reducing the ability to adapt in unexpected, adverse, and critical situations.

One of the main attractions of our social time is the possibility of creating super-human or transhuman minds with vastly superior intelligence capabilities. Currently the two main pathways to reach such a goal are the intelligence amplification of the human brain and the building of artificial general intelligence machines. The amplification of the human brain can be explored with bioengineering, genetic engineering, establishment of brain–computer interfaces, and mind uploading to computers. Artificial intelligence will play a crucial role in the development of artificial general intelligence machines capable of recursive self-improvement. After this first step, the emergence of artificial superintelligence, defined as an "intellect that greatly exceeds the cognitive performance of humans in virtually all domains of interest" (Bostrom 2014) becomes a likely possibility. The achievement of artificial superintelligence may or may not be associated with a technological singularity.

Transhumanism is the extreme expression of human enhancement. Journalist and writer Mark O'Connell defines it as a movement propelled by the "belief that we can and should eradicate ageing as a cause of death; that we can and should use technology to augment our bodies and our minds; that we can and should merge with machines, remaking ourselves, finally, in the image of our own higher ideals" (O'Connell 2017). No doubt this is a powerful belief, one that tends to render god an anachronistic figure of past times when humankind was still primitive and insufficiently ambitious. If transhumans start to emerge in relatively large numbers and ignore, dominate, or fight the more primitive humans, the end of *Homo sapiens* time may well be near. Recognition of humans and post-humans as two types of distinct "species" will necessarily involve a long and difficult process. In any case, the coexistence of the two may be harmonious and last a long time, as James Hughes argues when defending a democratic transhumanism (Hughes 2004). Or it may not. Even if it is harmonious, it is hard to imagine an outcome other than the progressive marginalisation of *Homo sapiens*, due to its inferior abilities and capacities, and its eventual extinction.

There is a wide range of studies and points of view about the ethical, social, political, and economic implications of human enhancement (Savulescu and Bostrom 2009; Agar 2010; Clarke et al. 2016; Roache and Savulescu 2016). The opinions can be classified into three main groups (Roache and Savulescu 2016). Transhumanists enthusiastically welcome opportunities to overcome biological limitations such as a limited healthy lifespan and the current average levels of intelligence. Bioliberals do not necessarily consider human enhancement technologies as good, but they are prepared to accept them as long as they serve a useful purpose and are safe and their use in society is not inherently unfair. Bioconservatives, such as Kass (2003) and Sandel (2004), oppose human enhancement essentially, because they consider that it destroys more worthy values than the benefits it promises to bring.

Human enhancement, superintelligence, and transhumanism are extremely interesting ideas and movements that help characterize the operative social time of the contemporaneous social generation. The operative social time of each generation has an intrinsic coherence in spite of the diversity of points of view that coexist. The present time is driven by a sacred belief in technology and its power to fulfil all human desires. It would be unjustifiable if we were not sufficiently brave and ambitious to reach the limits of perfection, intelligence, and pleasure and do away with time. It is difficulty to wait until all this becomes within reach. The singularity symbolizes the acceleration of operative social time. After the singularity transhumanism will be at hand and with it humans will get rid of time by eradicating "ageing as the cause of death". Time will then follow God in leaving the stage.

Coherence is also provided by the fact that human enhancement, superintelligence, and transhumanism is a program for the affluent individual, or rather for the exceptional affluent individuals that are better prepared to follow such stupendous avenues leading to a magnificent future. It is not a program with social structures in mind. Not a program to improve social relations, tolerance, justice, and human rights. It is hardly a program for humankind, when billions live only slightly above, at, or below the poverty line. Furthermore, there is also coherence in that the program needs nothing besides lots and lots of money, ingenuity, ambition, and self-interest, or, if preferred, egoism. It is a program that comforts and promotes the ego, confirming the superiority of the human being and its unquenchable desire to surpass itself. Finally, it takes place in well protected niches shielded from the turbulence of the present world. In the case of a nuclear war or an environmental collapse, it could probably be heroically pursued in one of the many underground bunkers that are proliferating in the most advanced economies.

One of the main defenders and promoters of transhumanism and the ICT path to human enhancement is Ray Kurzweil. Kurzweil believes that, in 2029, technological progress, particularly the development of artificial intelligence, will create computers that will be far more powerful than the human brain. During the 2030s, he foresees that some humans will become hybrid machines. A physical connection will be established between the brain's neocortex and the cloud and computers. This project will have to solve the challenge of building an interface between the system of neurons and synapses, where the electrochemical signals travel at around 100 m/s, and the electronic circuits that connect the brain to the computer and other digital systems, where signals travel at the speed of light, or around 300 million m/s. It is hoped that in the end man and machine will become one unit with the help of artificial intelligence.

Transhumanism is now part of American folklore. At the South by Southwest or SXSW annual festival held in Austin, Texas, which involves music festivals, films, and conferences, Kurzweil said in 2017: "I have set the date 2045 for the 'Singularity' which is when we will multiply our effective intelligence a billion-fold by merging with the intelligence we have created". It will be an opportunity to enhance humans. As well as being more intelligent, "we're going to be funnier, we're going to be better at music, we're going to be sexier".

Chapter 7
Anthropocene, Technosphere, Biosphere, and the Contemporary Utopias

7.1 The Great Acceleration and Its Consequences for the Earth System

Up to now, we have discussed some aspects of the history of human cultural evolution as regards human values and ideas, socio-economic and political movements, science, and technology. The resulting transformation in human societies and activities, in other words, in the human subsystem of the Earth system, has interfered with its biogeophysical subsystems and had a noticeable impact on them. Through an extensive and detailed range of observations, models, and reconstructions of the past, mainly since the end of the 19th century, science has made it possible to assess the characteristics, breadth, and intensity of local, regional, and global impacts on those subsystems.

It has already been mentioned that, depending on their origin, global changes may be natural or anthropogenic. The latter are often simply called global changes, when the context is clear enough to understand that they are anthropogenic in origin. Although it is sometimes a difficult process, it is now possible in many cases to identify and distinguish anthropogenic global changes from the natural variability in different biogeophysical subsystems arising from natural global changes. Both types of global change affect the Earth system, and therefore its biogeophysical and human subsystems, in different ways and at different spatial and temporal scales.

Natural or anthropogenic global changes in biogeophysical subsystems are also called global environmental changes. Turner et al. (1990) distinguished between systemic and cumulative change in global environmental changes. The latter comprises changes arising from the accumulation of localized and regional changes, but with a wide and dense distribution worldwide, or changes that have such an intensity in a few regions but generate a problem on a global scale, including situations in which they put a natural resource at risk on a global scale (Santos 2012).

Anthropogenic climate change, caused by greenhouse gas emissions into the atmosphere as a result of human activities, is the best-known example of a systemic global change. But there are other very important examples. The first systemic global change

© Springer Nature Switzerland AG 2021
F. Duarte Santos, *Time, Progress, Growth and Technology*, The Frontiers Collection,
https://doi.org/10.1007/978-3-030-55334-0_7

to be identified was depletion of stratospheric ozone in the Antarctic region caused by the worldwide use of chlorofluorocarbons (CFCs), known as the "ozone hole". A third systemic global change is the change in the Earth's albedo, the proportion of the Sun's radiation that is reflected by the Earth. Albedo change is caused by modifications to the Earth's surface, such as land use changes, including forestation, reforestation, deforestation, and desertification or by the extensive melting of sea ice in the Arctic.

The main examples of cumulative global changes are as follows:

- Progressive scarcity of water due to various causes, in particular overexploitation of water resources—rivers, lakes and aquifers.
- Degradation and destruction of a huge range of ecosystems, such as tropical moist forests, wetlands, and coral reefs.
- Deforestation.
- Degradation and destruction of habitats of endangered species.
- Loss of biodiversity.
- Degradation and loss of soils with the capacity to support agriculture.
- Desertification.
- Interference with the nitrogen and phosphorus cycles.
- Air, ocean, water, and soil pollution.
- Increase in the atmospheric concentration of anthropogenic aerosols.
- Increasing scarcity of some renewable and non-renewable natural resources.

These cumulative changes are being more widely observed and they are becoming more visible. They are among the greatest environmental challenges facing our civilisation.

The end of the Second World War marked the beginning of a period of acceleration in the pace of change in the human subsystem, which induced significant changes in the biogeophysical subsystems and in the interactions between all such subsystems. Many of the changes that manifested themselves globally had already been observed, but their variety, intensity, and pace increased significantly after 1950. This led to the proposal of the Great Acceleration concept (Hibbard et al. 2006; McNeill and Engelke 2016), as already mentioned. The term aims to capture the holistic, integrated, interdependent, accelerated, and intense character of global environmental changes caused by some human activities since the mid-20th century. There is a strong consensus in the scientific community that the last 50 years of the 20th century were the period during which the interference between human and biophysical subsystems underwent the fastest transformations (Steflen et al. 2004).

The path that led to the Great Acceleration from 1750 to 2000 has been documented in quantitative terms using twelve indicators related to socioeconomic transformations and twelve indicators related to transformations in the biogeophysical subsystems of the Earth system, published in a book by the International Geosphere–Biosphere Programme (IGBP) (Steffen et al. 2005). The relevant evolutionary graphs since the beginning of the Great Acceleration have become one of the most widely used and striking ways of demonstrating that we are in a new epoch of the Anthropocene.

Later, the graphs showing the indicators were updated for the period 1750–2010 (see Figs. 5.1–5.3), and some socioeconomic transformation indicators were broken down into the OECD countries, the so-called BRIC countries (Brazil, Russia, India, China, and South Africa), and other countries (Steffen et al. 2015). This breakdown is important to understand how the different pressures exerted on the Earth system by these three groups have developed. Most of world population growth since 1950 has been outside OECD countries, but most consumption of goods and services has been within the OECD. In 2010, OECD countries held 74% of world GDP, but only 18% of the population. All twelve socioeconomic transformation indicators show sharp and sometimes exponential growth from 1950 onwards. In some indicators, such as global population, the number of large dams, and water consumption, there has been a slowdown in growth in the most recent years. Global population grew between 1950 and 2010 at a rather higher annual rate than in the period between 1750 and 1950, but it has been falling in more recent times (Klein Goldewijk et al. 2010).

The study of the systemic and cumulative global change that the human subsystem has generated in biogeophysical subsystems has led scientists involved in the International Geosphere–Biosphere Programme to conclude that the Earth system is currently operating in a no-analogue state as regards its dynamics and functioning (Steflen et al. 2004). What this actually means is that some environmental indicators that characterize the state of the Earth system have moved outside the natural variation ranges of at least the last half million years.

7.2 Driving the Earth into No-Analogue States with Anthropogenic Climate Change. Do We Understand the Short and Long Term Consequences?

One emblematic indicator of the Earth system's new situation is the atmospheric concentration of CO_2, which in July 2019 was 411 ppmv (NASA 2019), higher than estimated values for at least the last 800 000 years, calculated from the composition of air trapped in small air bubbles in the Antarctic ice sheets (Petit et al. 1999; Siegenthaler 2005; Lüthi et al. 2008). Furthermore, the 411 ppmv value is likely to be higher than the estimated values of the atmospheric CO_2 concentration derived from alkenones, boron isotopes, and fossil leaf stomata preserved in layers of rock and sediment over the past two to three million years (WMO 2016; Martínez-Botí et al. 2015; Kürschner et al. 1996).

Analyses of air trapped in the ice cores extracted from the Antarctic ice sheets reveal that the concentration of CO_2 in the atmosphere fluctuated in cycles lasting around 100 000 years (see Fig. 7.1), related to the Milankovitch climate cycles (Milankovitch 1930), between around 180 ppmv during glacial periods and 280 ppmv in interglacial periods. In the transition from a glacial to an interglacial period the warming is not initiated by CO_2 emissions but by the astronomical forcing that drives the Milankovitch cycles. The warming causes the oceans to release more CO_2,

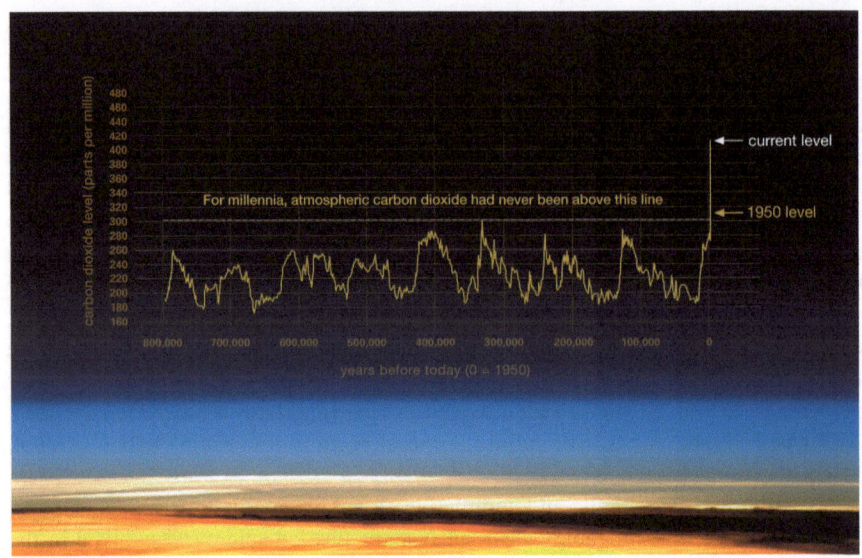

Fig. 7.1 Reconstruction of the global atmospheric CO_2 concentration for the past 800 000 years, obtained from the analysis of ancient air bubbles trapped in Antarctic ice cores. As explained in the text, there is an approximate correlation between the global CO_2 concentration and the global mean temperature of the atmosphere at the Earth's surface, which can be seen by comparing with the lower curve of Fig. 3.4. CO_2 concentration during the last 800 000 years has varied from about 180 ppmv in glacial periods to about 280 ppmv in interglacial periods. However, in the last 250 years the CO_2 concentration has increased significantly due to anthropogenic CO_2 emissions to the atmosphere, reaching 411 ppmv in July 2019 (NASA 2019). This value is likely to be higher than the estimated values of the atmospheric CO_2 concentration over the past two to three million years (Martínez-Botí et al. 2015; Kürschner et al. 1996). Data from Lüthi et al. (2008), Petit et al. (1999), and NOAA Mauna Loa CO_2 record. *Image credit* NASA

because warmer oceans hold less gas than colder oceans, and these atmospheric CO_2 emissions amplify the increase in global mean surface temperature.

The current CO_2 concentration is unambiguously outside the 180–280 ppmv range of variation due to anthropogenic emissions mostly from the intensive use of fossil fuels and land use changes, especially deforestation. CO_2 is an important indicator for the Earth system because it is one of the greenhouse gases naturally found in our atmosphere whose concentration has, as it was previously shown, a significant impact on the Earth's global climate.

In fact, the increase in CO_2 concentration by more than 46% since the Industrial Revolution is causing the atmospheric greenhouse effect to become more intense, which leads to an increase in the global mean temperature at the Earth's surface, a condition that is called global warming. The increase in the atmospheric concentration of all greenhouse gases with significant anthropogenic emissions, the most important of which in terms of radiative forcing (see Fig. 7.2) being CO_2, CH_4 and N_2O, causes a global climate change with other characteristics besides global warm-

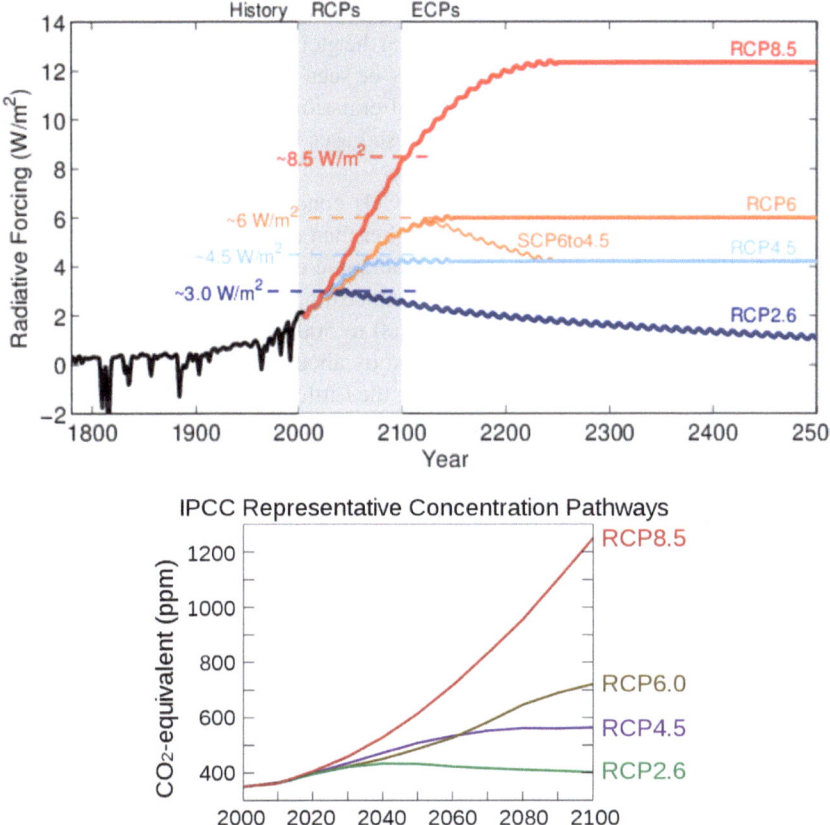

Fig. 7.2 Representative concentration pathways (RCP) used for climate model research for the IPCC Fifth Assessment Report (IPCC 2013). The RCPs are scenario trajectories of the radiative forcing until 2100. Radiative forcing is the net change in the energy balance of the Earth system due to some imposed perturbation, such as the increase in anthropogenic greenhouse emissions or a volcanic eruption. It is usually expressed in watts per square meter, averaged over a particular period of time, and quantifies the energy imbalance that occurs when the imposed change takes place. From the physical point of view, it is advantageous to consider radiative forcing scenarios instead of greenhouse gas concentration scenarios, because the former variable is the direct driver of climate change. *Upper*: Historical radiation forcing since 1765 up to 2000 (*black*), the RCP scenarios (RCP2.6, RCP4.5, RCP6.0 and RCP8.5) up to 2100, and the extended concentration pathways (ECP) for the period 2100–2500 (Meinshausen et al. 2011). The ECPs are simple extensions of the RCPs, based on the assumption of either smoothly stabilizing concentrations or maintaining constant emissions. The sharp minima in the observed historical data correspond to the effect of large volcanic eruptions. Of the four RCP scenarios, only the RCP2-6 scenario complies with the Paris Agreement objective of limiting the increase in the global mean temperature below 2 °C relative to pre-industrial global mean temperatures. Image adapted from Fig. 4 of Meinshausen et al. (2011). *Lower*: Greenhouse gas concentrations expressed in CO_2-equivalent ppmv for the four RCPs (IPCC 2013). Note that the concentration of anthropogenic greenhouse gases reaches a peak before 2100 only in the RCP2.6 scenario. Image produced by Efbrazil and licensed in Wikimedia Commons under the Creative Commons Attribution

ing, such as an increasing intensity and frequency of certain extreme weather events. Furthermore, it also has the effect of increasing the global mean sea level. The impacts of anthropogenic climate change can already be seen and will likely get worse over the coming decades and centuries if global emissions are not reduced. The total radiative forcing by all long-lived greenhouse gases in July 2019 corresponds to a CO_2-equivalent mole fraction of 497 ppmv.

The rate of change of the atmospheric CO_2 concentration since the Industrial Revolution is between 10 and 100 times higher than natural variations in CO_2 over the last 420 000 years (Falkowski 2000; Loulergue et al. 2008). The global mean surface temperature has increased by 4–7 °C since the Last Glacial Maximum 21 500 years ago. During the Holocene, from 10 000 to 5000 years ago, the global mean temperature was stable and then it decreased by about −0.7 °C, culminating in the coolest temperatures of the Holocene during the Little Ice Age, about 200 years ago (Marcott et al. 2013). The cooling was greater in the North Atlantic and may have been caused by a weakening of the meridional overturning circulation in the middle Holocene.

Since 1880, when reliable temperature records became available around the world, the yearly global mean land and ocean temperature has increased at an average rate of 0.07 °C per decade (IPCC 2014a), as a result of anthropogenic greenhouse gas emissions. Since 1981, the average rate of yearly temperature increase jumped to 0.17 °C per decade, more than twice the value (NOAA 2016). This latter rate of increase is roughly ten times faster than the average rate of global mean temperature increase in the transition from the last glacial period to the present interglacial.

The scientific progress in tracing the Earth's climate history has made it possible to conclude that the current rate of carbon emissions into the atmosphere in the form of CO_2 and CH_4 is the highest for the last 66 million years (Zeebe et al. 2016). Over this long period of time, the natural event that was most similar to the current situation, in terms of carbon emissions to the atmosphere, was the Paleocene–Eocene Thermal Maximum (PETM). However, we are currently emitting more than 9 GtC into the atmosphere every year, which is an emission rate almost ten times higher than at the start of the PETM. In the decade 2008–2017, just the global carbon emissions in the form of CO_2 emissions from burning fossil fuels and industrial processes (such as the manufacture of metal, cement, and chemical products) were estimated to be 9.4 ± 0.5 GtC yr^{-1} (Olivier et al. 2015; Le Quéré 2018).

Rather surprisingly this amount is about 0.8 GtC higher than the projection made for 2017 in the central, "business-as-usual" scenario for CO_2 emissions, meaning without additional mitigation measures, constructed by the Intergovernmental Panel on Climate Change (IPCC) in 1992 (Leggett et al. 1992). In other words, this means that all the mitigation measures that were pursued all over the world in the period 1992–2017 had a negligible effect when compared with the business as usual scenario made in the early 1990s by the IPCC.

The total amount of current anthropogenic emissions of CO_2 is greater, because the emissions from changes in land use also need to be taken into account, and they represent approximately a sixth of emissions from fossil fuel combustion and industrial processes (IPCC 2014a). Anthropogenic methane emissions into the atmosphere are difficult to measure and are currently estimated at 0.57 Gt CH_4 (IEA 2019a; Höglund-Isaksson 2012). About 40% of such emissions arise from natural sources and the remainder are from anthropogenic sources. Thus, human emissions of CO_2 contain far more carbon atoms than CH_4 emissions. The atmospheric concentration of methane reached 1859 ppbv in 2017 (WMO 2018), which is estimated to be around 2.5 times higher than at the beginning of the Industrial Revolution (Saunois et al. 2016).

These results show that current rate of anthropogenic carbon emissions into the atmosphere in the form of carbon dioxide and methane are higher than at the start of the PETM (Zeebe et al. 2016) by a factor of between 9.1 and 16.7. Considering the last 66 million years, one concludes that the Earth system is currently in a no-analogue state characterised by the most intense flux of carbon injection into the atmosphere in the form of CO_2 and CH_4 from organic sources. The evolution of this situation and its consequences are therefore more difficult to predict.

It is important to remember that current atmospheric CO_2 concentrations are much lower than during the initial stages of the Eocene, when concentrations were likely between 1000 and 2000 ppmv (Zachos et al. 2008; Anagnostou et al. 2016). If CO_2 emissions from fossil fuels and land use changes continue at current levels until 2100 and follow a logistic curve determined solely by fossil fuel reserves and resources, it is likely that humanity will have emitted about 5270 GtC between 1750 and 2400 (Caldeira and Wickett 2003). In this scenario, CO_2 concentrations would reach 1900 ppmv around the year 2300 and the surface ocean pH would be reduced by 0.7 units. This scenario is likely to be an event without precedent at least in the last 300 million years of the Earth system.

The faster the growth in CO_2 concentrations, the more pronounced the acidification of the ocean, especially in the layer closest to the surface. The acidification of surface waters in oceans is now 30 times higher than regional-level natural variation intervals (Friedrich et al. 2012). The same authors conclude that the acidification rate of the Atlantic and Pacific oceans currently observed is two orders of magnitude higher than the acidification during the transition from the last glacial period to the current interglacial period. The rapid variation in the ocean's acidity makes it difficult or impossible for the most vulnerable marine organisms to adapt. This is the case for phytoplankton and corals. Ocean acidification and increases in ocean water temperature will affect fishing and aquaculture, threatening food safety for many hundreds of millions of people. If it turns out to be impossible to achieve carbon neutrality at the end of the century, a situation in which anthropogenic CO_2 emissions are offset by natural or artificial carbon sinks, it is likely that surface water temperature will increase by 2–3 °C, the pH will fall by more than 0.2, and oxygen concentrations will fall by 2–4% (Mora et al. 2013). These values have no known precedent in the last 20 million years and they will imply a decrease in the ocean's productivity in terms of food resources. The authors point out that approximately 470–870 million of the poorest people in the world rely heavily on the ocean for food, jobs, and revenues and live in countries that will be most affected by simultaneous changes in ocean biogeochemistry induced by the current trends in anthropogenic greenhouse emissions (Mora et al. 2013). A more recent study (Moore et al. 2018b) concluded that potential fishery yields, constrained by lower trophic-level productivity, are likely to decrease by more than 20% globally and by nearly 60% in the North Atlantic. Furthermore, continued high levels of greenhouse gas emissions could have a long term effect in suppressing marine biological productivity for a millennium.

One of the most worrying aspects of this scenario is the destruction of marine ecosystems caused by ocean anoxic events, periods in which the concentration of oxygen in the ocean is very low. There are records in the Earth's history of great extinctions of marine organisms caused by oceanic anoxic events characterized by widespread distribution of marine organic matter-rich sediments, called black shales, and significant perturbations in the global carbon cycle (Gröcke et al. 2011). These perturbations are recorded globally in sediments as carbon isotope excursions and are mainly caused by high CO_2 emissions during phases of extensive volcanic activity. One example is the Toarcian Oceanic Anoxic Event (T-OAE), which took place 183 million years ago during the Early Jurassic (Al-Suwaidi et al. 2016). It caused a series of global environmental impacts, including the increase in the global mean surface temperature, ocean acidification, and anoxic events.

Studies of the effects of anoxia carried out in Ya Ha Tinda in the state of Alberta, Canada, can help us to understand how different marine ecosystems react to a pronounced climate change that has analogies to the current one (Martindale and Aberhan 2017). The increase in the atmospheric concentration of CO_2 in the T-OAE has been well documented through the associated carbon isotope excursion (Them et al. 2017), as in the PETM.

There are currently no oceanic anoxic events, but a growing number of ocean areas with low oxygen content, or hypoxic areas, have been observed since the 1960s. These are also called dead zones because most fish and other marine organisms cannot live there. The number of dead zones has doubled every decade since 1960 and there are now roughly 500, covering an area measuring 250000 km^2, mostly located in coastal areas and areas near estuaries (UNDP 2013). The coastal marine areas of countries with advanced economies are the most widely studied and it is there that the areas most affected by hypoxia are to be found, such as in the east and south coastal marine areas of the USA, the North and Baltic Seas in Europe, and the areas surrounding Japan (Diaz and Rosenberg 2008). It is very likely that the coastal marine areas of countries with emerging economies will be increasingly affected in the future if they do not address the causes of hypoxia. The damage presently caused to fishing and coastal tourism by dead zones is valued at tens of billions of dollars.

Hypoxia results from an excess of nutrients in the ocean, mostly due to the unsustainable use of fertilisers containing nitrogen and phosphorus in agriculture. When flowing into the sea, they cause the eutrophication of coastal ocean waters together with an increase in phytoplankton. Phytoplankton, through respiration, reduces the quantity of dissolved oxygen. Furthermore, when it falls to the ocean floor after dying, it is decomposed by bacteria that consume more oxygen dissolved in the water. The problem of spreading dead zones is being made worse by climate change because it gradually contributes to the hypoxia of oceans by way of several factors, especially by warming ocean water (Altieri and Gedan 2015). Furthermore, climate change increases eutrophication levels in lakes and rivers, not only due to the increase in the water temperature but also through changes in precipitation regimes, because precipitation tends to happen more frequently over short periods of time and to be stronger in those periods. There are signs that some marine regions of the world, such as the USA and the south, south-east, and east of Asia will be particularly affected (Sinha et al. 2017). The combination of climate change and eutrophication of coastal waters triggered by inputs of sewage, sludge, fertilizers, or other wastes containing nutrients such as nitrogen and phosphorus is a serious problem in the medium and long term. The solution is particularly difficult because the problem is diffuse and directly involves various socio-economical sectors, such as energy and agriculture, that are crucial to ensure economic growth.

The Earth system has begun an anthropogenic journey outside its interval of natural variability, in some cases reaching over the last millions to hundreds of millions of years. This new situation is reflected in a wide range of biogeophysical indicators, just a few of which, related to climate change, have been mentioned here. There are many others, however, such as those that involve a growing loss of terrestrial and marine biodiversity, disturbances in the nitrogen and phosphorus cycles, global water usage, air and ocean pollution, and desertification (Rockström et al. 2009a).

Curiously, what currently characterises the Earth's no-analogue state most distinctly is not, in most cases, the fact that the biogeophysical indicator values are outside the natural range of variability, but the fact that the rate of variation of those values is faster than it has been for tens of millions of years. The most significant aspect of the phenomenon is, once again, the rate of change and not the anomaly in the values themselves. The population of *Homo sapiens* has succeeded in contaminating their home planet with the increasing haste represented by the acceleration of operative social time. It is difficult to believe that we would not engulf other planets or even the whole universe if we could, in the same haste, forcing those, too, into no-analogue states and paths.

The evolution of the Earth system into a no-analogue state is irrelevant to planet Earth, but worrying to humans for two main reasons. The first is that the stronger and faster the forcing of Earth's biogeophysical subsystems, the more likely that the response will not follow a linear path, but instead generate an abrupt, non-linear change that is harder to model and predict, and which will be harder for humans to adapt to. The Earth system's natural abrupt changes occur over relatively short time intervals of tens to thousands of years and are relatively frequent (Steffen et al. 2005). The other reason is the possibility that the evolution of the no-analogue state lead the Earth system to an area of state space far from the area that characterises the Holocene.

The Holocene is an especially cherished period for *Homo sapiens* because it was at the start of that period that agriculture emerged and it saw the sequence of civilisations up to modernity, bringing us to the current age. These remarkable human developments benefited from the Holocene being an interglacial period, and therefore warm and wet, and favourable to global-scale agriculture and human activities in general. It should be remembered, however, that the *Homo* genus emerged in Africa roughly 2.6 million years ago, during the transition into a cooler climate caused by the glaciation of the Arctic. Since around 1.2 million years ago, relatively long glacial periods have alternated with shorter interglacial periods in cycles lasting roughly 100 000 years.

The last interglacial period before the Holocene lasted for approximately 15 000 years, and is known as the Eemian, as already mentioned, based on the study of sediments in the River Eem in Holland. Such studies enable us to better understand how the Earth system reacts to the forcing caused by an interglacial period. It may also improve our understanding of how the Earth system will react to the climate forcing caused by anthropogenic greenhouse gas emissions which, as we have seen, is producing changes at a faster pace than the astronomical forcing that generated transitions from glacial to interglacial periods. Studying this interglacial period may therefore be useful to better understand the Earth's response as greenhouse gases accumulate in the atmosphere.

The global mean surface temperature in the Eemian was only 1.5–2 °C higher than in the pre-industrial period (Masson-Delmotte et al. 2013; Hansen and Sato 2012; Hansen et al. 2016). Currently, the global mean surface temperature is already 1 °C higher than in that period. The Paris Agreement, reached in the UNFCCC negotiations in 2015, establishes that the global mean surface temperature should not rise by more than 2 °C relative to pre-industrial values. Thus, to comply with the Paris Agreement, the global temperature can only increase 1 °C more.

The important point here is that, in Polar Regions, the mean temperature tends to be higher than the global mean temperature due to positive feedback processes. The warming amplification in the Arctic and, to a lesser extent in the Antarctic, is the result of those regions' instability as they react to the astronomical forcing that caused the transition to the interglacial period or to the current forcing driven by the increasing emissions of greenhouse gases into the atmosphere. The present Arctic amplification has increased the mean surface temperature in the region to the north of the Arctic Circle by a factor of about two relative to the global mean temperature increase. One of the most important causes of such an amplification is the accelerated melting of large areas of sea ice. As a result the Arctic retains more heat, since the ocean surface reflects only 6% of incoming solar radiation, while sea ice reflects 50–70%.

In the Eemian the Arctic amplification was greater. Studies made with the objective of understanding the response of the Greenland ice sheet to the warmer climate of the Eemian concluded that the estimated surface temperature on the Greenland ice sheet was about 7.5 °C warmer compared to the last millennium (Dahl-Jensen et al. 2013). The warmer climate produced extensive melting of the Greenland ice sheets during the Eemian. Estimated elevations were about 130 m lower than at present (Dahl-Jensen et al. 2013).

Due to the melting in Greenland and Antarctica, the global mean sea level in the Eemian rose by at least 6–9.3 m compared to its level today (Kopp et al. 2009; Dutton and Lambeck 2012), although at the time CO_2 concentrations in the atmosphere were lower than 280 ppmv. The study of Eemian geological deposits in northern Europe lead to the conclusion that the Baltic Sea was then part of the much larger Eemian Sea, which was connected to the White Sea, one arm of the Barents Sea. Scandinavia was an island, and a significant part of northern Europe, including Holland, Denmark, and part of Germany, was flooded by the Eemian Sea.

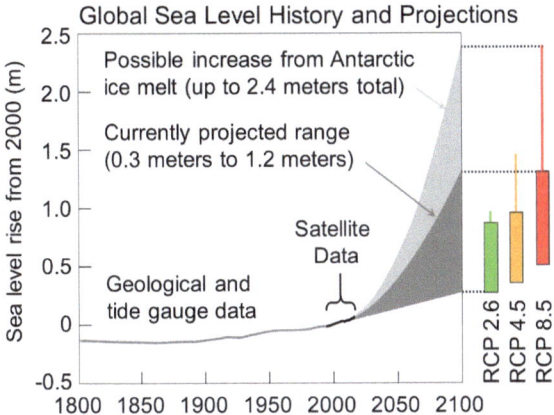

Fig. 7.3 Observed and projected changes in global mean sea level for 1800–2100. Measurements of global mean sea level using satellites started in 1993 and the obtained data agrees with tide gauge data. The boxes on the right show the likely range of change in sea level rise by 2100 relative to 2000 projected by three different RCPs, as described in the IPCC Fifth Assessment Report (IPCC 2013) (see Fig. 7.2). The lines above the RCP boxes show possible increases according to research done after 2013, based on the potential contribution to sea level rise from Antarctic ice sheet melt. Data from the US Global Change Research Program for the Fourth National Climate Assessment. Figure adapted from Sweet et al. (2017). Image produced by Efbrazil licensed in Wikimedia Commons under the Creative Commons Attribution

Around 20 000 years ago, in the preceding glacial period, global mean sea levels were around 120 m lower than they are today. In the last 11 700 years of the Holocene, the global mean sea level remained relatively stable, and since 1900 it has increased by about 16–21 cm, with about 7 cm since 1993, mainly as a result of anthropogenic climate change (USGCRP 2017). It is very likely that this rise was the fastest the Earth system has seen in the last 28 centuries (Kopp et al. 2016). Observations using tide gauges and satellites show that the annual sea level rise is increasing, having been on average 3.2 mm per year in the period 1993–2010 (IPCC 2013). The ocean has a very slow response to climate change, so the global mean sea level will continue to rise for many centuries and millennia (see Fig. 7.3). An important question is to know the future characteristics of that rise. Will it be roughly linear or exponential?

The observed rise in the global mean sea level results from atmospheric and ocean warming, which act to increase both the mass of the ocean, primarily through the melting of land ice in mountain glaciers and in the polar ice sheets of Greenland and Antarctica, and the volume of the ocean, primarily through thermal expansion. In the period 1971–1990, thermal expansion and melting of mountain glaciers outside the Polar Regions have been responsible for about 75% of the observed rise (IPCC 2013). The melting in the Polar Regions has contributed about 25% of that. However, in the future, the latter contribution will become dominant.

The ice sheets in Greenland and Antarctica contain the vast majority of the Earth's ice, and if they were to melt entirely, sea level would rise by more than 70 meters. The Arctic contains 3.1×10^6 km^3 of land ice, equivalent to 7.7 m of sea level rise, of which about 97% is stored in the Greenland Ice Sheet (Moon et al. 2018). Science has generally assumed that it is very difficult to perturb the equilibrium of the polar ice sheets, but more recent studies indicate a growing ice sheet imbalance, with more mass being melt than is being replaced by snowfall, especially in Greenland and West Antarctica. Studying how these ice sheets responded to the increase in global mean temperature during the Eemian may help us to project their future behaviour under anthropogenic climate change.

Analysis of what happened in the Eemian shows that it was the partial melting of the polar ice caps that contributed the most to the rise in global mean sea level in the later interglacial period. It is estimated that ice melting in Greenland and Western Antarctica would have contributed around 3.4 m and 2.5 m, respectively, while the other two factors previously mentioned would have contributed only 1 m (Clark and Huybers 2009). The reason for this behaviour is the instability of ice shelves in Greenland and in the Antarctic, particularly in the western region.

An ice shelf is a body of glacial ice that forms where a glacier or an ice sheet flows down to a coastline and floats, at least in part, in the ocean. When floating in the ocean, ice shelves are unstable because they have few points where they are supported by solid rock. This is the case for the Filchner, Ronne, and Larsen ice shelves in the Weddell Sea, the Pine Island and Thwaites glacier ice shelves in the Amundsen Sea, and the Ross ice shelf in the Ross Sea, all in western Antarctica.

The mechanism accelerating the melting of ice shelves in Antarctica is already being seen today. Melting gets faster due to contact between the lower part of the ice shelf and the subsurface waters of the ocean, which come from the Pacific and are getting warmer. These waters open up passageways beneath the shelves and force their grounding lines to retreat (DeConto and Pollard 2016). Moreover, the relatively warmer freshwater resulting from the melting generates a complex system of channels that end up creating fissures in the platform surface. Over time, the broadening and propagation of fissures cause large blocks of ice to separate and produce icebergs. The ice shelves and sea ice surrounding them act as buttresses that stop the ice sliding from Antarctica into the ocean. However, as the shelves disintegrate, the speed at which the ice above the global mean sea level moves towards the ocean increases, its melting speeds up, and the global mean sea level tends to rise exponentially. The disintegration processes can be seen, for example, in various sections of the Larsen ice shelf on the east coast of the Antarctic Peninsula. The Larsen A shelf disintegrated in 1995, Larsen B in 2002 (Rignot et al. 2019), and Larsen C, the largest, released a 5800 km^2 iceberg on 12 July 2017, and this reduces its shelf area by roughly 10%.

Data from the Gravity Recovery and Climate Experiment (GRACE) using a gravity measuring satellite reveal that Greenland and the Antarctic have been losing mass due to melting ice since 2002 at a yearly rate of 286 Gt and 127 Gt, respectively (GRACE 2019). Recent measurements show that, between 2012 and 2017, the average annual loss of ice in Antarctica was 219 Gt, which is approximately triple the 76 Gt measured in the previous decades 1992–2011 (Shepherd et al. 2018). In the periods 1992–1997, 1997–2002, 2002–2007, and 2007–2012, the average annual ice losses were 49, 38, 73, and 160 Gt, respectively. A study of the data obtained using several techniques reveals that there was an acceleration in melting that started in 2002 (Shepherd et al. 2018).

The identification and modelling of mechanisms that accelerate ice melting in Antarctica lead to the conclusion that there will be a doubling of ice mass loss in Greenland and Antarctica over the next 10–20 years (Hansen et al. 2016), which implies an exponential increase in global mean sea level. This means that the increase in global mean sea level may be greater than 1 m before the end of the century (DeConto and Pollard 2016). Controlling the rise of global mean sea level will depend on humankind's ability to reduce greenhouse gas emissions so that the global mean surface temperature does not rise more than 2 °C in relation to the pre-industrial period.

A realistic estimate of current climate change mitigation trends implies that it will be practically impossible to reach the Paris goal of a 2 °C cap and that the increase in the global mean temperature is likely to be between 3 and 5 °C. In this business as usual scenario, considerable financial resources need to be spent to avoid permanent flooding in large swathes of low-lying, populous coastal areas, including the Nile, Ganges–Brahmaputra, and Mekong deltas, as well as some of the coastal plains of northern Europe, the east coast of the USA, East Asia, and the Pacific, and some regions of Africa.

Humanity is faced with a new challenge that involves addressing and adapting to an increasingly serious situation that will slowly but relentlessly take place over the coming centuries and millennia (Santos 2014). The global mean sea level will continue to rise for many centuries, causing more frequent floods and submerging progressively larger areas of low-lying coastal areas and coastal systems, saltwater intrusion into surface waters and groundwater, increased erosion, and extremely negative social and economic impacts all over the world. The problem is aggravated by urbanisation, which is projected to lead to a situation where about two-thirds of the world population lives in cities, a high percentage being located in low-lying coastal zones. It is estimated that about 570 coastal cities with a total population of 800 million are vulnerable to a global mean sea level rise of 0.5 m by 2050 (C40 Cities 2019).

If one dares to take the long term view the situation looks even more challenging. According to Clark et al. (2016), it is very likely that, in the next 10 000 years, the global mean sea level will rise by between 25 and 52 m due to anthropogenic climate change. The two values correspond to scenarios with lower and higher greenhouse gas emission projections, respectively. Values of the global mean sea level rise below 25 m are very unlikely, due to the amount of greenhouse gases that have already been emitted and have accumulated in the atmosphere, given the slowness of the response of the hydrosphere and cryosphere to anthropogenic climate change forcing.

Only a tiny fraction of the human population knows about these projections, and if they were told, they would probably outrightly deny them or would consider them to be irrelevant and absurd because they refer to a time that is extremely far away. Furthermore, it is likely that some people believe that humankind will self-destruct before reaching the thirteenth millennium of the Common Era, so sea level rise projections for that time would be inconsequent. However, if we believe in the narrative of science and combine it with the sober expectation that *Homo sapiens* time will not end earlier than double the time in which we have benefited from the agricultural revolution, we come to the surprising realisation that humanity may witness a gradual rise in the global mean sea level of tens of metres over 10 000 years. It is an outlook we find very difficult to accept because it would reveal humankind's powerlessness in the face of a risk for which it is itself responsible. It is more likely that a response to that risk will be sought using large-scale climate geoengineering to cancel out some of the effects of anthropogenic climate change without, however, abandoning the current energy model based on fossil fuels. But will that be possible?

The comparison between what happened in the Eemian and the Earth system's current journey towards no-analogue states is likely to involve another violent aspect. There is evidence of very intense storms in the later part of the Eemian, at the close of the peak interglacial, when the global mean sea level rose by 6–9 m as major ice sheets melted or collapsed, while the Earth's atmosphere was less than 1 °C warmer than today (Hearty and Olson 2011; Hearty and Tormey 2017). Such evidence has been obtained from the physical characteristics of the geomorphology, stratigraphy, and sedimentology of the Eemian Marine Isotope Substage 5e sequence in the seashore of the Bahamas and Bermuda Islands. On the southern seashore of the Bermuda Islands, high waves and storm surges generated by hurricanes transported loose sediments onto the shore and piled up enough sand to bury established forests with trees that were 8–10 m tall (Hearty and Olson 2011). In the Eastern Bahamas, one finds sedimentary evidence of megaboulders transported by strong ocean waves on higher sea levels, lowland chevron storm beach ridges, and hillside run-up deposits on eolian ridges, which can be explained by very intense storms generating sustained long-period waves (Hearty and Tormey 2017). Hansen et al. (2016) have proposed that the very powerful storms at the interglacial–glacial transition ending Marine Isotope Substage 5e result from increases in atmospheric temperature gradients, eddy kinetic energy, and baroclinicity. During that period, the Southern Ocean surface was cooling, increasing ocean stratification, slowing deepwater formation, and increasing ice sheet melting. The global climate conditions at the end of the Eemian are far from being an analogue to the current anthropogenic forcing of the climate system. However, it is important to learn more about them, especially about extreme weather events, since the two periods share certain features.

It is very likely that storms will get stronger with anthropogenic climate change. Signs are currently emerging of more intense extreme weather events in the Northern hemisphere (Mann et al. 2017), as well as a higher percentage of very intense tropical cyclones (Elsner et al. 2008; Mendelsohn et al. 2012; Sobel et al. 2016). We still know little about these extreme storms, but they represent a high risk at present and especially in the future for human life and coastal assets in a world with a high population density in coastal zones.

Homo sapiens emerged around 200 000 years ago during the penultimate glacial period, and the beginning of our cultural evolution happened about 100 000 years ago with the emergence of symbolic thought and the capacity for spoken language, also during a glacial period. The relatively warm Holocene is an exception compared to the long glacial periods our species has lived through. Long-term climate models indicate that the Holocene interglacial period would have lasted for an abnormally long time, somewhere on the order of 60 000 years, compared to the duration of the Eemian and other, previous interglacial periods (Berger 2002), which typically lasted somewhere between 10 000 and 15 000 years. The cause of the different durations of the Eemian and the Holocene can be found in the variations in insolation at high latitudes of the Northern hemisphere, which are critical in determining the start of Milankovitch cycles. For example, at latitude 65°N, insolation will vary by only 25 W m^{-2} in the next 25 000 years, but at the start of the Eemian it varied by 110 W m^{-2} over 125 000 to 115 000 years (Berger 2002). What is behind this difference is the fact that the orbital eccentricity of the Earth is currently very low, which cushions the variations in insolation caused by the precession of the Earth's axis. In conclusion, the reason for the Holocene's climate stability is that we are close to reaching a minimum in the eccentricity of the ellipse traced out by the Earth in its orbit around the Sun.

Paleoclimatic models indicate that without anthropogenic interference in the climate, the current interglacial period would last roughly another 50 000 years, which is an exceptional situation within the context of the last million years (Ganopolski et al. 2016). With anthropogenic climate change, it is very likely that the interglacial period will last for at least 100 000 years, depending on the quantity of greenhouse gases that are emitted into the atmosphere and the response of polar ice to that forcing (Archer and Ganopolski 2005). Although there is a great deal of uncertainty about the duration of the anthropogenic extension of the current interglacial, it is well established that the Earth is on a trajectory of no-analogue states. This journey will have far-reaching consequences on living conditions for humankind and most living organisms in the biosphere.

7.3 Difficulties with the Definition of the Anthropocene as a Geological Epoch

Humans are interfering and changing the Earth system, leaving sustained marks on it. This interference is largely the result of three factors, whose interactions and feedback have a multiplying effect: the growth in global population, average per capita global consumption, and technological progress. These are the factors proposed by Holdren and Ehrlich in their well-known equation

$$I = PAT \, ,$$

where I represents human impact on the environment, P is the population, A is affluence, and T is technology (Holdren and Ehrlich 1974). The maximum impact that the environment can sustain is called the carrying capacity. If I is less than the carrying capacity then the environment can self-regenerate and the system is sustainable. If I exceeds the carrying capacity then the system is in an overshoot situation leading to more or less intense, persistent, or irreversible changes on a given time scale. Furthermore, overshoot tends to reduce the carrying capacity and may lead to various forms of collapse.

The Ehrlich–Holdren equation was presented at a time when negative predictions about the consequences for humankind of anthropogenic global environmental changes were beginning to be developed, and to reach the public media in the USA. However, there ensued a strong reaction denying such dire predictions, and techno-optimistic views about the future gained ground and became the dominant narrative in the country. In 2000, Ronald Bailey, the influential American libertarian science writer ridiculed the "infamous" Ehrlich–Holdren equation (Bailey 2000), pointing out that none of the dramatic impacts projected by ecologists since the 1970s had ever come true. On the contrary, Bailey considered that in 2000 the "the planet's future never looked better" (Bailey 2000). He describes himself as a "libertarian transhumanist", and belongs to the group of opinion makers that share the point of view that "technology will solve all environmental problems and that present environmental dilemmas are simply a necessary outcome of much needed economic growth" (Holt et al. 2009).

The use of an equation by environmental scientists to represent the drivers of the human impact on the environment was also criticised by a mathematician as "mathematical propaganda" (Koblitz 1981). A curious opinion when remembering that mathematics had first been used in the social sciences in the 18th century by Jacques Turgot and Nicholas de Condorcet, and that the field of social mathematics had been defined for the first time at the end of that century. To deny the legitimacy of using equations to describe environmental problems and their causes is probably another manifestation of the preoccupation in the USA that the treasured paradigm of "growth forever" might be endangered by environmental concerns seeking the respectability of mathematics.

The Ehrlich–Holdren equation represents an over-simplification of a complex global situation. The factors P, A, and T are not independent and it would therefore be more correct to say that I is a function of those three quantities. Human communities are very diverse around the world as regards their relation to and impact on the environment, and the equation is of course unable to account for this diversity and for the social, economic, political, and cultural influences that contribute to defining that impact. In spite of all these shortcomings the Ehrlich–Holdren equation identifies three of the essential drivers of the current global environmental crisis in a form that can be easily communicated.

It is consensual that humankind has an observable and ever greater impact on the environment at the local, regional, and global scales. This fact has been recognized since the end of the 19th century, when the Italian geologist Antonio Stoppani (1824–1891) spoke about a "new telluric force which in power and universality may be compared to the greater forces of earth," considering that it generated the "Anthropozoic era" (Crutzen 2002). It seemed therefore appropriate to identify a human-dominated geological epoch that would follow the Holocene, and this was called the Anthropocene by its proponents (Crutzen and Stoermer 2000; Steffen et al. 2011). The justification for this proposal was mostly based on environmental criteria, but there were also arguments that justified the identification of a new geological unit that could have various ranks: epoch, era, stage, or period.

From the point of view of geologists the term "Anthropocene" should only be used after being first formalised through the definition of a new stratigraphic unit with an identifiable beginning based on the human footprint on the Earth system. However, the word "Anthropocene" is already widely used in scientific books and articles and moreover in very diverse scientific and non-scientific environments. The word is frequently found in the media, in philosophical, political, and economic discussion groups, and its meaning is described in most dictionaries. Implicitly, it is the agreed way to refer to an epoch in which *Homo sapiens* is having an unprecedented impact on the Earth system, which is harmful to a large part of humankind and represents a high risk for its future.

The process to formally approve the use of the word "Anthropocene" as a geological epoch is the following, as described by a geologist: "The first step is to assemble a proposal to be submitted to the Subcommission of Quaternary Stratigraphy (SQS) and then to its parent body, the International Commission on Stratigraphy (ICS), for approval. If approved, the proposal can or cannot be ratified by the Executive Committee of the International Union of Geological Sciences (IUGS). If ratified, the new unit can be incorporated to the International Chronostratigraphic Chart (ICC), which is the international stratigraphic reference and is the basis for the Geological Time Scale (GTS), one of the great achievements of humanity, comparable to the periodic table of elements" (Rull 2018). The Anthropocene Working Group (AWG) was established in 2009, led by Jan Zalasiewicz, and since that time it has been working to produce a formal proposal to the ICS, defining the Anthropocene as a new epoch following the Holocene.

This project requires the identification of the Anthropocene from a stratigraphic viewpoint. Geological epochs are defined through precise boundaries left in accumulated sediments. For example, the transition from the Pleistocene to the Holocene is defined by the boundary between two layers in an ice core extracted from Greenland (Walker et al. 2009). After a long and intense debate, the AWG made a preliminary proposal to the International Geological Congress, at a meeting in Cape Town in August 2016, that the start of the Anthropocene epoch should be in the 1950s, when the first thermonuclear bomb tests were held. These blasts dispersed long-term radioisotopes in the atmosphere which, after some time, have been deposited in the soil and can be easily identified in stratigraphic analyses. They could be used to define the beginning of the Anthropocene (AWG 2019). However, the AWG has not yet made its final proposal to the SQS and the ICS.

Other ideas to define the start of the Anthropocene include the spread of agriculture, the Columbian Exchange of species between continents that followed the Age of Discovery (Crosby 1972), the Industrial Revolution, and the Great Acceleration (Waters et al. 2016). If the rapid transformations that are taking place in the Polar Regions are preferred, it is found that recent lake sediments in the Arctic have clear signatures of stratigraphic transitions from proglacial sediments to nonglacial organic matter, which could be used to demarcate the beginning of the Anthropocene.

Some scholars are of the opinion that it is not possible to define a single date for the start of the Anthropocene because of the diachronous character of the physical strata that reveal a human impact on the environment. The number of scientific papers and books that have been written by the AWG about the AWG's work and about defining the Anthropocene as a geological epoch is staggering in relation to the obvious human footprint on the global environment, although these things are very important to advance scientific knowledge. Nevertheless, after more than ten years of dedicated work by the members of the AWG, the success of their proposal is far from being guaranteed, as the AWG emphasizes in their last report (AWG 2019). If it is not successful some geologists consider that the term "Anthropocene" should be "removed from our dictionaries", implying that it should cease to be used (Rull 2018).

What are the reasons for the present situation, if there is such a large body of evidence for a stratigraphic basis for the Anthropocene (Waters et al. 2014, 2016)? The reason is likely to be that there is a strong opposition to the formal recognition of the Anthropocene as the geological epoch following the Holocene. This opposition is muted and scarcely visible, making it difficult to identify the true reasons behind the apparent deadlock. Glimpses of such reasons come mostly from American geologists, who point out that the project of defining the Anthropocene as a geological epoch is not genuinely scientific, but has political and environmental motivations (Finney and Edwards 2016; Edwards 2015).

The main reason why the Anthropocene cannot be identified as a geological epoch is that the "documentation and study of the human impact on the Earth system are based more on direct human observation than on a stratigraphic record" (Finney and Edwards 2016). We shall probably have to wait tens or hundreds of thousands of years, or maybe even millions of years, before geologists, by studying the stratigraphic record, will be able to identify a rock record that deserves to be considered a specific segment of geological time, distinct from the Holocene, according to the best procedures and the best current knowledge at that time, so that it can be included in the geological time scale. It will be utterly irrelevant whether that rock record does or does not carry the footprint of human activities. It is assumed implicitly that, whenever such a rock record is identified, humankind will continue to benefit from "growth forever", but will have managed to drastically reduce its impact on the environment. If that happens, there will be no ground to consider the Greek word *anthropos* to define the geological epoch following the Holocene. The Anthropocene will just have been a "holistic concept that involves time, place, human cultural attainment and dominance, and a variety of environmental effects" (Edwards 2015).

The main reason for the protracted process of trying to define a geological epoch that refers explicitly to the human impact on the environment is the cultural war to which Autin and Holbrook refer to in their article "Is the Anthropocene an issue of stratigraphy or pop culture?" (Autin and Holbrook 2012). They consider the Anthropocene to be an iconic term in pop culture which should not be elevated to the point of becoming evidence to change formal stratigraphic practice. The concept of the Anthropocene "is a focal point in the cultural wars over the recognition and interpretation of environmental process" (Autin and Holbrook 2012). Again one finds an opening to value and to incorporate those points of view in which the relevance of the environmental process is not recognized or is considered to be marginal.

Let us briefly review what the AWG considers in its preliminary proposal (AWG 2019) to be "the sharpest and most globally synchronous signals", for the definition of the beginning of the Anthropocene. Trinity, the first atomic bomb, was created by the Manhattan Project and blasted into the atmosphere in Alamogordo, New Mexico, USA, on 16 July 1945. The explosion launched into the air ash containing long-lived radioactive elements such as strontium-90 (^{90}Sr) and cesium-137 (^{137}Cs). These were only deposited locally in the soil due to the relatively low power of the device. The situation changed in 1954 when the USA launched Operation Castle. This involved detonating a series of thermonuclear bombs in the Bikini Atoll, one of the Marshall Islands in the Pacific. The first, called Castle Bravo, was detonated on 1 March 1954. It was predicted to have a power equivalent to 6 megatons of TNT but, due to miscalculations, it had a power of 15 megatons, around 1000 times the power of the Hiroshima and Nagasaki bombs.

The explosion created a mushroom-shaped cloud measuring 40 km in height and 100 km in diameter and generated residual radioactive material that turned into what is called nuclear fallout, formed of very small particles measuring between 10^{-7} m and 2×10^{-2} m in diameter. Part of this material fell directly back to the ground, but due to the strong upward air movements in the atmosphere generated by the explosion, the smaller particles went up into the stratosphere. Here they were dispersed by the general circulation of the atmosphere and eventually deposited on the Earth's surface over the following weeks, months, and years.

Nuclear fallout from the Castle Bravo test began by affecting the Rongelap, Rongerik, Alinginea, and Utirik atolls of the Marshall Islands, located to the east of Bikini, causing the people there to develop serious health problems (Simon et al. 2010). Sometime later, it contaminated practically the entire world, especially Australia, Japan, the United States of America, India, Europe, and almost all the islands of Oceania.

The main radiobiological danger of global nuclear fallout is that it contains radioisotopes that have long half-lives, such as ^{90}Sr and ^{137}Cs, which accumulate in the human body when people ingest food contaminated by the fallout. On 22 November 1954, the Soviet Union carried out its first fully successful thermonuclear test in the Semipalatinsk Test Site, also known as the Polygon, in the northeast steppe of today's Kazakhstan. The health impacts on the regional population of radiation exposure produced by the nuclear tests at the Polygon were very serious and are still being worked out (Yan 2019). On 8 November 1957 the United Kingdom made its first successful test on Christmas or Kiritimati Island, a coral atoll island in today's Republic of Kiribati in the Pacific. Most of the relevant data on the health impacts of the nuclear tests on the population of Christmas Island remains classified.

The fallout created by fission and fusion nuclear weapon testing in the atmosphere and in the ocean, with a total of roughly 520 bombs, has become a serious global problem that has increased radiobiological risk around the world. It has only been resolved by implementing agreements on nuclear weapons and tests among the superpowers, negotiations for which began in the 1980s.

The radioisotopes from nuclear fallout deposited in the environment by nuclear tests carried out in the atmosphere are one of the most significant signs of anthropogenic activity at the global level, and their beginning is clearly located in time and relatively easy to identify locally. This is the case for plutonium-239 (^{239}Pu), with a half-life of 24 100 years, produced in the explosion by the absorption of neutrons by uranium-238 (^{238}U), and carbon-14 (^{14}C), a naturally occurring element commonly found in the environment, with a half-life of 5730 years. The abundance of these two radioisotopes in the environment at global level began to rise in 1951 and reached a peak in 1964, known as the "bomb spike" (Hancock et al. 2014). Carbon-14 is continually produced in the upper troposphere and lower stratosphere through the collision of cosmic rays with ^{14}N nuclei in nitrogen molecules and then absorbed by living organisms through photosynthesis.

The bomb spike corresponded to an abnormal amount of ^{14}C in the troposphere, so all living beings, including humans, born around 1950, contain an excess of ^{14}C in their bodies compared with the natural level of this radioisotope. Carbon-14 is harmful to living organisms because it emits beta- radiation. Exactly how dangerous it is depends on the quantity of the radioisotope present in the organism. The chemist Linus Pauling (1901–1994), a pioneer in the movement to abolish nuclear weapons, was one of the few scientists who first studied and warned about the effects of the bomb spike on human health, doing so in an article published in 1958 (Pauling 1958).

7.4   The Materials of the Technosphere

Minerals and their distribution constitute one of the most extraordinary aspects of the human footprint on Earth. They are usually defined as crystalline compounds with a well-defined chemical composition and a specific crystal structure. Their diversification throughout the Earth's history started with photosynthesis by cyanobacteria, which began around 2.7 billion years ago and emitted huge quantities of oxygen into the atmosphere, finally changing its composition. This global change, known as the Great Oxidation Event or Great Oxygenation Event (GOE), began at the beginning of the Proterozoic, between 2.4 and 2.2 billion years ago. It gave rise to a series of chemical, physical, and geological processes that caused the appearance of many new minerals. It is estimated that around two thirds of the roughly 5000 species of minerals that naturally occur on Earth were created at that time (Hazen et al. 2008). It was then that the great iron oxide deposits were formed. They continue to feed a growing demand for iron to support economic growth throughout the world.

The first metals to be used by humans were found in an unreacted state and are called native metals. These include gold, silver, and copper. In the Neolithic, from about 7000 BC onwards, copper was used systematically to produce knives and sickles. As many other metals started to become useful, humankind became progressively responsible for the production of new minerals and for the distribution of chemical elements and minerals around the globe.

Aluminium, which was unknown in its pure metallic form before the 19th century, has seen a total production of around 500 Gt, 98% of which has occurred since 1950 with the beginning of the Great Acceleration (Zalasiewicz et al. 2014). Concrete, a composite of three basic components, water, aggregate (rock, sand, or gravel) and a cement, which is usually Portland cement today, was first used by the Egyptian and Greek civilisations, but more frequently by the Romans. Even today, we can see how the Romans used concrete in their buildings, by observing the Pantheon's dome and the Baths of Caracalla, both in Rome. Its use grew exponentially and reached a total amount estimated at about 500 000 Gt, enough to cover the whole of planet Earth with a layer of one kilo per m^2 (Waters et al. 2016).

The vast majority of minerals—iron, copper, mercury, nickel, zinc, lead, silver, gold, and many others—are being extracted from the subsoil in quantities that have on average grown exponentially since the mid-20th century to later be dispersed around the surface of the Earth for a wide range of purposes. Rare-earth elements are a set of seventeen chemical elements in the periodic table, which include the fifteen lanthanides, together with scandium and yttrium, which have similar chemical properties. These elements, which are relatively abundant but widely dispersed in the Earth's crust and rarely found concentrated in rare-earth minerals, were first explored from the end of the Second World War. Their dispersion, a consequence of their geochemical properties, makes their exploration more difficult and expensive.

Rare-earth elements have become very important for the current global economy because they are used in high-technology products like components of smartphones, digital cameras, computer hard disks, DVDs, flat screen televisions, electronic displays, rechargeable batteries, catalytic converters, magnets, fluorescent and light-emitting-diode (LED) lights, and many types of military hardware such as lasers, precision guided munitions, and guidance and control systems to steer missiles and bombs. The country with the highest reserves is China, which has also been the largest producer since the USA ceased to be the leading producer in the 1980s. China is carefully controlling its production of rare-earth elements to defend its geostrategic objectives.

A large proportion of intensively used metals are not recycled and often end up spread around the environment. Examples are platinum, rhodium, and palladium, used in catalytic converters of motor vehicles, which accumulate in soils near roads and motorways (Jarvis et al. 2001). The exploration of minerals is one of the human activities with the strongest negative impact on the environment, affecting water quality, destroying habitats for sometimes rare or endangered species, and profoundly changing the landscape in vast regions of the world. It is calculated that mineral extraction in mines around the world moves around 57 billion tonnes of sediment every year, approximately three times more than the total sediment transported every year by all the rivers in the world (Douglas and Lawson 2000).

Another aspect of the human intervention in the field of minerals is the creation of an enormous variety of synthetic crystalline compounds. It is calculated that human science and technology have already created around 180 000 new inorganic crystalline compounds, some probably new to the Solar System and even further afield. This anthropogenic diversification event took place over a period of time that is much shorter than the Great Oxygenation Event (Hazen et al. 2008). The International Mineralogical Association describes minerals simply as inorganic crystalline compounds generated in natural processes, including processes said to be natural but mediated by involuntary human intervention. There are 208 minerals in this latter category created by humankind's indirect action (Hazen et al. 2008). One famous example is calclacite, a very rare mineral discovered in 1959 on a limestone sample stored in the Brussels Museum of Natural History, where it was formed by the action of acetic acid released by the oak wood of the cabinet where the stone had been kept for many years.

Other materials used every day by almost all human beings are synthetic or semi-synthetic organic polymers, commonly called plastics. They are mostly derived from fossil fuels, but some come from renewable materials such as cellulose. The annual production of plastics has grown exponentially, having increased by a factor of around 20 since the 1960s to reach 311 million tonnes in 2014 (WEF 2016). It is estimated that global plastic production will reach 900 million tonnes each year in 2050, absorbing around 20% of all oil consumption. A recent estimate shows that globally around 8.3 billion tonnes of plastic have been produced since large-scale production began in the 1950s (Geyer et al. 2017). In 2015, 6.3 billion tonnes of plastic were thrown away as waste, of which 9% was recycled, 12% incinerated, and 79% accumulated in landfills, waste dumps, and in the environment. This accumulation in the environment will reach roughly 12 billion tonnes in 2050 (Geyer et al. 2017).

Plastics are dispersed in the terrestrial and marine environments, principally through rivers, lakes, and coastal areas in the form of both plastic objects and microplastics, which are produced directly, for example in cosmetics manufacturing, or may result from the fragmentation of larger pieces through the influence of UV radiation. Because plastics degrade from exposure to sunlight, they tend to last longer after being dumped into the ocean, where they are better protected from sunlight. Depending on the type of plastic, they are likely to remain in the ocean from decades to more than a thousand years, producing a cumulative problem with serious consequences for a wide range of living organisms, including fish, birds, turtles, and cetaceans. They frequently cause animals to die due to suffocation, ingestion, infection, or entanglement (Laist 1997; Derraik 2002).

Furthermore, plastics in the ocean create a new ecosystem of microbial communities that inhabit the biofilm formed on the outside of plastics, called the plastisphere. Microbes in the plastisphere can be transported long distances, making them potentially invasive species with unknown consequences for other marine ecosystems. The environmental impacts of this plastisphere are not yet fully understood (Zettler et al. 2013).

If measures are not taken to control the amount of plastic accumulated in the ocean, which is currently estimated at around 1.5 billion tonnes, plastics will exceed all fish, in terms of mass, in 2050 (WEF 2016). It is difficult to foresee all the implications of this interference with the ocean. One problem is that plastics contain various chemical substances called additives for enhancing polymer properties and prolonging their life, such as stabilisers, plasticisers, and pigments (Hahladakis et al. 2018). These substances produce waste management and pollution challenges and they likely have adverse effects for human health and for other living organisms as they spread through the biosphere over time.

The intensive use of fossil fuels since the beginning of the Industrial Revolution has inflicted many serious impacts on human health and the environment. Such externalities happen at each point in the fossil fuel life cycles, including the extraction process that can generate air, ocean, water, and soil pollution, and the transportation process, which can cause air pollution and dangerous accidents and spills. Burning fossil fuels leads to the emission of toxins, carbon dioxide, sulphur dioxide, nitrogen oxides, mercury, and particulate matter, and in particular black carbon, which have impacts on human health and the environment, especially on climate, as is well known.

Black carbon is composed of carbon particles resulting from the incomplete combustion of hydrocarbons and contains polycyclic aromatic hydrocarbons and heavy metals, which are toxic. Fossil fuel atmospheric emissions account for about 65% of the observed mortality caused by air pollution (Lelieveld et al. 2019). It is estimated that a global phasing out of fossil fuel use could avoid worldwide an excess mortality of about 3.61 million per year from outdoor air pollution.

Wildfires and more generally biomass burning occurs frequently around the world and produces aerosols and gases that also have harmful effects on health and the environment. Climate change is increasing the risk of wildfires particularly in Mediterranean type climates and in high latitude forests. A wide range of chemical compounds and aerosols are produced in the combustion of fossil fuels and biomass burning and some of them have a significant impact on the global climate. Such an impact is determined by the way they interact with electromagnetic radiation, or more specifically how they absorb and emit radiation at the different wavelengths of the electromagnetic spectrum. Some compounds cause positive radiative forcing at the boundary between the troposphere and the stratosphere, called the tropopause, and this tends to warm the troposphere. Others cause negative radiative forcing, which tends to cool the troposphere.

Anthropogenic aerosols are responsible for radiative forcing of the climate through multiple processes which can be grouped into two main types: aerosol–radiation interactions and aerosol–cloud interactions (IPCC 2013). Black carbon is particularly important because it is highly efficient at absorbing solar radiation, causing high positive radiative forcing. Its particles have a diameter less than about 2.5 micrometres. Around 40% of black carbon emissions come from burning fossil fuels, 20% from burning biofuels, and 40% from wildfires (Ramanathan and Carmichael 2008). Among fossil fuels, the main sources of black carbon are coal and diesel (Bond et al. 2007). Annual black carbon emissions have increased in a roughly linear way from approximately 1 billion tonnes in 1850 to 4.4 billion tonnes in 2000 (Bond et al. 2007).

The positive radiative forcing caused by black carbon emissions is currently highly significant when compared to the radiative forcing driven by the main greenhouse gases with anthropogenic emissions, viz., CO_2, CH_4, and nitrous oxide (N_2O). Radiative forcing by black carbon in 2011 was estimated at 0.4 Wm^{-2}, although with a large uncertainty, while forcing from the greenhouse gases mentioned, namely, CO_2,

CH_4, and N_2O, had values of $1.82\,Wm^{-2}$, $0.48\,Wm^{-2}$, and $0.17\,Wm^{-2}$, respectively (IPCC 2013). The black carbon radiative forcing estimate is not much lower than forcing caused by CH_4, which has anthropogenic emissions mostly from paddy fields where rice is grown, enteric fermentation emissions from ruminant animals raised for their meat and milk, biomass burning, and anaerobic decomposition of organic waste in landfills.

Black carbon also further intensifies another effect that speeds up anthropogenic climate change. The growing amount of particulate matter that circulates in the atmosphere, particularly dust from arid soils and deserts, black carbon, and other organic aerosols, ends up being deposited on the Earth's surface, particularly on mountain glaciers and the polar ice sheets, more in the Arctic due to the higher number of sources in the Northern hemisphere. This deposition generates a phenomenon known as "black snow", which causes a reduction in albedo (Dumont et al. 2014). At places where snow is darker due to the higher number of particles deposited, greater absorption of solar radiation increases the temperature and accelerates ice melting. It is very likely that the successive layers of black carbon and other organic aerosols of anthropogenic origin that are deposited on the Earth's surface will last for many centuries and millennia, leaving another stratigraphic Anthropocene marker. In conclusion, anthropogenic emissions of gases and aerosols have changed the composition of the atmosphere and have interfered with the climate system in a significant way. These global environmental changes are interfering with other global changes in ways that are still not entirely understood.

One of them is biodiversity loss. Humankind has destroyed part of the biodiversity of the biosphere by degrading or destroying many habitats and driving a significant number of species to extinction. Furthermore, the vital requirement of guaranteeing food security for a growing human population with increasing standards of living has been met by preferentially using a restricted number of genetically uniform, high-yield varieties of crops and livestock. Biodiversity for food and agriculture, also called agrodiversity, defined as the subset of biodiversity that contributes in one way or another to agriculture and food production, makes production systems and farmers more resilient to shocks and stresses, including climate change (FAO 2019a). However, there is evidence that the proportion of livestock breeds at risk of extinction is increasing and some 75% of plant genetic diversity has been lost as farmers worldwide are no longer cultivating their many local varieties and landraces. Nearly a third of ocean fish stocks are overfished and a third of freshwater fish species are considered threatened (FAO 2019a). Terrestrial ecosystems that deliver numerous services essential to food and agriculture are declining rapidly worldwide, including those that contribute to the supply of freshwater and provide habitat for species such as fish and pollinators.

Biodiversity at the ecosystem level is decreasing as a result of the crops and livestock that have been introduced around the world since the beginning of the Columbian Exchange and the worldwide proliferation of invasive species. While anthropogenic interference in the biosphere shows a tendency for biodiversity loss and therefore tends to make it more homogeneous, humans have increased diversity globally in the form of inorganic matter, particularly regarding crystalline compounds. The opposing tendencies of anthropogenic interference in the living and non-living realms reveals the nature, the likely future, and the risks of human progress.

The human subsystem of the Earth system is constantly generating significant amounts of material products to ensure the flows of energy and physical goods needed to guarantee the stability of the current financial and economic system. The overall set of these highly diverse and complex physical materials either in use or discarded, which are amassed in urban, rural, coastal, and maritime regions, constitutes the physical component of the human subsystem, or physical technosphere. Its total mass was recently estimated at 30 trillion tonnes (Zalasiewicz et al. 2017), which is around 10 000 times more than the total mass of human beings and 200 million times less than the mass of the Earth. There is therefore still a great deal of matter on Earth that humans may extract, remove, transform, use, and reject.

The artefacts produced in the human subsystem by the various contemporary and future technologies will eventually be discarded at the end of their useful life, but some will remain in the environment for long periods of time to form technofossils. These artefacts range in size from urban conglomerations, to buildings, cars, computers, other electronic devices, bottles, glasses, and cutlery, to nano-artefacts and anthropogenic particulate matter. The evolution and diversification of the artefacts belonging to the physical component of the human subsystem is much faster than that of living organisms in the biosphere. Technofossils are likely to increase in number and to diversify very rapidly because the technosphere is unable to recycle its artefacts, unlike the biosphere, which recycles almost all its constituent matter by recreating and maintaining complex organic systems (Waters et al. 2014). Although it is presently very difficult to evaluate the diversity of technofossils, it is likely that it will eventually become greater than that of fossils formed throughout the whole history of the biosphere. The human subsystem is carrying out the deepest transformation that the Earth system has had since the origin and flourishing of the biosphere.

7.5 Human Transformation of the Biosphere. Can We Avoid a Critical Transition?

An important question about the future is trying to foresee the states of the biosphere that could emerge from the activities of contemporaneous and future social generations and their relationship with the human subsystem. The biosphere, a concept introduced by the English geologist Eduard Suess (1831–1914) and by the Russian physicist Vladimir Vernadsky (1863–1945), is the set of living organisms, or biota, that interact with the Earth system and live in a thin layer that includes the atmosphere, hydrosphere, cryosphere, and upper part of the lithosphere. The combination of the biosphere and the physical systems that support it forms the global ecological system or ecosphere, as mentioned before.

Ecological systems at the local, regional, and global scales constitute examples of a dynamical system, a mathematical concept describing a system whose state, defined by a set of variables, evolves over time in state space according to some fixed rule. The concept has applications in many fields besides mathematics, such as physics, climate studies, chemistry, biology, medicine, engineering, finance, and economics. Ecological systems, as already mentioned, are also complex systems, since they are built from a large number of interacting subunits whose interactions result in some kind of collective behaviour.

Complex dynamical systems can undergo critical transitions where the system shifts abruptly and irreversibly from one stable dynamical regime to another at a critical threshold called the tipping point. The concept of critical transition is very general and can be applied to many natural and social phenomena such as the extinction of a biological species, the ecological collapse of a river or a lake, and the financial collapse of an institution. It is difficult to predict when such a tipping point will be reached but recent work has shown that in many cases it is possible to identify signs that it is approaching (Scheffer et al. 2009). One of these signs is the decrease in recovery rate to equilibrium, or a critical slowing down of the system's capacity to recover.

Analysis of fossil records shows that, since the emergence of life on Earth more than 3.77 billion years ago (Dodd et al. 2017), and possibly even 4.1 billion years ago (Bell et al. 2015), there has been a consistent trend towards increased biodiversity measured in terms of the number of phyla. However, the history of the biosphere is complex and increasingly difficult to decipher as we move away from the present.

The most recent global critical transition, and the one about which the most evidence and knowledge has been accumulated, happened during the passage from the last glacial period to the current interglacial period, which began around 12900 years ago and lasted for approximately 3300 years. The transition was accompanied by the extinction of some species, by speciation, and by changes and redistributions in biodiversity at the local and regional scales.

Over very long timescales during the Phanerozoic, which began 541 million years ago, the biosphere was deeply affected by five great mass extinctions events. At the beginning of the Cambrian, the first period of the Phanerozoic, the "Cambrian explosion" took place, one of the most extraordinary evolutionary events ever to occur in the history of life on Earth, during which most animal phyla emerged in a relatively short period of time of roughly 30 million years. Since that time, there have been oscillations in the number of genera and species caused by changes in the global climate, by the relative motions of the continents and changes in their configuration, by orogeny associated with plate tectonics, and by mass extinction events (Rohde and Muller 2005). These took place roughly 443, 359, 251, 200, and 65 million years ago at the end of the Ordovician, Devonian, Permian, Triassic, and Cretaceous, respectively. In all cases, the number of species fell by more than 75% in a time interval of 2 million years or less.

However, the biosphere always managed to regenerate itself and return to comparable or higher levels of biodiversity compared to those before over periods of time of at least 10 million years, by exploring new evolutionary paths. Most explanations for this long delay are based on environmental factors, but recent studies on evolution after the mass extinction at the end of the Cretaceous, using the morphological complexity of planktic foraminifers, suggest that evolution limits the speed of biodiversity recovery (Lowery and Fraass 2019). Research has shown that ecological complexity was recovered before the increase in the number of genera and species. The species that survived the impact of the asteroid bounced back rapidly to fill the available ecological niches, but further significant speciation had to wait for evolutionary innovations leading to the construction of a new morphospace. This is a much slower process that controls the speed of biological diversification after a mass extinction event. If the current pace of anthropogenic biodiversity loss is allowed to transform itself into the sixth mass extinction (Novacek 2001; Novacek and Cleland 2001), we already have an estimate of 10 million years as the time required to return to the levels of biodiversity at the beginning of the Holocene.

The formation of a new species is a relatively slow process that may take hundreds of thousands of years (Weir and Schluter 2007). Biological diversity finds its origin in a continuous process of adaptation and evolution through speciation and extinction mechanisms (Coyne and Orr 2004). It is thought that around 99% of the 4–10 billion species that are estimated to have appeared in the biosphere have already become extinct (Novacek 2001). All species become extinct sooner or later. Their duration is highly variable and depends on the taxonomic group (Lawton and McCredie 1995). Generally, extinction is caused by the inability to adapt to environmental changes that affect the habitat of the species. If the change happens more quickly than the species can evolve, the likely outcome is extinction.

Some species have a long duration, such as the cyanobacteria and their colonies, called stromatolites, which appeared 3.5 billion years ago and can still be found in areas of high salinity or in zones with high sediment influx. Another example are the ctenophora, also known as comb jellies, a phylum of invertebrate marine animals that live in the oceans around the world. Some of these are bioluminescent, with "combs" of cilia whose oscillatory motions allow them to move forward. They emerged more than 500 million years ago and continue to be abundant. Chimaeras are cartilaginous sea fish belonging to the Chimaeriformes order, close to sharks and rays, which emerged roughly 420 million years ago in the Silurian (Inoue et al. 2010). These have survived four mass extinctions and can still be found in the ocean, particularly in the deeper waters.

In this context, it is useful to define a background extinction rate, usually represented as the average number of extinctions per million species and per year (E/MSY) in normal situations, meaning outside the mass extinction events. To calculate the background rate, two difficult things need to be done: the total number of biological species must be estimated, and the overall number of extinctions per year established. The figures published for these variables are frequently different and research results are also far from convergent.

Work by taxonomists over the last 250 years or so has made it possible to identify and catalogue around 1.2 million species. Nonetheless, the total current number of species is much greater, estimated at around 8.7 million (Mora et al. 2011), of which 2.2 million are marine species. Many of these species are still unknown and will become extinct before we identify them. Some may have a potential contribution for human well-being that will be lost forever. Recent estimates place the background extinction rate between 0.023 and 0.135 E/MSY and the speciation rate between 0.05 and 0.2 new species per million species and per year (De et al. 2015). If we choose 0.1 E/MSY as a representative value for the background extinction rate, it can be concluded that the current extinction rates are 1000 times higher than natural background rates and are likely to become 10000 higher in the future (De et al. 2015; Pimm et al. 2014).

These results refer to global biodiversity. At the local and regional levels, there are cases where the number of species has increased due to the introduction of non-native species or to efforts to reintroduce and conserve endangered species. The increase in the overall extinction rate is mainly caused by human habitat fragmentation and destruction, which reduces the size of species populations, disrupts the structure of the species communities, and by the introduction of non-native species, also known as alien or exotic species, some of which become invasive. Deforestation in the tropics has been one of the major causes of current extinctions. But there are other causes such as hunting, illegal trade in endangered species, and pollution of the air, soils, terrestrial water resources, and the ocean through the introduction of unnatural chemicals, heavy metals, plastics, fertilizers, and synthetic pesticides and herbicides. Biodiversity decline is also caused by the introduction and proliferation of pathogenic agents into the environment, the disruption of biogeochemical cycles, and climate change.

In the past much attention has been given to biodiversity loss among terrestrial and marine vertebrates (Pimm and Raven 2000; Ceballos et al. 2017), but recently concern has also been extended to invertebrate taxa. It has been established that the biodiversity of insects, which constitute the largest group within the arthropod phylum, is threatened worldwide (Sánchez-Bayo and Wyckhuys 2019). Lepidoptera, Hymenoptera, and Coleoptera, especially dung beetles, are the most affected terrestrial taxa, while Odonata, Plecoptera, Trichoptera, and Ephemeroptera are the most affected aquatic taxa (Sánchez-Bayo and Wyckhuys 2019). At the same time some more adaptable generalist species of insects are occupying the vacant niches and becoming more abundant. The current rates of decline may lead to the extinction of 40% of the world's insect species over the next few decades.

In this context, it should be remembered that insect pollination plays a critical role in supporting global food security and is estimated to be responsible for about 35% of food supply. A decline in bee populations has been observed worldwide (EASAC 2015; Meixner and Le Conte 2016). In the apple and pear orchards of south-west China, the lack of bees to pollinate the trees has been temporarily solved through hand-pollination by farmers and their children. The need to solve the problem has already attracted investment for research projects that seek to create robotic bees or robobees (Potts et al. 2018; Jafferis et al. 2019; SSRG 2019), an example of the current effort to achieve weak sustainability by substituting human capital for natural capital.

The contemporary systemic and cumulative anthropogenic pressures on the biosphere have a higher intensity and rate of change than those that resulted from the last glacial–interglacial transition. Although it is likely that we are moving towards a global critical transition in the biosphere, there are as yet no reliable frameworks or criteria to forecast when it will begin and what particular characteristics it will have (Barnosky et al. 2012). There is a need to improve the detection of early warning signs of critical transitions on the global, regional, and local scales. Humankind is still a long way from having caused the extinction of 75% of species, but given the current pace it seems likely that the sixth mass extinction has indeed begun (Novacek and Cleland 2001; Barnosky et al. 2011). According to Barnosky et al. (2011), the current extinction rate is higher than those that caused the previous five mass extinctions and if it continues uncontrolled it could produce a mass extinction of similar magnitude in just three centuries. The percentages of threatened species for various groups are as follows: amphibians 41%, selected reptiles 34%, reef-forming corals 33%, mammals 25%, sharks, rays, and chimaeras 30%, birds 13%, conifers 34%, and selected dicots 36% (IUCN 2014). Tropical forests are likely to be the biome where the anthropogenic mass extinction caused by systematic deforestation over many decades will first be identified (Alroy 2017).

What should be done to prevent a planet-wide critical transition? Following the analysis of Barnosky et al. (2012), the most important action is "global cooperation to stem current global-scale anthropogenic forcings. This will require reducing world population growth and per-capita resource use; rapidly increasing the proportion of the world's energy budget that is supplied by sources other than fossil fuels while also becoming more efficient in using fossil fuels when they provide the only option; increasing the efficiency of existing means of food production and distribution instead of converting new areas or relying on wild species to feed people; and enhancing efforts to manage as reservoirs of biodiversity and ecosystem services, both in the terrestrial and marine realms, the parts of Earth's surface that are not already dominated by humans."

The concentration of terrestrial biomass in plants and animals useful for crop and livestock farming began with the Agricultural Revolution which first occurred in the Fertile Crescent of the Middle East, about 11 000 years ago. Now more than three quarters of the terrestrial biosphere has been transformed by human populations into anthropogenic biomes or anthromes (Ellis and Ramankutty 2008). Eighteen types of anthromes have been identified (Alessa and Chapiniii 2008), including dense urban areas, dispersed population settlements, croplands, rangelands, and forest plantations. About 43% of the Earth's land surface has been modified to be used for agriculture or for urban areas. The remaining area is fragmented by a growing global network of roads into roughly 600 000 fragments (Ibisch et al. 2016). More than half have an area of less than 1 km^2, only 7% have an area of more than 100 km^2, and only around a third are free of regular human usage for a variety of purposes, especially agriculture.

Land fragmentation harms ecosystem functions and services and reduces their biodiversity. If adequate ecological protection measures are not taken, the situation will get worse, since a 60% increase in the size of the road network is foreseen between 2010 and 2050 (Dulac 2013). In the ocean, it is calculated that around three quarters of continental platforms have suffered the effects of bottom trawling. Every year, humankind removes an average of 90 million tonnes of fish and other edible living organisms from the ocean legally and around 11–26 million illegally (FAO 2014; McCauley et al. 2015). On land, the analysis of paleobiological data from the youngest sedimentary record provide information on the status of species, communities, and biomes over the last few decades to millennia and on how they are responding to natural and anthropogenic environmental change (Kidwell 2015).

Another important feature of the current biosphere is the deep recomposition and restructuring of ecosystems and biomes resulting from the displacement of species outside their original native ranges caused by voluntary human action, as in the case of crops and livestock, or involuntary human action (Williams et al. 2015). This anthropogenic intervention has contributed to the naturalisation outside their native range of about 13 000 vascular plant species, representing 4% of the world's flora (van et al. 2015). In general, exotic species do not experience the constraints imposed on native species in land areas where they have been introduced and are cultivated by humans.

There have been many natural displacements and migrations of species in the history of life, but the mobility of terrestrial species was heavily conditioned by the relative position and separation between continents. The proliferation of non-native species throughout the world grew faster during the Great Acceleration and has caused a homogenisation of the biota at global scale that is likely to be unprecedented in the history of the biosphere due to its speed and magnitude. The biogeographical diversification of flora and fauna that began roughly 200 million years ago, when the continents that were united in the supercontinent Pangaea began to separate, is now being reversed.

Many non-native species are invasive and harm indigenous ones, thereby becoming factors of ecological change (Didham et al. 2005) that influence the evolutionary journey of the ecosystems in which they proliferate. In most cases, climate change is likely to favour the spread, establishment, and survival of invasive species (Bradley et al. 2010; Huang et al. 2011).

The most remarkable paradigm in the process of ecological change caused by the migration of species is that of *Homo sapiens* itself, which is a non-native species outside Africa that ended up invading all the continents and adapting to the full range of climates and biomes from hyper arid deserts to polar tundras, and from temperate climates to alpine environments and tropical rain forests. In this process, human beings took with them a wide variety of plants, crops, livestock, and domestic animals around the globe, allowing them to develop the hunting, fishing, farming, and forestry anthromes that are most useful and best suited to the new habitats, as well as many other species, some of which became invasive and had a significant impact on local biodiversity.

One famous example is the cat (*Felis silvestres catus*) and the mouse (*Rattus norvegicos*), particularly on islands (Medina et al. 2011). In the globalisation process, humankind also spread around the world a wide range of pathogenic agents, transported by people themselves and by the species that accompanied them, thereby causing serious negative impacts on human health and on the flora and fauna of the newly occupied regions.

Anthropogenic global changes are also inducing evolutionary changes in other species through changes in their environment, such as herbicide resistance in plants, pesticide resistance in insects, and changes in the growth rate of fishes caused by over-fishing. Another example in the human health domain is the evolution of antibiotic resistance in bacteria, which is a continuing healthcare risk on a global scale. Antibiotic resistance is a very good example of Darwinian evolution by natural selection. When an antibiotic kills most of the bacteria in an infected patient, a few individual bacteria that carry a gene that allows them to survive reproduce, passing that gene on to their descendants. In this way the patient remains with an infection that is now resistant to that particular antibiotic and can be passed on to other patients. The important point is that, in the 160 years or so since Darwin proposed his theory, our knowledge of evolution by natural selection, in particular in bacteria, has improved significantly, so medical researchers and doctors are better prepared to counteract the adaptation of bacteria to new antibiotics. The surprising issue in this regard, however, is the creationist opinion, particularly strong in the USA, which denies that antibiotic resistance in bacteria is an example of and follows the laws of Darwin's theory of evolution (Bergman 2003; Singh et al. 2016; Bohlin 2017).

7.6 Human Appropriation of Net Primary Productivity and Human Interference in the Carbon Cycle and in the Earth's Energy Balance

Material and energy flows in the biosphere have been modified by human activities. The biosphere can be conceptualized as a system of material and energy flows which, to a very good approximation, is materially closed but driven by the Sun's radiant energy and also by a relatively very small energy flux from the Earth's interior. Through photosynthesis, plants in terrestrial ecosystems, phytoplankton and algae in marine ecosystems, and other living organisms convert atmospheric CO_2 and water into organic carbon compounds and oxygen. The net primary productivity (NPP) of such living organisms, called photoautotrophs, is equal to the carbon taken up through photosynthesis, or gross primary production, minus the carbon in the form of CO_2 used in the process of respiration. For a given ecosystem, it is measured as the mass of carbon produced per square meter and per year ($g/m^2/year$). At the upper trophic level, heterotrophic organisms such as animals, fungi, and parasitic plants use the energy provided by the NPP. Net secondary productivity is the rate of energy consumption by the heterotrophs, minus the energy used in their respiration processes. The important point to emphasize here is that secondary productivity is limited by the NPP and there is energy loss during each transfer between trophic levels.

Values of the terrestrial NPP obtained by the MODIS spectroradiometer on NASA's Terra and Aqua satellites range from near zero grams of carbon per square meter per day to 6.5 grams per square meter per day. In some places the NPP has a negative value, which means that more carbon in the form of CO_2 is released to the atmosphere than absorbed by plants, because of plant decomposition or because respiration overpowers carbon absorption.

Furthermore, the data shows that the global terrestrial NPP has remained roughly constant for the last 30 years, with a yearly value close to 53.6 GtC (Ito 2011). This has happened despite efforts to raise yields in agriculture by increasing cropland areas, intensive farming, and in particular irrigation farming, by the growing use of synthetic fertilizers containing nitrogen and phosphorus, and furthermore by raising yields in forestry. Nevertheless, such increases in terrestrial NPP have been offset by the reduction in biomass as a result of anthropogenic land use changes, especially the felling of tropical rain forests and the degradation and destruction of habitats (Barnosky et al. 2012).

It is now possible to estimate the part of the NPP that is used for human consumption, called human appropriation of NPP, or HANPP (Haberl et al. 2007, 2014). The difference between the NPP and the HANPP determines the amount of energy available to transfer from the photoautotrophs, which are mostly plants and fungi, to other species besides *Homo sapiens* at other levels in the trophic webs of ecosystems. The increase in the HANPP has an effect on biodiversity, on the water and carbon cycles, and on the capacity of ecosystems to provide essential services to humans. The study of the HANPP improves our understanding of humankind's impact on terrestrial environments and how that impact is changing natural ecosystems.

Haberl et al. (2007) have presented a global assessment of the HANPP based on vegetation modelling, agricultural and forestry statistics, and geographical information on anthropogenic land use change, soil degradation, and impacts on ecosystems. It was estimated that in the year 2000, the HANPP amounted to about 15.6 GtC/year of an estimated global potential NPP (the plant cover that would prevail in the absence of human intervention) of 65.51 GtC/year, which represents 23.8% (Haberl et al. 2007). The HANPP had a 53% contribution from harvest, 40% from land-use-induced productivity changes, and 7% from human-induced vegetation fires, mostly in forests (Haberl et al. 2007). The fact that one biological species appropriates about one quarter of the Earth's potential NPP is remarkable. In spite of current and future efforts for agricultural intensification, it seems highly likely that cropland area will go on increasing over the coming decades to ensure the food security of a growing global population, and this implies a continuing increase in the HANPP.

The HANPP is used for food production and also to generate bioenergy. If all the above ground part of NPP, estimated to be about 30 GtC/year of biomass growth (Haberl et al. 2013), were to be converted into thermal energy through biomass combustion, the total resulting energy is estimated at 1110 EJ/year (one exajoule (EJ) equals 10^{18} joules). To establish this conversion, it is assumed that 1 kg of dry matter biomass is equivalent to 0.5 kg of carbon and has a calorific value of 18.5×10^6 J (Haberl et al. 2010).

Humans harvest about 230 EJ/year of biomass growth to produce food, fibre, and livestock feed, to maintain pastures for animal grazing, and to generate bioenergy, a large part of which comes from residues and waste flows (Haberl et al. 2013). To harvest this amount of energy in the form of biomass, humans affect or destroy about

70 EJ/year in the form of plant parts not harvested and left over and as burned biomass in anthropogenic vegetation fires. The remaining 810 EJ/year of potential energy from the terrestrial NPP cannot be easily exploited by humans because 48% grows in forest ecosystems and about 28% in national parks, conservation and wilderness areas, or cultivated ecosystems that are already heavily harvested (Haberl et al. 2013).

Anthropogenic global changes are interfering with the NPP in other ways besides those already mentioned. The anthropogenic increase in the CO_2 atmospheric concentration caused by the intensive use of fossil fuels and by land use changes increases the efficiency of photosynthesis, especially in C3 type plants, and also improves the efficiency of water use. This effect is called CO_2 fertilisation. Several experimental studies confirm that for certain plants, in conditions of nutrient abundance, the NPP increases by 20–25% when the CO_2 concentration is doubled compared to pre-industrial levels (Ainsworth and Long 2004; IPCC 2013). Nevertheless, in some ecosystems and for some types of plants, the CO_2 fertilisation effect is reduced or eliminated (Newingham et al. 2013; IPCC 2013). The main cause of this variability is likely to be the sensitivity of the CO_2 fertilisation effect to the abundance of nutrients in the soil, especially nitrogen and phosphorus. The NPP is conditioned not only by the amount of CO_2 present in the atmosphere, but also by the temperature and the availability of water and nutrients.

Anthropogenic climate change also interferes with the NPP through climate change itself. Various studies for Europe and China have shown that droughts reduce the terrestrial NPP at the regional level (Ciais et al. 2005; Lai et al. 2018). As regards temperature the results for China suggest that a 1.3 °C increase in the temperature stimulated a positive trend in the NPP at the national scale during the past 50 years (Lai et al. 2018). However, there was also an indication that the NPP tended to decrease when warming exceeded 2 °C. It is difficult at present to project quantitatively the effect of climate change on the global terrestrial NPP since it involves the aggregate effects of global mean temperature increase, more frequent droughts at the regional level, and CO_2 fertilization. Nevertheless, it is likely that if the global temperature increases above 2 °C, there will be an increasing tendency toward a reduced global NPP in terrestrial and marine ecosystems.

The NPP process is part of the carbon cycle, which describes the fluxes of carbon between the four main reservoirs of the Earth system: the atmosphere, hydrosphere, biosphere, and lithosphere. Since the Industrial Revolution, humans have been modifying the carbon cycle by converting the chemical energy contained in fossil fuels—coal, oil, and natural gas—into thermal energy. These fossil fuels are found in deposits in sedimentary rocks, and were formed from the fossilized remains of dead plants exposed to various conditions of heat and pressure in the Earth's crust over millions of years. Thus, fossil fuels have their origin in NPP processes that occurred millions of years ago. The burning of fossil fossils for energy generation and for other uses emits large quantities of CO_2 into the atmosphere, thus contributing to global climate change, and these emissions have been increasing over the last few decades. Abiogenic hypotheses for the origin of fossil fuels were proposed by various Russian geologists and by the Austrian-born American astrophysicist Thomas Gold (1920–2004) in the 20th century, but there is no scientific evidence to validate them, mainly because of the increasingly sophisticated understanding of the modes of formation of hydrocarbon deposits in nature (Glasby 2006).

Decadal global mean emissions of CO_2 from fossil fuel burning and industry, especially cement manufacturing, have increased from a value of 5.5 ± 0.3 GtC/year in the period 1980–1989 to 9.4 ± 0.5 GtC/year in the period 2008–2017, which represents an increase of about 70% in a period of less than 30 years (Le et al. 2018). Anthropogenic land use changes also give rise to CO_2 emissions but they were practically constant over that period of about two and a half decades, with a decadal global mean of 1.2 ± 0.7 GtC/year in the period 1980–1989 to 1.3 ± 0.7 GtC/year in the period 2007–2016 (Le et al. 2018). The large increase in total anthropogenic emissions in about 25 years is a surprising result in view of the scientific consensus (Cook et al. 2013) on the causality relationship between anthropogenic greenhouse gas emissions and climate change that has been established since the 1980s, in particular considering the creation of the IPCC in 1988. This means that the science of climate change and the warnings from scientists about the damaging effects of climate change have been largely ignored and have not led to climate action.

What happens with all the CO_2 that has been emitted into the atmosphere? The two main sinks are ocean uptake through CO_2 dissolution in sea water and land uptake through photosynthesis. The decadal global mean uptakes of those two sinks for the period 2007–2016 were 2.4 ± 0.5 GtC/year and 3.0 ± 0.8 GtC/year, respectively (Le et al. 2018). The global mean increase in atmospheric CO_2 during the same decade was 4.7 ± 0.1 GtC/year expressed in carbon atom mass. In conclusion, at present, about half of the CO_2 emitted into the atmosphere remains there. A large part of this will remain for centuries, increasing the Earth's greenhouse effect and consequently provoking global climate change.

Let us now analyse briefly the anthropogenic energy fluxes. The use of fossil fuels is by far largest source of primary energy for humankind. According to the IEA it represented 81.4% of the global primary energy supply in 2017 (IEA 2019a). Nuclear energy, biofuels plus waste, and hydro plus modern renewables, such as solar, wind, and geothermal, represented 4.9%, 9.4%, and 4.3% of the total, respectively. The global supply of all forms of primary energy in 2017 is estimated at 583.8 EJ (IEA 2019a). Biomass and waste, which are part of the HANPP, amount to 54.9 EJ. Thus the total amount of primary energy used in 2017 with its origin in the NPP, which includes fossil fuels and biomass plus waste, was 90.8% of the total, corresponding to 530 EJ.

It is curious to observe that in spite of all the current emphasis on technological development, nuclear and modern renewables represent only 9.2% of the global primary energy supply, amounting to 53.8 EJ. The likely explanation is that the fossil fuel industry has been able to supply the world with plenty of energy and become one of the most lucrative industries globally. This success has led to a lock-in situation as regards the transition from fossil fuels to renewable energies.

One attractive way to reduce global greenhouse gas emissions is to use biomass to convert it directly to thermal energy, to produce biochar or liquid biofuels, which have relative low cost and good storability and can easily be substituted for liquid fossil fuels. Estimates of global primary bioenergy that can be produced by 2050 range from 30 EJ/year to 1300 EJ/year (Haberl et al. 2010). However, with optimistic assumptions, the currently remaining land ecosystems that could be used to generate bioenergy imply an upper biophysical limit for primary bioenergy of about 190 EJ/year. To generate all this bioenergy, which is about triple the presently available bioenergy, it would be necessary to produce biomass outside croplands, wilderness, and natural forests (Haberl et al. 2013). Pasture lands, part of the remaining woodlands, savannahs, and tundra would have to be converted into an intense production of biomass to provide such an amount of energy. This program is not advisable since it would have a very harmful effect on biodiversity and would also constitute a risk for future food security. One of the greatest challenges regarding the HANPP is to find sustainable levels of bioenergy production that are compatible with biodiversity conservation and food security.

Humankind continuously requires a greater supply of primary energy for very diverse uses ranging from the energy sector, particularly for conversion into electricity, industry, agriculture, land, sea, and air transportation, infrastructures, urban areas, and rural settlements. The primary energy consumption, currently increasing at an annual rate of 1.5%, is likely to grow from 583.8 EJ/year in 2017 to 1000 EJ/year in 2050 (Moriarty and Honnery 2012). This amount of energy is still considerably smaller than the Earth system's main energy source, which is the total solar radiant energy that it absorbs in one year.

The electromagnetic energy flux received from the Sun at the average distance between the Earth and the Sun at the top of the atmosphere in an area of 1 m^2 perpendicular to the solar rays is 1366.1 Wm^{-2} and is called the solar constant (Fröhlich 2000). Solar radiant energy reaching the Earth system therefore has a power of 174PW (one petawatt is 10^{15} W) (Smil 2008). Because the Earth has an albedo of around 0.3, only 70% of solar radiation is absorbed, predominantly within the visible range. The total solar power absorbed by the Earth is therefore 121.8 PW. Over one year, this solar radiant power accumulates an energy of 3.841×10^6 EJ, which is about 6580 times more energy than the total primary energy consumed worldwide in 2017. In 2050, this ratio is projected to decrease to about 3841.

It is obviously impossible to use all this solar radiant energy as a primary energy source for humankind. There are many limiting factors, such as the diurnal variation of the solar radiant energy flux, the latitude effect, the local climate, and the intermittence caused by changes in the weather. Estimates of the global technical potential for solar energy use vary greatly, between 118 and 22 592 EJ/year (Moriarty and Honnery 2012). The upper limit is 22 times greater than the projected global energy supply for 2050.

So the Earth absorbs about 121.8 PW of the power from the Sun in one year. What happens to this energy? It has to be returned to outer space, otherwise the Earth's atmospheric temperature would increase continuously. In other words the Earth system is always moving towards an energy balance in which the incoming energy is equal to the outgoing energy. However, anthropogenic greenhouse gas emissions have displaced the Earth from its energy balance state, since they cause positive radiative forcing in the troposphere. The total anthropogenic radiative forcing in 2011 compared with 1750 is 2.29 (1.13–3.33) Wm^{-2} and has increased more rapidly since 1970 than in previous decades (IPCC 2013).

It is estimated that the Earth's climate system has gained 274 000 EJ of thermal energy during the period 1971–2010, due to the enhanced greenhouse effect caused by greenhouse gas emissions. This energy increase corresponds to a power rate of 0.213 PW in a linear fit to the annual values over that period. This power amounts to an annual energy increase of 6717 EJ, which is about 16.6 times greater than the mean annual global supply of all forms of primary energy used by humans in the period 1971–2017 (IEA 2019c). In other words, the mean global fossil fuel energy consumption in that period is the main cause for the large annual energy increase in Earth's climate system. It is the main cause in the sense that the radiative forcing of all well-mixed greenhouse gases with anthropogenic emissions arises mostly from CO_2. For instance, in 2011, the total radiative forcing from greenhouse gas emissions was 2.83 Wm^{-2}, while the radiative forcing from CO_2 emissions was 1.82 Wm^{-2}, which represents 64% of the total (IPCC 2013).

Where does all this excess heat go in the climate system? About 93% goes into ocean warming, most of it, 64%, to the upper layers between 0–700 m, 3% for melting Arctic sea ice, ice sheets, and mountain glaciers, 3% for warming the continents, and only 1% to warming the atmosphere (IPCC 2013). If the ocean covered much less than about 71% of the Earth's surface, the mean global temperature increase of the atmosphere would be considerably higher. The large ocean surface area has protected us from a much faster temperature increase in the atmosphere.

7.7 Can Biodiversity Loss Be Stopped?

There is currently widespread scientific consensus that humankind has profoundly transformed the biosphere and, through a range of different activities with local, regional, and global impacts, is driving it towards a critical transition, the onset time and detailed characteristics of which are still hard to foresee. As regards biodiversity, it is also consensual that the biosphere is moving towards the sixth mass extinction of species. The rising question is how humans are going to collectively react to these risks, which are far from being well defined, and what the future of the biosphere is in the Anthropocene.

To analyse these questions, it is useful to start by acknowledging that it is part of humankind's essential nature to use and exploit the useful resources in the biosphere at the local and regional levels, with no limitations except those that derive from short-term self-interest, and if necessary until those resources are exhausted. This acknowledgement, which may be uncomfortable for our image and self-esteem, does not apply to all people at all times but reflects the prevailing human behaviour up to now. The conclusion is reinforced by studying the biological evolution of our ancestors in the *Homo* genus. Can we change this dominant behaviour through cultural evolution? The future will answer this question.

Throughout history, *Homo sapiens* was often the victim of predation, but humans are also predators. A recent study makes a comparative analysis of the predatory behaviours of several animals and humankind, concluding that "humans function as an unsustainable 'super predator', which—unless additionally constrained by way of managers—will continue to change ecological and evolutionary processes globally" (Darimont et al. 2015). However, it is also important to acknowledge that cultural evolution has led humans to observe, study, and gain knowledge about the biosphere and to decipher the way it functions and evolves in great detail. It has also led to an appreciation and respect for nature and the extraordinary biodiversity of flora and fauna in the different world regions, to develop some remarkable nature conservation initiatives, especially in places that are biodiversity hotspots. Cultural evolution may therefore lead to a sustainable relation between humans and the biosphere.

An extensively analysed and debated example on the darker side of the problem is the destruction of megafauna in several world regions during the Late Pleistocene, the main cause of which was human action together with the climate change associated with the transition to the Holocene (Koch and Barnosky 2006). The human origin of the collapse in megafauna is particularly well documented in the case of Australia (van et al. 2017). A recent study that analysed the way the extinction of mammals evolved over the last 125 000 years concluded that it was selective regarding the size of the species' body and took place on all continents (Smith and Myers 2018). This selectiveness is another argument indicating that its main origin was the territorial expansion of the human population throughout the world in the Upper Pleistocene. *Homo sapiens* naturally preferred to kill species with larger body sizes since it tended to make the hunting more cost-effective in terms of effort. If the extinction rate trend is maintained, it is likely that the largest remaining mammal will be a domesticated species, probably the cow. The long term sustainability during many centuries in various forms of captivity of healthy populations of the large emblematic mammals, such as elephants, rhinoceroses, hippopotamuses, and many others is uncertain and far from being guaranteed.

As for the remaining wild mammals, it is known that their habits have changed profoundly so they can defend themselves from human predation. Human influence manifests itself in the spatial distribution and size of populations as well as in the time patterns of animal activities during the day and night. Gaynor et al. (2018) concluded, in a study involving 62 species on six continents, that mammals are becoming more nocturnal to avoid encounters with humans. The shift away from their natural patterns of activity has consequences for these mammals' fitness, capacity for survival, community interaction, and evolution.

The decreasing populations and the extinction of species that humankind has caused since the Pleistocene contrasts with the deep affection that humans have towards pets, particularly mammals such as cats and dogs. This dual behaviour is fully accepted as being quite natural and probably has the same nature and origin as the dual social behaviour that humans have when attacking mercilessly the members of a rival social group and defending unconditionally the members of their own social group. During the biological evolution of the hominids, there was a distinct advantage in being close to other animals, particularly mammals, living in the same environment since their presence was a sign of life and of useful natural resources such as water and edible plants. This attraction to life in the environment is an example of the biophilia hypothesis, according to which humans subconsciously seek to establish connections with other living organisms (Wilson 1984).

It is now very frequent for humans to develop very strong and maybe even excessive bonding with pets. The current sophistication of human–animal bonding is one of the outcomes of progress and economic prosperity, the pursuance of which has had the collateral effect of implying a profound and to a large extent irreversible transformation of the biosphere. The relationship between humans and their pets, which probably started with the domestication of dogs during the final Palaeolithic (Janssens et al. 2018), became much more complex and important for human well-being after the Agricultural Revolution. Frequently now, the relation with pets is also a statement about other human beings, especially for people who live isolated and forgotten by family and past friends and desperately in need of company and compassion. Pets can supply a kind of unconditional relationship that is usually impossible with people (Archer 1997). Dogs in particular bring security to their owners and are compassionate and understanding, providing irreplaceable company.

Given the importance of megafauna in the structure and function of ecosystems, their gradual disappearance has had a significant role in the evolution of terrestrial biodiversity over the last 125 000 years. Human behaviour as regards the biosphere at the local, regional, and global scales will certainly continue to evolve, but it is likely that the essential tendencies that have marked this behaviour in the past will remain the same. In other words, survival and economic prosperity are supreme overarching values that are placed above any ecosystem conservation actions that have no direct bearing on those human values in the short term. Medium and long term ecosystem conservation is recognized as good and defensible, but in practice as a very low global priority. The supremacy of human values, interests, and expectations will tend to get stronger and to manifest itself more firmly and forcibly in situations of danger that may be caused by decreasing natural resources, increasing climate change, and ecological breakdown.

Responses to the current situation are highly varied and range from triumphant techno-optimist points of view, which argue for continued transformation of the biosphere and for it to be fused with technology to benefit humankind, to the perspectives of those who actively engage in conservationist activities and environmentalist movements. There is an enormous diversity of positions between these two extremes. Within this large range of viewpoints, the platform for debate that is perhaps the most consensual and productive is finding out to what extent development, proliferation, and exploitation of farming and forestry anthromes will be able to offset the loss of services from "natural" ecosystems, many of which are seriously affected by overexploitation, pollution, and global anthropogenic changes, particularly climate change. The question can be reworded by asking whether or not future food security for a large portion of humanity will be jeopardised by the degradation and eventual collapse of "natural" ecosystem services.

We use scare quotes on "natural" because global anthropogenic changes have affected all biomes, ecosystems, and habitats in the biosphere to a certain extent, with more or less lasting effects. Pristine nature and ecosystems free from direct or indirect interference from human activities belong to the past, before the Age of Discovery, and have been replaced by disrupted natural ecosystems. Although many places in the world still look as though they are pristine, especially in photographs or films, a closer look in situ would show that they have all been impacted by global anthropogenic changes, particularly by climate change.

In the last few decades of the 20th century, the scientific community, and in particular biologists and Earth scientists, concerned with the risks of growing anthropogenic interference in the Earth system, especially the biosphere, and faced with the growing hegemonic value assigned to economic growth, began to devote more attention to assessing the ecosystem services and their economic value (Costanza et al. 1997). The Millennium Ecosystem Assessment was carried out from 2001 to 2005, under the aegis of the United Nations (MEA 2005). Its goal was "to assess the consequences of ecosystem change for human well-being and the scientific basis for action needed to enhance the conservation and sustainable use of those systems, so that they can continue to supply the services that underpin all aspects of human life". In that assessment, ecosystem services were classified into four large groups: supporting, provisioning, regulating, and cultural.

The first category represents services at the base of the biosphere subsystem that support the other groups of services. This includes NPP, soil formation and conservation, and preservation of nutrient cycles and habitats, including nurseries, temporary habitats for migratory species and pollination. These are fundamental services but their relationship with human needs is neither simple nor direct. The main provisioning services are the production of food, raw materials, biogenic minerals, energy in the form of biomass, medical resources, in particular genetic resources, especially for medicine and pharmacology, and therapeutic services, such as ecotherapy and therapies assisted by animals and by outdoor natural environments. Regulating services are generally more complex. The main services in this category help regulate the carbon and water cycles, provide waste decomposition, contribute to control water and air quality, contribute to control erosion and floods, provide biological control of pests, regulate prey populations through predation, and contribute to climate regulation, in particular by carbon sequestration. Finally, cultural services of ecosystems have a direct human relationship, although with highly varied expressions throughout the world, depending on the religion, culture, and state of socio-economic development of the populations. These services arise from the aesthetic values of ecosystems and their landscapes, seen as an integrated set of ecosystems, from their historical and identity values at the individual level and the community, society, and nation levels, and from their recreational and tourist values (Daniel et al. 2012; Cooper et al. 2016). The Millennium Ecosystem Assessment defines the cultural values of ecosystems as "nonmaterial benefits people obtain from ecosystems", accentuating the dichotomy between economic and non-economic values. They include cultural, spiritual, and recreational values, such as ecotourism and outdoor sports.

A more recent example of the strategy of highlighting the importance of the economic value of ecosystem services as a way of defending the sustainability of biological resources was the drawing up of a study called "The Economics of Ecosystems and Biodiversity" (TEEB 2010), presented in 2010 at the 10th Conference of the Parties of the United Nations Convention on Biological Diversity. At this meeting, a Strategic Plan for Biodiversity 2011–2020 and the Aichi Targets were approved. The Aichi Targets comprise 20 quantified targets for protecting biodiversity at global level to be achieved by 2020. Approving these objectives reveals that the scientific community has done high-quality work to alert governments, politicians, governmental institutions, non-governmental organisations, businesspeople, and the public in general about the dangerous situation the biosphere is in and the associated risks. Meeting the Aichi Targets would be a very important step towards the sustainability of ecosystem services and the preservation of biodiversity. However, assessments carried out in 2014 revealed that humankind was far from being able to achieve in 2020 what was approved in 2010 (Leadley et al. 2014; Tittensor et al. 2014; WWF 2018a).

Aichi Target number 5, at least halving the rate of loss of natural habitats, including forests, and significantly reducing degradation and fragmentation of those habitats will not be met. A 2016 WWF report (WWF 2016) presented the results of monitoring 14 152 populations of 3706 vertebrate species in several regions of the globe. The populations of these species fell by 58% on average between 1970 and 2012, primarily due to degradation and destruction of habitats, overexploitation of natural resources, pollution, and climate change. If current trends continue, the reduction in populations of vertebrates will reach 67% in 2020, compared with 1970 numbers. Falling populations are not only affecting vertebrates, but also many other phyla.

These warnings and many more that are not mentioned here have been unable to change the course of events. The more recent Report of the Plenary of the Intergovernmental Science-Policy Platform on Biodiversity and Ecosystem Services (IPBES) (IPBES 2019) indicates that only 4 of the 20 Aichi Biodiversity Targets are likely to be achieved in 2020. This is the first major Report since the Millennium Ecosystem Assessment of 2005 and introduces innovative ways of evaluating evidence, based in part and for the first time on indigenous and local knowledge. It states that "nature is declining globally at rates unprecedented in human history—and the rate of species extinctions is accelerating, with grave impacts on people around the world now likely. [...] The average abundance of native species in most major land-based habitats has fallen by at least 20%, mostly since 1900. More than 40% of amphibian species, almost 33% of reef forming corals and more than a third of all marine mammals are threatened. The picture is less clear for insect species, but available evidence supports a tentative estimate of 10% being threatened. At least 680 vertebrate species had been driven to extinction since the 16th century and more than 9% of all domesticated breeds of mammals used for food and agriculture had become extinct by 2016, with at least 1000 more breeds still threatened". The causes of the current trends are attributed in descending order to changes in land and sea use, direct exploitation of organisms, climate change, pollution, and invasive alien species.

The report also draws attention to the negative impacts that biodiversity loss and ecosystem degradation are having on human well-being, health, and food security at the global scale. It states that the "current negative trends in biodiversity and ecosystems will undermine progress towards 80% (35 out of 44) of the assessed targets of the Sustainable Development Goals, related to poverty, hunger, health, water, cities, climate, oceans, and land (SDGs 1, 2, 3, 6, 11, 13, 14, and 15)".

In the tropics from 1980 to 2000, 100 million hectares of land were converted to agriculture about half of it at the expense of undisturbed forests, mainly in Latin America and Southeast Asia. More than 33% of the world's land surface is now devoted to crop or livestock production. These changes are leading to the degradation or even to the collapse of some ecosystem services.

One worrying example is the decrease in insect populations and other pollinating animals that play a very important role in food production and the preservation of genetic biodiversity in plants of nutritional value. This decrease began to reveal itself at the start of the 20th century and is becoming dangerously worse in the 21st century (Kluser and Peduzzi 2007). Around 75% of cultivars on sale around the world rely on pollination. The annual global economic value was estimated at between $235bn and $577bn (IPBES 2016). In Europe, the monetary value of pollination has been estimated at €14.6bn (EASAC 2015). However, as already remarked, a decline is being observed in honey bee and wild bee populations and populations of other wild pollinators, particularly in north-west Europe, North America and parts of China. The diversity of pollinators is very important because it can improve crop yield or fruit quality and it is also good for natural vegetation (EASAC 2015). More than 40% of invertebrate pollinators, such as bees, wasps, and moths, and around 16.5% of vertebrate pollinators, such as bats and birds, are at risk of extinction (IPBES 2016). A recent study revealed that in the 27-year period between 1973 and 2000, flying insect populations in protected areas of Germany fell by 76%, with adverse effects on ecosystems, particularly in the medium and long term (Hallmann et al. 2017). The most likely causes for the decline in bee numbers are loss of habitat, the use of pesticides, particularly neonicotinoids (EASAC 2015), pollution, invasive species, pathogenic agents, and climate change.

The dominant discourse to avoid the risk of a collapse of ecosystem services has been to assign monetary value to its services and try to ensure this value is used in economic and political decision-making. The aim of the exercise is then to assess the monetary value of the damage to human society that arises from the degradation or partial incapacitation of the different types of biomes and ecosystems and the corresponding benefits of their being maintained and restored. Although estimates of these monetary values are difficult to make and are sometimes affected by large uncertainties, significant progress has been made (de Groot et al. 2012). Nevertheless, the integration and systematic use of these values in governance is very limited because the degradation or destruction of an ecosystem in a particular place is frequently the negative externality of an investment that generates immediate or short-term financial or economic benefit in the same region or elsewhere. The ecosystem service that is degraded or destroyed may only generate benefits in the medium and long-term.

Once again, the problem is time. The increasing acceleration of operative social time makes it harder or even impossible to opt for a benefit that continually manifests itself in the medium and long term instead of a concentrated, accessible benefit in the short term. Furthermore, the short-term benefit, the collateral consequence of which is the degradation or destruction of an ecosystem, generally benefits a restricted but powerful group of people, while the benefit from the ecosystem service benefits a much wider population, which in some cases includes all of humankind, in a diffuse and poorly documented way. Felling an area of 10000 hectares of tropical rain forest to provide pastures for cattle benefits investors and consumers after a short time, but it cancels the benefits of services provided by the forest ecosystem, specifically its contribution to the sustainability of the water and carbon cycles and to soil and biodiversity conservation. Since the farming and forestry anthrome that replaces the tropical rain forest is also an ecosystem, the monetary value of the services of the two ecosystems must be assessed and compared in the short, medium, and long term. This process must include, as far as the water and carbon cycles are concerned, accounting for the water consumption involved in cattle production and the CO_2 emissions connected to livestock farming, as well as CH_4 emissions from enteric fermentation in bovine animals.

Similar calculations should be made when planning to destroy a mangrove to build a prawn aquaculture unit, a highly profitable activity that proliferates in the coastal areas of Asia, particularly China and Thailand, and also in Brazil, Ecuador, and Mexico (Páez-Osuna 2001). Many other examples could be given of decisions regarding initiatives geared towards almost immediate economic growth, but that involve the degradation or destruction of ecosystems whose services carry benefits mostly in the medium and long term. In conclusion, the monetary assessment of ecosystem services is far from being enough to guarantee the sustainable exploitation of the biosphere's natural resources. Those most harmed by that unsustainable exploitation are poor people, especially in developing countries, whose food security depends on their agriculture and fishing practices, and more generally, future generations.

Despite these current difficulties, resulting from the prevailing mode of human behaviour, the monetary assessments of ecosystem services are important because they inform political decision-makers at different levels of governance, as well as businesspeople and civil society in general about the value of the different types of biomes and ecosystems. They may also serve to influence public opinion and public policy.

There are signs that the strategy to assure the sustainability of ecosystem services and to stop biodiversity loss is changing. Robert Watson, the IPBES Chair, when presenting the latest IPBES Global Assessment Report (IPBES 2019), after stating that it presents "an ominous picture" said that "it is not too late to make a difference, but only if we start now at every level from local to global". This is the typical posture of environmental scientists engaged in communicating the risks involved in the current overexploitation of natural resources, environmental degradation and pollution, and climate change. They know that they have to strike an optimistic note otherwise most of those who listen would just go away and stop listening. They don't want to receive the bad news of a less affluent future with few alternatives, especially when the proposed solutions bring into question their behaviour and values, and involve a change of attitude.

However, the most important part of Watson's talk came after that, when he said that "Through 'transformative change', nature can still be conserved, restored and used sustainably—this is also key to meeting most other global goals. By transformative change, we mean a fundamental, system-wide reorganization across technological, economic and social factors, including paradigms, goals and values." Transformative change is a concept that embraces Gandhi's words that "we must be the change we want to see happen in the world." It can be applied in the social, economic, and environmental domains and represents a systems approach to start a cultural transformation that will lead to a revolutionary change in society. The recognition that sustainability can only be achieved through some form of human transformative change is a positive development and has the advantage of acknowledging that it is a possible but very difficult task for humankind to achieve. Most probably the transformative change needed to achieve the Sustainable Development Goals by 2030 will require policies that involve a gradual but persistent departure from the current financial and economic system (UNRISD 2016). However, the current system is likely to react and oppose such a transformative change, arguing for a wide range of technological transformations that would replace natural capital by human capital. This course of action, based on the concept of weak sustainability, is likely to bring about a critical transition in the biosphere.

On the other hand, there are many people and thousands of institutions and organizations that believe in the need for a transformative change and have already started to work in that direction. They dedicate a significant part of their lives to sustainability and in particular to the conservation of terrestrial and marine biodiversity. Many of these efforts have had remarkable success and have managed to halt the extinction of several species. They are local and regional initiatives spread around the world, which still have limited expression at the global scale and will not be sufficient to reverse the dominant trend towards extinction. Nonetheless, they are fundamental examples that show that it is possible to meet the challenge facing us. They should be generously supported by governments, companies, and the general public. One of the most encouraging outcomes from the effort to conserve biodiversity in the 20th and 21st centuries is the growing number of protected areas, the total area of which is around 12% of terrestrial land. To maintain a high degree of conservation of flora and fauna in those areas, it is imperative to monitor and evaluate the level of protection that is actually enforced (Joppa et al. 2008).

From an aesthetic point of view, landscapes, plants, and animals have always provided inspiration for artists all over the world, especially in China and the West. Influenced by Buddhism and Taoism, landscape painting acquired great importance in China from the 8th century onwards, with artists like Wang Wei (699–759), and flourished in northern China in the period between the Tang and Song dynasties. Painters used the sumi-ê (ink wash) technique, a form of monochromatic painting focusing on representing essential features, in which drawing is combined with calligraphy. Jing Hao (855–915) and other famous artists sought to highlight the greatness and harmony of nature by painting rocky mountain landscapes, populated with isolated trees and rivers and confined in deep valleys, using strong, expressive brush strokes. In the south of China, Dong Yuan (934–962) and his disciples developed a more intimate style of landscape painting with undulating hills, lakes, and forests.

Several centuries later, Joachim Patinir (1475–1524), a Flemish Renaissance painter, was one of the first in the West to appreciate the beauty of landscapes, making them the main feature of his paintings. Pieter Bruegel the Elder (1525–1569) was deeply inspired by nature and became one of the first great masters of landscape painting, animated by scenes of everyday country life. Landscapes were gradually freed from human representation and other signs of their presence and soon became the only object of the painter's attention. Jacob von Ruisdael (1628–1682) and Caspar David Friedrich (1774–1840) have left us famous examples of this aesthetic journey. With Impressionism, the wild landscapes, or landscapes benignly transformed by man, became an excellent field of experimentation to achieve new aesthetic values. One of the impressionists' challenges was to represent the effects of light and colour on the landscape, the surface of water, the sea, snow, trees, and flowers. The aim was to paint "impossible things" in nature, as Claude Monet (1840–1926) said when he painted *The Boat* in 1887 (Musée Marmottan Monet, Paris). For many 20th century artists, the landscape was also one of the paths to abstraction, as was the case for Vassili Kandinsky (1866–1944).

Despite civilization's long journey towards cultural appreciation of landscapes, ecosystems, plants, and animals, we are now faced with the risk of a critical transition in the biosphere, probably accompanied by a mass extinction of species, the collapse of some ecosystems, and deep changes in our landscapes. What consequences will there be for humanity in this scenario? A first group of consequences relate to culture and heritage and include, among other things, the loss of essential and irreplaceable value that arises from the extinction of each biological species. A second group of consequences relates to well-being and human prosperity. As already mentioned, the current negative trends in biodiversity and ecosystems will undermine progress towards 80% of the assessed targets of the Sustainable Development Goals, related to poverty, hunger, health, and water, among others. A recent distressing trend is the continued growth in the number of undernourished people in the world in the period 2015–2018, jumping from 785.4 million in 2015 to 821.6 million in 2018 (WMO 2019).

7.8   Can We Develop a Better Technobiosphere

What if it turns out to be impossible to reduce the risk of some ecosystem services collapsing and a critical transition occurring in the biosphere, such as would threaten the well-being, economic prosperity, and food security for a large part of the global population? The most frequent answer is conceptual and based on the recognition that human societies have become a transforming force in the biogeophysical subsystems of the Earth system. Natural ecosystems have been replaced by anthroecosystems, meaning systems that arise from direct or indirect human action on the previous natural ecosystems. Natural ecosystems and landscapes free from direct human intervention are increasingly rare, but even these are subject to indirect global intervention, as already mentioned. The continuous and varied transformations in the biosphere that arise from direct human intervention are generating an anthropogenic biosphere or technobiosphere. This extraordinary human capacity to deeply transform the biosphere has been analysed in several approaches relating to biology, sociobiology, and social sciences, such as ecosystem engineering (Jones et al. 2010b), niche construction (Odling-Smee et al. 1996; Scott-Phillips et al. 2014), ecological heritage (Odling-Smee 2012), ultrasociality (Tomasello 2014), cultural evolution, and social change.

In his attempts to analyse and interpret the origins of the current situation, the US biologist Erle C. Ellis proposes the theory of anthroecology as a way of studying ecology in the new technobiosphere (Ellis 2015). The first step on this path is recognising the first law of the Anthropocene: "The ecological patterns, processes, and dynamics of the present day, deep past, and foreseeable future are shaped by human societies" (Ellis 2015). Note that the wording of this law does not distinguish between the different levels of intensity and breadth of human interference in the biosphere since the distant past and into the foreseeable future. It implies that there is nothing surprising or exceptional about the current rate of biodiversity loss at global level. It is assumed that ecological processes and dynamics have been and always will be moulded by human societies.

According to this first law of the Anthropocene, anthroecology describes the integration of human societies into an increasingly anthropogenic biosphere by way of a sociocultural process of niche building. The anthropogenic evolution of the biosphere will gradually lead it to be integrated into the human subsystem of the Earth system. In the most extreme limit, the biosphere will become part of the human subsystem.

The conceptual framework of anthroecology, by promoting techno-optimism, pretends to efface the risks associated with a critical transition in the biosphere, with the overexploitation of natural resources, with environmental degradation and pollution, and with climate change. In the new framework, the evolution of the technobiosphere is a dynamic process led by humankind to assure the continuation of its progress towards more well-being and global food security. From this point of view, it is not a priority to discuss potential risks, generate anxieties, and point out alternative paths that would in practice go against the overwhelming human quest for more economic prosperity and consumerism.

Since we are already in the Anthropocene, the best solution is to build an excellent Anthropocene supported by a technobiosphere that is also expected to be excellent. There are two main transformation factors in the evolutionary process towards the development of a better technobiosphere: direct human interference in the populations, genetic characteristics, and evolution of living organisms that are useful as sources of food and of other essential commodities, especially through genetic engineering and synthetic biology, and the generalized application of the other leading technologies, such as nanotechnology, big data, and artificial intelligence.

One of the main justifications to defend the construction of a new technobiosphere is that "we have the potential to create a much better planet than the one we are creating now" (Ellis 2018). However, by focusing on environmental limits, including limits on the global mean temperature increase, "instead of on the social strategies that enable environmental and social outcomes, we fail to engage the only force of nature that can help us: human aspirations for a better future" (Ellis 2018). This is a cultural point of view that arises from a culture with strong roots in the USA, a country of immigrants moved by a large horizon of expectations populated by religious freedom, economic prosperity, and a perception of limitless opportunities for a better future.

Creating a better planet involves costs, but in spite of all the strength that can be found in the human aspirations for a better future: "No better future will be possible if those most able to bear the costs—those who've benefited the most, the wealthy and the vested interests of this world—don't step up to pay for it" (Ellis 2018). Here one finds another difficulty related to human nature. Why should those that benefited the most, or in other words the ultra high-net-worth (UHNW) individuals and the politicians who serve them, pay for the costs of an expensive and demanding program that will benefit the majority of humankind but not themselves in particular in any significant way? The UHNW individuals and most politicians are not vulnerable to environmental degradation and pollution, or to the over-exploitation of natural resources, or to climate. In fact some of them benefit from the activities that are creating those side-effects. UHNW individuals don't require a liveable environment because they have the means to create one for themselves with all the required natural amenities. Some of them, aware of the dangers that the global anthropogenic changes are creating for humankind even now, but especially in the future, are contributing to the solution of the problem with their actions and their money, but the majority do not appear to feel concerned and are not prepared to intervene.

In conclusion, even those that propose the human construction of a "much better planet" acknowledge the difficulties ahead and identify solutions that require an unlikely change in the human behaviour of crucial social groups in society.

7.9 How to Assure Global Food Security up to 2100 and Beyond Using Genetic Engineering. Challenges, Successes, and Risks

The growth in global agricultural production since the Green Revolution in the mid-20th century has reduced hunger and undernourishment, increased food security, and significantly changed food diets worldwide. World meat production increased by a factor greater than four in the period 1961–2015. To meet the FAO scenario for 2050 of an average daily global demand of 3130 kcal per capita (FAO 2006), annual global agricultural production would need to grow by 70% between 2005–2007 and 2050 (Bruinsma 2011). One of the main intrinsic uncertainties in these types of scenarios is the future evolution of the human diet in the different regions of the world, especially the amount of meat included in diets and the type of meat, be it beef, pork, or poultry.

More recent scenarios, in which a global average daily supply of 3070 kcal per capita is forecast for 2050, involve an increase in global agricultural production of 60% between 2005–2007 and 2050, rather than 70% (Alexandratos and Bruinsma 2012). In this scenario there would be around 4% of the population of developing countries, corresponding to about 320 million people, who would remain chronically undernourished. This projection is considerably smaller than the current number of people in this situation, which, as already mentioned, has been rising since 2015, to reach about 10.7% of the global population in 2018. In the same scenario, there will be a slowdown in the growth of global agricultural production from 2050 onward, due to the deceleration in global population growth and the fact that it is projected that a growing part of the population will have medium or high food consumption per capita levels from 2050 onwards (Alexandratos and Bruinsma 2012).

Climate change is projected to have an adverse impact on world food security during the 21st century, more seriously if the global mean temperature increases above the Paris Agreement limit of 2 °C. Every aspect of food security will be affected, including the availability of food, access to food, the use of food, and price stability (FAO 2008). Developing countries are much more vulnerable to the impacts of climate change on food security than advanced economies, primarily due to the greater dependence of their economies on the agriculture sector. It is estimated that those countries would bear 75–80% of the costs of damage caused by climate change (WB 2010). In some regions of the world, particularly tropical and subtropical regions, agricultural production will tend to fall progressively as climate change becomes more intense, due mainly to the higher mean temperature, the higher frequency of extreme weather events, especially heat waves, droughts, and floods, and the greater interannual variability (IPCC 2014b). In some other regions, particularly at higher latitudes, agricultural production will tend to increase, although the mean increase per decade is likely to fall towards the end of the century (IPCC 2014b).

Extreme weather events reduce agricultural productivity. The 6% chance in each decade of extreme events reducing corn production simultaneously in the two main corn-producing regions in the world, the USA and China (Kent et al. 2017), also implies a risk for the world supply of corn. The consequences of that situation and of analogous situations for wheat and rice would be harmful for world food security. In the fishing sector, the migration of species to higher latitudes due to warming of the surface layers of the ocean will generate significant risks for food security in tropical countries. If the increase in average atmospheric temperature reaches more than 4 °C relative to the pre-industrial period, the risks for food security at the regional and global levels will be high (IPCC 2014b).

Growth in food production has been achieved in part by selecting species and hybridisation, and, more recently, using biotechnology. This includes a wide range of techniques for genetically manipulating organisms, including molecular marker-assisted breeding, the use of DNA or RNA modified in vitro, cell fusion of plant cells from different taxonomic families, and the use of recombinant DNA technology to create genetically-modified organisms (GMO) that produce genetically-modified crops. A GMO is an organism produced by genetic engineering that does not occur naturally by mating and/or natural recombination.

GMOs have been used to produce food since 1994, the first GM food being the Flavr Saver tomato, which never made it to the market. They are used to improve productivity, particularly through increased resistance to pathogenic agents and herbicides. Current GM soybeans, corn, cotton, potatoes, and wheat resist the herbicides that are sprayed on farms to kill weeds. The impact of GMO use on the environment and human health is likely to be significant but the medium and long term impact will probably take more time to be fully deciphered by science.

The harmfulness of such impacts has been debated, often passionately, in the last two decades (NRC 2010; Buiatti et al. 2013; Gilbert 2013; EASAC 2013; Oliver 2014; Perry et al. 2016). In 2015, the World Health Organisation's International Agency for Research on Cancer (IARC) classified glyphosate, a broad-spectrum systemic herbicide and crop desiccant, and the herbicides malathion and diazinon as probably carcinogenic to humans (IARC 2017). Glyphosate was created by the USA agrochemical and agricultural biotechnology corporation Monsanto, founded in 1901 and acquired by Bayer in 2018, and commercialized under the name Roundup. Monsanto then developed glyphosate-resistant Roundup Ready crop seeds. By planting glyphosate-resistant seeds and applying glyphosate when the plants are growing, farmers can plant the seeds in closer rows since they don't need to remove the weeds. In this way they can substantially increase their yields. The likely harmfulness of glyphosate for soil health and plant fecundity, and ultimately to agricultural sustainability, is currently a matter of active research (Yamada et al. 2009; Hagner et al. 2019). Glyphosate slows down amino acid synthesis, thereby inhibiting the protein synthesis necessary for plant growth.

Following the publication of the IARC report the Agency was targeted by attempts to discredit its work, particularly from the USA Congress. However, the intensity of the debate had no significant effect on the exportation of GMO seeds by major production companies (IAASTD 2009) and on its use in many countries over the world. Around 18 million farmers in 28 countries, including eight with advanced economies, use GMOs in their crops, primarily soya, corn, cotton, and rapeseed, covering a total area of around 450 million hectares. The largest GMO producers are the USA, Brazil, Argentina, Canada, and China. A group of 38 countries prohibit the growing of GMO but most countries import and consume them (SP 2015). Of the countries that ban GMO production, the one with the largest population is Russia.

Successive US governments have argued for the development and use of genetic engineering in agriculture, particularly the production and consumption of GMOs throughout the world, especially in less industrialised countries that are affected by food insecurity. This position is justified by the conviction that genetic engineering applied to agricultural production is vital for combating malnutrition and hunger and for ensuring food security under climate change, now and especially in the future. It is also considered essential to avoid situations of social and economic unrest and collapse, but also the terrorist movements and activities, which famine and extreme poverty contribute to create.

The humanitarian and security arguments come together and reinforce the economic and financial arguments that support the production of very lucrative products, since GMO seeds have to be purchased anew every year by farmers, unlike traditional seeds, which belong to the crop's life cycle. By controlling the access over seeds, Bayer, Monsanto, Syngenta, DuPont, and other corporations that produce GMOs do in practice control the lives of millions of farmers since seeds are the first link in the food chain. Many farmers fear the cycles of financial obligation in which they have become involved, and also the loss of control over local systems of food production.

By leading scientific research in genetic engineering applied to food production, the USA and some other countries, including Germany, Switzerland, Israel, Japan, and Australia, have created multinational corporations that dominate the global GMO market. US governments have politically supported their large pesticide and GMO corporations, by promoting genetic engineering in agriculture around the world, often in the face of opposition from public opinion and the governments of the countries targeted and to the detriment of other agricultural practices that are more sustainable at the local and regional levels and have a smaller environmental impact (FWW 2013). Monsanto, in particular, has used its large financial capabilities to hire lawyers to deliver lawsuits on small farmers that denounce the harmful effects of Roundup and to hire internet trolls to intimidate food consumers who question Roundup's health effects.

GMO seeds designed to create drought- and heat-resistant crops (Liang 2016) have also been produced by Monsanto and other GMO corporations and are presently used in various countries, especially in Africa. Drought is the largest climatic stress factor leading to reduced crop yields and is becoming more frequent due to climate change. In drylands the situation is aggravated by water scarcity because of water pollution and over-exploitation of water resources, especially aquifers. Furthermore, the problem becomes more difficult to solve because the human population in a large part of the world's drylands is increasing fast.

Drought is now and especially in the future the leading threat to global food security. Crops have different levels of natural resistance to droughts. The average global annual yield loss of maize that can be attributed to drought is about 15% and can reach 25% in Africa (Barker et al. 2010). The extra yields from drought-tolerant maize help to reduce significantly the number of people living in poverty in Africa and in other regions of the world (Gilbert 2014). Soya is more sensitive to drought and its annual yield can be reduced by approximately 40% (Clement et al. 2008; Liang 2016).

In Southern Africa, a region that is suffering consecutive droughts caused by climate change, some farmers have for decades been using seeds conventionally crossed and screened to identify the best gene patterns to be more tolerant to droughts. There is also evidence that drought-tolerant hybrids of traditional crops use water more efficiently. The development of a sustainable agriculture resilient to droughts that gives priority to local breeding of seeds and local knowledge, coupled with good farm practices and the use of crops adapted to the local climate and soil appears to be the best solution. More recently, drought tolerant GMO seeds produced by Monsanto have also become available and are being used, but seed sovereignty is now an important issue for many farmer groups in Africa. Furthermore, GMO seeds are currently lagging behind conventional breeding in efforts to create drought-resistant maize (Gilbert 2014).

Many politicians, businesspeople, and researchers think that conventional breeding cannot satisfy the global demand for food, both grain and animal protein, necessary to feed adequately a likely global human population of 9.7 billion people in 2050 (UNDESAPD 2017a) and that the large-scale use of GMO foods is unavoidable (Oliver 2014), while acknowledging in a more or less explicit way that there are still concerns as to the safety of GM crops for human consumption and the environment. In any case GMOs will not be able to solve the problem of long term global food security.

Several studies have concluded that to feed 9.7 billion people in 2050 (UNDE-SAPD 2017a), the global crop yield will have to double by that time. Yields have continued to grow in an approximately linear way at an annual mean rate of about 1.7% since 1960, due mostly to the development of synthetic fertilizers, hybrid breeding, and GMO seeds. However, to double the yield by 2050 it will be necessary to increase the annual rate from 1.7% to 2.2% in the next 30 years (Weber and Bar-Even 2019). Most of this increase could not come from expanding agricultural land, since the reserves of arable land are fast decreasing and the control of biodiversity loss requires the conservation of some natural habitats. Since the beginning of the Great Acceleration, about 30% of the world's arable land has become unproductive due to erosion, and topsoil is being lost 10–40 times faster than it is being replenished by natural processes (Laybourn-Langton et al. 2019). The only possible solution will be to increase the yield per unit area of land.

There are essentially two strategies to reach this objective: to increase planting density or to increase individual plant performance within a crop canopy. Regarding the first, humankind is already approaching a ceiling in intensely managed crop plantations (Mansfield and Mumm 2014). The strategy that is more promising in the medium and long term is to increase the performance of individual plants within a canopy and the most favourable way to achieve it is to improve photosynthesis in edible plants (Long et al. 2006).

Cyanobacteria were the first major producers of oxygen through photosynthesis, starting about 2.7 billion years ago. Later, around 1.25–0.75 billion years ago, red, brown, and green algae incorporated the photosynthesis mechanism and green algae evolved to create the vascular plants that are now essential for human life. The enzyme that catalyses the reaction in the Calvin–Benson photosynthesis cycle where atmospheric CO_2 is captured, to be converted later into sugar molecules, releasing oxygen, has been inherited from algae. It is called RuBisCO and is relatively inefficient. This is due to two things: the slowness of the process, since on average only three CO_2 molecules are converted per second, and the difficulty that RuBisCO has in distinguishing CO_2 molecules from O_2 molecules. Roughly 25% of the time, RuBisCo attaches O_2, in a process called oxygenation, instead of CO_2, in the process named carboxylation.

This problem was not as significant about one billion years ago because there was still little oxygen in the atmosphere at the time. Meanwhile, cyanobacteria have been able to adapt to the increasing percentage of molecular oxygen in the composition of the atmosphere that occurred up to the Carboniferous, when it reached a peak of about 30%, 300 million years ago (Berner 1999). They developed a photosynthesis mechanism that is more efficient than the one in vascular plants, in which RuBisCO is concentrated into carboxysomes that impede direct contact with O_2 in the air.

Human gene technology can take advantage of this adaptation by introducing the single-cell carbon concentration mechanism that cyanobacteria have developed to reduce oxygenation into the vascular plants that are essential for food production.

Recent molecular, biochemical, and physiological studies have unravelled the way the cyanobacteria operate to elevate CO_2 around RuBisCO molecules during photosynthesis. It is already possible to produce tobacco plants (*Nicotiana tabacum*), which are relatively easy to manipulate genetically, with part of their genetic photosynthesis mechanism replaced by the mechanism from the cyanobacterium *Synechococcus elongatus* PCC7942 (Lin et al. 2014). This is a first step towards increasing the efficiency of photosynthesis in crops. Other pathways to increase the efficiency of photosynthesis involve rerouting photorespiratory metabolism by introduction of synthetic bypasses and new ways that turn photorespiration into a process that captures carbon (Weber and Bar-Even 2019).

In general, plants directly exposed to solar radiation absorb more energy than they can use in photosynthesis. The excess energy is dissipated in the form of thermal energy by way of a non-photochemical quenching mechanism. However, plants take a long time to switch off their protective dissipation mechanism, typically many minutes, when, for example, a cloud passes overhead and they are left in the shade. The slowness of the response makes photosynthesis less efficient and may reduce by 20% the potential yield of crop plants (Kromdijk et al. 2016). About 50% of canopy carbon sequestration in crop plants occurs under light limitation and the efficiency of photosynthesis in the shade declines strongly with rapid light transitions caused by clouds and wind-driven movement of overshadowing leaves. Crop yields can therefore increase if genetic engineering succeeds in developing a more rapid adaptation to fluctuating light conditions by reducing the loss of excitation energy due to overprotection of plant photosystems (Weber and Bar-Even 2019). Kromdijk et al. (2016) have already succeeded in increasing the leaf carbon dioxide uptake and plant dry matter productivity by about 15% in a situation of fluctuating light in tobacco plants (*Nicotiana tabacum*), by transferring the genetic coding sequences of another plant.

Synthetic biology, which combines several sciences and technologies, including molecular biology, biophysics, biotechnology, computer sciences, and various emerging technologies such as big data and artificial intelligence will make it possible to build new biological systems and biological machines for a wide range of purposes. In 2010, a team of scientists led by the US geneticist and businessman J. Craig Venter managed to synthesise the genome of the *Mycoplasma mycoides* bacterium from its genetic code filed on a computer and introduce it into the *Mycoplasma capricolum* bacterium, whose DNA had previously been removed (Gibson et al. 2010). The new synthetic bacterium, which was given the name *Mycoplasma laboratorium*, began to live and to reproduce billions of times, controlled by the new genome.

This was one of the first steps towards creating synthetic life in a laboratory. In the future it is likely that it will become possible to build artificial multicellular organisms. The application of synthetic biology technologies and synthetic life in agriculture places humankind at the beginning of a process of creation of organisms that may be called superplants, supertrees, and superanimals, able to generate a future superagriculture.

The main justification for pursuing this path is to increase the likelihood of ensuring food security for a world population that is likely to continue to increase, although at a decreasing annual rate, until the end of the 21st century and probably beyond to reach eventually more than 11 billion people (UNDESAPD 2017a). All these human beings will hold a natural and legitimate expectation of having healthy and abundant food diets in a world that is likely to continue to be restricted by the overexploitation of natural resources, environmental pollution and degradation, and climate change. All these challenges have specific, differentiated, and interfering impacts in agricultural productivity, and most tend to be negative.

Will it be possible to continue achieving global food security in a relatively stable world up to 2100 and beyond? A very powerful driving force that contributes to reducing the risk of falling into a chaotic situation is that the applications of genetic engineering and synthetic biology in agriculture are likely to generate enormous profits for the businessmen and corporations that fund such applications and indirectly also the politicians and governments that support those businessmen and corporations. It is very likely that, in the future, food security for billions of people will be increasingly dependent on the large biotechnology corporations, which will strive to dominate the new food production markets.

It is impossible at this stage to foresee the impact that the global proliferation of superplant, supertree, and superanimal farming will have on human health and the environment. It is very likely that the environmental impact of increasing individual plant performance within a crop canopy will be less damaging than other strategies, in particular the generalized use of GMOs. The development of a superagriculture will not be hampered by the application of the precautionary principle regarding the medium and long term side effects on health and the environment. Development of superagriculture will be considered essential for human survival. It is also improbable that the risk of biodiversity loss and of endangering the remaining "natural" ecosystems by spreading the superagriculture beyond the farming and forestry anthromes will be a major obstacle to its development. Driven by a techno-optimist view, one may imagine that it will be possible to help the "natural" ecosystems resist the future global changes by using the new genetic technologies. The implementation of all these plans involves many dangerous risks, but food security is undeniably a crucial priority for humankind.

From an anthroecology perspective, all this is normal and even to some extent expected. The surprises that may emerge in this process will be controlled and managed to guarantee that the unquestionable goals of human well-being, economic prosperity, and economic growth can be met, at least for the part of humankind that already benefits from having achieved those goals. Many people and organisations do not share this one-dimensional and egoistic viewpoint and will continue to argue for other forms of prosperity (Jackson 2009), while trying to implement strong sustainability. These alternative views will continue to develop and likely become stronger through increasingly wider circles of acceptance worldwide.

The practical implementation of a superagriculture based on genetic engineering and synthetic biology is still a long way ahead. The forms it will take, the acceptance it may or may not find among people in countries with very diverse cultures and levels of development, especially in those with more intense food security problems, and the extent to which it will actually be able to solve the problems of food security, undernourishment, and famine are yet to be seen. On the other hand the way the biosphere will respond to such intensive and extensive transformations is also unknown and therefore a further potential risk for humankind.

There is a way, however, to contribute effectively to food security everywhere in the world but especially in countries with advanced economies that is accessible to all and has no damaging side effects on human health or the environment. According to the FAO (2019b) in 2011, about one third of the food produced globally per year for human consumption, which represents approximately 1.3 billion tonnes, gets lost or wasted. In industrialized countries, which have about 1.3 billion people, 17% of the global population, annual food losses and waste amounts to 680 billion US dollars, while the remaining part of the world population, about 6.4 billion, account for 310 billion US dollars. Per capita waste by consumers is between 95–115 kg a year in Europe and North America, while consumers in Sub-Saharan Africa and South and South-East Asia each throw away only 6–11 kg a year. If the developed countries continue to lead the world in the per capita value of food loss and waste, and if they do not manage to reduce it significantly, below the per capita value in developing countries, it will be much more difficult and likely impossible to ensure food security for everybody in the world during the 21st century and beyond.

7.10 Relationship Between Humans and the Technosphere and Biosphere

Global population growth, increasing energy consumption and use of natural resources, and the progress of science and technology will continue to extend and complexify the technosphere, and furthermore they will intensify and broaden its interference with the biosphere. As a result, it is likely that there will be a growing trend towards forms of fusion between the technosphere and the biosphere, which corresponds to the emergence of the technobiosphere. However, the technosphere and the biosphere have very different characteristics, especially regarding metabolism and their relationships with humans. The technosphere is a symbiotic system, built and kept running by humans for their benefit over many hundreds of years. Its evolution is determined by humans through social, political, and economic structures and activities and the application of science and technology over several time scales and space scales that range from the individual to families, communities, settlements, urban areas, and at the administrative level from municipalities to national governments and international organisations. The central objective of the technosphere is to extract energy from the environment to produce the work that is necessary for its functioning, which sustains the contemporaneous human civilization on a global scale.

The relationship between the technosphere and humans is a dynamic and reflexive one (Donges et al. 2017). Humans, by way of legislation, regulation, and social structures, are constantly trying to benefit, adapt, and react to the technosphere that they have created. They attempt to make it serve their multitudinous interests in the best way possible but without that imposing significant restrictions on their highly diverse and changeful liberties and behavioural styles. Nevertheless, due to the inertia, diversity, and the increasing complexity of its functional systems and structures at the national and global levels, the technosphere has acquired a semi-autonomous momentum with regard to humans, which has been described by Haff according to the following six rules (Haff 2014a):

> (1) The rule of inaccessibility, that large components of the technosphere cannot directly influence the behaviour of their human parts; (2) the rule of impotence, that most humans cannot significantly influence the behaviour of large technological systems; (3) the rule of control, that a human cannot control a technological system that expresses a larger number of behaviours than he himself; (4) the rule of reciprocity, that a human can interact directly only with systems his own size; (5) the rule of performance, that most humans must perform at least some tasks that support the metabolism of the technosphere; and (6) the rule of provision, that the technosphere must provide an environment for most humans conducive to their survival and function.

The last two rules are particularly important. The performance rule, establishes that humans are compelled to keep the systems of the technosphere functioning, for example, those that provide constant access to energy, food, transport, communications, and a wide range of other services, and material and immaterial goods. If such systems stopped functioning due to social upheaval, terrorism, wars, or economic collapse, the negative consequences for well-being and economic prosperity would be felt rapidly and would increase dangerously with the prolongation of the emergency.

The provision rule establishes that the technosphere must supply the essential conditions for the provision of goods and services that assure the increasing well-being and economic prosperity that most humans expect. The present growth model of the financial and economic system requires that the human desire for consumption of goods and services be boosted by all means available in order to maintain the sustainability of such a system. This process generates an unlimited positive feedback because science and technological innovation have a limitless capacity to create attractive new services and goods that boost the desire for consumption. Each step in this ongoing process extends and complexifies the technosphere.

The relationship between humans and the biosphere extends much longer in time, covers all of *Homo sapiens* time, and has a different kind of complexity. All members of the *Hominidae* family, whose evolution led to our species, have been part of the biosphere just like all species in the phylogenetic tree of life. The situation has changed because humankind has transformed and degraded a large part of the biosphere and therefore started to distinguish between the property of belonging to and the condition of depending on the biosphere. Now we depend primarily on the functioning of the technosphere, but we also depend on the biosphere, for instance, by its supply of raw materials, such as wood and biofuels, and in many other ways that we already mentioned in our brief description of ecosystem services in Sect. 7.7. Cutting away that dependence is impossible in the foreseeable future and reducing it is an additional risk for the future.

7.11   Energy Fluxes in the Biosphere and Technosphere, Energy and Exergy Efficiency, and Energy Sufficiency

To further analyse the relationship between humans and the biosphere and techno-sphere and its conditioning factors and limitations, one must consider the energy and material flows in the Earth system. Energy will be considered first, since energy supply is the foremost goal of the technosphere.

All subsystems of the Earth system, both biogeophysical and human, are open as regards energy and mass; in other words, there are energy and material flows between them. However, the Earth system is practically closed as regards material flows. The main material flows to and from outer space are relatively small: around 40 000 tonnes of cosmic dust falls to Earth each year (Zook 2001) and around 95 000 tonnes of hydrogen and 1600 tonnes of helium escape the Earth's gravitational field per year (Catling and Zahnle 2009).

In terms of energy, the Earth is open and receives a solar radiant energy flux of 174 PW, as already mentioned. There is another energy flux of thermal energy that is transferred from the Earth's interior through its surface, which is much less intense, amounting to 47 ± 2 TW (Davies and Davies 2010). This surface heat flux originates in the geothermal gradient inside the Earth, whose central core has temperatures of more than 5000 K.

Due to the energy conservation principle, the Earth radiates an energy flux into outer space with a mean value that tends to be equal to the incoming solar radiant energy flux. Around 30% of the incoming flux is reflected by the Earth into outer space and the remainder is absorbed and re-emitted, but in the form of electromagnetic radiation in the infrared range. As mentioned in Sect. 7.6, anthropogenic atmospheric emissions of greenhouse gases produce an energy imbalance in the infrared radiation at the tropopause that is causing a change in the global climate.

The outgoing infrared radiation is a "degraded" energy when compared with the incoming solar radiant energy, which has its greatest flux in the visible range of the electromagnetic spectrum. The number of photons, or quanta of light, in the visible range received from the Sun and absorbed by the Earth, each having a certain amount of energy, is much lower than the number of photons the Earth re-emits in the form of infrared radiation. The reason is that each photon in the visible range carries more energy than each photon in the infrared range, and the total absorbed energy is equal to the total re-emitted energy in order to keep energy balance. The energy of a photon in the visible range is around 25 times greater than the energy of a photon in the infrared range. The outgoing energy is intrinsically less ordered or degraded because it is distributed over more particles, and thus corresponds to more degrees of freedom.

From a physical perspective, this means that the entropy of infrared radiation emitted by the Earth is greater than the entropy of solar radiation absorbed by the Earth. It is the difference between the low entropy of the incoming flow of radiant energy and the relatively high entropy of the outgoing flow of radiant energy that makes it possible for there to be life on Earth.

In fact living organisms are open systems as regards energy. They maintain greater order than their surroundings by importing free energy in the form of food and exporting entropy to their environment as heat and waste (Schrödinger 1944). In the environment, one finds many other examples of dynamic self-maintaining open systems that import free energy and export less useful forms of energy, in particular heat, such as flames, forest fires, cyclones, and volcanic eruptions. While they persist as self-organizing and self-maintaining structures, they are far from thermodynamic equilibrium with their environment. When they decompose and lose their order, they reach thermodynamic equilibrium with their environment.

If the solar radiation coming towards the Earth were to stop, after a while the energy and material flows in the biogeophysical subsystems of the Earth system would also stop, and they would shift towards a state of thermodynamic equilibrium, a state of maximum entropy, a "dead" state with no movement or temperature gradients. In other words, there would be no free energy to allow life to continue. The water cycle would cease, there would be no more wind, precipitation, glaciers, ice sheets, or rivers, and living organisms would end up dying. The activity at the Earth's surface would only be driven by the differential gravitational forces that cause the tides, by seismic and volcanic events that depend on the heat generated in the radioactive decay of radioisotopes found inside the Earth, and by meteorites and other bodies that come from outer space and collide with the Earth.

The fall into a dead state through irreversible thermodynamic processes is avoided by the incoming solar radiation essentially through two main mechanisms: photosynthesis and the general circulation of the atmosphere. The latter results from the fact that the amount of solar radiant energy absorbed daily per square meter of the Earth's surface is greater in equatorial regions than at higher latitudes. The absorbed energy is redistributed over the Earth's surface by the general circulation of the atmosphere, which is responsible for the water cycle, and also by the ocean currents. About 75% of the latitudinal global heat transport is constituted by the air currents and 25% by the ocean currents.

Photosynthesis, on the other hand, is part of the carbon cycle, the biogeophysical cycle in which carbon is exchanged between the biosphere, atmosphere, hydrosphere, and lithosphere, and which directly or indirectly provides the energy necessary for living organisms, except for some bacteria that live on the ocean floor near hydrothermal vents and get their food by chemosynthesis of chemicals, such as methane and hydrogen sulphide. The chemical energy accumulated in the plant through photosynthesis is then consumed in a controlled way through respiration, a slow combustion process that keeps the organism in a stable living state and which involves organic matter and the atmospheric oxygen. Instead of consuming the accumulated energy in a controlled way through respiration to keep the organism alive, it is possible to return to the initial state of high entropy directly through the complete combustion of the plant, releasing CO_2, water vapour, and mineral waste into the environment.

In both autotrophs and heterotrophs, the vehicle for the flow of energy for the controlled consumption of the chemical energy that maintains their daily life is provided by ATP (adenosine triphosphate) molecules. Some organisms, besides consuming energy in respiration, emit radiant energy by way of bioluminescence, fluorescence, or phosphorescence as a result of several chemical and physical processes.

Let us now briefly consider the relationship between the technosphere and energy. Humans increased their total annual consumption of energy by a factor of more than 28 in a period of 217 years from 1800 to 2017, from 20.35 EJ to 583.8 EJ. This is an impressive achievement that lies at the heart of the social, economic, scientific, and technological progress made during that period of time. From the early beginning of the intensive use of fossil fuels, humankind was aware of the need to increase the efficiency of primary energy extraction and use. The increasingly efficient use of energy has made a decisive contribution to economic growth since the Industrial Revolution (Ayres 2009). Moreover, the greater efficiency has generated a positive feedback, which has increased global energy consumption. The British economist William Stanley Jevons (1835–1882), who pioneered the application of mathematics to economics, first noticed this effect, now called the Jevons paradox, when he observed that the increased efficiency in the use of coal led to an increased consumption of coal in various industries (Jevons 1865). Jevons was preoccupied with the gradual exhaustion of coal supplies from British mines. Of course, he could not foresee the future exponential increase in the global conversion and consumption of energy in the form of fossil fuels, but because of his concerns at the national level during his lifetime, he is considered to be the first economist who developed an ecological perspective of the economy in which human capital cannot always replace natural capital (Martínez-Alier 1987).

It is curious and significant that Jevons' observation about the relation between increased efficiency and increasing consumption came to be called a paradox, since in fact the underlying behaviour is an example of the essential human trait of consuming eagerly what is desirable and available, without thinking much about the medium and long term implications.

Energy efficiency is currently considered to be an essential condition to ensure economic growth and also to achieve both weak and strong sustainability. In physics, energy conversion efficiency is defined as the energy ratio between the useful output from an energy conversion machine and the input. From an economic point of view, energy efficiency is the ratio between an energy service output and an input of primary energy. The output may be evaluated in terms of performance, as for instance thermal comfort in a building, of services provided, as in the transport sector, and of all sorts of goods that require the use of energy, or simply of energy of a given form.

Better energy efficiency can be achieved at all points of the energy cycle from the energy required for the extraction or capture of primary energy sources to final energy consumption. Overall energy efficiency requires the shifting in time of energy conversion and consumption, during the day or even for longer periods of time, in order to diminish the use of energy without affecting the expected outputs.

The need to maintain an abundant supply of energy at the lowest possible price and to diminish the dependence on fossil fuels created an urgent need to improve all forms of energy efficiency. Regarding this process, in 1956, the Yugoslavian chemical engineer Zoran Rant (1904–1972) started to use the word "exergy" for a physical concept that was previously developed by the USA physicist Josiah Willard Gibbs (1839–1903).

To define exergy, first recall that the extent to which a certain form of energy can be converted into another form of energy depends on the nature of the energy that is converted. For instance, electrical energy can be completely converted into thermal energy, but the opposite is impossible: thermal energy cannot be completely converted into electricity. The same applies to the conversion of thermal energy into mechanical energy, which is a particularly important form of energy for humans. Exergy is a thermodynamic concept that measures the capacity of a given form of energy to be converted into mechanical energy, or in other words, to do work. It is defined as the maximum useful work that can be extracted from a system as it reversibly comes into equilibrium with its environment. The remaining energy is called anergy and corresponds to the residual thermal energy. Anergy is energy that has no capacity to do work, such as for instance the atmospheric heat in a parcel of air.

Exergy is a property of all material and energy flows, and depends upon the temperature, chemical composition, and electric potential of the thermodynamic system relative to its external environment. The contrast between the system and its environment defines the amount of exergy available. The exergy of the system increases with the difference in temperature between the system and its environment.

Exergy can be defined in a system that belongs to the biosphere and is a living organism or ecosystem, or belongs to the technosphere and is, for example, a motor vehicle, a boat, a plane, or a building. One can also consider exergy in more complex systems, such as an urban area, the economy of a country, or the global economy.

The design, evaluation, and optimisation of systems as regards their energy consumption is usually based on the concept of energy efficiency. This type of analysis can be complemented by the evaluation of the exergy efficiency, also known as second-law efficiency or rational efficiency. The objective is to evaluate how far the system is from an ideal system in which the processes are reversible, which means that the system is optimised from the perspective of the second law of thermodynamics (Bejan 2006). When speaking about the energy efficiency of an engine, one does not measure how much it approaches the ideal condition permitted by the second law of thermodynamics. That can only be done using exergy efficiency.

Exergy efficiency is a more suitable indicator for determining ways to facilitate economic growth by reducing energy consumption to the physically allowed minimum. One of the likely causes for the downturn in economic growth in the USA in the last 30 years was the slowdown in exergy efficiency growth, which had a mean annual rate of increase of 1.45% between 1950 and 1980, but only 0.41% between 1980 and 2010 (Laitner 2013). In the first period, average annual economic growth was 2.25%, while in the second it was 1.72%.

Exergy efficiency is still difficult to determine so most analyses are based on energy efficiency. The positive financial, operational, and societal benefits that result from improving energy efficiency are currently fully recognised by governments, businesspeople, international organizations, and non-governmental organizations. Furthermore, there are signs that society in general is beginning to be aware of those benefits and to participate in the effort to increase energy efficiency, especially in countries with advanced economies. The application of the emerging new technologies is likely to accelerate energy efficiency. Artificial intelligence will allow the optimization of energy use through real time analysis of supply and demand of electricity at the micro-grid level. It would be possible to alert people when energy is being wasted and to take action to reduce that waste. Furthermore, the use of micro-grids and distributed renewable energy sources will enable companies and prosumers to sell their excess energy with a high degree of autonomy and security. The management of their own supply of energy from renewable sources is being facilitated by the advances in energy storage as battery costs come down. Storage can help companies and prosumers to reduce the variance in electricity prices that results from surges in demand. An accurate record of the energy transactions between producers and consumers can be achieved with the help of blockchain technology. Energy efficiency in buildings can be further increased by the use of air- and ground-source heat pumps. Waste heat from factories and urban infrastructures can be used for space heat elsewhere in the urban area.

Is energy efficiency an adequate concept to assure that the global energy consumption in the medium and long term is environmentally sustainable at the global level? In other words, should people in advanced economies change their energy consumption habits and start consuming less energy motivated by considerations based on the advantages of greater energy equality in the world in order to achieve global sustainability? Some of those who think the answer is affirmative, have developed the concept of "energy sufficiency" or energy conservation (Princen 2005). The generalization of this concept leads to eco-sufficiency, which aims at reducing in absolute terms both the consumption of energy and natural resources and the generation of waste (Sachs 1993; Boulanger 2010). The crucial point behind energy sufficiency is to accept that there are environmental limits or planetary boundaries (Rockström et al. 2009a, b) that is advisable not to cross. By accepting these boundaries, there is an onus on affluent countries to bring about reductions in energy use so that developing countries can converge with them without increasing global energy use by too much (Darby and Fawcett 2018).

The question is whether to recognize or not that there are environmental limits that will be reached in the near future if we do not adopt a long-term sustainable use of energy and natural resources. The energy consumption of a house, building, town, or transport system built according to the principle of energy sufficiency would be lower in absolute terms than if the same infrastructures and services were planned to satisfy the stringent energy efficient specifications. Energy sufficiency implies a change in our attitude and behaviour driven by the objective of reaching long-term strong sustainability at the global level.

Energy sufficiency is also a way to avoid the Jevons paradox. More energy-efficient technologies usually tend to encourage more energy consumption, while energy sufficiency has the explicit voluntary objective of reducing energy consumption in absolute terms. Energy sufficiency involves, and benefits from, the application of the same technologies and procedures that are used to implement energy efficiency. The difference is that there is now a further requirement based on preventing the crossing of planetary boundaries. It will be very difficult to implement the practice of energy sufficiency worldwide, but on the other hand, controlling the growth of energy consumption is one of the most important conditions to ensure sustainability in the 21st century and beyond.

7.12 Metabolism of the Biosphere and Technosphere. Hopes for a Circular Economy and the Problem of Waste

Let us now briefly address the question of material flows and metabolism in the biosphere and technosphere. During the evolution of life the biosphere developed an almost perfect metabolism in the life cycles of different organisms. When an organism dies, the biosphere has other organisms that recover its materials and reinsert them in the web of life. The materials of the dead organisms are repeatedly upcycled and there is no loss of quality in the recycled products, in contrast with the technosphere which, by down-cycling, destroys the original value of materials. From the first unicellular organisms to humans, the biosphere has used cells as the building blocks of organisms of increasing complexity that colonize the ecosphere and recycle its material flows in a sustainable way by constantly adapting to a very diverse and changeable environment. For about four billion years, the biosphere did not accumulate what humans have come to designate as waste, because all the outputs of life were reused without loss of value.

In the long story of the evolution of life, there has been at least one notorious exception when the biosphere probably took some time to solve a relatively difficult problem. The episode occurred during the Carboniferous, between 358.9 and 298.9 million years ago, and is related to the beginning of the large scale formation of coal. The huge coal deposits from the Carboniferous result primarily from the emergence of lignin (Weng and Chapple 2010), suberin, cutin, and other complex organic polymers that made it possible to support plants above the ground, overcoming gravity and thereby starting the formation of forests. These forests were composed mostly of arborescent vascular plants called Lycophytes, related to mosses, Monilophytes, related to ferns, Equisetophytes related to horsetails, and Spermatophytes related to seed plants. Some of them reached heights of more than 50 m. The Carboniferous had a unique combination of very wet tropical conditions and extensive low lying areas covered by wetlands, swamps, lakes, and shallow coastal areas, which afforded ideal conditions for the development of luxuriant forests of vascular plants.

This was a time when tectonic plates were moving towards the progressive formation of the supercontinent Pangaea. The climate and orography created depositional systems for the dead trees that contributed to preventing their decomposition (Nelsen et al. 2016). Moreover, when lignin emerged as the solution for plants to overcome gravity, there were still no organisms capable of decomposing it when the plants died, some falling into swamps, where they were later covered by sediments and transformed into peat, lignite, and, finally, coal after hundreds of millions of years. Lignin is particularly difficult to rot and it took tens of millions of years for the evolution of white rot fungi that could achieve any substantial degradation of this molecule (Robinson 1990; Hibbett et al. 2016). Molecular clock analyses suggest that the origin of lignin degradation may have coincided with the sharp decrease in the rate of organic carbon burial around the end of the Carboniferous period (Floudas et al. 2012). Only in the Cenozoic did lignin-degrading organisms become sufficiently abundant to accelerate the land organic carbon cycle, which implied a subsequent decline in the primary terrestrial production of coals and kerogens.

These events in the Carboniferous had strong impacts on the composition of the atmosphere and on the climate. The high sequestration of atmospheric CO_2, without significant CO_2 emissions from the decomposition of the dead trees, decreased the atmospheric CO_2 concentration, which reached values of 100 ± 80 ppmv at the beginning of the Permian, giving rise to a glaciation that narrowly escaped from being a fully glaciated Snowball Earth state (Feulner 2017). Furthermore, the photosynthesis of the large Carboniferous forests increased the oxygen concentration in the atmosphere to values of the order of 30%, as already mentioned. The greater oxygen availability promoted gigantism in insects and amphibians and probably increased global biodiversity. However, it was a short term event, since it ended with the collapse of the Carboniferous rainforest about 305 million years ago, which resulted from the change to a much cooler and drier climate.

This extraordinary story contains a metaphor for our contemporary social time. Most of the coal that has been used as the primary energy source for the Industrial Revolution resulted from the sequestration of CO_2 by the forests of the Carboniferous. The CO_2 emitted from burning this coal was returned to the atmosphere and is driving anthropogenic climate change. If the biosphere had been able to react faster to solve the problem of the "wasted" lignin wood that was not decomposed, it is likely that no glaciation would have occurred at the end of the Carboniferous and the present anthropogenic climate change would have been weaker. Oil and gas are also contributing to climate change, but their origin bears no direct relation to the forests of the Carboniferous. They are mostly derived from dead marine micro-organisms that accumulate in sea-floor sediments, escape oxidation, and become buried beneath other sediments for hundreds of millions of years. Depending on the temperature and pressure conditions and the nature of the rocks, some of the deposits gave rise to oil, others to natural gas.

Unlike the biosphere, the technosphere generates large quantities of materials that have no direct or immediate human value and are called waste, mostly with low energy and high entropy. A large part of this waste builds up in the environment and is not reused. In the biosphere, reuse gives value to all kind of materials. It happens everywhere and tends to create a situation of equilibrium that benefits all living organisms. In the technosphere, solid and liquid waste together with aerosol and gas emissions are fast increasing and leading to a situation of growing disequilibrium that is having a negative impact on the environment and creating a risk for humans.

Risk is an anthropic concept, which is usually defined as something having a probability of damage, injury, liability, or loss in different domains such as health, food security, economy, finance, and insurance. The negative impacts of waste on the environment may generate a risk for humans, but only indirectly. Consider, for instance, the case of some biological species, with no direct value, that are in danger of disappearing because their habitats are being degraded or destroyed by human activities. To say that their extinction is a risk for humankind would involve attributing an economic value to those species, or to the ecosystem services in which they participate. As shown previously, this is a problematic issue where there is no consensus. If we imagine that humans were to disappear or that their population would drastically fall to a few thousand for some reason, most of the effects of previous human interference in the Earth system, such as waste, environmental degradation, and climate change, would be reversed. After some time, probably tens or hundreds of thousands of years, or even millions of years, only a few technofossils would remain. The biosphere would reuse a large portion of the accumulated human waste and the other biogeophysical subsystems would help redirect the Earth system towards a new state of equilibrium.

Humankind is increasingly using the Earth's natural resources and producing various kinds of solid and liquid wastes and emissions at a rate at which they cannot be renewed or sequestered. Resource accounting has usually been made through two indicators: the ecological footprint, which quantifies the demand for the biosphere's renewable natural capital; and biocapacity, which quantifies the Earth's supply of that capital. According to the Global Footprint Network, humankind has been using natural capital 1.7 times as fast as the Earth system is able to renew it (Lin et al. 2018). This situation causes greater scarcity and increases the price of some natural resources, the more vulnerable countries being those that have greater resource dependency. Some industries are already threatened by the difficulty to overcome the scarcity of resources. It can therefore be safely concluded that the future will be increasingly shaped by ecological constraints. Nevertheless policy decisions by governments and businesspeople are still often made under the assumption of limitless resources and ecosystem services. This approach will gradually change because trends in natural resource consumption are no longer immaterial for economic growth, not as they used to be after about 1820, when the major social and economic benefits of the Industrial Revolution began to be felt.

The economy of the industrialized countries during most of this period of about 200 years has been running on a one-way model of production and consumption. Goods manufactured from raw materials have been sold, used, and then incinerated or discarded as waste. With increasing industrialization around the world, a rising global population that entertains high expectations of well-being and economic prosperity, and the associated growing use of natural resource, it becomes impossible to reach sustainability with the linear business-as-usual model of production and consumption.

The circular economy concept addresses these challenges. It is based on practices that were intuitive to humans when they belonged to the biosphere. Humans were an integral part of the regenerative cycles of life, with needs comparable to those of their fellow primates. Now that humankind requires much more for its well-being, it is time to apply the essential principles of the metabolism of the biosphere to the technosphere. It is going to be a challenging problem. The biosphere has an autonomous evolution that has complexified its organisms and processes for about 4 billion years, including its recycling processes. However, the durations required by Darwinian evolution are incommensurate with the element of operative time of human life. The new metabolism in the technosphere is also different from that of the biosphere, because it involves only one biological species and depends primarily on its collective behaviour. Without a well organized and fluid collective global behaviour, it is likely to fail.

One of the pioneers of the circular economy was the Swiss architect Walter Stahel, who wrote a report to the European Commission in 1976 entitled "The potential for substituting manpower for energy" (Stahel and Reday-Mulvey 1976) and a few years later "The product life factor" where he defended the idea that the transition towards a sustainable society would require a closed loop economy, which was one of the names given initially to the circular economy (Stahel 1982).

Towards the end of the decade, two US environmental economists further developed the concept of a circular economy (Pearce and Turner 1989), also known as "cradle to cradle", as an alternative to the linear economy, known as "cradle to grave". The latter is characterised by the traditional economic model based on a "take-make-consume-throw away" approach to resources, with the waste dumped in the environment. Systems that use the circular economic model reuse, share, repair, refurbish, remanufacture, and recycle to create a closed-loop system that minimises the use of resource inputs and the creation of waste, pollution, and carbon emissions (Geissdoerfer et al. 2017). However, the circular economy still faces considerable challenges and at present only a few percentage points of the original product value is recovered after use. One of the major challenges of the circular economy is how to control the accumulation of waste, in particular plastics.

According to a 2012 World Bank study (WB 2012a), the amount of municipal solid waste (MSW) produced globally was about 1.3 billion tonnes. Around 44% of this waste came from OECD countries, but production was growing faster in other countries, particularly in the emerging economies. Based on global population estimates, it was projected that in 2025 the annual global production of MSW would hit 2.2 billion tonnes. From 1900 to 2000, the global urban population grew from 13% to 49%, while the solid urban waste produced increased tenfold, from 300 000 tonnes to 3 million tonnes per day. As the GDP per capita of a country grows, the inorganic component of urban solid waste, made up of materials such as plastics, paper, glass, aluminium, and many other metals, grows in relation to organic materials, making a circular economy more necessary than ever. This problem is less significant in rural regions, where food products are less packaged and food waste is comparatively lower. It is calculated that, on average, the amount of solid waste produced per capita in urban areas is four times higher than in rural areas.

According to a World Bank report from 2018 (WB 2018), the world was generating 2.01 billion tonnes of MSW annually, a value that should have been reached only in 2022, according to the 2012 estimate (Hoornweg and Bhada-Tata 2012). About one third of the MSW that is currently generated annually is not managed in an environmentally safe manner. There are very large variations in the per capita daily generation of MSW, ranging from 0.11 to 4.54 kg, with a global mean value of

0.74 kg. The contribution of the high-income countries to generating MSW has decreased from about 44% to 34% in the period 2012–2018, but these countries account for only 16% of the world's population.

Global MSW is projected to increase to 3.4 billion tonnes by 2050 (Kaza et al. 2018). Projections until the end of the century indicate that MSW generation in high-income countries will peak in mid-century, but the peak of global production may not be expected before 2100 (Hoornweg et al. 2015). It could be brought forward to 2075, but that depends on how citizens, governments, and businesspeople respond to the challenges of the quest for a circular economy in the next few decades. The timing of the peak waste will depend substantially on development in certain regions of the world, especially in the cities of Sub-Saharan Africa, where population growth rates are more than double those in the rest of the world.

More than half of the waste produced in the Middle East and North Africa, Sub-Saharan Africa, and South Asia is openly dumped or burned (Kaza et al. 2018). About 54% of the waste produced by upper-middle-income countries is left in land-fills. In high-income countries, 39% goes to landfills, 36% is recycled, and 22% is incinerated (Kaza et al. 2018). Landfills constitute an additional problem for the environment since globally they are responsible for about 11% of anthropogenic methane emissions into the atmosphere, resulting from the decomposition of organic waste under anaerobic conditions.

There are many solutions to the problems of recycling waste, but they usually require a considerable investment to address questions related to waste management, such as infrastructure, governance, financing, and capacity. Waste management is a costly activity that can be among the highest budget items for many municipalities. A recent interesting example was that China, which imported a large part of US waste, restricted imports of some recyclables such as mixed paper and most plastics from the USA. As a result, waste management companies told municipalities that there was no longer a market for their recycling, which implied that they would have to pay the additional costs to keep on recycling. Without a market, certain kinds of waste tend not to be recycled. Waste recycling is not the kind of business that can reward the investor with very large profits essentially because waste, instead of being part of the main goals of human life, is an unwanted by-product of consumption. However, uncollected waste and poorly disposed waste have negative impacts on health, economic prosperity, and the environment.

7.13 Perpetual Economic Growth

Economic growth as we know it was practically non-existent in the world until 1750 and has been increasingly robust and continuous for roughly the last 270 years, except for relatively short intervals of time. This period of global GDP growth has been a bonanza that has amazingly improved the living conditions of billions of people throughout the world by contributing to their health, economic prosperity, and well-being. In most economic and political circles, and generally speaking in public opinion, there is a predominant belief that such growth is a potentially perpetually achievable goal with innumerable social and political benefits (Toman et al. 1995; Drews and van den Bergh 2016). This culture is based on a conviction formally supported in the neoclassical theory of economic growth developed by Robert Solow and other economists. In this theory the economy is represented as a mechanical, circular, closed system where natural capital can always be replaced by human capital.

The aggregate production function includes only capital and work as factors of production and excludes natural resources. A literal interpretation of this functional relationship implies that human civilisation does not need natural resources. In Solow's words (Solow 1974): "If it is very easy to substitute other factors for natural resources, then there is in principle no 'problem'. The world can, in effect, get along without natural resources […]."

The Romanian American mathematician and economist Nicholas Georgescu-Roegen (1906–1994), one of the founders of ecological economics, underlined the incompatibility between Solow's point of view and the application of the laws of physics to the economy, particularly regarding the energy and material flows that run through it and enable it to operate (Georgescu-Roegen 1975). He also criticised the inability to deal with the problem of the intergenerational distribution of the Earth system's finite natural resources, some of which are not renewable. He called this "a dictatorship of the present over the future" (Georgescu-Roegen 1975). The criticisms went by unnoticed by the community of neoclassical economists and Solow ignored them, probably due to their irrelevance in the short term compared with the immediate benefits that strong economic growth was generating in the USA and other countries at the time.

However, a little later, Robert Solow and Joseph Stiglitz proposed a Cobb–Douglas type production function (Cobb and Douglas 1928) that explicitly includes natural resources. Production P is given by

$$P = K^a L^b R^c ,$$

where K is the capital stock, L is work, and R is the flow of natural resources. The exponents are positive and satisfy $a + b + c = 1$. The US ecological economist Herman Daly demonstrated that this function, which Georgescu-Roegen called the Solow–Stiglitz variant of the Cobb–Douglas function, was not accurate in its representation of the role natural resources play in the economy in the long term (Daly 1997b). It would be possible to increase P without limit while keeping L fixed, and also keeping R at an arbitrarily small value as long as K were to increase. However, that is not physically possible because an unlimited increase in K requires the use of growing amounts of energy and natural resources, for example, to make the machines used to produce goods and services. A production process in the economy always demands the use of quantities of energy and materials that cannot be entirely replaced by increasing other factors of production. This statement takes into account the fact that science, technology, and innovation make it possible to develop procedures, techniques, and materials that reduce the amount of energy and natural resources needed in the production processes. It also bears in mind that full implementation of a circular economy increases energy efficiency, reduces the amount of natural resources used in production and consumption, and reduces emissions and waste.

The problem of energy availability and finite natural resources is acknowledged, but intervenes in current economic discourse in only a rather marginal way. The important fact that stands out is that the neoclassical theories of economic growth continue to be applied successfully, strengthening the conviction that perpetual growth is within our reach. Following Daly's rather provocative article, Stiglitz (1997) felt the need to reply in a very short article to, according to him, minimise the consumption of natural resources, stating that: "In the end, we hope that we have made our essential points, using somewhat fewer trees and other resources than Daly did in his 15-page note." The answer began by acknowledging that Daly was rightly concerned about the wasteful use of natural resources and environmental degradation. However, Stiglitz believes that when resources are scarce, markets reflect that scarcity and economize their use. If that is not the case one is facing a market failure that can always be corrected. Regarding the viewpoints of Georgescu-Roegen that the law of thermodynamics establishes limits on economic growth, he stated that "No one, to our knowledge, is proposing repealing the laws of thermodynamics!"

The explanation for the misunderstanding is found when Stiglitz writes that the goal of neoclassical economists with their economic growth theories is to formulate analytical models that ensure sustainable economic growth "for the next 50–60 years". In the end, they try to ensure growth that is as strong and continuous as possible in the short and medium term—in Stiglitz's own terms, during the "intermediate run". If we go beyond the 50–60 year period and look for the limits of equations used in the models when the variable time continues to move forward indefinitely, results become absurd "but no one takes these limits seriously" (Stiglitz 1997). The equation that gives us the production function is only valid in the relatively short time of the intermediate run, but incorporates variables that are subject to the laws of physics, which are time independent.

The reason for favouring the intermediate run rather than the long term arises directly from the fact that the operative social time of the contemporary generation is focused on its own element of operative time. The spaces of experience and horizons of expectation of future generations are not relevant or barely relevant because they go beyond the operative time of the contemporary social generation. With the rise in life expectancy, especially in countries with advanced economies, a person at the start of their working life has an outlook of 40–50 years ahead of them. The economic models mentioned by Stiglitz are calibrated to satisfy the horizons of expectation of well-being, economic prosperity, and capacity to consume goods and services of that young age group, the importance of which is clear from a social, political, and economic standpoint.

Developing a theory of economic growth that includes three, four, or more social generations is not a priority because it creates the risk of impairing the contemporaneous generation. Doing so would require the long term planning of energy and natural resources consumption, atmospheric emissions, and waste disposal for a range of generations. Introducing these variables in the long term places limitations on current global consumption and on the present levels of economic prosperity so that successive social generations may have similar opportunities. It is therefore preferable to argue that very likely, in the future, science and technology will have solved the problems that now seem so intractable. Neoclassical economic theory acknowledges the over-exploitation of natural resources, pollution, emissions, and waste as market failures that should be corrected. However, in most cases, it proposes palliative corrections focused in the short and medium term period of 50–60 years that ignore the consequences of the long term trends.

In the mainstream neoclassical theory, economic growth is a vital policy objective that is essential to improve well-being, economic prosperity, and high levels of employment. Within this framework it becomes imperative that human ingenuity founded on science and technology will be able to guarantee the necessary and sufficient conditions to achieve perpetual economic growth so that all future generations will benefit from it. In practice, however, many people in the advanced economies and in many other developing economies feel that they are already living in a zero-growth economy. This contradiction between theory and experiment is putting pressure on neoclassical economics, but significant changes are likely to be very difficult to achieve.

In today's financial and economic environment there are many instances where a reasonable level of economic growth needs to be boosted by a strong growth in debt. Ultimately, however, it is the confidence in the future of the economy that justifies the systematic use of credit to promote economic productivity. The lender must have confidence in the borrower's ability to repay the loan received. Trust is essential to the sustainability of the economic and financial institutions.

When there is strong growth and the economy is flourishing most businesses rush to take extra debt in order to grow even more and increase profits, which is an essential form of human behaviour that capitalism tends to favour. In an economic downswing, there is a tendency to panic and pay off debt in order to avoid its consequences, and this, too, is an essential form of emotional human behaviour that can be damaging to the economy. This is one of the mechanisms that can lead to the recurrent financial and economic crises of capitalism.

Since the 2008–9 financial and economic crisis, most countries have increased their debt to levels never before reached. A 2016 IMF report covering 113 countries concluded that global nonfinancial debt, which includes debt incurred by governments, households, and businesses, had doubled since the beginning of the century and reached the record level of 152 trillion dollars in 2015, representing 225% of global GDP (IMF 2016a). In 2019 global debt reached 188 trillion dollars, which is 230% of the global GDP, while private debt made up almost two-thirds of the total debt (Georgieva 2019). This debt has partly been encouraged by the central banks, with quantitative easing and cuts in benchmark lending rates down to zero or even below zero to stimulate the economy by providing easy access to credit. In advanced economies, public debt in 2019 was at levels only seen since the Second World War, while emerging markets had debts at levels last seen only in the 1980s debt crisis. China at the end of 2019 had a total debt of 40 trillion US $. In low income countries, public debt has also increased fast in the period 2014–19.

There is no consensus about the value of the debt to GDP ratio at which the situation may be considered to be alarming, but it is known that it was the uncontrolled growth of private debt that led to the subprime crisis in the USA. High debt burdens involve the risk of leaving governments, households, and businesses vulnerable to a sudden tightening of financial conditions. The rise in private debt is not spread evenly throughout the world, and is more marked in countries with advanced economies and in some emerging economies such as China.

The most conspicuous leaders of private debt usually follow a trend of positive feedback in which the desire to increase relative economic prosperity persists or grows with it. Again we find the same essentially greedy form of human behaviour that was referred to above. There are many individual examples of exceptions to this trend, but they are not dominant at macroeconomic level.

The IMF, central banks, other regulatory institutions, and governments would not have to worry about debt if world GDP growth had been higher. One success story in which this mechanism worked well took place during *Les Trente Glorieuses*. The current problem is that, although GDP is growing in many countries, that growth is not believed to be high enough. According to OECD and World Bank reports (OECD 2017; WB 2019), global real GDP grew 2.806%, 2.482% 3.109%, and 2.974% in the years 2015, 2016, 2017, and 2018, respectively, which is lower than the average of about 4% in the two decades that preceded the 2008–9 crisis (WB 2019).

The likelihood of returning to averages of 4% primarily depends on being able to increase the productivity of the economy. Techno-optimists believe it can be achieved with the expected technological innovations of the Fourth Industrial Revolution, such as artificial intelligence, robotics, big data, nanotechnology, and synthetic biology. However, current indications don't allow much optimism. The expectations of well-being and economic prosperity are unlikely to be fulfilled by the Fourth Industrial Revolution. Robert Gordon has provided evidence that current technological innovations are giving a relatively limited boost to productivity growth (Gordon 2016).

A likely cause for the present situation is that demand-induced scarcity of natural resources caused by population growth and increasing levels of consumption, and supply-induced scarcity of natural resources caused by environmental degradation and climate change, create additional production and technological costs that constrain economic productivity growth (Heinberg 2011). These constraining factors are likely to become more forceful in the medium and long term. Governments, especially in the advanced economies, address this challenge by trying to decouple GDP growth from energy and natural resource consumption, mostly by encouraging the transition to a circular economy. This model is defined by the European Union action plan, as a process "where the value of products, materials and resources is maintained in the economy for as long as possible, and the generation of waste minimised […]" (EC 2015). In other words, the aim is to reduce the use of natural resources and to recycle waste as much as possible so that the cost of products, materials, and resources does not increase. Furthermore, it aims at reducing emissions, sources of pollution and the production of non-recycled waste in order to lessen the environmental impact of human activities. Maintaining price stability means creating conditions favourable to growth in economic productivity, which means making the economy more competitive. In short, this is a way of recognising that current constraints on the use of energy and natural resources tend to harm economic productivity growth.

When considering decoupling one must consider to what extent economic growth, defined as growth in GDP, can be decoupled from the consumption of energy and materials. Another type of decoupling that is more general and difficult to define precisely is the decoupling of economic growth from its environmental impacts. There is an enormous variety of types of human interference in the environment that interact among each other. These range from hunting and collecting fruit and plants, as practised by indigenous peoples in remote regions, such as the interior of the Amazon, to changes in land use, including the subsistence farming used extensively throughout the world, intensive agriculture whose area of production is growing fast, and deforestation. Other examples of interference include the many different forms of fishing, the exploitation of water resources, the exploitation of mineral resources, urbanisation, air, water, soil, and ocean pollution, biodiversity loss, and anthropogenic greenhouse gas emissions. The extension, intensity, and complexity of these anthropogenic impacts on the environment and the interactions and synergies between them are growing. For example, loss of biodiversity has many causes that interact with one another, including poaching, degradation, and destruction of habitats caused by land use changes and climate change. It is likely that climate change, from 2070 onwards in business-as-usual scenarios, will begin to put more pressure on the biodiversity of vertebrates than land use changes (Newbold 2018). Representing the complexity of all these interferences and pressures on the environment using quantitative indicators remains a challenge.

In the decoupling analysis, the global human population must be explicitly taken into account. Decoupling may happen in per capita decoupling indicators but even so global decoupling may not, due to the increase in population. The evolution of per capita indicators at country level is especially important to evaluate the extent to which a given country is progressing towards sustainable consumption and production.

Current natural resource scarcity, environmental degradation, and climate change are already giving signs that perpetual economic growth will not be possible. However, in the short term, and in the medium term of the next 50–60 years, there is a range of previously tested and innovative solutions for maintaining an economic growth that is as robust as possible. The emerging innovations that compete for media visibility include the circular economy, green growth (Smulders et al. 2014), the dematerialisation of the economy, and the decoupling of economic growth from the use of energy and natural resources (UNEP 2011). Addressing the question of perpetual economic growth is not a priority as long as it is possible to assure economic growth in the intermediate run.

There are economists who do not share the techno-optimist view that perpetual economic growth is feasible, but their capacity to influence economic policy is restricted. The Club of Rome, founded in 1968 at the *Accademia dei Lincei* in Rome, and presently composed of former heads of state, United Nations and government officials, politicians, diplomats, scientists, economists, and business leaders from many countries defends the position that economic growth cannot continue indefinitely because of resource depletion. They point out that, in spite of theoretical arguments based on the potential of technology and human ingenuity, the complexity and unpredictability of the real world and essential human behaviour and aspirations will play a major role in shaping the future pattern of natural resource overexploitation, environmental degradation, and climate change. Furthermore, they warn about the serious consequences of encouraging a belief in the feasibility of perpetual growth in a world with limited natural resources and energy (Meadows et al. 1972, 2004).

A recent assessment of the World3 model forecasts used by Meadows et al. (2004) shows that the "business as usual" scenario accurately reproduces the world's path since 1972 (Turner 2008), showing that it is unsustainable and dangerous. However, the ideas in *The Limits to Growth* continue to be disputed or ignored by mainstream neoclassical economic policies, by governments, and by the overwhelming majority of the human population. Most people are uninformed about the limits to economic growth. At best they consider it a depressing subject that provides no added value or enjoyment to life and about which they are unable to do anything.

There is a very strong tendency to attribute a stagnated or decreasing personal economic prosperity to deficient government policies at the local and national levels and to discard factors that have their origin in anthropogenic global changes. If the relevance of such drivers is acknowledged, people often assume that governments have the obligation and the power to solve the problem and assure well-being and economic prosperity for all citizens, without requiring much involvement from them.

More recently, however, as humankind has become increasingly confronted with the Earth's biogeophysical limits, some mainstream economists have been exploring the possibility of stable post-growth economic models that retain some of the key features of capitalism, but do not require continuous economic growth. The idea of a stationary or steady state economy is quite old and dates back to the concept of a stationary state proposed by Adam Smith (1776) and John Stuart Mill (1848). In 1930, Keynes predicted that a state of economic saturation would be reached within one century in which people would live in abundance with no further need for greed, exaction of interest, or love of money, which is an amazingly naive point of view about humans (Keynes 1930). In his own words (Keynes 1930): "Mankind would finally have solved 'the economic problem', that is, the struggle for existence."

In the second half of the last century, the steady state economy was no longer considered to be an inevitable feature of the medium or long term evolution of the capitalist economy, but a condition required by ecological sustainability. Beginning in the 1970s, Herman Daly followed Georgescu-Roegen's work and became the leading proponent of a steady state ecological economy that rejects capitalism (Daly 1980). Later, Europe saw the development of the less technological and more ideological and radical theory of degrowth, proposed by Serge Latouche (2007).

Still more recently, mainstream economists disenchanted with the excesses of capitalism and conscious of the need to protect Earth's limited natural resources have been exploring and defending a transition to a post-growth economy within the larger framework of capitalism (Barrett 2018). In the model used by Barrett, based on a non-linear dynamical system that incorporates Minsky's financial instability hypothesis, the end of growth would increase the wages share of output and would not exacerbate income inequalities (Barrett 2018). Furthermore, the model indicates that zero productivity growth is more likely to lead to long-term stability than positive productivity growth. That mainstream economists are becoming interested in exploring post-growth economies is an encouraging development that should help forge a strategy to reach sustainability.

The long term of future human social generations is a field that belongs almost exclusively to the academic world or to fiction. The likelihood of humankind becoming extinct would be much higher if other living organisms outside the artificial environments built and controlled by humans were to become progressively extinct. It is inconceivable that humans can survive if there is no biosphere free and independent from them. This scenario relates to situations in which astronauts currently live for many consecutive months in space and to those in which the first human colonies on the Moon, Mars, and other planets of the Solar System or beyond may be established. These colonies will basically require energy, water, oxygen, and food. The latter is likely to be the most challenging resource to obtain. The other planets of the Solar System, and in particular Mars, which is perhaps the most likely to receive a human colony, don't have a living soil full of microorganisms able to support plants. The conversion of surface materials such as dust and rocks to soil would be very difficult and expensive. The most likely solution would be to use cellular agriculture where food is derived from cells grown in laboratory dishes, for instance to produce algae, meat, and fish, to grow plants in tunnels illuminated by powerful LEDs, and to breed insects rich in proteins, such as crickets. All these foods are based on living organisms.

Without other living organisms, the survival of *Homo sapiens* is unimaginable. Our life is deeply dependent on photosynthesis in plants to convert solar radiant energy into chemical energy as well as maintaining the current concentration of oxygen and carbon dioxide in the atmosphere. Technology is unable to feed people without using living organisms, but it can help to increase efficiency in all stages of agricultural production.

The biosphere would persist without any problems in the absence of humans. We can imagine scenarios where the extinction of humans arises from widespread destruction such as in a global nuclear war. Even in that case, it is not very likely that all living organisms would die out. Some would survive and restart the evolutionary cycle of life by way of natural selection. A remarkable example of survival in extreme conditions can be found in several gingko trees (*Ginkgo biloba*) in Hiroshima after the "Little Boy" nuclear bomb was dropped by the USA on 9 August 1945. One of them was located at the Hosen-ji temple about 1100 metres from the place where the bomb fell, immediately killing roughly 140000 people and completely destroying almost all the buildings. In spring 1946, the tree grew new shoots and is still alive, now surrounded by the stairway of the rebuilt temple.

The gingko is a tree that has medicinal properties and its seeds are edible. It is considered sacred and a carrier of hope by the Chinese and Japanese. It is the only surviving species in a family of gymnosperms that emerged in the Permian around 270 million years ago. It is probably the oldest existing species of tree, and Darwin called it a "living fossil", as it has remained essentially unchanged for more than 200 million years. It almost became extinct during the last glacial period but it survived in refuges in the Dalou mountains in south-west China (Tang et al. 2012) and was preserved in eastern China for around 1000 years by monks in their temples (Shen et al. 2005). In autumn, its leaves turn a golden yellow colour and cover the ground around them. It is known that the tree can live for more than 1500 years but it could probably live even longer. It has an extraordinary resistance to insects and pathogenic agents—viruses, bacteria, and fungi—due to its large genome which has 10.6 thousand million base pairs. It includes a high percentage of repeated series that, over time, have created innovative and more efficient defence processes (Guan et al. 2016).

Without humans, the biosphere will continue to evolve for many hundreds of millions of years, as happened in the past. The likelihood of a biosphere without humans recreating something similar to the *Homo* genus or to *Homo sapiens* after some time from a non-specified starting situation, although probably catastrophic for living organisms, is vanishing small. It is practically impossible that a highly improbable or perhaps even unique event, such as the emergence of intelligent life on Earth and of species similar to *Homo sapiens*, be repeated (Sandberg et al. 2018).

The scenario of an Earth system without humans has been explored by several authors. Alan Weisman (2007) imagined a situation where humans disappeared from the Earth magically, leaving it as it was. Animals and plants would quickly adapt to the new situation and take over towns and cities, and all settlements and infrastructures would become unrecognisable within a few decades.

7.14 Benefits, Risks, and Uncertainties of Globalisation

Five processes can be identified in the human subsystem which, since the Enlightenment and the Industrial Revolution, have generated profound global socio-economic and cultural transformations. Furthermore, they have played a decisive role in anthropogenic interference in biogeophysical subsystems. These primary drivers are: science, through the increase in scientific knowledge and its applications, and technology, through the creation and use of new technologies; the development of new energy sources and systems, accompanied by growing energy consumption; global population growth; increased production and consumption of goods and services; and, finally, political, social, economic, financial, and institutional transformations, including education and vocational training.

These five drivers are intimately linked and form a complex changing system. Identifying and distinguishing them from one another is useful as a methodology to better understand their impact both on human cultural evolution and on the evolution of human interference in biogeophysical subsystems. As time went on these primary drivers generated secondary global processes with strong social, economic, financial, political, cultural, and environmental impacts. Among them globalisation, urbanisation, and migration stand out as three of the most important, and will probably remain so in the future. Let us start by considering globalisation.

The origin and history of use of the word reflect the emergence and development of deep transformations in the human subsystem during the twentieth century. The word "globalisation" only began to appear in dictionaries in the 1960s, precisely at the time when the Great Acceleration was maturing, but use of the word exploded in the 1980s and 1990s and then surprisingly stabilised and began to weaken in the 2000s. In economics, the globalisation of markets was forcefully defended by the US scholar Theodore Levitt at the beginning of the 1980s (Levitt 1983). The sudden interest in globalisation led to the publication of many articles in science and technology journals, together with books, articles, and news pieces published in the media, all attempting to analyse, understand, and debate the globalisation phenomenon and its effects on society. Interest in the new concept can be measured using several databases, including Ngram, Factiva, and the ISI Web of Knowledge (James and Steger 2014).

Only after the process of analysis and debate on globalisation began did other concepts emerge: proto-globalisation, to describe the Age of Discovery led by Europeans from the 15th century onwards, and modern globalisation, which was split into two phases: the first beginning in the 19th century and the second in the 20th century, with considerable economic, social, political, and environmental impacts (O'Rourke and Williamson 2002; Friedman 2005). The high point of the second globalisation occurred in the 1990s, after the collapse of the Soviet Union, when the former communist countries were becoming market economies, with the multilateral trade agreement of the Uruguay Round being concluded in 1993, after long negotiations.

Globalisation has been an extraordinarily efficient way of reducing poverty and increasing well-being and economic prosperity for hundreds of millions of people around the world, especially in countries with emerging economies, but also in less developed countries. It is one of the factors that has contributed most to economic convergence between countries with advanced economies and the rest of the world. It is estimated that about one billion people, mostly in Asia, have been pulled out of poverty as a result of globalisation. Globalisation is characterized by openness, competitiveness, and harmonisation (Oramah and Dzene 2019). The first property describes the faster pace at which legal, regulatory, and political obstacles to trade are removed. The growing competitiveness results from the greater ability of domestic producers to establish strong positions in international markets, and harmonisation describes the opportunity to increase the international standardisation of rules, procedures, and technologies. The world is becoming increasingly complex, interdependent, and interconnected, mainly because of the ongoing process of globalisation.

However, at the same time, globalisation has also increased inequalities at the global level. There have been winners and losers (Stiglitz 2002). Investment in less developed countries with cheap labour has increased employment and income. Countries with advanced economies have benefited from goods produced at low cost, greater access to developing markets, and increased economies of scale and scope. On the other hand, the relocation of production in industrial sectors that use mostly conventional technology to countries offering more favourable market conditions and the emergence of industries using new technologies with low employability have generated unemployment and stagnation or loss of income in countries with advanced economies. Labour market rigidities have constrained the migration of labour from the declining to the booming sectors of the economy, creating large income inequalities (Oramah and Dzene 2019). Furthermore, the acceleration in global integration brought by globalisation has increased national vulnerability to global financial and economic crises. One of the greatest challenges of globalisation is how to balance the benefits of global openness, interdependence, interconnectivity, and competitiveness with national politics and priorities.

Globalisation, besides its effects on socioeconomic and technological development, also has powerful impacts on the environment because it stimulates strong economic growth, particularly in some emerging economies. The rate and intensity of human interference with the environment and the exploitation of natural resources has increased with the second modern globalisation. Atmospheric greenhouse gas emissions have also grown significantly. In the period 1980–2002, China's CO_2 emissions increased from 1467 to 3694 million tonnes, at a rate of 8% per year. After China joined the WTO in 2002, emissions increased remarkably with a mean annual rate of 13% in the period 2002–2007, making China the world's top CO_2 emitter. By facilitating growth and consumerism, globalisation has led high-income and densely populated countries, especially Japan and European countries, to run up an ecological deficit with other regions in the world, since they live mostly on imported biocapacity.

A study on income distribution around the world between 1988 and 2008, carried out by the economists Branko Milanovic and Christoph Lakner (Milanovic 2012; Lakner and Milanovic 2016), shows how rising inequality was fuelled by stronger globalisation. Some of their conclusions were summarized in a graph of percentage income growth as a function of income that became known as the "elephant chart", because of the shape of the curve.

The global emerging middle class, situated between the 15th and 65th income percentiles, grew more than 55%, especially in some developing countries. In China, impoverished rural populations were lifted into the middle class. This group of people correspond to the graph's peak at the elephant's torso.

By contrast, the global upper middle class, especially in countries with advanced economies, situated between the 75th and 90th percentiles, has seen its income stagnate or falling. The near zero growth in income over two decades has very likely been one of the major causes for the development of populist movements and policies in the advanced economies, in particular impacting the outcome of the presidential election in the USA and the Brexit referendum in the UK, both in 2016.

However, the global elite, especially the top 1%, were winners. Their income rose by more than 60% between 1988 and 2008, capturing a large share of global income growth. They represent the elephant's raised trunk. Finally the greatest losers are the global extreme poor with income in the 5th percentile, who remained stuck in their profound, inhumane poverty during those 20 years. Some of them live in fragile, vulnerable, or dysfunctional states stuck in cycles of poverty and violence. They are Wells' "people of the abyss" and correspond to the elephant's slumped tail.

Some view the economic globalisation that began in the late 20th century mainly as a way for the "ruling elites" to promote the expansion of world markets for their own interests. Social movements have emerged as a reaction using various names such as, alter-globalisation, anti-globalisation, anti-corporate globalisation, and globalisation from below (Juris 2008). These movements support global dialogue and cooperation between people around the world, but they oppose the economic globalisation of the last decades because of its negative effects on justice, civil liberties, labour protection, indigenous cultures, the environment, sustainable use of natural resources, and climate change. Participants in such movements consider that multinational corporations have unregulated political power, exercised through socially unfair trade agreements and deregulated financial markets, which maximize their profit by eroding working conditions, reducing wages, and undermining the integrity of national legislative authority and sovereignty. However, the outcome of the anti-globalisation movements has been mostly rhetorical and had no effective impact on the progress of globalisation.

Nevertheless, since 2016, there was an upsurge in the backlash against globalisation, fuelled by a growing perception that globalisation causes job losses and depressed wages. This narrative has contributed to extraordinary political shifts in some advanced economies, especially in Anglo-Saxon countries (Stiglitz 2017).

The calls made by the IMF for "more redistribution" of the globalisation profits to promote the support for free trade were far from successful. This problem is particularly acute in the USA, where domestic fiscal and industrial policies have prevented the economic benefits of globalisation from being shared widely throughout society. Furthermore, the political system does not facilitate redistribution through taxes and high-quality public social services. Public spending currently represents around 37% of the USA's GDP, while that percentage is generally higher in Europe, reaching 50% in Sweden.

The long-standing consensus promoted by the West that trade liberalization is beneficial for development and economic growth around the world is now under attack by the USA. Its government has recently moved to implement protectionist measures by imposing tariffs on China and by threatening to do the same with the EU and other major trading partners. The trade war began in early 2018 when the USA imposed 25% tariffs on 50 billion US dollars' worth of Chinese goods. Such protectionist measures were welcomed by the workers of the former industrial heartlands that support the US President Donald Trump. They had been the victims of 40 years of deindustrialization, in part promoted by an unregulated globalisation process that mainly benefited the economic elites. However, trade wars based on zero-sum 18th century interventionist economic theory do not solve the problem, since they inevitably lead to a slowdown in global economic growth that will affect all countries, including the USA. In 2019, Christine Lagarde, then managing director of the IMF, warned that such trade wars will likely reduce global GDP by 0.5%, or 455 billion US dollars, in 2020.

The protectionist policies of the USA are unlikely to progress much further because, at a certain point, the economic elite that supports the dysfunctional US democracy will start arguing that there is no acceptable alternative to neoliberal globalisation. Up to now the trade war with China has been tolerated by the business oligarchs that constitute the core constituency of the US government because they have benefitted from generous tax cuts. But beyond certain limits, it will start to hurt Wall Street and the growth of financialization, or the growing scale and profitability of the finance sector, an outcome that would be anathema to the current global financial and economic system.

A more sustainable but less likely development of the present situation would be to transform the backlash against globalisation into a driver for a new world order focused more on solutions at the local level, stronger international cooperation, and a reformed international system that controls the power of the global financial and economic system through curbs on capital flows and a more strongly regulated international trade and global financial sector. These reforms would facilitate progress towards achieving the UN Sustainable Development Goals for 2030.

The current wave of protectionism will probably be short-lived. Globalisation is most likely an irreversible process that will develop into new forms of economic, financial, and cultural exchange and cooperation, better adapted to the future challenges facing humankind. In any case, the US economic hegemony will continue to be actively challenged by China through its determination to become the most powerful world economy. Its strategy is based on a growing network of investments and infrastructure projects, especially across Asia and Europe, in order to extend its influence around the world and provide a market for Chinese goods.

7.15 Urbanisation, Its Drivers, and Its Challenges

From the earliest city-states of Mesopotamia and Egypt, throughout history up to the 18th century, the majority of the human population lived from subsistence agriculture in rural areas. There was a rather stable equilibrium in the ratio between rural and urban populations since the economic activity in towns consisted mainly of trade in markets and small scale manufacturing. The industrialisation process started by the Industrial Revolution changed the status quo by significantly increasing the level of urbanisation. Urbanisation is defined as the increasing share of the population living in urban areas, which are nowadays defined differently in different countries, but are generally taken as areas that are more populous and dense than rural settlements (Tacoli et al. 2015). Urban population growth results from natural increase, migration, and reclassification or new national definitions of urban space. From the 18th century onwards, the expansion of European colonisation also began to create modern urban structures in Africa, the Americas, and Asia. In Europe, from the 19th century onwards, the urbanisation process accelerated with the large scale development of trade and industry, greater mobility provided by public transport, and the extraordinary expansion of transport based on internal combustion engines powered by gasoline or diesel, starting in the 20th century. It is estimated that, in 1800, only 6.2% of the world's population lived in settlements with more than 5000 inhabitants (Grauman 1976). In 1900, the percentage rose to 18%, increasing further to 31.4%

in 1950, 46.68% in 2000, and 58% in 2018, and it is forecast to reach 68% in 2050 (UNDESAPD 2019a).

There are significant differences in urbanisation levels in different regions of the world, with the highest values being 82% in North America, 81% in Latin America and the Caribbean, 74% in Europe, and 68% in Oceania, and the lowest values being 43% in Africa and 50% in Asia (UNDESAPD 2019a). This range of values is largely the result of strong correlations between urbanisation and levels of socioeconomic development, industrialisation, and demographics. Globalisation has also been a strong driver of urbanisation, especially in the emerging economies during the last 30 years. The model of food production based on intensive energy, land, and water use and on large greenhouse gas emissions that resulted from the Industrial Revolution has led to a global decline in the ratio of food producers to food consumers (Satterthwaite et al. 2010). This decline is strongly correlated with the decline in the global rural population and therefore anticorrelated with the level of urbanisation.

Rapid population growth and internal migration to urban areas in search of economic prosperity are main drivers in the urbanisation of a country's population. An important cause of migration to urban areas in developing countries is food insecurity in the rural areas. Rural–urban migration from mountainous areas where employment opportunities are scarce and subsistence agriculture prevails has become a worldwide trend in developing countries. However, these migrations do not ensure sustainable economic opportunities and food security (Tacoli et al. 2015). Hundreds of millions of urban dwellers face malnutrition or undernutrition that is more frequently caused by their lack of income than by a local or regional incapacity to produce food (Satterthwaite et al. 2010).

To determine whether and how starvation and famines relate to food supply it is necessary to understand the entitlement systems in the region where they occur (Sen 1981). Famines are often related to conflicts or disrupted societies where social links and solidarity break down and some groups actively or passively prevent others from accessing food supplies allowing a crisis to become a full-scale famine. After the famines in the Horn of Africa in 1984–1985 and 1992, North Korea in 1994–1998, and Somalia in 2011, it seemed that famines were eradicated in an age of declining poverty and hunger. However, in 2017, about 70 million people worldwide in unconnected countries such as Guatemala, Haiti, Nigeria, Malawi, Zimbabwe, South Sudan, Ethiopia, Somalia, Yemen, Syria, and Afghanistan needed food assistance in rural and urban areas. According to the Famine Early Warning System (FEWS) of the USAID organization, created in 1985, global food assistance needs in 2017 were unprecedented and represented an almost 50% increase compared the previous two years. Usually a famine has multiple causes, such as violent conflict, rapid inflation, and natural disasters, especially droughts and other weather-related extreme events that are becoming increasingly intensive and frequent due to climate change.

A crucial issue regarding the relation between agriculture and urbanisation is how the increasing and changing food demands from fast growing urban populations can be satisfied in a world progressively affected by climate change, while continuing to reduce both rural and urban poverty, ensure the sustainable use of water, combat desertification, halt and reverse land degradation, and halt biodiversity loss as prescribed by the UN Sustainable Development Goals 2, 6, and 15.

The rate of increase of urbanisation is particularly high in less developed countries, where it began to intensify in the 1950s, associated with some of the drivers of the Great Acceleration. According to United Nations projections, the global population in the period 2018–2050 will increase from 7.6 billion to 9.8 billion, while the global urban population will increase slightly more from 4.2 billion to 6.7 billion. Urbanisation levels will reach 56% in Africa and 64% in Asia in 2050 (UNDESAPD 2019a), and about 90% of the 2.5 billion increase in global urban population by 2050 will be in those two continents. In recent years, the urban population has declined in some cities located in the low-fertility countries of Europe and Asia, especially in countries where the national population is stable or contracting and the economy is stagnant.

While the global urban population is tending to increase in the 21st century, the global rural population has grown slowly since 1950. It is expected to reach a maximum of slightly more than 3.4 billion in the next few years and fall to 3.1 billion in 2050. The two countries with the largest rural populations are India and China, with 893 million and 578 million, respectively, in 2018. The size of these two populations indicates that they are likely to continue to contribute decisively to the economic growth of their countries through their continuing migration to the urban areas. The opposing trends in the size of the world's rural and urban populations coupled with an increasing global population represent one of the greatest sustainability challenges of the 21st century.

One characteristic of the global urbanisation process is the emergence of megacities. The number rose from two in 1950—New York and Tokyo—to four in 1975, with the addition of Shanghai and Mexico City, to 28 in 2014, of which only seven were in countries with advanced economies, and to 33 in 2018 (UNDESAPD 2018), and it is projected to reach 43 in 2030 (UNDESAPD 2018). The city with the largest population in 2018 was Tokyo, with 37.5 million, but the vast majority of the other megacities are in developing countries. Some sources indicate that the current number of megacities is considerably larger, of the order of 47, probably because there is no agreed standard criterion for determining the boundaries of a city for the purpose of determining its population. Some studies include only the population inside the city's administrative boundaries while others include the surrounding suburban communities. As regards the future, the uncertainties become larger. Nevertheless, a common feature of future scenarios is that the projected total population will grow in the largest cities. Projections based on the United Nations World Urbanisation Prospects indicate that the largest megacities in 2100 will be in Africa: Lagos, Nigeria, with 88.3 million, Kinshasa, Democratic Republic of the Congo, with 83.5 million, and Dar es Salem, Tanzania, with 73.7 million (Hoornweg and Pope 2017).

Managing large urban areas sustainably without compromising their residents' well-being will be one of the greatest challenges of the 21st century. The strong trend towards urbanisation has shifted a large part of the human sustainability challenges to the management of cities. However, sustainability is a much wider concept that involves all the other 16 Sustainable Development Goals, besides Goal 11 on sustainable cities and communities, including the goals regarding rural populations and the environment.

Cities have played a crucial role in the history of civilizations, including the advent of the modern age up to the present time. They are the privileged hubs of ideas, socioeconomic development, politics and government, trade, industrial development, financial activities, science, technology, culture, and the arts. They support the regional and national economies and they underpin a large part of the sustainability initiatives at the regional and global levels.

One of the main goals of cities is to enable people to progress socially, to support the creation of opportunities for employment and economic prosperity, and to provide the conditions for human well-being and healthy living. These objectives must be achieved without overstraining natural resources and with urban environmental pollution reduced to a minimum, especially air pollution. Furthermore, cities strive to reduce the production of non-recycled solid waste and the volume of greenhouse gas emissions to the atmosphere. Cities are expected to provide, in collaboration with national governments and through large technical investments and adequate institutional development, basic services to its residents, such as education and health care, to construct, develop, and maintain infrastructures, including access to clean water supplies, sanitation, energy systems, and green spaces, and to solve the problems of housing and transportation.

These objectives have led to the concept of a sustainable city, green city, or ecocity designed to achieve social, economic, and environmental sustainability (Register 1987). Ideally, a sustainable city should be able to feed itself with natural resources from the surrounding environment, to provide sustainably clean water to all residents, to achieve carbon neutrality through the use of renewable energy sources, and to implement energy efficiency programs. From the social point of view, a sustainable city should be democratic, equitable, socially cohesive, and diverse, while providing the conditions for a good quality of life.

In general, cities in advanced economies are significantly closer to the sustainable city model than cities in developing countries. In most cities of the latter, the sustainability challenges are difficult to address and solve successfully. Lagos, Nigeria, is a prime example of the problems facing such cities. Its population increased from about 325 000 in 1950 to 13.9 million in 2019 and is expected to become the largest world's urban population in 2100. Lagos is a major financial centre with the fourth highest GDP in Africa. It generates a quarter of Nigeria's GDP and in 2019 it had four billionaires and more than a thousand millionaires (Christian 2019), while being the fastest growing and most populous city of the African continent. It spreads chaotically over an area of 1171 km^2, and while some quarters of the town are very wealthy, about 66% of residents live in slums unconnected to piped water or to sanitary systems, some of them located around lagoons with no access to roads, electricity, or waste disposal, where crime is uncontrolled. The city's streets are choked with traffic and the main Olusosun dump site receives 10 000 tons of municipal solid waste daily and covers an area of about 40 hectares (Vidal 2018). All these problems coincide with the fact that Nigeria has emerged as the front runner in Africa in the forecast of the future increase in the number of millionaires, ahead of Egypt and Kenya (Christian 2019), making it a very attractive country for investment opportunities in Africa.

Clean water availability in cities will be one of the major problems of the 21st century and probably of the following centuries. According to the FAO Aquastat data, global internal freshwater resources (internal river flows and groundwater from rainfall) per capita have decreased by more than half in the period 1962–2014 from 13 402 m^3 to 6073 m^3. It is likely that fresh water demand will exceed supply by about 40% in 2030 due to climate change, population growth, and difficulties in finding alternative sources of water. Water reserves are increasingly contaminated by industrial pollutants and human waste, especially in medium- and low-income countries. In the first group of countries, less than one-third of wastewater is treated, while in the latter the share is much smaller. According to the WHO, at least 2 billion people globally use a drinking water source contaminated with faeces, and contaminated drinking water is estimated to cause 485 000 diarrhoeal deaths each year (WHO 2019b). Water-related tensions are on the rise between countries that share water resources and within countries between rural and urban communities, and among agricultural, industrial, and household consumers. Cooperative management agreements between countries exist only in about two-thirds of the world's transboundary rivers. Climate change is aggravating the scarcity of water resources in various regions of the world, especially in Mediterranean type climates, and in the drylands of the Middle East, parts of central and southern Asia, Sub-Saharan Africa, Southern Africa, and the Americas. In most of these regions groundwater is being explored unsustainably. It is estimated that 2 billion people live in countries experiencing high water stress (UN Water 2019).

Cities are becoming increasingly vulnerable to water scarcity, in part due to climate change and insufficient adaptation measures. One of the most visible cases of urban water crisis happened in Cape Town, South Africa, during the period 2015–2018, due to a severe drought that affected the Western Cape Province. The city was very close to "day zero", when the municipal water supply would be shut off. The water crisis had significant economic, health, and safety impacts in both the urban and rural areas. Economic losses were particularly severe in agriculture, estimated at Rand 5.9 billion, with a resulting 30 000 job losses and exports dropping 13–20% (WWF 2018b). Water scarcity is now the new normal with annual precipitation projected to decrease by 30% by 2050 relative to the last 30 years of the 20th century. Other cities that have experienced or have the risk of experiencing water crises include: São Paulo in Brazil, Mexico City in Mexico, Chennai and Bangalore in India, Cairo in Egypt, Istanbul in Turkey, Jakarta in Indonesia, Beijing in China, Melbourne in Australia, Moscow in Russia, and London in the UK (BBC 2018; Chapman 2019). India is a particularly vulnerable country, with 600 million people facing high to extreme water stress, where 70% of the water is contaminated and 75% of households do not have drinking water on premise (Kant 2018). After two consecutive years of weak monsoons, about 330 million people, a quarter of the country's population, is affected by a severe drought. Twenty-one major cities are at risk of reaching day zero water levels by 2020 (Kant 2018).

The human landscape of the world is gradually becoming more urban, a process that has consequences for every field of human activity, including the social, economic, political, cultural fields, as well as for people's health and safety. Regarding the relation between urbanisation and the environment, it is important to distinguish between the characteristics and quality of the urban environment, as well as its impact on human health, in particular, air quality and climate, and the specific impact that the urbanisation process has on the Earth's biogeophysical subsystems. In relation to the first aspect, urban air pollution is responsible for 4–7 million early deaths annually (WHO 2014; Landrigan et al. 2018). According to WHO data, nine out of ten people breathe air containing high levels of pollutants (WHO 2019a).

Concerning the climate, the approximately 2.3 billion people that represent the increase in urban population up to 2050 will be exposed to extreme heat risks resulting from the combination of the urban heat island effect and climate change. A recent study has shown that urban land areas are expected to expand by 0.6–1.3 million km^2, which corresponds to an increase of 78–171% over the urban footprint of 2015 (Huang et al. 2019). Urban expansion through 2050 will increase daily minimum and maximum temperatures by 0.5–0.7 °C on average during the summer, but it can go up by around 3 °C in some locations. This increase is about 39–64% of the greenhouse gas induced warming projected by a multi-model ensemble average in the IPCC RCP 4.5 scenario (see Fig. 7.2) (Huang et al. 2019). During winter, urban expansion warming will be weaker: on average 0.4–0.6 °C, up to about 2 °C in some cases, which is about 43–65% of the effect of global warming (Huang et al. 2019). The coupling between the urban heat island and global warming implies that about half of the future global urban population will experience extreme heat risks, especially in the lower latitudes and during heat waves, where temperatures are higher and the capacity to adapt is often lower.

Climate model simulations for the region around the Arabian Gulf under the business-as-usual scenario of future greenhouse gas concentrations, project wet-bulb temperatures, a combined measure of temperature and humidity, which are likely to approach and exceed the 35 °C limit of survivability for a fit human being (Pal and Eltahir 2016). The way to adapt to these climate extremes is to be outdoors only during short periods of time of less than six hours or to use special clothing for high temperature protection.

To evaluate the effect of urbanisation on biogeophysical subsystems, it is necessary to assess and compare the impacts produced by urban and rural populations. It is often argued that the concentration of people in urban areas increases efficiency in the management of food, energy, water, and other natural resources and leads to more sustainable consumption and production. However, the higher mean socio-economic development that people achieve, as they expected, by migrating from rural to urban areas, tends to increase their per capita energy and natural resource consumption. The main attraction that urban areas have for rural populations is the promise of affluence, the promise of economic prosperity, well-being, and easier access to the consumption of goods and services, which tends to generate consumerism.

A study of the energy and material flows through the world's 27 megacities as of 2010 showed a considerable variation in the energy and material flows among megacities (Kennedy et al. 2015). Rates between the lowest- and highest-consuming megacities differ by a factor of 28 for energy per capita, 23 for water per capita, 19 for waste production per capita, 35 for total steel consumption, and 6 for total cement consumption (Kennedy et al. 2015). Although there are exceptions, it is generally observed that cities with a higher percentage of wealthy people consume more material resources and ultimately discard more materials than poor cities. The per capita use of energy in New York is 24 times higher than in Kolkata, and production of solid waste more than 15 times higher (Kennedy et al. 2015). Smart urban policies, including a good and vast network of public transport, can reduce energy use, air pollution, and resource use.

On a per capita basis, people living in rural areas consume less energy and natural resources and emit smaller amounts of greenhouse gases than people living in megacities. Megacities have 6.7% of the world's population, but they consume 9.3% of global electricity and generate 12.6% of global waste (Kennedy et al. 2015). As regards emissions, all the world's urban areas accounted for 53.5% of the global population in 2014, but were responsible for 65% of global energy use and for 71–76% of global CO_2 emissions (Seto et al. 2014). However, per capita emissions in cities can be lower than the national per capita emissions, as for instance in New York relative to the USA (Dodman 2009). Economic prosperity in urban areas does not necessarily lead to larger per capita emissions. An example is given by Tokyo, which has lower per capita emissions than either Beijing or Shanghai.

Urban areas have a great potential to become more sustainable, but at present they are very far from reaching that goal at the global level. If consumption and production patterns remain unchanged, the social, economic, and environmental sustainability of the ongoing urbanisation process is likely to be unachievable. Furthermore, sustainability at the global level depends crucially on the capacity to reach that goal in future cities and megacities. To achieve sustainability, it is not enough to have marginal efficiency gains from incremental changes in society. It will be necessary to promote transformational changes.

Bioregionalism seeks a solution to some of the problems created by urbanisation by recovering some of the values of pre-industrial societies, encouraging self-sufficiency, using regional natural resources, and promoting social cohesion, sharing, and solidarity among members of the regional community. Many initiatives are currently being developed in the field of bioregionalism, focusing on the search for sustainability, but with differing visions and objectives. Most sustainable communities are planned and built to encourage social, economic, and environmental sustainability at local level (Phillips et al. 2013). Sometimes these communities are called transition towns or ecovillages, when there is a specific concern with minimising the community's ecological impact (Litfin 2014). Intentional communities form a wider group that also includes communities that have specific spiritual, political, and social objectives, as well as sustainability objectives.

Sustainable communities are spreading throughout the world and form a very promising sustainability movement at the local and regional levels. Their contribution to achieving sustainability in large urban areas and megacities is only indirect for the moment. The main challenge for the scalability of bioregionalism is finding a way to reduce the specific economic and social problems of high concentrations of people, including the consumption of energy, water, and other natural resources, crime, drugs, homelessness, human isolation, lack of solidarity and social cohesion, and the different forms of urban poverty.

7.16 Migrations

Before dealing with the matter of migration, it is appropriate to remember that, according to fossil records, long-distance running is a characteristic of the *Homo* genus that originated roughly 2 million years ago and likely played an important role in the evolution of the human body (Carrier et al. 1984; Bramble and Lieberman 2004; Pickering and Bunn 2007). Some mammals are sprinters, because they are four-legged and run faster, but *Homo sapiens* is able to run much longer distances than most mammals. A possible reason for this adaptation might have been that it allowed more meat to be incorporated into the human diet. Humans probably learned to recognise the flight of scavenger birds in the sky towards a recently deceased animal and follow them as quickly as possible to be the first to get to it. Another hypothesis, which applied before the use of slingshots and bows and arrows, would have been persistence hunting, which involves pursuing prey over large enough distances to exhaust them and kill them easily. This type of hunting by exhaustion was used by the San people of the Khoisan ethnic group that live in the Kalahari basin in Botswana, by the Tarahumara indigenous people in northern Mexico to hunt deer, by the Paiute and Navajo peoples in the USA to hunt antelope, and by Aboriginals in Australia to hunt kangaroos (Liebenberg 2006, 2008). Hunting by exhaustion was likely done during the hottest part of the day to tire the prey as quickly as possible.

Humans have remarkable characteristics for endurance running (defined as running distances longer than 5 km) in adverse temperature conditions using aerobic metabolism. In terms of distance and mean speed, these capabilities are only comparable to a few other mammals, such as wolves, hunting dogs, and hyenas, which are known habitually to run distances of 10–20 km a day (Lieberman et al. 2009). One reason why humans are so good at endurance running is the very efficient form of thermoregulation due to the characteristics of the sebaceous glands and the lack of fur. There are several other adaptations to endurance running in *Homo sapiens'* body, such as the nuchal ligament, which makes it possible to stabilise the position of the head in relation to the torso (Lieberman et al. 2009), the arch of the foot, which works like a spring when running, the elastic ligaments and tendons in the foot and leg, which allows storage of energy, a more developed gluteus maximus muscle, and parallel toes, shorter than in *Homo* genus ancestors, adapted to optimise running.

By running large distances, humans would have got used to "travelling" and discovering new places and environments. *Homo sapiens* was the only species in the *Homo* genus that took over the entire planet with a succession of migratory processes that began in Africa roughly 120 000 years ago and covered a very large part of the globe, including most of Europe, Asia, and the Americas, in the relatively short period of time of 60 000 years. No obstacles were able to stop human progress over tundras, steppes, sprawling plains, deserts, mountains, glaciers, deep valleys, winding rivers, dense forests and jungles, straits, and seas, in search of food, resources, and mythical abundance. The need to roam across long distances all over the world to search for food and resources is intrinsic to human nature. Later in the history of human civilization, that drive evolved into the need to travel, explore, conquer, and possess.

Running habits certainly underwent significant changes with the agricultural revolution, but migrations continued. The history of civilisations is essentially a history of migrations, conquests, and wars that bring and carry new myths, religions, languages, codes of behaviour and governance, weapons, clothing, artefacts, jewels, styles, design techniques, painting, sculpture and architecture styles, war, building, mobility, hunting and fishing technologies, and new agricultural practices.

Among many other examples, one finds the Neolithic expansion throughout the Mediterranean and Europe towards the west and the north, the Indo-European expansion from the Pontic-Caspian steppe to Western Europe and Siberia, the expansion of the Roman Empire, and the invasions from the north and east of Europe that came after the Romans. The Age of Discovery, considered to be a proto-globalisation, and the later colonisation of Africa, the Americas, and parts of Asia by Europeans accelerated migratory processes, including the profitable forced migration of the transatlantic slave trade.

There are several theories about types of migration (King 2012), although they can essentially be defined in relation to space, time, and the motivation to migrate. Space deals with the distance covered and whether or not the path crosses one or more borders. We have, then, external or international migration and internal migration. Time deals with the duration of the migration and whether it is temporary or permanent. A migration is normally considered to involve a time frame of more than a year. Finally, motivation deals with whether migrants are voluntary or forced.

The latter group includes refugee migrants, as covered by the United Nations Convention on the Status of Refugees, adopted in 1951. Refugee status depends on the recognition of a well-founded fear of being victims of persecution for reasons of race, religion, or political convictions. A person who, despite being considered as a refugee, does not have officially recognised refugee status is an asylum seeker. People who are displaced (internally or externally) due to conflicts, terrorism, famine, natural disasters, and environmental causes, including climate change, are also forced migrants. In developing countries, there are forced internal migrants due to development projects, such as mining operations or dam building. Locally, migration may also take place due to ethnic or socio-economic changes in the resident population, for example, gentrification driven by tourism.

In voluntary migration, migrants normally begin by analysing the "push and pull" factors and primarily have economic or political reasons or are simply driven by the quest for better climate, lower cost of living, and countries that provide better quality of life, especially in the case of retired people in some advanced economies. Conflicts, particularly in the Middle East and Africa, are currently destroying state economies and structures so seriously and deeply that people are forced to migrate for both economic and political reasons (Bommes et al. 2014).

Historically, since the 19th century, the greatest human migrations in the world have been the result of economic migrants shifting from rural areas to urban areas. This trend was described in 1885 by the German–British geographer Ernst Georg Ravenstein (1834–1913) (Ravenstein 1885, 1889) in the second and third laws of migration: "Most migration is from agricultural to industrial areas" and "Large towns grow more by migration than by natural increase". As already mentioned, rural migration is expected to continue feeding the strong urbanisation process of the 21st century.

The International Organization for Migration estimates that in 1965 there were 75 million migrants. This number increased to 152.5 million in 1990 and jumped to 257.7 million in 2017 (UNDESAPD 2017b). Refugees and asylum seekers accounted for about 10% of migrants in 2005. A large part of the increase from 1990 to 2017 happened in the later period of 2005 to 2017, when the annual increase was about 5.6 million, while in the former period of 1990 to 2005 it was 2.5 million. The international migrant stock grew at an average annual rate of 1.2% in the period 1990–2017, 2.4% in the period 2000–2010, and 2.3% in 2010–2017 (UNDESAPD 2017b). As a proportion of the world population, the international migrant stock is still quite small, but increased from 2.3% to 2.7% and 3.4% in the years 1965, 1990, and 2017, respectively. Of the 257.7 million international migrants in 2017, about 57% cent lived in the developed regions, while the developing regions hosted 43%. It is estimated that, from 2000 to 2014, migrants contributed between 40% and 80% to the growth in the workforce in countries in developed regions (Woetzel et al. 2016).

These numbers are only estimates and are affected by some uncertainty. The definition of what constitutes an international migrant varies from country to country, and in any given country, the number of migrants is always changing, in particular due to naturalization processes and mobility. Irregular, illegal, or unauthorized migrants include those that entered the country of residence without authorization, overstayed a visa or did not leave after being ordered to do so. The majority of the world's migrants arrive in their host country in a regular manner and only later become irregular due to administrative overstay. A migrant's irregular status is often the result of policies that tend to complicate the naturalization process using unclear or overly bureaucratic procedures, high visa renewal costs, language barriers, or thwarted access to legal aid, among other things. Although it is difficult to determine the number of illegal migrants, some authors claim that their number is increasing faster than the increase of legal migrants (King 2012).

In 2007, in view of the positive effects that resulted from the evolution in the overall number of international migrants, United Nations Secretary-General Ban Ki-moon, proposed that "we should welcome the dawn of the migration age", considering that it is likely to be the second stage of globalisation, the first being the movement of goods and capital (Ki-moon 2007). The United Nations reaffirmed that "migration is an engine of economic growth, innovation and sustainable development" (UN 2017), but most governments of countries with advanced economies were beginning to view migratory movements from poorer countries with growing mistrust and opposition. On 19 September 2016, the UN General Assembly approved the New York Declaration for Refugees and Migrants (UNHCR 2016), which expresses the political will of world leaders to save lives, protect rights, and share responsibilities around the world. It was also decided at that time to develop a Global Compact for Safe, Orderly and Regular Migration, which was adopted at a UN conference in Marrakech on 10–11 December 2018 and finally endorsed on the following 19 December at the UN General Assembly, with 152 countries in favour, 12 countries abstaining, and the Czech Republic, Hungary, Israel, Poland, and the USA voting against (UN 2019). According to the text of the resolution (UN 2019):

> Global Compact expresses our collective commitment to improving cooperation on international migration. Migration has been part of the human experience throughout history, and we recognize that it is a source of prosperity, innovation and sustainable development in our globalized world, and that these positive impacts can be optimized by improving migration governance. The majority of migrants around the world today travel, live and work in a safe, orderly and regular manner. Nonetheless, migration undeniably affects our countries, communities, migrants and their families in very different and sometimes unpredictable ways.

However, the global compact is (UN 2019):

> [...] a non-legally binding cooperative framework that recognizes that no State can address migration on its own because of the inherently transnational nature of the phenomenon. It requires international, regional and bilateral cooperation and dialogue. Its authority rests on its consensual nature, credibility, collective ownership, joint implementation, follow-up and review.

The total number of international migrants may be seen as revealing the "paradox of immobility", since, considering only "push and pull" factors arising from poverty and deep socioeconomic inequalities throughout the world, migratory flows would be expected to be much greater (Malmberg 1997). What is the reason for this paradox? Could it be that, despite the conflicts, the droughts, floods, and other extreme weather events, natural disasters, famine, malnutrition, poverty, and unemployment in many areas of the world, people who live there prefer not to migrate because of family ties, social and cultural connections, the well-being provided by their homes, or other reasons? Or are there many hundreds of millions of people wishing to migrate but who are stopped from doing so because poverty prevents them from obtaining the means to travel and to get the documents they need and pay for the journey? Or could it be that many, although they can afford to migrate, are stopped by national and international institutional barriers or forced to return? It is practically impossible to answer these questions fully and quantify such issues.

In the period between the end of the 19th century and the mid-20th century, there was more freedom for voluntary migration. Now, in the face of numerous immigration barriers in receiving countries, an increasing proportion are irregular migrants. A large part of those who leave Sub-Saharan Africa for Europe are rejected by authorities on their way to the north and return in terrible economic conditions or die (Sengupta 2016). Cape Verde, an archipelagic state off the west coast of Africa, has seen significant legal emigration to Brazil, the USA, and Europe throughout its history, and its economy has become dependent on remittances from emigrants. Cape Verdeans still wish to emigrate, but that has become difficult or impossible. The age of migration in Cape Verde has become the age of involuntary immobility (Carling 2002).

Global-level migratory pressure is growing, but the systems set up by destination countries to control or stop migration are increasingly powerful and efficient, especially in countries with advanced economies. Around the world, both developed and developing countries are creating various obstacles to receiving migrants, such as decreasing the number of refugees they accept, and denying asylum to those who might have been admitted in the past. Almost half of the international migrants are women, now mostly migrating on their own and not as family members.

Despite courageous proclamations by the United Nations, there is no sign that we are going into the "new migration age" that should mark the second stage of globalisation. On the contrary, globalisation has enabled the free movement of goods, capital, science, and some technologies, it has enabled prompt access to information and news at the global scale, it has created employment and economic prosperity, mostly in emerging economies, and it has boosted the formation, growth, and empowerment of multinational corporations. But in terms of labour, one of the pillars of productivity, economic growth, and sustainability, developed countries have been restricting the free movement of workers that emerges from globalisation. The stagnation or even the fall of immigration in countries with advanced economies is detrimental for their future economic sustainability and dynamism, because in most of these countries population ageing is decreasing the percentage of the total population that is of working age. Robotization and artificial intelligence will not be a substitute for a decreasing working-age population, because people are irreplaceable actors as regards scientific progress, technological innovation, and creativity in the arts.

As people have fewer and fewer children, the demographics of the future are becoming quite sobering, especially for the developed countries. In the European Union, a quarter of the population will be over 60 in 2020, and by 2050, that proportion will reach more than one-third. Projections for 2060 in the European Union indicate that the total population will be 517 million, 16 million higher than in 2010, but population decline is projected for about half of the EU member states, while the others will experience population growth (EC 2014). The working-age population, defined as those aged 15–64, is predicted to fall from 67% in 2010 to 56.2% in 2060. Over the next 50 years, the working-age population is expected to decline by nearly 42 million (EC 2014). The situation in Japan is more difficult. The population projections indicate that the percentage of people aged 65 years and older will increase from the current level of 28%–38% by 2050. In the USA the situation is slightly better than in the EU. The total population is projected to grow by 79 million, from 327 million in 2018 to about 406 million in 2060 (Vespa et al. 2018).

Ageing populations will strain various socioeconomic activities during the 21st century, especially welfare systems, which are likely to fall short on demand for health-care professionals, in particular medical doctors and nurses. However, the problem of population ageing is rapidly becoming global. According to UN data, in 2018, for the first time in history, people more than 64 years old worldwide outnumbered children under age five. Projections indicate that by 2050 there will be more than twice as many people above 64 as children under five. In the same year, the global number of people over 64 years old will also exceed the number of adolescents and young people aged 15–24 years (UNDESAPD 2019b). The global population with more than 64 years of age will increase from 665 million in 2020 to more than 1.9 billion in 2090, which represents 18% of the global population.

Economic migration has increasingly become a life strategy for many women and men that don't have employment opportunities and decent work in their own countries. However, due to a variety of country-specific cultural and religious characteristics, integration of migrants in some developed countries has been difficult or impossible to achieve. Poor integration of migrants in those countries hinders the recognition of their skills and experience, leads to unemployment, in particular for second generation migrants, and to high levels of discrimination and xenophobia.

Developed countries, especially those that have adverse demographics would benefit greatly from increasing the levels of immigration, preferentially of highly skilled immigrants. Although Japan has in the past almost entirely blocked immigration, it is now developing policies to attract immigration. Since 1985, the foreign share of the Japanese population has tripled, from 0.6 to 1.8%. The United States has had great success at attracting skilled immigrants. About 57% of immigrants who have patented an invention move to the USA, as do 41% of the global stock of college-educated migrants (Kenny 2019). Other countries, such as Canada, China, France, Germany, and the UK, have all recently reformed their laws, including the recognition of foreign diplomas for professional licensing purposes, in an attempt to attract entrepreneurs and experienced professionals. However, the highly skilled migrants represent only a very small part of current migration flows. What happens to all the others?

There is an increasing risk of moving towards a fragmented world, divided by walls and barriers. There are examples of such practices in the history of civilizations, but only now have they seen a global expression. One of the most notorious is the building by China's first emperor, Qin Shi Huang (259–210 BC), of the Great Wall of China, between 220 and 206 BC as a military defense against invasion by northern nomadic nations and to facilitate the establishment of border controls and the imposition of duties on goods transported along the Silk Road.

Although today's technologies to control people's mobility are becoming more and more sophisticated, it is still necessary to build walls. In the Middle East, walls are spreading significantly. Israel is a country separated from all its neighbours by different kinds of walls, from Lebanon to Syria, Jordan, the West Bank, the Gaza Strip, and Egypt. Kuwait has built a wall along the border with Iraq, Saudi Arabia on the borders with Iraq and Yemen, and Iran on the border with Pakistan. In Asia, India has built a wall on the border with Bangladesh, stretching over more than 3200 km, and is building another on the border with Myanmar. China has built a wall along the border with North Korea. In Europe, Austria, Bulgaria, and Hungary have built barriers on the borders with Slovenia, Turkey, Serbia, and Croatia, respectively. In Africa, Botswana has built a wall with Zimbabwe, and in North America, the USA is slowly implementing a project to extend the wall with Mexico along the entire border. Besides this short summary there are many others barriers around the world that are being planned or are already under construction.

Maritime borders are more difficult to control than land borders. Europe provides an excellent example of this difficulty. More than 2.5 million irregular migrants have crossed the Mediterranean towards Europe since the beginning of the 1970s (Fargues 2017). The increase in illegal migration was initially the result of introducing compulsory visas for migrant workers from North Africa and Turkey during the 1973 oil crisis, when unemployment in Europe rose. Migration followed three major routes in the Mediterranean, one to the east, one in the central area towards Italy, and another to the west. They were gradually controlled from east to west through agreements with the migrants' countries of origin or countries of departure along the Mediterranean coast. Entries into Europe via the Mediterranean that were not authorised by police authorities reached a high of 1 017 294 in 2015 and fell to 356 502 in 2016 and 127 991 in 2017 (Fargues 2017).

Access to Europe via the Mediterranean is the deadliest border in the world for migrants. Between 2000 and 2017, 33 761 migrants were found dead or declared missing (Fargues 2017). The dramatic wave of migration over the Mediterranean has had extensive political consequences in Europe by feeding populist, right-wing movements in several countries. Surveys indicate that, due to the 2015 crisis, migration became the largest concern for several countries in the European Union and contributed to influencing the results of elections, especially in Austria, Germany, Hungary, Italy, and France, not to mention the Brexit referendum in the United Kingdom.

Nonetheless, the humanitarian crisis is not in Europe but in countries outside Europe, in the Middle East, where civil war and terrorism in Syria have generated in 2018 about 5.6 million Syrian international migrants and 6.6 million internally displaced people, 2.98 million people in hard-to-reach besieged areas, and 13.1 million people in need of humanitarian assistance inside Syria (UNHCR 2018). In Africa, recurrent humanitarian crises have been caused by conflicts, terrorism, poverty, famine, and malnutrition, combined with the increasingly frequent adverse consequences of a warmer, drier climate.

Immigration across the Mediterranean to Europe is a dramatic problem that is far from being solved and will likely last for many decades. It has been difficult to establish a consensus as regards social services, security issues, deportation policies, and integration efforts for migrants in the European Union. The number of unauthorized migrants living in Europe reached a peak in 2016 and was estimated to be between 3.9 million and 4.8 million in 2017 (Connor and Passel 2019). These numbers represent less than 1% of Europe's total population of more than 500 million people living in the 28 European Union member states, and four European Free Trade Association (EFTA) countries, namely, Iceland, Liechtenstein, Norway, and Switzerland. In the summer of 2019 the Italian government started criminal proceedings against the German boat captain Pia Klemp and other crew members who have rescued thousands of migrants who were at risk in the Mediterranean. Some migrants are begin forcibly returned to Libya where they endure indefinite detention and torture or are forced to return to their original countries. Migrants coming from the Sub-Saharan countries have to overcome the days-long desert crossing to Libya where they are sometimes beaten, robbed, and detained for weeks by smugglers. Yet they keep trying to reach Europe because of the desperate situation in which they live.

In many countries with advanced economies, there is a significant number of people who view the influx of migrants and refugees as a dangerous existential threat, probably due to feelings of anxiety, insecurity, or fear. These feelings are fed and amplified by nativist organisations and political parties with agendas that go far beyond migration. The indoctrination processes are largely carried out using false or purposely distorted facts in the news media and social media.

For example, it is stated as a well established, uncontroversial fact that the arrival of migrants and refugees increases crime and terrorism, as well as being an additional burden on the economy and contributing to the destruction of the country's identity and civil, religious, and cultural values. A recent study on the refugee crisis in Germany in 2014 and 2015, when a total of about 1.7 million migrants entered the country, concluded that it had not caused an outbreak in crime (Gehrsitz and Ungerer 2017). Moreover, terrorist attacks by radical Islamic groups in Europe began long before the recent influx of refugees and migrants. The situation is similar in the USA, and it is common in some political camps to say that migrants that cross the Mexican border cause crime waves and that Muslims living in the USA are a danger to national security. However, studies have shown that migrants, both legal and illegal, have lower crime rates than American citizens born in the USA (Ghandnoosh and Rovner 2017). This type of argument, rather than soothing those who feel threatened, convinces them that they are reading fake news and strengthens their convictions that they themselves are in danger, along with the values and identity of their country. In most cases the nature of such convictions makes them irrationally impervious to communicated facts that contradict their convictions.

Migration is disruptive because it faces residents with human beings who have a different physical appearance, different social behaviour, a different culture, and who often also practise a different religion. Migrants may be seen as intruders who come to compete for common resources. In countries with advanced economies, they are often perceived as a symbol of today's socioeconomic changes and the new challenges of this century. Their physical presence exacerbates nostalgia for the past, which for many was a time of greater economic prosperity and well-being. It is as if the transformation that the world is undergoing with globalisation, the emergence of large and powerful economies, the growth in global population, and the numerous problems of less developed countries, especially poverty, conflict, and socioeconomic and institutional weaknesses, were coming into the cities, streets, and buildings where people live. Refugees vividly represent the fears that some nowadays share about the future. Through a psychological mechanism, these fears become a limiting factor in their capacity to accept the hospitality extended to migrants. Those who do not feel fearful are associated or identified by those who do so feel with dangerous intellectual elites that have hidden, twisted interests or are seen as naive and falsely convinced that there is a feasible path to a fairer, more equitable, and sustainable global world.

However, in some specific cases the concern with an immigration flow may be fully justified. An example is provided by the difficult relation that arises in some highly industrialized Western countries towards their Islamic communities. France, the Western country with the largest number of Muslims, estimates that there are between 5 and 6 million Muslims in a total population of 70 million, with only about one-third saying they are practising believers. The French Muslim population originated in migrations from North Africa, West Africa, and the Middle East that started in the late 1960s, mainly from Algeria. The French political establishment has made strenuous efforts to integrate the Islamic community into its social and political structures but has not fully succeeded. Instead, a large number of Muslim networks have succeeded in establishing ideological enclaves in some urban areas with religious communities that oppose in a more or less explicit way the liberal, democratic, and republican values of French society. These communities are proud of their *territoires conquis* (conquered territories) and reject integration because it would suppress their identity and would finally convert them into a French middle class which opposes the non-democratic and authoritarian Muslim regimes in North Africa and the Middle East countries they originate from (Rougier 2020). The majority of Muslim terrorists developed their vision of the world and their devotion to the Jihad against the Western unbelievers in just such *territoires conquis*.

The OECD has published reports since 2005 that contain detailed analyses of the different types of "fragility in achieving the aspirations of the 2030 Agenda, sustainable development and peace" (OECD 2018) and the best strategies and ways for overcoming them. The last report recalls that, in 2016, the number of forcibly displaced people was 65.6 million, the highest number since the end of the Second World War (OECD 2018). In the same year, 26 000 people died in terrorist attacks and 560 000 died due to violent conflicts. Most of these incidents took place in developing countries and were a consequence of their political, social, economic, environmental, and security weaknesses. This fragility arises from a multitude of factors that interact with one another, such as conflicts and violence, poverty and social and economic inequalities, dysfunctional government policies, growth in public debt, corruption, scarcity of some natural resources, degradation of the environment, and climate change. In 2030, if countries with advanced economies do not act by way of cooperation and public help policies and programmes for development, roughly the world's poorest 80% will be living in fragile countries and the number of forced migrants will continue to rise (OECD 2018).

Climate change is increasing the problems of the most vulnerable developing countries and thereby contributing to intensifying forced migration flows (Stapleton et al. 2017). Extreme meteorological events, the frequency and intensity of most of which is increasing due to climate change, are causing a growing number of internally displaced people. According to the Internal Displacement Monitoring Centre (IDMC 2017), such events caused 24 million internal displacements in 2016, 32 times the number of people displaced due to geophysical events in the Earth's interior, such as earthquakes, tsunamis, and volcanic eruptions, and three times more than those displaced due to conflicts. Climate change increases not only the frequency and intensity of some extreme meteorological events, known as sudden-onset climate-related hazards (Stapleton et al. 2017), but also causes slow-onset climate-related hazards, such as droughts, changes in precipitation patterns, desertification, melting glaciers and ice sheets, ocean acidification, and the rise in global mean sea level. These phenomena are having an increasing harmful impact on populations, especially in poorer countries, sometimes forcing people to migrate when living conditions become insufferable (Bremner and Hunter 2014).

A study of asylum applications to the European Union from 103 source countries in the period 2000–2014 found that temperatures higher than the moderate optimum of about 20 °C increased asylum applications in a nonlinear fashion, which implies an accelerated increase in the future if there is unabated climate change (Missirian and Schlenker 2017). Another study that used bilateral data on asylum seeking applications for 157 countries concluded that climate change, by increasing drought severity and the likelihood of armed conflict, played a significant role as an explanatory factor for asylum seeking in the period 2011–2015 (Abel et al. 2019). The projected increase in the flow of migrants towards OECD countries fleeing from the adverse impacts of a future intensification of climate change in their non-OECD countries may not happen or may be greatly mitigated because of the economic and financial capacity that these countries have to put up all sorts of defensive barriers, physical, administrative, and political, to block those flows. Climate migrations are more likely to happen within and between non-OECD countries.

A 2018 World Bank Report (Rigaud et al. 2018) analysed the impacts of climate change on migration flows in Sub-Saharan Africa, South Asia, and Latin America by 2050. Without robust adaptation and development measures, it predicts that just over 143 million people, or 2.8% of the population, will migrate internally within their countries to avoid the most serious impacts of climate change. Nonetheless, the number of migrants may be reduced to 31 million if there are sound mitigation measures at the global level and suitable adaptation measures at the regional and local levels.

With a world increasingly fragmented by permanent physical barriers, how will *Homo sapiens* satisfy his essential need to rove the Earth? Those who are barely surviving and would greatly benefit from migrating to regions and countries where they could find more opportunities to prosper and reach economic prosperity are mostly impeded from doing so. Those that live relatively well can satisfy their need to rove the Earth through tourism.

The vast majority of humankind have stopped being long-distance runners, but a significant and increasing part cover much longer distances, much more rapidly, sitting or lying comfortably on jet planes. According to a press release from the World Travel & Tourism Council dated 22 March 2018 (WTTC 2018), the travel and tourism sector is the one with the highest growth at global level, having grown 4.6% in 2017, roughly 45% more than the global economy, which grew by 3.165%. That year, it contributed $8.3 trillion to global GDP, which represents 10.4% of the total, and it provided about 20% of the new jobs created in the world.

The travel and tourism economic sector is the only large, global-level socioeconomic sector that does not select the places where it concentrates its activity on the basis of the affluence of the local people and their high purchasing power or on the value of the natural resources that may be exploited locally. The selection criteria are based on their cultural, ethnographic, artistic, architectural, and historical heritage values and also on the attractiveness and uniqueness of their environmental features, flora, fauna, and landscape. Places of touristic interest are found in all countries from the richest to the poorest.

It is likely, for that reason, that the industry is unusually explicit in its warnings about what it calls eco-limits and socio-economic megatrends (WTTC 2013). With regard to eco-limits, it believes that (WTTC 2013):

Resource constraints and the impact of climate change will continue to affect Travel & Tourism, in particular as most of the industry's growth is expected to take place in regions which are prone to climate change and resource scarcity. These impacts can also influence the demand for certain destinations particularly the ones in climate change hotspots.

Regarding socio-economic megatrends, it believes that (WTTC 2013):

Demographic change, increasingly complex patterns of labour migration, high unemployment especially among women and young people and implications of growing income inequality are aspects of a bigger picture that no industry, least of all Travel & Tourism, can afford to ignore. If ignored the risks to the short term and long-term business investment strategy will be high. Travel & Tourism businesses will be forced to potentially delay or cancel their investment and seek opportunities elsewhere. A social uprising could be detrimental to the perception of the destination but also to the tourism operator's products.

A world where the global mean temperature increased by more than 3 °C relative to the pre-industrial era due to anthropogenic climate change would not be the best one to continue developing the travel and tourist industry. This industry relies to a significant extent on sustainability, recognises its current and future challenges, and seeks to have a strategy for dealing with them.

A large part of tourist travel is made through air travel, which has gone through an exponential development, with passenger numbers rising from 576 to 1467 million, then to 4233 million in the years 1978, 1998, and 2018, according to the International Air Travel Association (IATA) statistics. The industry has the capacity to grow considerably in the future in part because it is estimated that only about 20% of the world population has ever flown in a plane. The rapid expansion of air travel results in part from the proliferation of low-cost airlines and a fast growing middle class around the world. Scenarios indicate that the number of civil aviation passengers will almost double in 2037, reaching 8.2 billion, which represents about 91% of the projected global human population in that year, but admittedly protectionism and trade wars pose a threat to such expectations (Garcia 2018). Tourists currently account for about 50% of air travellers.

It is estimated that 20% of total international travel is nature-based tourism where people enjoy hiking, camping and nature walks, wildlife viewing, botanical excursions, bird watching and many other outdoor activities (RA 2017). According to the Adventure Travel Trade Association, ecotourism had the highest level of client demand overall for travel activities in 2017, followed by cultural, environmentally sustainable, and hiking activities. The *Homo sapiens* continues to have the ability to be deeply attracted by nature and by the diversity and beauty of its life forms, ecosystems, and land and coastal zone landscapes.

A further example of this attraction is the increasing number of privately owned large land estates with gardens, fields, forests, and abundant wildlife to enjoy outdoor activities. To possess a private almost pristine island somewhere in the ocean is particularly cherished and a powerful sign of high socioeconomic and wealth status. A few examples among billionaires include: Dietir Mateschitz, Red Bull co-founder, owner of Laucala Island in the Republic of Fiji; Larry Ellison, executive chairman of Oracle, owner of 98% of Lanai Island in the Hawaii archipelago; John C. Malone, chairman of Liberty Media and Liberty Global, owner of Sampson Cay in the Bahamas and the brothers David and Frederick Barclay, owners of the Brecqhou Island in the Channel Islands. Some of these islands are a private retreat for the owner's family and friends while others accept visitors, usually for very high prices. There are still hundreds of islands around the world that present and future millionaires can buy, where they can enjoy the contact with nature in a secure private environment.

There is the danger that the drive to rove the Earth, and in particular to enjoy nature everywhere in the world where it is still in a relatively good state of conservation, may become a serious contributor to the unsustainability of the current development model. The rise in tourism and in particular ecotourism can improve the economy of developing countries and promote environmental protection. Nevertheless the negative impacts on the environment of the rise in tourism are increasing, especially as regards air travel, which is one of the more environmentally damaging activities that people can undertake. A return flight from London to New York generates about 986 kg of CO_2, which is more than the annual per capita emissions in most African countries. In 2018, CO_2 emissions from air travel accounted for about 2.5% of the global CO_2 emissions and they are predicted to triple by 2050. To accommodate the increase in air travel, it is necessary to redevelop many of the existing airports and to build new ones around the world. The building of a new airport is often a contentious project because of the large area of land that it occupies and of the negative environmental impacts that it generates, such as increased noise and air pollution, negative impacts on wildlife and biodiversity, and a greater risk of accidents.

Airports epitomize the evolution of operative social time since the beginning of the era of mass air travel which coincided with the start of the Great Acceleration. They represent the triumph of technology and consumerism and they exemplify the acceleration of operative social time. Time is the essence of air travel and is either short to encompass all that has and can be done in an airport or becomes extremely long due to a late connecting flight, a delayed or cancelled flight, a strike or some form of digital breakdown, or a security emergency. Security checks constitute a remarkable application of probability theory. The probability that the person in line in front of you plans to blow up the plane that he or she is going to board is infinitesimal, but nevertheless greater than zero. For each person, the security screening procedures appear absurd but relaxing them would dangerously increase that infinitesimal probability. People accept the screening checks with good temper to be immediately gratified with the opportunity to consume as much as possible during the remaining time in a labyrinth of duty-free shops. The risk of terrorism in airports is an unquestionable property of contemporary operative social time and its extinction has become unconceivable. Technology is unable to modify the laws of probability theory, but in the future it may shorten the time spent in the screening checks in airports. The screening equipment, which currently uses metal detectors, cabinet X-ray and backscatter X-ray machines, millimetre wave scanners, and more recently very sensitive infra-red cameras, will be using new technologies in the future.

According to a Global Data report, global airport construction projects are estimated at \$737.3 billion, with the Asia–Pacific region accounting for \$241.4 billion. The new projects include Al Maktoum Airport in Dubai, United Arab Emirates, Beijing Daxing International Airport in China, Mexico City International Airport in Mexico, Istanbul Ataturk Airport in Turkey, and Long Thanh International Airport in Long Thanh, Vietnam. Los Angeles International Airport and John F. Kennedy International Airport in the USA and Heathrow Airport near London in the UK will benefit from large redevelopment projects.

It is hoped that the concepts of sustainable mobility and sustainable tourism will be further developed and effectively applied in the future so that humankind can continue to rove the Earth safely and sustainably to enjoy a thriving biosphere.

7.17   Fossil Fuels and Climate Change

The history of intensive fossil fuel use by humankind and of anthropogenic climate change, for which fossil fuels are largely responsible, is one of the most symbolic examples of the lack of sustainability of the current world development model. Fossil fuels are largely responsible for the civilisation we enjoy today and, therefore, for the economic prosperity, well-being, and progress in the global mean value of quality of life indicators during most of the last 250 years. Nonetheless, the release of CO_2 into the atmosphere from burning fossil fuels increases the atmospheric greenhouse effect and contributes to anthropogenic climate change. There are other human activities that lead to climate change but CO_2 from fossil fuel usage is responsible for roughly 65% of the radiative forcing behind it (IPCC 2013). Land-use changes, particularly deforestation, produces CO_2 emissions that contribute 11% of the total radiative forcing, while all the other greenhouse gases with anthropogenic emissions, such as CH_4 and N_2O, contribute 24%.

We are facing an inadvertent consequence of fossil fuel use, acknowledged by science only since the end of the 19th century, which makes the climate more violent and destructive in the medium and long term and which causes profound transformations in other subsystems of the Earth system, the hydrosphere, cryosphere, and biosphere. There are not enough CO_2 sinks in the climate system for countries to continue to prospect for, exploit, and consume fossil fuels at the rate they do today. From this point of view, maintaining the atmospheric emissions of greenhouse gases at the same unabated level is an example of the tragedy of the commons (Paavola 2012). As we have seen, if climate change is not controlled, its effects will get gradually more harmful for most of humanity. We are still uncertain about the magnitude and the pace of evolution of such effects, but science already makes it possible to conclude unequivocally that they will be become more negative and dangerous the more greenhouse gases are emitted into the atmosphere and the longer the period of time over which emissions continue.

On the other hand, as already mentioned, the industry for exploring, extracting, exploiting, refining, and delivering the finished fossil fuels is the largest in the world in terms of economic and financial value and profitability. According to an analysis by the Climate Accountability Institute, the top 20 fossil fuel companies have contributed to 35% of all energy-related world emissions of carbon dioxide and methane, totalling 480 billion tonnes of CO_2 equivalent since 1965 (Taylor and Watts 2019). The state-owned company that is responsible for the largest volume of emissions is Saudi Aramco, which produced 4.38% of the global total. A relatively large number of countries owe their economic prosperity and wealth to fossil fuels. They are used throughout the world in practically all sectors of human activity, mainly for energy generation, but also in most industries, agriculture, air, sea, and land transport, buildings, and infrastructures. Over the last 250 years, they have become one of humankind's most valuable natural resources. Much of world history since the end of the Second World War has been determined by the geostrategic game of supply and demand for coal, oil, and natural gas among large producing and consuming countries. Fossil fuels make up roughly 80% of the world's primary energy sources (see Fig. 4.2). If we imagine a scenario in which the world stopped consuming fossil fuels suddenly without replacing them with another energy source, it would mean the collapse of today's civilisation.

When free markets do not maximise society's welfare, neoclassical economists consider that there is a market failure that can always be corrected by appropriate policy interventions. Thus climate change is an example of a market failure. The impacts of emissions fall mainly on people living in developing countries, on young people around the world, and on people yet to be born who have a relatively small or no responsibility at all for the emissions of greenhouse gases into the atmosphere. The adverse effects constitute a multigenerational externality, since they will take place for many centuries and are external to the market, in the sense that they will only affect it indirectly, which implies that there is no economic incentive to mitigate climate change. The motivation for policy interventions must therefore be largely ethical, based on intragenerational and intergenerational solidarity and justice. This plan has been very difficult to implement, due to the nature, time persistence, and history of the problem, and has provided the opportunity for some of the most surprising behaviour of humans in the operative social time of the modern age. Furthermore, it has revealed some essential but up to now unknown features of human nature.

The Middle East region is in many respects paradigmatic of the extraordinary issues related to climate change that emerged at the beginning of the 20th century. Since that time, it has acquired an exceptional visibility and importance in world history, primarily due to the large fossil fuel reserves it contains. The Western powers that discovered and began to exploit those reserves have been systematically intervening in the region since then to defend what they consider to be their strategic interests.

Following the First World War, Great Britain and France began to control the territories arising from the Sykes–Picot Agreement that partitioned the Ottoman Empire, which had stretched from Anatolia to Iran, the western banks of the Persian Gulf, and to Egypt. On 14 February 1945, on board the American cruiser Quincy, anchored in the Great Bitter Lake, close to the Suez Canal, President Franklin Delano Roosevelt (1882–1945) made a secret agreement with significant future implications

Fig. 7.4 Meeting between US President Franklin D. Roosevelt and King Ibn Saud of Saudi Arabia on board the US Navy heavy cruiser USS Quincy in the Great Bitter Lake, Egypt, on 14 February 1945. The King is speaking to the interpreter, Colonel William A. Eddy, and Fleet Admiral William D. Leahy, the president's Aide and Chief of Staff is at left. The meeting was secret because the war was not yet over and one of Roosevelt's objectives was to convince the King about the establishment of a Jewish State in Palestine, although he had pledged Winston Churchill that the USA would not intervene in territory controlled by the British. The agreement regarding the exploration of Saudi Arabia's large oil reserves in exchange for military protection consolidated US supremacy in the oil and natural gas industries and laid the foundations for the future wealth of Saudi Arabia. *Source* Naval History and Heritage Command, US Navy photo USA-C-54

with a remarkable man: Abd al-Aziz ibn Abd al-Rahman al-Faysal Al Saud (c. 1880–1953), better known in the Arab world as Abdul Aziz and in the West as Ibn Saud, a member of the House of al Saud and the founder of the Kingdom of Saudi Arabia (see Fig. 7.4). The agreement with the USA embraced Saudi Arabia within the American economic sphere of influence, provided security and military protection for the Al Saud's, including weapons, training, and a military base in Dhahran, in exchange for safe access to the oil reserves located in land controlled by Ibn Saud. Saudi Arabia soon became one of the first countries to be built almost exclusively upon the economic and financial importance of oil and natural gas.

At the Quincy meeting, Ibn Saud, when probed about Jewish migration to the Middle East after the atrocities of the Second World War, warned that he was against the formation of a Jewish state in Palestine because it would disregard the legitimate interests of the Arabs who lived there, and Roosevelt assured him that the USA would not allow such a state to be created. However, his successor, President Harry S. Truman (1884–1972), reversed that policy and the Western countries suggested dividing the "Territory under the British Mandate for Palestine" into two sovereign states, one Jewish and one Arab. This plan was adopted by the United Nations General Assembly as Resolution 181 (II) on 29 November 1947 (UN 1947), but is still not being implemented and probably never will.

The Al Saud's ruled parts of the Arabian Peninsula from 1744 up to 1881 when the Al Rashid clan, their main enemies, conquered Riyadh, forcing the poverty-stricken House of Al Saud into exile in Kuwait, while Ibn Saud was still an infant. In January 1902, there was a turn of fate when Ibn Saud supported by brothers, relatives, and a few dozen followers attacked Riyadh's clay and mud-brick al Masmak fort and killed Ibn Ajlan, the governor appointed by Ibn Rashid. After this victory and following about 30 years of fighting, Ibn Saud succeeded in uniting and controlling the competing families and tribes over much of central Arabia. During this time Ibn Saud cleverly decided to revive his dynasty's support for Wahhābism, which the House of Al Saud had embraced since the end of the 18th century. He knew that religious fanaticism could serve his ambition and founded a militantly religious brotherhood organization known as the Ikhwān, or the brethren. These islamicized Bedouins, converted into soldiers of Islam, were encouraged to fight and massacre their tribal Arab rivals.

Ibn Saud proclaimed himself king in 1926 and in 1932 renamed his kingdom as Saudi Arabia. The law and morals of the new kingdom were based on the fundamentalist Sharia law derived from the Quran, adopted in an uncodified form, which implies a considerable uncertainty in the content and scope of the laws, and on the medieval texts of the Hanbali School, the strictest traditionalist school of jurisprudence in Sunni Islam. At that time, Ibn Saud and his family lived frugally, practising the strict religious principles of the kingdom and rejecting all the luxuries of the modern world. Meanwhile, the Ikhwān, after defeating the internal tribal insurgents, engaged in cross-border raids into parts of Trans-Jordan, Iraq, and Kuwait to attack all non-Wahhabis, because they were infidels. Great Britain reacted to the attacks in Iraq and the pragmatic Ibn Saud turned against his own creation and imposed a defeat in the Battle of Sabilla in 29 March 1929, where the Ikhwān fought from the backs of camels or on foot, while the Saudis used British made machine guns.

Negotiations for an oil concession in the al-Ahsa eastern province began in March 1933 and led to the signing of the first oil contract between Saudi Arabia and an American Company, the American Standard Oil Co. (California), later called Chevron. On 3 March 1938, a Chevron-owned oil well in Dhahran drilled into what would be identified as the largest source of petroleum in the world. Dhahran is now the headquarters of the Saudi Arabian Oil Company or Saudi Aramco, which has both the world's second-largest proven crude oil reserves, estimated at more than 270 billion barrels, and second largest daily oil production. Furthermore, Saudi Aramco is the world's biggest energy company, generating more than a billion dollars a day in revenues.

Ibn Saud had a large family, with about three hundred wives and forty-three sons that survived infancy and an unknown number of daughters. In less than 30 years, Ibn Saud and his family went from being penniless to becoming one of the richest men in the world. His experience showed that oil was able to deliver almost immediate wealth on a scale never experienced before in the world. In view of the miracle produced by the black gold of the earth, people realized that if it could happen in Saudi Arabia, it would also take place elsewhere in the world, since the appetite for energy of an increasing global human population would be insatiable.

In spite of being extremely rich, Ibn Saud continued to adhere to the frugality of his previous life, prescribed by the strict religious principles of Wahhābism, but he was very fearful that such huge amounts of money would corrupt his family and alienate them from their religion. In the end some form of coexistence was established between Wahhābism and the conspicuous consumption and luxurious way of life of the Saudi Royal family, estimated to have now about 15 000 members. The present king is Salman bin Abdulaziz Al Saud, one of Ibn Saud's son's, and the presumptive heir apparent is his son Crown Prince Mohammad bin Salman who is the first deputy prime minister of Saudi Arabia and holds much of the executive power in the country.

Wahhābism is a fundamentalist Sunni movement founded by Muhammad ibn Abd al-Wahhab (1703–1792), who argued for a return to the practices of the Prophet Muhammad's initial community, his first successors and caliphs. Later, in the 1920s and in the context of reactions to the collapse of the Ottoman Empire, the Salafi movement emerged in Egypt. Salaf is a word that comes from the Arabic "salaf" which means "pious forefathers", and is also a fundamentalist version of Sunnism. Under the powerful influence of Ibn Saud, Salafism took on its modern form, converging with Wahhābism, which was declared the official religion of the kingdom of Saudi Arabia (Zaman 2012). With the protection offered by the USA to cope with the more populous neighbouring Arab countries and with the huge profits accumulated from oil exports, some Saudis developed an active policy of proselytism for Wahhābism in the 1960s, both at home and abroad, and thereby became the greatest vector of expansion and strengthening of Salafi jihadist movements around the world.

Saudi Arabia is a totalitarian state with an absolute monarchy where the monarch is both the head of state and government and accumulates judicial functions. As an enforcer of divine law, he has the right to enact or amend laws and regulations by royal order. As would be expected, Saudi Arabia is among the ten countries with the lowest democracy index as defined by the Economist Intelligence Unit. Gender equality is very limited, since Saudi women must have a male legal guardian who makes a number of decisions on their behalf. The regime employs strict censorship of the media and uses the police and security forces to repress any form of political dissent, in particular the ideologies that inspired the Arab Spring (Raposo 2017).

With its peculiar political regime, Saudi Arabia has been very convenient for the defence of the West's interests in the Middle East, and particularly for those of the USA and other Anglo-Saxon countries. Basically, Saudi Arabia is a country run by an extremely rich, extended family that is a friend of the USA and owns huge quantities of oil. Other countries in the region with large reserves of oil and natural gas have a much longer, more complex, and richer cultural, social, economic, and political history and have therefore been much more difficult to manipulate and control. The symbiosis between Saudi Arabia and US interests have been carefully preserved since the 1945 meeting at Great Bitter Lake.

The value of oil and the other fossil fuels comes well before the sporadic, volatile, and opportunistic Western preoccupation with defending and promoting human rights and democracy around the world. The success of oil and natural gas in transforming the poverty-stricken tribes of Jazirat al-Arab, or the peninsula of the Arabs, into rich countries, with the exception of Yemen, may help us to understand the reaction to climate change of most of the business people who benefit from fossil fuels the world over. How can some scientists have the arrogance to forcibly advise that the continuing use of fossil fuels is changing the global climate and is harmful for humankind? How can they suggest that the world should practically stop using fossil fuels just to avoid a climate change that is slow and will more greatly affect future generations than the contemporaneous generation, and even then mainly poor people in underdeveloped countries? Is it worthwhile changing the habits of about 20% of the world population currently enjoying well-being and economic prosperity? The world economic and financial elites are safely protected from the adverse effects of global climate change. However, they consider that they are vulnerable to the economic and financial risk of an energy transition from fossil fuels to renewable energies.

To better understand the complexity of the current situation, let us look in more detail at the problem of mitigating climate change. Despite the efforts made under the aegis of the UNFCCC since it came into force on 21 March 1994, the global energy model has not changed since then. Fossil fuels have continuously provided about 80% of the global primary energy sources in the following quarter of a century, up to 2019. About 90% of the anthropogenic CO_2 emissions arise from fossil-fuel burning for energy generation and industrial processes and from cement production, while land-use changes, especially deforestation, release the remaining 10%.

The Intergovernmental Panel on Climate Change (IPCC) started to produce scenarios of greenhouse gas emissions in its first assessment report, published in 1990 (IPCC 1990). These scenarios are important because climate change is a slow process relative to the element of operative time of human life. This means that responses to climate change must last as long as many of those elements and therefore that their evaluation for decision-making will involve medium and long term scenarios. Emission scenarios provide essential inputs to the climate models which are used to produce future climate scenarios and allow evaluation of the future impacts of climate change. The first IPCC assessment report presented four CO_2 emission scenarios (IPCC 1990). The scenario labelled A, named "business as usual", is a "non-intervention scenario" (IPCC 1994), since it does not explicitly include mitigation measures. In this scenario, fossil fuels would continue to be used as if there were no climate change. The other three other scenarios, labelled B, C, and D, are "intervention scenarios", since they "incorporate a progressive penetration of controls on greenhouse gas emissions", or in other words, explicit mitigation measures (IPCC 1990). As expected, the annual global CO_2 emissions of the intervention scenarios are predicted to be much lower than those of the non-intervention scenario.

In 2014, global CO_2 emissions from burning fossil fuels and industrial processes were estimated at 35.7 $GtCO_2$ (Olivier et al. 2015), as already mentioned. This amount is about 6 $GtCO_2$ more than the projection of the 1990 IPCC non-intervention scenario A, and about 13 $GtCO_2$ more than the projection of the highest of the three intervention scenarios. The mitigation measures that the authors of the first IPCC report predicted to be possible to implement have not yet been realized. Furthermore, the CO_2 emissions in that year were considerably greater than in the predicted non-intervention scenario. As regards fossil fuels, the world kept on using them as if the adverse side-effect of climate change did not exist.

Intervention scenarios were excluded from all the following IPCC emission scenarios, namely the IS92 series (Leggett et al. 1992), the 2000 Special Report on Emissions Scenarios (SRES) (IPCC 2000) and the Shared Socioeconomic Pathways (SSPs) (O'Neill et al. 2014; Riahi et al. 2017), although in this case some scenarios are differentiated by the degree of investment in carbon-intensive sources or low-carbon energy sources. The reason for this limitation is rather obscure, but is likely related to the backstage pressure that the fossil fuel lobby exerted on the IPCC proceedings, mainly through the USA.

The discrepancy between the business as usual scenario and the observed volume of emissions continued in the updated IPCC IS92 series of scenarios published in 1992. The global CO_2 emissions from burning fossil fuels and industrial processes observed in 2014 is roughly 2.9 $GtCO_2$ higher than the forecast for 2014 in the IS92a central scenario in a range of six "business-as-usual" scenarios for CO_2 emissions, drawn up 22 years ago by the IPCC (Leggett et al. 1992). The various IS92 scenarios are built around different assumptions regarding the main drivers that determine the volume of anthropogenic greenhouse gas emissions: population trends, GDP growth, primary energy use, energy mix, and technological changes.

The third series of IPCC scenarios, known as SRES, were used to implement public adaptation and mitigation policies around the world during more than a decade. They were constructed with Terms of Reference imposed by the IPCC that required that "none of the scenarios in the set includes any future policies that explicitly address climate change, although all scenarios necessarily encompass various policies of other types" (IPCC 2000). In a footnote on page 23, it is further explained that (IPCC 2000):

> For example, no scenarios are included that explicitly assume implementation of the emission targets in the UNFCCC and the Kyoto protocol.

In other words, the IPCC report authors were prevented from analysing, constructing, and discussing in their report an emission scenario that would satisfy the emission target of the UNFCCC, which is the main goal of the UNFCCC. In fact, Article 2 of the Convention states that (UNFCCC 1992):

> The ultimate objective of this Convention and any related legal instruments that the Conference of the Parties may adopt is to achieve, in accordance with the relevant provisions of the Convention, stabilization of greenhouse gas concentrations in the atmosphere at a level that would prevent dangerous anthropogenic interference with the climate system. Such a level should be achieved within a time frame sufficient to allow ecosystems to adapt naturally to climate change, to ensure that food production is not threatened and to enable economic development to proceed in a sustainable manner.

It is worth mentioning that the Convention has a wide legitimacy, since it has at present 197 parties to it.

Let us now return to the evolution since 2014 of CO_2 emissions from burning fossil fuels and industrial processes. During the period 2013–2016, these emissions were practically constant, a very welcome development that was considered to announce the beginning of the decrease in global emissions that will be essential to achieve the stabilization of the CO_2 concentration in the atmosphere, and subsequently the stabilization of the mean global surface temperature. Meanwhile, in those three years, global GDP grew 8.2% and this was tentatively interpreted as the possibility of decoupling emissions from economic growth. Some immediately warned that this conclusion was a dangerous myth. Most of the decoupling was attributed to the transition in China from coal to other fossil fuels and renewables after 2013. In the period 2000–2013, China's use of coal increased at an average annual rate of 8%.

Taking into account the evolution of greenhouse gas emissions over the last three decades and the global inertia regarding the decarbonisation of the different socio-economic sectors, particularly the energy, transport, and agriculture sectors, it is very unlikely that the 2 °C goal in the Paris Agreement will be achieved. To justify this assertion, one should remember that the mean global surface temperature has already increased 1 °C since the pre-industrial period, so we can only increase it by another 1 °C to meet the agreement. Given that there is an approximately linear relationship between cumulative CO_2 emissions and the rise in global mean surface temperature (Matthews et al. 2009), it is possible to estimate the maximum amount of CO_2 which, starting at a particular time, can be emitted while still maintaining a high probability of not going beyond 2 °C.

This amount of CO_2 is called the carbon budget (Meinshausen et al. 2009). It is estimated that between the pre-industrial period and 2015, 2050 $GtCO_2$ were emitted into the atmosphere (Rogelj et al. 2016) and that, to have a 66% chance of not going beyond a 2 °C rise in mean global surface temperature, we have a carbon budget of 1200 $GtCO_2$ from 2015 onwards (Friedlingstein et al. 2014). According to the Fifth IPCC Report (IPCC 2013), the carbon budget for the starting period of 2011–2015 is between 870 $GtCO_2$ and 1240 $GtCO_2$. The lower amount gives a higher chance of not going beyond 2 °C. If we go over the higher amount, a temperature increase above 2 °C becomes very likely. Although there are several estimates, we can conclude that the carbon budget is roughly 1000 $GtCO_2$.

One of the reasons why it is unlikely that the Paris Agreement will be met is the US$7.2 trillion investment in power plants planned over the next ten years, most of which are to be run on coal and natural gas (Pfeiffer et al. 2018). It is estimated that emissions from fossil fuel power plants that are currently operating represent an excess of 240 $GtCO_2$ during their lifetime with regard to the carbon budget corresponding to a global mean surface temperature increase of 1.5–2 °C. Emissions from the plants that have already been planned and are in the pipeline represent an additional excess of 270 $GtCO_2$. Put another way, if today no more power plants were built, we would still need to close roughly 20% of global power plants running on fossil fuels to be able to meet the Paris Agreement (Pfeiffer et al. 2018).

It should be noted, however, that despite this situation in the energy sector, it is still possible to meet the Paris Agreement if there is, particularly in the countries that are the biggest CO_2 emitters—China, the USA, the EU, India, and Russia—sufficient social mobilisation and political will to make the necessary energy transition (Rogelj et al. 2018). This social mobilisation does not yet exist, however. The issue of climate change is, for most people, diffuse, complex, and affected by great uncertainties, partly due to the generalized effort in misinformation, disinformation, and manipulation. Moreover, the idea that a solution involving an energy transition would be harmful to a country's economic growth and therefore to people's economic prosperity is widely promoted.

Many believe that continuing to defend the possibility of not going over the 2 °C goal is crucial to raise awareness in societies about the urgency of the problem and to create an incentive to be more ambitious in emissions reduction, but either they do not know how or they are powerless to overcome the practical difficulties required to achieve it. Others, on the other hand, believe that this objective is counter-productive because it is a fiction that harms both mitigation and adaptation efforts (Nordhaus 2018). The main problem is that decarbonising the world economy does not favour the prevailing utopia of growth, technology, and egoism.

How much coal, oil, and natural gas can we use without violating the Paris Agreement? If we used all the reserves of fossil fuels currently known and economically feasible to exploit, usually simply called reserves, CO_2 emissions would total around 2900 GtCO$_2$ (McGlade and Ekins 2015). In conclusion, to meet the Paris Agreement, we need to leave roughly 2/3 of fossil fuel reserves in the ground. Using all known reserves, including those that it is not economically feasible to exploit, known as resources, would lead to a total of 18 000 GtCO$_2$ in emissions (Tokarska et al. 2016). This would cause an increase in the mean global surface temperature of 6.4–9.5 °C and an increase of 14.7–19.5 °C in the Arctic, due to the warming amplification in the region (Tokarska et al. 2016). In this scenario, atmospheric concentrations of CO_2 in 2300 would be roughly 2000 ppmv, which is roughly five times the mean concentration of 411 ppmv measured in July 2019 (NOAA 2019). Climate change caused by this scenario would be much stronger, and its impacts on the economy, human economic prosperity and well-being, and the biosphere would be devastating, primarily because of the speed with which this change would occur. The last time the Earth's atmosphere saw comparable CO_2 concentrations was during the last part of the Eocene, between 45 and 34 million years ago.

What is the current situation regarding prospecting for new fossil fuel deposits? In practice, companies and interested governments continue to prospect for new deposits, particularly oil and natural gas deposits, as if there were no Paris Agreement, although many parliaments of the respective countries have ratified it. One of the potential dangers of this inconsistency is the possibility that a rapid evolution of low-carbon technologies in the energy and transport sectors quickly reduce demand for fossil fuels and lead to most of these reserves becoming stranded assets. A recent simulation of the consequences of current trends in the transition to low-carbon technologies continuing until 2050 leads to the conclusion that the macroeconomic impact of stranded fossil fuel assets would be a carbon financial bubble that would lead to a discounted global wealth loss of between US$1 trillion and US$4 trillion (Mercure et al. 2018). According to this study, there will be winning and losing countries due to the macroeconomic impact of the carbon bubble. The winners will be countries that import fossil fuels, such as China and the EU, and the losers will be producing countries, whose economies heavily rely on those fuels, such as Russia, the USA, Canada, and some Middle Eastern countries.

Today, when a country prospects for and finds fossil fuels in its territory that are economically attractive to exploit, it is welcomed with great joy by the government and people of that country. The reasoning that leads to this conclusion is extremely simple. If other countries have been blessed with the luck of finding abundant fossil fuels in their territories that have given them greater prosperity and wealth, why should such an opportunity be rejected? If the country already exploits fossil fuels, it would be incomprehensible for the majority of people not to continue benefiting from prospection and exploitation.

There are several suppositions in this argument. Not all countries that have large reserves of fossil fuels, particularly oil and natural gas, are effectively able to benefit from those resources. It is the so-called "oil curse" or more generally the "resource curse" that can often affect a country's governance negatively and lead to weaker economic growth than in countries with less natural resources. There is evidence that oil tends to make authoritarian regimes more durable, increase certain types of corruption, and help trigger violent conflict in low- and middle-income countries (Ross 2015). Furthermore, the assumption that the economic value of fossil fuel reserves will increase or remain stable indefinitely, in other words, that the assets will never depreciate or become stranded is unlikely. It is this assumption that, as we have already seen, is being called into question by the development of low-carbon technologies, particularly those connected to renewable energies. Finally, we cannot ignore all the other negative externalities of fossil fuel exploitation and use, which include the emission of large quantities of CO_2 into the atmosphere and, particularly in the case of power plants, the contribution to air pollution, which is estimated to cause 4–7 million early deaths each year, as already mentioned (WHO 2014; Landrigan et al. 2018).

Another important aspect to bear in mind when investing in prospection and eventually exploitation of fossil fuels is the increase in exploitation costs, particularly in the case of oil and natural gas. In the past, the costs of coal extraction were low because mines had thicker and longer seams and layers close to the surface. However, these types of mineral deposits are now less abundant so extraction has become harder and more expensive. We are still far from peak coal, peak oil, and peak gas, but there is a need to exploit non-conventional fossil fuels, also known as "extreme energy", because conventional fossil fuels are becoming relatively scarce. Extreme energy includes tar sands in the state of Alberta in Canada, from which oil is extracted, tight oil and shale gas in the USA, which today cover most of the country's consumption, coal bed methane extraction in Australia, and deepwater offshore oil reserves in the Gulf of Mexico and South Atlantic along the coast of Brazil. Exploitation of all these forms of extreme energy involves high environmental risks (EEI 2018).

Another example of an extreme fossil energy is the exploitation of offshore oil in the Arctic by Russia, which began in 2013 with the Prirazlomnoye well in the Pechora Sea in highly challenging environmental conditions. It is estimated that unconfirmed reserves of oil and natural gas in the Arctic represent about 20% of the world total of that type of reserve (EY 2013). To benefit from this enormous potential wealth of extreme energy, Russia is developing underwater prospection and exploitation technologies under sea ice using robotic submarines. It is not only Russia that is interested in exploiting oil and natural gas in the Arctic, but also Norway, Denmark, the USA, and Canada. All these countries intend to compete with one another for exploitation in the Arctic, each one fearing that only the others will benefit from those resources. Highly adverse environmental conditions, which make prospection and exploitation relatively expensive and subject to greater risks, and the market volatility of fuel prices do nothing to deter them from their intentions. A positive factor from the point of view of enabling this plan of action is anthropogenic climate change which, by increasing the global mean surface temperature and causing the Arctic sea ice to melt, gradually generates easier exploitation conditions, thereby boosting the profitability of the investment. It is noted that the exploitation of oil and natural gas in the Arctic generates serious environmental risks and will likely create irreversible pollution and the degradation or destruction of Arctic ecosystems that are unique in the world (NOAA 2016).

It is likely that other forms of extreme fossil energy will be discovered and used. A recent example of a new source of natural gas production is underground coal gasification and the exploitation of continental and coastal oceanic natural gas hydrate deposits. According to I. Schindler and J. Schindler, the higher costs of exploiting fossil fuels, coal and oil in particular, were in part responsible for the slower growth in global energy consumption per capita in the period 2011–2017, and this tendency contributed to limiting global economic growth (Schindler and Schindler 2018).

It is significant that the argument for following a behaviour consistent with the Paris Agreement is normally ignored by the countries that signed or ratified it, when they decide to start or continue to prospect for and explore coal, oil, or natural gas in their onshore or offshore deposits. This attitude shows that not even the perspective of a progressively more violent and destructive climate, nor the volatility of prices and the possibility of stranded fossil fuel assets, nor the negative environmental impacts of using fossil fuel energy, especially air pollution caused by coal-fired power stations, can surpass the short-term expectation of increased economic growth and economic prosperity. Humankind is dangerously addicted to fossil fuels.

7.18 The Energy Transition from Fossil Fuels to Renewable Energy. Hopes and Doubts

In spite of the reluctance of the fossil fuel industry to fully adhere to the Paris Agreement goal of not going above 2 °C in mean global temperature, the official language of the United Nations is that sustainable energy is central to the success of the UN Sustainable Development Agenda for 2030. The Sustainable Development Goal 7 has three main targets: ensure affordable, reliable, and universal access to modern energy services; increase substantially the share of renewable energy in the global energy mix; and double the global rate of improvement in energy efficiency (UNDESAPD 2018). A global energy transition away from fossil fuels to low-carbon solutions, which will be enabled mainly through technological innovation in renewable energy generation (Gielen et al. 2019) and energy efficiency, is urgently needed. Along this path a recent positive trend is that solar photovoltaics and wind power have benefited from falling costs and higher competitiveness with other primary energy sources.

Several studies have shown that it is technically possible to make an energy transition to 100% renewable energy system at national and global levels (Steinke et al. 2013; Mathiesen et al. 2015; Jacobson et al. 2015, 2017a, 2018; Barbosa et al. 2017; Gulagi et al. 2017; Jacobson et al. 2017b; Ram et al. 2018). To achieve this goal all types of renewables and energy storage technologies will need to be used to deal with the problems of intermittency and energy system reliability. During fluctuations in energy demand it will be necessary to use pumped storage systems, various types of batteries, power-to-hydro, power-to-heat, power-to-mobility, and power-to-liquid systems (Ram et al. 2018). It is possible to build and implement, well before the end of the century, safe energy networks at national, regional, and global level that are 100% renewable, mainly from wind, hydro, and solar sources, but also from geothermal, bioenergy, and tidal energy, that would be able to meet energy demands in all sectors, including the electricity sector, transportation, industry, and thermal energy for heating and cooling. With sufficiently high investment at national level around the world, such a program could also be in place by the middle of the century.

The initial study by Mark Jacobson for the USA (Jacobson et al. 2015) caused a strong reaction from other American researchers who claimed it contained significant shortcomings in the analysis (Clack et al. 2017). They disagreed that it would be possible to provide "low-cost solutions to the grid reliability problem with 100% penetration of wind, water and solar power across all energy sectors in the continental United States between 2050 and 2055", with only electricity and hydrogen as energy carriers (Clack et al. 2017). The main controversial issue in Jacobson's 2015 plan for the USA, which involves a 25-fold increase in renewable energy use over 35 years, is the dependence on hydroelectric energy during the transition, something which requires additional high investments.

Jacobson protested (Jacobson et al. 2017b) and took his critical colleagues to court, but eventually withdrew his lawsuit. The episode gave a negative image of scientists exploring the possibility of increasing the penetration of renewable energy sources. Christopher Clack and the article's other authors (Clack et al. 2017) acknowledge that the transition from fossil fuels to renewables is possible and desirable but they believe that it is not advisable to depend totally on renewables, and that it is safer to keep a small fraction of other primary energy sources in the energy mix.

The concern that, due to the intermittency of some renewable energies, such as solar and wind, renewables would be unable to meet baseload 24-hour per day demand is quite widespread but unfounded. Some renewable energies are as reliable for baseload energy as fossil fuels, such as bio-electricity generated by burning biomass from crop residues, plantation forests, or other sources, concentrated solar thermal power with thermal storage, for instance in molten salt, and geothermal power. These renewables can avoid balancing intermittent power sources with storage or new transmission (Matek and Gawell 2015).

Some authors (Heuberger and Mac Dowell 2018) have highlighted the practical difficulties involved in implementing a global transition to a system of 100% or close to 100% renewable energies. The transition to renewable energies requires a significant social and political mobilisation. People's trust in the feasibility of the transition requires the development of technical solutions that guarantee the reliability and operability of energy generation and supply systems throughout the transition process.

The important conclusion is the technical and economic feasibility of decarbonising the economy at the national, regional, and global levels by the mid-21st century using mostly current science and technology, striving for greater energy efficiency and using 100% or a high percentage, around 80%, renewable energies (Ram et al. 2018; Jacobson et al. 2018). This is good news for all those who recognise the climate change problem and its progressively harmful consequences. To achieve a probability greater than 66% of meeting the Paris Agreement goal of 2 °C, net zero global anthropogenic greenhouse gas emissions must be achieved by 2080–2100, coupled with a 40–70% reduction by 2050 relative to 2010 levels (IPCC 2014c). Net zero greenhouse gas emissions means that all anthropogenic emissions are offset by the natural sinks of those gases.

The energy transition is harder in sectors that involve infrastructures with high power consumption. This is true for the industrial sector and especially for the blast furnaces for iron and steel production and the Portland cement kilns, which need temperatures above 1500 °C. It is possible to operate these furnaces with renewable energy by using concentrated solar power (Fernández-González et al. 2018) and high-capacity energy storage systems, although this would require high initial investments. Another way of running heavy industry with renewable energy is by first using electricity to produce hydrogen by way of water hydrolysis or by producing synthetic methane. In Sweden, the company Vattenfall is building the first blast furnaces for steel production using hydrogen from renewable sources.

Transport is another problematic sector and the technologies needed to make the transition have not yet been fully developed. There has been significant progress in electric automobiles and sales are growing exponentially, with China leading the way. Several countries have planned to ban the future sale of passenger vehicles with gasoline, liquefied petroleum gas, or diesel internal combustion engines before 2050. These include India, China, Japan, South Korea, and various EU countries, Denmark, Sweden, Germany, France, the Netherlands, Spain, Portugal, and Costa Rica in Central America.

The decarbonisation of the maritime and aviation transport sectors is considerably more difficult. Technologies based on renewable energies are less developed but it is foreseen that the sale of fully carbon free boats and planes will begin before the middle of the century. The Frenchman Victorien Erussard's Energy Explorer is the first hydrogen boat that uses exclusively solar and wind energy as its primary energy sources (EE 2017). Decarbonized hydrogen is obtained from the hydrolysis of seawater performed with electricity from wind and solar photovoltaic power obtained on board. During its inaugural voyage, which began in 2017 and is set to last six years, the technologies will be optimised, which should help with the process of decarbonising maritime transportation.

According to the International Civil Aviation Organization (ICAO), fossil fuel consumption in air travel grew from 280 billion liters of kerosene in 2013 to 356 billion liters in 2018, a growth of 27%. The resulting greenhouse gas emissions increased by 26%, from 710 million tonnes in 2013 to 895 million tonnes in 2018, which represents about 2.4% of global anthropogenic CO_2 emissions (Timperley 2019). Emissions will continue to grow with the number of air passengers expected to reach 8.2 billion in 2037, as already mentioned.

In 2005, the EU launched the world's first large scale greenhouse emissions trading scheme (ETS) and, in 2008, legislation was enacted to extend the scheme to the aviation sector. The EU ETS Directive 2008/101/EC of the European Parliament and of the European Council of 19 November 2008 was intended to cap aviation emissions of CO_2 in 2012 at 97% of the average in 2004–2006 and at 95% in each year from 2013 to 2020, from all flights within airports in the European Economic Area (EEA), from flights departing from airports in the EEA to third countries and, if not exempted through delegated legislation, from incoming flights to airports in the EEA from third countries (the two latter categories covered by derogation) (EEUETS 2020).

The airline industry and some countries, especially USA, Russia, China, and India reacted adversely to the proposed regulation and argued that the EU could not legally regulate flights when they were not in EU skies. Both the USA and China passed legislation prohibiting their airlines from participating in the aviation EU ETS. China went further ahead and threatened to cancel $60 billion in plane orders from Airbus. The European Commission backed down and froze the ETS aviation legislation, claiming that the unilateral measures were necessary due to the resistance of other countries in adopting joint mitigation measures within ICAO.

Partly as a response to the pressure exerted by the EU, an agreement on the Carbon Offsetting and Reduction Scheme for International Aviation (CORSIA) was finally reached at the 39th session of the ICAO Assembly on 26 October 2016. The agreement was much acclaimed and aviation was hailed as the first industry sector to adopt a global market-based measure scheme to respond to the problem of CO_2 emissions from fossil fuel use. However, CORSIA, a UN deal based on the "aspirational goal" of making all growth in international flights after 2020 "carbon neutral", is not aligned with the Paris Agreement. At best it can only "modestly reduce" the net climate impact of international aviation up to 2035, according to the International Council on Clean Transportation (Timperley 2019). The CORSIA voluntary requirements only apply for operators with international emissions above 10 000 tCO_2 per year, which implies that, for instance, most of the world's private jets are exempt.

CORSIA's current measures include primarily offsetting emissions through the process of an airline purchasing emission units, equivalent to its offsetting requirements, and alternative fuels mostly biofuels and especially palm oil, which is an "unsustainable source", since it is a leading driver of deforestation, biodiversity loss, and human rights abuse in South-East Asia countries. The Finnish company Neste, the largest producer of aviation biofuels, uses mostly palm oil in the form of a derivative called palm fatty acid distillate and plans to concentrate its aviation biofuel production in Singapore, at the centre of the world's largest palm oil producing region. This is an example of a contribution to solving the climate change crisis at the cost of aggravating environmental degradation and destruction. The amount of unsustainable biofuels needed to support the fast-growing air travel would have very serious implications for the conservation of forests and biodiversity.

A better way to decarbonize the aviation sector is to develop new advanced technologies. Electric aircraft and hybrid hydrogen fuel cell-electric aircraft are being developed. The first prototype of a hydrogen-powered plane was the Tu-155 made by Russian manufacturer Tupolev. However, the use of hydrogen in aircraft raises problems that are difficult to solve, such as those that result from the low energy density of the fuel, the weight of containers, safety, and relatively high costs. Electric aircraft are a more promising technology, but this works only for small and medium-sized electric aircrafts. More than 70 aviation companies are planning to use electric aircraft by 2024 (Fourtané 2019), but larger aircraft may have to continue relying on the fossil fuel-based jet engine. In the N+3 Project, NASA has set aggressive fuel burn, noise, and CO_2 emission reduction goals for a new generation of aircraft that could become operational by 2035 (Ashcraft et al. 2011; Jagtap 2019).

Various institutions have developed detailed models for the global energy transition to renewable energies. The International Renewable Energy Agency (IRENA), has shown that it is possible to meet the 2 °C target if the contribution of renewable energies to the global primary energy supply is increased from around 15% in 2015 to roughly two-thirds in 2050 (IRENA 2018). However, it points out that this transition is not enough. It is also necessary to reduce the energy intensity of the global economy by two-thirds by 2050, which means that, despite the predicted growth in world population, the global supply of primary energy in 2050 would be a little lower than in 2015. Renewable energies would constitute 60% or more of final energy consumption in many countries in 2050 (IRENA 2018). In the period 2015–2050, they could increase from 7% to 67% in China and from 17% to 70% in the EU. The International Energy Agency has also presented a detailed model for a global energy transition that would achieve carbon neutrality in 2060, limiting the global mean temperature increase to 1.75 °C by 2100 (IEA 2017b).

The transition to renewable energy and to greater energy efficiency can potentially achieve 90% of the required carbon emission reductions that are needed to achieve the 2 °C goal of the Paris Agreement. However, the present pace and scale of that transition is not in line with the climate target. What are the main obstacles to implementing this plan?

The first problem is that it requires a large capital expenditure implying significant upfront borrowing. Later, the positive outcome is that the fuel costs of renewable energies are essentially free. Studies indicate that in the more favourable energy transition scenarios the main energy source is solar, accounting for 50% of the total, followed by hydro and wind energy, and a smaller contribution from geothermal, marine, and tidal energies. All these sources of energy should be used to optimise the transition, although the mix varies according to the conditions and opportunities that are specific to each country or region.

A second problem associated with the energy transition is that it will require far-reaching changes and investment, not only in the energy system, but also in transportation, industry, agriculture, buildings, infrastructure, and services. The transformational changes across all sectors of human activity required by the energy transition would likely demand the world's largest ever peacetime investment (Cooper 2019).

A third problem is that the transformational changes needed to implement the energy transition involve social and professional training changes related to unemployment and new employment opportunities, new types of services, and new business models. There will be significant job losses in the conventional energy sector, but more employment opportunities in renewable energy industry services, research, and technology. There is also a social cost in the sense that many people are against any change in their habits and routines that does not benefit them directly and in an obvious way, and moreover they are suspicious of a transition that is initially so expensive and disruptive. All the noise, misinformation, and disinformation that surrounds the climate change issue enhances their rejection to change. Even more so, they are opposed to any change that would imply any cost that would come out of their budget.

Many people recognize the risks of climate change but the prevailing culture is that the government must be able to solve the problem without any additional cost to their budget, especially if it turns out to be through taxation. They react as if the origin of climate change had nothing to do with each of them. The legitimacy of governments is to a large extent based on the premise that it is always able to provide the conditions for an increasing and unlimited economic prosperity for its citizens.

On the other hand, many people are fully aware of the origin of climate change and of its risks and are willing to promote the energy transition even if it involves some costs for them, but they represent a small fraction of the total population. The increase of such a fraction depends on extensive awareness-raising and social mobilisation regarding energy and climate change matters and sustainability at the global level. Without that, it will be very difficult or even impossible to change our individual, collective, and institutional behaviour.

The fourth problem is where to find the investment necessary to implement the energy transition. From the financial point of view climate change brings three kinds of risk: physical risk, which is the name adopted in this context for the risk associated with the harmful physical impacts that climate change imposes on companies and on their assets, the risks involved in the mitigation process, which are called energy transition risks, and liability risks that arise from people or businesses seeking compensation for losses they may have suffered from the physical or transition risks. Liability risks are those that are more directly related to the need for information disclosures about assessing the future impacts of climate change on an existing or a proposed business. Furthermore, they incorporate more explicitly the question of time and the long term nature of climate change risks. Future generations are likely to be much more adversely affected by climate change than the contemporary generation, and may ask who is responsible for the impacts that they will suffer.

The transition is viewed as presenting significant risks to the current economic and financial system due to its disruptive effect in almost all economic sectors and industries, rendering companies, business models, and some economic sectors obsolete, producing financial dislocations and losses in asset values, like for instance fossil fuel deposits. The Economist Intelligence Unit estimated the value at risk, as a result of climate change, to the total global stock of manageable assets as ranging from $4.2 trillion to $43 trillion between now and 2100 (Gardner 2015). The value at risk measures the loss a portfolio may experience, within a given time horizon, at a particular probability, and the stock of manageable assets is defined as the total stock of assets held by non-bank financial institutions. Although most people would likely agree with the statement that the most important climate change risks are those that directly involve the human population, its health, food security, and well-being, and result from the harmful impacts of sea level rise and extreme events, such as heat waves, droughts, floods, stronger tropical and extra-tropical cyclones, and forest fires, the fact is that the proposed solution, namely the energy transition, is considered by the financial system as a high risk. With less egoism and more compassion around the world, and greater adherence to truthful information about the science of climate change, that high risk would be considered of less importance and easily overcome.

The large-scale and long-term nature of climate change is an unusual problem for the current economic and financial system, which implies that investment decisions in this field are particularly difficult to make. A contribution to solving this problem was provided in 2007 by the formation of the Climate Disclosure Standards Board, an international consortium of business and environmental NGOs that "works to provide decision-useful information about material climate change and natural capital to markets via mainstream corporate reports" (CDSB 2019). In April 2015, the G20 Finance Ministers and Central Bank Governors asked the Financial Stability Board to review how the financial sector can provide the information needed by investors, lenders, and insurance underwriters to appropriately assess and price climate-related risks and opportunities. Following this initiative, a Task Force on Climate-Related Financial Disclosures (TCFD) was established in December of 2015, when the Paris Agreement was reached at the UNFCCC Conference of Parties (COP) 21, chaired by the American businessman and politician Michael Bloomberg.

The task force was charged to consider "the physical, liability and transition risks associated with climate change and what constitutes effective financial disclosures across industries". The final report of the TCFD was presented on 15 June, 2017 (TCFD 2017) and its recommendations were incorporated in an updated version of the CDSB Framework (CDSB 2019) in 2018. In June 2019 the TCFD published its second status report where it states that climate-related disclosure is becoming mainstream: "Nearly 800 public- and private-sector organizations have announced their support for the TCFD and its work, including global financial firms responsible for assets in excess of $118 trillion" (TCFD 2019). However, the TCFD also manifests its concern that not enough companies are disclosing decision-useful climate-related financial information. A minority of companies disclosed forward-looking climate targets or the resilience of their strategies under different climate-related scenarios, including the 2 °C Paris Agreement target. The TCFD considers that progress must be accelerated and that "today's disclosures remain far from the scale the markets need to channel investment to sustainable and resilient solutions, opportunities, and business models" (TCFD 2019).

In spite of the positive efforts to provide better information many companies and organizations still perceive the implications of climate change to be long-term and uncertain, therefore, not necessarily relevant to decisions made today. Furthermore, they are also influenced by the prevailing misinformation and disinformation regarding the nature and severity of the impacts of anthropogenic climate change on the human population around the world. These shortcomings have implied that carbon reporting often fails to capture the nature and severity of all climate change impacts and risks, and this tends to slow down or prevent investments in the low-carbon economy, even though the current low interest rates are particularly favourable (Cooper 2019). In fact the competitiveness of renewable energy depends strongly on the cost of borrowing, which means that the present time is especially advantageous. The frequent impasses at the UN climate change negotiations also act as a brake on such investments because they increase the uncertainty about the regulatory and fiscal policies at the global and national levels.

The establishment of and the work done by the CDSB and TCFD constitute very important positive steps in the path to achieve the energy transition in time to avoid a "dangerous anthropogenic interference with the climate system". Nevertheless, the International Energy Agency (IEA 2019c) stated in 2019 that "the momentum behind clean energy technologies is not enough to offset the effects of an expanding global economy and growing population". Carbon dioxide emissions are set to rise up until 2040, which implies that the world will fall far short of shared sustainability goals. This kind of dysfunctionality of the current economic and financial system, apparently convinced that it is possible to move simultaneously in opposite directions, is well exemplified by the fossil fuel industry. According to a Carbon Tracker report to keep emissions within the international climate target of the Paris Agreement and to protect shareholder value, the major world's oil and gas companies would have to cut their combined production by 35% before 2040 (CT 2019), which is very unlikely. Instead, global natural gas consumption is predicted to rise by 40% up to 2050 (IEA 2019c).

Natural gas is the fossil fuel with the highest increase in annual rate of consumption, having increased by 4.6% in 2018, and it is likely to continue that way for several decades. The trend results to a large extent from the new strategy of the oil and gas companies, especially in the USA, which is to remarket themselves as crucial contributors to the eradication of the problem of climate change, which they have been systematically dismissing or denying, by promoting natural gas as the solution. In fact natural gas emits 50–60% less CO_2 when combusted in a new and efficient natural gas power plant compared with emissions from a typical new coal power plant, and about 15–20% less greenhouse gases than the tailpipe emissions from a modern internal combustion engine vehicle (UCS 2014). In other words, when generating the same energy unit through fossil fuel combustion, natural gas produces less CO_2 than oil and oil less than coal. Nevertheless, burning natural gas produces about 500 grams of CO_2 per kW h. Thus, by substituting natural gas for coal or oil,

CO_2 emissions are reduced but not eliminated. It is therefore a fallacy to state that by switching to natural gas we are solving the problem of climate change. In spite of these facts it is estimated that five oil companies—ExxonMobil, BP, Chevron, Royal Dutch Shell, and ConocoPhillips—are actively advertising the benefits of increasing the use of natural gas. In the period 1986–2015 they have spent at least $3.6 billion on advertising to promote their company's position with regard to environmental issues, especially climate change (Brulle et al. 2020). These campaigns are often characterized by misinformation and disinformation on the role of fossil fuels in climate change and about the trends in CO_2 emissions around the world.

Mike Coffin, oil and gas analyst and author of the CarbonTracker report (Coffin and Grant 2019), said: "If companies and governments attempt to develop all their oil and gas reserves, either the world will miss its climate targets or assets will become 'stranded' in the energy transition, or both. The industry is trying to have its cake and eat it—reassuring shareholders and appearing supportive of Paris, while still producing more fossil fuels". These developments indicate that the current economic and financial system does not look upon the Paris Agreement temperature target as a realistic goal to be reached, but just as an annoyance that provokes an unwelcome pressure on business activity around the world.

The fifth problem that affects the energy transition stems from the geostrategic competition between the major world powers. In 2018, the USA began setting tariffs and other trade barriers on China to counteract what it considered China's "unfair trade practices". This form of protectionism, motivated largely by the USA's fear that China may become the world's most powerful economy, creates uncertainty, hurts businesses, and increases the risk of slowing down the global economy. Up to the end of 2019, the USA imposed tariffs on about $360 billion of Chinese goods and China has retaliated by targeting about $110 billion of USA products. China's economic growth in 2019 is at its slowest in nearly 30 years, which is likely to be partly a consequence of the trade war.

In view of this situation and taking into account the fact that economic growth is the paramount objective of government, China decided to boost the economy and create new jobs by restarting the exploration of coal mines and building new coal-fired power plants. Cheap and abundant energy is one of the surest ways to revitalize the economy and to achieve greater GDP growth. Coal is relatively cheap and there is still plenty of coal in the Shaanxi province of Central China. It is estimated that China still has a reserve to production ratio of 38 years (BP 2019). However, the reasons for the revival of coal in China are more complex than just the trade war with the USA.

The country increased its coal-fired power by 38 GW, 35 GW, and 25.5 GW in the years 2016, 2017, and 2018, and in 2019 coal plants with a total power of 148 GW are either being built or starting to be built. In part ,this upsurge results from the effective application of a well-known but relatively expensive technology that reduces the air pollution caused by coal power plants. Since 2016, coal operators have been required to install smokestack scrubbers to remove SO_2, N_2O, and other hazardous gases, and presently about 80% of China's coal plants have scrubbers. As already mentioned, China's CO_2 emissions rose sharply in the period 2002–2007, reaching a peak in 2013 and then declining slightly in the period 2014–2016, mainly because the government was very concerned with air pollution and had declared a "war on pollution". Because of the new scrubbers, the air quality in many Chinese cities, including Beijing, improved significantly between 2013 and 2017. The government, less preoccupied with the air pollution created by coal power plants, allowed the construction of more plants. The upsurge is also related to a power struggle with the provincial governments, which are preoccupied with the social costs of the energy transition, particularly as regards jobs in the coal producing regions. Another debate that is going on in China is how well renewables can supply the baseload power to support large scale vehicle electrification. The dominant opinion is that the best cost-effective solution is not concentrated solar power or other advanced technologies, but a coal-fired power baseload of about 1400 GW.

China intends to continue its investment in renewable energies and has a plan to reduce coal power in the north-west of the country by about a quarter to bring down overcapacity and global carbon emissions (Shepherd 2019). Zhao Yingmin, China's vice environment minister, said in November 2019: "We continue to work hard to advance the fight against climate change, but on the other hand, we are indeed facing multiple challenges such as developing the economy, improving the people's livelihoods, eliminating poverty and controlling pollution" (Stanway 2019). In 2013, the whole Chinese population gained access to electricity. The per capita use of energy increased by an amazing factor of 2.48 in the period 2000–2014, but in 2014 it was still about three times less than in the USA (WB 2020). On the other hand, China's CO_2 emissions increased by 1.7% and 2.3% in 2017 and 2018, which represents a trend incompatible with the Paris Agreement goal of 2 °C. A UNEP report published at the end of 2019 indicates that the world needs to cut CO_2 emissions by 7.6% a year over the next decade in order to fulfil that goal (UNEP 2019).

7.19   The Long-Term Future of Increasing Energy Demand

The success of the present economic and financial system presupposes a robust annual global GDP growth, usually recommended to be well above 3%, which it has been possible to achieve with increasing levels of global energy consumption, although only closer to 3% since 1975. The increase in consumption results only in part from the continuous global population growth, since global energy per capita has also generally been increasing, but at slower rates during some periods, as already mentioned. Thus, the sustainability of the success of the present economic and financial system depends on the continuous availability of increasing amounts of primary energy sources. Assuming that the present system endures without major changes this century and beyond, what are those energy sources going to be?

A distinctive feature of the current situation is that the global energy system is a two-speed system where advanced industrialized economies had a mean energy consumption per capita in the period 1970–2020 about 5 times higher than the rest of the world, according to IEA statistics. The gap between the two groups of countries is slowly decreasing, because the per capita consumption in the advanced economies increases at a rate that is about three times lower than the rest of the world. A significant point, however, is that the energy consumption per capita in the advanced economies continues to increase. Can we expect a saturation in the amount of energy consumption per capita in the advanced economies in the future? The most likely answer is that we cannot, although on average the population in these countries already enjoys a relatively high level of well-being and economic prosperity compared with the other countries. Such a saturation would represent a very significant advance towards sustainability and could be reached by applying the principle of energy sufficiency.

To discuss the long term future of humankind's energy sources, it is convenient to revisit the various types of energy flux in the Earth system. We have seen that the total power received from the Sun in the form of radiant energy in the visible domain and absorbed by the Earth system amounts to 121.8 PW, which is of the order of 10^{17} W. When there is an energy balance, the same amount of power is returned to outer space as radiant energy in the infrared domain. In fact, as also shown previously, there is an imbalance caused by the anthropogenic accumulation of greenhouse gases in the atmosphere, which creates a positive radiative forcing. This implies that the amount of outgoing radiant energy is slightly less than the amount of incoming radiant energy. The mean difference in power for the period 1971–2010 was estimated at 0.213 PW by the IPCC (2013), and is responsible for the observed global warming and all its other consequences. This power, of the order of 10^{14} W, delivers an energy of 0.213 PJ per second, which is greater by a factor of 3.38 than the energy of 0.063 PJ delivered by the bomb "Little Boy" dropped on Hiroshima.

What are the other energy fluxes on Earth? Global photosynthesis converts a fraction of the absorbed radiant energy, viz., 121.8 PW, into a global NPP of 104.9 GtC per year, with 46.2% contributed by the oceans and 53.8% contributed by land (Field 1998). This carbon is stored on land as wood and fibres of trees and plants, and in the ocean it is fast recycled into the food chain. It represents the annual growth of organic matter, which is of course essential for life in general and for human life in particular. The rate of energy storage by global photosynthesis can be estimated as the calorific power of all carbon produced in one year by photosynthesis, which amounts to a power of 0.122 PW, using the previously mentioned conversion formula. Estimates based on the photochemical reactions involved in photosynthesis give values of the same order of magnitude (Dimitriev 2013), so the value usually accepted for the rate of energy storage by global photosynthesis is 10^{14} W.

As already mentioned we use, or destroy in the process of using, about 300 EJ per year of the energy stored by terrestrial photosynthesis for food, pastures, bioenergy, including wasted biomass and forest fires (Haberl et al. 2013), corresponding to an energy power of 9.51×10^{-3} PW, which is of the order of 10^{12} W. The part of that energy used currently for bioenergy amounts to about 54.9 EJ per year, or 1.74×10^{-3} PW. Furthermore, it has been proposed that 190 EJ per year, or 6.02×10^{-3} PW, is an upper biophysical limit for primary bioenergy (Haberl et al. 2013).

Let us now consider the human use of global primary energy, which as previously mentioned was estimated at 583.8 EJ in 2017, and corresponds to a power of 1.85×10^{-2} PW. This power is of the order of 10^{13} W and about 5.4 times smaller than the power of the global photosynthesis. However, humankind's power demand is on the way to exceeding the power of global photosynthesis next century. Primary energy consumption has been increasing at a mean annual rate of 1.5%, although in 2018 it increased by 2.3%, the fastest rate of the last decade. If this rate of 1.5% is maintained, global primary energy consumption will reach about 1000 EJ in the middle of the century, as already noted (Moriarty and Honnery 2012). If the annual growth rate of 1.5% is maintained for 113 years, which is less likely, the global consumption of primary power will reach 10^{14} W, the power of global photosynthesis, in 2130.

So the question is: what primary energy sources will be used up to that time and beyond? The largest energy sources that humankind has been using throughout history are derived from the net primary productivity (NPP) either in the form of biomass or fossilized biomass or fossil fuels. In 2017, the percentage of the global energy consumption involving the NPP in the present time or in the distant past was 90.8%. This situation reveals the deep conservatism of the current large scale energy technologies around the world and the reluctance to believe in and invest in technological innovation in this area. The reason for this "lock in" are the considerable vested interests of the fossil fuel industry in the present status quo. Thus the global dependence on fossil fuels is likely to be maintained for many decades in spite of the climate change crisis.

The fossil fuel infrastructure for production and consumption continues to expand and it will take a long time to replace it by the infrastructure of a new energy paradigm. A revealing example of the underlying trends in the present situation can be found in the transportation sector. Global annual electric car sales are expected to increase from 2 million in 2018 to 20 million vehicles by 2030, and nearly 7% of the global car fleet is expected to be electric in that year (IEA 2019d). At the same time, the sales of internal combustion engine cars declined in 2018 and 2019. This might be taken as a sign that CO_2 emissions from cars are peaking and set to start decreasing. In fact, this is untrue because consumers are buying ever larger, more powerful and less fuel-efficient cars, known as Sport Utility Vehicles (SUVs). The global number of SUVs has duplicated in the past decade and around 40% of present car sales are SUVs. This trend reveals that a significant fraction of global consumers continue to follow the culture that more power is an imperative for progress and that there is no alternative to progress and unlimited economic growth.

The most likely future outcome is that the global share of modern renewables, particularly solar and wind, will continue to increase, but at a pace unable to satisfy the demand from the growing economies of developing countries and from increasing global population. This scenario implies that global CO_2 emissions will only peak after 2040, a situation that could be avoided by using the well-known technology of carbon capture and sequestration (CCS) on a large scale around the world; this is where the CO_2 emissions from fossil fuel power plants, and particularly from coal-fired power plants, are captured and injected deep into the ground to prevent the gas from being released into the atmosphere (IPCC 2014c). However, the implementation of this technology is not commercially viable in the absence of a carbon tax or adequate incentives because it increases the price of electricity. Capturing and liquefying CO_2 requires a lot of energy and increases the fuel requirements of a coal-fired power plant by 25–40%. The cost of electricity from a new power plant with CCS is estimated to be 21–91% higher than from a conventional power plant (BGS 2020). In spite of this problem, when CCS is combined with burning biomass, the power plant has the advantage that it generates negative CO_2 emissions. Solar and wind energy are rapidly becoming cost-competitive as regards conventional coal power plants. In the USA, 74% of the coal power plants provide electricity that's more expensive than wind or solar (Gimon et al. 2019). Keeping the coal in the ground is by far the most economical way of reducing CO_2 emissions and preventing the air pollution that results from its use.

With the continuing use of fossil fuels their readily available reserves will inevitably become scarce. There are large uncertainties in the estimates of global fossil fuel reserves and resources because of the difficulty in foreseeing future technological innovations that would allow increased extraction from the ground without significantly reducing the energy return on energy invested (EROEI). The most frequently mentioned estimate of about 5000 GtC (Caldeira and Wickett 2003; Tokarska et al. 2016) is considered to be at the lower end of the range of estimates (IEA 2013). Total fossil fuel reserves have been estimated to lie in the range 930–1954 GtC and resources in the range 8527–13 648 GtC (Grubler et al. 2014).

To estimate how long fossil fuels will last it is also necessary to take into account the present and future rates of production and consumption. Assuming that all the estimated reserves and resources of fossil fuels are burned with CO_2 emissions following the IPCC IS92a scenario (Leggett et al. 1992) in the period 2000–2100 and that they subsequently follow a logistic function (Caldeira and Wickett 2003), the emissions peak would be reached around 2150. Fossil fuels would be exhausted between 2400 and 2500 and the CO_2 atmospheric concentration would peak in 2300 at about 2000 ppmv, a value that last occurred only between 50 and 40 million years ago. However, it is likely that in this scenario fossil fuels would last much longer since their consumption rate would tend to decrease with time as a result of the slowing down of the global economy due to the environmental disruption caused by intense anthropogenic climate change.

Taking into account only the global fossil fuel reserves (which include the deposits that are currently known to be economically feasible to exploit) and at the present rates of production, oil will run out in 53 years, natural gas in 54, and coal in 110 (IEA 2015). Technological innovations may extend these limits considerably, but the other problem is that there has been a global secular trend of decreasing EROEI for all fossil fuels, in particular for oil and natural gas. This trend is consistent with the fact that the deposits that are easily exploited have become more difficult to find, whence investors are now going after more difficult deposits and forced to back "extreme energy", which decreases the profitability of the investment.

In the USA, the EROEI for finding oil and gas decreased exponentially from 1200 in 1919 to 5 in 2007 (Guilford et al. 2011). The EROEI for production in the oil and gas industry was about 22 in the mid-1950s and decreased to 11 in the mid- to late 2000s (Guilford et al. 2011). Around the world, most easy oil has already been extracted. The oil reserves remaining to be exploited are difficult to reach since they are located in ultra-deep deposits off-shore, in the Arctic, in tar sands, or in shale rocks and they are usually difficult to refine. It now takes more energy and therefore more investment to extract fossil fuels from the ground, and this trend tends to increase the costs of energy and food at the global level. Furthermore, this tendency has the potential to significantly slow down global economic growth.

Technological innovations have allowed new oil sources to be explored, but the creeping scarcity problem is far from being resolved. The Hubbert peak prediction for oil in the USA, obtained using the oil production data up to 1965, failed completely because oil shale was first produced in large quantities in 2008. Due to the shale revolution, the USA is now leading global oil and natural gas production, with record growth in 2018. Nevertheless, the country is still exposed to what happens in global oil markets, in part because most of its refineries are designed to refine the heavy crude from Venezuela, Nigeria, and the Middle East but not the light sweet crude produced domestically from shale rocks. The USA continues to import oil in part because of the distorted capabilities of its refineries. Still, the refineries are hesitant to make the necessary but expensive changes in the equipment because the shale oil boom is expected to level off within the next few years.

From the financial point of view, the USA's shale oil boom has failed to significantly increase corporate profits, and shale producers have created a large debt to finance the required large production investments. It is estimated that shale oil companies have more than $200 billion of debt maturing over the next four years, with about $40 billion at the beginning of 2020 (FT 2020). A balance must be found between production and returns, and this is likely to reduce the levels of production. It is important to realize that this situation arises even though the oil and gas production boom has been aided by tax provisions and other subsidies that support private investment in infrastructure for oil exploration and development (Erickson et al. 2017). Federal tax laws enable oil and gas producers to deduct capital expenditures faster, or at higher levels, than standard tax accounting rules would typically allow, and this boosts investment returns. With oil prices at $50 per barrel, about half of the yet-to-be-developed shale oil deposits in the USA will only become profitable with federal subsidies, potentially increasing US oil production by almost 20 billion barrels over the next few decades (Erickson et al. 2017).

When the fossil fuel EROEI is estimated at the final energy stage, at the time when energy enters the economy, instead of being measured at the primary energy stage, the values obtained are much closer to those of renewable energies than previously expected (Brockway et al. 2019). Furthermore, the fossil fuel EROEI at the final stage is generally decreasing and could decline precipitously in the near future (Brockway et al. 2019), while the renewable EROEI is increasing, and is likely to continue increasing in the mid- to long term. A further important aspect to take into account when deciding to invest in renewable energies or shale oil and gas is to consider the health and environmental impacts of the two energy options. The former energy sources either have no discernible impacts or negligible impacts compared with those of the latter (Dimanchev et al. 2019).

For instance, fracking often releases toxic chemicals into water and air (Jackson et al. 2014). Air and water quality dynamics around fracking sites are not yet fully understood, but cumulative health impacts of fracking operations for nearby residents and workers have begun to be identified. Drilling a single well requires the use of between 3–6 million gallons of water posing significant risks to local water supplies, especially in drought-prone regions (Kondash and Vengosh 2015). In some southern states of the USA, such as Arkansas and Texas, researchers have found water levels dangerously depleted due to water withdrawals for fracking operations.

Furthermore, the treatment and disposal of the large volume of chemical-laden flowback water and brine-laden wastewater from fracking poses serious problems. In the period 2011–2016 when shale oil and gas exploration experienced a rapid growth, the water use per well increased up to 770%, while flowback and wastewater volumes generated within the first year of production increased up to 1440% (Kondash et al. 2018). This trend creates significant challenges regarding the water supply for fracking and the treatment of the increasingly larger volumes of wastewater produced. Besides the impact on water resources, fracking operations and the deep-well injection of their wastewaters are known to cause earthquakes (Foulger et al. 2018).

There is now evidence that people living closer than 3 km from fracking sites suffer negative health impacts. These include a greater incidence of low-birth weight babies as well as significant declines in several other measures of infant health. There is strong evidence for negative health effects of in utero exposure to fracking within 1 km of fracking sites (Currie et al. 2017). About 29 000 of the nearly 4 million annual US births occur within 1 km of an active fracking site (Currie et al. 2017). Most likely, people living in the affected regions are becoming increasingly aware of these risks, but it is also likely that there is much misinformation and disinformation about the subject. Meanwhile, some US states have banned fracking based on its negative environmental and health impacts. These include New York State, Maryland, and Vermont. Although there is shale gas in Europe, large scale exploration has not yet started and some countries have banned fracking, such as France and Germany.

It may be expected that, when shale oil and gas replace coal, the climate change impact will be smaller because CO_2 emissions from coal are significantly higher than those from oil and gas. However, this conclusion may not be entirely correct because of the fugitive CH_4 emissions at the fracking sites. Global CH_4 emissions rose steeply in the last decades of the 20th century and then levelled off. The Paris Agreement was built upon the assumption of stable atmospheric methane concentrations in the 21st century. Unexpectedly, around 2016, emissions started to rise again for no apparent reason. In 2018, global methane concentrations reached a new maximum, after the second highest year-on-year jump in the last 20 years. The new trend in methane emissions is a further threat to the Paris Agreement. In 2019, evidence began to accumulate that the recent increase in methane emissions results predominantly from shale oil and gas production, especially from shale gas (Howarth 2019) in the USA. The country accounts for 89% of the global shale gas production, while Canada accounts for most of the remaining part. It has been estimated that fracking operations leak, vent, or flare between 2% and 6% of the natural gas extracted.

In view of all the shortcomings associated with shale oil and gas exploitation it might be thought that any reasonable outlook and responsible long-term planning would advise against investing in shale oil and gas, and instead advocate investment in renewable energies. Why is this not happening in the USA? The reason is largely cultural, since historically the economic supremacy of the USA and the role of the dollar as the world's dominant currency has been deeply dependent on fossil fuels. This dependence is part of the country's identity and it will take a long time to change it. Oil exploration and production is deeply rooted in the history of the country and it is believed that it will continue to support the USA's economic and military supremacy in the world, with no end in sight. People who believe in this narrative are prepared to endure great sacrifices to pursue the dream of America greatness.

They are prepared to support a dream that imperils the health of US citizens by allowing businesses to increasingly pollute the air and water resources in order to promote economic growth. Not all Americans believe in this narrative and many oppose it for various reasons, in particular the increasing human and environmental risks, the financial costs of extreme energy, and the risk of climate change, but currently that is the narrative that supports current US domestic and international energy policy. Support for the opposing narratives is strongly correlated with the political polarization between the Republican and Democratic parties and the final future outcome will depend on the functionality of the democratic institutions.

The power of the USA in the Middle East has been increasing since the time of the momentous meeting between Franklin Roosevelt and Ibn Saud at Great Bitter Lake in 1945. Since the beginning of the Iraq war in 2003, the USA has controlled most Middle East oil both in Iraq, which has some of the world's largest oil reserves, and in Saudi Arabia, considered to be one of its most precious allies. The use by Iraq of its own oil revenues is controlled by the Federal Reserve Bank in New York and ultimately by the government of the USA. In fact, the USA could easily implode the Iraq economy with unilateral sanctions that would cut Iraqi government access to its account at the Reserve Bank. The USA's power to impose sanctions on various countries around the world depends on the American dollar functioning as the world's reserve currency. This control over Middle East oil provides the USA with a strong dominating influence over China, Europe, India, and Japan, which are all dependent on oil from the Middle East. With the shale oil boom, the USA no longer depends directly on Middle East oil, but its importance, instead of decreasing, has in fact increased because the outstanding goal is to preserve the world energy paradigm based on fossil fuels for as long as possible. Oil is to a large extent the source of American power and world dominance. By inundating their domestic market with shale oil, the USA expects oil to go on being economically competitive so that the transition to other energy sources, in particular renewable energies, will become less attractive and more difficult to implement.

If the current energy policy of the USA prevails globally, supported by other fossil fuel producing countries, the full transition away from fossil fuels is likely to be delayed past the Paris 2 °C threshold, and humankind will be confronted with dangerous anthropogenic interference in the climate system. The USA will be deeply affected by climate change, along with all other countries in the world. However, being an advanced economy, it counts on having a much stronger capacity to adapt to the inclemency of the new climate than most other countries, especially those developing countries that are particularly vulnerable to climate change. In fact, their greater capacity for adaptation will tend to increase their economic and military power relative to other countries, which is a good example of the application of the prevailing principle of rational egoism. In this scenario, it is likely that the impacts of natural resource overexploitation and climate change will lead to regional economic slowdowns and upheavals and that the demand for energy will drop significantly. On the other hand there are signs that the climate crisis is generating political pressure on the major banks to assess climate risks to their fossil fuel investments and the financial system as a whole. The next ten years will be crucial for the major oil and gas corporations to decide how to position themselves as regards climate change liability risks resulting from its stranded assets.

In any case the energy transition will be most likely based on renewable energies, since the capacity of nuclear fission energy to contribute significantly to a carbon neutral world is undermined by the fact that nuclear fission power plants have become costlier compared with other competing energy technologies, especially in Western countries (MITEI 2018). The economic and financial conditions for the development of fission nuclear power plants are considerably more favourable in Russia, China, India, South Korea, and other countries. State-owned nuclear companies in Russia and China are now offering emerging countries attractive finance and fuel services for the building of nuclear power plants. Nuclear fission energy provides an important contribution to decreasing the global dependence on fossil fuels. France and Sweden have achieved a rapid decarbonisation of their electricity supply through the building of nuclear power plants. In these countries, electricity is supplied to the consumer with less than a tenth of the world average CO_2 emissions per kWh. Both fossil fuels and nuclear energy involve risks, but it is very difficult to compare them quantitatively and in a way that generates consensus because of their different nature, people's different perception of the risk, and their deeply different relationship with time. Furthermore, the science-based assessment of the risks involved in the two energy sources requires quite different levels of scientific literacy to be made understandable to all citizens.

A better alternative to decarbonize the global economy is potentially provided by nuclear fusion reactions, a process similar to the one that gives the Sun and the stars their very large power output. Nuclear fusion reactors are likely to provide nearly inexhaustible energy since they are based on the deuterium–tritium (D-T) fusion reaction. Deuterium, ^2H, is an isotope of hydrogen with a natural abundance of 0.0156% in the hydrogen contained on Earth's water molecules. This means that in every cubic metre of seawater there are 33 grams of deuterium. Tritium, ^3H, is a fast-decaying isotope of hydrogen that can be produced in a nuclear reaction involving neutrons and lithium. Since fusion produces neutrons, the reaction can take place inside a reactor with lithium on its interior walls. Land-based lithium resources are estimated to be sufficient to operate fusion power plants for more than 1000 years and ocean lithium resources are practically inhexaustible.

Temperatures of some tens of millions of degrees Celsius are required to initiate fusion reactions in a plasma containing deuterium and tritium, and the energy needed to heat the plasma up to those temperatures is considerable. When the fusion reactions start, they rapidly produce more energy than the energy used to heat the plasma. If these fusion reactions can be controlled and sustained then a process called controlled thermonuclear fusion becomes possible. This is probably the most important goal in energy technology since the discovery of the atomic nucleus and the strong nuclear force. In spite of all the effort and investments made, controlled fusion has not yet been fully achieved. The longest contained fusion reaction to date lasted for just 70 seconds and was made by the Korea Superconducting Tokamak Advanced Research reactor in 2016.

As regards the history of nuclear fusion, the first patent of a thermonuclear reactor was produced in the UK by the Atomic Energy Authority and presented on 28 April 1946. Independently, the Russian physicists Igor Tamm (1895–1971) and Andrei Sakharov (1921–1989) proposed in 1951 the Tokamak (a Russian acronym for "toroidal chamber with magnetic coils") to confine the hot plasma within powerful magnetic fields, a design that has since become the leading candidate for a commercial fusion reactor. Man-made uncontrolled fusion was realized when the USA detonated the Castle Bravo thermonuclear bomb in the Bikini Atoll. In this kind of nuclear device the temperature needed to start the fusion reaction in the bomb is obtained with the energy released by a nuclear fission explosion that compresses and heats the fusion fuel. Humankind has been trying to achieve controlled thermonuclear fusion for 74 years without success and it will probably take several more decades to realize a practically useable energy source based on the D-T nuclear fusion reaction (Ongena 2016).

While waiting to evaluate the feasibility of using thermonuclear fusion reactors to supply energy on a large scale at the global level, the safest and most economical option to decarbonize the world economy is to develop renewable energies. The question then arises of whether or not renewable energies will be able to satisfy the expected energy demand by the world economy in the foreseeable future. The scientific consensus is that they will, although there are various social, economic, financial, and technological challenges that must be addressed and resolved to reach that objective. In fact, the global energy transition coupled with a significant increase in energy efficiency requires a transformative social change with the explicit goal of social justice and sustainability.

In the case of fossil fuels, a distinction is made between reserves and resources. Renewable energies are of a different nature to fossil fuels and nuclear energy, but they all represent some form of power, or in other words, an annual energy flow that is likely to be available indefinitely in *Homo sapiens* time, if harvested in a sustainable way. The total natural flows of growing biomass, solar, wind, hydro, geothermal, and ocean energy are usually called theoretical potentials. The technical potential of a renewable energy reflects the possible degree of use determined by physical, in particular thermodynamic, geographic, technological, and social limitations, without considering economic feasibility (Grubler et al. 2014). The part of the technical potential that is economically feasible to use, i.e., cost effective with current technology and costs of production, is called the economic potential, which is the equivalent of a reserve for fossil fuels. The global theoretical, technical, and economic potentials of renewable energy sources have been analysed and discussed by various authors (Miller et al. 2011; Johansson et al. 2012; Moriarty and Honnery 2012; Marvel et al. 2013; Grubler et al. 2014; Korfiati et al. 2016).

The Global Energy Assessment of 2012 concluded that the renewable energy resource base is potentially sufficient to meet a global demand 10 to 100 times greater than the current global energy demand (Johansson et al. 2012). The renewable energy sources that have the greatest technical potential are: direct solar radiant energy with an estimated potential power between 1.97 PW and 8.90 PW; ocean energy with a power between 101 TW and 328 TW; wind energy with a power between 39.0 TW and 70.2 TW, and geothermal power with a power between 25.3 TW and 48.2 TW. Using the GEA Pathways, which are normative scenarios that illustrate possible transition pathways towards a more sustainable future, the economic potential of all renewable energies in 2100 ranges from about 11 TW to 79 TW.

Wide-ranging technological innovations and more efficient manufacturing procedures have made solar and wind energy the leaders of the energy transition. According to Bloomberg New Energy Finance "wind and solar energy are now the least expensive forms of power in two-thirds of the world" (BNEF 2019). In 2017, the global renewable primary energy power was estimated at 2.55 TW, which is about 13.8% of the global energy power of 18.51 TW. Assuming that global primary energy power continues to grow at an annual rate of 1.5%, it will reach 63.7 TW by the end of the century. If all this energy is provided by renewable energy sources, it will be close to the upper range of the renewable energy GEA Pathways. The jump from 2.55 TW to 63.7 TW would be extremely difficult to achieve. Without a significant improvement in global energy efficiency it will become increasingly difficult to make a full transition from fossil fuels to renewable energies, as previously emphasized. Moriarty and Honnery (2012) consider that a global shift to renewables depends on a reduction in overall energy use. If the sustainability goal is not fully accepted around the world, there is the risk of responding to the energy demand by returning to the use of coal, which is the fossil fuel with longest lasting reserves and the most cost-efficient exploitation. In conclusion, there are large uncertainties regarding the way humankind is going to develop and satisfy its energy needs up to the end of the century and especially beyond. In any case, it is well known that energy has been a determining factor for the human civilization since the Industrial Revolution, and it will continue to be.

The issue of energy is not limited to the question of environmental security and global availability in the future, but involves other crucial challenges. One of the most important is the question of energy equity, which can be defined as ensuring that all people around the world have access to the level of energy needed to provide for their security and well-being (Moss et al. 2011). The world is far from reaching this objective, although Target 7.1 of Sustainable Development Goal number 7 is: "By 2030, ensure universal access to affordable, reliable and modern energy services".

This goal does not specify the minimum per capita consumption, although it is known that, below a certain value, well-being and economic prosperity decline sharply. This question is closely related to the issue of energy poverty, which has very different expressions and consequences around the world, namely in the least developed countries and in countries with advanced industrialized economies. In the latter group of countries, recent energy poverty is partly the result of large economic inequalities and the economic slowdown in those countries, which in turn finds among its causes the increasing cost of fossil fuel exploitation. At the same time, it is precisely in such countries that energy sufficiency should start to be gradually implemented.

Day et al. (2016) analysed energy poverty on the global scale within the framework of developing the essential individual capabilities proposed by Amartya Sen (1999) and others. Energy poverty is very far from being eradicated in developing countries. Nevertheless, the number of people without access to electricity has fallen, and dropped below 1.1 billion for the first time in 2017 (IEA 2017a). It must be noted, however, that due to the relatively high fertility rates in communities without access to electricity, the total number of people in that situation in 2016 was higher than in the year 2000. Lack of access to electricity and modern energy sources makes it notably more difficult to escape from poverty, particularly in rural areas. These are much worse affected than urban areas, since roughly 80% of people who do not have access to electricity live in rural areas. About two billion people do not have cooking facilities based on modern energy services, forcing them to use animal waste and other toxic alternatives as cooking fuel.

There are also positive achievements. A good example is that, since the year 2000, roughly 1.2 billion people have gained access to electricity, 500 million of these being in India, which is a huge success. The most problematic region of the world continues to be Sub-Saharan Africa where, nonetheless, the electrification rate has outstripped population growth since 2014 (IEA 2017a). Roughly 71% of the 1.2 billion people who have gained access to electricity since 2000 have done so using fossil fuels: 45% coal, 19% oil, and 7% natural gas. Renewable energies are managing to increase their contribution to the electrification effort in rural areas thanks to their falling cost and innovative financing initiatives. Roughly 6% of new access in the last five years has been achieved using decentralised renewable systems (IEA 2017a), a model that is very important for achieving environmental security. The fossil fuel industry views target 7 of the Sustainable Development Goals mostly as a business opportunity for consuming fossil fuels. The use of decentralized renewable energy systems and the promotion of energy efficiency is not their priority.

The deep energy inequalities around the world have harmful consequences for the global mitigation of climate change. Reducing the CO_2 emissions of emerging and less-developed countries can slow down the economic growth that they require to converge with advanced economies. The fast-growing economies of Asia have been the largest contributors to the global increase in CO_2 emissions in recent years. Instead of criticizing such countries, the advanced economies should feel an ethical obligation to help them make the transition to renewable energies without jeopardizing their right to benefit from economic growth. Operational costs for renewable energy are decreasing but the construction of the generating facilities requires access to long-term credit that is difficult to obtain in many developing countries. Advanced economies could play a crucial role in helping to finance the energy transition in those countries. This role is well recognised in the international climate negotiations under the UNFCCC and in particular in the Paris Agreement. The Green Climate Fund was established in 2010 within the framework of the UNFCCC as an operating entity of the Financial Mechanism to assist developing countries in adaptation and mitigation. It became operational in 2015 and by the end of 2019 it had allocated about $5.2 billion to 111 climate projects in 99 countries, addressing both mitigation and adaptation. However, the original pledges made by the advanced economies totalled $10.4 billion. The most important contributors include France, Germany, Japan, Norway, Sweden, and the UK, while the governments of the USA and Australia have refused to contribute to the fund. Advanced economies could also contribute to mitigation by phasing out their coal power plants, which have an average age of 45 years, while in Asia it is only 12 years (Rogoff 2020). A world mired in protectionism and trade wars is incapable of seeing the mutual advantages that will come economic convergence of the emerging and less-developed countries with the advanced economies, and it will become much more vulnerable to climate change.

7.20   Controlling the Weather to Dominate the World. The Background to Geoengineering

The fossil fuel industry and the economic, and the financial system that supports it, are now often acknowledging publicly when pressed by people concerned with the climate crisis that the risk of climate change is real and that the risk warrants action. However, in other circumstances, they also claim that reducing greenhouse gases is too expensive and that the developing world needs fossil fuels to combat and eradicate poverty. One of the most likely escapes from this *impasse* is a solution blessed by the conviction that human technology will be able to solve all the ensuing human problems.

The technological solution to climate change that might allow the world's dependence on fossil fuels to perdure is geoengineering or climate engineering (RS 2009; IPCC 2012; NRC 2015a, b; Blackstock and Low 2018). On 20 July 2009, the American Meteorological Society made a political declaration on geoengineering of the Earth system (AMS 2009) in which it sets out three proactive strategies to reduce the risks of climate change: (1) mitigation, (2) adaptation, and (3) geoengineering, defined as "deliberately manipulating physical, chemical, or biological aspects of the Earth system", with the aim of neutralising the consequences of growing greenhouse gas emissions.

In 1965, 44 years earlier and also in the USA, a report by the Presidential Science Advisory Committee to President Lyndon B. Johnson (1908–1973) acknowledged the risks of climate change caused by growing global CO_2 emissions from burning fossil fuels, and suggested, as a response, the adoption of "countervailing climatic changes" able to offset and perhaps cancel out the effects of the former (PSAC 1965). Specifically, it suggested spreading floating particles in the ocean that reflect solar radiation and increase the Earth's albedo by roughly 1% with an annual cost estimated at $500 million. A suggestion filled with technological hope and hubris. The official position in the US Congress and government, since the time when climate change entered the political agenda, has been to view the use of fossil fuels as unquestionable and irreversible because they provide abundant energy at affordable prices, make a powerful contribution to economic growth, and are one of the main sources of American power and supremacy in the world. Furthermore, making an energy transition to control climate change would be an affront to free enterprise and it would represent unacceptable state interference in the private sector.

David Keith, Gordon McKay Professor of Applied Physics in the School of Engineering and Applied Sciences and Professor of Public Policy in the Harvard Kennedy School, both at Harvard University, is one of the most outspoken defenders of geoengineering. He was one of the first scientists to include geoengineering in a wider, more ambitious project, stating that the evolution of our technological society will inevitably lead to the intentional transformation of nature at planetary level (Keith 2010). Nonetheless, Keith also says, "I urge caution" (Keith 2010). According to him, geoengineering is the "intentional, large scale manipulation of the environment".

To understand the motivation and hope involved in geoengineering it is worth analysing them in an historic context by revisiting a much older quest to modify the weather. Indeed, this idea of modifying the weather has a long history (Fleming 2006) which began in the 1830s in the USA when an economic depression and crisis in agricultural production meant wheat had to be imported from Europe. James P. Espy (1785–1860), an American meteorologist, developed a theory of storms, published under the title *The Philosophy of Storms* (Espy 1841), based on which he suggested artificially increasing precipitation by lighting large forest fires to produce upward movements in the air. He spent several years unsuccessfully requesting authorisation from the USA Congress to start forest fires along a 600-mile north–south line along the Rocky Mountains in the west of the USA (Fleming 2010). Finally, in 1849, Espy set fire to 12 acres of pine forest in Fairfax County, Virginia, and waited for rain that never came (Moore 2015). After this failure, the newspapers stopped talking about Espy's theories for artificially producing rain and the matter was no longer discussed.

Inside a cloud, when water vapour saturation is reached, water droplets form on condensation nuclei, which are small particles of about 0.2 μm, or one hundredth the size of a water droplet, and when the temperature falls below freezing, ice crystals form on crystallization nuclei. In 1911, the German geophysicist Alfred Wegener (1880–1930) showed for the first time that precipitation from a cloud is triggered by the instability of the thermodynamic state resulting from the co-existence of liquid water and ice in a cloud (Wegener 1911). In 1928, Tor Bergeron (1891–1977), a Swedish meteorologist of the Bergen School of Meteorology in Norway, proposed that clouds would produce precipitation if they contained the right mixture of ice crystals and supercooled water drops (Bergeron 1928). Ten years later, Walter Findeisen (1909–1945), a German meteorologist contributed to the previous work of Wegener and Bergeron and built a cloud chamber where he carried out experiments that confirmed the theories that had been proposed (Findeisen 1938). What is now called the Wegener–Bergeron–Findeisen process refers to the rapid conversion of liquid to ice that may occur when supercooled droplets and ice crystals co-exist in a cloud, a conversion that depends on the difference in saturation vapour pressure over liquid and ice surfaces. This Wegener–Bergeron–Findeisen process may abruptly transform non-precipitating liquid clouds into heavily precipitating ice clouds, changing the clouds' radiative properties and profoundly impacting both weather and climate. The discovery of this process led to the conclusion that the introduction of a substance that acts as ice nuclei in a supercooled cloud can convert the supercooled water droplets into ice, thereby increasing the probability of rain or snow.

In 1930, the Dutch meteorologist August Veraart (1881–1947) was the first to be convinced that he could manipulate clouds by changing their thermal structure through seeding them with dry ice, a form of solid CO_2. He went ahead and made seeding experiments from a small airplane flying over the Zuider Zee, and this did indeed produced small amounts of rain. He was also convinced that it would be possible to dissolve clouds and produce more sunshine by carrying out the dry ice seeding in the early morning. His experiments are reported in a book with the English title *More Sunshine in the Cloudy North, More Rain in the Tropics* (Veraart 1931), but his work and his enthusiastic reports were not considered to have sound scientific foundations. They were not accepted by the Dutch scientific community and they were subsequently forgotten. Nevertheless, Veraart's achievements were mentioned in the popular scientific literature of his time as the first sign of progress in human control of the weather.

Just after the end of the Second World War, in 1946, the self-taught chemist and meteorologist Vincent Schaefer (1906–1993), the physical chemist Bernard Vonnegut (1914–1997), and the chemist, physicist, and engineer Irving Langmuir (1881–1957), who won the 1932 Nobel Prize in Chemistry, all three American, carried out the first large-scale experiments in which clouds were artificially seeded with dry ice and silver iodide, the aim being to produce abundant precipitation (Fleming 2006). Following the success achieved in the first flights to produce artificial rain, Langmuir became excited and imagined that it would be possible to control and modify weather, not only at the local level, but also at the global level, for example, diverting the paths of tornados and tropical cyclones. He also dreamt of modifying the climate to transform arid regions into luxuriant lands capable of sustaining agriculture and producing abundant food.

As would be expected given the essential characteristics of human nature, the first impulse following a promising and potentially powerful scientific discovery was to develop technologies to manipulate the weather and the climate for military purposes, specifically in the Cold War, which was just getting under way in the 1950s. Irving Langmuir and the American engineer and science administrator Vannevar Busch (1890–1974), who played a crucial role in persuading US governments to undertake the Manhattan Project and set up the National Science Foundation, quickly convinced military personnel that the offensive potential of manipulating weather and the climate was huge. In 1947, General George C. Kenney, US commander of the Strategic Air Command, stated: "The nation that first learns to plot the paths of air masses accurately and learns to control the time and place of precipitation will dominate the globe" (Fleming 2006).

However, knowledge of atmospheric dynamics and meteorology was still very limited. Only after further advances in the physics of weather and climate and the development of numerical weather prediction supported by the first computers, particularly ENIAC and EDVAC in the USA (Edwards 2010), was it finally realised that projects for military domination based on the paths of air masses would be impossible to achieve. One of the main initial contributors to the progress of scientific knowledge in the physics of the atmosphere was the Hungarian-born mathematician and physicist John von Neumann (1903–1957), together with the circle of mathematicians and physicists who worked with him. They contributed decisively to the development of weather and climate models that use numerical methods to solve the Navier–Stokes equations, the fundamental partial differential equations that describe the flow of incompressible fluids. In 1950 a highly simplified approximation to the set of equations that govern the dynamics of the atmosphere, including the Navier–Stokes equations, was solved numerically to produce the first weather forecasts based directly on those equations (Charney et al. 1950).

A very important initiative regarding the human knowledge of the Earth system was the International Geophysical Year of 1957–1958, agreed and jointly funded by the USA and the USSR. The results of the research projects that started at that time further revealed the complexity of the Earth system to the scientists and governments of both countries, and highlighted the importance of maintaining its current state of equilibrium for the well-being and economic prosperity of all countries. But the 1950s were also a time of the intensifying Cold War and on 1 November 1955 the Vietnam War started, lasting until the fall of Saigon on 30 April 1975.

In March 1971, Jack Anderson (1922–2005), the well-known American journalist and pioneer of investigative journalism, revealed in his Washington Post column of 18 March 1971, under the code name Intermediary-Compatriot, a covert weather warfare program run by the US military in South-East Asia. It was later revealed that this program, called Operation Popeye, actually started on 20 March 1967 (USS 1974). The news of the covert operation further entered public awareness through an article in the New York Times, published on 3 July 1972 (Hersh 1972). Two days later, on 5 July, Operation Popeye ended. This episode presents an example of the advantages of a functioning democracy with a free press, whereby it is possible to inform citizens about controversial activities, giving them an opportunity to react and influence their voting choices. Nevertheless, the development of controversial covert actions harmful for people, directly or indirectly through their effect on the environment, and their power to inflict injuries and suffering is far from being prevented just by democracy and free speech.

In 1966, the US Secretary of Defense Robert McNamara expressed doubts about the possibility of a victory in the Vietnam War. One of the reasons was the difficulty in preventing the movement of troops and equipment in the Ho Chi Minh Trail network through Laos and Cambodia from North Vietnam towards South Vietnam. This situation was probably one of the main reasons for creating Operation Popeye, which had the objective of increasing rainfall during the southwest monsoon season, from April to September, in order to flood the trails, cause landslides along roadways, and wash out river crossings (USS 1974). The cover operation was based on previous experience gained in the use of lead iodide and silver iodide deployed by aircraft on tropical cumulus clouds and hurricanes in Okinawa, Guam, Philippines, Texas, and Florida in a program called Project Stormfury carried out by the United States Navy and Air Force and by NOAA. The technical aspects of Operation Popeye were approved by the Pentagon's Research and Engineering Office and by Donald F. Hornig, Special Assistant to the President of the United States for Science and Technology. After the publication of the New York Times article (Hersh 1972), the USA Secretary of Defense Melvin Laird was forced to testify before Congress in 1972 and categorically denied that the Pentagon had conducted rainmaking warfare operations in South-East Asia. However, on 28 January 1974, a private letter from Melvin Laird was leaked to the press where he admitted that his 1972 testimony to Congress on weather warfare had been false.

In October 1966 the operation was tested in the Se Kong river valley in Sekong Province of Laos by personnel of the Naval Ordnance Test Station in China Lake, California, now called the Naval Air Weapons Station. In 82% of the 58 cloud seeding experiments performed, the clouds were reported to produce rain within a brief period after seeding. These results were considered to be sufficiently encouraging for the US Department of Defense to start an operational program. During the five years that the program lasted, 2602 secret cloud seeding missions were flown from a US Air Force base in Thailand, targeting territories in Cambodia, Laos, and Vietnam without the knowledge or authorization of the Thai, Laotian, or Cambodian governments. It was the first known weather warfare program in military history.

Operation Popeye did not change the course of the Vietnam War and there was disagreement within the Pentagon about the efficacy of cloud seeding to modify the weather. At the secret Senate hearing on 20 March 1974, Lt. Col. Ed Soyster, representing the Joint Chiefs of Staff, was optimistic about the outcome of Operation Popeye, while Dennis J. Doolin, Deputy Assistant Secretary of Defense, considered that the amount of additional rain was small, of the order of 10% (USS 1974). The latter interpretation is likely to be closer to the truth because, already in 1972, the USA decided to discontinue the use of weather modification in warfare and to ensure that its adversaries would do the same. In fact, "in a statement before the Subcommittee on Oceans and International Environment of the Senate Foreign Relations Committee on 26 July 1972, the US Government renounced the use of climate modification techniques for hostile purposes, even if the development of such techniques proves to be feasible in the future" (USACDA 1974).

Following that statement, the US Senate and the House of Representatives held hearings during the same year of 1972, and the Senate adopted a resolution the year after, calling for an international agreement "prohibiting the use of any environmental or geophysical modification activity as a weapon of war". In the summit meeting in Moscow in July 1974, the US President Richard Nixon and the USSR General Secretary Leonid Brezhnev formally agreed to hold bilateral discussions on how to bring about "the most effective measures possible to overcome the dangers of the use of environmental modification techniques for military purposes". Finally, the

Environmental Modification Convention, formally the Convention on the Prohibition of Military or Any Other Hostile Use of Environmental Modification Techniques opened for signature on 18 May 1977 in Geneva and entered into force on 5 October 1978. An environmental modification technique is defined in the convention as any technique for changing—through the deliberate manipulation of natural processes—the dynamics, composition or structure of the Earth, including its biota, lithosphere, hydrosphere, and atmosphere, or of outer space. The agreement explicitly encourages peaceful applications of environmental modification.

In fact the "manipulation of natural processes" is practised daily around the world as a means of sustaining all kinds of human activities from farming to eating, living and working in houses, buildings, and cities, and travelling. Anthropogenic climate change provides an example of an unintentional manipulation of the natural climate that results mainly from the type of primary energy sources used by humankind. The constant manipulation of the environment can be reduced to a minimum when people do care about the environment and are aware of the need to achieve sustainability in all its dimensions—social, economic, and environmental. The US initiative to propose and sign the Environmental Modification Convention is likely to have resulted mostly from being convinced that weather and climate warfare is relatively ineffective and powerless and less from concerns about the harmful effects of environmental modification. Today, the Convention has been almost forgotten and at best lies dormant.

Nevertheless, the widespread conviction that it was possible to modify and control weather for military purposes remained firmly present in the US Air Force until the end of the 20th century and most probably beyond. One study by military researchers published in 1996 stated that "in 2025, US aerospace forces can 'own the weather' by capitalizing on emerging technologies and focusing development of those technologies to war-fighting applications" (House et al. 1996). This article had surprising consequences, since it was one of the sources of inspiration for environmental journalist William Thomas, who first popularised the idea, in 1999, that contrails (condensation trails) left behind by jet planes were actually chemtrails (chemical trails) formed of toxic products launched from planes by the US Air Force to control the weather through geoengineering (Thomas 1999, 2004). According to some versions of the chemtrail narrative, military and commercial airliners are deliberately spraying some kind of mixture of toxic chemicals in the USA and possibly across the globe with various goals, ranging from large-scale weather modification to mass population or mind control, and constitutes one of the largest covert operations ever attempted in the world (Tingley and Wagner 2017).

Thomas also based his opinions on a military facility constructed in Gakona, Alaska, dedicated to implementing the High Frequency Active Auroral Research Program (HAARP) with the purpose of studying the ionosphere and investigating the potential for developing ionospheric enhancement technology for radio communications and surveillance (USAF 1993), which started in 1990 and was closed in 2015. Major aspects of the program were kept secret for alleged reasons of "national security", making it impossible for scientists to assess the program's potential impacts on people and on the environment. HAARP has been the origin of many conspiracy theories, which range from being capable of modifying weather, causing earthquakes, droughts, storms, and floods, disabling satellites and exerting mind control over people. From what is known about the HAARP facility and its program there is no scientific basis to believe in such an array of effects.

Thomas' opinions were also influenced by a patent called Stratospheric Welsbach Seeding for Reduction of Global Warming, presented on 23 April 1990 by the inventors David B. Chang and I-Fu Shih, with the Hughes Aircraft Company as assignee. This patent constitutes a fascinating story about the relation between science and the environment over the last few decades. The objective of the patent was very ambitious, namely to solve the energy radiation imbalance on the tropopause caused by the accumulation of greenhouse gases in the troposphere. To achieve this goal it proposed to seed the atmosphere with 10 to 100 μm metal oxide particles, such as thorium dioxide and aluminium oxide, from airplanes at altitudes between 7 and 13 km. These oxides have high emissivity in the visible and far-infrared ranges of wavelengths, but lower in the near-infrared. The inventors were convinced that the oxides would absorb the thermal radiation emitted by the Earth's surface, which contains mainly far-infrared, and instead of re-emitting it in the same range of wavelengths, they would re-emit the absorbed energy in the visible region of the spectrum.

They believed in a simplistic and erroneous analogy with the physical mechanism involved in the incandescent mantle, invented by the Austrian physicist and chemist Carl Auer von Welsbach (1858–1929). In fact the oxide particles would absorb and re-emit the thermal radiation emitted by the ground and therefore act as greenhouse gases. They would not re-emit in the visible region of the spectrum because they would not have the energy available for that, being in thermodynamic equilibrium with their environment. For the proposed patent to work, at least three basic laws of physics, namely, Kirchhoff's and Planck's laws of radiation and the second law of thermodynamics, would have to be violated. Surprisingly the USA Patent Office granted the patent on 26 March 1991. Nothing came out of the Welsbach patent except that it helped fuel conspiracy theories about sinister forces spraying the atmosphere with toxic substances to modify the weather and the climate or even to harm people.

A 3015-subject survey conducted in three Anglo-Saxon countries—the USA, the UK, and Canada—concluded that 16.6% of the respondents believed that there was a secret government program that used aircraft to put harmful chemicals into the air (Mercer et al. 2011). This large-scale atmospheric program is commonly referred to as chemtrails or covert geoengineering. There are several websites, such as *Geoengineering Watch* and *Global Sky Watch* that purport to show evidence of the covert geoengineering narrative (Allgaier 2019), which is particularly developed in the USA, the UK, and Canada and is spread almost evenly across the political spectrum (Tingley and Wagner 2017). It would be helpful to research the reasons why the covert geoengineering narrative is more developed and popular in Anglo-Saxon countries than in others. Curiously, the covert geoengineering narrative is often associated with scepticism about anthropogenic climate change, a point of view that is relatively more common in Anglo-Saxon countries, particularly in the USA, Canada, and Australia. According to a nationally representative 1000-subject survey carried out during the 2016 USA Senate elections as part of the Cooperative Congressional Election Study, 10% of the people questioned believed the covert geoengineering theory to be entirely true and 20–30% thought it was somewhat true (Tingley and Wagner 2017). Previously, in 2000, the US Environmental Protection Agency (EPA), the Federal Aviation Administration (FAA), the National Aeronautics and Space Administration (NASA), and the National Oceanic and Atmospheric Administration (NOAA) emphatically refuted the foundations of the theory (EPA 2000), but believers' reaction was to view this rebuttal as further proof of a conspiracy.

Rose Cairns (Cairns 2016) has pointed out that the covert geoengineering narrative contains perspectives and implications that are important for the emerging policies on climate geoengineering implementation and cannot be dismissed as merely paranoid or pathological. In other words, the chemtrails theory may be a reaction narrative to geoengineering *avant la lettre*. She asks if it is more irrational to believe that the climate is being controlled by powerful elites or to believe that the powerful actors currently discussing how to manipulate the global climate will finally do it (Cairns 2016). Cairns also considers that the association of the covert geoengineering narrative with extreme forms of anthropogenic climate change scepticism may limit the possibilities for critical engagement with other trends of environmental discourse.

Scientists in Anglo-Saxon countries have been concerned with the covert geoengineering narrative and addressed the problem from the scientific point of view by surveying two groups of experts, one on atmospheric chemistry with knowledge of condensation trails and another on geochemistry working on atmospheric deposition of dust and pollution to know their opinions (Shearer et al. 2016). Results of the study show that 76 of the 77 scientists interviewed said they had not encountered evidence of a secret large-scale atmospheric program, and that the data cited as evidence could be explained through other factors, including well-understood physics and chemistry associated with aircraft contrails and atmospheric aerosols (Shearer et al. 2016).

Interestingly, the authors explicitly say in the paper that their goal is not to try to convince the believers that their covert geoengineering narratives are false but to establish a source of peer-reviewed science that can inform governance and the public discourse (Shearer et al. 2016). Somehow they are concerned about the power of the covert geoengineering narrative, but they acknowledge their powerlessness to stop it. The gap between peer review science and the covert geoengineering narrative is taken as unbridgeable. Sociologists that endeavour to understand the essence of the covert geoengineering narrative refuse to abandon the hope of establishing a rational dialogue with the believers, which, however, is unlikely to be based on science. Here we are entering the contemporary science–policy–society interfaces that justify the concepts of "extended facts" and "extended peer community", although these tend to devalue the methodology of science.

Meanwhile, weather modification continues to be attempted around the world and new techniques are being developed. In spite of all the hopes and dedicated work, the practical results have been modest. The use of hail cannons by farmers to destroy hailstones with explosives or with sound blasts, a practice that started in the 19th century, continues to be used, although there are no scientific reasons to prove their efficacy (Wieringa and Holleman 2006). In 2003, the US National Research

Council concluded that "there is still no convincing scientific proof of the efficacy of intentional weather modification efforts" (NRC 2003). Cloud seeding to increase precipitation continues to be used in the USA and in other countries, with rather modest results. For instance, the California Department of Water Resources has estimated a 4% annual precipitation increase attributable to the combined state seeding projects (Hunter 2007). Research on cloud seeding conducted in South Africa and Mexico has shown that seeding warm rain clouds with salt particles, known as hygroscopic seeding, is more effective than seeding cold rain clouds with silver iodide. It must be noted that silver iodide is toxic and can cause temporary incapacitation or residual injury to humans and other mammals with intense or chronic exposure. However, studies have shown negligible environmental and health impacts (WMA 2009), because the amount of silver iodide released in cloud seeding operations is usually very small, of the order of one gram per square kilometer.

China appears to be the country that most firmly believes in the benefits to be gained from weather modification. Chinese research in the field began in 1958 and is still being fast developed. The government-run Weather Modification Program launches thousands of specially designed rockets and artillery shells containing silver iodide into clouds every year in an attempt to manipulate weather conditions. Run by the Weather Modification Department, a division of the Chinese Academy of Meteorological Sciences, the program employs and trains about 40 000 people across China, some of them farmers, 3000 of whom work with 7000 canons and 4687 rocket launchers. Chinese authorities managed to divert rain from Beijing during the opening ceremony of the 2008 Olympics by firing 1100 silver iodide rockets into the skies. It is now normal practice to try to ensure clear skies on public holidays in Beijing, such as the National Day in 1 October, by seeding clouds in the preceding days so that the rain produced in this way disperses the pollution.

China is deploying a massive weather modification network for the Tibetan Plateau to channel more rainfall into China's northern regions, with expected additional amounts in the range of 5–10 billion m^3 per year (ETC 2018a). The plan proposed by the Aerospace Science and Technology Corporation and called "Sky River" is to cover an area of about 500 000 km^2 with tens of thousands of fuel burning chambers that blast silver iodide into the atmosphere. One of the main problems with these large-scale weather modification infrastructures is the effect they have in other regions and ultimately on the general circulation of the atmosphere. By diverting the rain to Northern China, Tibet will become drier, affecting the livelihoods of people living there. Because of the general circulation of the atmosphere, it is simply impossible to contain large-scale weather modification to a limited area. If a given country or region achieves the desired effect in its territory, it is very likely to produce undesirable effects elsewhere.

Meteorologists have also long been considering the possibility of applying weather modification techniques to divert, tame, or even destroy tornados and tropical cyclones. So why have they not succeeded? A tropical cyclone is a rapidly rotating storm system with a diameter between 100 and 2000 km, characterized by a deep low-pressure centre, strong winds, and a spiral arrangement of thunderstorms that produce heavy rains. It functions approximately as a heat machine, obtaining its energy from the thermal energy of the warm ocean waters near the surface through evaporation and releasing the energy through the latent heat of water vapour condensation into rain in the clouds close to the eyewall and in the rainbands, and the ascending outflow of warm air towards the cold upper levels of the troposphere. The total rate of energy release by a tropical cyclone depends mostly on the amount of rain that is produces. Assuming that the cyclone produces 1.5 cm per day of rain inside a circle of radius 665 km (Gray 1981), a simple calculation shows that it will release an energy of 5.2×10^{19} J per day or a power of 6.0×10^{14} W, which corresponds to 600 TW. Other estimates range from 50 to 200 TW (NOAA 2020). Less than 10% of a tropical cyclone's power is converted into the kinetic energy of the generated winds. How do these numbers compare with other energy indicators essential to humankind?

As previously mentioned the total solar radiant power absorbed by Earth is 121 800 TW, the mean rate of thermal energy increase in the climate system due to the accumulation of greenhouse gases in the atmosphere is 213 TW, and the global rate of primary energy consumption in 2017 was 18.5 TW. During the mean tropical cyclone lifetime of 5–8 days, a tropical cyclone has a power about 30 times greater than the current global rate of primary energy consumption by humankind.

To weaken a tropical cyclone it would be necessary to significantly increase the barometric pressure inside the eye, which would imply adding about half a billion tons of air for an eye of radius 20 km. It is difficult to imagine how such a displacement of air could be achieved. Trying to destroy a tropical cyclone with an atomic bomb would be a complete failure and would cause a major human disaster. The pulse of high pressure, or shock wave, produced by the explosion would travel away faster than the speed of sound but would be unable to increase the barometric pressure in the cyclone's eye. On the other hand the radioactive fallout released from the bomb would be scattered widely by the powerful cyclone winds and further away to populated areas by the trade winds.

Another possibility would be to weaken the developing tropical cyclone in its early stages before it became very powerful. One could try to change the sea surface temperature by coating the ocean with a thin layer of biodegradable oil in order to reduce evaporation. The problem, however, is that only about 6% of the tropical disturbances evolve to become tropical cyclones and it is very difficult to identify those that will grow to full power. The energy consumed trying to stop all these disturbances from growing in various spots across the ocean far away from land is impracticable. Many other ideas are being proposed, but the sheer size and energy of tropical cyclones and the instability of the weather and size of the ocean environment where tropical cyclones develop make it very difficult to intervene.

The situation as regards controlling tornados has similarities to that of tropical cyclones. A tornado is a violently rotating column of air that is temporarily in contact with both the ground and a cumulonimbus cloud, and it can be very destructive. tornados have much less total energy than tropical cyclones and their energy is almost exclusively kinetic energy, while in tropical cyclones most of the energy is thermal. The destructive power of a tornado results from the concentration of kinetic energy in a small moving area. Because of this concentration, tornados have the highest energy density of nature's storms, about six times higher than that of tropical cyclones. Estimates of the total kinetic energy of tornados in the USA have concluded that half of all tornados have total kinetic energy exceeding 62.1 GJ, a quarter exceeding 383.2 GJ, and 1% exceeding 31.9 TJ (Fricker and Elsner 2015). To calculate the power of a tornado while it is potentially destructive, one has to know how long it is connected with the ground, which is variable, ranging from a few seconds to three hours or more. Assuming a mean value of 15 minutes, the power of a tornado, while it is potentially destructive, ranges from 4×10^{-4} to 1.5×10^{-2} TW, which is ten thousand to a million times weaker than the power of a tropical cyclone.

An infrastructure able to weaken or prevent the development of tornados must be scientifically well justified, and its efficacy should be previously tested before implementation. Furthermore, its practicality in terms of the required technology, investment, and cost–benefit analysis must be assessed. The instantaneous state of the atmosphere or, in other words, the weather in a given location is variable over short time intervals of the order of minutes and highly non-uniform across space, especially in unstable atmospheric conditions. This means that it is an environment where it is difficult to convincingly demonstrate that the outcome of any weather modification effort aimed at preventing the development of tornados is in fact the result of human intervention and not the result of the natural evolution of the state of the atmosphere. Many ideas are being put forward to prevent tornados from developing, such as identifying the critical regions of a thunderstorm mesoclyne that are likely to lead to the formation of a tornado and heating them with a microwave beam coming from large solar arrays on satellites, or using laser beams to heat the desired regions.

For the techno-optimists, it is deeply disappointing that up to now technology has been unable to produce more significant advances in weather modification and in the control and weakening of potentially harmful extreme weather events. The atmosphere has turned out to be much more complex, changeable, unpredictable, and powerful than initially expected. Despite the difficulties encountered in modifying and taming the weather, there is a persistent hope that technology will eventually succeed and offer humans a chance to "own the weather".

7.21 Climate Geoengineering. Carbon Dioxide Removal

There are fundamentally two types of climate geoengineering or climate intervention actions with the objective of countervailing anthropogenic climate change. One is called carbon dioxide removal (CDR) or negative emissions, the aim of which is to remove CO_2 directly from the atmosphere using natural sinks or chemical engineering processes (IPCC 2012; NRC 2015a). The other type of geoengineering is called solar radiation management (SRM) or albedo modification, which consists of purposely modifying the energy balance in the atmosphere in order to reduce or cancel out part of the climatic consequences of the current anthropogenic climate change according to a specific metric such as the surface air temperature, precipitation, or others (see Fig. 7.5) (IPCC 2012; NRC 2015b; Reynolds 2019).

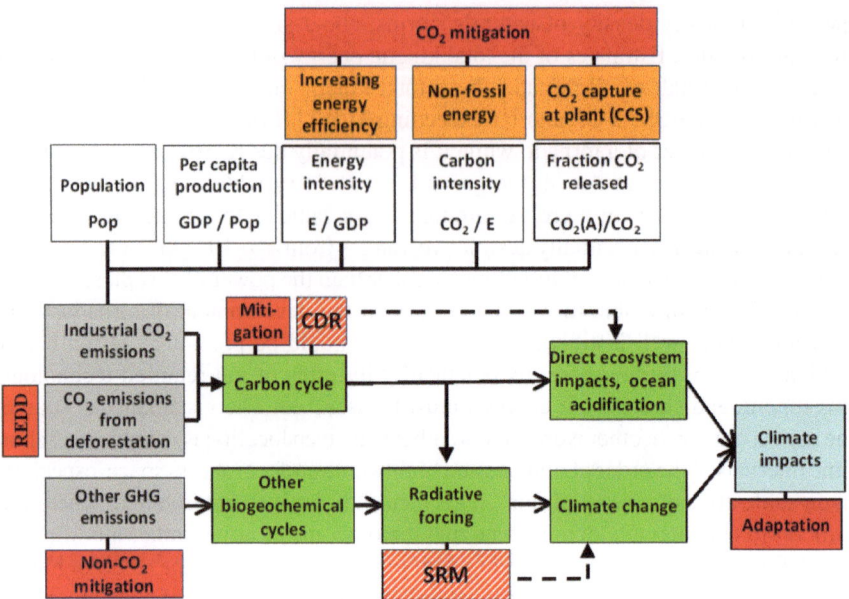

Fig. 7.5 Diagram illustrating the relations between the human, socioeconomic, and climate systems and the processes of mitigation, adaptation, and geoengineering in its two forms, carbon dioxide removal (CDR) and solar radiation management (SRM) (IPCC 2012). The first four white boxes on top represent the Kaya identity, which is an example of the equation $I = PAT$. According to the Kaya identity, the total carbon dioxide emissions can be expressed as the product of four factors: human population, GDP per capita, energy intensity (energy consumed per unit of GDP), and carbon intensity (CO_2 emissions per unit of energy consumed). Carbon capture and sequestration (CCS) adds a fifth box on the right since with that technology only a fraction of the CO_2 emissions are released into the atmosphere. CDR reduces the amount of CO_2 in the atmosphere and therefore reduces the anthropogenic interference in the carbon cycle. SRM reduces radiative forcing without mitigation, but the direct impacts of high CO_2 atmospheric concentrations persist, especially ocean acidification. *Credit* Figure 1.1 of reference IPCC (2012)

Recently, a third type of geoengineering for polar regions is receiving attention, which aims to slow down the melting of glaciers, ice shelves, and ice sheets, particularly in Antarctica (Moore et al. 2018a). Geoengineering of glaciers is primarily the result of concern about the global mean sea level change, which in 2100 is likely to be about 1 m higher relative to the pre-industrial level (Jevrejeva et al. 2016). This rise means that some of the world's population that lives in low-lying coastal areas, i.e., some 0.5–5% of the global population, will see floods and loss of land with global costs estimated at $50 trillion per year. To avoid these impacts, coastal protection barriers will need to be built, the cost and maintenance of which are estimated to be tens of billions of dollars a year. In this scenario, according to Moore et al. (2018a), geoengineering of glaciers becomes competitive.

The methods proposed consist of using concrete structures on the seabed to avoid some of the more stable ice shelves from sliding into the ocean and melting.

Another proposal involves building underwater embankments at some fjords to stop the ocean's warmer waters from penetrating the fjords and causing the submerged bases of the ice shelves to melt. These proposals are rather far-fetched because of the high cost of building such large-scale structures in the adverse environmental conditions that prevail in Antarctica. The fact that they are published in a respectable peer-review science journal can be interpreted as revealing some desperation about the inability to control mean sea level rise through mitigation. Geoengineering of polar glaciers raises many environmental and political questions that would have to be dealt by the Antarctic Treaty System, which regulates governance of Antarctica and protects its ecosystems.

In the Arctic, US researchers have started to perform small-scale geoengineering experiments to slow the melting process of the sea ice in the summer months using bright floatable materials in a granular form, such as small hollow glass spheres to reflect more sunlight. Ice911 is a non-profit organization founded in 2008 that has carried out this type of experiment in the Arctic. The projects currently have small budgets of the order of $100 000 dollars, but they have the ambition of building up to millions of dollars.

There are six main CDR processes (EASAC 2018). Two of these, afforestation or reforestation to increase biological CO_2 sequestration and agricultural practices and soil management that increase the capacity to retain carbon in the soil, can be considered mainstream mitigation measures. The third process, already mentioned, is carbon capture and sequestration (CCS) at fossil fuel power plants, particularly coal plants, and the use of bioenergy with carbon capture and sequestration or BECCS. CCS technologies for power plants have already been developed in Europe and the USA, but are not being implemented because their use increases the cost of the electricity generated (EASAC 2018). Incentives will need to be created to apply them and develop feasible business models. The problem is finding industrial applications to reuse the gigantic amounts of CO_2 that will be produced during the CCS processes and pay for the costs involved in the capture and sequestration (GCCSI 2011).

Curiously, the largest current industrial use of CO_2 is enhanced oil recovery (EOR), which is the extraction of crude oil from an oil field that cannot be extracted otherwise. EOR is currently a very important technique in global oil production, as production from mature fields is declining and new fields are harder to find. The commonest approach to EOR is gas injection and the fluid most frequently used is CO_2, because it reduces the viscosity of the oil and is less expensive than natural gas, liquefied petroleum gas, and nitrogen. The total amount of CO_2 sequestered in projects of CO_2 capture, utilization, and sequestration is estimated at 2209 Mt in 2028 (Tcvetkov et al. 2019).

The CO_2 emissions from the conventional cement industry are about 5–8% of the global CO_2 emissions. Various technologies are being developed to reduce emissions in this industry, such as CO_2 uptake using cement-based materials. The carbonation reaction that occurs between the reactive compounds from cement-based materials can fix CO_2 in the form of thermodynamically stable carbonates. This CO_2 uptake can be used as a way to implement carbon capture, utilization and sequestration (Jang et al. 2016). The percentage of cement in concrete can vary greatly, from as low as

30% to as high as 100%. Different concrete mixes are chosen to impart different properties to the final concrete. Injecting CO_2 during the mixing process increases the strength of concrete, allowing for less cement to be used in the concrete mix. Ordinary concrete cures are done by reacting with water over the course of several weeks. Concrete can also be produced by making it react with a water–CO_2 solution and curing in 24 hours. During this curing process, the concrete can sequester up to 300 kg of CO_2 per ton of cement used in the concrete (Rissman 2018). Application of the various technologies that are being developed may potentially make cement production carbon neutral by 2050 (Rissman 2018).

The three remaining CDR processes involve technologies that are clear examples of geoengineering. The first is iron fertilization of the ocean to stimulate oceanic carbon sequestration through the growth of phytoplankton in iron-limited, high-nutrient, low-chlorophyll regions. Since 1990, researchers around the world have conducted 13 major iron-fertilization experiments in the open ocean. It has been difficult to estimate reliably the additional amount of CO_2 that the fertilized phytoplankton extracts from the atmosphere and deposits in the deep ocean after dying. Meanwhile scientists have warned that ocean fertilization is likely to have potential adverse side-effects, such as creating toxic algal blooms, deep-water oxygen depletion, alteration of distant food webs, and deep-water ocean acidification (Williamson et al. 2012).

In view of these concerns, the United Nations Convention on Biological Diversity decided, on 30 May 2008, to establish a moratorium on large-scale experiments (Tollefson 2008). Five years later, the Convention adopted rules for evaluating the ocean fertilization experiments. In the same year Parties to the London Protocol to the London Convention on the Prevention of Marine Pollution by Dumping of Wastes and Other Matter unanimously adopted a resolution prohibiting all ocean fertilization activities other than "those for legitimate scientific research." This resolution was followed by another in 2010 that set out the conditions defining what constitutes "legitimate scientific research."

The largest documented ocean fertilization experiment was carried out in 2012, after the moratorium was passed, by US businessman Russ George, who launched about 100 tonnes of iron sulphate from a fishing boat 200 nautical miles west of the islands of Haida Gwaii, formerly the Queen Charlotte Islands, an archipelago approximately 45–60 km off the northern Pacific coast of Canada. Russ George started by convincing the local council and the president of the Haina nation, Guujaaw, to pursue iron fertilization as an ocean restoration initiative that would boost salmon populations, which are of great value to the local populations. He also hoped to make a large profit from selling lucrative carbon credits based on the amount of CO_2 that would be sequestered in the ocean.

According to Russ George, the iron dumping on the ocean caused an algae bloom over a $10\,000\ km^2$ area. His team of unidentified scientists analysed the result of the experiment and allayed "all the possible fears that have been raised about ocean fertilization" (Lukacs 2012). However, there is no published scientific evidence that the experiment worked. In fact the existence of the experiment was first reported by Silvia Ribeiro of the international technology watchdog ETC Group, who pointed out its possible harmful side-effects. After hearing this story, Guujaaw said (Lukacs

2012): "The village people voted to support what they were told was a 'salmon enhancement project' and would not have agreed if they had been told of any potential negative effects or that it was in breach of an international convention".

Proposals for iron ocean fertilization are still being developed around the world. The published plan for the Korean Iron Fertilization Experiment in the Southern Ocean (KIFES) is particularly detailed (Yoon et al. 2018). More recently, it was proposed to use biogenic poorly crystalline Fe oxides produced by chemosynthetic iron-oxidizing bacteria (Emerson 2019). Upon drying, these oxides produce a fine powder that could be dispersed over the ocean at altitude by aircraft to increase the wind-driven aeolian dust that is a primary iron source to the open ocean. As the energy transition becomes increasingly more difficult and climate change more menacing, and in spite of all the possible negative side-effects, iron fertilization is likely to be more frequently used.

The second geoengineering process is to capture atmospheric CO_2 by large-scale enhancement of rock weathering (Strefler et al. 2018), a natural process that slowly removes CO_2 from the atmosphere through its reaction with rainwater to form carbonic acid, which in turn reacts with the silicates and carbonate rocks producing calcium and carbonate ions. These ions travel in the streams and rivers and after reaching the ocean are used by marine calcifying organisms to build their shells and skeletons. Finally, the calcium carbonate contained in shells is deposited on the ocean floor after the marine organisms die and is subducted beneath a tectonic plate, forming again silicate rocks and releasing the CO_2 back into the atmosphere through volcanoes and the surrounding vents. These emissions complete the carbonate–silicate geochemical cycle. CO_2 weathering is intensified when the global climate is warmer and therefore globally rainier. The resulting higher CO_2 rate of removal from the atmosphere decreases the greenhouse effect and cools the climate, which means that this CO_2-induced negative feedback mechanism acts as a remarkable long-term stabilization of the Earth's surface atmospheric temperature.

The acceleration of weathering reactions of minerals that consume CO_2 can be achieved by pulverizing and distributing large amounts of crushed silicate minerals on the land surface, a process called enhanced weathering. One of the most favourable minerals for that purpose is olivine, because it is widely available and reacts easily with the atmospheric CO_2 dissolved in rain water (Schuiling and Tickell 2010). The acceleration of weathering processes, by increasing the concentration of bicarbonate ions in the ocean, increases ocean alkalinity, and this can potentially reduce the ocean acidification caused by CO_2 anthropogenic emissions (Renforth and Henderson 2017). Currently, about a quarter of CO_2 emissions are absorbed by the oceans, and these contain more carbon than soils, plants, animals, and the atmosphere combined. It has been estimated that more than 1.5 mole of carbon is removed from the atmosphere for every mole of Mg and Ca dissolved in the ocean from weathered silicate minerals and 0.5 mole from weathered carbonate minerals (Renforth and Henderson 2017). These processes represent a mean carbon uptake in the form of CO_2 of 0.13 GtC, which is about 18 times smaller than the mean annual carbon uptake by the oceans through CO_2 dissolution (Le et al. 2018). Various technologies have been proposed for ocean alkalinity enhancement, such as accelerated weathering of limestone, electrochemically promoted weathering, and ocean liming, but there are still many unanswered technical, economic, environmental, social, and ethical questions, as is common in geoengineering.

Finally, the sixth and final CDR process is the direct capture of CO_2 from the air using chemical methods (Socolow et al. 2011), which is sometimes called "industrial air capture" when carried out on a large scale. This technology is promising but has higher costs than capturing CO_2 from point sources with relatively high concentrations of the gas, such as coal power plants and cement production plants.

The first commercial facility for direct capture from the air began operating in Switzerland in 2017. In Canada, a company called Carbon Engineering, founded by David Keith, has developed a process called "air to fuels" which produces fossil fuels, such as petrol, diesel, or jet fuel, by direct capture of CO_2 from the air using renewable energies. Part of the CO_2 captured may be sequestered in underground deposits or at a second stage used to produce liquid fossil fuels by way of a thermo-catalytic reaction with hydrogen obtained through the hydrolysis of water using renewable electricity. This is a way of harnessing sequestered CO_2 but using those fuels finishes up releasing CO_2 back into the atmosphere again. Most of the planned commercial applications of direct air capture are for production of liquid fuel or enhanced oil recovery.

CDR processes will very likely be able to remove CO_2 at local levels in amounts of the order of tens to hundreds of millions of tonnes. However, separately or together, it is very unlikely that they will be able to remove the gigatonnes of carbon needed to achieve carbon neutrality at the global scale in time to avoid going beyond the 2 °C limit of the Paris Agreement (EASAC 2018). Moreover, there are significant uncertainties about the technological, economic, and environmental feasibility of CDR processes (Smith et al. 2016; Field and Mach 2017).

A 2011 estimate (House et al. 2011; Realff and Eisenberger 2012) gauges the total costs involved in direct capture of CO_2 from the air at 1000 dollars per tonne. Considering the current value of annual global CO_2 emissions from fossil fuels and industrial processes, which reached 36.6 ± 1.8 $GtCO_2$ in 2018 (Jackson et al. 2019), it can be concluded that capturing a significant fraction of that amount every year will be a financially demanding and problematic proposal to put into practice. On the other hand, the most favourable scenarios, among the 900 analysed by the IPCC (IPCC 2014c), indicate that BECCS could be used at the end of the century to extract 12 $GtCO_2$ from the air every year. However, this programme would put the world's food security at risk, because around 1.2 billion hectares would need to be used, which is more than 25% of the world's total arable land and up to 8% of Earth's land area (Field and Mach 2017).

7.22 Solar Radiation Management Geoengineering

SRM geoengineering directly interferes with the climate system to reduce global warming without needing to reduce anthropogenic greenhouse gas emissions and therefore without needing to make an energy transition, at least so fast as required to avoid a dangerous interference in the climate system. SRM is a category of geo-engineering that is more frequently addressed and thoroughly discussed because it is considered to be more attractive from the economic viewpoint, and more effective and promising than CDR. The deployment of SRM could be achieved rapidly, it would produce more significant and faster results than large-scale CDR, and its proponents consider that it would involve less cost to the current economic and financial system than either mitigation or large-scale CDR. However, even in Anglo-Saxon countries, SRM geoengineering is mainly discussed in elite circles and in the publications of USA conservative and libertarian Think Tanks (Collomb 2019), but seldom appears in daily newspapers or in the social media, which means that few people except physical and social science scientists, economists, and investors are familiar with its complexities and side-effects.

It is a technological fix that essentially involves increasing the percentage of solar radiant energy reflected by the Earth. Humankind would therefore have a climate system in which the atmospheric greenhouse effect would continue to intensify but would absorb less energy from the Sun. Essentially the same idea was suggested to the president of the USA in 1965 (PSAC 1965). The global, regional, and local consequences of this dual interference in the Earth system are naturally more difficult to model and project into the future.

However, it is clear that SRM geoengineering does not resolve the effects on the climate system that arise directly from the increase in the concentration of atmospheric CO_2, which is the case of ocean acidification and the serious impacts that this is having on marine and coastal ecosystems and fishing (Mora et al. 2013). Another direct consequence of the higher atmospheric CO_2 concentration is that it causes a faster growth rate in crops but, at the same time, the higher growth rate favours more the production of carbohydrates and less the production of nutrients that are especially important for human health. It has been found that many food crops that were grown under a CO_2 concentration of 550?ppm have protein, iron, and zinc contents that are reduced by 3–17% compared with current conditions (Smith and Myers 2018). This reduction could cause an additional 175 million people to be zinc deficient and an additional 122 million people to be protein deficient by 2050.

Several SRM geoengineering proposals have emerged over time based in outer space or on the Earth system. In 1997, the US Lawrence Livermore National Laboratory proposed a range of space solutions (Teller et al. 1997), including placing a space sunshade at the inner Lagrange point L1, located roughly 1.5 million kilometres from Earth. This sunshade could be formed of many small disks able to reflect into outer space around 2% of the solar radiation that reaches the Earth, but the cost of this project is too high. On Earth, it has been suggested that the terrestrial albedo could be raised by increasing the reflectivity of low clouds over the ocean by injecting sea water into the atmosphere (Latham et al. 2008) and also increasing the reflectivity of the terrestrial surface, for example by painting the roofs of urban buildings and other infrastructures white.

The most promising method, which is attracting the most research and investment, is to launch large amounts of sulphate aerosols into the atmosphere by injecting precursors, such as sulphur dioxide (SO_2), hydrogen sulphide (H_2S), and sulphuric acid (H_2SO_4), launched by plane, balloon, or artillery (Rasch et al. 2008; Robock et al. 2009). This kind of project can be viewed as simulating the effects of a powerful volcanic eruption which, by expelling large quantities of ash and aerosols into the stratosphere, reflects solar radiation, causes global dimming, and temporarily lowers the mean global temperature of the atmosphere at the surface.

A well-known example from history was the gigantic eruption of the Mount Tambora volcano on the island of Sumbawa in Indonesia, which reached its climax on 10 April 1815. It is estimated that the eruption produced a fall in global mean surface temperature between 0.4 °C and 0.7 °C, although meteorological data is very scarce and highly uncertain (Stothers 1984). The following year, 1816, became known as the "year without a summer" or the "poverty year" in Europe due to the destruction of crops, causing the "last great subsistence crisis in the Western world" (Post 1977). More recently, a network of meteorological observation incomparably better than the one in place at the time of the Tambora eruption recorded a decrease in mean global temperature of the atmosphere at the surface of roughly 0.5 °C as a result of the eruption of the Mount Pinatubo volcano in the Philippines on 15 June 1991 (Parker et al. 1996).

The first suggestions of climate geoengineering using aerosol injection into the atmosphere were made by the Russian climatologist Mikhail Budyko (1920–2001), one of the main founders of modern climate theory (Budyko 1956). He was one of the first to calculate the effects of anthropogenic climate change and to warn about the potential adversity of its impacts (Budyko 1974, 1977). Later, aerosol geoengineering began to be seen as a possible solution to climate change and it has been estimated that the annual costs of a 1% reduction in the mean amount of global solar radiation that reaches the troposphere using aerosols would be $100 billion (PPIGW 1992). Further studies have shown that it is possible to use aerosols that are more efficient in scattering light per unit of mass and reduce the costs so that, according to David Keith, they would be "within reach of the world's richest individuals or private foundations" (Keith 2010).

In 2016, Paul Crutzen, a well-known researcher in atmospheric chemistry and winner of the Nobel Prize in Chemistry, concerned with the global deadlock on international climate change mitigation policy negotiations, wrote an article in which he proposes cooling the planet by injecting sulphate aerosols into the stratosphere (Crutzen 2006). However, Crutzen emphasised that geoengineering should only be used if human inability to carry out mitigation led to a dangerously disruptive situation. The article was very well received by most techno-optimists and viewed with great caution by others, particularly those with extensive knowledge of the climate system. In practice, its main effect was to boost a surge in research and in the number of published scientific papers on geoengineering (Oldham et al. 2014).

Injecting aerosols into the stratosphere interferes with the climate system in several ways. The simplistic view of this plan for anthropogenic interference is that, by increasing albedo in a controlled way, all that happens is that the mean global temperature of the atmosphere at the surface falls by an amount set in advance to cancel the expected mean global temperature increase caused by the anthropogenic greenhouse gas emissions over a certain interval of time since the pre-industrial period, for example, $2\,°C$, and all the rest remains unchanged in the climate system. This is not the case, because the meteorological variables that characterize the state of the atmosphere are highly correlated and not independent from each other.

Stratospheric aerosol geoengineering modifies the physics and chemistry of the atmosphere and interferes in several aspects of the climate system in a way that is different from the intensification of the greenhouse effect. As a result, the second interference does not simply cancel out the first; it may countervail some of its effects, but it generates new ones, such as reducing the concentration of stratospheric ozone in the case of sulphate aerosols, altering the global hydrological cycle, reducing the mean global precipitation by about 4.5%, and diminishing the amplitude of the seasonal cycle of temperature at high-latitude locations (Bala et al. 2008; Tilmes et al. 2008, 2013; Keith et al. 2016; Jiang et al. 2019). Climate modelling shows that stratospheric aerosol geoengineering generates significant reductions in monsoon precipitation in Africa and Asia (Bala et al. 2008; Tilmes et al. 2013; Keller et al. 2014; ETC 2018b), which are essential in sustaining the lives of billions of people.

The reason why injecting sulphate aerosols into the stratosphere leads to a decrease in global mean precipitation is a robust scientific conclusion that can be understood using a simple argument based on the surface energy budget. The negative change in the mean temperature of the atmosphere at the surface that results from an increase in the Earth's albedo has the same absolute value as the positive change resulting from an increase in greenhouse atmospheric concentration with the same nominal radiative forcing. However, the response of the land surface evaporation, and therefore of the precipitation, is quite different. In the albedo case, the land surface is affected by a decrease in incident shortwave radiation and also by a decrease in incident longwave radiation from a cooler atmosphere, while in the greenhouse effect case, it is only affected by an increase in the incident longwave radiation from a warmer atmosphere.

In the model used by Bala et al. (Bala et al. 2008), the hydrological sensitivity, defined as the percentage change in global mean precipitation per degree of warming, is 2.4%/K for solar forcing, but only 1.5%/K for CO_2 forcing. Reducing the amount of solar radiation that penetrates the troposphere reduces global precipitation and changes the atmospheric heating patterns over the globe. These are dominated by the latent heat release related to the (Trenberth and Stepaniak 2004; Trenberth and Dai 2007) flow of energy through the climate system between the incoming and outgoing radiation.

Recalling that stratospheric aerosol geoengineering events and large volcanic eruptions are similar as regards their effect on the Earth's albedo, scientists have studied the latter in great detail in order to better understand the potential effects that would result from implementing the former. The global precipitation and streamflow records from 1950 to 2004 were examined to identify the effects of the volcanic eruptions from three large tropical eruptions—Agung in May 1963, El Chichón in April 1982, and Pinatubo in June 1991 (Trenberth and Dai 2007). It was necessary first to disentangle the effects of the volcanic eruptions from those of the ENSO events.

The eruption that is best documented and produced the largest amount of aerosols was the Mount Pinatubo eruption. It is estimated that it injected 20 Mt of sulphur dioxide into the atmosphere, and that the combined effect with other aerosols and ash produced an increase in albedo of about 0.007, corresponding to an additional reflection of solar radiant energy of 2.5 Wm^{-2} over two years (Wielicki 2005). Following the eruption of Mount Pinatubo, there was a substantial decrease in precipitation over land and a record decrease in runoff and river discharge into the ocean from October 1991 to September 1992 (Trenberth and Dai 2007). Trenberth and Dai conclude (Trenberth and Dai 2007) that "creating a risk of widespread drought and reduced freshwater resources for the world to cut down on global warming does not seem like an appropriate fix. Our results suggest that considerable caution should be used regarding any intentional human intervention in the climate system that we do not fully understand."

The results regarding the climate changes induced by stratospheric aerosol geoengineering obtained with different climate models have been compared and they agree reasonably well. Jones et al. (2010a) concluded that the consequences of continuous injection of SO_2 into the lower stratosphere using the Hadley Centre's HadGEM2-AO and the Goddard Institute for Space Studies ModelE climate models led to broadly similar geographic distributions of near-surface mean air temperature and mean precipitation. According to the projections made by current climate models, it is not possible at either global or regional level to restore the mean temperature of the atmosphere and the precipitation conditions that prevailed in the pre-industrial period using stratospheric aerosol geoengineering to countervail anthropogenic climate change. In theory, geopolitical agreements would need to be reached on optimising temperature and precipitation at regional level (Ricke et al. 2010).

The effects stratospheric aerosol geoengineering on agricultural productivity have been a topic of increasing interest. Interestingly, some research has manifestly tried to emphasize positive outcomes while others have pointed out to the uncertainties associated with the problem. Pongratz et al. (2012) find that aerosol geoengineering in a high-CO_2 climate "generally causes crop yields to increase, largely because temperature stresses are diminished while the benefits of CO_2 fertilization are retained". Precipitation changes are devalued because it is considered that most of the climate-induced yield reductions result from the increase in mean global temperature and "precipitation effects in individual regions are cancelled out when averaged over latitude bands". The disruption across many countries, especially in the tropical regions, that would result from precipitation effects in individual regions are not addressed. Xia et al. (Xia et al. 2016) emphasize that enhanced diffuse radiation, combined with atmospheric cooling, changes in soil water content, and total solar radiation reduction, significantly increases plant photosynthesis rates in temperate and tropical regions, although it reduces rates in high latitude and mountain regions. According to the models used, the net effect is found to increase the global land carbon sink at an annual rate of 3.8 GtC (Xia et al. 2016).

More recently, Proctor et al. (Proctor et al. 2018) estimated global agricultural effects of aerosol geoengineering using El Chichón and Mount Pinatubo volcanic eruptions as a proxy. They find that the effect of the stratospheric aerosols on crop yields is negative for both C4 (maize) and C3 (soy, rice, and wheat) crops. Furthermore, the application of the yield model to an experiment in aerosol geoengineering leads to projected damage from scattering of sunlight that is roughly equal in magnitude to the benefits from lowering the mean global temperature (Proctor et al. 2018). The net effect of geoengineering would attenuate little of the global agricultural damage from the climate change resulting from anthropogenic greenhouse gas emissions.

The eventual benefits that may result from implementing stratospheric aerosol geoengineering on a large scale do not arise immediately but only after some years. Moreover, geoengineering, once started, must be pursued continuously as long as the anthropogenic emissions of greenhouse gases are kept at values close to the current level. Climate models show that if geoengineering is removed, the mean global temperature of the atmosphere would increase rapidly and many ecosystems could be negatively affected by the high rates of temperature change (Jones et al. 2010a; Ross and Matthews 2009). Ross and Matthews conclude that "climate engineering in the absence of deep emissions cuts could arguably constitute increased risk of dangerous anthropogenic interference in the climate system under the criteria laid out in the United Nations Framework Convention on Climate Change" (Ross and Matthews 2009).

7.23 The Future of Geoengineering

After the publication of Paul Crutzen's article (Crutzen 2006), research on geoengineering has moved forward significantly, but there are still many uncertainties about how it impacts the climate system in conjunction with the impacts resulting from continuing anthropogenic emissions of large volumes of greenhouse gases. The other thing we don't know enough about are the unintended side-effects on human societies and activities, in particular on health, agriculture, and the economy of countries that may be diversely affected around the world.

Large-scale CDR processes implemented with existing technology will be unable to fulfil the 2 °C goal of the Paris Agreement if the consumption of fossil fuels is maintained at the current levels (USGCRP 2017). The combination of mitigation and large-scale CDR suffers from the fact that the latter requires a long implementation period. To make this choice would require a previous economic and financial evaluation, including a comparative cost–benefit analysis between mitigation measures and CDR deployment, including the optimum scale of such a deployment. Furthermore, to be effective, CDR would have to carry on until the emissions from fossil fuels are significantly reduced.

Unlike CDR, SRM using aerosol injections into the stratosphere can be put into place quickly and produce a fall in the mean global temperature of the atmosphere at the surface within a few years. Another important feature of this form of SRM is that it is relatively cheap compared to CDR. With today's technology, a 1 Wm^{-2} reduction in radiative forcing can be achieved with an emission rate of one million tonnes of sulphur per year into the stratosphere using balloons, planes, or artillery shells with an annual cost of a few billion dollars (USGCRP 2017).

Such an amount is easily within reach for a country, a billionaire, or a group of concerned billionaires and private foundations. In other words, an enlightened group of New Republic members, using H.G. Wells' metaphor, could "save" the fossil fuel industry and the interests of its leaders and shareholders, not to mention the economic models of the countries that most benefit from it, while at the same time presenting themselves as potentially saving humankind from dangerous interference with the climate system. Furthermore, geoengineering will create new business models and profit opportunities for the very same industries and corporations that have contributed decisively to the climate crisis. All this with a minimum amount of intervention by ostensible governments, to use Wells' language once again, so as not to create regulatory impediments and complex bureaucracies.

The uncertainties about the magnitude, severity, and regional distribution of stratospheric aerosol geoengineering side-effects are unlikely to be fully resolved before actually deploying the technology and studying its effects (Robock et al. 2010). Deploying large-scale stratospheric aerosol geoengineering is a planetary-scale geophysics experiment analogous to the increase of greenhouse gas emissions that has been happening since the Industrial Revolution. The detailed consequences for humankind of conducting both climate interventions and their timing are unknown. The first one was unintentional. The second would result from the human refusal to address the causes of the first and would be fully conscious and premeditated. Furthermore, the second experiment must last until greenhouse gas emissions from the use of fossil fuels are significantly reduced, which likely means more than a century.

It is noted that the most favourable commercially available technologies for mitigation of climate change have been known for quite a long time, but have not been implemented worldwide on a large scale. A well-known set of mitigation measures that could have been implemented up to the mid-21st century to avoid a doubling of the pre-industrial concentration of CO_2 was proposed in 2004 by Pacala and Socolow (Pacala 2004). They identified and described 14 mitigation options using technologies that have been accessible since the beginning of the present century.

Unsurprisingly, the favoured technology of stratospheric aerosol geoengineering, which will be abbreviated from now onwards simply as SRM geoengineering, has generated various currents of opinion. These currents manifest themselves mostly in technology, economics, and politics, but also, curiously, in peer review science, albeit in more subtle forms. The advocates of implementing SRM geoengineering and those that are actually planning its deployment are mostly from Anglo-Saxon countries (Collomb 2019). In fact, the largest part of the scientific, technological and political discussion regarding SRM geoengineering has been taking place in Anglo-Saxon countries, in a relatively open and transparent way. The subject is also addressed in China and Russia, although in these countries there is less information and transparency and no public discussion of the political implications of SRM geoengineering (Huttunen et al. 2015). In continental Europe, research on geoengineering is developing, but plans for performing SRM geoengineering experiments in the stratosphere are less developed and less discussed than in the USA.

Politically, SRM geoengineering suffers from the fact that the two sides of the political spectrum tend to reject it for quite different reasons. Those on the centre and left tend to reject it because they favour an energy transition away from fossil fuels to solve the problem of anthropogenic climate change, and they consider that geoengineering involves unnecessary risks for humankind and conflicts with the Sustainable Development Goals (Schneider 2019). Those on the right and extreme right who have indulged in climate change scepticism and defending the incapacity of science, and in particular of climate models, to project the future of the Earth's climate have become trapped in their own narrative. If climate change is not a problem that requires a solution, why try to combat its effects through geoengineering? Negating science has a price because science is a coherent edifice built by the scientific method. By discriminating the truthfulness of scientific results according to the benefit that one can derive from each of them destroys the usefulness of science. As already mentioned, many people, especially in Anglo-Saxon countries, believe in a covert geoengineering narrative, according to which geoengineering is already taking place and harming people (Cairns 2016). Scientists trying to disclaim this narrative reluctantly admit that their effort further convinces the believers that they are right (Shearer et al. 2016).

The outcome of systematically promoting climate change scepticism for decades is that it becomes difficult to organize and boost support for research on SRM geoengineering. This type of research is presently supported vigorously by a small number of scientists, mostly from the USA and the UK, such as David Keith in Harvard University and Ken Caldeira from the Carnegie Institution for Science at Stanford University, but they complain about a lack of support from the federal government. In a joint paper, the two scientists begin by stating that "like it or not, a climate emergency is a possibility, and geoengineering could be the only affordable and fast-acting option to avoid a global catastrophe." They finish by proposing a research plan in three phases, involving an annual funding of 30 to 100 million US dollars in the second phase that would culminate in the development of a deployable SRM geoengineering system (Caldeira and Keith 2010).

In general, ecomodernists defend SRM geoengineering. David Keith is one of those who signed the Ecomodernist Manifesto. There are many other examples of the connection between ecomodernism and SRM geoengineering. For instance, Steven Pinker, who unequivocally sides with the ecomodernists (Collomb 2019), finishes up supporting SRM geoengineering (Pinker 2018). For its advocates, ecomodernism is the way to include the imperative of progress in the realm of the environment. Since people, following their basic instincts, are not attracted by "limits to growth," the only way forward is to believe in the imperative of progress and promote the politics of the possible in order to lead them to higher levels of enlightenment. Enlightening people about the dangers of not recognizing the limits to growth damages the economy and does not promote growth. Or, making the circular argument clearer, there are no limits to growth, because growth has no limits.

Those scientists that defend the need for SRM geoengineering seize every opportunity to argue for its deployment even if only in a limited way. An example is given by a group of USA researchers, including David Keith (MacMartin et al. 2018), who argue that a limited deployment of SRM geoengineering in addition to insufficient global mitigation measures could help keep global mean temperature less than 1.5 °C above the pre-industrial level in an overshoot scenario relative to the Paris Agreement that would otherwise peak near 3 °C. It is difficult to imagine how to perform the required research on SRM geoengineering that it would be advisable to do before deployment and still be able to start a limited deployment in time to achieve the 1.5 °C goal, since this requires immediate action.

The IPCC Special Report on Global Warming of 1.5 °C (IPCC 2018) dismisses SRM geoengineering in their list of measures to limit the rise in global mean temperature because they "face large uncertainties and knowledge gaps as well as substantial risks and institutional and social constraints to deployment related to governance, ethics, and impacts on sustainable development. They also do not mitigate ocean acidification."

Another group of US researchers, also involving David Keith (Irvine et al. 2019), claim that, although SRM geoengineering may increase climate risk for some regions, especially as regards the decrease in precipitation, concerns about the quantitative extent of the regional inequality of adverse impacts of SRM geoengineering may be overstated.

In recent years, David Keith has developed extensive arguments (Keith 2017) about the advantages of carrying out SRM geoengineering experiments. He acknowledges the difficulties and dangers of SRM geoengineering, but argues that "the Earth is already so transformed by human actions that it is, in effect, a human artefact" (Keith 2010). It is therefore preferable to use geoengineering rather than running the risk of not being able to mitigate climate change. Together with Lizzie Burn and Selena Wallace, he currently leads the Harvard Solar Engineering Research Program at Harvard University, partly funded by billionaires and private foundations, such as Bill Gates and the William and Flora Hewlett Foundation.

One of his colleagues, Frank Keutsch leads the Stratospheric Controlled Perturbation Experiment (SCoPEx), which began in 2017 and is due to run to 2024 with a budget of about $20m. The Harvard team will be the first in the world to conduct a SRM geoengineering experiment in the stratosphere. The experiment will release small plumes of calcium carbonate, each of around 100 grams, from a steerable balloon 20 km above the World View Spaceport in Arizona, or somewhere else in the US southwest, and then observe the optical properties of the aerosol formed by the dispersed particles. Sulphates are not used because of their negative effect on the stratospheric ozone (Tilmes et al. 2008). It should be noted that limited local experiments are unable to fully test stratospheric aerosol geoengineering (Robock et al. 2010). The intensity and the adversity of the side-effects can only be tested with a full-scale implementation.

The researchers complain that they are "funded by a hodgepodge of private funds and the redirection of federal research grants" (Caldeira and Keith 2010) and point out that the National Science Foundation, the Department of Energy, and the National Aeronautics and Space Administration would like to initiate research funding, but they cannot act without political cover from the Federal Government.

It must be noted that SCoPEx is a type of experiment that is banned by the moratorium on geoengineering experiments passed on 29 October 2010 by 193 member countries of the United Nations Convention on Biological Diversity (ETC 2010). However, the USA refused to ratify the convention, and the only requirement in the country if one wishes to perform such experiments is to previously report them to the National Oceanic and Atmospheric Administration.

There is currently a clear preoccupation among US scientists engaged in geoengineering research and particularly in SRM geoengineering about the lack of guidance on the subject by the US government. In fact, there is at present no national or global governance framework for geoengineering technologies except for the above-mentioned moratoria imposed by the Convention on Biological Diversity, the London Protocol to the London Convention on the Prevention of Marine Pollution by Dumping of Wastes and Other Matter, and the restrictions on using geoengineering for military purposes imposed by the Convention on the Prohibition of Military or Any Other Hostile Use of Environmental Modification Techniques. The resulting void implies that the practices of scientific research and intellectual property acquisition are *de facto* shaping the development of the field of geoengineering (Oldham et al. 2014). Meanwhile, bibliometric monitoring of geoengineering research publications and patenting activity is considered to be useful for the long awaited future governance of climate engineering (Oldham et al. 2014).

The position of the US Congress regarding federal mitigation measures is well illustrated by the Climate Security Act proposed by Senators Joseph Lieberman and John Warner on 18 October 2007 and supported by Senator Barbara Boxer, which involved the establishment of a cap and trade system for greenhouse gas emissions, analogous to the one implemented in the EU. After long debates, Republican senators managed to have the proposal rejected on 6 June 2008, arguing that the law would damage the country's economy.

A few days earlier, on 3 June, the influential Republican politician Newt Gingrich, former Speaker of the House of Representatives, wrote a piece on his blog defending geoengineering called (Gingrich 2008): "Stop the Green Pig: Defeat the Boxer–Warner–Lieberman Green Pork Bill Capping American Jobs and Trading America's Future". In his text he begins by theorizing as follows: "I've long maintained that there are two ways to protect the environment: There is the old-style, command and control, government regulation and litigation approach. Or there is the new way which uses technology, innovation and incentives to protect our environment." He defends the modern technological way, a precursor of ecomodernism. Further along in the text he writes: "Geoengineering holds forth the promise of addressing global warming concerns for just a few billion dollars a year. Instead of penalizing ordinary Americans, we would have an option to address global warming by rewarding scientific innovation. Our message should be: Bring on the American Ingenuity. Stop the green pig. Let your Senator know today".

The country most likely to begin a large-scale injection of aerosols into the stratosphere over its territory is the USA, because of its greater motivation, its know-how on the subject, and the power of the fossil fuel industries, in particular the oil and natural gas industries. The initiative would end up having a global effect, like a large volcanic eruption, and would affect other countries to different extents. Some of these countries would probably also feel free to use geoengineering to modify the climate and meet their specific domestic climate interests. A dangerous, uncontrolled race would likely start. Researchers from the USA and the UK have already addressed the concept of "counter-geoengineering" as a means of stopping or countervailing unilaterally deployed geoengineering (Parker et al. 2018). In other words, they have addressed the possibility that climate geoengineering could create international conflicts and become a weapon to be used in such conflicts. Reaching a worldwide consensus on intervention in the climate system using SRM geoengineering acceptable for all countries will be very difficult to achieve, largely because there are significant differences in the nature and intensity of the adverse side-effects at regional level around the world (Ricke et al. 2010; Robock et al. 2010; Schellnhuber 2011; Hamilton 2013; Hulme 2014; Kreuter 2015; Schneider 2019).

Meanwhile, study centres dedicated to the issue of governing geoengineering are beginning to emerge, such as the Carnegie Climate Governance Initiative (C2G), previously called Carnegie Climate Geoengineering Governance Initiative (C2G2), a non-profit organization based in New York. Its executive director is Janos Pasztor, who handled climate issues under former UN secretary-general Ban Ki-moon. C2G states that "it is not for or against the research, testing, or potential use of any proposed method or technology" for climate governance. It considers that "urgent action is needed at all levels to reduce emissions and adapt to climate impacts, and that an honest and open conversation is needed around all potential measures to manage the increasing risks of climate change". Pasztor goes around the world talking to high-level government officials and says that, as far as SRM geoengineering is concerned (Tollefson 2018): "They need to understand the risks—not just the risks of doing it, but also the risks of not understanding and not knowing".

The Solar Radiation Management Governance Initiative is an international NGO-driven project that is contributing to building the capacity of developing countries to evaluate SRM geoengineering. Their stated official position is one of neutrality regarding how SRM should be governed or whether it should ever be used. They do capacity building through outreach meetings in developing and emerging economies, where they invite scientists from advanced economies that are engaged in SRM geoengineering to report on their research. All these organizations are very careful to reiterate their neutrality, but the justification for their existence relies on a non-vanishing probability that SRM geoengineering will eventually be deployed on a large-scale. As the probability increases, their visibility and their funding also tends to increase. Their admirable and commendable work presupposes that SRM geoengineering will happen one day.

The question of geoengineering governance was addressed at a multilateral level at the fourth United Nations Environment Assembly (UNEA-4) held in Nairobi, Kenya, in March 2019. Switzerland, with the support of Burkina Faso, Federated States of Micronesia, Georgia, Liechtenstein, Mali, Monaco, Montenegro, New Zealand, Niger, Republic of Korea, and Senegal proposed a resolution on "Geoengineering and its governance" motivated by its belief in multilateralism and concerned about the ethical and social questions associated with geoengineering and the environmental and geopolitical risks that would result if a country were to initiate unilateral deployment. The document was very far from proposing a robust, broad, and multilateral geoengineering governance mechanism. It had a very limited ambition and essentially called for the preparation of an assessment of the status of geoengineering technologies, including CDR and SRM management, but it was withdrawn due to the lack of consensus.

From the accounts of the meeting, it is very difficult, perhaps impossible, to fully understand the hidden political motivations of the countries that were against the proposal. One clear outcome was that the positions of the USA and Saudi Arabia were aligned in their strong opposition to the document (Jinnah and Nicholson 2019). Some countries were concerned that the UNEA would have the power to make governance recommendations about geoengineering while others were concerned that the UNEA would weaken the existing international efforts to control some geoengineering initiatives under the Convention on Biological Diversity and the London Protocol. Most countries were concerned that geoengineering would undermine mitigation, and defended the idea that the text should explicitly recall the primacy of reducing emissions, while others opposed such a text.

The USA and Saudi Arabia left the impression that they will oppose any regulation or recommendation that acknowledges the urgency of addressing global climate change. The advocacy groups that were present at Nairobi called for a prohibition of geoengineering and pointed out that SRM geoengineering is a dangerous expedient that the fossil fuel corporations are planning to deploy at the right time to keep their businesses.

Among scientists there are also currents of opinion that are against the deployment of SRM geoengineering essentially because they value the potential dangers of trying deliberately to change the climate to countervail anthropogenic climate change and consider that the best option to solve the problem is to move away from fossil fuels (Robock et al. 2009, 2010; Schellnhuber 2011; Hamilton 2013; Hulme 2014; Kreuter 2015; Schneider 2019). Some foundations such as the Center for International Environmental Law and the Heinrich Boell Foundation have promoted information and research about the risks of SRM geoengineering (Muffe et al. 2019).

A frequently used argument to defend SRM geoengineering is that it reduces climate damage if mitigation proves insufficient to limit climate change. However, if it is actually deployed and starts to cool the climate system, there is no guarantee at all that greenhouse gas emissions from fossil fuels would be further reduced. Thus, in practice, SRM geoengineering is likely to allow for the business-as-usual reliance on fossil fuels. Consequently, SRM geoengineering is set to start new businesses and profit opportunities for industries and corporate interests that have systematically delayed by various means the transition to renewable energies. It would most likely extend and secure the ongoing concentration of power and control by the large multinational technological corporations.

In fact, one of the arguments for the deployment of SRM geoengineering points out that it is "more consistent with individual liberty than are the greenhouse gas controls" and it "would require no government-imposed lifestyle changes" (Lane 2006). Diana Furchtgott-Roth of the US Manhattan Institute points out the economic disadvantages of emissions reductions: "The approach would raise energy prices and costs of production, suppress wage and employment growth, and drive up prices of houses, home heating and cooling, cars, and other manufactured goods by raising production costs. It is a recipe for economic drag" (Furchtgott-Roth 2009). In view of these negative impacts on consumer capitalism, she advocates SRM geoengineering as the preferred alternative over mitigation.

From the point of view of climate science, SRM geoengineering would certainly involve very dangerous additional anthropogenic interference in the climate system. John Schellnhuber considers that SRM potentially means "mutually assured destruction" (MAD), an expression used for the military intimidation strategy to describe a widespread conflict using nuclear weapons that would guarantee the mutual extermination of the parties.

Our civilization has been profoundly marked by the success of fossil fuels to provide accessible and abundant energy at affordable prices for about two centuries. Oil has been particularly powerful in shaping world history, including the Bretton Woods System, the oil crises of the 1970s, the long ongoing wars in the Middle East, the financial and economic crisis of 2008, besides many other events (Auzanneau 2015).

The Anglo-Saxon countries, especially the USA and the UK, have been extremely shrewd at taking advantage of the power that came from the exploration and commercialization of fossil fuels, and in particular oil. An example of this shrewdness was the As-Is Agreement of August 1928, a secret pact between the major oil companies of the time. It is interesting to briefly revisit the circumstances under which the agreement was reached, since they reveal the inner workings of the element of operative social time of the fossil fuel world energy hegemony or simply the operative social time of fossil fuels.

The setting for the agreement was the austere Achnacarry Castle, in the Scottish West Highlands, the seat of the chiefs of Clan Cameron, that had been rented by Sir John Cadman, the head of the Anglo-Persian Oil Company that later gave rise to British Petroleum. He invited his three greatest rivals, Henri Deterding, head of Royal Dutch Shell, William Mellon, of Gulf Oil, and Walter Teagle, head of Standard Oil to a pheasant hunt. It is impossible to reconstruct accurately all the details of the fortnight that the group of oil businessmen spent dining, shooting, fishing, and doing business in the lands of Sir Ewen Cameron of Lochiel (1629–1719), in the Lochaber region of Scotland, whom the historian Lord Macaulay (1800–1859) described as the "Ulysses of the Highlands". The operative social time was more relaxed and they probably had the opportunity to enjoy lavish meals of smoked wild salmon, roast grouse, and venison served with the finest wines (Stevenson and Singh 2017).

The purpose of the As-Is Agreement was to establish a cartel that would guarantee profits by controlling the world oil market, excluding the USSR. The members of the cartel agreed on fixing production, prices, and the division of markets in each geographical area, which was illegal because it breached anti-trust laws. For instance, the crude from East Texas would be implacably cut from 98 cents a barrel to just 10 cents a barrel, forcing many wildcatters and smaller oil companies out of business (Stevenson and Singh 2017). A mechanism called "phantom freight" for rigging prices was created to build a protective dike around the US oil industry (Auzanneau 2015). It was probably one of the most efficient, far-reaching, and amoral agreements ever established by a group of respectable businessmen. It was endorsed on 17 September 1928, without a single signature.

In doing so they were firmly convinced that they were supremely responsible. The text of the Achnacarry agreement states that "the petroleum industry has not of late years earned a return on its investment sufficient to enable it to continue to carry in the future the burden and responsibilities placed upon it in the public's interest" (Auzanneau 2015). The As-Is Agreement assured that the major Anglo-Saxon companies would benefit from an almost eternal source of capital given by the inexhaustible abundance of black gold, without the dispensable risk of destructive competition.

If one wants to try to foresee the future of humankind, one must look at the future of the Earth's climate system over the next few centuries. To do that, it is essential to look at the future of the fossil fuel industry, and in particular the oil and gas industries. Finally, the best way to do that is to reconstruct and understand the outstanding successes of their past and the culture of power and hubris that was developed and became well established and ingrained in people's minds over more than a century. The leaders of the fossil fuel industry remain convinced of their supreme responsibility when arguing that the only way to reduce poverty in the world is to promote economic growth in developing countries by providing them with abundant fossil fuels, especially oil and natural gas. Regarding SRM geoengineering, there are many voices from the fossil fuel industry that support its deployment because it would slow down climate change and give a breathing-space to reduce world poverty in developing countries with the help of fossil fuels, without forcing them to adapt to a changed climate for several decades.

Many people, organizations, institutions, and governments that accept and follow science-based policies defend an energy transition away from fossil fuels and oppose the use of SRM geoengineering. The final outcome depends on how long the operative social time of fossil fuels will last: 30, 50, 100 years or more? The complexity of the human subsystem and of its interactions with the other subsystems of the Earth system make it impossible to make reliable predictions. The uncertainties are very great indeed.

The USA, Saudi Arabia, Russia, and other coal, oil, and gas producing countries will do everything in their power to extend as long as possible the supremacy of fossil fuels in the global energy mix, but their economic power insofar as it is derived from fossil fuels is in decline. The main reason for their decline is that they are selectively negating or omitting the results of science. In fact, they disregard the evidence and the advice from science that the world must decarbonize its economy to prevent the worst impacts of climate change now and especially in the future. In fact, most countries, whether fossil fuel producers or not, tend to disregard scientific advice when it does not conform to their economic and financial vision of the future and to their short term interests. They profess their unquestionable belief in technology, but they reject the limits that science makes so clear for sustaining the well-being of humankind in the Earth system.

7.24   The COVID-19 Pandemic and Its Global Impacts

The global crisis that resulted from the COVID-19 pandemic, which had its origin in Wuhan, China, at the end of 2019, is throwing light on or exacerbating most of the major challenges facing humankind in the 21st century that have been addressed in the present book. The pandemic is a zoonosis, a type of infectious disease caused by a pathogen such as a bacterium, virus, or prion that is naturally transmitted between vertebrate animals and people. Various studies have shown that zoonotic pathogens are the most likely source of emerging and re-emerging infectious diseases (Wool-house and Gowtage-Sequeria 2005; Kemunto et al. 2018). The analysis of a database of 335 emerging infectious diseases that appeared between 1940 and 2004 shows that 60.3% are zoonoses and that 71.8% of these originated in wildlife. It also showed that they have a tendency to increase over time (Jones et al. 2008).

The risk to human health posed by zoonoses started to increase with the Agricultural Revolution through the closer contact with animals and the higher human population density in cities. Since that time humankind has become acquainted with the unpredictable appearance of deadly and devastating new infectious diseases. It took a long time for science to discover their origin, how they propagate, and and how to cure them. It is usually difficult to reconstruct the origin and history of the pathogen responsible for a zoonosis. There is uncertainty about the animal or animals from which it started to spread because the pathogen can mutate many times before it reaches humans and becomes able to penetrate human cells, but also because the pathogen may frequently be found in different species. We now know that the bubonic plague came from rats and prairie dogs to fleas and then to humans, while smallpox came from cows, measles from dogs, rabies from vampire bats to cattle and then humans, influenza from horses and pigs and also from wild birds before being transmitted to domestic fowl, Lyme disease from deer mice to ticks to humans, West Nile virus from birds to mosquitos to humans, Nipah virus from bats and pigs, HIV from chimpanzees, Ebola from the African fruit bat and also from chimpanzees, gorillas, and baboons, and Zika fever from non-human primates. Some of the recent zoonoses that have rapidly increased in incidence or geographical range include the Rift Valley fever, pandemic influenza H1N1 2009, yellow fever, avian Influenza H5N1 and H7N9, West Nile virus, severe acute respiratory syndrome SARS-CoV, and Middle East respiratory syndrome coronavirus MERS-CoV (WHO 2020). It is estimated that about one billion cases of illness and millions of deaths occur every year globally from zoonoses (WHO 2020).

The major pandemics of the 20th century were originated by influenza viruses of animal origin, although most new strains arise in human populations. The so-called "Spanish flu", caused by an A(H1N1) virus, very likely of avian origin, is estimated to have been responsible for 20–50 million deaths in the period 1918–1919, while the "Asian flu", caused by an A(H2N2) virus in 1957–58, and the "Hong Kong flu", caused by an A(H3N2) virus in 1968, were responsible for an estimated 1–4 million deaths each.

The present COVID-19 pandemic is caused by a SARS coronavirus that has been named SARS-CoV-2. Coronaviruses are spherical, 120 to 160 nm in diameter (1 nanometre is 10^{-9} metre), with an outer envelope bearing 20-nm-long club-shaped projections that collectively resemble a crown. These spikes attach to a protein called ACE2 on the surface of cells. This is a protein that plays a role in regulating blood pressure and sets off chemical changes that effectively fuse the membranes around the cell and the virus together, allowing the virus's RNA to enter the cell. The virus then takes over the cell's protein-making machinery to translate the cell's RNA into new copies of the virus. In a few hours a single cell can be forced to produce tens of thousands of new virions that go about reproducing the process of virion production. The most common symptoms include fever, cough, tiredness, shortness of breath or difficulty breathing, muscle aches, sore throat, and loss of taste or smell. These lead in some cases, especially in elderly people or people with health problems or immunologic deficiencies, to pneumonia, multi-organ failure, heart problems, blood clots, acute kidney injury, and additional viral and bacterial infections and death. In May 2020 there were still many unknowns about how the virus attacks the human immunologic system and the epidemiology of diseases it causes.

In mid-May 2020, approximately six months after the first cases of SARS-CoV-2 infection, about 5 million people had been infected globally and there had been over 320 000 deaths. During that time the pandemic had reached all countries, while the most affected were the USA, Russia, and some European countries, especially the UK, Italy, Spain, and France. The only effective measure so far found to slow down the spread of the disease has been the implementation of social distancing measures, which have had a devastating effect of the economy, especially through the reduction of industrial activity, commerce, travel, and tourism. The impacts are leading to a deep global recession which, according to an IMF report of April 2020, will shrink global GDP in 2020 by 3%, 6.1% in advanced economies and 1% in emerging market and developing economies, relative to 2019, and produce a negative per capita GDP growth in 90% of all countries in 2020 (IMF 2020).

The first characterization and full-length genome sequencing of the new coronavirus that started the epidemic of acute respiratory syndrome in Wuhan in December 2019, later named SARS-CoV-2, was published in Nature on 3 February 2020 (Zhou et al. 2020b). It was shown that the virus shares a 79.6% sequence identity to SARS-CoV, and is 96% identical at the whole-genome level to a coronavirus found in a *Rhinolophus affinis* bat sampled in the Yunnan province of China. Later it was reconfirmed that SARS-CoV-2 is not a laboratory construct or a purposefully manipulated virus (Andersen et al. 2020). Although SARS-CoV-2 is likely to have its origin in bats, it is still unknown how it was able to move to humans, possibly through another host, and undergo mutations leading to molecular changes, most likely in its spike glycoprotein, so that it was able to enter human cells. A coronavirus found in the *Rhinolophus malayanus* bat in 2020 is also very similar in most genomic regions to SARS-CoV-2, as well as to a coronavirus found in *Manis javanica*, the Malayan pangolin, which is illegally imported into the Guangxi and Guangdong provinces of southern China (Zhou et al. 2020a).

The observed increase in zoonoses since the beginning of the Great Acceleration is very likely a result of the environmental crisis and of the associated cultural changes in some regions of the world. The ancestral relationship between people and wildlife that can still be observed in a few indigenous groups has been deeply transformed by modernity and more than two centuries of global economic progress. In advanced economies, there is usually an effort to preserve some natural habitats. However, in emerging and developing economies, most natural habitats, except those included in protected areas, are being systematically degraded or destroyed by logging, mining, and other forms of natural resource extraction, agriculture, new settlements, roads, and growing urban areas. This organized trespassing, viewed by most as an imperative of progress, thrusts wildlife and people into new forms of contact that creates a new cultural relation. While the eating of wild animals by indigenous people follows cultural paradigms that evolved through millennia, the new consumption of wild animals by non-indigenous people has little or no memory of the past and is guided by the rules of shifting circumstances and those of the market.

In China, there was widespread famine in rural areas during the years 1959–1961 at the time of Mao Zedong's Great Leap Forward. People were forced to eat anything available including all sorts of wild animals. This experience led in the late 1970s to the lucrative establishment of a massive program of wildlife farming where the bushmeat was primarily consumed in rural communities by relatively poor people. With time, this habit evolved, and bushmeat is now consumed mostly by the middle class as a luxury product. The risk of transmission of pathogens to humans increases when wildlife animals are stressed and immunosuppressed by being farmed in small cages and sent to wet markets where animals of different species in piled up cages are likely to exchange pathogens through their excretions, before being slaughtered to satisfy customers who like to eat freshly killed meat. Throughout this cruel process, the wild animals are frequently in close contact with humans, which increases the risk of pathogen transmission. The SARS-CoV coronavirus responsible for the SARS epidemic that started in Guangdong province in China in 2002 is very likely to have originated in horseshoe bats and spread to humans either directly or through animals held in Guangdong wet markets. SARS-CoV-like viruses were isolated from Himalayan, or masked, palm civets (*Paguna larvata*) and other animals such as the raccoon dog (*Nyctereutes procyonoides*), found live in one of those wet markets (Guan 2003). The civets are eaten mostly in the autumn and winter because they are believed to help people better withstand the cold weather.

On 21 December 2019, about two thirds on the patients with viral pneumonia in a Wuhan hospital were found to have some relation to the Wuhan Seafood Wholesale Market, which includes a wet market of wild animals. In social networks people were advised not to go there. One possibility is that SARS-CoV-2 was able to jump from bats to some wild animal sold in the market, possibly the pangolin, one of the most illegally traded mammals in the world. On 24 February, China's Standing Committee of the National People's Congress, issued a decree entitled *Comprehensively prohibiting the illegal trade of wild animals, eliminating the bad habits of wild animal consumption, and protecting the health and safety of the people.* This decision could potentially have far reaching consequences by impairing or debilitating the multibillion-dollar trade of protected wild species, but it is too early to say what may happen in the future. Chinese authorities have stated that they closed down some 20 000 wildlife farms and that more than 2 550 people have been accused of wildlife crimes. Meanwhile, it is known that in April 2020 some wet markets reopened, but there is no reliable information about the sale of wildlife.

Wet markets continue to be found in many Asian and African countries, posing a significant risk for human health at the global scale. In North Sulawesi, Indonesia, snake and bat meat are often sold in supermarkets (Paddock and Sijabat 2020). Many markets in the islands of Java, Sumatra, Bali, and Sulawesi sell wildlife for food and for other uses. Song bird species captured in nearby forests are sold as ornamental pets or to enter into singing competitions, in a thriving illicit trade that is threatening wild populations. Local authorities resist closing these markets because they provide traditional food and very valuable income for those in greatest need. In Africa, primate bushmeat, the source of the simian immunodeficiency virus (SIV) which mutated into human HIV, continues to occur, and urban bushmeat markets are presently growing in Africa.

Unless humankind decides to effectively mitigate the environmental crisis by finding more sustainable pathways for our relation with the biosphere, the risk of pandemic diseases like COVID-19 will persist and is likely to increase. A further factor that is likely to aggravate the risk of future pandemic diseases is the current climate change crisis. Climate change shapes the biogeographical distribution of species, and constitutes an additional factor that is liable to disrupt their habitats and force some of them to migrate. In particular, it changes the geographical areas where vector-borne diseases are found. The ongoing increase in global mean temperature, the changes in regional precipitation patterns and the more frequent extreme weather events, together with changes in other environmental and social determinants, affect the development and dynamics of the disease vectors and of the pathogens they carry. This process is creating greater uncertainty and risk regarding the spread of vector-borne diseases such as dengue fever, malaria, Lyme disease, Zika and West Nile virus, schistosomiasis, leishmaniasis, Chagas disease, and African trypanosomiasis (Campbell-Lendrum et al. 2015).

The COVID-19 crisis is inflicting substantial and often dramatic economic and social shocks as global production, consumption, and employment levels drop abruptly. Thus it is also having a strong impact on the other two main current global crises, which have a socioeconomic nature: the debt crisis and the crisis that results from increasing socioeconomic inequalities at the global level. The COVID-19 pandemic will aggravate the debt crisis that has been looming in the last few years. To equilibrate the risk between public health and economic growth, populations were temporarily ordered to stay at home, while schools, offices, and factories were closed or limited their activities, most businesses came to a temporary standstill, road traffic was reduced to a minimum, and air traffic decreased by about 60%. This dramatic slowdown forced governments and central banks to deliver massive fiscal and monetary stimulus into the economy. We have little idea how this surge in public and private debt will affect economic growth in the coming years. The combination of defaults and bankruptcies and the high levels of public debt are likely to lead to a weaker recovery than the one that followed the 2008–2009 financial and economic crisis. Furthermore, the recovery in advanced economies is likely to be hampered by the increase in public spending needed to assure universal health care and social security for ageing societies debilitated by the pandemic.

The COVID-19 pandemic is also expected to exacerbate socioeconomic inequalities at the global level. There are various factors that contribute to this outcome. The living conditions of a large part of the population in the emerging and developing economies is likely to get worse. For hundreds of millions of people living in export-led emerging and developing economies the loss of income due to the global economic crisis is likely to mean severe poverty, famine, and forced migration. According to the International Labour Organisation, over 60% of the global workforce, which corresponds to about two billion people, hold informal jobs, meaning that they mostly lack social protections, workers' rights, and decent working conditions. The socioeconomic conditions of this large part of humankind are likely to worsen.

The pandemic is accelerating the tendencies for de-globalisation, protectionism, and fragmentation that were already apparent before the crisis. The USA on one side and China, Russia, and Iran on the other are likely to enter a period of open trade war and economic decoupling that may evolve into some new and dangerous form of Cold War. Other countries are likely to respond by adopting policies for progressive self-reliance, trying to shield their businesses and workers from the recurring global crises. Various governments are already imposing protectionist measures in the pharmaceutical, medical equipment, and food sectors as a response to the crisis. These tendencies are likely to lead to tighter restrictions on the movement of goods, services, capital, and labour at the global level. Technology, data, and information exchanges will become more protected and difficult. The re-shoring of production from low-cost regions to higher-cost advanced economy markets will tend to accelerate the Fourth Industrial Revolution, which will increase the precariat and working poor, and encourage populist and nativist political movements.

Five interrelated and concurrent global crisis have been briefly identified. It is debatable whether they are crises and if they can be considered catastrophic crises resulting from catastrophic risks or even whether they are existential risks (Leslie 1996; Posner 2004; Smil 2008; Bostrom 2002, 2013). Three are related to the environment and are more likely to be considered as catastrophic. The COVID-19 crisis has a limited time frame, while the other two environmental crises are progressing with no end in sight. The other two are the developing socioeconomic crises related to the current globalized economic and financial system. It has been argued in this book that these two crises are closely related the two long-term environmental crises. The fact that humankind has currently to deal with five simultaneous and interrelated crises is significant and should be a matter for thought.

Curiously, this simultaneity is reminiscent of the worldview of the future expressed in 1998 by Guillaume Faye (1949–2019), a far right French journalist and writer, advocate of identitarianism and one of the main theoreticians of the Nouvelle Droite, in his book *L'Archéofuturisme* (Faye 1998). There he writes that "a series of 'dramatic lines' are drawing near: like the tributaries of a river, they will converge in perfect unison at the breaking point (between 2010 and 2020), plunging the world into chaos." In 2004, under the pseudonym of Guillaume Corvus, he wrote a book entitled *La convergence de catastrophes* (Faye 2004), in which he analyses the actual and potential risks facing Western civilisation, including among others the environmental crisis, climate change, immigration, terrorism, and pandemics, and he prognosticates that they will soon combine to overturn the present order.

The concept of archaeofuturism describes a world divided into an economic bloc that continues to develop and depend on science and technology and another bloc that reverts to "archaic values" and solutions (Faye 1998):

> The techno-scientific portion of humanity would have no right to intervene in the affairs of the neo-medieval communities that form the majority of the population, nor – most importantly – would it in any way be obliged to 'help' them.

Although it is a simplistic view of the world, one finds here another form of the deep division of humankind into two groups that H. G. Wells called the New Republic and their followers and the People of the Abyss. This tendency for social and economic apartheid is a real risk that must be avoided. However, if the four long-term crises are allowed to persist and become stronger their combined effect will become very difficult to address and humankind will tend to segregate part of itself. What would be left of the values of the Enlightenment, in particular the egalitarian values of equal political, economic, and civil rights for all people, human perfectibility, and progress?

The COVID-19 crisis could and should be seized as an opportunity to explore and pursue new pathways for sustainability. Temporarily at least, it has alleviated some of the human pressures on the environment and the resulting effects have been surprising and inspiring for many people around the world. The significantly reduced economic and industrial activities and the lower mobility and travelling have temporarily reduced emissions of carbon dioxide, nitrogen oxides, and related ozone formation, and also emissions of particulate matter. Urban areas have become less polluted, skies have been noticeably clearer, and animals have come out to visit our habitats. Some people may have acquired a renewed reverence for the environment. But there have also been negative environmental consequences related to a higher volume of unrecyclable waste, lower maintenance of natural ecosystems, illegal deforestation, fishing, and wildlife hunting.

Climate change is one of the areas where there are opportunities for change since the crisis is likely to have long-lasting effects on the global energy system. One of the main questions in May 2020 was how the fossil fuel industry will resist the crisis and how it will emerge from it. Coal is likely to be affected in the long-term. With decreasing demand for electricity, many utilities cut back on coal first, because it is more expensive than natural gas, wind, and solar. Coal power stations had to close in several countries during the crisis. In the EU, imports of coal for thermal power plants have plunged by almost two-thirds in recent months to reach lows not seen in

30 years. In the USA the fall in electricity demand is likely to decrease coal's share of electricity generation from 50% a decade ago to about 10%. The peak of global coal consumption is likely to have passed.

The situation regarding oil is more complex. The COVID-19 pandemic has reduced oil demand by about 30%, in particular as a result of the reductions in road transport and air travel. Oil production was slow to adapt to such a decrease, leading the US benchmark West Texas Intermediate futures contracts for May to be priced negatively at −40 US dollars a barrel on 20 April. The next day, the International benchmark Brent crude oil settled at \$19.33 per barrel, but after one month it recovered to values around \$35.

To assess the future of the oil industry Wood Mackenzie's uses three scenarios (WM 2020). In the "full recovery" scenario, oil demand is restored to previous levels and continues to increase up to the mid-2030s, before beginning to decline slowly by the end of the decade. In the "go it alone" scenario characterized by a strong de-globalisation process, global oil consumption is likely to reach peak consumption before 2030. The "greener growth scenario" is the only one that includes the possibility of an effective implementation of the energy transition in the EU and the USA. In this scenario, oil demand is predicted to stabilise in the 2020s and to fall significantly in the 2030s.

A report from McKinsey (Barbosa et al. 2020), points out that, under most scenarios, "oil and natural gas will remain a multi-trillion market for decades". It emphasizes that the industry has a "proud history of bold structural moves, innovation, and safe and profitable operations in the toughest conditions". The oil industry was able to restructure itself after the 1980s oil glut following the 1970s oil crises and to recreate the opportunity for highly lucrative investment. In the period 1990–2005, the total return to shareholders in the oil and gas industry, except refining and marketing companies, exceeded the total return of the S&P 500 index during the same period. However, it has underperformed that index in the past 15 years (Barbosa et al. 2020). The proud history of bold structural moves that McKinsey refers to surely includes the move by Exxon, the most powerful private corporation of the USA, to deny previous research results obtained by the corporation linking fossil fuels and climate change (Santos 2020), when in 1984 Lee R. Raymond, elected to the Board of Directors, became one of the leading architects of climate change denial in the USA and one of the most powerful forces against climate change mitigation in the world. Finally, the 19-year tenure of Raymond as the lead independent director of JPMorgan Chase, the world's largest lender to the fossil fuel industry, was announced on 1 May 2020, satisfying demands for change led by New York City's comptroller Scott Stringer and climate change action groups.

The effect of the COVID-19 crisis on climate change is measurable but very small compared with what is needed to reach atmospheric concentration levels of greenhouse gases that do not imply a dangerous anthropogenic interference on the climate system. The reduction in global CO_2 emissions in 2020 relative to 2019 is estimated by the IEA at about 8%, corresponding to about 2.6 billion $GtCO_2$ (IEA 2020), but at this time there is still uncertainty about these numbers. It must be realized, however, that the climate change driver is the concentration of atmospheric

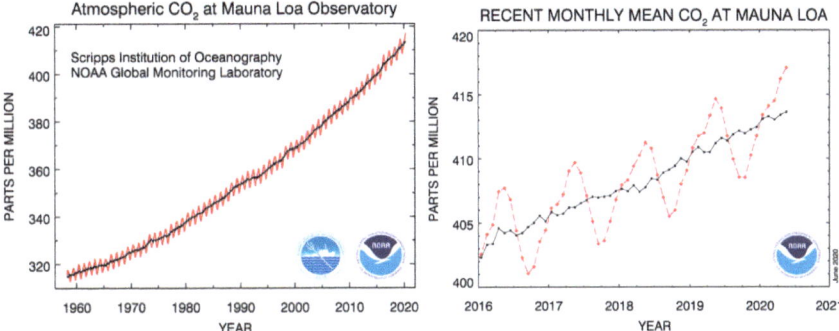

Fig. 7.6 Mean monthly CO_2 atmospheric concentration measured at Mauna Loa Observatory, Hawaii, as the mole fraction in dry air. Measurements at Mauna Loa, at an altitude of 3400 m, in the northern subtropics of the central Pacific were first carried out by the remarkable American scientist Charles David Keeling (1928–2005) of the Scripps Institution of Oceanography in March of 1958 with funding from the International Geophysical Year of 1957–1958, at a facility of the National Oceanic and Atmospheric Administration (NOAA) (Keeling et al. 1976). NOAA started its own CO_2 measurements in May of 1974, and they have run in parallel with those made by the Scripps Institution of Oceanography since then. *Left*: The full record of combined Scripps data and NOAA data from 1958 to June 2020. Notice that the growth in the atmospheric CO_2 concentration is faster than linear. However, to control climate change, the concentration must reach a peak, decrease, and stabilize at about 350 ppmv. *Right*: Monthly mean values centered on the middle of each month (*red dashed line*) and a moving average of seven adjacent seasonal cycles centered on the month to be corrected (*black line*). The seasonal cycle with higher CO_2 concentration in the months of April and May results from the asymmetry between the northern and southern hemispheres as regards biological sequestration due to a larger land mass in the former hemisphere. *Credit* Earth System Research Laboratories, Global Monitoring Laboratory, National Oceanic and Atmospheric Administration (https://www.esrl.noaa.gov/gmd/ccgg/trends/)

CO_2. As long as emissions are large the CO_2 concentration will keep increasing. It is like the water level in a swimming pool. If we reduce by 8% the water flow into the pool from a tap, the water level in the pool continues to increase although at a slower pace. In the case of the atmosphere, it would mean that CO_2 concentrations would increase in 2020 relative to 2019 by 2.48% instead of the expected 2.80% in the absence of the crisis, which is a very small effect (Betts et al. 2020). Atmospheric CO_2 concentrations measured at the Mauna Loa observatory in Hawaii have increased from an annual average of 316 ppmv in 1959 to 411 ppmv in 2019 (see Fig. 7.6). To control climate change, CO_2 concentrations have to stabilise and for that to happen it is necessary to reduce drastically the emissions until they become close to zero. Only a few decades after the concentration stabilizes will the mean global temperature begin to stabilize. According to a UN Environment Programme 2019 report (UNEP 2019), global greenhouse gas emissions must decrease annually by 7.6% each year between 2020 and 2030 in order to comply with the 1.5 °C temperature goal of the Paris Agreement.

The COVID-19 crisis presents a sobering perspective of what has to be done to meet the Paris Agreement and of how far humankind is from achieving that goal.

Although we are all facing a problem that is very difficult to solve, it is still possible to reach the goals of the Paris Agreement. The price for negating the challenge or avoiding any attempt to address it decisively will be a progressively dangerous climate change crisis that will put billions of people at risk, especially those that are most vulnerable, which usually implies that they are also those who have contributed the least to the problem.

Once more and finally, one returns to the question of segregation. Personally, I stand for the core values of the Enlightenment, for the defence of the scientific method, for the prevalence of ethical values over self-interest, for human rights, democracy, and human progress, for the sustainability of an inclusive civilization that welcomes freedom, diversity, cooperation, solidarity, and creativity, protects the integrity of the biosphere, and respects the Earth system as its home.

Chapter 8
Conclusions

8.1 Knowing the Facets of Time

The main lines of the role played by science and technology in the history of civilizations up to the present time have been succinctly presented. The myth that has prevailed over all the changes in myths, religions, and religious practices and over all social and political transformations, and is now flourishing, is technology. Technology is the current magical companion of *Homo sapiens*, able to fulfil all his wildest dreams of well-being, entertainment, and economic prosperity and all the excesses that make life exciting and worth living.

Time is different from all the other constraints that science and technology have helped humankind to overcome. Time remains, to our chagrin, unconquerable. Technology is utterly unable to change or improve its nature for our benefit. Happily, we dominate space and speed, and we dream of colonizing Mars and exploring the Solar System and eventually beyond, as fast as possible. Our angry reaction to the impregnability of time is to use it as much as possible and to extend the life that gives access to it as long as possible. Humankind has learned about how to use time ever more efficiently, and about experiencing the thrill of time acceleration. The practice of the maxim that there is no time to lose, that time is money, and that there is always something more to do than simply being, is widespread and rarely discussed. Nevertheless, we cannot help feeling that time is never enough to appease or satisfy all our expectations. Instead of using time greedily as if it is a rare resource, we would do better to recover peace and learn to enjoy the flow of time at our leisure.

The historical narrative and analysis of the advent of progress, growth, and technology developed in this book is guided and supported by the concepts of operative social time, operative social time structure, the element of operative social time, and historical time, and also by their relationship with the concepts of psychological time, biological time, and physical time. The long and continuing interaction between the two groups of concepts has contributed decisively to influencing and characterizing our long cultural evolution, which can be traced back to the Upper Palaeolithic, when *Homo sapiens* developed a more complex social behaviour that shaped cognitive and

© Springer Nature Switzerland AG 2021
F. Duarte Santos, *Time, Progress, Growth and Technology*, The Frontiers Collection,
https://doi.org/10.1007/978-3-030-55334-0_8

symbolic thought and symbolic representation. The process of adapting to external cycles and a constantly and deeply changing environment has played a decisive role in the biological evolution of hominids that led to the emergence of *Homo sapiens*. The contrast between the uncertainty of everyday human life and the superhuman regularity of the external time cycles of the Sun, Moon, and planets was a very fertile mystery that influenced various civilizations, especially the Egyptians and the Maya. The latter went as far as admitting that time had a cyclical nature that was a reflection of the cycles observed by the Maya in life and in the terrestrial and cosmic environment. More importantly, the cycles of time enabled them to peer into the future and help make the right decisions. The fascinating initial steps in the use and interpretation of the concepts of time have marked the cultural evolution of humans.

We do not know how long our species *Homo sapiens* will live but it will be practically impossible for it to last forever. There is, then, an element of operative social time—*Homo sapiens* time—that includes and surpasses all the other elements of operative social time. One of the main difficulties in the analysis of *Homo* and *Homo sapiens* time is our inability to imagine periods of time that last hundreds of thousands of years and the impossibility of gaining access to and analysing the relevant events and processes that took place within them and had a crucial influence on our evolution. Some of the modes for the termination of *Homo sapiens* time induced by natural environmental transformations or by causes related to human behaviour, including the advent of transhumanism, were addressed.

The concepts of time that have been introduced and analysed also play a crucial role in understanding the impacts of progress, growth, and technology on sustainability. It is argued that the contrast and resulting tension between the cyclical nature of the element of generational social time and the uniqueness of the element of time that is our lifetime is an essential challenge of the human condition. To what extent should we appreciate and enjoy something that is unique, personal, and cannot be repeated—our lifetime as an element of operative time—and, on the other, value our active participation in the social generational cycle by helping it to continue in a sustainable way? What is the compatibility of personal interests, behaviour, and lifestyle with the interests of future generations, including our own descendants? The permanent tension between these two types of driving forces is responsible for shaping the human approach to sustainability and intergenerational justice.

Sustainability also depends crucially on intertemporal choices that involve decisions with consequences having repercussions in the future. This type of decisions, which are very common in everyday life, force us to imagine the future and make comparisons between the costs and benefits that take place at different times. They are personal decisions in which time, the way it is perceived, and our awareness of the finiteness of our own lifetime play an essential role. Intertemporal decisions are also made collectively, for instance by governments, institutions, and organizations at the local, national, and international levels in all fields of action, including social, political, economic, financial, and environmental, in which case they have a much broader influence on sustainability. Choosing to obtain valuable sustainable benefits that are distant in time and having the patience to wait for them, rather than gain immediate access to unsustainable benefits now, is one of the hardest exercises for human willpower.

Most analyses of intertemporal choice in the context of economics are based on the model of discounted utility, where it is assumed that people assess the gains and losses resulting from intertemporal choices in accordance with an exponential time-discounting law, just as the financial markets assess gains and losses over time. This model implies a rate of time preference or time discounting rate independent of time. Deviations from the discounted utility model are often interpreted in neoclassical economics as anomalous patterns of economic behaviour. However, the rate at which people discount future rewards is not constant but declines with the length of the delay.

The rate of time preference is an important characteristic of operative social time. People who have a high rate of time preference tend to favour their interests and well-being more in the present or very short term, while people who have a low rate of time preference assign relatively greater importance to their interests and well-being in a more distant future. It is argued that the tendency for an increase in the rate of time preference that has been documented in various domains, such as eating habits, personal savings, and consumer habits, is likely to have its origin in the acceleration of operative social time associated with modernity, and more recently with the current economic and financial model centred on fostering consumerism.

Economic intertemporal choices are determined by the time-discounting rate that makes it possible to convert cost and benefit flows with a future economic value into the equivalent value in the present. There are a wide range of economically important projects and plans for which the most suitable time-discounting rate needs to be determined. On the one hand, there are private, corporate projects that are generally short in duration, lasting less than one social generation, and social projects, which are often run by governments or private funding organizations that frequently have time frames spanning several social generations. Thus, there is a need to calculate corporate and social rates of time discounting. In general, corporate rates are higher than social rates because people are most interested in their prosperity, well-being, and quality of life in the short term of their lives and are more averse to risk during their lifetimes, while society as a whole tends to have a time horizon that reaches further into the future over the medium and long term. The value chosen for the social discounting rate is systematically used in public policy and governance to calculate the amount of future investment in health, education, research, and development, as well as investment in protecting the environment, achieving the sustainable use of natural resources, and combating climate change, which implies that it is an instrument for intergenerational equity and justice.

In Ramsey's theory of optimal long-term economic growth based on a constant social time discount rate, time discounting results from expectations of perpetual economic growth based on scientific and technological progress. In a world where future generations will certainly benefit from growing economic prosperity, current utilities will gradually lose their capacity to attract and satisfy. The search for the compatibility between the principles of intergenerational justice and economic intertemporal choices has led to the development of several utility optimisation criteria and models, such as the discounted utilitarianism model. There are numerous studies that attempt to adapt the current economic system to deal with long-term issues by way of models for calculating social discount rates in the intergenerational context. A constant social time discount rate overly devalues what will happen in the distant future. The use of social time discount rates that fall over time, or soft discounting, in long-term public policy is better adapted to the present uncertainty about the future.

Climate change intergenerational justice is especially sensitive to the value chosen for the social time discount rate used in investing in climate change mitigation, a choice that has generated long and unresolved controversies. A high discount rate corresponds to a low social cost of carbon and implies that there is not much need to mitigate climate change now because future generations will have better ways to deal with the problem with the help of forthcoming technological progress. A low discount rate is favoured by ethical considerations of intergenerational justice based on the severity of the future impacts of unmitigated climate change.

The importance of time has increased very quickly since the beginning of the 16th century, with the firm resolve to measure physical time and to use clock time to regulate transport and progressively everyday life. Operative social time gained more economic and financial value and the temporal resolution of the operative time structure increased significantly. Modernity is to a large extent the history of time.

With the theory of relativity the concept of physical time evolved away from Newton's absolute time and became disconnected from psychological time. But technology ensures that relativistic physical time is also relevant in everyday life, although in a much more subtle way. Atomic clocks in GPS satellite run fast by on average roughly 38 000 nanoseconds per day relative to atomic clocks on the ground, and this is incompatible with the precision required by a GPS positioning system. This time dilation is a relativistic effect resulting from the relative speed of satellites to atomic clocks on the ground and from the fact that the atomic clocks on satellites are located at a greater distance from the Earth's centre, where the Earth's gravitational field is weaker. If this time difference were not continually corrected, the GPS system would become useless. This is an example of a technological application that depends on scientific knowledge in ways that are disregarded by the end users. There are many other examples in our operative social time where science slips into the background and is devalued. This devaluation may convince people that it is possible to preselect whatever in science is useful for their own interests and negate or omit whatever is unfavourable for them. That is dangerous because science is a coherent structure built by the scientific method, whose conclusions and predictions are universal and potentially falsifiable. If one accepts the narrative of science, one has to accept all its outcomes with impartiality.

Physical time, like all the other concepts of time, is an inexhaustible source of new ideas and knowledge. It retains the same mystery and elusiveness that it had when the Sumerians, Egyptians, Chinese, and others started to invent devices to measure it. Our knowledge has evolved remarkably and we are now reaching a point in which the incompatibilities between quantum mechanics and the theory of relativity are becoming clearer and require a solution, especially as regards time. The problem of non-locality associated with quantum entanglement is that it apparently clashes with the theory of relativity because the instantaneous connectivity between particles separated in space is incompatible with the speed of light being the maximum speed at which matter and all forms of information can travel. Up until now, it has been possible to include the theory of relativity in quantum mechanics by making quantum mechanics covariant, but what probably needs to be done is to start with relativistic spacetime and find a way to include quantum mechanics in it. This program will probably involve the discovery of a new structure of spacetime that includes the possibility of retrocausality. There is going to be an end to physical time which entails also an end to all other facets of time.

It is impossible to retrace the steps in the development of human lifetime awareness and how life became an element of operative social time. Nevertheless, we may attempt to glimpse the initial forms of that awareness by observing how chimpanzees, bonobos, and other animals behave when they face the death of a member of their group. In the case of *Homo sapiens*, a possible conjecture is that the awareness of the finiteness of life and all its psychological consequences started to evolve more rapidly with the emergence of symbolic thought and spoken language acquired about 100000 years before present (BP). Probably the main defence that was developed against death anxiety were some primitive forms of repression and denial. The denial of death became progressively one of the most powerful expressions of the psychological ego's defence mechanism of denial. Nowadays, denialism is a fairly common form of psychological strategy that is unconsciously used to protect from the feelings of anxiety which arise because we feel threatened by very diverse risks.

Humans have been affected by many catastrophes since prehistoric times. At present we are continually confronted with catastrophic risks, which can be regional or global. The former include natural disaster risks, such as floods, droughts, tropical cyclones, intense extra-tropical storms, earthquakes, tsunamis, volcanic eruptions, and socioeconomic and technological risks such as regional famines, epidemics, terrorism, large migrations, wars, and potential regional effects of ill-judged application of the emerging technologies. Global catastrophic risks seriously threaten the well-being and economic prosperity of humankind on a global scale and are considerably more challenging because their mitigation depends on the cooperative and coordinated response of all the countries affected. Global catastrophic risks can be natural, such as large asteroid collisions, supervolcanoes, and natural pandemic diseases. They can also be anthropogenic, such as a generalized nuclear war followed by a nuclear winter, accidental or engineered misuse of emerging technologies, such as nanotechnology or artificial intelligence, engineered pandemic diseases, extreme anthropogenic climate change, climate geoengineering impacts that result from the failure to countervail anthropogenic climate change, undetermined forms of global ecological collapse, critical transitions in the biosphere, extreme consequences from overexploitation of natural resources coupled with increasing demand, worldwide social and political disruption and tyrannies, and severe global economic crises generated by the inner workings of the current economic and financial system, by an escalating trade war, or by concurrence with the preceding drivers of risk.

Some of these global catastrophic risks, because they can be perceived as having the potential to imperil humankind in a long-lasting or even irreversible way, are called existential risks. It is important to realize that there are unknown catastrophic risks that do not fit entirely into the preceding list and also that may be perceived as existential risks. An increasingly common way to react to the anxiety generated by global anthropogenic catastrophic risks is to understate or deny them, or to defer actions that would reduce the risks.

The oldest known records of the human awareness of the transition from life to an afterlife is the use of graves by *Homo sapiens* from the Middle Palaeolithic period, about 80 000 to 100 000 years BP, already reflecting the emotional importance of the person who died. Evidence for the use of graves by *Homo neanderthalensis* is rare and up to now has appeared only at the end of their existence, about 50 000 years BP, at a time when miscegenation between Neanderthals and humans probably began.

The study of the development of myths, the symbolic narratives about the origins, nature, and future of the world and humans, is a way to decipher the evolution of the concept of time. According to Michael Witzel, world mythology systems can be divided into two main groups: the Laurasian group, which includes mythologies from North Africa, Eurasia, and the Americas, and the Gondwanan group, which includes mythologies from Sub-Saharan Africa, the Andaman Islands, Papua New Guinea, and Australia. While the former group of myths has a structured narrative in which the Universe has a beginning and includes the origins and development of the human presence within it, most Gondwanan mythologies describe a time-less Universe without a well-defined beginning, in which the emergence of humans and their forms of cultural expression happens through a forest of tales unrelated in time. Over the last 3000 years, many of the characteristics that distinguish Laurasian mythologies have been adopted and reformulated by some of the major world religions, especially Zoroastrianism, Judaism, Christianity, and Islam. The Abrahamic religions replaced Laurasian polytheism with a monotheist framework, but retained the linear time narrative of a Universe created by God and a final demise with the promise of a paradise. In all these narratives, the element of operative time of the human life became a metaphor for the element of operative time of the Universe.

The records left by the ancient civilisations enable us to retrace the evolution and increasing complexity of the concepts of life and afterlife. Egyptians were fascinated by the enigma of the end of life from the very beginnings of their civilisation, and they constantly sought to reinterpret and revere it. This essential concern can be seen in all aspects of social, political, and religious life and, above all, in the funeral rituals and in their tombs, which they called "Houses of Eternity". The latter ranged from the simplest mastaba to the Pyramid of Khufu. According to Hecataeus of Abdera, a Pyrrhonist philosopher who was active in the 4th century B.C., the Egyptians gave the time they spent living very low value and called the houses for the living "dosshouses". They placed "the most value on the time after death, during which the memory of virtue will preserve them" (Assmann 1984). Especially in the Old Kingdom, the Houses of Eternity and the temples were usually built in stone or were cut in the rock, while the houses and palaces of the Pharaohs were built in air-dried loam bricks.

The Dharmic religions of the Indian subcontinent developed more complex concepts of life and the afterlife by introducing a cyclic and a cosmic dimension, where the identity or soul of each human being, of each animal or plant, of each stone, river, or mountain is equally valued. The Abrahamic religions are structured in belief systems that encourage humans to follow the right path towards salvation along a linear and unrepeatable time, while in Dharmic religions, humans are invited to accept a much more complex and contradictory world, where life is dominated by continuous cycles of birth, life, death, and rebirth until one reaches some form of deliverance.

Faced with the impossibility of achieving immortality, humans, after the beginning of the triumphal development of modern science and technology that started about 1820, have become increasingly accustomed to rely on both to extend healthy life expectancy as much as possible. Absolute immortality goes beyond the framework of biological life as we know it on Earth. Conditional immortality or ammortality, a situation in which the organism's lifetime is not limited by ageing or by disease, but can potentially terminate due to an unavoidable external cause, is compatible with biology. Most humans strive to reach amortality but they reserve the concept of eternity for the realm of religion. Eternity may be the quality of that which is outside time (timelessness) or the property of lasting forever (everlastingness). The latter meaning implies an infinite time and is therefore unrelated to physical time.

Unicellular organisms have a form of conditional immortality because the original organism reproduces through cell division to produce clones that only stop living if they are destroyed by an accidental external factor. The new cells produced in binary fission are not exact copies of the parent cell because the division process generates malformations, which accumulate in the proteins. However, the situation is very different in multicellular organisms that are subject to a programmed aging process at cellular level or apoptosis, which eventually leads to death. The differentiation between germ cells and somatic cells is ultimately responsible for the ageing processes. The fact that the separation of these two types of cells takes place at the start of the embryo's growth ensures that genetic or regulatory modifications in somatic cells that occur during the development process do not have consequences for the cells involved in the sexual reproduction.

Sexual reproduction provides evolutionary advantages because it leads to a greater genetic diversity and therefore higher adaptive capacity in a continually changing environment. There is a trade-off between the much more advanced capacity for evolution and diversification of species provided by sexual reproduction, leading to an enormous range of increasingly complex organisms including intelligent beings, and ageing followed by death. It is a small price to pay for the remarkable human capacities, but also one that is deeply perturbing and conditions life.

We are still far from fully understanding the cellular, genetic, and molecular processes that lead to ageing, and, in particular, the reasons why multicellular organisms have quite different lifespans. Ageing is mostly the result of an accumulation of somatic damage caused by the organism's decreasing investment in maintaining and repairing DNA. Longevity is regulated by a range of genes that control maintenance and repair actions. Nature's strategy has been to allow this damage to accumulate without being duly repaired, which inexorably leads to the decline and death of the organism. On the other hand, it has invested in the renewed vigour and adaptation capacity brought by descendants generated through sexual reproduction.

But nature's pursuit of survival is unbounded. Some multicellular organisms have devised ways to circumvent apoptosis to keep alive. The jellyfish *Turritopsis dohrnii* and some other cnidarian species reproduce sexually through fertilisation of eggs in the marine aquatic environment to generate larvae that swim freely and then attach themselves to the seabed and produce a colony of polyps that eventually detach themselves to become adult jellyfish. However, if the jellyfish faces adverse environmental factors or senescence, it can invert its development process and return to the juvenile benthic polyp stage. The jellyfish is capable of ontogeny reversal using a process of cellular transdifferentiation to return to its juvenile form, a cycle which, by being repeated indefinitely, delivers conditional immortality.

Humans are unable to use inverted metamorphosis but try to prolong their lives with the help of medicine and the most advanced science and technology. This situation coupled with the weakening belief that human life is an unalienable gift from God has changed the way people deal with death, making euthanasia increasingly popular, especially in countries with advanced economies.

8.2 Origins of *Homo Sapiens* Time and the Acceleration of Time

The emergence of the *Homo* genus, 3 to 2 million years BP, and the emergence of *Homo sapiens* much later, 3 to 2 hundred thousand years BP, is a perennial subject of fascination. What circumstances led to it in the context of the Darwinian evolution of primates? What are the essential characteristics of our biological species? Does unravelling our deep past and inquiring about its main drivers tells us something about our future? Is there an essential coherence within *Homo sapiens* time?

A first crucial aspect of our evolution is that primates found their adaptive niche in trees. Statistical methods applied to a model of speciation and molecular-clock studies indicate that primates started to diverge from other mammals around 81.5 million years BP in the Late Cretaceous. The oldest know fossil of a primate lived more recently, about 55 million years BP, and was an early member of the tarsier lineage named *Archicebus achilles* (Ni et al. 2013). It was a very small tree-dwelling animal weighing about 20–30 grams that lived in the tropical forests of China.

The early primates had three major characteristics that resulted from their adaptation to arboreal living: hands and feet adapted to life in the trees, with nails instead of claws; front-facing stereoscopic vision in which the images provided by each eye overlap, resulting in the capacity for depth perception; and a higher level of encephalisation, the relationship between brain mass and the animal's body mass, compared to other families of animals. These characteristics were the result of adaptation to the increased diversity of forests of angiosperms during the Upper Cretaceous and interestingly the latter would later became critical for the *Homo* genus. Front-facing vision made it easier to locate and consume insects and small plant elements, particularly the fruit which was becoming abundant and diverse at the time.

Angiosperms, plants with flowers and fruit, emerged about 160 million years BP, and dispersed quickly across continents, overtaking forests of conifers, dominating global flora and becoming a great evolutionary conquest. Their flowers have the crucial advantage of allowing animals to participate and facilitate the reproduction process, the success of which is the central aim of Darwinian evolution. Thus the evolution of our primate ancestors is inextricably linked with flowering trees.

The second crucial development in the history of our prehistoric ancestors is the long-term global climate change initiated after the Eocene Climatic Optimum, 54–48 million years BP, when the mean global temperature of the atmosphere peaked at values about 9–14 °C higher than today. There were no polar ice caps and the atmospheric CO_2 concentration was likely higher than 1000 ppmv. After that time, the Earth's climate started to cool, changing from a greenhouse to an icehouse in about 50 million years. Instead of being continuous, the cooling was quite complex, showing various oscillations on time-scales of a few million years.

The cooling trend had its origin in the lithosphere through a reduction in magmatic activity and in tectonic plate movements that reduced the global heat transport by ocean currents from the equatorial regions to the high latitudes. Large ice sheets started to form in Antarctica at about 34 million years BP but only about 3 million years BP in the Arctic. All living species were forced to adapt to these changes, which implied significant and long-lasting impacts on their evolution strategies. Some evolved through an adaptive radiation where organisms diversify rapidly to take advantage of the new environmental niches created by the changing climate. An interesting example is the radiation of mammals during the Eocene Climatic Optimum into two new orders: Artiodactyla, which includes the pig, oxen, sheep, and goats, and Perissodactyla, which includes the horse and donkey, all animals that were much later on domesticated by humans. Plants adapted to the colder and drier climate by the development of grasslands dominated by Gramineae at the expense of a reduction of forested areas.

The increasing seasonality and loss of forest habitats resulting from the cooling climate implied that, at the end of the Miocene, about 10 million years BP, apes became extinct in Eurasia and restricted to Africa. The first fossils of bipedal hominins that have been found in Africa are from the late Miocene, about 7 million BP, while during the Pliocene, hominins experienced an adaptive radiation. At the end of the Pliocene, about 3 million years BP, the global climate suffered a new transition becoming cooler, drier, and more variable, progressively dominated by the Milankovitch cycles (see Fig. 3.4), as a result of the formation of ice sheets in the northern hemisphere. The emergence of the *Homo* genus occurs precisely at this time when the hominins had to adapt to a more adverse and challenging climate, which was responsible for replacing large swathes of forest by grasslands and savannahs, with open spaces where it was difficult to compete in the search for food. This adaptation is very likely to have benefited from the intensive encephalisation that is one of the most distinctive morphological features in the evolution of the *Homo* genus. This implies that global climate change played an important role in the emergence of the *Homo* genus and consequently of *Homo sapiens*. Adaptation to the arrival of a more adverse climate benefited from the enhanced complexity of social intragroup and intergroup relations and strategies.

Fossil records indicate that *Homo sapiens* emerged in Africa and until recently it was thought that they appeared in East Africa around 200 000 years ago. However, fossils of individuals morphologically very close to *Homo sapiens*, dated to 315 000 years BP, were discovered in Morocco, North Africa, in 2017. It is more likely that present-day *Homo sapiens* results from interbreeding between human groups across Africa that emerged in different locations and were linked by migrations. Interestingly, 315 000 years BP was a time close to the third interglacial period before the present one, the Holocene, when the global climate was warmer and wetter and the Sahara had rivers, lakes, and savannah. It is very likely that global climate changes associated with the approximately 100 000 year cycle of glacial and interglacial periods influenced human dispersal and evolution in Africa and elsewhere. Studies based on the analysis of maternal L0 mitochondrial DNA have indicated that the primordial population of *Homo sapiens* had its origin about 200 000 years BP in southern Africa in a region that is now northern Botswana and migrated during the last interglacial period 130 000–110 000 years BP because the increased humidity opened green corridors to the northeast and to the southeast (Chan et al. 2019). There is much that remains unknown about the origins of *Homo sapiens* and about the influence the climate had on its evolution, migrations, and global dispersal. One of the biggest challenges is to correlate the global climate change data with terrestrial records of climate change at the local and regional level where hominin fossils are found.

In the Holocene, and particularly during the historical period, it is less difficult to reconstruct the effect of climate changes on human activities and development. Africa is a revealing example because of the records left by the Egyptian civilization. It is well known that the intensity of the monsoon in Africa varies with a 26 000 year cycle associated with the precession of the Earth's axis. Due to this motion, the monsoon was intense in the period 11 000–6000 years BP and most of the Sahara desert was transformed into a savannah. However, 5500–6000 years BP, the wet period ended and the Sahara returned to being a desert in just 1000 years. This climate change was likely to have caused the migration of people that lived in the Sahara to the Nile valley, where they would later develop the Egyptian civilisation. The same climate event affected other regions and contributed to the collapse of the Ubaid culture, the first stage of the Sumer civilisation, which lasted from c. 8500 to 5800 years BP in Mesopotamia and was replaced by the Uruk culture, a more evolved stage of the same civilisation.

Later, around 4200 years BP, a new climate event, which has little known origins, produced droughts in various regions of the world with impacts that are relatively well documented since various civilizations had already appeared. In Egypt, around the year 2150 BC, and for two or three decades, the Nile floods fell drastically, sands invaded part of the river valley, the Faiyum Oasis dried up, the soils of the delta deteriorated, and famine spread throughout the country, paralysing the political institutions and sowing chaos. After some time, the reaction to the deep social and economic crisis was the emergence of a new political vision characterised by greater sensitivity to social issues, mercy, and compassion. It was probably one of the first instances in history where a strong government based on a highly centralised hierarchy, adopted, albeit in an embryonic way, social concepts of equality that involved the pharaoh protecting the weakest and poorest in society, especially in times of adversity. The droughts that occurred about 4200 years BP also contributed to the fall of the Akkadian Empire in Mesopotamia, the decline of the Harappan civilisation in the Indus valley, the disappearance of the Liangzhu Culture in China, and the emergence of the Motillas Culture in Central Spain. There are many other examples of climate influencing the rise and fall of civilizations, such as the end of the Classic Maya in Central America between 750 and 900 AD. In all the above cases there were certainly various non-physical factors that contributed to the civilizational upheavals, but climate was surely one of the drivers.

A crucial process in *Homo sapiens* time was the development of symbolic representation. Symbolic representation in the form of writing, visual arts, and all types of digital visual communication is so much taken for granted nowadays that we tend to forget the time when humans had yet to create it. Symbolic representation is one of the most important building processes of operative social time and especially of multigenerational operative social time because of its capacity to promote intergenerational communication. The oldest consistent records that we have of symbolic representation are engraved ochre stones and an engraved bone dated between 100 000 and 70 000 years BP, found in the Blombos Cave in the southern tip of South Africa, near Cape Agulhas. The collection of human artefacts found in the cave is very rich and includes also beads made from shells, bone instruments, and stone tools that are very advanced for the time. All these items are probably the first visible signs of symbolic thought and a proxy for the emergence of the capacity for spoken language in humans. The next step in symbolic representation is the representation of figures of therianthropes and animals interacting, dating from the Upper Palaeolithic, 43 900 years BP, found in a cave in the island of Sulawesi in Indonesia (Aubert et al. 2019), and the more recent figurative representations found in Eurasian caves dating from about 40 000 years BP onwards. We will never fully understand the symbolic meaning of the parietal art that began to flourish in Europe, from the Urals to the Iberian Peninsula, about 37 000 years BP. Nevertheless, we understand that it had a social purpose and represented a world view that was transmitted to successive social generations over about 22 000 years, shaping a multigenerational operative social time identifiable by that cultural canon. During those 22 000 years, there were evolutions in stone tool technology, hunting and fishing implements, adornments, and probably in many other human activities, but the essential characteristics of rock art remained the same.

Later, time began to accelerate with the Agricultural Revolution which began in five independent centres of domestication in the early Holocene epoch—the Near East, China, southwestern Mexico, northwestern South America, and southwestern Amazonia (Lombardo et al. 2020)—and operative social time changed profoundly. With the development of agriculture based on the domestication and use of some plants and animals, food surpluses were generated, part of the population was freed from strictly agricultural activity, and a social process of division of labour began, which created an increasing specialisation of human activities, new occupations and professions, and, eventually, new forms of social stratification. Division of labour increased efficiency in the production of goods and services in the agricultural society, and it created more opportunities to develop new technologies. In turn, these technologies helped boost efficiency and diversity in the production of goods and services. Such processes, many of which were amplified by positive feedback, created surpluses of different types—not only food—and therefore greater wealth that was distributed in society according to the relative power of the various social actors and groups, creating further inequalities.

The Agricultural Revolution represents the beginning of a process that led to the main ancient civilisations, in which the most diverse technologies were discovered, where mathematics, philosophy, and primordial forms of the natural and social sciences began to flourish. The abundance created by the Agricultural Revolution enabled the remarkable development of the arts: architecture, visual arts, theatre, dance, music, literature, and poetry. The same abundance eventually led to an increase in human population, to the militarisation of societies, to wars and large scale battles, and later on to widespread degradation of the environment and the overexploitation of natural resources. After the Industrial Revolution, the latter trends became more serious and climate change emerged.

Since the Agricultural Revolution, humanity has become accustomed to surpluses in the production of goods and has been able to use them to increase the power of a few powerful groups in society, but also to increase economic prosperity, well-being, and quality of life for a large part of the human population. All these changes have exacerbated the expression of some essential traits of human nature that have been present since the beginning of *Homo sapiens* time. Social stratification and the resulting social and economic inequalities were strongly present in hunter–gatherer societies, but the surpluses created by increased productivity in the Agricultural Revolution opened new corridors for the intensification and diversification of inequalities.

The Agricultural Revolution created the possibility of experiencing increasingly differentiated and luxurious lifestyles, new opportunities for social mobility, professional diversification, unlimited enrichment, rapid ascent to power and political power over increasingly larger populations, huge military victories, large land conquests, and lucrative pillaging. We may curb the impulses that drive us to indulge in individual or collective behaviours that are damaging to society, but we cannot suppress them. We will have to go on learning to live with them until the end of *Homo sapiens* time.

The Industrial Revolution that emerged in Great Britain in the 18th century immensely increased efficiency in the production of goods and services by mechanisation in the new factory system that gradually replaced the domestic system. This transition to the intensive use of machines was made possible by the ability to convert the chemical energy of coal into mechanical energy, provided by James Watt's steam engine, but it implied a much bigger energy consumption per capita. Between 1820 and 2017 the diversification and global expansion of this chain of processes meant that the world energy annual consumption per capita rose roughly from 21 to 77 GJ ($1 \text{ GJ} = 10^9$ J). The greater supply of goods and services induced population growth and increased consumption of goods and services, leading to a significant increase in GDP. During the same period 1820–2017, the world population increased by a factor of 7.1, and world GDP at constant prices and per capita increased by a factor of 14.2, although in a highly unequal way between different regions and countries. The success of the new development model depends critically on the availability of increasing supplies of energy and continuous economic growth.

The Industrial Revolution, modern science and technology, and the extraordinary growth of economies all over the world promoted by capitalism has produced a vast range of social, political, ethical, and cultural changes. Family relations, class structure, and society have been reformed, gender equality has become accessible and improved significantly, a large part of the rural population has moved to cities, leading to the global phenomenon of urbanisation, extreme poverty has been progressively eradicated across the globe, and mobility has increased, considerably facilitating contact between people with different cultural, political, and religious backgrounds. Migration, segregation, and integration issues have become increasingly important. A growing number of countries have adopted democratic forms of government, human rights have been proclaimed and applied to a growing part of humankind. Democracy, economics, and demographics are core issues now and especially for the future. Natural resources, environment, climate change, and sustainability have become emerging issues, embarrassing and often controversial.

One of the prices that humankind had to pay to benefit from the successes of the Industrial Revolution was the overwhelming increase in the power of time. The prototypes of current operative social time structures were the assembly lines of the first factories of the Industrial Revolution. Division of labour in production forced the introduction of strict working hours that all workers had to follow. If some did not comply with them, it had a multiplicative effect on the assembly line and reduced production per unit of time, i.e., it reduced labour productivity, leading down the line to reduced profits. First factories and then schools, hospitals, public administration, and trade all adopted working hours that had to be followed rigorously. The town itself began to operate as a metaphor for the public clock installed in the church bell tower or in the clock tower. An ever increasing number of people adopted the belief that one should not "waste time", but instead make full use of every minute or else some irretrievable opportunity might be lost.

As the Industrial Revolution progressed people were immersed in an increasing density of experiences, perceptions, feelings, actions, activities, commitments, deadlines, events, and all sorts of changes per unit time, which amounts to an acceleration of operative social time. In other words, the tempo of life has been ever increasing since then, except in a few very abnormal periods, such as in the confinement period of the 2020 COVID-19 pandemic. This acceleration is closely linked to increases in speed in the most diverse activities: in human performance, in all lines of business and professional activity, in all three economic sectors—extracting raw materials, processing them into products, and providing goods and services –, in the communication of all kinds of news through news media and social media, in the access to information and knowledge, and in all forms of transportation and travel. On longer temporal and spatial scales, it relates to a faster pace of social, cultural, political, economic, and technological changes, and finally to the acceleration of historical time. In the past, the time scale for decisive societal changes was generally longer than human lives, so adaptation to change often took place over several generations. Currently, the time scale for change is frequently shorter than a human lifetime, so this requires training for constant adaptation throughout one's life.

Time plays a crucial role in the way humankind deals with the highly disruptive impacts that the contemporary financial and economic system, based as it is on ever increasing consumerism, has on the environment and on natural resources. The events, activities, experiences, and expectations experienced, shared, and communicated in the social context, which structure one's operative social time, are mostly located within the element of operative time of one's own life. The central role played by this element of operative time, our own lifetime, is in fact a form of time discounting and is characterised by the personally adopted rate of time discount. What is thought will happen to human societies and to the environment beyond each of our own lives, whether they are expectations, forecasts, or projections based on scenarios, has a much lower value than what is expected or imagined will happen during our lives. In the latter case, events and situations have a direct personal value, whereas in the former they have only an indirect value based on different forms of family and social solidarity, involving our descendants, other members of the family, friends, and people who are from the same area or nationality, and also on solidarity with humankind considered as a whole, in some infrequent cases. Naturally, all forms of solidarity tend to disappear as we imagine social generations in a progressively more distant future.

The problem is that the biogeophysical subsystems of the Earth system where anthropogenic interference is creating human risks have a response time that is much longer than the human life expectancy. From a human point of view, it is very different if we can foresee that a risk will manifest itself severely in the coming hours, days, and months, or in the next few decades or centuries. In the former case, we are faced with a risk associated to an event that, if it takes place, is very likely to affect us and our family. In this situation, the response to the risk is driven by an instinct for survival and to protect our lives and the lives of those who are close to us and that we especially care for. We share our concern with all those who are subjected to the same risk and we are willing to help them. The situation changes when the risk only becomes severe in the long term, for example, in time horizons of 50, 100, or more years, as is the case for very severe anthropogenic climate change. In this case, the harmful impacts make relatively slow progress, leading people to believe that they are unlikely to significantly affect their own lives, well-being, or economic prosperity. However, future generations, including our descendants, will very probably be more severely affected. The justification for acting in the present to reduce future risk in the long term is no longer our own survival, but solidarity between generations, which is an ethical and moral issue.

8.3 Science, Progress, and Economic Growth

The second half of the 18th century, when the Industrial Revolution was beginning to emerge, was a very productive time from the conceptual point of view. Jacques Turgot and Nicolas de Condorcet, both of whom were Encyclopaedists, were the first to analyse and explicitly advocate the idea of progress. In 1750, in his *Discours sur le progrès successifs de l'esprit humain*, Turgot said: "The total mass of the human race, by alternating between calm and agitation, good and bad, marches always, however slowly, towards greater perfection." He acknowledged that progress in the sense of human perfectibility is not continuous, or the same for all, but was statistical in nature, as we would say today. Turgot had a linear concept of historical time in which the human race, like an individual, advances from its infancy towards greater perfection, opposed to the cyclical "time of nature".

Condorcet, a protégé and close friend of Turgot, was the main promoter of the ideology of progress in his famous *L'esquisse d'un tableau historique des progrès de l'esprit humain*, written under dramatic circumstances, precisely when history started to seriously question the hopes of the Encyclopaedists. The book constitutes a formulation of the ideology of progress within the framework of an historical analysis and reveals Condorcet's unbreakable belief in the perfectibility of man. For Condorcet, human perfectibility could always be surpassed and had only physical limits. In fact, he wrote that "the perfectibility of man is absolutely indefinite; that the progress of this perfectibility, henceforth above the control of every power that would impede it, has no other limit than the duration of the globe upon which nature has placed us". Condorcet's belief was that progress could be planned and secured across the whole world by using reason, developing the sciences, perfecting moral ideas, and implementing human rights. In fact, more than 200 years later, a large segment of humankind in various regions of the world still live in a mental framework comparable to those of pre-Enlightenment times. Claude Lévi-Strauss, the French anthropologist, who worked among indigenous groups in Brazil, noted that the word "anthropology" should be changed to "entropology"—the study of the homogenization of human life around the world. In this sense, Condorcet's plan for progress is the ultimate and definitive step in entropology.

Louis de Bonald, a contemporaneous critic of Condorcet, continued to defend the apocalyptic salvation announced by Catholicism and believed that science was usurping it, forgetting the brutal, everlasting realities of passion, conflict, and human violence. Thomas Malthus preferred to stress the idea that no matter what, misery was an integral and unstoppable part of societies, stating that "no possible form of society could prevent the almost constant action of misery upon a great part of humankind, if in a state of inequality, and upon all, if all were equal". David Hume was less categorical about progress, pointing out that the mutual dependence of political and intellectual development implied that progress would require political stability.

In spite of the criticisms and different points of view of many thinkers such as Kant, Schopenhauer, Marx, Engels, and Nietzsche, the ideology of progress based on the principles of the Enlightenment and liberated from religious constraints has been increasingly adopted and the belief that modern times are clearly superior to previous ones has been encouraged and widely accepted. People began to be convinced that historical time is changeable and that an evolution towards a better future driven by progress can be achieved. Before, historical time was relatively slow because horizons of expectation were very limited. The main expectation about the future was eschatological and it consisted of an apocalyptic prediction of the Last Judgement. The better future promised by progress came associated with an acceleration of time. Since the beginning of the 19th century, the acceleration of time has intensified, pushed forward by economic growth based on the growing supply of accessible energy and expanding commodity markets, greater scientific knowledge, and growing technological innovation.

In Judeo-Christian religious doctrines, historical time is linear, irreversible, and finite because there will be a final day after which the chosen will receive the eternal joy for which man was created. With the modern faith in progress, historical time has remained linear and is supposed to lead progressively to ever better times, but now with the condition that its irreversibility must be guaranteed by human actions. In other words, humans must actively and continuously engage in the generation of progress.

How do these two perspectives of historical time blend with the narrative of science? The first one is outside the narrative of science because it relies on a salvation action at the end of time by supernatural powers whose existence the scientific method has been unable to prove. However, those that believe in or practise the scientific method in their professions and accept the narrative of science may also believe in the historical time perspective offered by the Judeo-Christian doctrines or in historical time perspectives offered by other religions. The belief in God and in other supernatural beings continues to be strongly rooted either as faith or in a subconscious way, and plays a vital role in the minds of people from the contemporaneous social generation. Some consider that this belief is in contradiction with the methodology of science but contradictions are one of the most essential and powerful drivers of *Homo sapiens* time, although we don't like to acknowledge that we entertain them. Contradictions inhabit the human mind and may be a source of creativity. In classical logic, they must be eradicated, but in everyday life they constitute a vehicle to interpret the absurdity and nonsense that one often finds in the world and in one's personal experiences. Although permanent consistency is not part of our nature, contradictions are potentially dangerous.

Blending the second perspective of historical time with the science narrative should be straightforward because it originated in the recognition that science and technology coupled with the power of capitalism have the capacity to promote progress indefinitely by improving human health and well-being and raising economic prosperity. However, one finds contradictions once again. Before addressing them, one has to further analyse the concept of progress.

Progress manifests itself in different areas of human activity and takes different forms, including human, social, economic, political, ethical, and moral progress, as well as the progress of science and the progress of technology. These forms are always changing with historical time, but the arrow of operative social time has become firmly supported by the arrow of progress. Currently, the dominant standpoint is that among all the above forms of progress there is an essential and overarching one that conditions all the others, which is economic growth, usually measured by GDP growth, at national or global level. One of the most distinctive features of contemporary operative social time is the exclusive focus on economic growth as a measure of progress. Human, ethical, and moral progress is occasionally an interesting subject for debate, but of limited practical consequence.

It has frequently been pointed out that GDP is an economic indicator that does not reflect levels of poverty and social exclusion, wealth distribution, education, and other aspects of human and social development, environmental conservation, and questions regarding sustainability, but attempts to agree and start using alternative indicators have so far failed to be put into practice. In the present economic and financial system, income distribution inequalities have been aggravated for decades by the deterioration of worker's bargaining power and by a rise in profits as a share of total income, with the rate of return on capital being higher than the rate of economic growth. Furthermore, with the emergence of the new technologies of the Fourth Industrial Revolution, inequality is worsening because capital is becoming a frequent substitute for all but very highly skilled labour. The last four decades have witnessed a decoupling between productivity growth and employment growth in countries with advanced economies, which implies that salaries have less capacity to redistribute wealth. These new developments have led to the expansion of the gig economy, to a larger precariat, and to a distribution of wealth that may become incompatible with social inclusion and democracy, meaning that they are contrary to human, social, and political progress.

A further perturbing aspect is that the model used to secure economic progress is perpetual economic growth, which science says will eventually become incompatible with maintaining the health of the Earth's environment, the sustainability of natural resources, and the capacity to recycle rejected material flows. The way out has been to formulate analytical models that ensure sustainable economic growth for the "intermediate run", meaning the next 50–60 years. The reason for this preference is very simply that the operative social time of the contemporary generation is focused on its own element of operative time. What happens after should gradually become the concern of the following social generations as has happened in the past. In fact, in spite of various warnings, in particular those of the Club of Rome that started in 1972, continuous economic growth has been achieved since the 18th century, with the exception of periods of crisis.

Science, nevertheless, has been very explicit in saying that we are now rapidly approaching quantitative planetary boundaries, which when crossed, increase the risk of generating large-scale abrupt or irreversible environmental changes that will be harmful for humankind as a whole, although in differentiated ways. In fact, some of these planetary boundaries have already been crossed, such as the increasing atmospheric concentration of greenhouse gases that controls climate change, the lost integrity of the biosphere and the uncontrolled perturbation of some biogeochemical flows. The hazardous and destructive effects of this global interference are already affecting communities around the world, especially in developing countries. Science is saying explicitly that the present situation is not a recurrent one that *Homo sapiens* has previously encountered in his time but a truly new and exceptional one. The Earth system is now operating in a no-analogue state as regards its dynamics and functioning, which means that some environmental indicators that characterize its state have moved outside the natural ranges of variation of at least the last half a million years.

In spite of the damaging effects on human life of environmental pollution and degradation, natural resource overexploitation, the incapacity to manage and recycle the flow of waste, climate change, and scientific warnings about the future dangers of our interference with the Earth system, there is no visible political will around the world to effectively adapt the current economic and financial system at the national and global levels to the need of achieving sustainability. Furthermore, in some countries, particularly in the USA's, conservative and libertarian circles, often associated with certain religious movements, there is an attempt to distinguish between "solid science" and "politicized science", the latter being the one that warns about the dangers of environmental degradation and especially about climate change. Politicized science should be rejected because it hinders economic growth. Science is no longer viewed as a provider of truthful and useful evidence to inform policy but as one of the many inputs into production which can in some cases hinder profitability. Profitability is the overriding objective of the multinational corporations that free-ride the market place, supported by their respective government institutions.

The modern perspective of historical time that owes so much to science and technology has become so powerful and presumptuous that it now tends to selectively reject the advice of science. This rejection, besides being a conceptual contradiction, is a dangerous reaction of negation. It is an ominous sign that to guarantee the progress that is supposed to secure the irreversibility of modern historical time, science must be selectively rejected, especially when it addresses the present and future consequences for humankind of the anthropogenic interference on the Earth system. The tendency to refuse selectively certain scientific results and advice also manifests itself in the rejection of evidence-based policies and in spurning experts, especially when their advice is not aligned with the professed world view. Very often such world views are constructed in a process of negation or systematic misunderstanding of contemporaneous problems, a process that can be fuelled by a cultural reaction of disenchantment and anger, by economic dislocation, and by a nostalgia for the past. The systematic discrediting of experts that form the elites creates an inexhaustible supply of charlatans and is leading to the decline of democracy.

The narrative of science says that humankind is already suffering the destructive effects of the perpetual economic growth model and advises that it is highly recommendable to adapt the model to the new challenges of sustainability if humankind wants historical time to remain linear and lead to better times. By refusing to accept the narrative of science, but at the same time being forced to acknowledge external changes that are impossible to fully understand without the help of science, leads to a dangerous flourishing of conspiracy theories and to political polarization.

Climate change provides a clear example of the risks of unsustainability. It has been estimated, using a data set of 174 countries over the period 1960–2014, that unmitigated climate change, represented by the RCP 8.5 scenario corresponding to a mean increase in average global temperature of 0.04 °C per year, will lead to reductions in global real GDP per capita of 0.8, 2.51, and 7.22% in 2030, 2050, and 2100, respectively (Kahn et al. 2019). Abiding by the Paris Agreement, represented by the RCP 2.6 scenario corresponding to a mean increase in average global temperature of 0.01 °C per year, will reduce the loss of GDP to 1.07% in 2100. Unmitigated climate change will have long-lasting adverse effects on most economic sectors, employment, and labour productivity, affecting all countries, although in differentiated ways.

A unilateral transition to a more sustainable model is unlikely, due to the extreme economic competition between countries, especially between the major economies. There is a lock-in situation regarding the present economic and financial system that is supported by the majority of the human population. The system has been very successful in convincing people that there is no other credible alternative model and that the present model is the only one that can provide well-being and economic prosperity. Economic growth will not necessarily provide further well-being and economic prosperity to a significant part of the global population, but humankind as a whole is stuck on a "treadmill" of production and consumption (Gould et al. 2004) that generates excessive environmental risks. Incremental changes to the business-as-usual models are being implemented in the framework of various national and international initiatives, in particular the UN SDGs, but the trends regarding the goals that involve natural systems continue to be negative. This failure is leading the UN SDG community and other international organizations to consider promoting transformational changes that would tackle the root causes of the present situation (Díaz et al. 2019). Another approach is to develop transition movements where communities seek to establish sustainable ecovillages and transition towns and territories. These communities have been successful in reaching their objectives at local level and constitute a good example, but the main challenge that remains is their scalability to large urban areas at national and global level.

The present status quo is likely to change slowly but progressively as successive social generations begin to feel the economic insecurity. Generation Z will mostly enter the work force at the beginning of the serious economic recession caused by the COVID-19 pandemic which began in 2020, after having already witnessed the negative impacts of the financial and economic crisis of 2008–2009. An analogous situation looks likely to be repeated for generation Alpha, the demographic cohort succeeding generation Z, with birth years in the period between the early 2010s to the mid-2020s. Generation Alpha will begin life with a greater uncertainty about the future than did generation Z or generation Y at their age. Generation Y found themselves in a worse situation than generation X as regards improving their economic prosperity relative to their parents, although they had on average a better education and better health.

8.4   Technology and Democracy

There is another crucial player in the debate about progress, which is technology. Technology is much older than science and precedes *Homo sapiens* time. As already mentioned, there is significant consensus that the emergence of the *Homo* genus was very likely a form of adaptation to a drier and colder climate that spread more savannahs grasslands in some regions of Africa. This event happened at the time when the production of sharp-edged stone tools of the Oldowan culture started around 2.6 million BP. These earliest tools, which were discovered in 1964 in association with fossils looking more likely *Homo* than *Australopithecus*, reveal that those who produced them, had a good knowledge of stone-fracture mechanics and that they knew how to extract flakes from the cores of particular stones. They had learned a specific lithic technology that was going to evolve extraordinarily through the Palaeolithic, Mesolithic, and Neolithic periods, the latter beginning just 12 000 years BP. More recently, in 2015, a discovery was made near Lake Turkana in Kenya of simpler 3.3 million-year-old stone artefacts in spatiotemporal association with hominin fossils (Harmand et al. 2015). Surely, many more discoveries will be made in the future that will provide further light on the interconnectedness between the origin of the *Homo* genus time, the first usages of technology, and the repercussions on the African climate of a global climate change. However, it is fair to say that primitive forms of technology have been closely linked with the *Homo* genus from its first existence. The only remaining species of the *Homo* genus now firmly believes that technology will always lighten up his future and rescue him from his adventures and excesses.

Technology progressed steadily but relatively slowly from the beginning of the Neolithic up to the 16th century, with major inventions such as irrigation in Mesopotamia about 6000 BC, metallurgy, sailing, paper production, gunpowder, windmills, the mechanical clock, the oceanic carrack, and printing in 1455. Then came the Renaissance, which created the conditions at the end of the 16th century for the emergence of what we now call science, although at the time it was called natural philosophy. *Scientia* originally meant just knowledge and until the eighteenth century included also theology. Some of the most important contributors to the process that led to the establishment of the methodology of science up to the end of the 17th century were Leonardo da Vinci, Nicolaus Copernicus, Andreas Vesalius, Tycho Brahe, Francis Bacon, Galileo Galilei, Johannes Kepler, René Descartes, and Isaac Newton. Science, however, was practised away from society in the seclusion of observatories, laboratories, libraries, and academies, and therefore remained rather mysterious to society as a whole. The public spectacle of science began in the 18th century, in the Age of Enlightenment in the streets of several European cities, especially London and Paris. Inventors, science communicators, and enthusiasts offered demonstrations and experiments in the streets, squares, markets, and cafés that amazed an audience eager to experience new sensations, enjoy themselves, and be surprised and dazzled. There were long queues of people waiting to take part in Abbé Jean-Antoine Nollet's experiment, where they could experience an electric discharge, which was called a *sensation scientifique*.

The invention of the steam engine, which played a crucial role in the development of the Industrial Revolution, was made by Thomas Newcomen, an inventor, and perfected in the period 1763–1775 by James Watt, an inventor and instrument maker who worked as a technician in the laboratories of the University of Glasgow. So the invention did not arise from an application of what we now call scientific knowledge and information but by the successful exercise of practical mechanical skills. Only later, in 1824, did the French mechanical engineer Sidi Carnot's (1796–1832) theoretical discussion on the efficiency of an idealized steam engine initiate the development of the science of classical thermodynamics. The invention of the steam engine was what we would call today an outstanding technological innovation, but when it occurred the word "technology" was very uncommon in Great Britain. Instead, people referred to "technics" for the useful arts as an antonym to the performing and fine arts. Only after 1820 did the word "technology" come into more frequent use, acquiring its present meaning in the second half of the 20th century, especially with the fast development of information and communications technologies (ICT).

The First Industrial Revolution was mainly characterized by the beginning of the intensive use of coal, the invention of the steam engine and other powerful new machines, such as the spinning jenny used in textile mills and the cotton engine, and the development of new technologies such as a way to make puddled iron with better properties than pig iron and the new road-building technology developed by John McAdam. The Second Industrial Revolution, which started in the latter half of the 19th century and lasted up to the beginning of the Digital Revolution, was mostly based on the use of electricity and made it possible to develop production lines in factories. It led to successive outbursts of remarkable inventions, such as the telegraph, photography, telephones, water turbines, phonographs, electric light bulbs, cinema, electric motors, dishwashers, aircraft, photoelectric cells, plastics, washing machines, television, refrigerators, antibiotics, radar, photocopiers, satellites, lasers, microwave ovens, nuclear power plants, and many others. Many of these inventions were based on modern science. It became increasingly clear that the synergy between science and technology was very creative and could be used to improve human well-being and economic prosperity.

It can also be used or developed specifically for the exercise of the darker impulses of human nature, such as violent aggression, war, human oppression and persecution, and the promotion of vice. In the military domain, the main goal of applying science and technology is simply to produce new weapons more deadly and destructive than those that one's adversaries are able to build. In the civil domain the success of a technology depends on its capacity to render people addicted to the products it helps to produce. The range of products is unlimited, from smartphones, tablets, and SUVs to semisynthetic opioids, which have been responsible for tens of thousands of fatal overdoses annually in the USA in recent years. The addictive power of technologies is enhanced by using the most advanced techniques and the most advanced science in various domains, for instance frontier knowledge in human psychology. The countries that first benefited from the Second Industrial Revolution witnessed a profound change in their way of life and a general improvement in the well-being of their citizens. Then the new economy and the wave of new lifestyles propagated progressively around the world. Intellectuals, writers, and philosophers tried to imagine how the world would be once it had been utterly transformed by the overwhelming power of scientific and technological progress. One of the more interesting insights is from Herbert George Wells' non-fiction bestseller *Anticipations of the Reaction of Mechanical and Scientific Progress upon Human Life and Thought*, published in 1901. According to Wells, nation-states would disappear and be substituted by a great federation of English-speaking peoples using English as a universal language and guided by the New Republic, an educated class of unprecedented people who would influence and control the apparatus of their ostensible governments. The New Republic is reminiscent of the contemporary economic and financial elites who run the large multinational corporations and share a disproportionate part of the world's economic, financial, and political power. Wells was also a visionary when he wrote that "the peasant of today will be represented tomorrow by the people of no account whatever, the classes of extinction, the People of the Abyss", whom we now recognize in the precariat and the working poor, left behind by the inexorable Fourth Industrial Revolution.

Nation-states have not disappeared as Wells forecasted about 120 years ago, but the English-speaking world, comprising Great Britain, together with its former colonies and the imperial possessions of the Victorian era British Empire, and the USA, a former colony from the pre-Victorian era, is now the dominating economic and military power structure in the world. These countries do not constitute a unified nation or a federation, but they are a rather loose family that know well how to take advantage of their common language and pragmatically unite when the circumstances so require. The English-speaking peoples have been guided by the ideas of Adam Smith and uncountable followers, by the success of the First Industrial Revolution, by the early firm belief in science and technology, by utilitarianism, by the systematic adherence to parliamentary democracy, by a pragmatic rule of the Empire summarized in the motto under which the British army went to war in 1914—For God, King, and Country –, and by US exceptionalism and the inevitability of its superpower status. Their contribution to contemporary civilization is undeniable and in certain specific ways they represented the civilizational values of the West and have often assumed its leadership.

However, the world is fast transforming and there are many pressing old and new challenges that should be urgently addressed and require new solutions. The question that one may ask is whether the English-speaking world will be able and willing to lead to a safer and more sustainable world future or whether it will just be part of the problem. Various indicators point to a decline of US influence in the world. The same applies to Great Britain, which is currently dominated by an ideological impulse to move away from Europe and build a stronger English-speaking world with the USA, Australia, Canada, New Zealand, India, Malaysia, Singapore, and various countries in Africa. A further crucial issue is to identify the possible alternatives. Surely, there will be no power vacuum. There is fierce competition from China and other emergent economies to gain increasing prominence and economic power. The European Union spends most of its energy consolidating itself, and although it constitutes a remarkable achievement, especially as regards peace, human rights, freedom of movement and residence for its citizens, scientific and cultural development, environmental protection, and an ambitious pioneering world level program of science-based sustainability, it has no superpower ambitions.

China is an old civilisation that started with the Shang dynasty in the 16th century BC, and is very proud of its identity and achievements. The country went through a painful period of decline during the Qing dynasty caused in part by the colonialism and imperialism of the Western powers and later by the intervention of Russia and Japan. That period ended with what the Chinese call the century of humiliation from 1839 to 1949. On 1 October 2019, on a high balcony overlooking Tiananmen Square, President Xi Jinping said: "On this spot, seventy years ago, Comrade Mao Zedong solemnly declared to the world the establishment of the People's Republic of China. That great event thoroughly transformed China's tragic fate, ending more than a century of poverty, weakness, and bullying." There is no doubt that China wants to be the world's strongest economic power and to surpass the West precisely in the domains that created its unsurpassable power—science and technology, the ideology of progress, and the current economic and financial system. Its primary goal is to copy and adopt the West's flawed paradigm of perpetual economic growth. In this process it will go its own way, since China has a different culture and different values from the West, which will endure for a long time yet.

China adopted an authoritarian political system based on the communist party and wants to prove that its system is superior to Western democracies, especially as regards achieving robust economic growth and increasing the economic prosperity of its citizens. Thus China pursues the objective of patiently extending its model all over the world, just as the West used to do so vigorously in the past. Democratic values, in particular political freedom of expression and dissent, protection of minorities, diversity, and popular sovereignty are ruthlessly suppressed in China. As long as the vast majority of its citizens are satisfied, finding that their expectations of well-being, increasing economic prosperity, and growing consumerism are being fulfilled, they may consider political freedoms to be a secondary issue that may be forgotten. The open-ended question is which of the two political regimes, democracy or the Chinese authoritarian regime, will be the most resilient to the recurrent crises that are occurring as a result of an unsustainable global economic and financial system. A more fundamental question is whether democracy will be able to adapt to the powerful forces that are trying to perpetuate the present system or whether it is bound to decline under such pressures. Science plays a crucial role in the elucidation of these questions and on the outcome of these ongoing processes. All political regimes that selectively negate science are likely to be less resilient to crisis. Moreover, democracies that selectively negate science tend to decline and eventually adopt authoritarian forms of government. Those political regimes that strive to maintain a thriving democracy are more likely to accept the advice of science, adopt evidence-based policies, and make a future transition to a sustainable economic and financial system. Science should be sufficient to convince humankind that it needs to reverse its course of action. If it turns out that it isn't, humankind will learn the hard way how to reach sustainability. Western values will decline and authoritarian and dehumanised values will take their place, while dystopian societies will proliferate.

Various writers in the first half of the 20th century, particularly the Russian writer Yevgeny Zamyatin and the English writers Herbert Wells, Aldous Huxley and George Orwell, used dystopian novels to explore the consequences of totally ordered and controlled utopian societies, coupled to or more often created by the power of technology, which became increasingly autonomous and ruled the course of history. They were probably motivated by the hope of a change in direction and just tried to convey their concerns in an attractive way to a wide audience. Later in 1959, Martin Heidegger recognized the inevitability of the technological dominance of historical time when he wrote (Heidegger 1959):

> In all areas of his existence, man will be encircled ever more tightly by the forces of technology. These forces, which everywhere and every minute claim, enchain, drag along, press and impose upon man under the form of some technological contrivance or other – these forces [...] have moved long since beyond his will and have outgrown his capacity for decision.

Sixty years later Heidegger's forecast has been entirely vindicated, but societies, especially in the advanced economies, are still not entirely convinced about the supreme admirableness of the contemporaneous model. Dystopian literature has become increasingly popular in the past few decades, particularly since the financial and economic crisis of 2008–2009, and mostly among young people, creating a whole new business for movies and merchandise. It is difficult to fully understand this tendency, but it likely reveals the need to find analogies of today's operative social time in fictional narratives of various coherent forms of dystopia. The renewed success of such novels implies that they offer a model and possibly a foretaste for some of today's programmatic worldviews.

The Third Industrial Revolution began in the 1960s and was mostly based on the Digital Revolution and the rapid development of ICT. The extraordinary success of digitised information communication resulted from the development by Claude Shannon of a Boolean algebra of electronic circuits that could be used to transmit any type of information electronically—documents, books, images, or music. Information theory unified the issue of communication and storage of digitised information in an increasingly diverse range of devices and technological media. Today, one can access the news, social media, online lectures, courses and meetings, music, games, videos, and films in personal computers and mobile phones. All this bounty is instantaneously and constantly accessible and has become inseparable from billions of people's everyday lives. The internet platforms are now the map and the clock, the printing press and the typewriter, the filming and photographic camera, the calculator, telephone, sound recorder, music player, radio, and TV for billions of people. Since the 1980s, we have been witnessing a transition from the role of being a citizen to the role of being a digital citizen.

Humanity is now navigating in two parallel worlds, the physical world of the Earth system, where we live as a biological organism, and the virtual digital world of ICT, where the mind spends an increasing amount of time. The way towards the future is to integrate these parallel realms as far as possible with a wide variety of devices and structures located at different points in physical space and virtually mapped in digital space so that we can extend our presence and power over the physical world. The relationship between physical space and time in the real-time virtual digital world is regulated by the speed of light, making it almost time-coincident with the physical world in the Earth system and in the surrounding outer space where satellites operate.

ICT digital communication and media devices are changing social interactions and the processes and structure of mental activities. Face-to-face encounters are now frequently considered to be time consuming and inefficient for many objectives. Generation Z teenagers have been using video call applications, in particular Zoom, for years and call themselves "Zoomers". With the COVID-19 pandemic, hundreds of millions have become Zoomers by using various video platforms socially in high schools and universities and in all sorts of professional activities. Air travel will be reduced in the future, specifically for face-to-face meetings. One of the main outcomes of the COVID-19 pandemic is an increasing use of modern digitally based strategies in most fields of activity, especially in medicine.

The compulsive and irresistible use of digital devices and platforms is essentially driven by a desire for instant gratification as often and intensely as possible. One of the cognitive processes that is most affected is attentional control because the internet and social media platforms are specialized in providing information and many forms of entertainment in an unceasing atomized flow, leaving little or no time for concentration. The relatively long periods of attention fragmentation and scattering are transforming operative time structures around the world. Idleness, tedium, and solitude continue to exist but we may now have the feeling that they can be chased away by overloading time. Much more diverse experiences have been created. All were potentially possible and just needed technology's remarkable capacity for innovation to appear, flourish, and culturally transform us. Digital information devices have broadened human private and social abilities and experiences, in particular the capacity to communicate, stay informed, be entertained, be alienated, and be exploited. Each person's digital world has become a refuge in which virtual time and space are firmly under their control.

The Information Age and its ICT devices have produced a deep transformation in societies and in the political, economic, and financial systems of all countries, significantly contributing to the globalisation process. The new global digital world, and the connectivity that provides access to it, have made it possible to improve health conditions in many regions of the globe, prevent and better manage natural disasters, boost social inclusion and cohesion, improve access to education and professional training, promote access to information, knowledge, and active citizenship, and enable civil society movements and organisations in social, political, economic, and environmental fields.

At the same time, the intensive use of digital devices is likely to change our memory and thinking habits and maybe also the way we think. There are various research studies on the subject, emphasizing either positive or negative impacts, but there is a notorious lack of scientific consensus, and positions tend to become rather polarized and emotional. The most frequent conclusion is that the effort to understand what is happening to humans is futile because there's no going back from the progressive atomization of our operative social time by a constant flood of information. Probably, due to brain neuroplasticity, humans will become much better at the modes of thinking involving surfing, skimming, scanning, and multitasking, while some of the other modes will tend to be atrophied.

Digitalization opens up new opportunities to reach the UN's Sustainable Development Goals, but it can also generate risks for humankind. ICT, in cooperation with emergent technologies and various social sciences, such as big data, data science and artificial intelligence, psychology, political science, and marketing, have been improving methodologies and procedures to successfully use the internet and social media for abusive personal data collection and to disinform, misinform, manipulate, and deceive citizens. Authoritarian states excel in using such methods to recruit new supporters, undermine the democratic debate, and attack and black out their opponents. The same techniques, usually in more sophisticated form, are also used in democratic countries, particularly by foreign players and interests, to influence people's political opinions, actions, and voting intentions, using Facebook, Twitter, and other digital platforms. An example of these emerging techniques was provided by unacknowledged Russian government interference in the 2016 US presidential election through the Internet Research Agency, based in Saint Petersburg, and other institutions. The same digital means of communication are also used by corporate interests, especially in the USA, to make an assault on specific areas of science considered by some lobbies, corporations, news media, politicians, and opinion makers to be "politicised" science.

The internet is used today in almost all countries, democratic or non-democratic. China has the largest online community in the world and the percentage of people using the internet is around 60%. At the global scale, the political use of the internet is low compared with other uses, such as online music, films, games, and sports. In countries where there is no internet censorship, citizens are free to criticize their country's political institutions and leaders, exerting an increasing pressure for the adoption of reforms. On the other hand, in non-democratic countries, citizens that spend a large amount of time in the internet enjoying its social and entertainment content tend to be less concerned with living under authoritarian political conditions as long as the regime offers some measure of well-being and economic prosperity (Stoycheff et al. 2016). The internet's democratic potential can be boosted by building demand for internet freedom, but unregulated internet freedom is also used to attack democracy. The access that digital platforms have to hundreds of millions of people's personal data, travelling movements, and political, ideological, and economic preferences, as well as other information, is a potential threat to the future of democracy.

Digital platforms could be an instrument for actively promoting democracy, but instead they are mostly used for optimising both individual self-interests, in the form of online consumption of goods and services, and the profits and economic power of the digital technological corporations, start-ups, and entrepreneurs that support them. They are not prepared to promote human and social development and innovation. Initiatives focused on these objectives may appear in the digital platforms, but they remain outside their main consumer culture. Solidarity-based economic experiments that present themselves as alternatives to the current economic and financial system have great difficulty in penetrating the digital platforms and are not considered to be sufficiently attractive or are outright rejected.

Finally, the Fourth Industrial Revolution is an extension of the third but differs from it in terms of the new opportunities and plans for automation, interoperability, the use of intelligent technical support systems, decentralised decision-making, and information exchange in production technologies for goods and services. It is marked by the development of emerging technologies and the production of new products and services. In the Fourth Revolution, information technology, which automates business and office processes, will be coupled with operational technology in the internet of things so that industrial and operating processes in factories, infrastructures, and homes can be fully automated.

The main areas of activity and emerging technologies that will support the Fourth Revolution are ICT, artificial intelligence, robotics, the internet of things, big data, additive manufacturing, including 3D printing, rapid prototyping and direct digital manufacturing, blockchain, digital cryptocurrencies, autonomous vehicles, drone technologies, precision agriculture, nanotechnology, genetic engineering, synthetic biology, and geoengineering. The new 5G technology, the fifth generation in mobile network technology, is essential for the success of the Fourth Industrial Revolution since it is designed to connect virtually everyone and everything, including people, objects, devices, offices, businesses, markets, infrastructures, and factories. By operating at a much higher frequency than the four preceding generations, 5G has faster download speeds, reduced congestion, and lower latency. The main driver behind this accelerated program is the military, economic, and technological competition between the major world powers, especially between the USA and China.

Artificial intelligence (AI) will likely come in four waves, named by the American computer scientist Kai-Fu Lee as Internet AI, Business AI, Perception AI, and Autonomous AI. With the fourth wave, machines will finally become entirely autonomous. Some hope that these machines will eventually be transformed into human-like entities with selected powers superior to those of humans, an exploit that will give their creators the unique feeling of superiority that results from overshadowing humans. Others are fully convinced that science and the emergent technologies of the Fourth Industrial Revolution will eventually deliver the unlimited extension of healthy human life, or ammortality, which fulfils a dream that is central to the human condition. Finally, other researchers are using the same tools to achieve human enhancement, superintelligence, and transhumanism. The achievement of these plans is probably far away in time, but they are moved by powerful and unquenchable dreams.

Three hundred years after the beginning of the Enlightenment, humankind continues to be fascinated by the wonders of science and technology, just as thousands of Parisians watched in amazement a balloon with flammable air floating up into the air with Jacques Charles and Nicolas-Louis Robert aboard at the Jardin des Tuileries on 1 December 1783. Now we have the International Space Station, the promise of establishing a one million strong colony on Mars, the unbelievable progress promised by AI, the hope that synthetic biology will enhance humans and make them almost disease-free. Furthermore, they could be designed to reproduce the most popular and desirable stereotypes. This perpetual wonderment with technology has been a distinguishing characteristic of human time since the First Industrial Revolution. But besides producing feelings of awe, technology also is also useful and can solve pressing human problems such as ensuring the increase in agricultural productivity that would satisfy the food demands of more than ten billion people that are likely to inhabit the Earth by the end of the century.

AI surveillance technology is being deployed around the world to monitor and track citizens for security, economic, and political objectives, some of which are lawful while others violate human rights. Smart city and safe city platforms, smart policing, and facial recognition systems, are being fast developed in many countries, especially in China. It is estimated that in 2019 more than 75 countries out of 176 worldwide were actively using AI surveillance technologies. The social credit system that was initially promoted by private corporations in 2014 is now being used by the Chinese government to register all aspects of life to judge citizen's behaviour and political trustworthiness. If, for instance, someone writes about censorship and government corruption, they are likely to be blacklisted and lose some rights, such as the right to buy property, make a loan, or undertake certain kinds of travel, without any previous warning or official notification. The extension of this system so that the government can control the whole population represents a frightening dystopian future.

The COVID-19 pandemic has given the opportunity for the largest extension of state power over its citizens since the Second World War, through the application of the most advanced surveillance techniques to contain the disease. The success of China, South Korea, and Taiwan in controlling the spread of the virus faster than Western countries is largely due to the digital surveillance measures adopted, in particular the technologies of contact tracing. It is well-known that states increase their power in crises. The question is: will they give it up when the crisis ends? States' routine access to citizen's medical and electronic records may be a consequence of the 2020 pandemic. A potentially more dangerous result of the COVID-19 pandemic and similar crises is the risk of abuse of power by political leaders, which can seriously endanger democracy, as has happened in Hungary.

States also increase their economic power in violent crises. In the USA, Congress passed an initial $2 trillion stimulus package, representing 10% of GDP, to soften the economic blow of the COVID-19 pandemic for workers and businesses. The European Central Bank created a quantitative easing (QE) program worth 750 billion euros until the end of 2020 to buy government and corporate debt from countries across the Eurozone, representing 6% of its GDP. A large number of central banks at global level announced analogous measures to protect their economies. This generalized response underscores an underlying tendency for a transition from the capitalist market economy to an economy managed by the central banks, which has been apparent since the 2008–2009 financial and economic crisis. Between the two crises there was hope for normalization of global monetary policies but it is now likely that it will take some years to abandon very low interest rates and large scale QE. "QE infinity" is now a likely outcome. The danger is that, as happened in 2008, the liquidity is used to generate money through financial capitalism, in particular rewarding shareholders through stock-buyback schemes, instead of being directed toward good investment opportunities in the productive economy. The other danger is that while central banks must extend loans to businesses, this emergency happens at a time when the global private debt, which is the driving force behind the global debt, reached historical heights. According to a 2019 IMF report, global private debt has tripled since 1950 and the debt of the emerging economies has exceeded that of the advanced economies (IMF, 2019). Global debt has reached $184 trillion, which is equivalent to a global debt of $86 000 per capita, more than 2.5 times the mean income per capita. Global debt levels are now well above those at the time of the 2008 financial crash, which according to the IMF increases the risk of another crash.

The massive fiscal and monetary stimulus that governments and central banks have released to support their economies also shows that, in a time of crisis, the economic and financial system adopts Keynesian economics and temporarily puts aside the neoliberal roadmap. Countries that follow neoliberal economic policies, such as the USA, social democracies, particularly in Europe, and the single-party communist state in China, have all adopted the same type of large scale government spending policies. However, the prevailing system believes that robust economic growth must be rapidly reinstalled, and that can only be achieved by returning to strong neoliberal policies.

The COVID-19 pandemic confronts the emerging and developing economies with much more difficult and dramatic problems than the advanced economies. The former countries cannot replicate the solution of the latter by creating stimulus packages through massive QE. Furthermore, for hundreds of millions of people living in export-led emerging and developing economies, the loss of income due to the global economic crisis may mean severe poverty, famine, and forced migration. Advanced economies should show the utmost solidarity by announcing a stay on developing and emerging economies' debt service. In many of these countries, if their government chooses to continue paying their foreign creditors, that will likely mean starvation for many of their citizens.

The characteristics and inner-workings of the current economic and financial system make the situation of the emerging and developing economies particularly dramatic in a deep global crisis such as the one created by the COVID-19 pandemic. The G20 nations are already committed to a moratorium on $20 billion of the private and bilateral debt of some of those countries. Furthermore, the IMF and the World Bank, supported by the G20, have pledged grants to cover debt payments and plan to issue "pandemic bonds".

The problem is, however, that since the Bretton Woods agreements the dollar has become the world's reserve currency, which implies that the world's financial system is largely under the control of the Federal Reserve of the USA and functions as an instrument to secure the geopolitical supremacy of the USA. In the Bretton Woods Conference, John Maynard Keynes forcefully proposed the creation of an International Currency Union based on a new currency called bancor, but that was not accepted by the USA because it wanted to give disproportionate power to the Federal Reserve at the time of the Cold War. Times have changed and there are many proposals to end the centrality of the dollar in the global financial system, but they have all been unsuccessful.

Currently about 60% of all world's central bank reserves are held in US dollars and the Federal Reserve can discretionarily increase or decrease the flow of dollars. Furthermore, the superabundance of dollar-dominated debt in the world gives the Federal Reserve an inordinate power to control the world economy. Its first priority in the COVID-19 crisis is to keep the current economic and financial system in good health, which involves providing direct lending to overleveraged corporations that have benefitted from loans with low interest rates and quantitative easing since the last crisis in 2008–2009.

Meanwhile, indebted countries from emerging and developing economies receive only temporary loans that often come attached with conditions regarding their development model. The Federal Reserve chooses in accordance with its own criteria the countries for which it creates swap lines with their central banks, to make dollars less scarce. For instance, the economic and financial system favours those countries willing to sell off their drilling rights to the large multinational fossil fuel companies. Some of these countries, such as Mozambique, are perfectly aware that they will suffer the adverse climate change impacts resulting from their contribution to the continuing use of fossil fuels, but they desperately need to increase their debt-servicing revenue. The present crisis reveals the huge fragility and risk affecting most of the human population, particularly those living in the emerging and developing economies, and the need to implement a reform of the current economic and financial system based on the principles of human rights, justice, solidarity, and ethics. Without that kind of reform it will be increasingly difficult to avoid dramatic social and economic situations across the world that are likely to run out of control. David Beasley, the executive director of the United Nations World Food Program warned the Security Council on 21 April 2020 that: "We are not only facing a global health pandemic but also a global humanitarian catastrophe."

The COVID-19 crisis serves as a rehearsal of what is likely to happen as the environmental and climate change crises continue to unfold. The difference is that in these two crises there is no hope of finding a fix like a vaccine for the coronavirus that acts relatively rapidly. Effective responses are long overdue and whenever humankind decides to use them they will only be able to act slowly over many decades. The longer we wait to respond resolutely, the longer it will take to recover some sort of equilibrium with the environment.

8.5   Adapting to or Building the Future

Humankind is now following an unsustainable path encircled by five main types of strongly interrelated and dangerous crises: the COVID-19 pandemic, the debt crisis, the crisis that results from increasing socioeconomic inequalities at the global level, the environmental crisis, and the climate change crisis. The first is a sudden crisis that is progressing very rapidly and, at the time of writing, is still developing and not yet fully understood by science, while the other four are progressing in slow motion and are quite well understood by science. Time forces an immediate response to the first and has delayed effective solutions to the others, apparently indefinitely.

Many authors have warned that humankind's interference with the biosphere is forcing it into a global critical transition that would put ecosystem services at risk with dramatic consequences for human well-being and prosperity. The current species extinction rate is higher than those that caused the previous five mass extinctions and if it continues uncontrolled it may produce a comparable mass extinction. The percentage of threatened species for various groups are: amphibians 41%, selected reptiles 34%, reef-forming corals 33%, mammals 25%, sharks, rays, and chimaeras 30%, birds 13%, conifers 34%, and selected dicots 36%. Tropical forests are likely to be the biome where the anthropogenic mass extinction caused by systematic defor-estation over many decades will first be identified. To prevent a planetary-scale critical transition, the most important actions are (Barnosky et al. 2012):

> [...] global cooperation to stem current global-scale anthropogenic forcings. This will require reducing world population growth and per-capita resource use; rapidly increasing the propor-tion of the world's energy budget that is supplied by sources other than fossil fuels while also becoming more efficient in using fossil fuels when they provide the only option; increasing the efficiency of existing means of food production and distribution instead of converting new areas or relying on wild species to feed people; and enhancing efforts to manage as reservoirs of biodiversity and ecosystem services, both in the terrestrial and marine realms, the parts of Earth's surface that are not already dominated by humans.

There are many reasons why it is dangerous for humankind and also disgraceful that unlimited human greed, egoism, and hubris is destroying tropical forests, the biome that supported the emergence of our primate ancestors.

Climate change has been extensively addressed in the main text of this book, so it will be referred to here in a very succinct way. The climate change crisis results to a large extent from humankind's strong dependence on fossil fuels. Fossil fuel global energy consumption as a percentage of the total has oscillated between 78.72 and 94.63% in the period 1960–2018 and stayed systematically close to 80% in the period 1990–2018 (see Fig. 4.2) (WB, 2020a). The persistence of this high-level dependence over more than half a century in spite of the high visibility of the climate change issue, of the risk it represents for humankind, at present and especially in the future, and of the persistent call for action from countless scientists, engineers, economists, politicians, and activists the world over, clearly shows that the current economic and financial system is not interested in solving this problem. We know much more now about the science of climate change and about its present, middle, and long term impacts on the various socioeconomic sectors and biogeophysical subsystems than in the 1980s when climate change was first unequivocally observed across the globe and the IPCC was founded in 1988. Furthermore, the way to mitigate climate change so that it has a minimum impact on economic growth at national and global level is well known. On the other hand, it is well established that there will be a slowing down of the global economy if the business-as-usual emissions scenario continues to be followed. None of these arguments has had any noticeable influence on the attitude of the majority of the fossil fuel corporations and industry and on the governments that support them. On the contrary, they insist in prospecting more coal, oil, and natural gas deposits, often with the support of government subsidies, and to explore and sell as much fossil fuels as possible. The warning that by continuing prospecting for fossil fuel deposits, especially oil and natural gas, corporations risk creating a large amount of stranded assets and eventually a carbon bubble, is disregarded.

At the same time, the fossil fuel lobby has excelled in the art of creating doubts or fully denying that climate change is to a large extent caused by anthropogenic CO_2 emissions. Another frequent endeavour is to convince governments, politicians, and citizens that the continuous use of fossil fuels is the only way available to increase economic growth and economic prosperity and to reduce poverty, especially in the developing countries. This program of misinformation, disinformation, and manipulation made through publications and reports and on the social platforms, with the support of generous funding from the fossil fuel industry, has been very successful and has managed to confuse a significant fraction of the world's citizens, especially in the USA, which derives much of its economic and military power from oil and natural gas. Meanwhile some renewable energies have become cheaper and economically competitive with fossil fuels. A slow global energy transition from fossil fuels towards renewable energies has started and is progressing.

This transition does not mean that the extremely important contribution that fossil fuels gave to the Industrial Revolution and to its success in creating the contemporaneous civilization is not fully acknowledged. The point is that a harmful side-effect of burning fossil fuels has been identified, namely the emission of CO_2 in large quantities, which increases the atmospheric greenhouse effect and produces climate change. There are of course other anthropogenic emissions of greenhouse gases, but CO_2 contributes about 65% of the anthropogenic radiative forcing. So, in conclusion, it is time to modernise the global energy paradigm and start effectively decarbonising the world economy. Most of the large fossil fuel corporations, especially those that produce oil and natural gas, in cooperation with the governments, banks, and financial institutions that support them, do not accept this argument and they will make every possible effort to continue prospecting, exploring, and selling fossil fuels. The behaviour of the fossil fuel industry and of the associated governments is currently one of the best examples of rational egoism applied to world affairs.

In spite of all the negotiating effort that has been patiently made over more than three decades by many hundreds of negotiators, scientists, and engineers to reach a world consensus on climate change mitigation, the UNFCCC Paris Agreement will not be fulfilled. However, UNFCCC negotiations will proceed in the hope that climate change mitigation measures will keep increasing and that they will continue to become more effective. There is a great hope that there will be no symmetry between the beginning and end of the *Homo* genus time as regards global climate change. The symmetry will only prevail if humankind lets anthropogenic climate change develop out of control, allowing it to evolve into an existential threat. It is more likely that humans are entering a period of many hundreds of year of anthropogenic climate change, in which crises will be more frequent and the well-being of a large part of humankind will be seriously degraded.

As happens with all crises, the present five may bring the opportunity to start rebuilding a more inclusive and sustainable world, causing them to recede into oblivion. Essentially everything that should be learned about how to find and follow a new globally sustainable pathway is already known. New research and development on that subject is always desirable, necessary, and important, but that is not the main obstacle.

The main problem is the lack of collective global willingness to change direction because of long-standing opposing worldviews. The COVID-19 crisis, instead of opening new avenues for the convergence of those worldviews, is more likely to entrench them. The currently stronger worldview, or the worldview of the status quo, is founded on the belief that the present economic and financial system is the best one to improve well-being and economic prosperity around the world. Its defenders may concede that small incremental changes could be made to improve the system without putting it at risk, but deeper transformational changes are considered to be very risky and fully rejected. This is essentially the worldview of the global financial and economic corporate elites that control an inordinate part of the world economy and are the ones that most benefit from the system. The same worldview is also shared by many politicians, but unlike the corporate elites, instead of active players they just support and benefit from the system. The world view of the status quo is defended by most politicians in the emerging economies, where the outstanding priority is to improve the well-being and economic prosperity of its citizens in order to reach the standards of the advanced economies. In the advanced economies that worldview is shared with nuances by a political spectrum that goes from neoliberals, conservatives, and libertarians in the USA to most political parties that have governmental responsibilities in Europe and elsewhere. In fact, the strongest driver of the status quo is the large differences in economic prosperity between the advanced economies and the emergent and developing economies that urges them to bridge the gap as fast as possible, before it is too late. The solution to this problem would be proactive economic convergence of the advanced economies towards the emergent and developing economies. This programme, however, is contrary to the principle of rational egoism and individual self-interest.

The fact is that a majority of people around the world have identified consciously or subconsciously with the principle of the inevitability of progress based on perpetual economic growth. This conviction is supported by an unfaltering belief that technology will always provide further economic prosperity and free humankind from the adverse impacts of all five present crises and all future crises. Furthermore, one of the defining features of operative social time since the beginning of the Great Acceleration is that the extraordinary development of science and technology coupled with the application of emergent social sciences has succeeded in captivating and attracting humans to indulge in boundless consumerism, often with borrowed money, or in other words to embrace a lifestyle of opulence that helped them climb the social status ladder. The vast majority of people do not enjoy opulence, but that is the dream they entertain. All these beliefs in the irreplaceable power of economic growth are strongly related to or rooted in personal and collective behaviour that is very common in countries with advanced economies, increasingly common in emerging economies, and fast penetrating less developed countries. The notion of prosperity that is closest to this behaviour is naturally economic prosperity, which opens the doors directly to the delights of consumerism.

Economic prosperity is a modern construction that tends to rank people on a one-dimensional scale that is socially damaging because it is based on comparing what a few have and others don't. The one-dimensionality of the scale promotes the stairway effect and functions as a social trap where there are no limits to poverty and wealth. It encourages consumerism and a society dominated by greed and egoism. That is not surprising because neoclassical economics is based on individualism and more specifically on the assumption that the economy must be guided by the optimization of individual self-interest, which follows directly from the practice of rational egoism. However, the 20th century went much further and elevated egoism to the rank of the most important and productive of human qualities. One of the most visible champions of the new trend was the Russian émigré Ayn Rand, who fiercely and successfully defended the practice of a form of normative ethical–rational egoism in the USA under the pretence that it was a philosophical theory called Objectivism. Her efforts had an enduring influence on the ideological and moral grounding of the new neoliberal policies and made it possible to erroneously present them as an ethical system in which successful business people ensure an amazing economic progress for society as a whole and should therefore be venerated. Attempting to deny our imminently social nature by ignoring or rejecting actions that do not bring individual self-interest, such as cooperation, mutual help, solidarity, and active compassion, is a form of perversity that leads to social exclusion and dystopia. It means ethical and moral retrogression instead of progress. Is that the type of world that we want for us and for humankind? If citizens were to be confronted with this question, the answer would probably be "No", but that answer is inconsistent with the observation that rational egoism is increasingly accepted and diffused around the world as an inevitable behavioural style.

Life satisfaction does not depend exclusively on economic prosperity, interpreted as opulence or utility, but also involves social capital, which describes the quality of the social environment and the community, especially access to education and health, social, cultural, and psychological factors related to a shared identity, inclusion, trust, solidarity, compassion, integration, and participation in the community. There are other forms of prosperity besides economic prosperity. Sustainable prosperity emerges from developing the capabilities to flourish in a multi-dimensional social, economic, and environmental space guided by the objective of achieving sustainability at the local, national, and global levels. Sustainable prosperity does not exclude economic values, but it replaces opulence by utility and conditions utility to the main goal of sustainability. This option guarantees the social, economic and environmental sustainability of a civilisation that respects freedom, human rights, tolerance, solidarity, justice, peace, and social inclusion in the medium and long term.

Besides the status quo there are many alternative world views, but they all have the disadvantage that they are almost entirely rejected by the global financial and economic corporate elites that benefit from and control an undue part of the world economy. Furthermore, because of their great diversity, they have difficulty in converging to a common position that would present a strong, unified, and clear alternative capable of attracting a majority of citizens. Politically, most of the alternative world views are supported by a wide spectrum of political parties on the left, including the green parties. In non-democratic regimes, the alternative world views are not represented politically and are very often forced to be clandestine. The world views that oppose the status quo have done remarkable work to promote sustainability in many countries, especially at the local, regional, and national levels, but at the global level there is no significant progress because the opposing forces are dominant at that level. This situation aggravates the problem since a significant part of the unsustainability of human civilization has a global nature. This results from the relatively long human cultural evolution in the Holocene, especially in the period following the Great Acceleration.

The way the economic and financial system counteracts the alternative world views is through a "post-ideological" program based on the deployment of digital technologies. In the case of physical problems, such as climate change, instead of addressing and solving the cause of the problem, countervailing actions are deployed, supported by digital technologies, which are supposed to offset or cancel part of the negative impacts. In the case of dangerous contemporaneous behavioural, socioeconomic, and political tendencies, the post-ideological program, instead of addressing the core issues behind such tendencies, uses the digital technologies to build a virtual framework that provides individuals with the erroneous feeling that the harmful effects of such tendencies have been eliminated or are in the process of being eliminated. The origin of this trend that believes in "solving" complex social, economic, and environmental problems by circumventing, reformulating, and reinterpreting them with ICT has been attributed to Silicon Valley and represents an extension of solutionism.

Solutionism is an ideology devoid of ethical and moral dimensions, in which it is believed that all human everyday problems can somehow be solved by the right algorithms, codes, AI applications, automation processes, and robots (see Fig. 5.7) (Morozov 2013). Some of these technologies are very popular, such as the self-tracking gadgets that can be used to stimulate people to have healthier living habits, such as exercise more, eat diversified foods, monitor the blood pressure and other health indicators, practise safer driving habits, and have more stable financial situations. This digitalisation of human life opens a very profitable market for ICT start-ups and corporations. The next step for solutionism is to address and "solve" the global crises of income inequalities, environmental pollution, degradation and destruction, and climate change, which in fact have no solution compatible with the current economic and financial system. A possible way to give people the feeling that inequalities can be reduced by technological means is to expand the use of the blockchain technology that makes it possible to securely inventory, track, subdivide, and transfer wealth over the Internet, underpins the cryptocurrencies, in particular bitcoin and ethereum among many others, and serves as a vehicle for a sharing economy. The application of solutionism to the environmental crisis is essentially based on the ecomodernist view that the Earth system is now essentially a human artefact so that there are no grounds for any concern about further human interference. Finally, the climate change crisis can be solved by using blockchain and AI to optimise CDR and SRM geoengineering (Lockley et al. 2019).

What then is the solution that addresses the problems themselves? Here, we are once again confronted with one of the essential characteristics of our phylogenetic evolution up to and after the emergence of the *Homo* genus, namely, group-structured sociality. The origin of our direct ancestor's sociality began a long time ago when primates began banding together as loose groups of both sexes about 52 million BP at the anthropoid root (Shultz et al. 2011). Violent forms of adaptive aggressive strategies in intergroup relations can still be observed today in chimpanzees and bonobos, which belong to the hominin tribe just like *Homo sapiens* (Wilson et al. 2014). Killing members of another social group is an adaptive strategy by which killers increase their fitness by gaining access to territory, food, and mates. The natural selection pressures that led to the emergence of *Homo sapiens* created a complex social behaviour, structured in relatively small groups that simultaneously developed highly protective and cooperative intragroup strategies and highly competitive and aggressive intergroup strategies.

This essential form of behaviour is reflected and can be identified in all human social groups although it acquires different expressions that depend on kinship structures, the degree of social and economic development of the group's population, and the origin, nature, and function of the group. In a broad sense social groups include the tribes, clans, and chiefdoms of anthropology, the many types of social groups that emerge in contemporaneous societies with different natures, sizes, identities, goals, and functions, and also the sovereign states, federations, and political and economic unions of states, in which a large majority consciously share and protect the same cultural and historical identity. Loyalty to the social group, in particular as regards behaviour and way of thinking, and discrimination or animosity towards the members of other social groups is frequently called tribalism. Tribalism and tribal bias is deeply rooted in our cognitive system and is common to us all (Clark et al. 2019). Recently, the word "tribalism" has acquired a slighting and depreciative meaning by being used as synonymous with strong and destructive political polarization, especially in the USA, where there are symptoms of a faltering democracy. The extraordinary richness of the world's history, cultures, and artistic forms of expression, and of the world's national heritage sites, is a cherished and unique legacy of our very fruitful and creative group-based structured sociality, or tribalism in *sensu lato*, to use a simpler expression. However, our tribalism has great difficulty in dealing with global problems, particularly with global environmental changes, which require the ascendancy of the one-tribe-on-the-planet frame of mind, or in other words, the prevalence of humankind's interest over the diverse interests of each of humankind's "tribes". It is an analogous situation to that described by the tragedy of the commons, but applied to the whole of humanity.

In the contemporaneous operative social time most people in the world define their identity in terms of nationality. The utmost priority of a nationality is to protect the country itself, to maximise its autonomy and security, to develop its wealth and the well-being and economic prosperity of its citizens, to secure access or have control over crucial natural resources, and to develop its regional or global influence, its prestige, and its cultural identity. As has always happened in history, there are plenty of conflicts between countries and groups of countries. The most common in our operative social time are over monetary policy, trade, economic and political influences over other countries, natural resources, environmental issues, and over religious issues, especially when coupled with territorial and geopolitical conflicts. In many cases countries have been unable to resolve such conflicts at the negotiating table, but did not choose to resort to warfare given its immense costs. Nevertheless, the "War on Terror" conducted by the USA in the Middle East and in other regions of the world since 2001, stands out as a notable exception.

The relatively recent and rampant development and popularity of rational egoism has acted as an insurance against large scale warfare between major world powers and against world wars, since they would impair any hope of increasing well-being and economic prosperity across all economic brackets, including the upper 1%. Pyrrhic wars have become incompatible with the decline of all values that do not have an economical connotation. Nevertheless, some level of violent conflict is beneficial for the more powerful countries to assess the efficiency and sophistication of their weapons and to maintain a lucrative military–industrial complex. Recurring crises may progressively destroy economic values and replace them by other, more ideological values, which would tend to increase the probability of large scale warfare.

Meanwhile, conflicts have evolved in the direction of disabling all forms of cooperation between conflicting countries, in particular by denouncing military agreements, such as nuclear arms treaties, socioeconomical and cultural agreements, and the use of various forms of sanction. The preferred way of inflicting harm on the adversary is by imposing embargoes and economic sanctions, restrictions on economic assistance, financial restrictions, financial and travel restrictions on people individually, but also increasing tariffs on imported goods, which starts trade wars, as recently happened between the USA and China. These sanctions and restrictions are additional to those related to countering terrorism and international criminal organizations. At present the US Treasury Department, Commerce Department, and State Department lists embargoes against 30 countries or territories. Instead of resolving conflicts through negotiations or war, the current tendency is to use a panoply of economic and financial instruments aimed at disrupting, disabling, and eventually destroying the economies of adversaries in the hope that they will be forced to change their behaviour.

Solving the environmental and climate change crises is extremely difficult because sovereign states do not abdicate from giving priority to their national interests over humankind's interest, especially if the level of global geopolitical tension or confrontation is high, as is presently the case between the USA and China. The surge of globalisation in the late 20th century was mainly driven by the ruling elites of the economic and financial system with the objective of increasing profits by promoting the expansion of world markets and lowering labour costs. Protectionist measures in the form of a trade war imposed by the USA on China and on other countries started in 2018, less than two years before the COVID-19 crisis. The geopolitical outcomes of the new crisis are likely to be growing discord between the major powers, acceleration of the decline of the USA, an increasing nativism and protectionism, and a resurgence of the effort to promote greater self-sufficiency against the trend of recurring crises. However, it must be remembered that the "every country for itself" policies led in the past to a global economic crisis and that the reversal of globalisation would create a more fragmented, unstable, violent, and dangerous world. Citizens, non-governmental organizations, businessmen, and governments must develop all possible initiatives to increase international cooperation between people and states globally and in all aspects of human life and activities, especially at the cultural, scientific, and environmental levels.

So the question is: how can we achieve global sustainability and solve the environmental and climate change crises? Hope should remain intact and strong after we acknowledge the fundamental human shortcomings in dealing with global problems. The logic of our previous successes, based on the belief in progress, science, and technology, requires us to rise to the challenge and progress further. The challenge is to know whether progress is only possible within the framework of the multi-tribal viewpoint or whether it can be extended to the framework of the one-tribe-on-the-planet viewpoint. In the former case, humankind will be forced to adapt to a future shaped by an outdated framework. In the latter case, humankind will be constructing its future with new patterns of behaviour based on a new framework.

To reach the sustainability goal there are three crucial measures that must be undertaken. The first is to recognize that global problems are very difficult to solve, but the wisdom gained by knowing ourselves better is essential if we are to find workable solutions. The second is to progressively adopt the one-tribe-on-the-planet viewpoint framework so that globalisation, instead of being reversed, is developed as a privileged driver to reach sustainability through transformational changes guided by the UN Sustainable Development Goals for 2030. The third regards governance in democratic countries. The discourse of democracy's political leaders must undergo a change and start frankly addressing two crucial issues.

The first is the incompatibility between our civilization model and perpetual economic growth that implies uncontrolled natural resource depletion and dangerous environmental changes. The second is the urgent need to reverse the process of increasing inequalities worldwide. These issues are not of paramount importance to the electorates of almost the whole spectrum of political parties, because they are not directly related to the self-interests of most citizens, or indeed to their nationalistic preoccupations. However, since they are essential for peace and sustainability, party leaders should base their political discourse more on science and ethics and on the scientific understanding of the global changes that are affecting all world citizens, and refrain from paraphrasing what the majority of their specific electorate wants them to say. Some of the politicians who address these challenges will not be elected, but eventually they will be respected and supported. Without this courageous ethical attitude and vision, democracies will not be part of the solution to the current problems of inequality or the environmental and climate change crises that are threatening humankind, leaving ample space for the proliferation of authoritarian regimes. Without a renewal and adaptation of democracies to the new world challenges, authoritarian regimes are likely to become dominant. Humankind is confronted with the choice of adapting to the adverse future that it has created or to rebuild a new, more propitious future by a process of self-regeneration. Technological success has erroneously convinced us that we can disdain our essential forms of sociality. The self-interest cult leads to dystopia. Self-interest should be replaced by social self-regeneration.

The crucial condition for the success of this plan is to decelerate operative social time and to be as much as possible at peace with time. Deceleration does not impair progress but it changes the emphasis on its various forms, giving prominence to human and social progress. It also favours the development of other forms of prosperity besides economic prosperity. Deceleration is compatible with the human essential characteristics and there are many examples of contemporaneous decelerated operative social time at the individual and collective levels scattered throughout the world. If deceleration at the global level turns out to be impossible then there will be no way to reach any form of sustainable development. In this scenario humankind will be assailed by continuing crises that will result from persisting with impracticable perpetual economic growth models coupled with the continuing overexploitation of natural resources, environmental degradation and pollution, biodiversity loss, and climate change. Those crises would inevitably increase poverty and inequalities, decrease social inclusiveness, well-being, and economic prosperity around the world, and that would become too high a price to pay for not having been able to find a sustainable pathway into the future. The notion of progress in its various forms would be impaired and eventually outmoded. Human wisdom and foresight may prevent that outcome. Time resolves all human issues and uncertainties.

References

Abel GJ, Brottrager M, Crespo Cuaresma J, Muttarak R (2019) Climate, conflict and forced migration. Glob Environ Chang 54:239–249. https://doi.org/10.1016/j.gloenvcha.2018.12.003

Aboujaoude E (2011) Virtually You: The Dangerous Powers of the E-Personality. W. W. Norton & Company

Abramovitz M (1956) Resource and Output Trends in the United States Since 1870. Am Econ Rev 46:5–23

Adams FC, Laughlin G (1997) A dying universe: the long-term fate and evolution of astrophysical objects. Rev Mod Phys 69:337–372. https://doi.org/10.1103/RevModPhys.69.337

Addis DR, Wong AT, Schacter DL (2007) Remembering the past and imagining the future: Common and distinct neural substrates during event construction and elaboration. Neuropsychologia 45:1363–1377. https://doi.org/10.1016/j.neuropsychologia.2006.10.016

Ade PAR, Aghanim N, Armitage-Caplan C, et al (2014) Planck 2013 results. XVI. Cosmological parameters. Astron Astrophys 571:A16. https://doi.org/10.1051/0004-6361/201321591

AF (2019) Mapping Waorani Territory. In defence of a forest homeland, a culture, a way of life. In: Amaz. Front. https://www.amazonfrontlines.org/chronicles/mapping-waorani/

Agar N (2010) Humanity's End. The MIT Press

Aghion P, Howitt P (1992) A Model of Growth Through Creative Destruction. Econometrica 60:323. https://doi.org/10.2307/2951599

Ainslie G (1975) Specious reward: A behavioral theory of impulsiveness and impulse control. Psychol Bull 82:463–496. https://doi.org/10.1037/h0076860

Ainslie G, Haslam N (1992) Hyperbolic discounting. In: Choice over time. Russell Sage Foundation, New York, NY, US, pp 57–92

Ainsworth EA, Long SP (2004) What have we learned from 15 years of free-air CO_2 enrichment (FACE)? A meta-analytic review of the responses of photosynthesis, canopy properties and plant production to rising CO_2. New Phytol 165:351–372. https://doi.org/10.1111/j.1469-8137.2004.01224.x

Al-Suwaidi AH, Hesselbo SP, Damborenea SE, et al (2016) The Toarcian Oceanic Anoxic Event (Early Jurassic) in the Neuquén Basin, Argentina: A Reassessment of Age and Carbon Isotope Stratigraphy. J Geol 124:171–193. https://doi.org/10.1086/684831

Alesina A, Di Tella R, MacCulloch R (2004) Inequality and happiness: are Europeans and Americans different? J Public Econ 88:2009–2042. https://doi.org/10.1016/j.jpubeco.2003.07.006

Alessa L, Chapiniii F (2008) Anthropogenic biomes: a key contribution to earth-system science. Trends Ecol Evol 23:529–531. https://doi.org/10.1016/j.tree.2008.07.002

© Springer Nature Switzerland AG 2021   583
F. Duarte Santos, *Time, Progress, Growth and Technology*, The Frontiers Collection,
https://doi.org/10.1007/978-3-030-55334-0

Alexander S, Ussher S (2012) The Voluntary Simplicity Movement: A multi-national survey analysis in theoretical context. J Consum Cult 12:66–86. https://doi.org/10.1177/1469540512444019

Alexandratos N, Bruinsma J (2012) World agriculture towards 2030/2050. The 2012 revision, ESA Working Paper No. 12-03. Rome

Allgaier J (2019) Science and Environmental Communication on YouTube: Strategically Distorted Communications in Online Videos on Climate Change and Climate Engineering. Front Commun 4. https://doi.org/10.3389/fcomm.2019.00036

Allison G (2017) Destined for War: Can America and China Escape Thucydides's Trap? Houghton Mifflin Harcourt

Almond GA, Chodorow M, Pearce RH (1981) Progress and Its Discontents. Bull Am Acad Arts Sci 35:4. https://doi.org/10.2307/3823284

Alroy J (2017) Effects of habitat disturbance on tropical forest biodiversity. Proc Natl Acad Sci 114:6056–6061. https://doi.org/10.1073/pnas.1611855114

Alstadsæter A, Johannesen N, Zucman G (2017) Tax Evasion and Inequality. Cambridge, MA

Alter A (2017) Irresistible: The Rise of Addictive Technology and the Business of Keeping Us Hooked. Penguin Press

Altieri AH, Gedan KB (2015) Climate change and dead zones. Glob Chang Biol 21:1395–1406. https://doi.org/10.1111/gcb.12754

Alvaredo F, Chancel L, Piketty T, et al (2018) World Inequality Report 2018

Amnesty International (2019) Laws designed to silence: The global crackdown on civil society organizations. London

AMS (2009) AMS Policy Statement on Geoengineering the Climate System. Boston, MA

Anagnostou E, John EH, Edgar KM, et al (2016) Changing atmospheric CO_2 concentration was the primary driver of early Cenozoic climate. Nature 533:380–384. https://doi.org/10.1038/nature17423

Andersen H, Grush R (2009) A Brief History of Time Consciousness: Historical Precursors to James and Husserl. J. Hist. Philos. 47:277–307

Andersen KG, Rambaut A, Lipkin WI, et al (2020) The proximal origin of SARS-CoV-2. Nat Med 26:450–452. https://doi.org/10.1038/s41591-020-0820-9

Andy (2019) How many satellites orbiting the Earth in 2019? In: Pixalytics Ltd. https://www.pixalytics.com/satellites-orbiting-earth-2019/

Appenzeller T (2012) Human migrations: Eastern odyssey. Nature 485:24–26. https://doi.org/10.1038/485024a

Archer D, Ganopolski A (2005) A movable trigger: Fossil fuel CO_2 and the onset of the next glaciation. Geochemistry, Geophys Geosystems 6. https://doi.org/10.1029/2004GC000891

Archer J (1997) Why do people love their pets? Evol Hum Behav 18:237–259. https://doi.org/10.1016/S0162-3095(99)80001-4

Arciga J (2019) Anti-Vaxxer Larry Cook Has Weaponized Facebook Ads in War Against Science. In: Dly. Beast. https://www.thedailybeast.com/anti-vaxxer-larry-cook-has-weaponized-facebook-ads-in-war-against-science

Arlidge J (2009) I'm doing "God's work". Meet Mr Goldman Sachs. In: The Times. https://www.thetimes.co.uk/article/im-doing-gods-work-meet-mr-goldman-sachs-zflqc78gqs8

Asafu-Adjaye J, Blomqvist L, Brand S, et al (2015) An Ecomodernist Manifesto

Ashby N (2002) Relativity and the Global Positioning System. Phys Today 55:41–47. https://doi.org/10.1063/1.1485583

Ashcraft SW, Padron AS, Pascioni KA, et al (2011) Review of Propulsion Technologies for N+3 Subsonic Vehicle Concepts

Aspa J (2019) 8 Anti-aging Stocks in the US. In: Invest. News Netw. https://investingnews.com/daily/life-science-investing/genetics-investing/anti-aging-companies-invest/

Assmann J (1984) Death and Salvation in Ancient Egypt, 2005th edn. Cornell University Press

Associated Press (2019) Opioid strong enough to sedate elephants on rise in Ohio, coroners warn. In: The Guardian. https://www.theguardian.com/us-news/2019/feb/09/ohio-opioid-crisis-carfentanil-elephant-tranquilizer-warning

Atran S (2006) The moral logic and growth of suicide terrorism. Wash Q 29:127–147. https://doi. org/10.1162/wash.2006.29.2.127

Aubert M, Lebe R, Oktaviana AA, et al (2019) Earliest hunting scene in prehistoric art. Nature 576:442–445. https://doi.org/10.1038/s41586-019-1806-y

Ausubel JH, Frosch RA, Herman R (1989) Technology and Environment: An Overview. National Academies Press, Washington, D.C.

Autin WJ, Holbrook JM (2012) Is the Anthropocene an issue of stratigraphy or pop culture? GSA Today 60–61. https://doi.org/10.1130/G153GW.1

Auzanneau M (2015) Or noir: la grande histoire du pétrole, La Découverte. Chelsea Green Publishing (2018)

AWG (2019) Results of binding vote by AWG. In: Anthr. Work. Gr. http://quaternary.stratigraphy. org/working-groups/anthropocene/

Ayres R, Warr B (2009a) The Economic Growth Engine. Edward Elgar Publishing

Ayres RU, Warr B (2009b) Energy Efficiency and Economic Growth: the 'Rebound Effect' as a Driver. In: Energy Efficiency and Sustainable Consumption. Palgrave Macmillan UK, London, pp 119–135

Ayyadevara S, Tazearslan Ã, Bharill P, et al (2009) Caenorhabditis elegans PI3K mutants reveal novel genes underlying exceptional stress resistance and lifespan. Aging Cell 8:706–725. https:// doi.org/10.1111/j.1474-9726.2009.00524.x

Backhouse RE, Boianovsky M (2016) Secular stagnation: The history of a macroeconomic heresy. Eur J Hist Econ Thought 23:946–970. https://doi.org/10.1080/09672567.2016.1192842

Bacon F (1627) New Atlantis. no publisher given

Bacon F (1620) Novum Organum. Rés (1978)

Bae CJ, Douka K, Petraglia MD (2017) On the origin of modern humans: Asian perspectives. Science 358:eaai9067. https://doi.org/10.1126/science.aai9067

Bailey R (2000) Earth Day, Then and Now. The planet's future has never looked better. Here's why. In: Reason Found. https://reason.com/2000/05/01/earth-day-then-and-now-2/

Bala G, Duffy PB, Taylor KE (2008) Impact of geoengineering schemes on the global hydrological cycle. Proc Natl Acad Sci 105:7664–7669. https://doi.org/10.1073/pnas.0711648105

Baltimore D, Berg P, Botchan M, et al (2015) A prudent path forward for genomic engineering and germline gene modification. Science 348:36–38. https://doi.org/10.1126/science.aab1028

BAML (2015) Robot Revolution - Global Robot & AI Primer

Bar-Yosef O (2002) The Upper Paleolithic Revolution. Annu Rev Anthropol 31:363–393. https:// doi.org/10.1146/annurev.anthro.31.040402.085416

Barbosa F, Bresciani G, Graham P, et al (2020) Oil and gas after COVID-19: The day of reckoning or a new age of opportunity? In: McKinsey Co. https://www.mckinsey.com/industries/oil-and-gas/ our-insights/oil-and-gas-after-covid-19-the-day-of-reckoning-or-a-new-age-of-opportunity

Barbosa L de SNS, Bogdanov D, Vainikka P, Breyer C (2017) Hydro, wind and solar power as a base for a 100% renewable energy supply for South and Central America. PLoS One 12:e0173820. https://doi.org/10.1371/journal.pone.0173820

Barker T, Campos H, Cooper M, et al (2010) Improving Drought Tolerance in Maize. In: Plant Breeding Reviews. John Wiley & Sons, Inc., Oxford, UK, pp 173–253

Barnosky AD, Hadly EA, Bascompte J, et al (2012) Approaching a state shift in Earth's biosphere. Nature 486:52–58. https://doi.org/10.1038/nature11018

Barnosky AD, Matzke N, Tomiya S, et al (2011) Has the Earth's sixth mass extinction already arrived? Nature 471:51–57. https://doi.org/10.1038/nature09678

Barrett AB (2018) Stability of Zero-growth Economics Analysed with a Minskyan Model. Ecol Econ 146:228–239. https://doi.org/10.1016/j.ecolecon.2017.10.014

Barth E, Bryson A, Davis J, Freeman R (2014) It's Where You Work: Increases in Earnings Dispersion across Establishments and Individuals in the U.S. Cambridge, MA

Barzashka I (2013) Are Cyber-Weapons Effective? RUSI J 158:48–56. https://doi.org/10.1080/ 03071847.2013.787735

Barzilai N, Crandall JP, Kritchevsky SB, Espeland MA (2016) Metformin as a Tool to Target Aging. Cell Metab 23:1060–1065. https://doi.org/10.1016/j.cmet.2016.05.011

Basset NA (2015) Neanderthals and Modern Behaviour: Did they bury their dead? UMASA J 33:1–14

Bateman J (2018) Why China is spending billions to develop an army of robots to turbocharge its economy. In: EDGE

Baudelaire C (1863) Le Peintre de la vie moderne. Fayard/Mille et une nuits, 2010

Baudelaire C (1857) Les Fleurs du mal. Poulet-Malassis et De Broise, Alençon

Bauman Z (2000) Liquid Modernity. Polity Press

Bauman Z (2017) Retrotopia. Polity Press

BBC (2018) The 11 cities most likely to run out of drinking water – like Cape Town. In: Br. Broadcast. Corp. https://www.bbc.com/news/world-42982959

Beavan C (2009) No Impact Man: The Adventures of a Guilty Liberal Who Attempts to Save the Planet, and the Discoveries He Makes About Himself and Our Way of Life in the Process. Farrar, Straus and Giroux

Beck U (1986) Risk Society: Towards a New Modernity. SAGE Publications, 1992

Becker E (1973) The denial of death. Free Press

Beerling DJ, Lomax BH, Royer DL, et al (2002) An atmospheric pCO_2 reconstruction across the Cretaceous-Tertiary boundary from leaf megafossils. Proc Natl Acad Sci 99:7836–7840. https://doi.org/10.1073/pnas.122573099

Beets SD (2015) BB&T, Atlas Shrugged, and the Ethics of Corporation Influence on College Curricula. J Acad Ethics 13:311–344. https://doi.org/10.1007/s10805-015-9244-4

Behrensmeyer AK (2006) Climate Change and Human Evolution. Science 311:476–478. https://doi.org/10.1126/science.1116051

Bejan A (2006) Advanced Engineering Thermodynamics. Wiley

Bekelman JE, Li Y, Gross CP (2003) Scope and Impact of Financial Conflicts of Interest in Biomedical Research. JAMA 289:454. https://doi.org/10.1001/jama.289.4.454

Bell B (1970) The Oldest Records of the Nile Floods. Geogr J 136:569. https://doi.org/10.2307/1796184

Bell EA, Boehnke P, Harrison TM, Mao WL (2015) Potentially biogenic carbon preserved in a 4.1 billion-year-old zircon. Proc Natl Acad Sci 112:14518–14521. https://doi.org/10.1073/pnas.1517557112

Bell JS (1964) On the Einstein–Podolsky–Rosen paradox. Phys Phys Fiz 1:195–200. https://doi.org/10.1103/PhysicsPhysiqueFizika.1.195

Bentham J, Harrison R (1776) A Fragment on Government. Cambridge University Press (1988), Cambridge

Berger A (2002) CLIMATE: An Exceptionally Long Interglacial Ahead? Science 297:1287–1288. https://doi.org/10.1126/science.1076120

Berger A, Crucifix M, Hodell DA, et al (2016) Interglacials of the last 800,000 years. Rev Geophys 54:162–219. https://doi.org/10.1002/2015RG000482

Berger RM, Kelly JJ (1993) Social Work in the Ecological Crisis. Soc Work 38:521–526. https://doi.org/10.1093/sw/38.5.521

Bergeron T (1928) Über die dreidimensional verknüpfende Wetteranalyse (I). Geofys Publ 5:1–111

Bergman J (2003) Does the acquisition of antibiotic and pesticide resistance provide evidence for evolution? J Creat 17:26–32

Berman M (1982) All that is solid melts into air: the experience of modernity. Simon and Schuster, New York

Bernal JD (1939) The social function of science. George Routledge

Bernat JL (2006) The Whole-Brain Concept of Death Remains Optimum Public Policy. J Law, Med Ethics 34:35–43. https://doi.org/10.1111/j.1748-720X.2006.00006.x

Berner RA (1999) Atmospheric oxygen over Phanerozoic time. Proc Natl Acad Sci 96:10955–10957. https://doi.org/10.1073/pnas.96.20.10955

Berns GS, Laibson D, Loewenstein G (2007) Intertemporal choice – toward an integrative framework. Trends Cogn Sci 11:482–488. https://doi.org/10.1016/j.tics.2007.08.011

Bernstein PL (1998) Against the Gods: The Remarkable Story of Risk. John Wiley & Sons, Inc.

Betts R, Jones C, Jin Y, et al (2020) Analysis: What impact will the coronavirus pandemic have on atmospheric CO_2? In: Carbon Br. https://www.carbonbrief.org/analysis-what-impact-will-the-coronavirus-pandemic-have-on-atmospheric-co2

Bezanson A (1922) The Early Use of the Term 'Industrial Revolution'. Q J Econ 36:343. https://doi.org/10.2307/1883486

BGS (2020) The cost of CCS. In: Br. Geol. Surv. https://www.bgs.ac.uk/discoveringGeology/climateChange/CCS/TheCostofCSS.html

Bhatia KT, Vecchi GA, Knutson TR, et al (2019) Recent increases in tropical cyclone intensification rates. Nat Commun 10:635. https://doi.org/10.1038/s41467-019-08471-z

BIEN (2019) Basic Income European Network. In: Basic Income Eur. Netw. https://basicincome.org/

Binder CR, Hinkel J, Bots PWG, Pahl-Wostl C (2013) Comparison of Frameworks for Analyzing Social-ecological Systems. Ecol Soc 18:art26. https://doi.org/10.5751/ES-05551-180426

Biro D, Humle T, Koops K, et al (2010) Chimpanzee mothers at Bossou, Guinea carry the mummified remains of their dead infants. Curr Biol 20:R351–R352. https://doi.org/10.1016/j.cub.2010.02.031

Blackburn EH, Epel ES, Lin J (2015) Human telomere biology: A contributory and interactive factor in aging, disease risks, and protection. Science 350:1193–1198. https://doi.org/10.1126/science.aab3389

Blackburn R (1988) The overthrow of colonial slavery, 1776–1848. Verso

Blackstock JJ, Low S (eds) (2018) Geoengineering our Climate? Ethics, Politics, and Governance. Routledge

Blagoev B, Muhr SL, Ortlieb R, Schreyögg G (2018) Organizational working time regimes: Drivers, consequences and attempts to change patterns of excessive working hours. Ger J Hum Resour Manag Zeitschrift für Pers 32:155–167. https://doi.org/10.1177/2397002218791408

Blaylock J, Smallwood D, Kassel K, et al (1999) Economics, food choices, and nutrition. Food Policy 24:269–286. https://doi.org/10.1016/S0306-9192(99)00029-9

BNEF (2019) New Energy Outlook 2019. In: Bloom. New Energy Financ. https://about.bnef.com/new-energy-outlook/

Boehm A-M, Khalturin K, Anton-Erxleben F, et al (2012) FoxO is a critical regulator of stem cell maintenance in immortal Hydra. Proc Natl Acad Sci 109:19697–19702. https://doi.org/10.1073/pnas.1209714109

Bogdanov A (1922) Tektologiya: Vseobschaya Organizatsionnaya Nauka in 3 volumes. Essays in Tektology: The General Science of Organization, translated by George Gorelik, Seaside, California, Intersystems Publications (1980), Berlin and Petrograd-Moscow

Bohlin G (2017) Evolving germs – Antibiotic resistance and natural selection in education and public communication. Linköping University

Boix C (2010) Origins and Persistence of Economic Inequality. Annu Rev Polit Sci 13:489–516. https://doi.org/10.1146/annurev.polisci.12.031607.094915

Bommes M, Fassmann H, Sievers W (2014) Migration from the Middle East and North Africa to Europe: Past Developments, Current Status, and Future Potentials. Amsterdam University Press

Bonald L (1796) Observations sur un ouvrage posthume de M. de Condorcet. In: Théorie du pouvoir politique et religieux dans la société civile. Oeuvres de M. Bonald (1843), Paris

Bond TC, Bhardwaj E, Dong R, et al (2007) Historical emissions of black and organic carbon aerosol from energy-related combustion, 1850–2000. Global Biogeochem Cycles 21:n/a-n/a. https://doi.org/10.1029/2006GB002840

Bosch S van den, Rotmans J (2008) Deepening, Broadening and Scaling up: a Framework for Steering Transition Experiments. Dutch Research Institute for Transitions (DRIFT), Delft, Rotterdam

Bostrom N (2002) Existential Risks: Analyzing Human Extinction Scenarios and Related Hazards. J Evol Technol 9

Bostrom N (2013) Existential Risk Prevention as Global Priority. Glob Policy 4:15–31. https://doi. org/10.1111/1758-5899.12002

Bostrom N (2014) Superintelligence: Paths, Dangers, Strategies. Oxford University Press

Boulanger P-M (2010) Three strategies for sustainable consumption. SAPIENS 3:1–10

Bourke AFG (2011) The validity and value of inclusive fitness theory. Proc R Soc B Biol Sci 278:3313–3320. https://doi.org/10.1098/rspb.2011.1465

Boutang Y (1998) De l'esclavage au salariat. Economie historique du salariat bride. Presses Universitaires de France, Paris

Boyce CJ, Brown GDA, Moore SC (2010) Money and Happiness. Psychol Sci 21:471–475. https:// doi.org/10.1177/0956797610362671

Boyer P (2008) Evolutionary economics of mental time travel? Trends Cogn Sci 12:219–224. https:// doi.org/10.1016/j.tics.2008.03.003

BP (2018) Statistical Review of World Energy, 67th Edition

BP (2019) Statistical Review of World Energy, 68th Edition

Bradley BA, Blumenthal DM, Wilcove DS, Ziska LH (2010) Predicting plant invasions in an era of global change. Trends Ecol Evol 25:310–318. https://doi.org/10.1016/j.tree.2009.12.003

Bragg W (1936) The Progress of Physical Science. In: Jeans J (ed) Scientific Progress. George Allen and Unwin, London

Bramble DM, Lieberman DE (2004) Endurance running and the evolution of Homo. Nature 432:345–352. https://doi.org/10.1038/nature03052

Brand S (2009) Whole Earth Discipline: An Ecopragmatist Manifesto. Viking Penguin

Braudel F (1985) La Dynamique du Capitalisme. Paris

Bremner J, Hunter L (2014) Migration and the Environment. Popul Bull Popul Ref Bur 69:12

Breuil H (1952) Quatre cents siècles d'art pariétal. Les cavernes ornées de l'âge du Renne. Montignac

Bricker VR (1982) The Origin of the Maya Solar Calendar. Curr Anthropol 23:101–103. https:// doi.org/10.1086/202782

Brico E (2019) Addiction is a disease. But an opioid executive wants you to think it's a crime. In: Vox. https://www.vox.com/first-person/2019/4/10/18304572/addiction-opioid-purdue-pharma-sackler-family

Brigida A-C (2018) Nearly 60% of migrants from Guatemala's Dry Corridor cited climate change and food security as their reason for leaving. In: Glob. Cathol. Clim. Mov. https://catholicclimatemovement.global/nearly-60-of-migrants-from-guatemalas-dry-corridor-cited-climate-change-and-food-security-as-their-reason-for-leaving/

Brockman J (2011) How is the Internet Changing the Way You Think?: The Net's Impact on Our Minds and Future. Atlantic Books

Brockway PE, Owen A, Brand-Correa LI, Hardt L (2019) Estimation of global final-stage energy-return-on-investment for fossil fuels with comparison to renewable energy sources. Nat Energy 4:612–621. https://doi.org/10.1038/s41560-019-0425-z

Brodwin E (2019) The founder of a startup that charged $ 8,000 to fill your veins with young blood says he's shuttering the company and starting a new one. In: Bus. Insid. https://www. businessinsider.sg/young-blood-transfusions-ambrosia-shut-down-2019-6

Brondsted HV (1969) Planarian Regeneration. Elsevier

Brook BW, Barnosky AD (2012) Quaternary Extinctions and Their Link to Climate Change. In: Saving a Million Species. Island Press/Center for Resource Economics, Washington, DC, pp 179–198

Brooks AS, Yellen JE, Potts R, et al (2018) Long-distance stone transport and pigment use in the earliest Middle Stone Age. Science 360:90–94. https://doi.org/10.1126/science.aao2646

Brooks H (1994) The relationship between science and technology. Res Policy 23:477–486. https:// doi.org/10.1016/0048-7333(94)01001-3

Brooks N (2012) Beyond collapse: climate change and causality during the Middle Holocene Climatic Transition, 6400–5000 years before present. Geogr Tidsskr J Geogr 112:93–104. https:// doi.org/10.1080/00167223.2012.741881

Broome J (1992) Counting the Cost of Global Warming. White Horse Press, Cambridge

Broomfield M (2012) Support to the identification of potential risks for the environment and human health arising from hydrocarbons operations involving hydraulic fracturing in Europe

Brosius J (2003) From Eden to a hell of uniformity? Directed evolution in humans. BioEssays 25:815–821. https://doi.org/10.1002/bies.10313

Bruinsma J (2011) The resource outlook to 2050: By how much do land, water use and crop yields need to increase by 2050? In: Looking ahead in World Food and Agriculture: Perspectives to 2050. Food and Agriculture Organization of the United Nations (FAO), Rome, pp 10–12

Brulle RJ (2014) Institutionalizing delay: foundation funding and the creation of U.S. climate change counter-movement organizations. Clim Change 122:681–694. https://doi.org/10.1007/s10584-013-1018-7

Brulle RJ, Aronczyk M, Carmichael J (2020) Corporate promotion and climate change: an analysis of key variables affecting advertising spending by major oil corporations, 1986–2015. Clim Change 159:87–101. https://doi.org/10.1007/s10584-019-02582-8

Bruns S, Gross C, Stern D (2013) Is There Really Granger Causality Between Energy Use and Output? E.ON Energy Research Center, Future Energy Consumer Needs and Behavior (FCN)

Brynjolfsson E (1993) The productivity paradox of information technology. Commun ACM 36:66–77. https://doi.org/10.1145/163298.163309

Buckley BM, Anchukaitis KJ, Penny D, et al (2010) Climate as a contributing factor in the demise of Angkor, Cambodia. Proc Natl Acad Sci 107:6748–6752. https://doi.org/10.1073/pnas.0910827107

Budyko MI (1956) The Heat Balance of the Earth's Surface, Leningrad (Translated by Nina A. Stepanova). U.S. Department of Commerce (1958), Washington D.C.

Budyko MI (1974) Climate and Life. Academic Press

Budyko MI (1977) Climatic changes. American Geophysical Union, Minnesota

Buell L (2015) Enough is enough? In: Syse K, Mueller M (eds) Sustainable consumption and the good life. Routledge, pp 7–26

Buiatti M, Christou P, Pastore G (2013) The application of GMOs in agriculture and in food production for a better nutrition: two different scientific points of view. Genes Nutr 8:255–270. https://doi.org/10.1007/s12263-012-0316-4

Burckhardt J (1929) Judgments on History and Historians. Liberty Fund Inc. (1999)

Burger J, Escartin M, Ponce N, Lanza LV (2016) Agua Zarca Hydroelectric Project: Independent fact finding mission – Report and Recommendations

Burgess AG, Burgess JP (2011) Truth. Princeton University Press, Princeton

Butler J (1726) Fifteen Sermons Preached at the Rolls Chapel. Botham

C40 Cities (2019) Staying afloat: the urban response to sea level rise. In: C40 Cities Clim. Leadersh. Group, Inc. https://www.c40.org/other/the-future-we-don-t-want-staying-afloat-the-urban-response-to-sea-level-rise

Cáceres B (2015) Berta Caceres acceptance speech, 2015 Goldman Prize ceremony. In: Goldman Environ. Prize. https://www.youtube.com/watch?v=AR1kwx8b0ms

Cadena A, Giraut J, Grosman N, Vaz A de O (2019) Unlocking the economic potential of Central America and the Caribbean. In: McKinsey Co. https://www.mckinsey.com/featured-insights/americas/unlocking-the-economic-potential-of-central-america-and-the-caribbean/

Cai W, Borlace S, Lengaigne M, et al (2014) Increasing frequency of extreme El Niño events due to greenhouse warming. Nat Clim Chang 4:111–116. https://doi.org/10.1038/nclimate2100

Cai W, Wang G, Santoso A, et al (2015) Increased frequency of extreme La Niña events under greenhouse warming. Nat Clim Chang 5:132–137. https://doi.org/10.1038/nclimate2492

Cairns R (2016) Climates of suspicion: 'chemtrail' conspiracy narratives and the international politics of geoengineering. Geogr J 182:70–84. https://doi.org/10.1111/geoj.12116

Caldeira K, Keith D (2010) The Need for Climate Engineering Research. Issues Sci Technol 27:57–62

Caldeira K, Wickett ME (2003) Anthropogenic carbon and ocean pH. Nature 425:365–365. https://doi.org/10.1038/425365a

Callenbach E (1975) Ecotopia. Banyan Tree Books

Callenbach E (1981) Ecotopia Emerging. Banyan Tree Books

Campbell-Lendrum D, Manga L, Bagayoko M, Sommerfeld J (2015) Climate change and vector-borne diseases: what are the implications for public health research and policy? Philos Trans R Soc B Biol Sci 370:20130552. https://doi.org/10.1098/rstb.2013.0552

Canterbery ER (2006) Alan Greenspan. The oracle behind the curtain. World Scientific

Cantril H (1965) The pattern of human concerns. Rutgers University Press, New Brunswick

Carleton TA, Hsiang SM (2016) Social and economic impacts of climate. Science 353:aad9837–aad9837. https://doi.org/10.1126/science.aad9837

Carling J (2002) Migration in the age of involuntary immobility: Theoretical reflections and Cape Verdean experiences. J Ethn Migr Stud 28:5–42. https://doi.org/10.1080/13691830120103912

Carr N (2018) Is Google Making Us Stupid? In: The Atlantic, 15 August

Carr N (2010) The Shallows: What the Internet Is Doing to Our Brains. W.W. Norton Company

Carrier DR, Kapoor AK, Kimura T, et al (1984) The Energetic Paradox of Human Running and Hominid Evolution [and Comments and Reply]. Curr Anthropol 25:483–495. https://doi.org/10.1086/203165

Cascio J (2009) Get Smarter. In: Atl. https://www.theatlantic.com/magazine/archive/2009/07/get-smarter/307548/

Case A, Deaton A (2015) Rising morbidity and mortality in midlife among white non-Hispanic Americans in the 21st century. Proc Natl Acad Sci 112:15078–15083. https://doi.org/10.1073/pnas.1518393112

Catling DC, Zahnle KJ (2009) The Planetary Air Leak. Sci Am 300:36–43. https://doi.org/10.1038/scientificamerican0509-36

Cavaliere A, De Marchi E, Banterle A (2014) Healthy–unhealthy weight and time preference. Is there an association? An analysis through a consumer survey. Appetite 83:135–143. https://doi.org/10.1016/j.appet.2014.08.011

CBS (2010) The Cost of Dying: End-of-Life Care. In: CBS. https://www.cbsnews.com/news/the-cost-of-dying-end-of-life-care/. Accessed 8 Aug 2010

CDC (2019) National Center for Health Statistics. In: Centers Dis. Control Prev. https://www.cdc.gov/nchs/

CDSB (2019) CDSB Framework for reporting environmental & climate change information. Advancing and aligning disclosure of environmental information in mainstream reports

Ceballos G, Ehrlich PR, Dirzo R (2017) Biological annihilation via the ongoing sixth mass extinction signaled by vertebrate population losses and declines. Proc Natl Acad Sci 114:E6089–E6096. https://doi.org/10.1073/pnas.1704949114

Cederström C, Spicer A (2017) Desperately Seeking Self-Improvement: A Year Inside the Optimization Movement. OR Books

CEIC (2016) Global Database. In: CEIC. https://www.ceicdata.com/en/products/global-economic-database

Chabris CF, Laibson DI, Schuldt JP (2008) Intertemporal Choice. In: Palgrave Macmillan (eds) The New Palgrave Dictionary of Economics. Palgrave Macmillan UK, London, pp 1–8

Chan EKF, Timmermann A, Baldi BF, et al (2019) Human origins in a southern African palaeo-wetland and first migrations. Nature 575:185–189. https://doi.org/10.1038/s41586-019-1714-1

Chapman W (2019) 10 Cities Most At Risk of Running Out of Water. In: .S. News World Rep. https://www.usnews.com/news/cities/slideshows/10-cities-most-at-risk-of-running-out-of-water

Charney JG, Fjörtoft R, von Neumann J (1950) Numerical Integration of the Barotropic Vorticity Equation. Tellus 2:237–254

Chase K (2003) Firearms: A Global History to 1700. Cambridge University Press

Chauvet J-M, Deschamps EB, Hillaire C (1995) La grotte Chauvet à Vallon-Pont-d'Arc. Editions du Seuil

Chernyshevsky N (1863) What Is to Be Done? Cornell University Press (1989)

Chesney R, Citron D (2019) Deepfakes and the New Disinformation War The Coming Age of Post-Truth Geopolitics. In: Foreign Aff. https://www.foreignaffairs.com/articles/world/2018-12-11/deepfakes-and-new-disinformation-war

Chichilnisky G (1996) An axiomatic approach to sustainable development. Soc Choice Welfare 13:231–257. https://doi.org/10.1007/BF00183353

Christenhusz MJM, Byng JW (2016) The number of known plants species in the world and its annual increase. Phytotaxa 261:201. https://doi.org/10.11646/phytotaxa.261.3.1

Christensen C (2003) The Innovator's Solution: Creating and Sustaining Successful Growth. Harvard Business School Press

Christensen CM (1997) The Innovator's Dilemma: When New Technologies Cause Great Firms to Fail. Harvard Business Review Press

Christian A (2019) Nigeria Is Making More Millionaires Than Ever. In: Weetracker. https://weetracker.com/2019/02/02/nigeria-is-making-more-millionaires-than-ever/

Ciais P, Reichstein M, Viovy N, et al (2005) Europe-wide reduction in primary productivity caused by the heat and drought in 2003. Nature 437:529–533. https://doi.org/10.1038/nature03972

Clack CTM, Qvist SA, Apt J, et al (2017) Evaluation of a proposal for reliable low-cost grid power with 100% wind, water, and solar. Proc Natl Acad Sci 114:6722–6727. https://doi.org/10.1073/pnas.1610381114

Clark CJ, Liu BS, Winegard BM, Ditto PH (2019) Tribalism Is Human Nature. Curr Dir Psychol Sci 28:587–592. https://doi.org/10.1177/0963721419862289

Clark PU, Huybers P (2009) Interglacial and future sea level. Nature 462:856–857. https://doi.org/10.1038/462856a

Clark PU, Shakun JD, Marcott SA, et al (2016) Consequences of twenty-first-century policy for multi-millennial climate and sea-level change. Nat Clim Chang 6:360

Clarke S, Savulescu J, Coady T, et al (eds) (2016) The Ethics of Human Enhancement. Oxford University Press

CLB (2019) Migrant workers and their children. In: China Labour Bull. https://clb.org.hk/content/migrant-workers-and-their-children

Clement M, Lambert A, Herouart D, Boncompagni E (2008) Identification of new up-regulated genes under drought stress in soybean nodules. Gene 426:15–22. https://doi.org/10.1016/j.gene.2008.08.016

Clottes J, Lewis-Williams D (2001) Les chamanes de la préhistoire. Transe et magie dans les grottes ornées. Texte intégral, polémique et réponses. La Maison des Roches, Paris

Coady D, Parry I, Sears L, Shang B (2017) How Large Are Global Fossil Fuel Subsidies? World Dev 91:11–27. https://doi.org/10.1016/j.worlddev.2016.10.004

Cobb CW, Douglas PH (1928) A Theory of Production. Am Econ Rev 18:139–165

Cobham A, Janský P (2017) Global distribution of revenue loss from tax avoidance: Re-estimation and country results. UNU-WIDER

Coffin M, Grant A (2019) Balancingthe Budget: Why deflating the carbon bubble requires oil and gas companiesto shrink

Cohen L (2004) A Consumers' Republic: The Politics of Mass Consumption in Postwar America. J Consum Res 31:236–239. https://doi.org/10.1086/383439

Collier P (2018) The Future of Capitalism: Facing the New Anxieties. Harper

Collins R (2004) Lenski's Power Theory of Economic Inequality: A Central Neglected Question in Stratification Research. Sociol Theory 22:219–228. https://doi.org/10.1111/j.0735-2751.2004.00213.x

Collomb J-D (2019) US Conservative and Libertarian Experts and Solar Geoengineering: An Assessment. Eur J Am Stud 14:. https://doi.org/10.4000/ejas.14717

Colton J (1982) Foreword. In: Almond A, Chodorow M, Pearce RH (eds) Progress and its discontents. University of California Press, Berkeley, pp ix–xii Columbus C (1492) The Journal of Christopher Columbus (during his First Voyage, 1492–93). Hakluyt Society (2010)

Comte A (1852) Catéchisme positiviste, Carilian-G. Paris

Conard NJ (2009) A female figurine from the basal Aurignacian of Hohle Fels Cave in southwestern Germany. Nature 459:248–252. https://doi.org/10.1038/nature07995

Conboy IM, Conboy MJ, Wagers AJ, et al (2005) Rejuvenation of aged progenitor cells by exposure to a young systemic environment. Nature 433:760–764. https://doi.org/10.1038/nature03260

Condorcet J-A-N de C (1795) Esquisse d'un tableau historique des progrès de l'esprit humain. Librairie philosophique J. Vrin (1970)

Condorcet J-A-N de C, Sieyès E-J, Duhamel J-M (1793) Journal d'instruction sociale, par les citoyens Condorcet, Sieyès et Duhamel. [1 June–6 July 1793.]. Impr. des sourds-muets, Paris

Condorcet N (1805) Eléments du calcul des probabilités et son application aux jeux de hasard, à la loterie, et aux jugemens des hommes. éditions Royez, an XIII, réédition par l'IREM de Paris VII, 1986

Connor P, Passel JS (2019) Europe's Unauthorized Immigrant Population Peaks in 2016, Then Levels Off. In: Pew Res. Cent. https://www.pewresearch.org/global/2019/11/13/europes-unauthorized-immigrant-population-peaks-in-2016-then-levels-off/

Conrad P (1999) Modern Times, Modern Places: How life and art were transformed in a century of revolution, innovation and radical change. Alfred A. Knopf, New York

Cook J, Nuccitelli D, Green SA, et al (2013) Quantifying the consensus on anthropogenic global warming in the scientific literature. Environ Res Lett 8:024024. https://doi.org/10.1088/1748-9326/8/2/024024

Cooper D (2019) Low rates provide a historic opportunity to tackle climate change. In: Financ. Times. https://www.ft.com/content/c752698c-200c-11ea-92da-f0c92e957a96

Cooper N, Brady E, Steen H, Bryce R (2016) Aesthetic and spiritual values of ecosystems: Recognising the ontological and axiological plurality of cultural ecosystem 'services.' Ecosyst Serv 21:218–229. https://doi.org/10.1016/j.ecoser.2016.07.014

COPINH (2017) Carta firmada por el Consejo Indígena, el Consejo de Ancianas-os y mas de 150 miembras-os de la comunidad indígena lenca de Río Blanco. In: COPINH. http://copinhonduras.blogspot.com/p/blog-page_31.html

Corvus G (2004) La convergence des catastrophes. Vérité

Costanza R (1980) Embodied Energy and Economic Valuation. Science 210:1219–1224. https://doi.org/10.1126/science.210.4475.1219

Costanza R, D'Arge R, de Groot R, et al (1997) The value of the world's ecosystem services and natural capital. Nature 387:253–260. https://doi.org/10.1038/387253a0

Costanza R, Hart M, Talberth J, Posner S (2009) Beyond GDP: The Need for New Measures of Progress. Pardee Pap No 4, Bost Pardee Cent Study Longer-Range Futur

Coyne JA, Orr HA (2004) Speciation. Sinauer Associates, Inc., Sunderland, Massachusetts

Cranford R (1995) Criteria for Death. In: Reich W (ed) Encyclopedia of Bioethics, 2nd edn. Macmillan, New York, pp 529–534

Crittenden AN, Schnorr SL (2017) Current views on hunter-gatherer nutrition and the evolution of the human diet. Am J Phys Anthropol 162:84–109. https://doi.org/10.1002/ajpa.23148

Crivelli E, de Mooij R, Keen M (2016) Base Erosion, Profit Shifting and Developing Countries. Finanz Public Financ Anal 72:268–301

Crosby AW (1972) The Columbian Exchange: Biological and Cultural Consequences of 1492. Praeger Publishers (2003)

Crutzen PJ (2002) Geology of mankind. Nature 415:23–23. https://doi.org/10.1038/415023a

Crutzen PJ (2006) Albedo Enhancement by Stratospheric Sulfur Injections: A Contribution to Resolve a Policy Dilemma? Clim Change 77:211–220. https://doi.org/10.1007/s10584-006-9101-y

Crutzen PJ, Stoermer EF (2000) The 'Anthropocene.' Glob Chang Newsletters 41:17–18

Cullen HM, DeMenocal PB, Hemming S, et al (2000) Climate change and the collapse of the Akkadian empire: Evidence from the deep sea. Geology 28:379. https://doi.org/10.1130/0091-7613(2000)28<379:CCATCO>2.0.CO;2

Currie J, Greenstone M, Meckel K (2017) Hydraulic fracturing and infant health: New evidence from Pennsylvania. Sci Adv 3:e1603021. https://doi.org/10.1126/sciadv.1603021

Cyranoski D (2019) The CRISPR-baby scandal: what's next for human gene-editing. Nature 566:440–442. https://doi.org/10.1038/d41586-019-00673-1

Dabla-Norris E, Kochhar K, Suphaphiphat N, et al (2015) Causes and Consequences of Income Inequality: A Global Perspective. International Monetary Fund (IMF)

Dahl-Jensen D, Albert MR, Aldahan A, et al (2013) Eemian interglacial reconstructed from a Greenland folded ice core. Nature 493:489–494. https://doi.org/10.1038/nature11789

Dai A (2013) Increasing drought under global warming in observations and models. Nat Clim Chang 3:52–58. https://doi.org/10.1038/nclimate1633

Dalio R (2019) Why and How Capitalism Needs to Be Reformed (Parts 1 & 2). In: Bridg. Assoc. L.P. https://www.linkedin.com/pulse/why-how-capitalism-needs-reformed-parts-1-2-ray-dalio/

Daly HE (1980) Economics, Ecology, Ethics: Essays Toward a Steady-state Economy. W. H. Freeman

Daly HE (1993) Valuing the Earth: Economics, Ecology, Ethics, 2nd edn. MIT Press

Daly HE (1997a) Beyond Growth: The Economics of Sustainable Development. Beacon Press, Boston

Daly HE (1997b) Georgescu-Roegen versus Solow/Stiglitz. Ecol Econ 22:261–266. https://doi.org/10.1016/S0921-8009(97)00080-3

Daniel TC, Muhar A, Arnberger A, et al (2012) Contributions of cultural services to the ecosystem services agenda. Proc Natl Acad Sci 109:8812–8819. https://doi.org/10.1073/pnas.1114773109

Darby S, Fawcett T (2018) Energy sufficiency: an introduction. Concept paper

Darimont CT, Fox CH, Bryan HM, Reimchen TE (2015) The unique ecology of human predators. Science 349:858–860. https://doi.org/10.1126/science.aac4249

Darwin C (1871) The Descent of Man, and Selection in Relation to Sex. John Murray

Darwin CR (1859) On the origin of species by means of natural selection. John Murray, London

Dasgupta PS, Maler KG, Barrett S (1999) Intergenerational Equity, Social Discount Rates and Global Warming. In: Portney PR, Weyant JP (eds) Discounting and Intergenerational Equity. Resources for the Future, Washington, DC, pp 51–78

Davies JH, Davies DR (2010) Earth's surface heat flux. Solid Earth 1:5–24. https://doi.org/10.5194/se-1-5-2010

Davis K, Moore WE (1945) Some Principles of Stratification. Am Sociol Rev 10:242. https://doi.org/10.2307/2085643

Dawkins R (1976) The Selfish Gene. Oxford University Press, Oxford

Day R, Walker G, Simcock N (2016) Conceptualising energy use and energy poverty using a capabilities framework. Energy Policy 93:255–264. https://doi.org/10.1016/j.enpol.2016.03.019

De Grey A, Rae M (2007) Ending Aging: The Rejuvenation Breakthroughs that Could Reverse Human Aging in Our Lifetime. St. Martin's Press

de Groot R, Brander L, van der Ploeg S, et al (2012) Global estimates of the value of ecosystems and their services in monetary units. Ecosyst Serv 1:50–61. https://doi.org/10.1016/j.ecoser.2012.07.005

de Magalhães JP, Stevens M, Thornton D (2017) The Business of Anti-Aging Science. Trends Biotechnol 35:1062–1073. https://doi.org/10.1016/j.tibtech.2017.07.004

De Vos JM, Joppa LN, Gittleman JL, et al (2015) Estimating the normal background rate of species extinction. Conserv Biol 29:452–462. https://doi.org/10.1111/cobi.12380

Deane P (1965) The first industrial revolution. Cambridge University Press, Cambridge

DeConto RM, Pollard D (2016) Contribution of Antarctica to past and future sea-level rise. Nature 531:591–597. https://doi.org/10.1038/nature17145

Demoule J-P (2017) Les dix millénaires oubliés qui ont fait l'Histoire. Fayard

Derraik JG (2002) The pollution of the marine environment by plastic debris: a review. Mar Pollut Bull 44:842–852. https://doi.org/10.1016/S0025-326X(02)00220-5

Diaz del Castillo B (1632) Historia verdadera de la conquista de la Nueva España. 1800 (Printed for J. Wright, Piccadilly, by John Dean, High Street, Congleton)

Diaz RJ, Rosenberg R (2008) Spreading Dead Zones and Consequences for Marine Ecosystems. Science 321:926–929. https://doi.org/10.1126/science.1156401

Díaz S, Settele J, Brondízio ES, et al (2019) Pervasive human-driven decline of life on Earth points to the need for transformative change. Science 366:eaax3100. https://doi.org/10.1126/science.aax3100

Dibble HL, Aldeias V, Goldberg P, et al (2015) A critical look at evidence from La Chapelle-aux-Saints supporting an intentional Neanderthal burial. J Archaeol Sci 53:649–657. https://doi.org/10.1016/j.jas.2014.04.019

Diderot D, D'Alembert J le R (eds) (1751) Encyclopédie, ou dictionnaire raisonné des sciences, des arts et des métiers. André le Breton, Michel-Antoine David, Laurent Durand and Antoine-Claude Briasson

Didham RK, Tylianakis JM, Hutchison MA, et al (2005) Are invasive species the drivers of ecological change? Trends Ecol Evol 20:470–474. https://doi.org/10.1016/j.tree.2005.07.006

Dieckmann H (1951) Les Contributions de Diderot a la "Correspondance Littéraire" et a l' "Histoire Des Deux Indes." Rev Hist Litt Fr 51:417–440

Dimanchev EG, Paltsev S, Yuan M, et al (2019) Health co-benefits of sub-national renewable energy policy in the US. Environ Res Lett 14:085012. https://doi.org/10.1088/1748-9326/ab31d9

Dimitriev OP (2013) Global Energy Consumption Rates: Where is the Limit? Sustain Energy 1:1–6

Dobbs R, Koller T, Ramaswamy S, et al (2015) Playing to win: the new global competition for corporate profits

Dodd MS, Papineau D, Grenne T, et al (2017) Evidence for early life in Earth's oldest hydrothermal vent precipitates. Nature 543:60–64. https://doi.org/10.1038/nature21377

Dodge R, Daly A, Huyton J, Sanders L (2012) The challenge of defining wellbeing. Int J Wellbeing 2:222–235. https://doi.org/10.5502/ijw.v2i3.4

Dodman D (2009) Blaming cities for climate change? An analysis of urban greenhouse gas emissions inventories. Environ Urban 21:185–201. https://doi.org/10.1177/0956247809103016

Dong X, Milholland B, Vijg J (2016) Evidence for a limit to human lifespan. Nature 538:257–259. https://doi.org/10.1038/nature19793

Donges JF, Lucht W, Müller-Hansen F, Steffen W (2017) The technosphere in Earth System analysis: A coevolutionary perspective. Anthr Rev 4:23–33. https://doi.org/10.1177/2053019616676608

Douglas I, Lawson N (2000) The Human Dimensions of Geomorphological Work in Britain. J Ind Ecol 4:9–33. https://doi.org/10.1162/108819800569771

Douglas KM, Sutton RM (2011) Does it take one to know one? Endorsement of conspiracy theories is influenced by personal willingness to conspire. Br J Soc Psychol 50:544–552. https://doi.org/10.1111/j.2044-8309.2010.02018.x

Douglas RG, Woodruff F (1981) Deep-sea benthic foraminifera. In: The Oceanic Lithosphere (The Sea, Vol. 7). Wiley-Interscience, New York, pp 1233–1327

Dowden B (2009) The Metaphysics of Time: A Dialogue (New Dialogues in Philosophy). Rowman & Littlefield Publishers

Doyle JR (2013) Survey of time preference, delay discounting models. Judgm Decis Mak 8:116–135

Drews S, van den Bergh JCJM (2016) Public views on economic growth, the environment and prosperity: Results of a questionnaire survey. Glob Environ Chang 39:1–14. https://doi.org/10.1016/j.gloenvcha.2016.04.001

Drucker PF (1993) Post-Capitalist Society. Harper Collins Publishers

Drutman L (2015) The Business of America Is Lobbying. Oxford University Press

Duhigg C (2016) Smarter Faster Better: The Secrets of Being Productive. Random House

Dulac J (2013) Global land transport infrastructure requirements. Estimating road and railway infrastructure capacity and costs to 2050. Paris

Dumont M, Brun E, Picard G, et al (2014) Contribution of light-absorbing impurities in snow to Greenland's darkening since 2009. Nat Geosci 7:509–512. https://doi.org/10.1038/ngeo2180

Durán D (1581) Historia de las Indias de Nueva España y islas de Tierra Firme, 1867th edn. Imp. de J.M. Andrade y F. Escalante, México

Durkheim É (1912) Les formes élémentaires de la vie religieuse, cinquième. Les Presses Universitaires de France

Durkheim É (1897) Le suicide. Étude de sociologie, 2nd edn. Les Presses Universitaires de France, Paris

Dutton A, Lambeck K (2012) Ice Volume and Sea Level During the Last Interglacial. Science 337:216–219. https://doi.org/10.1126/science.1205749

EASAC (2015) Ecosystem services, agriculture and neonicotinoids

EASAC (2013) Planting the future: opportunities and challenges for using crop genetic improvement technologies for sustainable agriculture

EASAC (2018) Negative emission technologies: What role in meeting Paris Agreement targets?

Eberstadt N (2017) Our Miserable 21st Century. In: Commentary. https://www.commentarymagazine.com/articles/nicholas-eberstadt/our-miserable-21st-century/

EC (2015) COM(2015) 614 final: Closing the loop – An EU action plan for the Circular Economy. Brussels

EC (2014) Population ageing in Europe. Facts, implications and policies. Publications Office of the European Union, Bruxelles

Eco U, Culler J, Rorty R, Brooke-Rose C (1992) Interpretation and Overinterpretation. Cambridge University Press

Edgar RS, Green EW, Zhao Y, et al (2012) Peroxiredoxins are conserved markers of circadian rhythms. Nature 485:459–464. https://doi.org/10.1038/nature11088

Edwards D (2011) I'm Feeling Lucky: The Confessions of Google Employee Number 59. Houghton Mifflin Harcourt

Edwards L (2015) What Is the Anthropocene? Eos (Washington DC) 95. https://doi.org/10.1029/2015EO040297

Edwards PN (2010) A Vast Machine. Computer Models, Climate Data, and the Politics of Global Warming. MIT Press

EE (2017) Energy Explorer. In: Energy Explor. https://www.energy-observer.org/

EEI (2018) Extreme Energy Initiative. In: Extrem. Energy Initiat. https://extreme-energy-initiative.blogs.sas.ac.uk/

EEUETS (2020) Aviation in the EU ETS. In: Emiss. https://www.emissions-euets.com/carbon-market-glossary/965-aviation-in-the-eu-ets

Ehninger D, Neff F, Xie K (2014) Longevity, aging and rapamycin. Cell Mol Life Sci 71:4325–4346. https://doi.org/10.1007/s00018-014-1677-1

Einstein A (1916) Die Grundlage der allgemeinen Relativitätstheorie. Ann Phys 354:769–822. https://doi.org/10.1002/andp.19163540702

Einstein A (1915) Die Feldgleichungen der Gravitation. In: Albert Einstein: Akademie-Vorträge. Wiley, pp 88–92

Einstein A (1905a) Ist die Trägheit eines Körpers von seinem Energieinhalt abhängig? Ann Phys 323:639–641. https://doi.org/10.1002/andp.19053231314

Einstein A (1905b) Zur Elektrodynamik bewegter Körper. Ann Phys 322:891–921. https://doi.org/10.1002/andp.19053221004

Elgin D (1981) Voluntary Simplicity. Harper Paperbacks (1998)

Elliott B (2016) Natural Catastrophe: Climate Change and Neoliberal Governance. Edinburgh University Press

Ellis EC (2015) Ecology in an anthropogenic biosphere. Ecol Monogr 85:287–331. https://doi.org/10.1890/14-2274.1

Ellis EC (2018) Science Alone Won't Save the Earth. People Have to Do That. In: New York Times. https://www.nytimes.com/2018/08/11/opinion/sunday/science-people-environment-earth.html

Ellis EC, Ramankutty N (2008) Putting people in the map: anthropogenic biomes of the world. Front Ecol Environ 6:439–447. https://doi.org/10.1890/070062

Elsner JB, Kossin JP, Jagger TH (2008) The increasing intensity of the strongest tropical cyclones. Nature 455:92–95. https://doi.org/10.1038/nature07234

Eltahir EAB, Wang G (1999) Nilometers, El Niñ o, and climate variability. Geophys Res Lett 26:489–492. https://doi.org/10.1029/1999GL900013

Eltis D, Richardson D (eds) (2008) Extending the Frontiers: Essays on the New Transatlantic Slave Trade Database. Yale University Press

Emerson D (2019) Biogenic Iron Dust: A Novel Approach to Ocean Iron Fertilization as a Means of Large Scale Removal of Carbon Dioxide From the Atmosphere. Front Mar Sci 6:. https://doi.org/10.3389/fmars.2019.00022

Emiliani C (1955) Pleistocene Temperatures. J Geol 63:538–578. https://doi.org/10.1086/626295

Enerdata (2017) Global Energy Statistical Yearbook 2017. In: Enerdata. https://yearbook.enerdata.net/total-energy/world-energy-intensity-gdp-data.html. Accessed 21 Jun 2017

Engels F (1845) The Condition of the Working Class in England. Penguin Classics (2009)

EPA (2000) Aircraft Contrails Factsheet. Environmental Protection Agency (EPA)

Erickson P, Down A, Lazarus M, Koplow D (2017) Effect of government subsidies for upstream oil infrastructure on U.S. oil production and global CO_2 emissions. Seattle, WA

ESA (2019) Space debris by the numbers. In: Eur. Sp. Agency. https://www.esa.int/Safety_Security/Space_Debris/Space_debris_by_the_numbers

Espy JP (1841) The philosophy of storms. Charles C. Little and James Brown, Boston

Essen L, Parry JVL (1955) An Atomic Standard of Frequency and Time Interval: A Cæsium Resonator. Nature 176:280–282. https://doi.org/10.1038/176280a0

ETC (2018a) China's Plan to Seed Himalayan Clouds Is Geoengineering – Unintentional or Otherwise. In: ETC Gr. https://www.etcgroup.org/content/chinas-plan-engineer-himalayan-clouds-geoengineering-unintentional-or-otherwise

ETC (2018b) Solar Radiation Management Geoengineering and Climate Change: Implications for Africa

ETC (2010) The geoengineering moratorium under the UN Convention on Biological Diversity

Etzioni A (1999) Voluntary Simplicity: Characterization, Select Psychological Implications, and Societal Consequences. In: Essays in Socio-Economics. Studies in Economic Ethics and Philosophy. Springer, pp 1–26

Evans NP, Bauska TK, Gázquez-Sánchez F, et al (2018a) Quantification of drought during the collapse of the classic Maya civilization. Science 361:498–501. https://doi.org/10.1126/science.aas9871

Evans WN, Lieber EMJ, Power P (2018b) How the Reformulation of Oxycontin Ignited the Heroin Epidemic. Rev Econ Stat Forthcoming

EY (2013) Arctic Oil and Gas. In: Ernst & Young (EY). https://www.ey.com/Publication/vwLUAssets/Arctic_oil_and_gas/%24FILE/Arctic_oil_and_gas.pdf

Falkowski P (2000) The Global Carbon Cycle: A Test of Our Knowledge of Earth as a System. Science 290:291–296. https://doi.org/10.1126/science.290.5490.291

Fanelli D (2018) Opinion: Is science really facing a reproducibility crisis, and do we need it to? Proc Natl Acad Sci 115:2628–2631. https://doi.org/10.1073/pnas.1708272114

FAO (2015) Global Forest Resources Assessment 2015. Rome

FAO (2014) The State of World Fisheries and Aquaculture. Rome

FAO (2006) World agriculture: towards 2030/2050 – interim report. Prospects for food, nutrition, agriculture and major commodity groups. Rome

FAO (2008) Climate change and food scarcity: a framework document. Rome

FAO (2019a) The State of the World's Biodiversity for Food and Agriculture. Rome

FAO (2019b) Save Food: Global Initiative on Food Loss and Waste Reduction. In: Food Agric. Organ. United Nations. http://www.fao.org/save-food/resources/keyfindings/en/

Fargues P (2017) Four Decades of Cross-Mediterranean Undocumented Migration to EuropeA Review of the Evidence. Geneva, Switzerland

Fava GA (2016) The Hidden Costs of Financial Conflicts of Interest in Medicine. Psychother Psychosom 85:65–70. https://doi.org/10.1159/000442694

Faye G (1998) L'Archéofuturisme: Techno-science et retour aux valeurs ancestrales. L'ncre, Paris

Fernández-González D, Prazuch J, Ruiz-Bustinza Í, et al (2018) Iron Metallurgy via Concentrated Solar Energy. Metals (Basel) 8:873. https://doi.org/10.3390/met8110873

Ferriss T (2016) Tools of Titans: The Tactics, Routines, and Habits of Billionaires, Icons, and World-Class Performers. Houghton Mifflin Harcourt

Ferry J-M (1995) L'allocation universelle: pour un revenu de citoyenneté. Les Éditions du Cerf

Feulner G (2017) Formation of most of our coal brought Earth close to global glaciation. Proc Natl Acad Sci 114:11333–11337. https://doi.org/10.1073/pnas.1712062114

Fickling D (2019) China Could Outrun the U.S. Next Year. Or Never. In: Bloom. Opin. www.bloombergquint.com/opinion/will-china-overtake-u-s-gdp-depends-how-you-count

Field CB (1998) Primary Production of the Biosphere: Integrating Terrestrial and Oceanic Components. Science 281:237–240. https://doi.org/10.1126/science.281.5374.237

Field CB, Mach KJ (2017) Rightsizing carbon dioxide removal. Science 356:706–707. https://doi.org/10.1126/science.aam9726

Findeisen W (1938) Kolloid-meteorologische Vorgänge bei Niederschlagsbildung. Meteorol Zeitschrift 24:443–454. https://doi.org/10.1127/metz/2015/0675

Finney SC, Edwards LE (2016) The "Anthropocene" epoch: Scientific decision or political statement? GSA Today 26:4–10. https://doi.org/10.1130/GSATG270A.1

Flaherty MG (1991) The Perception of Time and Situated Engrossment. Soc Psychol Q 54:76–85. https://doi.org/10.2307/2786790

Fleming JR (2006) The pathological history of weather and climate modification: Three cycles of promise and hype. Hist Stud Phys Biol Sci 37:3–25. https://doi.org/10.1525/hsps.2006.37.1.3

Fleming JR (2010) Fixing the Sky. The Checkered History of Weather and Climate Control. Columbia University Press

Fletcher G (2015) The Routledge Hanbook of Philosophy of well-being. Routledge

FLI (2016) Autonomous Weapons: an Open Letter from AI & Robotics Researchers. In: Futur. Life Inst. https://futureoflife.org/open-letter-autonomous-weapons/

Florence CS, Zhou C, Luo F, Xu L (2016) The Economic Burden of Prescription Opioid Overdose, Abuse, and Dependence in the United States, 2013. Med Care 54:901–906. https://doi.org/10.1097/MLR.0000000000000625

Florio J (1598) A worlde of wordes, or Most copious, and exact dictionarie in Italian and English. Arnold Hatfield for Edw. Blount, London

Floudas D, Binder M, Riley R, et al (2012) The Paleozoic Origin of Enzymatic Lignin Decomposition Reconstructed from 31 Fungal Genomes. Science 336:1715–1719. https://doi.org/10.1126/science.1221748

FMO (2016) Frequently Asked Questions on The Agua Zarca Run-of-the-River Hydroelectric Generation Project

Foa RS, Mounk Y (2016) The Danger of Deconsolidation: The Democratic Disconnect. J Democr 27:5–17

Fohlen C (1971) Qu'est-ce que la révolution industrielle? Éditions Robert Laffont, Paris

Fontenelle BLB (1686) Entretien sur la pluralité des mondes. Nabu Press (2011), Paris

Forster P (2005) Evolution: Did Early Humans Go North or South? Science 308:965–966. https://doi.org/10.1126/science.1113261

Foulger GR, Wilson MP, Gluyas JG, et al (2018) Global review of human-induced earthquakes. Earth-Science Rev 178:438–514. https://doi.org/10.1016/j.earscirev.2017.07.008

Fourastié J (1979) Les Trente Glorieuses: Ou la révolution invisible de 1946 à 1975. Fayard

Fourtané S (2019) The Future of Aviation: Electric Airplanes Will Decarbonize the Aviation Industry. In: Interes. Eng. Inc. https://interestingengineering.com/the-future-of-aviation-electric-airplanes-will-decarbonize-the-aviation-industry

Fraisse P (1984) Perception and Estimation of Time. Annu Rev Psychol 35:1–37. https://doi.org/10.1146/annurev.ps.35.020184.000245

Francis (2015) Encyclical Letter Laudato Si' of the Holy Father Francis on Care for our Common Home. Vatican Press, Rome

Frederick S, Loewenstein G, O'Donoghue T (2003) Time discounting and time preference: A critical review. In: Time and decision: Economic and psychological perspectives on intertemporal choice. Russell Sage Foundation, New York, NY, US, pp 13–86

French HW (2017) Everything Under the Heavens: How the Past Helps Shape China's Push for Global Power. Alfred A. Knopf

Freud S (1901) The Psychopathology of Everyday Life, 1914th edn. The Macmillan Company, New York

Frey CB, Osborne M (2013) The Future of Employment. Oxford Martin School, University of Oxford, Oxford, UK

Fricker T, Elsner JB (2015) Kinetic Energy of Tornadoes in the United States. PLoS One 10:e0131090. https://doi.org/10.1371/journal.pone.0131090

Friedlingstein P, Andrew RM, Rogelj J, et al (2014) Persistent growth of CO_2 emissions and implications for reaching climate targets. Nat Geosci 7:709–715. https://doi.org/10.1038/ngeo2248

Friedman M (1962) Capitalism and Freedom. University of Chicago Press

Friedman TL (2005) The World Is Flat: A Brief History of the Twenty-first Century. Farrar, Straus and Giroux

Friedman WJ (1982) The Developmental Psychology of Time. Academic Press

Friedrich T, Timmermann A, Abe-Ouchi A, et al (2012) Detecting regional anthropogenic trends in ocean acidification against natural variability. Nat Clim Chang 2:167–171. https://doi.org/10.1038/nclimate1372

Froehlich A (ed) (2019) Space Security and Legal Aspects of Active Debris Removal. Springer International Publishing, Cham

Fröhlich C (2000) Observations of Irradiance Variations. Space Sci Rev 94:15–24. https://doi.org/10.1023/A:1026765712084

FSB (2014) Global Shadow Banking Monitoring Report 2014

FT (2020) America cannot escape oil price volatility. The shale revolution alone cannot deliver true energy independence. In: Financ. Times. https://www.ft.com/content/07dd54e4-2e3b-11ea-bc77-65e4aa615551

FTC (2016) Federal Trade Commission Cigarette Report for 2013

Fu Q, Li H, Moorjani P, et al (2014) Genome sequence of a 45,000-year-old modern human from western Siberia. Nature 514:445–449. https://doi.org/10.1038/nature13810

Fuentes-Nieva R, Galasso N (2014) Working for the few. Political capture and economic inequality

Fukuyama F (1992) The End of History and the Last Man. Avon Books, Inc, New York

Funtowicz SO, Ravetz JR (1993) Science for the post-normal age. Futures 25:739–755. https://doi.org/10.1016/0016-3287(93)90022-L

Furchtgott-Roth D (2009) Climate Change: Another Option. In: Manhattan Inst. https://www.manhattan-institute.org/html/climate-change-another-option-2232.html

FWW (2013) Biotech Ambassadors. How the U.S. State Department Promotes the Seed Industry's Global Agenda

Galilei G (1610) Sidereus Nuncius. Tommaso Baglioni, Venice

Galor O, Moav O (2007) The Neolithic Revolution and Contemporary Variations in Life Expectancy. Brown University, Department of Economics

Ganopolski A, Winkelmann R, Schellnhuber HJ (2016) Critical insolation–CO_2 relation for diagnosing past and future glacial inception. Nature 529:200–203. https://doi.org/10.1038/nature16494

Gantner J (1958) Leonardos Visionen von der Sintflut und vom Untergang der Welt. Geschichte einer Künstlerischen Idee. Francke, Bern

Gao X, Zhang X, Yang D, et al (2010) Revisiting the origin of modern humans in China and its implications for global human evolution. Sci China Earth Sci 53:1927–1940. https://doi.org/10.1007/s11430-010-4099-4

Garcia M (2018) Air Travel Projected To Double In 20 Years, But Protectionism Poses Threat. In: Forbes. https://www.forbes.com/sites/marisagarcia/2018/10/24/iata-raises-20-year-projections-to-8-2-billion-passengers-warns-against-protectionism/#22249df5150f

Gardner B (2015) The Cost of Inaction: Recognising the Value at Risk from Climate Change. In: Econ. Intell. Unit. https://eiuperspectives.economist.com/sustainability/cost-inaction

Gardner MN, Brandt AM (2006) "The Doctors' Choice Is America's Choice". Am J Public Health 96:222–232. https://doi.org/10.2105/AJPH.2005.066654

Gascon M, Triguero-Mas M, Martínez D, et al (2016) Residential green spaces and mortality: A systematic review. Environ Int 86:60–67. https://doi.org/10.1016/j.envint.2015.10.013

Gately I (2001) Tobacco: A Cultural History of how an Exotic Plant Seduced Civilization. Grove Press

Gaynor KM, Hojnowski CE, Carter NH, Brashares JS (2018) The influence of human disturbance on wildlife nocturnality. Science 360:1232–1235. https://doi.org/10.1126/science.aar7121

GCCSI (2011) Accelerating the uptake of CCS: Industrial use of captured carbon dioxide

GDB, Afshin A, Forouzanfar MH, et al (2017) Health Effects of Overweight and Obesity in 195 Countries over 25 Years. N Engl J Med 377:13–27 (2017). https://doi.org/10.1056/NEJMoa1614362

Gehrsitz M, Ungerer M (2017) Jobs, Crime, and Votes: A Short-Run Evaluation of the Refugee Crisis in Germany

Geissdoerfer M, Savaget P, Bocken NMP, Hultink EJ (2017) The Circular Economy – A new sustainability paradigm? J Clean Prod 143:757–768. https://doi.org/10.1016/j.jclepro.2016.12.048

Gell-Mann M, Ruhlen M (2011) The origin and evolution of word order. Proc Natl Acad Sci 108:17290–17295. https://doi.org/10.1073/pnas.1113716108

Gellman B, Poitras L (2013) U.S., British intelligence mining data from nine U.S. Internet companies in broad secret program. In: Washington Post. https://www.washingtonpost.com/investigations/us-intelligence-mining-data-from-nine-us-internet-companies-in-broad-secret-program/2013/06/06/3a0c0da8-cebf-11e2-8845-d970ccb04497_story.html

Georgescu-Roegen N (1971) The Entropy Law and the Economic Process. Harvard University Press, Cambridge, MA

Georgescu-Roegen N (1975) Energy and Economic Myths. South Econ J 41:347–381. https://doi.org/10.2307/1056148

Georgieva K (2019) How to Use Debt Wisely. In: Int. Monet. Fund. https://www.imf.org/en/News/Articles/2019/11/07/sp110719-how-to-use-debt-wisely

Geyer R, Jambeck JR, Law KL (2017) Production, use, and fate of all plastics ever made. Sci Adv 3:e1700782. https://doi.org/10.1126/sciadv.1700782

Ghandnoosh N, Rovner J (2017) Immigration and Public Safety. Washington, DC

GHWBI (2014) State of Global Well-Being. 2014 Country Well-Being Rankings. Gallup Healthways Well-Being Index (GHWBI)

Gibson DG, Glass JI, Lartigue C, et al (2010) Creation of a Bacterial Cell Controlled by a Chemically Synthesized Genome. Science 329:52–56. https://doi.org/10.1126/science.1190719

Giddens A (1999) Risk and Responsibility. Mod Law Rev 62:1–10. https://doi.org/10.1111/1468-2230.00188

Gielen D, Boshell F, Saygin D, et al (2019) The role of renewable energy in the global energy transformation. Energy Strateg Rev 24:38–50. https://doi.org/10.1016/j.esr.2019.01.006

Gilbert N (2013) Case studies: A hard look at GM crops. Nature 497:24–26. https://doi.org/10.1038/497024a

Gilbert N (2014) Cross-bred crops get fit faster. Nature 513:292–292. https://doi.org/10.1038/513292a

Gilens M, Page BI (2014) Testing Theories of American Politics: Elites, Interest Groups, and Average Citizens. Perspect Polit 12:564–581. https://doi.org/10.1017/S1537592714001595

Gimon E, O'boyle M, T.M.Clack C, Mckee S (2019) The coal cost crossover: economic viability of existing coal compared to new local wind and solar resources. San Francisco, CA

Gingerich PD (2006) Environment and evolution through the Paleocene–Eocene thermal maximum. Trends Ecol Evol 21:246–253. https://doi.org/10.1016/j.tree.2006.03.006

Gingrich N (2008) Stop the Green Pig: Defeat the Boxer–Warner–Lieberman Green Pork Bill Capping American Jobs and Trading America's Future. In: Hum. Events. https://humanevents.com/2008/06/03/stop-the-green-pig-defeat-the-boxerwarnerlieberman-green-pork-bill-capping-american-jobs-and-trading-americas-future/

Giubilini A, Sanyal S (2016) Challenging Human Enhancement. In: The Ethics of Human Enhancement. Oxford University Press, pp 1–24

Giustina M, Versteegh MAM, Wengerowsky S, et al (2015) Significant-Loophole-Free Test of Bell's Theorem with Entangled Photons. Phys Rev Lett 115:250401. https://doi.org/10.1103/PhysRevLett.115.250401

Glasby GP (2006) Abiogenic Origin of Hydrocarbons: An Historical Overview. Resour Geol 56:83–96. https://doi.org/10.1111/j.1751-3928.2006.tb00271.x

Gleick J (2011) The Information: A History, A Theory, A Flood. Pantheon Books

Global Witness (2016) Exposing the Truth: Annual Report 2016. In: globalwitness.org. https://www.globalwitness.org/en/about-us/exposing-truth/

Global Witness (2017) Honduras the deadliest place to defend the planet. In: globalwitness.org. https://www.globalwitness.org/documents/18804/English_Honduras_full_report_single_v6.pdf

Godwin F (1638) The man in the moon: or a discourse of a voyage thither by Domingo Gonsales, the speedy messanger. Scolar Press (1971), Menston, England

Góis D de (1566) Crónica do Felicíssimo Rei D. Manuel. Edies Vercial (2010), Lisboa

Gold R (2019) PG&E: The First Climate-Change Bankruptcy, Probably Not the Last. In: Wall Str. J. https://www.wsj.com/articles/pg-e-wildfires-and-the-first-climate-change-bankruptcy-11547820006

Goldberg S (2017) Surveillance without Borders: The "Traffic Shaping" Loophole and Why It Matters. In: Century Found. https://tcf.org/content/report/surveillance-without-borders-the-traffic-shaping-loophole-and-why-it-matters/

Gollier C (2002) Discounting an uncertain future. J Public Econ 85:149–166. https://doi.org/10.1016/S0047-2727(01)00079-2

Gordon R (2000) Interpreting the "One Big Wave" in U.S. Long-Term Productivity Growth. Cambridge, MA

Gordon RJ (2016) The Rise and Fall of American Growth. Princeton University Press

Gordon RJ (2015) Secular Stagnation: A Supply-Side View. Am Econ Rev 105:54–59. https://doi.org/10.1257/aer.p20151102

Gosseries A (2008) Theories of intergenerational justice: a synopsis. SAPIENS 1:63–74

Gould KA, Pellow DN, Schnaiberg A (2004) Interrogating the Treadmill of Production. Organ Environ 17:296–316. https://doi.org/10.1177/1086026604268747

Goulder LH, Williams RC (2012) The choice of discount rate for climate change policy evaluation. Clim Chang Econ 03:1250024.https://doi.org/10.1142/S2010007812500248

Graafland J, Compen B (2015) Economic Freedom and Life Satisfaction: Mediation by Income per Capita and Generalized Trust. J Happiness Stud 16:789–810. https://doi.org/10.1007/s10902-014-9534-3

Graafland J, Lous B (2018) Economic Freedom, Income Inequality and Life Satisfaction in OECD Countries. J Happiness Stud 19:2071–2093. https://doi.org/10.1007/s10902-017-9905-7

GRACE (2019) Measuring Earth's Surface Mass and Water Changes. In: Gravity Recover. Clim. Exp. https://grace.jpl.nasa.gov/

Granger GG (1956) La mathématique sociale du marquis de Condorcet. Presses Universitaire de France, Paris

Granjon M-C (1982) Les interventions des Etats-Unis en Amérique centrale (1885–1980): le poids du passé. Polit étrangère 47:297–308. https://doi.org/10.3406/polit.1982.3129

Grauman J V. (1976) Orders of magnitude of the world's urban population in history. United Nations Popul Bull 8:39

Gray C (1995) The World Equity Budget or Living on about $142 per Month. In: Anderson D (ed) Downward Mobility for Conscience's Sake: Ten Aotobiographycal Sketches. Tom Paine Institute, pp 97–124

Gray WM (1981) Recent advances in tropical cyclone research from Rawinsonde composite analysis. Colorado

Greenhow C, Chapman A, Marich H, Askari E (2017) Social media and social networks. In: The SAGE Encyclopedia of Out-of-School Learning. SAGE Publications

Greenspan A (2007) The age of turbulence. Penguin Press

Gregg R (1936) The Value of Voluntary Simplicity. Visva-Bharati Q

Griffiths JG (1980) The origins of Osiris and his cult. E.J. Brill

Grimal P (1985) The dictionary of classical mythology. Blackwell, Oxford, England; New York, NY

Gröcke DR, Hori RS, Trabucho-Alexandre J, et al (2011) An open ocean record of the Toarcian oceanic anoxic event. Solid Earth 2:245–257. https://doi.org/10.5194/se-2-245-2011

Gronau I, Hubisz MJ, Gulko B, et al (2011) Bayesian inference of ancient human demography from individual genome sequences. Nat Genet 43:1031–1034. https://doi.org/10.1038/ng.937

Grosberg RK, Strathmann RR (2007) The Evolution of Multicellularity: A Minor Major Transition? Annu Rev Ecol Evol Syst 38:621–654. https://doi.org/10.1146/annurev.ecolsys.36.102403.114735

Grossman G, Helpman E (1991) Innovation and Growth in the Global Economy. MIT Press, Cambridge

Grow M (2008) U.S. Presidents and Latin American Interventions: Pursuing Regime Change in the Cold War. University Press of Kansas

Grubler A, Nakicenovic N, Pachauri S, et al (2014) Energy Primer. Laxenburg, Austria

Grullon G, Larkin Y, Michaely R (2015) The Disappearance of Public Firms and the Changing Nature of U.S. Industries. SSRN Electron J. https://doi.org/10.2139/ssrn.2612047

GSMA (2016) The Mobile Economy Report: 2016

Gu G, Dickens GR, Bhatnagar G, et al (2011) Abundant Early Palaeogene marine gas hydrates despite warm deep-ocean temperatures. Nat Geosci 4:848–851. https://doi.org/10.1038/ngeo1301

Guan R, Zhao Y, Zhang H, et al (2016) Draft genome of the living fossil Ginkgo biloba. Gigascience 5:49. https://doi.org/10.1186/s13742-016-0154-1

Guan Y (2003) Isolation and Characterization of Viruses Related to the SARS Coronavirus from Animals in Southern China. Science 302:276–278. https://doi.org/10.1126/science.1087139

Guerrera F, Baer J, Braithwaite T (2009) Goldman apologises for role in crisis. In: Financ. Times. https://www.ft.com/content/782afd66-d3bd-11de-8caf-00144feabdc0

Guilaine J (2017) Les Chemins de la protohistoire: Quand l'Occident s'éveillait (7000–2000 avant notre ère). Odile Jacob

Guilford MC, Hall CAS, O'Connor P, Cleveland CJ (2011) A New Long Term Assessment of Energy Return on Investment (EROI) for U.S. Oil and Gas Discovery and Production. Sustainability 3:1866–1887. https://doi.org/10.3390/su3101866

Guillemette Y, Turner D (2018) The Long View: Scenarios for the World Economy to 2060. OECD Economic Policy Papers, No. 22, OECD Publishing, Paris

Guinot B (2011) Solar time, legal time, time in use. Metrologia 48:S181–S185. https://doi.org/10.1088/0026-1394/48/4/S08

Gulagi A, Choudhary P, Bogdanov D, Breyer C (2017) Electricity system based on 100% renewable energy for India and SAARC. PLoS One 12:e0180611. https://doi.org/10.1371/journal.pone.0180611

Gusmão B de (1709) Reproduction facsimilé d'un dessin à la plume de sa description et de la pétition adressée au Jean V. (de Portugal) en langue latine et en écriture contemporaine (1709) retrouvés récemment dans les archives du Vatican du célèbre aéronef de Bartholomeu Lou. 17

Guterres A (2019) #WorldRefugeeDay. In: twitter. https://twitter.com/antonioguterres/status/1141526080169185280

GWI (2018) Wellness Now a $4.2 Trillion Global Industry – with 12.8% Growth from 2015-2017. In: Glob. Wellness Inst. https://globalwellnessinstitute.org/press-room/press-releases/wellness-now-a-4-2-trillion-global-industry/

Gyngell C, Douglas T, Savulescu J (2017) The Ethics of Germline Gene Editing. J Appl Philos 34:498–513. https://doi.org/10.1111/japp.12249

Haberl H, Beringer T, Bhattacharya SC, et al (2010) The global technical potential of bio-energy in 2050 considering sustainability constraints. Curr Opin Environ Sustain 2:394–403. https://doi.org/10.1016/j.cosust.2010.10.007

Haberl H, Erb K-H, Krausmann F (2014) Human Appropriation of Net Primary Production: Patterns, Trends, and Planetary Boundaries. Annu Rev Environ Resour 39:363–391. https://doi.org/10.1146/annurev-environ-121912-094620

Haberl H, Erb K-H, Krausmann F, et al (2013) Bioenergy: how much can we expect for 2050? Environ Res Lett 8:031004. https://doi.org/10.1088/1748-9326/8/3/031004

Haberl H, Erb KH, Krausmann F, et al (2007) Quantifying and mapping the human appropriation of net primary production in earth's terrestrial ecosystems. Proc Natl Acad Sci 104:12942–12947. https://doi.org/10.1073/pnas.0704243104

Hafele JC, Keating RE (1972) Around-the-World Atomic Clocks: Predicted Relativistic Time Gains. Science 177:166–168. https://doi.org/10.1126/science.177.4044.166

Haff P (2014a) Humans and technology in the Anthropocene: Six rules. Anthr Rev 1:126–136. https://doi.org/10.1177/2053019614530575

Haff PK (2014b) Technology as a geological phenomenon: implications for human well-being. Geol Soc London, Spec Publ 395:301 LP – 309. https://doi.org/10.1144/SP395.4

Hagner M, Mikola J, Saloniemi I, et al (2019) Effects of a glyphosate-based herbicide on soil animal trophic groups and associated ecosystem functioning in a northern agricultural field. Sci Rep 9:8540. https://doi.org/10.1038/s41598-019-44988-5

Hahladakis JN, Velis CA, Weber R, et al (2018) An overview of chemical additives present in plastics: Migration, release, fate and environmental impact during their use, disposal and recycling. J Hazard Mater 344:179–199. https://doi.org/10.1016/j.jhazmat.2017.10.014

Haight A (1935) Banned Books: Informal Notes on Some Books Banned for Various Reasons at Various Times and in Various Places. R. R. Bowker company, New York

Hajdu T, Hajdu G (2014) Reduction of Income Inequality and Subjective Well-Being in Europe. Econ Open-Access, Open-Assessment E-Journal 8:1. https://doi.org/10.5018/economics-ejournal.ja.2014-35

Hallmann CA, Sorg M, Jongejans E, et al (2017) More than 75 percent decline over 27 years in total flying insect biomass in protected areas. PLoS One 12:e0185809. https://doi.org/10.1371/journal.pone.0185809

Hamermesh DS (2019) Spending Time: The Most Valuable Resource. Oxford University Press

Hamilton C (2013) Earthmasters. The Dawn of the Age of Climate Engineering. Yale University Press

Hamilton WD (1964a) The genetical evolution of social behaviour. I. J Theor Biol 7:1–16. https://doi.org/10.1016/0022-5193(64)90038-4

Hamilton WD (1964b) The genetical evolution of social behaviour. II. J Theor Biol 7:17–52. https://doi.org/10.1016/0022-5193(64)90039-6

Hancock GJ, Tims SG, Fifield LK, Webster IT (2014) The release and persistence of radioactive anthropogenic nuclides. Geol Soc London, Spec Publ 395:265–281. https://doi.org/10.1144/SP395.15

Handberg-Thorsager M (2008) Stem cells and regeneration in planarians. Front Biosci Volume:6374. https://doi.org/10.2741/3160

Hansen AH (1938) Full recovery or stagnation? W.W. Norton, New York

Hansen J, Sato M, Hearty P, et al (2016) Ice melt, sea level rise and superstorms: evidence from paleoclimate data, climate modeling, and modern observations that 2°C global warming could be dangerous. Atmos Chem Phys 16:3761–3812. https://doi.org/10.5194/acp-16-3761-2016

Hansen JE, Sato M (2012) Paleoclimate Implications for Human-Made Climate Change. In: Climate Change. Springer Vienna, Vienna, pp 21–47

Harari Y (2015) Homo Deus: A brief history of tomorrow. Harvill Secker

Harari YN (2014) Sapiens: A Brief History of Humankind. Harvill Secker

Hardin G (1968) The Tragedy of the Commons. Science 162:1243–1248. https://doi.org/10.1126/science.162.3859.1243

Hardoon D (2017) An Economy for the 99%: It's time to build a human economy that benefits everyone, not just the privileged few

Hargreaves S (2013) North Dakota grows five times faster than nation. In: Cable News Netw. https://money.cnn.com/2013/06/06/news/economy/north-dakota-economy/index.html

Harmand S, Lewis JE, Feibel CS, et al (2015) 3.3-million-year-old stone tools from Lomekwi 3, West Turkana, Kenya. Nature 521:310–315. https://doi.org/10.1038/nature14464

Harvey CM (1994) The reasonableness of non-constant discounting. J Public Econ 53:31–51. https://doi.org/10.1016/0047-2727(94)90012-4

Harvey D (1990) The condition of postmodernity. An enquiry into the origins of cultural change. Wiley-Blackwell

Hassan FA (1997) Nile Floods and Political Disorder in Early Egypt. In: Dalfes HN, Kukla G, Weiss H (eds) Third Millennium BC Climate Change and Old World Collapse. Springer-Verlag Berlin Heidelberg, pp 1–23

Hassan R (2009) Empires of speed: time and the acceleration of politics and society. Brill, Leiden, The Netherlands

Hayden BY (2016) Time discounting and time preference in animals: A critical review. Psychon Bull Rev 23:39–53. https://doi.org/10.3758/s13423-015-0879-3

Hayek FA (1944) The Road to Serfdom. University of Chicago Press

Hazen RM, Papineau D, Bleeker W, et al (2008) Mineral evolution. Am Mineral 93:1693–1720. https://doi.org/10.2138/am.2008.2955

Heal GM (2005) Intertemporal welfare economics and the environment. In: Maler K-G, Jeffrey V (eds) Handbook of Environmental Economics, Vol. 3. North-Holland, Amsterdam, pp 1105–1145

Hearty PJ, Olson SL (2011) Preservation of trace fossils and models of terrestrial biota by intense storms in mid-last interglacial (MIS 5c) dunes on Bermuda. Palaios 26:394–405. https://doi.org/10.2110/palo.2010.p10-132r

Hearty PJ, Tormey BR (2017) Sea-level change and superstorms; geologic evidence from the last interglacial (MIS 5e) in the Bahamas and Bermuda offers ominous prospects for a warming Earth. Mar Geol 390:347–365. https://doi.org/10.1016/j.margeo.2017.05.009

Hegel GWF (1830) Lectures on the Philosophy of World History. Cambridge University Press (2012)

Heidegger M (1927) Being and Time, 1962 edit. SCM Press, London

Heidegger M (1959) Discourse on Thinking. trans. by John M. Anderson & E. Hans Freund; introd. by John M. Anderson, Harper & Row (1966), New York

Heilbron JL (1999) Electricity in the 17th & 18th Centuries: A Study in Early Modern Physics. Dover Pubns

Heinberg R (2011) The End of Growth: Adapting to Our New Economic Reality. Clairview Books

Heine H (1869) Letzte Gedichte Und Gedanken. Hoffmann und Campe, Hamburg

Heine H (1854a) Vermischte Schriften. Facsimile Publisher (2019)

Heine H (1854b) Lutetia, The works of Heinrich Heine. Translated by C. G. Leland (1893), London

Heine H (1855) Lutèce. Lettres sur la vie politique, artistique et sociale de la France. Michel Lévy Frères Éditeurs, Paris

Held D, McGrew A (2001) The Oxford Companion to Politics of the World, 2nd edn. Oxford University Press (2004)

Heller AC (2009) Ayn Rand and the World She Made. Knopf Doubleday Publishing Group

Helliwell J, Layard R, Sachs J (eds) (2017) World Happiness Report 2017. Sustainable Development Solutions Network, New York

Hénin P-Y (1983) L'impact macro-économique d'un choc pétrolier. Rev économique 34:865–896. https://doi.org/10.3406/reco.1983.408744

Henry JS (2012) The Price of Offshore Revisited: New Estimates for "Missing" Global Private Wealth, Income, Inequality, and Lost Taxes

Hensen B, Bernien H, Dréau AE, et al (2015) Loophole-free Bell inequality violation using electron spins separated by 1.3 kilometres. Nature 526:682–686. https://doi.org/10.1038/nature15759

Henshilwood CS (2002) Emergence of Modern Human Behavior: Middle Stone Age Engravings from South Africa. Science 295:1278–1280. https://doi.org/10.1126/science.1067575

Henshilwood CS, D'Errico F, Watts I (2009) Engraved ochres from the Middle Stone Age levels at Blombos Cave, South Africa. J Hum Evol 57:27–47. https://doi.org/10.1016/j.jhevol.2009.01.005

Henson RG (1988) Butler on Selfishness and Self-Love. Philos Phenomenol Res 49:31. https://doi.org/10.2307/2107991

Hepburn C (2007) Valuing the far-off future: discounting and its alternatives. In: Atkinson G, Dietz S, Neumayer E (eds) Handbook of Sustainable Development. Edward Elgar, Cheltenham, UK, pp 109–124

Hersh SM (1972) Rainmaking Is Used As Weapon by U.S. In: New York Times. https://www.nytimes.com/1972/07/03/archives/rainmaking-is-used-as-weapon-by-us-cloudseeding-in-indochina-is.html

Hershkovitz I, Weber GW, Quam R, et al (2018) The earliest modern humans outside Africa. Science 359:456–459. https://doi.org/10.1126/science.aap8369

Heuberger CF, Mac Dowell N (2018) Real-World Challenges with a Rapid Transition to 100% Renewable Power Systems. Joule 2:367–370. https://doi.org/10.1016/j.joule.2018.02.002

Hibbard KA, Crutzen PJ, Lambin EF, et al (2006) Decadal-scale interactions of humans and the environment. In: Costanza R, Graumlich LJ, Steffen W (eds) Sustainability or Collapse?: An Integrated History and Future of People on Earth (Dahlem Workshop Reports). MIT Press, Cambridge, Massachusetts

Hibbett D, Blanchette R, Kenrick P, Mills B (2016) Climate, decay, and the death of the coal forests. Curr Biol 26:R563–R567. https://doi.org/10.1016/j.cub.2016.01.014

Hilbert M, Lopez P (2011) The World's Technological Capacity to Store, Communicate, and Compute Information. Science 332:60–65. https://doi.org/10.1126/science.1200970

Hilgevoord J, Atkinson D (2011) Time in Quantum Mechanics. Oxford University Press

Hill E, Ma L (2017) Shale Gas Development and Drinking Water Quality. Am Econ Rev 107:522–525. https://doi.org/10.1257/aer.p20171133

Hirschhorn N, Initiative WTF (2005) The tobacco industry documents: what they are, what they tell us, and how to search them: a practical manual

Hobbes T (1642) De Cive. Clarendon Press (1983)

Hobsbawm E (1968) Industry and Empire: The Birth of the Industrial Revolution. Weidenfeld & Nicholson

Hoff MD, McNutt JG (1995) The Global Environmental Crisis: Implications forSocial Welfare and Social Work. J Sociol Soc Welf 22

Hoffmann DL, Angelucci DE, Villaverde V, et al (2018a) Symbolic use of marine shells and mineral pigments by Iberian Neandertals 115,000 years ago. Sci Adv 4:eaar5255. https://doi.org/10.1126/sciadv.aar5255

Hoffmann DL, Standish CD, García-Diez M, et al (2018b) U-Th dating of carbonate crusts reveals Neandertal origin of Iberian cave art. Science 359:912–915. https://doi.org/10.1126/science.aap7778

Höglund-Isaksson L (2012) Global anthropogenic methane emissions 2005–2030: technical mitigation potentials and costs. Atmos Chem Phys 12:9079–9096. https://doi.org/10.5194/acp-12-9079-2012

Holdren JP, Ehrlich PR (1974) Human Population and the Global Environment: Population growth, rising per capita material consumption, and disruptive technologies have made civilization a global ecological force. Am Sci 62:282–292

Holt R (2016) What now for science? Science 354:947–947. https://doi.org/10.1126/science.aal4180

Holt RPF, Pressman S, Spash CL (eds) (2009) Post Keynesian and Ecological Economics Confronting Environmental Issues

Holton GJ (1993) Science and Anti-Science. Harvard University Press, Cambridge, Massachusetts; London, England

Hoornweg D, Bhada-Tata P (2012) What a waste. A global review of solid waste management. Washington, DC

Hoornweg D, Bhada-Tata P, Kennedy C (2015) Peak Waste: When Is It Likely to Occur? J Ind Ecol 19:117–128. https://doi.org/10.1111/jiec.12165

Hoornweg D, Pope K (2017) Population predictions for the world's largest cities in the 21st century. Environ Urban 29:195–216. https://doi.org/10.1177/0956247816663557

Horgan J (1996) The End Of Science: Facing The Limits Of Knowledge In The Twilight Of The Scientific Age. Helix Books

Houdin J-P, Brier B (2008) Le secret de la Grande Pyramide. Fayard

House KZ, Baclig AC, Ranjan M, et al (2011) Economic and energetic analysis of capturing CO_2 from ambient air. Proc Natl Acad Sci 108:20428–20433. https://doi.org/10.1073/pnas.1012253108

House TJ, Near JB, Shields WB, et al (1996) Weather as a force multiplier: Owning the weather in 2025

Houston A (1748) Advice to a Young Tradesman, Written by an Old One (21 July 1748). In: Franklin: The Autobiography and Other Writings on Politics, Economics, and Virtue. Cambridge University Press (2004), pp 200–202

Howarth RW (2019) Ideas and perspectives: is shale gas a major driver of recent increase in global atmospheric methane? Biogeosciences 16:3033–3046. https://doi.org/10.5194/bg-16-3033-2019

Howell RT, Howell CJ (2008) The relation of economic status to subjective well-being in developing countries: A meta-analysis. Psychol Bull 134:536–560. https://doi.org/10.1037/0033-2909.134.4.536

Hoy RC (1994) Parmenides' Complete Rejection of Time. J Philos 91:573–598. https://doi.org/10.2307/2941069

Hoynes HW, Rothstein J (2019) Universal Basic Income in the US and Advanced Countries. National Bureau of Economic Research, Inc

HRW (2018) Heed the Call: A Moral and Legal Imperative to Ban Killer Robots. In: Hum. Rights Watch. https://www.hrw.org/report/2018/08/21/heed-call/moral-and-legal-imperative-ban-killer-robots

Hsiang SM, Burke M, Miguel E (2013) Quantifying the Influence of Climate on Human Conflict. Science 341:1235367–1235367. https://doi.org/10.1126/science.1235367

Hsieh PN, Zhou G, Yuan Y, et al (2017) A conserved KLF-autophagy pathway modulates nematode lifespan and mammalian age-associated vascular dysfunction. Nat Commun 8:914. https://doi.org/10.1038/s41467-017-00899-5

Huang D, Haack RA, Zhang R (2011) Does Global Warming Increase Establishment Rates of Invasive Alien Species? A Centurial Time Series Analysis. PLoS One 6:e24733. https://doi.org/10.1371/journal.pone.0024733

Huang K, Li X, Liu X, Seto KC (2019) Projecting global urban land expansion and heat island intensification through 2050. Environ Res Lett 14:114037. https://doi.org/10.1088/1748-9326/ab4b71

Hubble E (1929) A relation between distance and radial velocity among extra-galactic nebulae. Proc Natl Acad Sci 15:168–173. https://doi.org/10.1073/pnas.15.3.168

Hublin J-J, Ben-Ncer A, Bailey SE, et al (2017) New fossils from Jebel Irhoud, Morocco and the pan-African origin of Homo sapiens. Nature 546:289–292. https://doi.org/10.1038/nature22336

Huff CD, Xing J, Rogers AR, et al (2010) Mobile elements reveal small population size in the ancient ancestors of Homo sapiens. Proc Natl Acad Sci 107:2147–2152. https://doi.org/10.1073/pnas.0909000107

Hughes J (2004) Citizen Cyborg: Why Democratic Societies Must Respond To The Redesigned Human Of The Future. Westview Press

Hulme M (2014) Can Science Fix Climate Change?: A Case Against Climate Engineering. Wiley

Hume D (1777) Of the Rise and Progress of the Arts and Sciences. In: Copley S, Edgar A (eds) Selected Essays. Oxford University Press (1993), pp 56–77

Hume R (1921) The Thirteen Principal Upanishads. Oxford University Press, London, Edinburgh, Glasgow, New York, Toronto, Melbourne, Cape Town, Bombay

Hunter SM (2007) Optimizing cloud seeding for water and energy in California. A Pier-Final Project Report. Denver, Colorado

Huntington SP (1991) The Third Wave: Democratization in the Late Twentieth Century. University of Oklahoma Press

Huttunen S, Skytén E, Hildén M (2015) Emerging policy perspectives on geoengineering: An international comparison. Anthr Rev 2:14–32. https://doi.org/10.1177/2053019614557958

Huxley A (1932) Brave New World. Chatto & Windus, United Kingdom

Huybregts MAC (Riny) (2017) Phonemic clicks and the mapping asymmetry: How language emerged and speech developed. Neurosci Biobehav Rev 81:279–294. https://doi.org/10.1016/j.neubiorev.2017.01.041

Huygens C (1657) De ratiociniis in ludo aleae. Ex officinia J. Elsevirii

Hyam R (2002) Britain's Imperial Century, 1815–1914. A Study of Empire and Expansion. Palgrave Macmillan UK, London

IAASTD (2009) Agriculture at a crossroads. International Assessment of Agricultural Knowledge, Science and Technology for Development. International Assessment of Agricultural Knowledge, Science and Technology for Development (IAASTD), United Nations Environmental Programme (UNEP)

IARC (2017) Some Organophosphate Insecticides and Herbicides IARC Monographs on the Evaluation of Carcinogenic Risks to Humans Volume 112. Lyon, France

Ibisch PL, Hoffmann MT, Kreft S, et al (2016) A global map of roadless areas and their conservation status. Science 354:1423–1427. https://doi.org/10.1126/science.aaf7166

IBM (2012) Demystifying Big Data: Decoding The Big Data Commission Report. In: Int. Bus. Mach. Corp. https://www-304.ibm.com/events/wwe/grp/grp004.nsf/vLookupPDFs/TimPaydos%27Presentation/%24file/TimPaydos%27Presentation.pdf

IDC (2014) The digital universe of opportunities: rich data and the increasing value of the internet of things. In: Int. Data Corp. https://www.emc.com/leadership/digital-universe/2014iview/index.htm

IDMC (2017) Global Internal Displacement Database. In: Intern. Displac. Monit. Cent. http://www.internal-displacement.org/database

IEA (2013) Resources to Reserves, Oil, Gas and Coal Technologies for the Energy Markets of the Future

IEA (2015) World Energy Outlook 2015. Paris

IEA (2017a) World Energy Outlook 2017. Paris

IEA (2017b) Energy Technology Perspectives. In: Int. Energy Agency. https://www.iea.org/topics/energy-technology-perspectives

IEA (2017c) Energy Access Outlook 2017. OECD

IEA (2019a) Methane Tracker. In: Int. Energy Agency. https://www.iea.org/reports/methane-tracker-2020/interactive-country-and-regional-estimates

IEA (2019b) OECD total primary energy supply (TPES) by source, 1971-2018. In: Int. Energy Agency. https://www.iea.org/data-and-statistics/charts/oecd-total-primary-energy-supply-tpes-by-source-1971-2018

IEA (2019c) World energy balances 2019 edition database documentation

IEA (2019d) World Energy Outlook 2019. Paris

IEA (2020) Global Energy Review 2020: The impacts of the Covid-19 crisis on global energy demand and CO_2 emissions

IHME (2018) Findings from the Global Burden of Disease Study 2017

Illge L, Schwarze R (2006) A Matter of Opinion: How Ecological and Neoclassical Environmental Economists Think about Sustainability and Economics. DIW Berlin, German Institute for Economic Research

ILO (2015) World Employment and Social Outlook – trends 2015

ILO (2016) Department of Statistics. In: Int. Labour Organ. http://www.ilo.org/stat/lang--en/index.htm

ILO (2018) World Employment Social Outlook – 2018. Greening with Jobs. Geneva

Imbach P, Beardsley M, Bouroncle C, et al (2017) Climate change, ecosystems and smallholder agriculture in Central America: an introduction to the special issue. Clim Change 141:1–12. https://doi.org/10.1007/s10584-017-1920-5

Imbach P, Chou SC, Lyra A, et al (2018) Future climate change scenarios in Central America at high spatial resolution. PLoS One 13:e0193570. https://doi.org/10.1371/journal.pone.0193570

IMF (2016a) World Economic Outlook. International Monetary Fund (IMF)

IMF (2016b) Fiscal Monitor. World economic and financial surveys, 0258-7440. Washington, DC

IMF (2017a) World Economic Outlook. International Monetary Fund (IMF)

IMF (2017b) Real GDP growth. In: Int. Monet. Fund. https://www.imf.org/external/datamapper/ NGDP_RPCH@WEO/OEMDC/ADVEC/WEOWORLD?year=2017

IMF (2018) Global financial stability report. A decade after the global financial crisis. Are we safer?

IMF (2020) Chapter 1: Global Prospects and Policies. In: World Economic Outlook, April 2020: The Great Lockdown. International Monetary Fund (IMF), p 25

Innes DC (1992) Francis Bacon, Christianity and the Hope of Modern Science. Boston College

Inoue JG, Miya M, Lam K, et al (2010) Evolutionary Origin and Phylogeny of the Modern Holocephalans (Chondrichthyes: Chimaeriformes): A Mitogenomic Perspective. Mol Biol Evol 27:2576–2586. https://doi.org/10.1093/molbev/msq147

Ioannidis JPA (2005) Why Most Published Research Findings Are False. PLoS Med 2:e124. https:// doi.org/10.1371/journal.pmed.0020124

IPBES (2016) The assessment report of the Intergovernmental Science-Policy Platform on Biodiversity and Ecosystem Services on pollinators, pollination and food production. Secretariat of the Intergovernmental Science-Policy Platform on Biodiversity and Ecosystem Services, Bonn, Germany

IPBES (2019) Global assessment report on biodiversity and ecosystem services of the Intergovernmental Science-Policy Platform on Biodiversity and Ecosystem Services. IPBES secretariat, Bonn, Germany

IPCC (1990) Climate change: The IPCC scientific assessment. Cambridge University Press, Cambridge, New York, Port Chester, Melbourne, Sydney

IPCC (1994) Climate Change 1994: Radiative Forcing of Climate Change and An Evaluation of the IPCC IS92 Emission Scenarios. Cambridge University Press, Cambridge

IPCC (2000) Special Report on Emissions Scenarios: A special report of Working Group III of the Intergovernmental Panel on Climate Change. Cambridge University Press, Cambridge, UK

IPCC (2012) Meeting Report of the Intergovernmental Panel on Climate Change Expert Meeting on Geoengineering. IPCC Working Group III Technical Support Unit, Potsdam Institute for Climate Impact Research, Potsdam, Germany

IPCC (2013) Climate Change 2013: The Physical Science Basis. Contribution of Working Group I to the Fifth Assessment Report of the Intergovernmental Panel on Climate Change. Cambridge University Press, Cambridge, United Kingdom and New York, NY, USA

IPCC (2014a) Climate Change 2014: Synthesis Report. Contribution of Working Groups I, II and III to the Fifth Assessment Report of the Intergovernmental Panel on Climate Change. IPCC, Geneva, Switzerland

IPCC (2014b) Climate Change 2014: Impacts, Adaptation, and Vulnerability. Part A:Global and Sectoral Aspects. Contribution of Working Group II to the FifthAssessment Report of the Intergovernmental Panel on Climate Change. Cambridge University Press, Cambridge, United Kingdom and New York, NY, USA

IPCC (2014c) Climate Change 2014: Mitigation of Climate Change. Contribution of Working Group III to the Fifth Assessment Report of the Intergovernmental Panel on Climate Change. Cambridge University Press, Cambridge, United Kingdom and New York, NY, USA

IPCC (2018) Global Warming of 1.5C. An IPCC Special Report on the impacts of global warming of 1.5C above pre-industrial levels and related global greenhouse gas emission pathways, in the context of strengthening the global response to the threat of climate change,. In Press

IRENA (2018) Global energy transformation: A roadmap to 2050 (2018 edition). Abu Dhabi

Irvine P, Emanuel K, He J, et al (2019) Halving warming with idealized solar geoengineering moderates key climate hazards. Nat Clim Chang 9:295–299. https://doi.org/10.1038/s41558-019-0398-8

Isasi R, Kleiderman E, Knoppers BM (2016) Editing policy to fit the genome? Science 351:337–339. https://doi.org/10.1126/science.aad6778

Istvan Z (2013) The Transhumanist Wager. Futurity Imagine Media

Ito A (2011) A historical meta-analysis of global terrestrial net primary productivity: are estimates converging? Glob Chang Biol 17:3161–3175. https://doi.org/10.1111/j.1365-2486.2011.02450. x

IUCN (2014) The IUCN Red List of Threatened Species. In: Int. Union Conserv. Nature's. http://www.iucnredlist.org/

Iyengar S, Massey DS (2019) Scientific communication in a post-truth society. Proc Natl Acad Sci 116:7656–7661. https://doi.org/10.1073/pnas.1805868115

Jackson RB, Friedlingstein P, Andrew RM, et al (2019) Persistent fossil fuel growth threatens the Paris Agreement and planetary health. Environ Res Lett 14:121001. https://doi.org/10.1088/1748-9326/ab57b3

Jackson RB, Vengosh A, Carey JW, et al (2014) The Environmental Costs and Benefits of Fracking. Annu Rev Environ Resour 39:327–362. https://doi.org/10.1146/annurev-environ-031113-144051

Jackson T (2009) Prosperity without growth? The transition to a sustainable economy. Sustainable Development Commission

Jacobson MZ, Delucchi MA, Cameron MA, Frew BA (2015) Low-cost solution to the grid reliability problem with 100% penetration of intermittent wind, water, and solar for all purposes. Proc Natl Acad Sci 112:15060–15065. https://doi.org/10.1073/pnas.1510028112

Jacobson MZ, Delucchi MA, Bauer ZAF, et al (2017a) 100% Clean and Renewable Wind, Water, and Sunlight All-Sector Energy Roadmaps for 139 Countries of the World. Joule 1:108–121. https://doi.org/10.1016/j.joule.2017.07.005

Jacobson MZ, Delucchi MA, Cameron MA, Frew BA (2017b) The United States can keep the grid stable at low cost with 100% clean, renewable energy in all sectors despite inaccurate claims. Proc Natl Acad Sci 114:E5021–E5023. https://doi.org/10.1073/pnas.1708069114

Jacobson MZ, Delucchi MA, Cameron MA, Mathiesen B V. (2018) Matching demand with supply at low cost in 139 countries among 20 world regions with 100% intermittent wind, water, and sunlight (WWS) for all purposes. Renew Energy 123:236–248. https://doi.org/10.1016/j.renene.2018.02.009

Jafferis NT, Helbling EF, Karpelson M, Wood RJ (2019) Untethered flight of an insect-sized flapping-wing microscale aerial vehicle. Nature 570:491–495. https://doi.org/10.1038/s41586-019-1322-0

Jagtap SS (2019) Systems Evaluation of Subsonic Hybrid-Electric Propulsion Concepts for NASA N+3 Goals and Conceptual Aircraft Sizing. Int J Automot Mech Eng 16:7259–7286. https://doi.org/10.15282/ijame.16.4.2019.07.0541

Jallon B, Napolitano U, Boutté F (eds) (2017) Paris Haussmann: A Model's Relevance. Park Books; Bilingual edition

James I. (1604) A counter-blaste to tobacco. Rodale Press, London (1954)

James P, Steger MB (2014) A Genealogy of 'Globalization': The Career of a Concept. Globalizations 11:417–434. https://doi.org/10.1080/14747731.2014.951186

James W (1918) The Principles of Psychology, Volume 1 (of 2). Henry Holt and Company, New York

Jamieson KH (2018) Crisis or self-correction: Rethinking media narratives about the well-being of science. Proc Natl Acad Sci 115:2620–2627. https://doi.org/10.1073/pnas.1708276114

Jancovici J-M (2012a) Les limites énergétiques de la croissance. Debat 171:80. https://doi.org/10.3917/deba.171.0080

Jancovici J-M (2012b) Quelques réflexions sur la transition énergétique. In: jancovici.com. https://jancovici.com/transition-energetique/choix-de-societe/quelques-reflexions-sur-la-transition-energetique/

Jang JG, Kim GM, Kim HJ, Lee HK (2016) Review on recent advances in CO_2 utilization and sequestration technologies in cement-based materials. Constr Build Mater 127:762–773. https://doi.org/10.1016/j.conbuildmat.2016.10.017

Janssens L, Giemsch L, Schmitz R, et al (2018) A new look at an old dog: Bonn-Oberkassel reconsidered. J Archaeol Sci 92:126–138. https://doi.org/10.1016/j.jas.2018.01.004

Jarvis KE, Parry SJ, Piper JM (2001) Temporal and Spatial Studies of Autocatalyst-Derived Platinum, Rhodium, and Palladium and Selected Vehicle-Derived Trace Elements in the Environment. Environ Sci Technol 35:1031–1036. https://doi.org/10.1021/es0001512

Jauss HR, Benzinger E (1970) Literary History as a Challenge to Literary Theory. New Lit Hist 2:7–37. https://doi.org/10.2307/468585

Jègues-Wolkiewiez C (2000) Lascaux, vision du ciel des Magdalénien, Arte preistorica e tribale. Conservazione e salvaguardia dei messaggi. Symp 2000 d'Art Rupestre, Val Camonica

Jevons WS (1865) The Coal Question: An Inquiry concerning the Progress of the Nation, and the Probable Exhaustion of our Coal-mines. Macmillan

Jevrejeva S, Jackson LP, Riva REM, et al (2016) Coastal sea level rise with warming above 2°C. Proc Natl Acad Sci 113:13342–13347. https://doi.org/10.1073/pnas.1605312113

Jiang J, Cao L, MacMartin DG, et al (2019) Stratospheric Sulfate Aerosol Geoengineering Could Alter the High-Latitude Seasonal Cycle. Geophys Res Lett 46:14153–14163. https://doi.org/10.1029/2019GL085758

Jinnah S, Nicholson S (2019) The hidden politics of climate engineering. Nat Geosci 12:876–879. https://doi.org/10.1038/s41561-019-0483-7

Johansson TB, Nakicenovic N, Patwardhan A, Gomez-Echeverri L (eds) (2012) Global Energy Assessment (GEA) - Toward a Sustainable Future. Cambridge University Press, Cambridge

Jones A, Haywood J, Boucher O, et al (2010a) Geoengineering by stratospheric SO 2 injection: results from the Met Office HadGEM2 climate model and comparison with the Goddard Institute for Space Studies ModelE. Atmos Chem Phys 10:5999–6006. https://doi.org/10.5194/acp-10-5999-2010

Jones CG, Gutiérrez JL, Byers JE, et al (2010b) A framework for understanding physical ecosystem engineering by organisms. Oikos 119:1862–1869. https://doi.org/10.1111/j.1600-0706.2010.18782.x

Jones CI (1995a) Time Series Tests of Endogenous Growth Models. Q J Econ 110:495–525. https://doi.org/10.2307/2118448

Jones CI (1995b) R & D-Based Models of Economic Growth. J Polit Econ 103:759–784. https://doi.org/10.1086/262002

Jones KE, Patel NG, Levy MA, et al (2008) Global trends in emerging infectious diseases. Nature 451:990–993. https://doi.org/10.1038/nature06536

Jönsson KI, Rabbow E, Schill RO, et al (2008) Tardigrades survive exposure to space in low Earth orbit. Curr Biol 18:R729–R731. https://doi.org/10.1016/j.cub.2008.06.048

Joordens JCA, D'Errico F, Wesselingh FP, et al (2015) Homo erectus at Trinil on Java used shells for tool production and engraving. Nature 518:228–231. https://doi.org/10.1038/nature13962

Joppa LN, Loarie SR, Pimm SL (2008) On the protection of "protected areas." Proc Natl Acad Sci 105:6673–6678. https://doi.org/10.1073/pnas.0802471105

Josephus F (1961) Jewish Antiquities, Loeb Class. Harvard University Press

Juris JS (2008) Networking Futures The Movements against Corporate Globalization. Duke University Press

Kaeberlein M (2013) Longevity and aging. F1000Prime Rep 5:. https://doi.org/10.12703/P5-5

Kaeberlein M, Rabinovitch PS, Martin GM (2015) Healthy aging: The ultimate preventative medicine. Science 350:1191–1193. https://doi.org/10.1126/science.aad3267

Kagermann H, Wahlster W, Helbig J (2013) Securing the future of German manufacturing industry: Recommendations for implementing the strategic initiative INDUSTRIE 4.0, Final report of the Industrie 4.0 Working Group. Office of the Industry-Science Research Alliance, Frankfurt

Kahn ME, Mohaddes K, Ng RNC, et al (2019) Long-Term Macroeconomic Effects of Climate Change: A Cross-Country Analysis

Kahneman D, Deaton A (2010) High income improves evaluation of life but not emotional well-being. Proc Natl Acad Sci 107:16489–16493. https://doi.org/10.1073/pnas.1011492107

Kalenscher T, Pennartz CMA (2008) Is a bird in the hand worth two in the future? The neuroeconomics of intertemporal decision-making. Prog Neurobiol 84:284–315. https://doi.org/10.1016/j.pneurobio.2007.11.004

Kang X, He W, Huang Y, et al (2016) Introducing precise genetic modifications into human 3PN embryos by CRISPR/Cas-mediated genome editing. J Assist Reprod Genet 33:581–588. https://doi.org/10.1007/s10815-016-0710-8

Kant A (2018) Composite Water Management Index (CWMI). A National Tool for Water Measurement, Management and Improvement. In: NITI Aayog. https://niti.gov.in/writereaddata/files/new_initiatives/presentation-on-CWMI.pdf

Kant I (1784) Idea for a Universal History with a Cosmopolitan Purpose. In: Reiss H, Nisbet HB (eds) Political Writings. Cambridge University Press (1991), Cambridge

Kass L (2003) Ageless bodies, happy souls: biotechnology and the pursuit of perfection. In: The New Atlantis. Spring, pp 9–28

Kaza S, Yao LC, Bhada-Tata P, Van Woerden F (2018) What a Waste 2.0: A Global Snapshot of Solid Waste Management to 2050. World Bank Group

Keefe PR (2017) The family that built an empire of pain. In: New Yorker. https://www.newyorker.com/magazine/2017/10/30/the-family-that-built-an-empire-of-pain

Keeley B (2015) Debate the Issues: Complexity and Policy making. Economic Co-operation and Development (OECD)

Keeling CD, Bacastow RB, Bainbridge AE, et al (1976) Atmospheric carbon dioxide variations at Mauna Loa Observatory, Hawaii. Tellus 28:538–551. https://doi.org/10.3402/tellusa.v28i6.11322

Keen S (2011) Debunking Economics – Revised and Expanded Edition: The Naked Emperor Dethroned? Zed Books; Expanded, Revised edition

Keith D (2010) Engineering the Planet. In: Schneider S, Mastrandrea M (eds) Climate Change Science and Policy by S. Schneider and M. Mastrandrea. Island Press, Washington DC, p 49

Keith DW (2017) Toward a Responsible Solar Geoengineering Research Program. Issues Sci Technol 33

Keith DW, Weisenstein DK, Dykema JA, Keutsch FN (2016) Stratospheric solar geoengineering without ozone loss. Proc Natl Acad Sci 113:14910–14914. https://doi.org/10.1073/pnas.1615572113

Keller DP, Feng EY, Oschlies A (2014) Potential climate engineering effectiveness and side effects during a high carbon dioxide-emission scenario. Nat Commun 5:3304. https://doi.org/10.1038/ncomms4304

Kelley CP, Mohtadi S, Cane MA, et al (2015) Climate change in the Fertile Crescent and implications of the recent Syrian drought. Proc Natl Acad Sci 112:3241–3246. https://doi.org/10.1073/pnas.1421533112

Kelly R (1883) The Alternative: A Study in Psychology. Macmillan, 1882. J Ment Sci 29:271–278. https://doi.org/10.1192/bjp.29.126.271

Kemunto N, Mogoa E, Osoro E, et al (2018) Zoonotic disease research in East Africa. BMC Infect Dis 18:545. https://doi.org/10.1186/s12879-018-3443-8

Kennedy CA, Stewart I, Facchini A, et al (2015) Energy and material flows of megacities. Proc Natl Acad Sci 112:5985–5990. https://doi.org/10.1073/pnas.1504315112

Kennedy JF (1961) Address in New York City to the National Association of Manufacturers (496), December 5, 1961

Kennett DJ, Breitenbach SFM, Aquino V V., et al (2012) Development and Disintegration of Maya Political Systems in Response to Climate Change. Science 338:788–791. https://doi.org/10.1126/science.1226299

Kennett JP, Stott LD (1991) Abrupt deep-sea warming, palaeoceanographic changes and benthic extinctions at the end of the Palaeocene. Nature 353:225–229. https://doi.org/10.1038/353225a0

Kenny C (2019) The Real Immigration Crisis The Problem Is Not Too Many, but Too Few. In: Foreign Aff. https://www.foreignaffairs.com/articles/2019-11-11/real-immigration-crisis

Kent C, Pope E, Thompson V, et al (2017) Using climate model simulations to assess the current climate risk to maize production. Environ Res Lett 12:054012. https://doi.org/10.1088/1748-9326/aa6cb9

Kepler J (1634) Kepler's Somnium: The Dream, Or Posthumous Work on Lunar Astronomy. Courier Dover Publications (1967), New York

Kerkhove R (2006) Dark Religion? Aztec Perspectives on Human Sacrifice. In: Hartney C, McGarrity A (eds) The Dark Side: Proceedings of the Seventh Australian and International Religion, Literature and the Arts Conference. University of Sydney RLA Press, Sydney, pp 136–160

Kerr JFR, Wyllie AH, Currie AR (1972) Apoptosis: A Basic Biological Phenomenon with Wide-Ranging Implications in Tissue Kinetics. Br J Cancer 26:239–257. https://doi.org/10.1038/bjc.1972.33

Keynes JM (1923) A Tract On Monetary Reform. Macmillan and Co., limited, London

Keynes JM (1933) Essays in Persuasion. W. W. Norton & Co., New York

Keynes JM (1930) Economic Possibilities for Our Grandchildren. In: Essays in Persuasion. Palgrave Macmillan UK (2010), London, pp 321–332

Khanna P, Francis D (2016) These 25 Companies Are More Powerful Than Many Countries. In: Foreign Policy's. https://foreignpolicy.com/2016/03/15/these-25-companies-are-more-powerful-than-many-countries-multinational-corporate-wealth-power/

Ki-moon B (2007) We should welcome the dawn of the migration age. In: The Guardian https://www.theguardian.com/commentisfree/2007/jul/10/comment.globalisation

Kibble A, Cabianca T, Daraktchieva Z, Gooding T, Smithard J, et al (2014) Review of the Potential Public Health Impacts of Exposures to Chemical and Radioactive Pollutants as a Result of the Shale Gas Extraction Process. Public Health England

Kidd B (1894) Social Evolution. Cambridge University Press (2009)

Kidd D, McIntosh K (2016) Social Media and Social Movements. Sociol Compass 10:785–794. https://doi.org/10.1111/soc4.12399

Kidwell SM (2015) Biology in the Anthropocene: Challenges and insights from young fossil records. Proc Natl Acad Sci 112:4922–4929. https://doi.org/10.1073/pnas.1403660112

Kiernan VG (2005) America, the new imperialism: from white settlement to world hegemony. Verso

Kim HL, Ratan A, Perry GH, et al (2014) Khoisan hunter-gatherers have been the largest population throughout most of modern-human demographic history. Nat Commun 5:5692. https://doi.org/10.1038/ncomms6692

Kim S, Hwang J, Lee D (2008) Prefrontal Coding of Temporally Discounted Values during Intertemporal Choice. Neuron 59:161–172. https://doi.org/10.1016/j.neuron.2008.05.010

King R (2012) Theories and typologies of migration: an overvew and a primer. Malmö, Sweden

Kinzer S (2006) Overthrow: America's Century of Regime Change from Hawaii to Iraq. Henry Holt and Company

Kirkwood TBL, Austad SN (2000) Why do we age? Nature 408:233–238. https://doi.org/10.1038/35041682

Klein Goldewijk K, Beusen A, Janssen P (2010) Long-term dynamic modeling of global population and built-up area in a spatially explicit way: HYDE 3.1. The Holocene 20:565–573. https://doi.org/10.1177/0959683609356587

Kluser S, Peduzzi P (2007) Global Pollinator Decline: A Literature Review. UNEP/GRID-Europe

Knoll A., Javaux E., Hewitt D, Cohen P (2006) Eukaryotic organisms in Proterozoic oceans. Philos Trans R Soc B Biol Sci 361:1023–1038. https://doi.org/10.1098/rstb.2006.1843

Koblitz N (1981) Mathematics as Propaganda. In: Mathematics Tomorrow. Springer New York, New York, NY, pp 111–120

Koch PL, Barnosky AD (2006) Late Quaternary Extinctions: State of the Debate. Annu Rev Ecol Evol Syst 37:215–250. https://doi.org/10.1146/annurev.ecolsys.34.011802.132415

Koke S, Grebing C, Frei H, et al (2010) Direct frequency comb synthesis with arbitrary offset and shot-noise-limited phase noise. Nat Photonics 4:462–465. https://doi.org/10.1038/nphoton.2010.91

Komlos J, Smith PK, Bogin B (2004) Obesity and the rate of time preference: is there a connection? J Biosoc Sci 36:209–219. https://doi.org/10.1017/S0021932003006205

Kondash A, Vengosh A (2015) Water Footprint of Hydraulic Fracturing. Environ Sci Technol Lett 2:276–280. https://doi.org/10.1021/acs.estlett.5b00211

Kondash AJ, Lauer NE, Vengosh A (2018) The intensification of the water footprint of hydraulic fracturing. Sci Adv 4:eaar5982. https://doi.org/10.1126/sciadv.aar5982

Kooijmans L, Smirnov Y, Solecki RS, et al (1989) On the Evidence for Neandertal Burial. Curr Anthropol 30:322–330. https://doi.org/10.1086/203747

Kopp RE, Kemp AC, Bittermann K, et al (2016) Temperature-driven global sea-level variability in the Common Era. Proc Natl Acad Sci 113:E1434–E1441. https://doi.org/10.1073/pnas.1517056113

Kopp RE, Simons FJ, Mitrovica JX, et al (2009) Probabilistic assessment of sea level during the last interglacial stage. Nature 462:863–867. https://doi.org/10.1038/nature08686

Korfiati A, Gkonos C, Veronesi F, et al (2016) Estimation of the global solar energy potential and photovoltaic cost with the use of open data. Int J Sustain Energy Plan Manag 9:17–29. https://doi.org/10.5278/ijsepm.2016.9.3

Koselleck R (2002) The practice of conceptual history: timing history, spacing concepts. Stanford University Press, Stanford

Koselleck R, Tribe K (2004) Futures Past. Columbia University Press

Kossin JP (2018) A global slowdown of tropical-cyclone translation speed. Nature 558:104–107. https://doi.org/10.1038/s41586-018-0158-3

KPCB (2013) Internet trends D11 Conference. In: D11 Conference. Kleiner Perkins Caufield & Byers, p 117

Krasnova IN, Cadet JL (2009) Methamphetamine toxicity and messengers of death. Brain Res Rev 60:379–407. https://doi.org/10.1016/j.brainresrev.2009.03.002

Kreft S, Eckstein D, Melchior I (2016) Global Climate Risk Index 2017: Who Suffers Most From Extreme Weather Events? Weather-related Loss Events in 2015 and 1995 to 2015. Germanwatch e.V., Bonn, Berlin

Kreuter J (2015) echnofix, Plan B or Ultima Ratio? A Review of the Social Science Literature on Climate Engineering Technologies

Krimsky S (2003) Science in the Private Interest: Has the Lure of Profits Corrupted Biomedical Research? Rowman & Littlefield Publishers

Kromdijk J, Glowacka K, Leonelli L, et al (2016) Improving photosynthesis and crop productivity by accelerating recovery from photoprotection. Science 354:857–861. https://doi.org/10.1126/science.aai8878

Krugman P (2007) The Conscience of a Liberal. W.W. Norton Company

Kubota S (2011) Repeating rejuvenation in Turritopsis, an immortal hydrozoan (Cnidaria, Hydrozoa). Biogeography 101:1–3

Kumar M, Babaei P, Ji B, Nielsen J (2016) Human gut microbiota and healthy aging: Recent developments and future prospective. Nutr Heal Aging 4:3–16. https://doi.org/10.3233/NHA-150002

Kürschner WM, van der Burgh J, Visscher H, Dilcher DL (1996) Oak leaves as biosensors of late neogene and early pleistocene paleoatmospheric CO_2 concentrations. Mar Micropaleontol 27:299–312. https://doi.org/10.1016/0377-8398(95)00067-4

Kurzweil R (1999) The Age of Spiritual Machines. Viking Press

Kushner D (2013) The Real Story of Stuxnet How Kaspersky Lab tracked down the malware that stymied Iran's nuclear-fuel enrichment program. In: IEEE Spectr. https://spectrum.ieee.org/telecom/security/the-real-story-of-stuxnet

La Mettrie JO (1747) L'Homme Machine. Gallimard (1999), Paris

Lafer G (2017) The One Percent of Solution. Cornell University Press, Ithaca, NY

Lai C, Li J, Wang Z, et al (2018) Drought-Induced Reduction in Net Primary Productivity across Mainland China from 1982 to 2015. Remote Sens 10:1433. https://doi.org/10.3390/rs10091433

Laist DW (1997) Impacts of Marine Debris: Entanglement of Marine Life in Marine Debris Including a Comprehensive List of Species with Entanglement and Ingestion Records. In: J.M. C, D.B. R (eds) Marine Debris. Springer Series on Environmental Management. Springer, New York, NY, pp 99–139

Laitner JA (2013) Linking Energy Efficiency to Economic Productivity: Recommendations for Improving the Robustness of the U.S. Economy. Washington, DC

Lakner C, Milanovic B (2016) Global Income Distribution: From the Fall of the Berlin Wall to the Great Recession. World Bank Econ Rev 30:203–232. https://doi.org/10.1093/wber/lhv039

Lander ES (2016) The Heroes of CRISPR. Cell 164:18–28. https://doi.org/10.1016/j.cell.2015.12.041

Landrigan PJ, Fuller R, Acosta NJR, et al (2018) The Lancet Commission on pollution and health. Lancet 391:462–512. https://doi.org/10.1016/S0140-6736(17)32345-0

Lane L (2006) Strategic Policy Options for the Bush Administration Climate Policy. AEI Press, Washington D.C.

Langley C, Parkinson S (2009) Science and the corporate agenda. The detrimental effects of commercial influence on science and technology. Folkestone, UK

Langs R (2004) Death Anxiety and the Emotion-Processing Mind. Psychoanal Psychol 21:31–53. https://doi.org/10.1037/0736-9735.21.1.31

Lanphier E, Urnov F, Haecker SE, et al (2015) Don't edit the human germ line. Nature 519:410–411. https://doi.org/10.1038/519410a

Larrasoaña JC, Roberts AP, Rohling EJ, et al (2003) Three million years of monsoon variability over the northern Sahara. Clim Dyn 21:689–698. https://doi.org/10.1007/s00382-003-0355-z

Las Casas B de (1552) Brevísima relación de la destrucción de las Indias. ANTIGONA

Latham J, Rasch P, Chen C-C, et al (2008) Global temperature stabilization via controlled albedo enhancement of low-level maritime clouds. Philos Trans R Soc A Math Phys Eng Sci 366:3969–3987. https://doi.org/10.1098/rsta.2008.0137

Latouche S (2007) Petit traité de la décroissance sereine. Fayard/Mille et une nuits, Paris

Lawrance EC (1991) Poverty and the Rate of Time Preference: Evidence from Panel Data. J Polit Econ 99:54–77. https://doi.org/10.1086/261740

Lawton JH, McCredie R (eds) (1995) Extinction Rates. Oxford University Press

Laybourn-Langton L, Rankin L, Baxter D (2019) This is a crisis. Facing up to the age of environmental breakdown (Initial report). London

Le Quéré C, Andrew RM, Friedlingstein P, et al (2018) Global Carbon Budget 2017. Earth Syst Sci Data 10:405–448. https://doi.org/10.5194/essd-10-405-2018

Leadley PW, Krug CB, Alkemade R, et al (2014) Progress towards the Aichi Biodiversity Targets: An Assessment of Biodiversity Trends, Policy Scenarios and Key Actions. Secretariat of the Convention on Biological Diversity

Lee K-F (2018) AI Superpowers: China, Silicon Valley, and the New World Order. Houghton Mifflin Harcourt, New York

Leggett J, Pepper WJ, Swart RJ (1992) Emissions scenarios for the IPCC: an update. In: Houghton JT, Callander BA, Varney SK (eds) Climate Change 1992: The Report to the IPCC Scientific Assessment. Cambridge University Press, New York, Victoria, pp 69–95

Lehner M (1997) The Complete Pyramids. Solving the Ancient Mysteries. Thames & Hudson Ltd, London

Lehto RH, Stein KF (2009) Death Anxiety: An Analysis of an Evolving Concept. Res Theory Nurs Pract 23:23–41. https://doi.org/10.1891/1541-6577.23.1.23

Leifer MS, Pusey MF (2017) Is a time symmetric interpretation of quantum theory possible without retrocausality? Proc R Soc A Math Phys Eng Sci 473:20160607. https://doi.org/10.1098/rspa.2016.0607

Lelieveld J, Klingmüller K, Pozzer A, et al (2019) Effects of fossil fuel and total anthropogenic emission removal on public health and climate. Proc Natl Acad Sci 116:7192–7197. https://doi. org/10.1073/pnas.1819989116

Lenin VI (1902) What Is to Be Done? Penguin Classics (1988)

Lenski G (2005) Ecological-Evolutionary Theory: Principles and Applications. Paradigm Publishers

Lenski GE (1966) Power and privilege: a theory of social stratification. McGraw-Hill

Leonhardt D (2017) Our Broken Economy, in One Simple Chart. In: New York Times. www.nytimes. com/interactive/2017/08/07/opinion/leonhardt-income-inequality.html. Accessed 7 Aug 2017

Lepousez V, Gassiat C, Ory C, et al (2017) Climate Risk Impact Screening: the methodological guidebook. Paris

Leroi-Gourhan A (1975) The Flowers Found with Shanidar IV, a Neanderthal Burial in Iraq. Science 190:562–564. https://doi.org/10.1126/science.190.4214.562

Leslie J (1996) The end of the world: the science and ethics of human extinction. Routledge, London

Levine RV, Norenzayan A (1999) The Pace of Life in 31 Countries. J Cross Cult Psychol 30:178–205. https://doi.org/10.1177/0022022199030002003

Levitsky S, Ziblatt D (2018) This is how democracies die. In: The Guardian. www.theguardian. com/us-news/commentisfree/2018/jan/21/this-is-how-democracies-die

Levitt T (1983) Globalization of Markets. Harv Bus Rev 10

Levitz S (2018) We Owe Central American Migrants Much More Than This. In: Intelligencer. https://nymag.com/intelligencer/2018/06/we-owe-central-american-migrants-much-more-than-this.html

Lewis J (2018) Economic Impact of Cybercrime – No Slowing Down. Santa Clara, CA

Liang C (2016) Genetically Modified Crops with Drought Tolerance: Achievements, Challenges, and Perspectives. In: Drought Stress Tolerance in Plants, Vol 2. Springer International Publishing, Cham, pp 531–547

Liang P, Xu Y, Zhang X, et al (2015) CRISPR/Cas9-mediated gene editing in human tripronuclear zygotes. Protein Cell 6:363–372. https://doi.org/10.1007/s13238-015-0153-5

Liebenberg L (2006) Persistence Hunting by Modern Hunter–Gatherers. Curr Anthropol 47:1017–1026. https://doi.org/10.1086/508695

Liebenberg L (2008) The relevance of persistence hunting to human evolution. J Hum Evol 55:1156–1159. https://doi.org/10.1016/j.jhevol.2008.07.004

Lieberman DE, Bramble DM, Raichlen DA, Shea JJ (2009) Brains, Brawn, and the Evolution of Human Endurance Running Capabilities. In: The First Humans – Origin and Early Evolution of the Genus Homo. Vertebrate Paleobiology and Paleoanthropology. Springer, Dordrecht, pp 77–92

Lieberthal K, Jisi W (2012) Addressing US-China strategic distrust. The Brookings Institution, Washington DC

Lin D, Hanscom L, Murthy A, et al (2018) Ecological Footprint Accounting for Countries: Updates and Results of the National Footprint Accounts, 2012–2018. Resources 7:58. https://doi.org/10.3390/resources7030058

Lin MT, Occhialini A, Andralojc PJ, et al (2014) A faster Rubisco with potential to increase photosynthesis in crops. Nature 513:547–550. https://doi.org/10.1038/nature13776

Lind RC (1982) A Primer on the Major Issues Relating to the Discount Rate for Evaluating National Energy Options. In: Lind RC (ed) Discounting for Time and Risk in Energy Policy. Resources for the Future, Washington, DC, pp 21–94

Linden N, Popescu S, Short AJ, Winter A (2009) Quantum mechanical evolution towards thermal equilibrium. Phys Rev E 79:061103. https://doi.org/10.1103/PhysRevE.79.061103

Lindley D (1993) The end of physics: the myth of a unified theory. Basic Books

Lindner AB, Madden R, Demarez A, et al (2008) Asymmetric segregation of protein aggregates is associated with cellular aging and rejuvenation. Proc Natl Acad Sci 105:3076–3081. https://doi. org/10.1073/pnas.0708931105

Lipset SM (1997) American Exceptionalism: A Double-Edged Sword. W. W. Norton & Company

Lisiecki LE, Raymo ME (2005) A Pliocene-Pleistocene stack of 57 globally distributed benthic δ 18 O records. Paleoceanography 20. https://doi.org/10.1029/2004PA001071

Litfin KT (2014) Ecovillages: Lessons for Sustainable Community. Polity Press

Lobell JA (2012) New Life for the Lion Man. Archaeology 65

Locke J (1690) An Essay Concerning Human Understanding, 1997th edn. Penguin Books, London

Lockley A, Mi Z, Coffman D (2019) Geoengineering and the blockchain: Coordinating Carbon Dioxide Removal and Solar Radiation Management to tackle future emissions. Front Eng Manag 6:38–51. https://doi.org/10.1007/s42524-019-0010-y

Loki R (2019) Indigenous peoples go to court to save the Amazon from oil company greed. In: Salon.com, LLC. http://www.salon.com/2019/04/12/indigenous-peoples-go-to-court-to-save-the-amazon-from-oil-company-greed_partner/

Lombardo U, Iriarte J, Hilbert L, et al (2020) Early Holocene crop cultivation and landscape modification in Amazonia. Nature. https://doi.org/10.1038/s41586-020-2162-7

Long SP, Zhu X-G, Naidu SL, Ort DR (2006) Can improvement in photosynthesis increase crop yields? Plant, Cell Environ 29:315–330. https://doi.org/10.1111/j.1365-3040.2005.01493.x

Loulergue L, Schilt A, Spahni R, et al (2008) Orbital and millennial-scale features of atmospheric CH_4 over the past 800,000 years. Nature 453:383–386. https://doi.org/10.1038/nature06950

Lowery CM, Fraass AJ (2019) Morphospace expansion paces taxonomic diversification after end Cretaceous mass extinction. Nat Ecol Evol 3:900–904. https://doi.org/10.1038/s41559-019-0835-0

Lucas RE, Schimmack U (2009) Income and well-being: How big is the gap between the rich and the poor? J Res Pers 43:75–78. https://doi.org/10.1016/j.jrp.2008.09.004

Luhrs J (1997) The Simple Living Guide: A Sourcebook for Less Stressful, More Joyful Living. Broadway Books

Lukacs M (2012) World's biggest geoengineering experiment "violates" UN rules. In: The Guardian. www.theguardian.com/environment/2012/oct/15/pacific-iron-fertilisation-geoengineering

Lüthi D, Le Floch M, Bereiter B, et al (2008) High-resolution carbon dioxide concentration record 650 000–800 000 years before present. Nature 453:379–382. https://doi.org/10.1038/nature06949

Lyotard J-F (1979) La Condition postmoderne: Rapport sur le savoir. Les Éditions de Minuit

Ma H, Marti-Gutierrez N, Park S-W, et al (2017) Correction of a pathogenic gene mutation in human embryos. Nature 548:413–419. https://doi.org/10.1038/nature23305

MacAskill W (2015) Doing Good Better: Effective Altruism and a Radical New Way to Make a Difference. Guardian Faber Publishing

Macaulay V (2005) Single, Rapid Coastal Settlement of Asia Revealed by Analysis of Complete Mitochondrial Genomes. Science 308:1034–1036. https://doi.org/10.1126/science.1109792

Mackay C (1854) Memories of extraordinary popular delusions: and the madness of crowds, Volumes 1–2. Routledge, London

MacLeod KG, Quinton PC, Seplveda J, Negra MH (2018) Postimpact earliest Paleogene warming shown by fish debris oxygen isotopes (El Kef, Tunisia). Science 360:1467–1469. https://doi.org/10.1126/science.aap8525

MacMartin DG, Ricke KL, Keith DW (2018) Solar geoengineering as part of an overall strategy for meeting the 1.5C Paris target. Philos Trans R Soc A Math Phys Eng Sci 376:20160454. https://doi.org/10.1098/rsta.2016.0454

Maddison A (2006) The World Economy. OECD

Maddison A (2007a) Contours of the World Economy 1-2030 AD. Oxford University Press

Maddison A (2007b) Chinese economic performance in the long run. Organisation for Economic Co-operation and Development (OECD)

Majkić A, D'Errico F, Stepanchuk V (2018) Assessing the significance of Palaeolithic engraved cortexes. A case study from the Mousterian site of Kiik-Koba, Crimea. PLoS One 13:e0195049. https://doi.org/10.1371/journal.pone.0195049

Malmberg G (1997) Time and Space in International Migration. In: Hammar T, Brochmann G, Tamas K, Faist T (eds) International Migration, Immobility and Development. Multidisciplinary Perspectives. Routledge, Berg, Oxford, pp 21–48

Malthus T (1798) An Essay on the Principle ofPopulationAn Essay on the Principle of Population, as itAffects the Future Improvement of Societywith Remarks on the Speculations of Mr. Godwin,M. Condorcet, and Other Writers. J. Johnson, London

Mann ME, Rahmstorf S, Kornhuber K, et al (2017) Influence of Anthropogenic Climate Change on Planetary Wave Resonance and Extreme Weather Events. Sci Rep 7:45242. https://doi.org/10.1038/srep45242

Mansfield BD, Mumm RH (2014) Survey of Plant Density Tolerance in U.S. Maize Germplasm. Crop Sci 54:157–173. https://doi.org/10.2135/cropsci2013.04.0252

Mantoux PJ (1906) La révolution industrielle au XVIIIème siècle, essai sur les commencements de la grande histoire moderne en Angleterre. Thèse 1906, révisée 1928, éd. Genin (1959), Paris

Manyika J, Lund S, Chui M, et al (2017) Jobs lost, jobs gained: What the future of work will mean for jobs, skills, and wages

Marcott SA, Shakun JD, Clark PU, Mix AC (2013) A Reconstruction of Regional and Global Temperature for the Past 11 300 Years. Science 339:1198–1201. https://doi.org/10.1126/science.1228026

Marshack A (1964) Lunar Notation on Upper Paleolithic Remains. Science 146:743–745. https://doi.org/10.1126/science.146.3645.743

Marshall A (2019) Louvre Removes Sackler Family Name From Its Walls. In: New York Times. https://www.nytimes.com/2019/07/17/arts/design/sackler-family-louvre.html

Martindale RC, Aberhan M (2017) Response of macrobenthic communities to the Toarcian Oceanic Anoxic Event in northeastern Panthalassa (Ya Ha Tinda, Alberta, Canada). Palaeogeogr Palaeoclimatol Palaeoecol 478:103–120. https://doi.org/10.1016/j.palaeo.2017.01.009

Martínez-Alier J (1987) Ecological Economics: Energy, Environment and Society. Basil Blackwell, Oxford

Martínez-Botí MA, Foster GL, Chalk TB, et al (2015) Plio-Pleistocene climate sensitivity evaluated using high-resolution CO_2 records. Nature 518:49–54. https://doi.org/10.1038/nature14145

Martínez DE (1998) Mortality Patterns Suggest Lack of Senescence in Hydra. Exp Gerontol 33:217–225. https://doi.org/10.1016/S0531-5565(97)00113-7

Marvel K, Kravitz B, Caldeira K (2013) Geophysical limits to global wind power. Nat Clim Chang 3:118–121. https://doi.org/10.1038/nclimate1683

Marx K (1867) Das Kapital. Erster Band. Buch I: Der Produktionsprocess des Kapitals. Verlag Von Otto Meissner, Hamburg

Marx K (1885) Das Kapital. Erster Band. Buch II: Der Produktionsprocess des Kapitals. Verlag Von Otto Meissner

Marx L, Mazlish B (1998) Introduction. In: Marx L, Mazlish B (eds) Progress: Fact Or Illusion? University of Michigan Press, pp 1–7

Mason DJ (2018) Forced Separation of Children from Parents: Another Consideration. JAMA 320:963. https://doi.org/10.1001/jama.2018.12154

Mason P (2015) PostCapitalism: A Guide to Our Future Hardcover. Allen Lane

Masson-Delmotte V, Schulz M, Abe-Ouchi A, et al (2013) Information from Paleoclimate Archives. In: Stocker TF, Qin D, Plattner G-K, et al. (eds) Climate Change 2013: The Physical Basis, Contribution of Working Group I to the Fifth Assessment Report of the Intergovernmental Panel on Climate Change. Cambridge University Press, Cambridge, United Kingdom

Masters RK, Reither EN, Powers DA, et al (2013) The Impact of Obesity on US Mortality Levels: The Importance of Age and Cohort Factors in Population Estimates. Am J Public Health 103:1895–1901. https://doi.org/10.2105/AJPH.2013.301379

Matek B, Gawell K (2015) The Benefits of Baseload Renewables: A Misunderstood Energy Technology. Electr J 28:101–112. https://doi.org/10.1016/j.tej.2015.02.001

Mathews JT (1989) Redefining Security. Foreign Aff 68:162.https://doi.org/10.2307/20043906

Mathiesen BV, Lund H, Connolly D, et al (2015) Smart Energy Systems for coherent 100% renewable energy and transport solutions. Appl Energy 145:139–154. https://doi.org/10.1016/j.apenergy.2015.01.075

Matthews D (2015) I spent a weekend at Google talking with nerds about charity. I came away ...worried. In: Vox. https://www.vox.com/2015/8/10/9124145/effective-altruism-global-ai

Matthews HD, Gillett NP, Stott PA, Zickfeld K (2009) The proportionality of global warming to cumulative carbon emissions. Nature 459:829–832. https://doi.org/10.1038/nature08047

Mbaye S, Badia MM (2019) New Data on Global Debt. In: Int. Monet. Fund Blog. https://blogs.imf.org/2019/01/02/new-data-on-global-debt/

Mcbrearty S, Brooks AS (2000) The revolution that wasn't: a new interpretation of the origin of modern human behavior. J Hum Evol 39:453–563. https://doi.org/10.1006/jhev.2000.0435

McCauley DJ, Pinsky ML, Palumbi SR, et al (2015) Marine defaunation: Animal loss in the global ocean. Science 347:1255641–1255641. https://doi.org/10.1126/science.1255641

McClure SM (2004) Separate Neural Systems Value Immediate and Delayed Monetary Rewards. Science 306:503–507. https://doi.org/10.1126/science.1100907

McCown T. (1937) Excavations of Mugharet es-Skhül. In: Garrod D, Bate DMA (eds) The Stone Age of Mount Carmel, Vol.1: Excavations in the Wady el-Mughara. Clarendon, Oxford, pp 91–107

McDonald D (2015) The Firm: The Inside Story of McKinsey: The World's Most Controversial Management Consultancy. Oneworld Publications

McGlade C, Ekins P (2015) The geographical distribution of fossil fuels unused when limiting global warming to 2°C. Nature 517:187–190. https://doi.org/10.1038/nature14016

McInerney FA, Wing SL (2011) The Paleocene-Eocene Thermal Maximum: A Perturbation of Carbon Cycle, Climate, and Biosphere with Implications for the Future. Annu Rev Earth Planet Sci 39:489–516. https://doi.org/10.1146/annurev-earth-040610-133431

McKinsey (2016) Income inequality: why so many households are not advancing. In: McKinsey Co. https://www.mckinsey.com/featured-insights/employment-and-growth/income-inequality-why-so-many-households-are-not-advancing#

McKnight SA (2006) The Religious Foundations of Francis Bacon's Thought. University of Missouri Press

McNeill JR, Engelke P (2016) The Great Acceleration: An Environmental History of the Anthropocene Since 1945. Harvard University Press, Cambridge, Massachusetts

McTaggart JE (1908) The Unreality of Time. Mind 17:457–474

MEA (2005) Ecosystems and Human Well-being: Synthesis. Island Press, Washington, DC

Meadows DH, Randers J, Meadows DL (2004) Limits to Growth: The 30-Year Update, 3rd edn. Chelsea Green Publishing

Meadows LH, Meadows DL, Randers J, Behrens III WW (1972) The Limits to Growth. Universe Books, New York

Medina FM, Bonnaud E, Vidal E, et al (2011) A global review of the impacts of invasive cats on island endangered vertebrates. Glob Chang Biol 17:3503–3510. https://doi.org/10.1111/j.1365-2486.2011.02464.x

Meek R (1963) The economics of physiocracy. Harvard University Press, Cambridge

Meinshausen M, Meinshausen N, Hare W, et al (2009) Greenhouse-gas emission targets for limiting global warming to 2°C. Nature 458:1158–1162. https://doi.org/10.1038/nature08017

Meinshausen M, Smith SJ, Calvin K, et al (2011) The RCP greenhouse gas concentrations and their extensions from 1765 to 2300. Clim Change 109:213–241. https://doi.org/10.1007/s10584-011-0156-z

Meixner MD, Le Conte Y (2016) A current perspective on honey bee health. Apidologie 47:273–275. https://doi.org/10.1007/s13592-016-0449-3

Mendelsohn R, Emanuel K, Chonabayashi S, Bakkensen L (2012) The impact of climate change on global tropical cyclone damage. Nat Clim Chang 2:205–209. https://doi.org/10.1038/nclimate1357

Mercer AM, Keith DW, Sharp JD (2011) Public understanding of solar radiation management. Environ Res Lett 6:044006. https://doi.org/10.1088/1748-9326/6/4/044006

Mercure J-F, Pollitt H, Viñuales JE, et al (2018) Macroeconomic impact of stranded fossil fuel assets. Nat Clim Chang 8:588–593. https://doi.org/10.1038/s41558-018-0182-1

Mersenne M (1636) Harmonie universelle, contenant la théorie et la pratique de la musique. Paris

Michaels D (2008) Doubt is Their Product: How Industry's Assault on Science Threatens Your Health. Oxford University Press

Milankovitch M (1930) Mathematische Klimalehre und Astronomische Theorie der Klimaschwankungen Volume 1 de Handbuch der Klimatologie Parte 1,Volume 1 de Handbuch der Klimatologie: In fünf Bänden. Gebrüder Borntraeger, Berlin

Milanovic B (2012) Global Income Inequality by the Numbers: in History and Now. An Overview

Mill JS (1848) Principles of Political Economy. John W. Parker, London

Miller LM, Gans F, Kleidon A (2011) Estimating maximum global land surface wind power extractability and associated climatic consequences. Earth Syst Dyn 2:1–12. https://doi.org/10.5194/esd-2-1-2011

Milman O (2016) Trump to scrap Nasa climate research in crackdown on 'politicized science.' In: The Guardian. https://www.theguardian.com/environment/2016/nov/22/nasa-earth-donald-trump-eliminate-climate-change-research

Missirian A, Schlenker W (2017) Asylum applications respond to temperature fluctuations. Science 358:1610–1614. https://doi.org/10.1126/science.aao0432

MITEI (2018) The Future of Nuclear Energy in a Carbon-Constrained World. An Interdisciplinary MIT Study. Massachusetts

Mithen S (1996) The prehistory of the mind. A search for the origins of art, religion and science

Miyagawa S, Lesure C, Nóbrega VA (2018) Cross-Modality Information Transfer: A Hypothesis about the Relationship among Prehistoric Cave Paintings, Symbolic Thinking, and the Emergence of Language. Front Psychol 9:. https://doi.org/10.3389/fpsyg.2018.00115

Mobus GE, Kalton MC (2015) Principles of Systems Science. Springer New York, New York, NY

Moffitt TE, Arseneault L, Belsky D, et al (2011) A gradient of childhood self-control predicts health, wealth, and public safety. Proc Natl Acad Sci 108:2693–2698. https://doi.org/10.1073/pnas.1010076108

Mokyr J (1990) The Lever of Riches: Technological Creativity and Economic Progress. Oxford University Press

Montgomery D (1989) The Fall of the House of Labor: The Workplace, the State, and American Labor Activism, 1865–1925. Cambridge University Press

Moon T, Ahlstrøm A, Goelzer H, et al (2018) Rising Oceans Guaranteed: Arctic Land Ice Loss and Sea Level Rise. Curr Clim Chang Reports 4:211–222. https://doi.org/10.1007/s40641-018-0107-0

Moore GE (1965) Cramming more components onto integrated circuits, Reprinted from Electronics. Electronics 38:114–117

Moore JC, Gladstone R, Zwinger T, Wolovick M (2018a) Geoengineer polar glaciers to slow sea-level rise. Nature 555:303–305. https://doi.org/10.1038/d41586-018-03036-4

Moore JK, Fu W, Primeau F, et al (2018b) Sustained climate warming drives declining marine biological productivity. Science 359:1139–1143. https://doi.org/10.1126/science.aao6379

Moore P (2015) The Weather Experiment: The Pioneers Who Sought to See the Future. Farrar, Straus and Giroux

Mora C, Tittensor DP, Adl S, et al (2011) How Many Species Are There on Earth and in the Ocean? PLoS Biol 9:e1001127. https://doi.org/10.1371/journal.pbio.1001127

Mora C, Wei C-L, Rollo A, et al (2013) Biotic and Human Vulnerability to Projected Changes in Ocean Biogeochemistry over the 21st Century. PLoS Biol 11:e1001682. https://doi.org/10.1371/journal.pbio.1001682

More T (1516) Utopia. More, Habsburg Netherlands

Moreno MM, Enrich LB de L, Tejado J del P, Sierra JM (2014) Los primeros aprovechamientos de aguas subterráneas en la Península Ibérica. Las motillas de Daimiel en la Edad del Bronce de La Mancha. Boletín Geológico y Min 125:455–474

Morgan PJ, King RW, Shapiro II (1985) Length of day and atmospheric angular momentum: A comparison for 1981–1983. J Geophys Res 90:12645. https://doi.org/10.1029/JB090iB14p12645

Morgan TH (1898) Experimental studies of the regeneration of Planaria maculata. Arch für Entwickelungsmechanik der Org 7:364–397. https://doi.org/10.1007/BF02161491

Moriarty P, Honnery D (2012) What is the global potential for renewable energy? Renew Sustain Energy Rev 16:244–252. https://doi.org/10.1016/j.rser.2011.07.151

Morozov E (2013) To Save Everything, Click Here: The Folly of Technological Solutionism. Public Affairs, New York

Moskalev A, Aliper A, Smit-McBride Z, et al (2014) Genetics and epigenetics of aging and longevity. Cell Cycle 13:1063–1077. https://doi.org/10.4161/cc.28433

Moss J, McMann M, Rae J, et al (2011) Energy Equity and Environmental Security

Mounteney J, Bo A, Oteo A (2016) The internet and drug markets. Luxembourg

Mourre V, Villa P, Henshilwood CS (2010) Early Use of Pressure Flaking on Lithic Artifacts at Blombos Cave, South Africa. Science 330:659–662. https://doi.org/10.1126/science.1195550

Muffe C, Feit S, Fuhr L, et al (2019) Fuel to the Fire: How Geoengineering Threatens to Entrench Fossil Fuels and Accelerate the Climate Crisis

Mumford L (1934) Technics and civilization. Harcourt, Brace and Co, New York

Munich RE (2019) Climate Change and Natural Disasters Overview. In: Munich RE. https://www.munichre.com/topics-online/en/climate-change-and-natural-disasters.html

Murray C (2006) In Our Hands: A Plan To Replace The Welfare State. Aei Press

Mystakidou K, Parpa E, Tsilika E, et al (2005) The Evolution of Euthanasia and Its Perceptions in Greek Culture and Civilization. Perspect Biol Med 48:95–104. https://doi.org/10.1353/pbm.2005.0013

Naghavi M, Marczak LB, Kutz M, et al (2018) Global Mortality From Firearms, 1990–2016. JAMA 320:792. https://doi.org/10.1001/jama.2018.10060

Nagy B, Farmer JD, Bui QM, Trancik JE (2013) Statistical Basis for Predicting Technological Progress. PLoS One 8:e52669. https://doi.org/10.1371/journal.pone.0052669

Nagy B, Farmer JD, Trancik JE, Gonzales JP (2011) Superexponential long-term trends in information technology. Technol Forecast Soc Change 78:1356–1364. https://doi.org/10.1016/j.techfore.2011.07.006

NASA (2019) Carbon Dioxide. In: Natl. Aeronaut. Sp. Adm. https://climate.nasa.gov/vital-signs/carbon-dioxide/

NASEM (2017a) Valuing Climate Changes. National Academies Press, Washington, D.C.

NASEM (2017b) Human Genome Editing. National Academies Press, Washington, D.C.

NBER (2018) US Business Cycle Expansions and Contractions. In: Natl. Bur. Econ. Res. https://www.nber.org/cycles

NCHS (2018) United States, 2017: With special feature on mortality. Hyattsville, MD

Nelsen MP, DiMichele WA, Peters SE, Boyce CK (2016) Delayed fungal evolution did not cause the Paleozoic peak in coal production. Proc Natl Acad Sci 113:2442–2447. https://doi.org/10.1073/pnas.1517943113

Nelson C (2006) Thomas Paine: Enlightenment, Revolution, and the Birth of Modern Nations. Viking Books, New York

Neuberg GW (2009) The Cost of End-of-Life Care. Circ Cardiovasc Qual Outcomes 2:127–133. https://doi.org/10.1161/CIRCOUTCOMES.108.829960

Neumayer E (2004) The environment, left-wing political orientation and ecological economics. Ecol Econ 51:167–175. https://doi.org/10.1016/j.ecolecon.2004.06.006

Newbold T (2018) Future effects of climate and land-use change on terrestrial vertebrate community diversity under different scenarios. Proc R Soc B Biol Sci 285:20180792. https://doi.org/10.1098/rspb.2018.0792

Newingham BA, Vanier CH, Charlet TN, et al (2013) No cumulative effect of 10 years of elevated [CO_2] on perennial plant biomass components in the Mojave Desert. Glob Chang Biol 19:2168–2181. https://doi.org/10.1111/gcb.12177

Newton I (1687) Philosophiæ naturalis principia mathematica. Jussu Societatis Regiæ ac Typis Joseph Streater ..., Londini

Ni X, Gebo DL, Dagosto M, et al (2013) The oldest known primate skeleton and early haplorhine evolution. Nature 498:60–64. https://doi.org/10.1038/nature12200

Nietzsche F (1882) The Gay Science. Cambridge University Press (2001)

Nietzsche F (1883) Thus Spoke Zarathustra. Cambridge University Press (2006), Cambridge

Nietzsche F (2003) Nietzsche: Writings from the Late Notebooks (Cambridge Texts in the History of Philosophy). Cambridge University Press

Nilsson M, Fielden FJ (1920) Primitive time-reckoning; a study in the origins and first development of the art of counting time among the primitive and early culture peoples. Lund: C.W.K. Gleerup

Nisbet RA (1980) History of the Idea of Progress. Transaction Publishers

NOAA (2016) Effects of Oil and Gas Activities in the Arctic Ocean: Final Environmental Impact Statement

NOAA (2019) Trends in Atmospheric Carbon Dioxide. In: ESRL's Glob. Monit. Lab. Natl. Ocean. Atmos. Adm. https://www.esrl.noaa.gov/gmd/ccgg/trends/global.html

NOAA (2020) Frequently Asked Questions. In: Natl. Ocean. Atmos. Adm. (NOAA), Hurric. Res. Div. https://www.aoml.noaa.gov/hrd/tcfaq/C5c.html

Nordhaus T (2018) The Two-Degree Delusion The Dangers of an Unrealistic Climate Change Target. In: Foreign Aff. https://www.foreignaffairs.com/articles/world/2018-02-08/two-degree-delusion

Nordhaus T, Shellenberger M (2004) The death of environmentalism. Global warming in a post-environmental world

Nordhaus T, Shellenberger M (2007) Break Through: From the Death of Environmentalism to the Politics of Possibility. Houghton Mifflin Harcourt

Novacek MJ (2001) The Biodiversity Crisis Losing What Counts. The New Press

Novacek MJ, Cleland EE (2001) The current biodiversity extinction event: Scenarios for mitigation and recovery. Proc Natl Acad Sci 98:5466–5470. https://doi.org/10.1073/pnas.091093698

Nowak MA, Tarnita CE, Wilson EO (2010) The evolution of eusociality. Nature 466:1057–1062. https://doi.org/10.1038/nature09205

NPL (2005) News from the National Physical Laboratory. Metromnia Wintr 2005:4

NRC (2010) Impact of Genetically Engineered Crops on Farm Sustainability in the United States. National Academies Press, Washington, D.C.

NRC (2015a) Climate Intervention: Carbon Dioxide Removal and Reliable Sequestration, National Academies of Sciences. National Research Council (NRC), National Academies Press, Washington, D.C.

NRC (2015b) Climate Intervention: Reflecting Sunlight to Cool Earth. National Research Council (NRC), National Academies Press, Washington, D.C.

NRC (2003) Critical Issues in Weather Modification Research. National Research Council (NRC), National Academies Press, Washington, D.C.

O'Connell M (2017) To be a Machine: Adventures Among Cyborgs, Utopians, Hackers, and the Futurists Solving the Modest Problem of Death. Doubleday

O'Neill BC, Kriegler E, Riahi K, et al (2014) A new scenario framework for climate change research: the concept of shared socioeconomic pathways. Clim Change 122:387–400. https://doi.org/10.1007/s10584-013-0905-2

O'Rourke KH, Williamson JG (2002) When did globalisation begin? Eur Rev Econ Hist 6:23–50. https://doi.org/10.1017/S1361491602000023

Odling-Smee FJ, Laland KN, Feldman MW (1996) Niche Construction. Am Nat 147:641–648. https://doi.org/10.1086/285870

Odling-Smee J, Laland KN (2011) Ecological Inheritance and Cultural Inheritance: What Are They and How Do They Differ? Biol Theory 6:220–230. https://doi.org/10.1007/s13752-012-0030-x

Odum AL (2011) Delay discounting: Trait variable? Behav Processes 87:1–9. https://doi.org/10.1016/j.beproc.2011.02.007

OECD (2013) OECD Guidelines on Measuring Subjective Well-being. Organization for Economic Co-operation and Development (OECD)

OECD (2016a) Making Cities Work for All: Data and Actions for Inclusive Growth. Organization for Economic Co-operation and Development (OECD) Publishing, Paris

OECD (2016b) Unemployment rate. In: Organ. Econ. Co-operation Dev. https://data.oecd.org/unemp/unemployment-rate.htm

OECD (2016c) Inclusive Framework on Base Erosion and Profit Shifting. In: Organ. Econ. Co-operation Dev. http://www.oecd.org/ctp/beps/

OECD (2017) Interim Economic Outlook

OECD (2018) States of Fragility 2018. Organization for Economic Co-operation and Development

Ofcom (2018) The Communications Market 2018

Ohler N (2015) Der totale Rausch Drogen im Dritten Reich. Kiepenheuer & Witsch Gmbh

Oldham P, Szerszynski B, Stilgoe J, et al (2014) Mapping the landscape of climate engineering. Philos Trans R Soc A Math Phys Eng Sci 372:20140065. https://doi.org/10.1098/rsta.2014.0065

Oliver MJ (2014) Why we need GMO crops in agriculture. Mo Med 111:492–507

Olivier JGJ, Janssens-Maenhou G, Muntean M, Peter JAHW (2015) Trends in global CO_2 emissions: 2015 Report. PBL Netherlands Environmental Assessment Agency

Ongena J (2016) Nuclear fusion and its large potential for the future world energy supply. Nukleonika 61:425–432. https://doi.org/10.1515/nuka-2016-0070

Oramah B, Dzene R (2019) Globalisation and the Recent Trade Wars: Linkages and Lessons. Glob Policy 10:401–404. https://doi.org/10.1111/1758-5899.12707

Oreskes N (2018) Beware: transparency rule is a Trojan Horse. Nature 557:469–469. https://doi.org/10.1038/d41586-018-05207-9

Oreskes N, Conway EM (2010) Merchants of Doubt. Bloomsbury Press

Orwell G (1949) 1984. Secker & Warburg, London

Osborne P (1992) Modernity Is a Qualitative, Not a Chronological, Category: Notes on the Dialectics of Differential Historical Time. In: Barker F, Hulme P, Iversen M (eds) Postmodernism and the Re-reading of Modernity. Manchester University Press, p 322

Oshio T, Kobayashi M (2010) Income inequality, perceived happiness, and self-rated health: Evidence from nationwide surveys in Japan. Soc Sci Med 70:1358–1366. https://doi.org/10.1016/j.socscimed.2010.01.010

Ossiander M, Riemensberger J, Neppl S, et al (2018) Absolute timing of the photoelectric effect. Nature 561:374–377. https://doi.org/10.1038/s41586-018-0503-6

Österberg J (1988) Self and Others. Springer Netherlands, Dordrecht

Otañez MG, Mamudu HM, Glantz SA (2009) Tobacco Companies' Use of Developing Countries' Economic Reliance on Tobacco to Lobby Against Global Tobacco Control: The Case of Malawi. Am J Public Health 99:1759–1771. https://doi.org/10.2105/AJPH.2008.146217

Owen R (1836) The book of the new moral world, containing the rational system of society, founded on demonstrable facts, developing the constitution and laws of human nature and of society. E. Wilson, London

Owens J (1978) The Doctrine of Being in the Aristotelian Metaphysics: A Study in the Greek Background of Mediaeval Thought, 3rd edn. Pontifical Institute of Mediaeval Studies, Toronto, Canadá

Paavola J (2012) Climate Change: The Ultimate Tragedy of the Commons? In: Cole DH, Ostrom E (eds) Property in Land and Other Resources. Lincoln Institute for Land Policy, pp 417–433

Pacala S (2004) Stabilization Wedges: Solving the Climate Problem for the Next 50 Years with Current Technologies. Science 305:968–972. https://doi.org/10.1126/science.1100103

Paddock RC, Sijabat DM (2020) Where Bats Are Still on the Menu, if No Longer the Best Seller. In: New York Times. https://www.nytimes.com/2020/05/13/world/asia/coronavirus-bats-market-Indonesia.html

Páez-Osuna F (2001) The Environmental Impact of Shrimp Aquaculture: Causes, Effects, and Mitigating Alternatives. Environ Manage 28:131–140. https://doi.org/10.1007/s002670010212

Pagani M, Huber M, Liu Z, et al (2011) The Role of Carbon Dioxide During the Onset of Antarctic Glaciation. Science 334:1261–1264. https://doi.org/10.1126/science.1203909

Page DN (1976) Particle emission rates from a black hole: Massless particles from an uncharged, nonrotating hole. Phys Rev D 13:198–206. https://doi.org/10.1103/PhysRevD.13.198

Paine T (1797) Agrarian Justice. In: The Origins of Universal Grants. Palgrave Macmillan UK (2004), London

Paine T (1776) Common Sense. Applewood Books (2002)

Paine T (1791) Rights of Man. Grove Press (2008)

Paine T (1792) Letter addressed to the addressers, on the late proclamation

Paine T (1794) The age of reason; being an investigation of true and fabulous theology. Broadview Press Ltd (2011)

Pal JS, Eltahir EAB (2016) Future temperature in southwest Asia projected to exceed a threshold for human adaptability. Nat Clim Chang 6:197–200. https://doi.org/10.1038/nclimate2833

Parfrey LW, Lahr DJG (2013) Multicellularity arose several times in the evolution of eukaryotes (Response to DOI 10.1002/bies.201100187). BioEssays 35:339–347. https://doi.org/10.1002/bies.201200143

Parker A, Horton JB, Keith DW (2018) Stopping Solar Geoengineering Through Technical Means: A Preliminary Assessment of Counter-Geoengineering. Earth's Future 6:1058–1065. https://doi.org/10.1029/2018EF000864

Parker AG, Goudie AS, Stokes S, et al (2006) A Record of Holocene Climate Change from Lake Geochemical Analyses in Southeastern Arabia. Quat Res 66:465–476. https://doi.org/10.1016/j.yqres.2006.07.001

Parker DE, Wilson H, Jones PD, et al (1996) The Impact of Mount Pinatubo on World-Wide Temperatures. Int J Climatol 16:487–497. https://doi.org/10.1002/(SICI)1097-0088(199605)16:5%3C487::AID-JOC39%3E3.0.CO;2-J

Parker J (1999) Spendthrift in America? On Two Decades of Decline in the U.S. Saving Rate. Cambridge, MA

Patterson S (2010) The Quants: How a New Breed of Math Whizzes Conquered Wall Street and Nearly Destroyed It. Crown

Pauling L (1958) Genetic and Somatic Effects of Carbon-14: This by-product of nuclear-weapon testing may do more genetic and somatic damage than has been supposed. Science 128:1183–1186. https://doi.org/10.1126/science.128.3333.1183

Pavlis EC, Sindoni G, Paolozzi A, et al (2017) El Niño effects on earth rotation parameters from LAGEOS and LARES orbital analysis. In: 2017 IEEE International Conference on Environment and Electrical Engineering and 2017 IEEE Industrial and Commercial Power Systems Europe (EEEIC / I&CPS Europe). IEEE, pp 1–6

Peabody JB (ed) (1973) John Adams: A Biography in His Own Words. New York, Newsweek

Pearce D, Groom B, Hepburn C, et al (2003) Valuing the future: Recent advances in social discounting. World Econ 4:121–141

Pearce DW, Turner RK (1989) Economics of Natural Resources and the Environment, 1989th, The edn.

Peres S (2016) Saving the gene pool for the future: Seed banks as archives. Stud Hist Philos Sci Part C Stud Hist Philos Biol Biomed Sci 55:96–104. https://doi.org/10.1016/j.shpsc.2015.09.002

Perlow LA (2012) Sleeping with Your Smartphone: How to Break the 24/7 Habit and Change the Way You Work. Harvard Business Review Press

Perrault C (1688) Parallèle Des Anciens Et Des Modernes. Tome 2 (Éd.1688–1697). Hachette Groupe Livre (2012)

Perry ED, Ciliberto F, Hennessy DA, Moschini G (2016) Genetically engineered crops and pesticide use in U.S. maize and soybeans. Sci Adv 2:e1600850. https://doi.org/10.1126/sciadv.1600850

Petit JR, Jouzel J, Raynaud D, et al (1999) Climate and atmospheric history of the past 420,000 years from the Vostok ice core, Antarctica. Nature 399:429–436. https://doi.org/10.1038/20859

Pettitt P (2002) When burial begins. Br Archaeol 8–13

Pfeffer J, Carney DR (2018) The Economic Evaluation of Time Can Cause Stress. Acad Manag Discov 4:74–93. https://doi.org/10.5465/amd.2016.0017

Pfeiffer A, Hepburn C, Vogt-Schilb A, Caldecott B (2018) Committed emissions from existing and planned power plants and asset stranding required to meet the Paris Agreement. Environ Res Lett 13:054019. https://doi.org/10.1088/1748-9326/aabc5f

Philibert C (2006) Discounting the Future. In: Pannell DJ, Schilizzi SGM (eds) Economics and the Future. Edward Elgar Publishing

Phillips R, Seifer B, Antczak E (2013) Sustainable Communities: Creating a Durable Local Economy (Earthscan Tools for Community Planning). Routledge

Pickering TR, Bunn HT (2007) The endurance running hypothesis and hunting and scavenging in savanna-woodlands. J Hum Evol 53:434–438. https://doi.org/10.1016/j.jhevol.2007.01.012

Pigou AC (1932) The economics of welfare, 4th edn. Macmillan, London

Piguet M-F (2008) Individualisme: origine et réception initiale du mot, Œuvres et critiques, Revue internationale d'étude de la réception critique des uvres littéraires de langue française. Narr Fr Attempto Verlag XXXIII:39–60

Pike AWG, Hoffmann DL, Pettitt PB, et al (2017) Dating Palaeolithic cave art: Why U–Th is the way to go. Quat Int 432:41–49. https://doi.org/10.1016/j.quaint.2015.12.013

Piketty T (2013) Le capital au xxie siècle. Seuil, Paris

Piketty T, Saez E (2013) Top Incomes and the Great Recession: Recent Evolutions and Policy Implications. IMF Econ Rev 61:456–478. https://doi.org/10.1057/imfer.2013.14

Piketty T, Saez E, Zucman G (2016) Distributional National Accounts: Methods and Estimates for the United States. Cambridge, MA

Pilcher J (1994) Mannheim's Sociology of Generations: An Undervalued Legacy. Br J Sociol 45:481. https://doi.org/10.2307/591659

Pimm SL, Jenkins CN, Abell R, et al (2014) The biodiversity of species and their rates of extinction, distribution, and protection. Science 344:1246752–1246752. https://doi.org/10.1126/science.1246752

Pimm SL, Raven P (2000) Extinction by numbers. Nature 403:843–845. https://doi.org/10.1038/35002708

Pimpaneau J (1989) Histoire de la littérature chinoise. Éditions Philippe Picquier, Arles Paris

Pinker S (2011) The Better Angels of Our Nature. New York, NY: Viking

Pinker S (2018) Enlightenment Now: The Case for Reason, Science, Humanism, and Progress. Viking

Piraino S, Boero F, Aeschbach B, Schmid V (1996) Reversing the Life Cycle: Medusae Transforming into Polyps and Cell Transdifferentiation in Turritopsis nutricula (Cnidaria, Hydrozoa). Biol Bull 190:302–312. https://doi.org/10.2307/1543022

Pitney N (2016) Elon Musk to Address 'Nerd Altruists' at Google HQ. In: Huffington Post, 16 July 2015

Pokharia AK, Agnihotri R, Sharma S, et al (2017) Altered cropping pattern and cultural continuation with declined prosperity following abrupt and extreme arid event at 4,200 yrs BP: Evidence from an Indus archaeological site Khirsara, Gujarat, western India. PLoS One 12:e0185684. https://doi.org/10.1371/journal.pone.0185684

Polo M, Pisa R da (1350) Les voyages de Marco Polo. Available at National Library of Sweden

Pongratz J, Lobell DB, Cao L, Caldeira K (2012) Crop yields in a geoengineered climate. Nat Clim Chang 2:101–105. https://doi.org/10.1038/nclimate1373

Popper KR (1959) The logic of scientific discovery. Julius Springer, Hutchinson & Co

Porter B (2006) Empire and Superempire: Britain, America and the World. Yale University Press

Posner RA (2004) Catastrophe: Risk and Response. Oxford University Press, New York

Post JD (1977) The last great subsistence crisis in the Western World. Johns Hopkins University Press, Baltimore

Potts SG, Neumann P, Vaissière B, Vereecken NJ (2018) Robotic bees for crop pollination: Why drones cannot replace biodiversity. Sci Total Environ 642:665–667. https://doi.org/10.1016/j.scitotenv.2018.06.114

Powell A, Shennan S, Thomas MG (2009) Late Pleistocene Demography and the Appearance of Modern Human Behavior. Science 324:1298–1301. https://doi.org/10.1126/science.1170165

PPIGW (1992) Policy Implications of Greenhouse Warming Mitigation, Adaptation, and the Science Base. National Academies Press, Washington, D.C.

Prasad V (2005) Dinosaur Coprolites and the Early Evolution of Grasses and Grazers. Science 310:1177–1180. https://doi.org/10.1126/science.1118806

PRC (2010) Future of the Internet IV. In: Pew Res. Cent. https://www.pewresearch.org/internet/2010/02/19/future-of-the-internet-iv-2/

PRC (2013) Global Indicators Database. In: Pew Res. Cent. https://www.pewresearch.org/global/database/indicator/3/survey/15/

PRC (2017) The future of truth and misinformation online. In: Pew Res. Cent. http://www.pewinternet.org/2017/10/19/the-future-of-truth-and-misinformation-online/

PRC (2016) The Politics of Climate. In: Pew Res. Cent. http://www.pewinternet.org/2016/10/04/the-politics-of-climate/

Princen T (2005) The Logic of Sufficiency. The MIT Press

Proctor J, Hsiang S, Burney J, et al (2018) Estimating global agricultural effects of geoengineering using volcanic eruptions. Nature 560:480–483. https://doi.org/10.1038/s41586-018-0417-3

Proctor RN (2012a) Golden Holocaust. Origins of the Cigarette Catastrophe and the Case for Abolition. University of California Press

Proctor RN (2012b) The history of the discovery of the cigarette–lung cancer link: evidentiary traditions, corporate denial, global toll: Table 1. Tob Control 21:87–91. https://doi.org/10.1136/tobaccocontrol-2011-050338

Przybylski AK, Mishkin AF (2016) How the quantity and quality of electronic gaming relates to adolescents' academic engagement and psychosocial adjustment. Psychol Pop Media Cult 5:145–156. https://doi.org/10.1037/ppm0000070

PSAC (1965) Restoring the Quality of Our Environment, Report of The Environmental Pollution Panel, Presidential Science Advisory Committee, Executive Office of the President. Presidential Science Advisory Committee (PSAC), Washington DC

PwC (2015) The Sharing Economy. Consumer Intelligence Series. https://www.pwc.fr/fr/assets/files/pdf/2015/05/pwc_etude_sharing_economy.pdf

PwC (2012) The evolution of video gaming and content consumption. In: Price waterhouse Coopers (PwC). https://www.pwc.com/sg/en/tice/assets/ticenews201206/evolutionvideogame201206.pdf

PwC (2017) The Long View: How will the global economic order change by 2050?

Quiles A, Valladas H, Bocherens H, et al (2016) A high-precision chronological model for the decorated Upper Paleolithic cave of Chauvet-Pont d'Arc, Ardèche, France. Proc Natl Acad Sci 113:4670–4675. https://doi.org/10.1073/pnas.1523158113

Quinn WH (1992) A study of Southern Oscillation – related climatic activity for A.D. 622 – 1990 incorporating Nile River flood data. In: Diaz HF, Markgraf V (eds) El Niño: Historical and paleoclimatic aspects of the Southern Oscillation. Cambridge University Press, pp 119–149

Quinn TR, Tremaine S, Duncan M (1991) A three million year integration of the earth's orbit. Astron J 101:2287. https://doi.org/10.1086/115850

Quinones S (2015) Dream land. The true tale of America's opiate epidemic. Bloomsbury Pres

Quoidbach J, Dunn EW, Petrides KV, Mikolajczak M (2010) Money Giveth, Money Taketh Away. Psychol Sci 21:759–763. https://doi.org/10.1177/0956797610371963

RA (2017) Market Research on Improving Connectivity of Sustainable Tourism Operations in Ecuador and Peru to the EU Marketplace. Berlin, Germany

Rae J, Bonar J, Mixter CW (1906) The Sociological Theory of Capital, Being a Complete Reprint of the New Principles of Political Economy, 1834. Econ J 16:97. https://doi.org/10.2307/2221147

Ram M, Bogdanov D, Aghahosseini A, et al (2018) Global Energy System based on 100% Renewable Energy: Energy Transition in Europe Across Power, Heat, Transport and Desalination Sectors. LUT University and Energy Watch Group, Lappeenranta, Berlin

Ramanathan V, Carmichael G (2008) Global and regional climate changes due to black carbon. Nat Geosci 1:221–227. https://doi.org/10.1038/ngeo156

Ramsey FP (1928) A Mathematical Theory of Saving. Econ J 38:543. https://doi.org/10.2307/2224098

Rand A (1943) Fountainhead. Plume, New York

Rand A (1957) Atlas Shrugged. Dutton (1992)

Rand A (1964) The virtue of selfishness. Penguin

Randers J (2012) 2052: A Global Forecast for the Next Forty Years. Chelsea Green Publishing

Raposo J (2017) Violence In Eastern Saudi Arabia: The Latest Manifestation Of Saudi Totalitarianism. In: Organ. World Peace. https://theowp.org/reports/violence-in-eastern-saudi-arabia-the-latest-manifestation-of-saudi-totalitarianism/

Rappenglück M (2004) A Palaeolithic Planetarium Underground – The Cave of Lascaux (Part 1). Migr Diffus an Int J 5:93–119

Rasch PJ, Tilmes S, Turco RP, et al (2008) An overview of geoengineering of climate using stratospheric sulphate aerosols. Philos Trans R Soc A Math Phys Eng Sci 366:4007–4037. https://doi.org/10.1098/rsta.2008.0131

Ravenstein EG (1885) The Laws of Migration. J Stat Soc London 48:167. https://doi.org/10.2307/2979181

Ravenstein EG (1889) The Laws of Migration. J R Stat Soc 52:241–301. https://doi.org/10.1111/j.2397-2335.1889.tb00043.x

Ravetz JR (1997) The science of 'what-if?' Futures 29:533–539. https://doi.org/10.1016/S0016-3287(97)00026-8

Ravosa MJ, Dagosto M (eds) (2007) Primate Origins: Adaptations and Evolution. Springer US, Boston, MA

Rawls J (1971) A Theory of Justice, 2nd edn. Oxford University Press, Oxford, UK

Rawls J (2001) Justice as Fairness. Harvard University Press, Cambridge, MA

RDC (2018) A Rising Tide? In: Reflective Democr. Campaign. https://wholeads.us/2018-report/

Realff MJ, Eisenberger P (2012) Flawed analysis of the possibility of air capture. Proc Natl Acad Sci 109:E1589–E1589. https://doi.org/10.1073/pnas.1203618109

Rees M (2013) Denial of Catastrophic Risks. Science 339:1123–1123. https://doi.org/10.1126/science.1236756

Reeves M, Püschel L (2015) Die Another Day: What Leaders Can Do About the Shrinking Life Expectancy of Corporations. In: Bost. Consult. Gr. https://www.bcg.com/en-ao/publications/2015/strategy-die-another-day-what-leaders-can-do-about-the-shrinking-life-expectancy-of-corporations.aspx

Register R (1987) Ecocity Berkeley: Building Cities for a Healthy Future. North Atlantic Books

Reich D (2018) Who We Are and How We Got Here: Ancient DNA and the New Science of the Human Past. Pantheon

REN21 (2018) Renewables 2018 Global Status Report. Paris

Rendu W, Beauval C, Crevecoeur I, et al (2014) Evidence supporting an intentional Neandertal burial at La Chapelle-aux-Saints. Proc Natl Acad Sci 111:81–86. https://doi.org/10.1073/pnas.1316780110

Renforth P, Henderson G (2017) Assessing ocean alkalinity for carbon sequestration. Rev Geophys 55:636–674. https://doi.org/10.1002/2016RG000533

Renne PR, Deino AL, Hilgen FJ, et al (2013) Time Scales of Critical Events Around the Cretaceous–Paleogene Boundary. Science 339:684–687. https://doi.org/10.1126/science.1230492

Retallack GJ (2001) Cenozoic Expansion of Grasslands and Climatic Cooling. J Geol 109:407–426. https://doi.org/10.1086/320791

Reynolds JL (2019) The Governance of Solar Geoengineering. Cambridge University Press

Riahi K, van Vuuren DP, Kriegler E, et al (2017) The Shared Socioeconomic Pathways and their energy, land use, and greenhouse gas emissions implications: An overview. Glob Environ Chang 42:153–168. https://doi.org/10.1016/j.gloenvcha.2016.05.009

Rice A, Ortiz LV (2019) The McKinsey way to save an island. Why is a bankrupt Puerto Rico spending more than one billion dollars on expert advice? In: Intell. Vox Media, Inc. https://nymag.com/intelligencer/2019/04/mckinsey-in-puerto-rico.html

Ricke KL, Morgan MG, Allen MR (2010) Regional climate response to solar-radiation management. Nat Geosci 3:537–541. https://doi.org/10.1038/ngeo915

Riess AG, Filippenko A V., Challis P, et al (1998) Observational Evidence from Supernovae for an Accelerating Universe and a Cosmological Constant. Astron J 116:1009–1038. https://doi.org/10.1086/300499

Rigaud KK, de Sherbinin A, Jones B, et al (2018) Groundswell: Preparing for Internal Climate Migration. Washington, DC

Rignot E, Mouginot J, Scheuchl B, et al (2019) Four decades of Antarctic Ice Sheet mass balance from 1979–2017. Proc Natl Acad Sci 116:1095–1103. https://doi.org/10.1073/pnas.1812883116

Riley JC (2005) Estimates of Regional and Global Life Expectancy, 1800–2001. Popul Dev Rev 31:537–543. https://doi.org/10.1111/j.1728-4457.2005.00083.x

Risse M (2018) Human Rights and Artificial Intelligence: An Urgently Needed Agenda. Carr Center for Human Rights Policy, Harvard Kennedy School

Rissman J (2018) Cement's role in a carbon-neutral future. https://energyinnovation.org/wp-content/uploads/2018/11/The-Role-of-Cement-in-a-Carbon-Neutral-Future.pdf

Ritchie H, Roser M (2015) Energy. In: Our World Data. https://ourworldindata.org/energy

Roache R, Savulescu J (2016) Enhancing Conservatism. In: The Ethics of Human Enhancement. Oxford University Press, pp 145–159

Robinson JM (1990) Lignin, land plants, and fungi: Biological evolution affecting Phanerozoic oxygen balance. Geology 18:607. https://doi.org/10.1130/0091-7613(1990)018%3C0607:LLPAFB%3E2.3.CO;2

Robinson N (2019) Rich white men rule America. How much longer will we tolerate that? In: The Guardian. https://www.theguardian.com/commentisfree/2019/may/20/rich-white-men-rule-america-minority-rule

Robock A, Bunzl M, Kravitz B, Stenchikov GL (2010) A Test for Geoengineering? Science 327:530–531. https://doi.org/10.1126/science.1186237

Robock A, Marquardt A, Kravitz B, Stenchikov G (2009) Benefits, risks, and costs of stratospheric geoengineering. Geophys Res Lett 36:L19703. https://doi.org/10.1029/2009GL039209

Rockström J, Steffen W, Noone K, et al (2009a) Planetary Boundaries: Exploring the Safe Operating Space for Humanity. Ecol Soc 14:art32. https://doi.org/10.5751/ES-03180-140232

Rockström J, Steffen W, Noone K, et al (2009b) A safe operating space for humanity. Nature 461:472–475. https://doi.org/10.1038/461472a

Rogelj J, den Elzen M, Höhne N, et al (2016) Paris Agreement climate proposals need a boost to keep warming well below 2°C. Nature 534:631–639. https://doi.org/10.1038/nature18307

Rogelj J, Popp A, Calvin KV, et al (2018) Scenarios towards limiting global mean temperature increase below 1.5°C. Nat Clim Chang 8:325–332. https://doi.org/10.1038/s41558-018-0091-3

Rogoff K (2020) We must tackle global energy inequality before it's too late. In: The Guardian. www.theguardian.com/business/2020/jan/06/global-energy-inequality-tax-emissions-co2

Rohde RA, Muller RA (2005) Cycles in fossil diversity. Nature 434:208–210. https://doi.org/10.1038/nature03339

Romer PM (1990) Endogenous Technological Change. J Polit Econ 98:S71–S102

Rosa H (2013) Social Acceleration. Columbia University Press, New York Chichester, West Sussex

Rosenberg M, Markoff J (2016) The Pentagon's "Terminator Conundrum": Robots That Could Kill on Their Own. In: New York Times. https://www.nytimes.com/2016/10/26/us/pentagon-artificial-intelligence-terminator.html

Ross A, Matthews HD (2009) Climate engineering and the risk of rapid climate change. Environ Res Lett 4:045103. https://doi.org/10.1088/1748-9326/4/4/045103

Ross ML (2015) What Have We Learned about the Resource Curse? Annu Rev Polit Sci 18:239–259. https://doi.org/10.1146/annurev-polisci-052213-040359

Rougier B (2020) Les territoires conquis de l'islamisme. PUF Roux

Rousseau J-J (1755) Discours sur l'origine et les fondements de l'inégalité parmi les hommes. Marc Michel Rey, Amsterdam

Roux F, Guyonvarc'h C-J (1969) No Title. In: Dictionnaire des Symboles. p 335

RS (2009) Geoengineering the climate: Science, governance and uncertainty. London

Rubenstein JC (2016) The Lessons of Effective Altruism. Ethics Int Aff 30:511–526. https://doi.org/10.1017/S0892679416000484

Rull V (2018) What If the 'Anthropocene' Is Not Formalized as a New Geological Series/Epoch? Quaternary 1:24. https://doi.org/10.3390/quat1030024

Ryan RM, Deci EL (2001) On Happiness and Human Potentials: A Review of Research on Hedonic and Eudaimonic Well-Being. Annu Rev Psychol 52:141–166. https://doi.org/10.1146/annurev.psych.52.1.141

Sachs J, Kotlikoff L (2012) Smart Machines and Long-Term Misery. Cambridge, MA

Sachs JD (2015) The Age of Sustainable Development. Columbia University Press

Sachs W (1993) Die vier E's: Merkposten für einen massvollen Wirtschaftsstil. Polit Ökologie 11:69–72

Sachsenmaier D (2011) Global Perspectives on Global History: Theories and Approaches in a Connected World. Cambridge University Press

Saez E, Zucman G (2014) Wealth Inequality in the United States since 1913: Evidence from Capitalized Income Tax Data. Cambridge, MA

Saint-Simon H de (1824) Catéchisme des industriels, Deuxième cahier. In: Œuvres de Saint-Simon et d'Enfantin. 47 vols. Paris, v. 37, Paris

Saltelli A (2018) Why science's crisis should not become a political battling ground. Futures 104:85–90. https://doi.org/10.1016/j.futures.2018.07.006

SAMHSA (2018) Key Substance Use and Mental Health Indicators in the United States: Results from the 2017 National Survey on Drug Use and Health

Samuelson PA (1937) A Note on Measurement of Utility. Rev Econ Stud 4:155–161. https://doi.org/10.2307/2967612

Sánchez-Bayo F, Wyckhuys KAG (2019) Worldwide decline of the entomofauna: A review of its drivers. Biol Conserv 232:8–27. https://doi.org/10.1016/j.biocon.2019.01.020

Sánchez-Quinto F, Lalueza-Fox C (2015) Almost 20 years of Neanderthal palaeogenetics: adaptation, admixture, diversity, demography and extinction. Philos Trans R Soc B Biol Sci 370:20130374. https://doi.org/10.1098/rstb.2013.0374

Sánchez Alvarado A (2012) Q&A: What is regeneration, and why look to planarians for answers? BMC Biol 10:88. https://doi.org/10.1186/1741-7007-10-88

Sandberg A, Drexler E, Ord T (2018) Dissolving the Fermi Paradox. https://arxiv.org/abs/1806.02404

Sandel MJ (2004) The Case Against Perfection. In: The Atlantic, April 2004

Sanger DE (2012) Confront and Conceal: Obama's Secret Wars and Surprising Use of American Power. Crown Publishing Group

Santos FD (2016) Space systems for sustainable development on Earth. In: Williamson R, Morris L (eds) Space for the 21st Century: Discovery, Innovation, Sustainability (Aerospace Technology Working Group) (Volume 5). CreateSpace Independent Publishing Platform

Santos FD (2012) Humans on Earth. Springer Berlin Heidelberg, Berlin, Heidelberg

Santos FD (2014) Vulnerability, impacts and adaptation of coastal zones to global change. In: Monaco A, Prouzet P (eds) Vulnerability of Coastal Ecosystems and Adaptation. Wiley-ISTE, pp 131–168

Santos FD (2020) Climate Change in the XXIst and Following Centuries: A Risk or a Threat? In: Jodelet D, Vala J, Drozda-Senkowska E (eds) Societies Under Threat. A Pluri-Disciplinary Approach. Springer, pp 143–155

Santos HC, Varnum MEW, Grossmann I (2017) Global Increases in Individualism. Psychol Sci 28:1228–1239. https://doi.org/10.1177/0956797617700622

Sardar Z (2010) Welcome to postnormal times. Futures 42:435–444. https://doi.org/10.1016/j.futures.2009.11.028

Sarton G (1936) The Study of the History of Science. Harvard University Press, Wisconsin

Satterthwaite D, McGranahan G, Tacoli C (2010) Urbanization and its implications for food and farming. Philos Trans R Soc B Biol Sci 365:2809–2820. https://doi.org/10.1098/rstb.2010.0136

Saturno WA, Stuart D, Aveni AF, Rossi F (2012) Ancient Maya Astronomical Tables from Xultun, Guatemala. Science 336:714–717. https://doi.org/10.1126/science.1221444

Saunois M, Bousquet P, Poulter B, et al (2016) The global methane budget 2000–2012. Earth Syst Sci Data 8:697–751. https://doi.org/10.5194/essd-8-697-2016

Savage I (2013) Comparing the fatality risks in United States transportation across modes and over time. Res Transp Econ 43:9–22. https://doi.org/10.1016/j.retrec.2012.12.011

Savulescu J, Bostrom N (eds) (2009) Human Enhancement. Oxford University Press

Schacter DL, Addis DR, Hassabis D, et al (2012) The Future of Memory: Remembering, Imagining, and the Brain. Neuron 76:677–694. https://doi.org/10.1016/j.neuron.2012.11.001

Scheffer M, Bascompte J, Brock WA, et al (2009) Early-warning signals for critical transitions. Nature 461:53–59. https://doi.org/10.1038/nature08227

Schellnhuber HJ (2011) Geoengineering: The good, the MAD, and the sensible. Proc Natl Acad Sci 108:20277–20278. https://doi.org/10.1073/pnas.1115966108

Schellnhuber HJ, Crutzen PJ, Clark WC, Hunt J (2005) Earth System Analysis for Sustainability. Environ Sci Policy Sustain Dev 47:10–25. https://doi.org/10.3200/ENVT.47.8.10-25

Schindler I, Schindler J (2018) Strategies for an Economy Facing Energy Constraints. In: Burlando R, Tartaglia A (eds) Physical Limits to Economic Growth Perspectives of Economic, Social, and Complexity Science. Routledge

Schneider J (2020) Interactive Extra-solar Planets Catalogue. In: exoplanet.eu. http://exoplanet.eu/catalog/

Schneider L (2019) Fixing the Climate? How Geoengineering Threatens to Undermine the SDGs and Climate Justice. Development 62:29–36. https://doi.org/10.1057/s41301-019-00211-6

Schneider SM (2012) Income Inequality and its Consequences for Life Satisfaction: What Role do Social Cognitions Play? Soc Indic Res 106:419–438. https://doi.org/10.1007/s11205-011-9816-7

Schoeninger MJ (1981) The Agricultural "Revolution": Its Effect on Human Diet in Prehistoric Iran and Israel. Paléorient 7:73–91. https://doi.org/10.3406/paleo.1981.4288

Schopenhauer A (1851) Additional Remarks on the Doctrine of the Suffering of the World. Parerga and Paraliponema: Short Philosophical Essays, Oxford

Schröder K-P, Connon Smith R (2008) Distant future of the Sun and Earth revisited. Mon Not R Astron Soc 386:155–163. https://doi.org/10.1111/j.1365-2966.2008.13022.x

Schrödinger E (1944) What Is Life? Cambridge University Press (1992), Cambridge

Schuiling RD, Tickell O (2010) Enhanced weathering of olivine to capture CO_2. J Appl Geochemistry 12:510–519

Schumpeter J (1942) Capitalism, Socialism and Democracy. Harper and Brother, New York

Schwab K (2016) The Fourth Industrial Revolution. World Economic Forum

Schwalm CR, Anderegg WRL, Michalak AM, et al (2017) Global patterns of drought recovery. Nature 548:202–205. https://doi.org/10.1038/nature23021

Schweitzer Y, Mendelboim A, Rosner Y (2017) Suicide Attacks in 2016: The Highest Number of Fatalities. Inst Natl Secur Stud 887

Scott-Phillips TC, Laland KN, Shuker DM, et al (2014) The niche construction perspective: A critical appraisal. Evolution (N Y) 68:1231–1243. https://doi.org/10.1111/evo.12332

Scott JC (2017) Against the Grain: A Deep History of the Earliest States. Yale University Press, New Haven, London

Seligman M (2007) What You Can Change and What You Can't: The Complete Guide to Successful Self-Improvement. Nicholas Brealey Publications

Seligman MEP (2002) Authentic happiness: Using the new positive psychology to realize your potential for lasting fulfillment. Free Press

Seligman MEP (2011) Flourish: a visionary new understanding of happiness and well-being. Free Press, New York

Sen A (1984) The Living Standard. Oxf Econ Pap 36:74–90. https://doi.org/10.1093/oxfordjournals.oep.a041662

Sen A (1981) Poverty and Famines: An Essay on Entitlement and Deprivation. Clarendon Press, Oxford

Sen A (1999) Development as Freedom. Oxford University Press

Sengupta S (2016) Heat, hunger and war force Africans onto 'road of fire.' In: TAIPEI TIMES. http://www.taipeitimes.com/News/editorials/archives/2016/12/20/2003661556

Seto KC, Dhakal S, Bigio A, et al (2014) Human Settlements, Infrastructure and Spatial Plan. In: Edenhofer O, Pichs-Madruga R, Sokona Y, et al. (eds) Climate Change 2014: Mitigation of Climate Change. Contribution of Working Group III to the Fifth Assessment Report of the Intergovernmental Panel on Climate Change. Cambridge University Press, Cambridge, United Kingdom and New York, NY, US, pp 923–1000

Shakespeare W (1623) Mr. William Shakespeare's Comedies, Histories, & Tragedies. Printed by Isaac Iaggard, and Ed. Blount, 1623, London

Shalm LK, Meyer-Scott E, Christensen BG, et al (2015) Strong Loophole-Free Test of Local Realism. Phys Rev Lett 115:250402. https://doi.org/10.1103/PhysRevLett.115.250402

Shamosh NA, DeYoung CG, Green AE, et al (2008) Individual Differences in Delay Discounting. Psychol Sci 19:904–911. https://doi.org/10.1111/j.1467-9280.2008.02175.x

Shannon CE (1948) A Mathematical Theory of Communication. Bell Syst Tech J 27:379–423. https://doi.org/10.1002/j.1538-7305.1948.tb01338.x

Shannon CE, Weaver W (1949) The Mathematical Theory of Communication. University of Illinois Press, Urbana, Illinois

Sharer RJ, Traxler LP (2005) The Ancient Maya, 6th edn. Stanford University Press

Sharkey A (2019) Autonomous weapons systems, killer robots and human dignity. Ethics Inf Technol 21:75–87. https://doi.org/10.1007/s10676-018-9494-0

Sharot T, Korn CW, Dolan RJ (2011) How unrealistic optimism is maintained in the face of reality. Nat Neurosci 14:1475–1479. https://doi.org/10.1038/nn.2949

Shaver R (2015) Egoism. In: Stanford Encycl. Philos. (Spring 2015 Ed. Edward N. Zalta. https://plato.stanford.edu/archives/spr2015/entries/egoism/

Shaver R (2019) Egoism. In: Stanford Encycl. Philos. (Spring 2019 Ed. Edward N. Zalta. https://plato.stanford.edu/archives/spr2019/entries/egoism/

Shearer C, West M, Caldeira K, Davis SJ (2016) Quantifying expert consensus against the existence of a secret, large-scale atmospheric spraying program. Environ Res Lett 11:084011. https://doi.org/10.1088/1748-9326/11/8/084011

Sheehan IS (2014) Are suicide terrorists suicidal? A critical assessment of the evidence. Innov Clin Neurosci 11:81–92

Shen L, Chen X-Y, Zhang X, et al (2005) Genetic variation of Ginkgo biloba L. (Ginkgoaceae) based on cpDNA PCR-RFLPs: inference of glacial refugia. Heredity (Edinb) 94:396–401. https://doi.org/10.1038/sj.hdy.6800616

Shepherd A, Ivins E, Rignot E, et al (2018) Mass balance of the Antarctic Ice Sheet from 1992 to 2017. Nature 558:219–222. https://doi.org/10.1038/s41586-018-0179-y

Shepherd C (2019) China aims to reduce coal power production. In: Financ. Times. https://www.ft.com/content/c31ff98e-1589-11ea-9ee4-11f260415385

Sherman R (2017) Uneasy Street: The Anxieties of Affluence. Princeton University Press

Shorrocks A, Davies J, Lluberas R (2018) Global Wealth Report 2018. Credit Suisse AG

Shorrocks A, Davies JB, Lluberas R, Koutsoukis A (2016) Global Wealth Report 2016. Credit Suisse AG

Shulgin A, Shulgin A (1991) Pihkal: A Chemical Love Story. Transform Press

Shulgin A, Shulgin A (1997) Tihkal: A Continuation. Transform Press

Shultz S, Opie C, Atkinson QD (2011) Stepwise evolution of stable sociality in primates. Nature 479:219–222. https://doi.org/10.1038/nature10601

Sibree J (1896) Madagascar before the conquest. London

Sidgwick H (1874) The Methods of Ethics. Hackett Publishing (1981)

Siegenthaler U (2005) Stable Carbon Cycle-Climate Relationship During the Late Pleistocene. Science 310:1313–1317. https://doi.org/10.1126/science.1120130

Simmons K, Wike R, Oates R (2014) People in Emerging Markets Catch Up to Advanced Economies in Life Satisfaction

Simon SL, Bouville A, Land CE, Beck HL (2010) Radiation Doses and Cancer Risks in the Marshall Islands Associated with Exposure to Radioactive Fallout from Bikini and Enewetak Nuclear Weapons Tests: Summary. Health Phys 99:105–123. https://doi.org/10.1097/HP.0b013e3181dc523c

Singer P (2009) The Life You Can Save: Acting Now to End World Poverty. Random House

Singer P (2015) The Most Good You Can Do: How Effective Altruism Is Changing Ideas About Living Ethically. Yale University Press

Singh N, Sit MT, Chung DM, et al (2016) How Often Are Antibiotic-Resistant Bacteria Said to "Evolve" in the News? PLoS One 11:e0150396. https://doi.org/10.1371/journal.pone.0150396

Sinha E, Michalak AM, Balaji V (2017) Eutrophication will increase during the 21st century as a result of precipitation changes. Science 357:405–408. https://doi.org/10.1126/science.aan2409

Smil V (2008) Global Catastrophes and Trends: The Next Fifty Years. The MIT Press

Smith A (1759) The Theory of Moral Sentiments, Indianapol. Glasgow Edition of the Works and Correspondence of Adam Smith

Smith A (1776) An Inquiry into the Nature and Causes of the Wealth of Nations, Vol. I and II. Liberty Fund, Indianopolis

Smith A, Anderson J (2014) Aaron Smith and Janna Anderson. In: Pew Res. Cent. https://www.pewresearch.org/internet/2014/08/06/future-of-jobs/

Smith CB (2004) How the Great Pyramid Was Built. Smithsonian Institution Press

Smith EA, Hill K, Marlowe FW, et al (2010) Wealth Transmission and Inequality among Hunter–Gatherers. Curr Anthropol 51:19–34. https://doi.org/10.1086/648530

Smith JM (1964) Group Selection and Kin Selection. Nature 201:1145–1147. https://doi.org/10.1038/2011145a0

Smith JM (1998) The origin of altruism. Nature 393:639–640. https://doi.org/10.1038/31383

Smith MR, Myers SS (2018) Impact of anthropogenic CO_2 emissions on global human nutrition. Nat Clim Chang 8:834–839. https://doi.org/10.1038/s41558-018-0253-3

Smith P, Davis SJ, Creutzig F, et al (2016) Biophysical and economic limits to negative CO_2 emissions. Nat Clim Chang 6:42–50. https://doi.org/10.1038/nclimate2870

Smith PK, Bogin B, Bishai D (2005) Are time preference and body mass index associated? Econ Hum Biol 3:259–270. https://doi.org/10.1016/j.ehb.2005.05.001

Smulders S, Toman M, Withagen C (2014) Growth theory and "green growth." Oxford Rev Econ Policy 30:423–446. https://doi.org/10.1093/oxrep/gru027

Smyth J (2015) Rio Tinto shifts to driverless trucks in Australia. In: Financ. Times. https://www.ft.com/content/43f7436a-7632-11e5-a95a-27d368e1ddf7

Sobel AH, Camargo SJ, Hall TM, et al (2016) Human influence on tropical cyclone intensity. Science 353:242–246. https://doi.org/10.1126/science.aaf6574

Socolow R, Desmond M, Aines R, et al (2011) Direct Air Capture of CO_2 with Chemicals

Solow RM (1987) We'd better watch out. In: New York Times Book Review. New York Times, New York, p 36

Solow RM (1956) A Contribution to the Theory of Economic Growth. Q J Econ 70:65. https://doi.org/10.2307/1884513

Solow RM (1957) Technical Change and the Aggregate Production Function. Rev Econ Stat 39:312. https://doi.org/10.2307/1926047

Solow RM (1974) The Economics of Resources or the Resources of Economics. Am Econ Rev 64:1–14

Sorokin PA, Merton RK (1937) Social Time: A Methodological and Functional Analysis. Am J Sociol 42:615–629

Sovová T, Kerins G, Demnerová K, Ovesná J (2017) Genome Editing with Engineered Nucleases in Economically Important Animals and Plants: State of the Art in the Research Pipeline. Curr Issues Mol Biol 21:41–62

SP (2015) GM Crops Now Banned in 39 Countries Worldwide – Sustainable Pulse Research. In: Sustain. Pulse. https://sustainablepulse.com/2015/10/22/gm-crops-now-banned-in-36-countries-worldwide-sustainable-pulse-research/

Specter M (2009) Denialism: How Irrational Thinking Hinders Scientific Progress, Harms the Planet, and Threatens Our Lives. Penguin Press, United States

Spencer H (1857) Progress: Its Law and Cause. In: Essays: Scientific, Political, and Speculative: Volume I. D. Appleton and Company, New York, pp 8–62

Spinoza B (1677) Ethics, Demonstrated in Geometrical Order. CreateSpace Independent Publishing Platform (2015)

Spratt DJ, Dunlop I (2018) What lies beneath: The understatement of existential climate risk

SSRG (2019) The Robobee Project. In: Self-Organizing Syst. Res. Gr. https://ssr.seas.harvard.edu/robobee-project

Stadler J, Matthews J, Raju R, et al (2018) New visionaries and the Chinese Century: Billionaires insights 2018

Stahel W (1982) The product life factor. In: An Inquiry Into the Nature of Sustainable Societies: The Role of the Private Sector. pp 72–96

Stahel WR, Reday-Mulvey G (1976) Jobs for tomorrow: the potential for substituting manpower for energy. Vantage Press (1981), New York

Standing G (2011) The Precariat The New Dangerous Class. Bloomsbury Academic

Standing G (2014) A Precariat Charter. From Denizens to Citizens. Bloomsbury Academic

Stanley J-D, Krom MD, Cliff RA, Woodward JC (2003) Short contribution: Nile flow failure at the end of the Old Kingdom, Egypt: Strontium isotopic and petrologic evidence. Geoarchaeology 18:395–402. https://doi.org/10.1002/gea.10065

Stanway D (2019) In China, coal creeps back in as slowing economy overshadows climate change ambitions. In: Reuters. https://www.reuters.com/article/us-climate-change-china-coal/in-china-coal-creeps-back-in-as-slowing-economy-overshadows-climate-change-ambitions-idUSKBN1Y60NU

Stapleton SO, Nadin R, Watson C, Kellett J (2017) Climate change, migration and displacement: the need for a risk-informed and coherent approach

Statista (2018) Global No.1 Data Platform. https://www.statista.com/. Accessed 16 Nov 2018

Staubwasser M, Sirocko F, Grootes PM, Segl M (2003) Climate change at the 4.2 ka BP termination of the Indus valley civilization and Holocene south Asian monsoon variability. Geophys Res Lett 30:. https://doi.org/10.1029/2002GL016822

Steffen W, Broadgate W, Deutsch L, et al (2015) The trajectory of the Anthropocene: The Great Acceleration. Anthr Rev 2:81–98. https://doi.org/10.1177/2053019614564785

Steffen W, Grinevald J, Crutzen P, McNeill J (2011) The Anthropocene: conceptual and historical perspectives. Philos Trans R Soc A Math Phys Eng Sci 369:842–867. https://doi.org/10.1098/rsta.2010.0327

Steffen W, Sanderson A, Tyson P, et al (2005) Global Change and the Earth System. Springer Berlin Heidelberg, Berlin, Heidelberg

Steffens G (2018) Changing climate forces desperate Guatemalans to migrate. In: Natl. Geogr. Soc. https://www.nationalgeographic.com/environment/2018/10/drought-climate-change-force-guatemalans-migrate-to-us/

Steflen W, Andreae MO, Bolin B, et al (2004) Abrupt Changes: The Achilles' Heels of the Earth System. Environ Sci Policy Sustain Dev 46:8–20. https://doi.org/10.1080/00139150409604375

Steinke F, Wolfrum P, Hoffmann C (2013) Grid vs. storage in a 100% renewable Europe. Renew Energy 50:826–832. https://doi.org/10.1016/j.renene.2012.07.044

Stephenson FR, Morrison L V., Smith FT (1995) Long-term fluctuations in the Earth's rotation: 700 BC to AD 1990. Philos Trans R Soc London Ser A Phys Eng Sci 351:165–202. https://doi.org/10.1098/rsta.1995.0028

Stern DI (2011) The role of energy in economic growth. Ann N Y Acad Sci 1219:26–51. https://doi.org/10.1111/j.1749-6632.2010.05921.x

Sterner T (1994) Discounting in a world of limited growth. Environ Resour Econ 4:527–534. https://doi.org/10.1007/BF00691927

Stevenson S, Singh T (2017) The Course of History: Ten Meals That Changed the World. Birlinn Ltd

Stieglitz S, Dang-Xuan L (2013) Social media and political communication: a social media analytics framework. Soc Netw Anal Min 3:1277–1291. https://doi.org/10.1007/s13278-012-0079-3

Stiglitz JE (2019) People, Power, and Profits: Progressive Capitalism for an Age of Discontent. W. W. Norton & Company

Stiglitz JE (2017) Globalization and Its Discontents Revisited: Anti-Globalization in the Era of Trump. W. W. Norton & Company

Stiglitz JE (1997) Georgescu-Roegen versus Solow/Stiglitz. Ecol Econ 22:269–270. https://doi.org/10.1016/S0921-8009(97)00092-X

Stiglitz JE (2002) Globalization and Its Discontents. W. W. Norton & Company

Stockhammer E, Yilmaz D (2015) Alternative economics. A new student movement. Radic Philos jan/fev:

Stocks T (2015) Red Roses and Slain Dragons. In: James Stock. Co. https://docplayer.net/5662945-Red-roses-and-slain-dragons-a-presentation-by-tim-stocks-chairman-james-stocks-co-and-partner-taylor-wessing-llp.html

Storr W (2017) Selfie: How We Became So Self-Obsessed and What It's Doing to Us. Picador

Stothers RB (1984) The Great Tambora Eruption in 1815 and Its Aftermath. Science 224:1191–1198. https://doi.org/10.1126/science.224.4654.1191

Stoycheff E, Nisbet EC, Epstein D (2016) Differential Effects of Capital-Enhancing and Recreational Internet Use on Citizens' Demand for Democracy. Communic Res 009365021664464. https://doi.org/10.1177/0093650216644645

Strefler J, Amann T, Bauer N, et al (2018) Potential and costs of carbon dioxide removal by enhanced weathering of rocks. Environ Res Lett 13:034010. https://doi.org/10.1088/1748-9326/aaa9c4

Summers LH (2014) U.S. Economic Prospects: Secular Stagnation, Hysteresis, and the Zero Lower Bound. Bus Econ 49:65–73. https://doi.org/10.1057/be.2014.13

Sweet WV, Horton R, Kopp RE, et al (2017) Ch. 12: Sea Level Rise. Climate Science Special Report: Fourth National Climate Assessment, Volume I. Washington, DC

Swan TW (1956) ECONOMIC GROWTH and CAPITAL ACCUMULATION. Econ Rec 32:334–361. https://doi.org/10.1111/j.1475-4932.1956.tb00434.x

Szpunar KK, Watson JM, McDermott KB (2007) Neural substrates of envisioning the future. Proc Natl Acad Sci 104:642 LP – 647. https://doi.org/10.1073/pnas.0610082104

Tacoli C, McGranahan G, Satterthwaite D (2015) Urbanisation, rural–urban migration and urban poverty. London

Tan TCJ, Rahman R, Jaber-Hijazi F, et al (2012) Telomere maintenance and telomerase activity are differentially regulated in asexual and sexual worms. Proc Natl Acad Sci 109:4209–4214. https://doi.org/10.1073/pnas.1118885109

Tang CQ, Yang Y, Ohsawa M, et al (2012) Evidence for the persistence of wild Ginkgo biloba (Ginkgoaceae) populations in the Dalou Mountains, southwestern China. Am J Bot 99:1408–1414. https://doi.org/10.3732/ajb.1200168

Tang L, Zeng Y, Du H, et al (2017) CRISPR/Cas9-mediated gene editing in human zygotes using Cas9 protein. Mol Genet Genomics 292:525–533. https://doi.org/10.1007/s00438-017-1299-z

Tatar M, Post S, Yu K (2014) Nutrient control of Drosophila longevity. Trends Endocrinol Metab 25:509–517. https://doi.org/10.1016/j.tem.2014.02.006

Tatomir A, McDermott C, Bensabat J, et al (2018) Conceptual model development using a generic Features, Events, and Processes (FEP) database for assessing the potential impact of hydraulic fracturing on groundwater aquifers. Adv Geosci 45:185–192. https://doi.org/10.5194/adgeo-45-185-2018

Tavaré S, Marshall CR, Will O, et al (2002) Using the fossil record to estimate the age of the last common ancestor of extant primates. Nature 416:726–729. https://doi.org/10.1038/416726a

Taylor M, Watts J (2019) Revealed: the 20 firms behind a third of all carbon emissions. In: The Guardian. https://www.theguardian.com/environment/2019/oct/09/revealed-20-firms-third-carbon-emissions

TCE (2017) The Christian Egoist. In: Christ. Egoist. http://www.thechristianegoist.com/category/christian-egoism/

TCFD (2017) Recommendations of the Task Force on Climate-related Financial Disclosures (Final Report)

TCFD (2019) Task Force on Climate-related Financial Disclosures: Status Report

Tcvetkov P, Cherepovitsyn A, Fedoseev S (2019) The Changing Role of CO_2 in the Transition to a Circular Economy: Review of Carbon Sequestration Projects. Sustainability 11:5834. https://doi.org/10.3390/su11205834

TEEB (2010) The Economics of Ecosystems and Biodiversity: Mainstreaming the Economics of Nature: A synthesisof the approach, conclusions and recommendations of TEEB

Teller E, Wood L, Hyde R (1997) Global warming and ice ages: I prospects for physics- based modulation of global change. United States

Testart A (2012) Avant l'histoire: L'évolution des sociétés, de Lascaux à Carnac. Gallimard

Testart A (2016) Art et religion de Chauvet à Lascaux. Gallimard

Testart A (1982) Les chasseurs-cueilleurs ou l'origine des inégalités. Société d'Ethnographie (Université Paris X-Nanterre), Paris

The Economist (2017) What people want at the end of life. In: Econ. https://www.economist.com/graphic-detail/2017/04/27/what-people-want-at-the-end-of-life. Accessed 27 Apr 2017

The Worldwatch Institute (2008) State of The World 2008: Innovations for a Sustainable Economy. W. W. Norton & Company

Them TR, Gill BC, Caruthers AH, et al (2017) High-resolution carbon isotope records of the Toarcian Oceanic Anoxic Event (Early Jurassic) from North America and implications for the global drivers of the Toarcian carbon cycle. Earth Planet Sci Lett 459:118–126. https://doi.org/10.1016/j.epsl.2016.11.021

Thomas W (1999) Contrails: Poison From the Sky. In: Citizens a Democr. Renaiss. https://www.renaissance.cyberjournal.org/1999/03/04/poison-in-the-skies-haarp-speculation/

Thomas W (2004) Chemtrails Confirmed. Bridger House Publishers

Thoreau DH (1854) Walden or Life in the Woods. Ticknor and Fields, Boston

Tierney J (2007) A Survival Imperative for Space Colonization. In: New York Times. https://www.nytimes.com/2007/07/17/science/17tier.html

Tilmes S, Fasullo J, Lamarque J-F, et al (2013) The hydrological impact of geoengineering in the Geoengineering Model Intercomparison Project (GeoMIP). J Geophys Res Atmos 118:11,036-11,058. https://doi.org/10.1002/jgrd.50868

Tilmes S, Muller R, Salawitch R (2008) The Sensitivity of Polar Ozone Depletion to Proposed Geoengineering Schemes. Science 320:1201–1204. https://doi.org/10.1126/science.1153966

Timperley J (2019) Airlines around the world have recently begun to monitor their CO_2 emissions as part of a UN climate deal. In: Carbon Br. https://www.carbonbrief.org/corsia-un-plan-to-offset-growth-in-aviation-emissions-after-2020

Tinari P (2005) China Trade: The Art and Commerce of Tobacco. In: Duke Mag. https://alumni.duke.edu/magazine/articles/china-trade

Tingley D, Wagner G (2017) Solar geoengineering and the chemtrails conspiracy on social media. Palgrave Commun 3:12. https://doi.org/10.1057/s41599-017-0014-3

Tittensor DP, Walpole M, Hill SLL, et al (2014) A mid-term analysis of progress toward international biodiversity targets. Science 346:241–244. https://doi.org/10.1126/science.1257484

Tocqueville A (1840) De la Démocratie en Amérique, Volume 2. Pagnerre Éditeur, Paris

Tokarska KB, Gillett NP, Weaver AJ, et al (2016) The climate response to five trillion tonnes of carbon. Nat Clim Chang 6:851–855. https://doi.org/10.1038/nclimate3036

Tollefson J (2008) UN decision puts brakes on ocean fertilization. Nature 453:704–704. https://doi.org/10.1038/453704b

Tollefson J (2018) First sun-dimming experiment will test a way to cool Earth. Nature 563:613–615. https://doi.org/10.1038/d41586-018-07533-4

Toman MA, Pezzey J, Krautkraemer J (1995) Neoclassical economic growth theory and 'sustainability'. In: Bromley DW (ed) Handbook of Environmental Economics. Blackwell, Oxford, pp 139–165

Tomasello M (2014) The ultra-social animal. Eur J Soc Psychol 44:187–194. https://doi.org/10.1002/ejsp.2015

Tomasky M (2019) Is America Becoming an Oligarchy? Growing inequality threatens our most basic democratic principles. In: New York Times. https://www.nytimes.com/2019/04/14/opinion/america-economic-inequality.html

Tooley M (2000) Time, Tense, and Causation. Oxford University Press, Oxford

Toye R (2010) Churchill's Empire: The World That Made Him and the World He Made. Henry Holt and Company

Toynbee A (1884) Lectures on the Industrial Revolution in England. Cambridge University Press (2011), Cambridge

Trainer T (2015) The case for simplicity. http://simplicitycollective.com/the-case-for-simplicity-by-ted-trainer

Trenberth KE, Dai A (2007) Effects of Mount Pinatubo volcanic eruption on the hydrological cycle as an analog of geoengineering. Geophys Res Lett 34:. https://doi.org/10.1029/2007GL030524

Trenberth KE, Hoar TJ (1996) The 1990-1995 El Niño-Southern Oscillation Event: Longest on Record. Geophys Res Lett 23:57–60. https://doi.org/10.1029/95GL03602

Trenberth KE, Stepaniak DP (2004) The flow of energy through the earth's climate system. Q J R Meteorol Soc 130:2677–2701. https://doi.org/10.1256/qj.04.83

Trivers RL (1971) The Evolution of Reciprocal Altruism. Q Rev Biol 46:35–57. https://doi.org/10.1086/406755

Turgot ARJ (1750) Tableau philosophique des progrès successifs de l'esprit humain. In: Schelle G (ed) Œuvres de Turgot et documents le concernant. 5 volumes, Paris: Felix Alcan, 1913–1923

Turkle S (2011) Alone Together: Why We Expect More from Technology and Less from Each Other. Basic Books

Turner BL, Kasperson RE, Meyer WB, et al (1990) Two types of global environmental change. Glob Environ Chang 1:14–22. https://doi.org/10.1016/0959-3780(90)90004-S

Turner G (2008) A comparison of The Limits to Growth with 30 years of reality. Glob Environ Chang 18:397–411. https://doi.org/10.1016/j.gloenvcha.2008.05.001

Turner SK (2018) Constraints on the onset duration of the Paleocene–Eocene Thermal Maximum. Philos Trans R Soc A Math Phys Eng Sci 376:20170082. https://doi.org/10.1098/rsta.2017.0082

Tverberg G (2018) Our Energy Problem Is a Quantity Problem. In: Our Finite World. https://ourfiniteworld.com/2018/05/30/our-energy-problem-is-a-quantity-problem/. Accessed 30 May 2018

Tverberg GE (2012) Oil supply limits and the continuing financial crisis. Energy 37:27–34. https://doi.org/10.1016/j.energy.2011.05.049

UCS (2012) Heads They Win,Tails We Lose: How Corporations Corrupt Science at the Public's Expense. Cambridge, MA; Washington, DC; Berkeley, CA; Chicago, IL

UCS (2014) Environmental Impacts of Natural Gas. In: Union Concerned Sci. https://www.ucsusa.org/resources/environmental-impacts-natural-gas

UN (1947) Resolution 181(II) on the Future government of Palestine, adopted 29 November 1947

UN (2002) United Nations treaties and principles on outer space. United Nations Publication, Vienna, Austria

UN (2017) Making migration work for all. Report of the Secretary-General. A/72/643

UN (2019) Resolution adopted by the General Assembly on 19 December 2018

UN Dispatch (2019) US Ceases Cooperation With UN Human Rights Special Rapporteurs. In: UN Dispatch. https://www.undispatch.com/us-ceases-cooperation-with-un-human-rights-special-rapporteurs/

UN Water (2019) United Nations Water. In: United Nations Water. https://www.unwater.org/

UNCTAD (2012) World Investment Report 2012: Towards a New Generation of Investment Policies

UNCTAD (2013) Global Value Chains and Development

UNDESAPD (2017a) World Population Prospects: The 2017 revision. In: United Nations Dep. Econ. Soc. Aff. Popul. Div. https://population.un.org/wpp/Publications/

UNDESAPD (2017b) International Migration Report 2017: Highlights (ST/ESA/SER.A/404). New York

UNDESAPD (2018) The World's Cities in 2018 – Data Booklet (ST/ESA/ SER.A/417)

UNDESAPD (2019a) World Urbanization Prospects: The 2018 Revision. New York

UNDESAPD (2019b) World Population Prospects 2019. In: United Nations Dep. Econ. Soc. Aff. Popul. Div. https://population.un.org/wpp/

UNDP (2013) Ocean hypoxia – "Dead Zones": Issue Brief

UNEP (2011) Decoupling natural resource use and environmental impacts from economic growth, A Report of the Working Group on Decoupling to the International Resource Panel. United Nations Environment Programme (UNEP), Paris

UNEP (2019) Emissions Gap Report 2019. Nairobi

UNFCCC (1992) United Nations Framework Convention on Climate Change. In: United Nations Framew. Conv. Clim. Chang. https://treaties.un.org/

UNGA (2010) Report of the Scientific and Technical Subcommittee on its forty-seventh session, held in Vienna from 8 to 19 February 2010. Vienna, Austria

Unger SH (1994) Controlling Technology: Ethics and the Responsible Engineer. John Wiley & Sons

UNHCR (2016) New York Declaration for Refugees and Migrants. In: United Nations High Comm. Refug. New York Declaration for Refugees and Migrants

UNHCR (2018) Syria emergency. In: United Nations High Comm. Refug. http://www.unhcr.org/syria-emergency.html

UNITED (2018) About the 'List of Deaths.' In: UNITED Intercult. Action. http://unitedagainstrefugeedeaths.eu/about-the-campaign/about-the-united-list-of-deaths/

UNODC (2017) World Drug Report 2017

UNRISD (2016) Policy Innovations for Transformative Change. Implementing the 2030 Agenda for Sustainable Development. Geneva, Switzerland

USACDA (1974) Arms Limitations Agreements, US Arms Control and Disarmament Agency, 1974 Report

USAF (1993) Final Environmental Impact Statement. High Frequency Active Auroral Research Program

USDA (2018) International Macroeconomic Data Set. In: United States Dep. Agric. Econ. Res. Serv. www.ers.usda.gov/data-products/international-macroeconomic-data-set/. Accessed 10 Aug 2018

USDHEW (1964) Smoking and Health: Report of the Advisory Committee to the Surgeon General of the Public Health Service. United States Department of Health, Education and Welfare (USDHEW)

USDHHS (2019) Adolescents and Tobacco: Trends. In: U.S. Dep. Heal. Hum. Serv. https://www.hhs.gov/ash/oah/adolescent-development/substance-use/drugs/tobacco/trends/index.html

USGCRP (2017) Climate Science Special Report: Fourth National Climate Assessment, Volume I. Washington, DC

USS (1974) Weather Modification, United States Senate, Secret hearing held on March 20 and made public on May 19. In: United States, Congress, Senate, et al. (eds) Weather Modification: Hearings, Ninety-third Congress, Second Session ... January 25 and March 20. U.S. Government Printing Office, pp 87–123

Vaesen K, Collard M, Cosgrove R, Roebroeks W (2016) Population size does not explain past changes in cultural complexity. Proc Natl Acad Sci 113:E2241–E2247. https://doi.org/10.1073/pnas.1520288113

Van den Bergh J, Antal M (2014) Evaluating Alternativesto GDP as Measures of Social Welfare/Progres. WWWforEurope Work Pap 56 van der Kaars S, Miller GH, Turney CSM, et al (2017) Humans rather than climate the primary cause of Pleistocene megafaunal extinction in Australia. Nat Commun 8:14142. https://doi.org/10.1038/ncomms14142

van Kleunen M, Dawson W, Essl F, et al (2015) Global exchange and accumulation of non-native plants. Nature 525:100–103. https://doi.org/10.1038/nature14910

van Noort RBJC (1991) The Present Environmental Crisis. Stud Environ Sci 45:3–14. https://doi.org/10.1016/S0166-1116(08)70353-4

Van Parijs P (1995) Real Freedom for All, What (if Anything) Can Justify Capitalism? Oxford University Press

Vandermeersch B (1970) Une sépulture moustérienne avec offrandes découverte dans la grotte de Qafzeh. Comptes Rendues Hebd des Séances l'Academie des Sci 270:298–301

Veblen T (1899) The Theory of the Leisure Class: An Economic Study of Institutions. Macmillan

Veraart AW (1931) Meer Zonneschijn in Het Nevelig Noorden, Meer Regen in de Tropen. N. V. Seyffardt's Boek en Muziekhandel, Amsterdam

Verdoux P (2009) Transhumanism, Progress and the Future. J Technol Evol 20:49–69

Vermeij GJ (1994) The Evolutionary Interaction Among Species: Selection, Escalation, and Coevolution. Annu Rev Ecol Syst 25:219–236. https://doi.org/10.1146/annurev.es.25.110194.001251

Vespa J, Armstrong DM, Medina L (2018) Demographic Turning Points for the United States: Population Projections for 2020 to 2060, Current Population Reports. Washington, DC

Vidal J (2015) Malawi's forests going up in smoke as tobacco industry takes heavy toll. In: The Guardian. https://www.theguardian.com/global-development/2015/jul/31/malawi-tobacco-industry-deforestation-chinkhoma

Vidal J (2018) The 100 million city: is 21st century urbanisation out of control? In: The Guardian. https://www.theguardian.com/cities/2018/mar/19/urban-explosion-kinshasa-el-alto-growth-mexico-city-bangalore-lagos

Vijg J, Kennedy BK (2016) The Essence of Aging. Gerontology 62:381–385. https://doi.org/10.1159/000439348

Vinge V (1993) The Coming Technological Singularity: How to Survive in the Post-Human Era. In: Landis GA (ed) Vision-21: Interdisciplinary Science and Engineering in the Era of Cyberspace. NASA Publication CP-10129, pp 11–22

Vitali S, Glattfelder JB, Battiston S (2011) The Network of Global Corporate Control. PLoS One 6:e25995. https://doi.org/10.1371/journal.pone.0025995

Voltaire (1759) Candide, ou l'Optimisme. Cramer, Marc-Michel Rey, Jean Nourse, Lambert, and others, France

Vrba E (1995) On the connections between paleoclimate and evolution. In: Vrba ES, Denton GH, Partridge TC, Burckle LH (eds) Paleoclimate and Evolution, with Emphasis on Human Origins. Yale University Press, New Haven, pp 24–48

WA (2019) WorldAtlas. In: worldatlas.com(WA). https://www.worldatlas.com/

Wade L (2018) Feeding the gods. Science 360:1288–1292. https://doi.org/10.1126/science.360.6395.1288

Walker M, Johnsen S, Rasmussen SO, et al (2009) Formal definition and dating of the GSSP (Global Stratotype Section and Point) for the base of the Holocene using the Greenland NGRIP ice core, and selected auxiliary records. J Quat Sci 24:3–17. https://doi.org/10.1002/jqs.1227

Walsh B (2008) Heroes of the Environment 2008. In: Time. http://content.time.com/time/specials/packages/article/0,28804,1841778_1841816_1843874,00.html

Walsh MW (2018) McKinsey Advises Puerto Rico on Debt. It May Profit on the Outcome. In: New York Times. https://www.nytimes.com/2018/09/26/business/mckinsey-puerto-rico.html?partner=rss%26emc=rss

Walters J (2019) Sackler family want to settle opioids lawsuits, lawyer says. In: The Guardian. https://www.theguardian.com/us-news/2019/apr/24/sackler-family-opioid-lawsuit-settle

Wang G, Cai W, Gan B, et al (2017) Continued increase of extreme El Niño frequency long after 1.5°C warming stabilization. Nat Clim Chang 7:568–572. https://doi.org/10.1038/nclimate3351

Wang Y, Hekimi S (2015) Mitochondrial dysfunction and longevity in animals: Untangling the knot. Science 350:1204–1207. https://doi.org/10.1126/science.aac4357

Warraich H (2017) Modern death. How medicine changed the end of life, 1st edn. St. Martin's Press

Waters CN, Zalasiewicz J, Summerhayes C, et al (2016) The Anthropocene is functionally and stratigraphically distinct from the Holocene. Science 351:aad2622–aad2622. https://doi.org/10.1126/science.aad2622

Waters CN, Zalasiewicz JA, Williams M, et al (2014) A stratigraphical basis for the Anthropocene? Geol Soc London, Spec Publ 395:1–21. https://doi.org/10.1144/SP395.18

WB (2010) World development report 2010. Development and climate change. Washington D.C

WB (2012) Information and communications for development 2012: maximizing mobile

WB (2015) WB Update Says 10 Countries Move Up in Income Bracket. In: World Bank. http://www.worldbank.org/en/news/press-release/2015/07/01/new-world-bank-update-shows-bangladesh-kenya-myanmar-and-tajikistan-as-middle-income-while-south-sudan-falls-back-to-low-income

WB (2018) Life expectancy at birth, total (years). In: World Bank. https://data.worldbank.org/indicator/SP.DYN.LE00.IN

WB (2019) Gross domestic product 2018, PPP. In: World Bank. https://databank.worldbank.org/data/download/GDP_PPP.pdf

WB (2020) Energy use (kg of oil equivalent per capita) - United States, China. In: World Bank. https://data.worldbank.org/indicator/EG.USE.PCAP.KG.OE?locations=US-CN

WCED (1987) Our Common Future. World Commission on Environment and Development. Oxford University Press, Oxford

Wealth-X (2017a) World Ultra Wealth Report 2017

Wealth-X (2017b) Billionaire census 2017

Wealth-X (2019a) World Ultra Wealth Report 2019

Wealth-X (2019b) Billionaire census 2019

Weber APM, Bar-Even A (2019) Update: Improving the Efficiency of Photosynthetic Carbon Reactions. Plant Physiol 179:803–812. https://doi.org/10.1104/pp.18.01521

Weber M (1905) The Protestant Ethic and the Spirit of Capitalism. Unwin Hyman, London & Boston

Weber M (1922) Wirtschaft und Gesellschaft: Grundriss der verstehenden Soziologie, 5th edn. Johannes Winckelmann

WEF (2016) The New Plastics Economy. Rethinking the future of plastics

Wegener A (1911) Thermodynamik der Atmosphäre. J.A. Barth, Leipzig

Weiner A (2016) How Bad Writing Destroyed the World: Ayn Rand and the Literary Origins of the Financial Crisis. Bloomsbury Academic

Weir JT, Schluter D (2007) The Latitudinal Gradient in Recent Speciation and Extinction Rates of Birds and Mammals. Science 315:1574–1576. https://doi.org/10.1126/science.1135590

Weisman A (2007) The world without us. Thomas Dunne Books

Weitzman ML (1994) On the "Environmental" Discount Rate. J Environ Econ Manage 26:200–209. https://doi.org/10.1006/jeem.1994.1012

Weitzman ML (1998) Why the Far-Distant Future Should Be Discounted at Its Lowest Possible Rate. J Environ Econ Manage 36:201–208. https://doi.org/10.1006/jeem.1998.1052

Wells HG (1901) Anticipations of the Reaction of Mechanical and Scientific Progress: Upon Human Life and Thought. Dover Publications (1999)

Wells HG (1905) A Modern Utopia. Chapman & Hall

Wells HG (1918) In the Fourth Year: Anticipations of World Peace. Chatto & Windus, Macmillan Inc., Reino Unido

Weng J-K, Chapple C (2010) The origin and evolution of lignin biosynthesis. New Phytol 187:273–285. https://doi.org/10.1111/j.1469-8137.2010.03327.x

Wesselink A, Hoppe R (2011) If Post-Normal Science is the Solution, What is the Problem?: The Politics of Activist Environmental Science. Sci Technol Hum Values 36:389–412. https://doi.org/10.1177/0162243910385786

West SA, Griffin AS, Gardner A (2007) Social semantics: altruism, cooperation, mutualism, strong reciprocity and group selection. J Evol Biol 20:415–432. https://doi.org/10.1111/j.1420-9101. 2006.01258.x

WFP-EU (2017) Saving Lives. Changing Lives: WFP-EU 2017 Partnership Repor. Brussels, Belgium

WFP (2017) Food Security and Emigration. Why people flee and the impact on family members left behind in El Salvador, Guatemala and Honduras. Research report

Whillans AV, Weidman AC, Dunn EW (2016) Valuing Time Over Money Is Associated With Greater Happiness. Soc Psychol Personal Sci 7:213–222. https://doi.org/10.1177/1948550615623842

White HB (1968) Peace Among the Willows. The Political Philosophy of Francis Bacon. Springer Netherlands, Dordrecht

White L (1949) The science of culture. A study of man and civilization. Grove Press, New York

White MP, Alcock I, Grellier J, et al (2019) Spending at least 120 minutes a week in nature is associated with good health and wellbeing. Sci Rep 9:7730. https://doi.org/10.1038/s41598-019-44097-3

White TD, Asfaw B, Beyene Y, et al (2009) Ardipithecus ramidus and the Paleobiology of Early Hominids. Science 326:64–64, 75–86. https://doi.org/10.1126/science.1175802

Whiten A (2009) The identification of culture in chimpanzees and other animals: from natural history to diffusion experiments. In: Laland KN, Galef BG (eds) The question of animal culture. Harvard University Press, Cambridge, MA, pp 99–124

Whitrow GJ (1989) Time in History: Views of Time from Prehistory to the Present Day. Oxford University Press

WHO (2002) Smoking Statistics. In: World Heal. Organ. http://www.wpro.who.int/mediacentre/factsheets/fs_20020528/en/

WHO (2008) WHO Report on the Global Tobacco Epidemic, 2008. The MPOWER package. Geneva, Switzerland

WHO (2009) Global Health Risks

WHO (2014) 7 million premature deaths annually linked to air pollution. In: World Heal. Organ. https://www.who.int/mediacentre/news/releases/2014/air-pollution/en/

WHO (2016a) Tobacco. In: World Heal. Organ. http://www.who.int/mediacentre/factsheets/fs339/en/

WHO (2016b) Various facts about tobacco consumption. In: World Heal. Organ. http://www.wpro.who.int/mediacentre/factsheets/fs_201203_tobacco/en/

WHO (2018) Global Health Observatory (GHO) data. In: World Heal. Organ. https://www.who.int/gho/mortality_burden_disease/life_tables/situation_trends_text/en/

WHO (2019a) Ten threats to global health in 2019. In: World Heal. Organ. https://www.who.int/news-room/feature-stories/ten-threats-to-global-health-in-2019

WHO (2019b) Drinking-water. In: World Heal. Organ. https://www.who.int/en/news-room/factsheets/detail/drinking-water

WHO (2020) Zoonotic disease: emerging public health threats in the Region. In: East Mediterr. Reg. Off. World Heal. Organ. www.emro.who.int/about-who/rc61/zoonotic-diseases.html

Wielicki BA (2005) Changes in Earth's Albedo Measured by Satellite. Science 308:825–825. https://doi.org/10.1126/science.1106484

Wieringa J, Holleman I (2006) If cannons cannot fight hail, what else? Meteorol Zeitschrift 15:659–669. https://doi.org/10.1127/0941-2948/2006/0147

Wilford JN (2009) Flutes Offer Clues to Stone-Age Music. In: New York Times. https://www.nytimes.com/2009/06/25/science/25flute.html

Wilkinson R, Pickett K (2010) The Spirit Level: Why Equality is Better for Everyone. Penguin Books

Williams GE (2000) Geological constraints on the Precambrian history of Earth's rotation and the Moon's orbit. Rev Geophys 38:37–59. https://doi.org/10.1029/1999RG900016

Williams M, Zalasiewicz J, Haff P, et al (2015) The Anthropocene biosphere. Anthr Rev 2:196–219. https://doi.org/10.1177/2053019615591020

Williamson P, Wallace DWR, Law CS, et al (2012) Ocean fertilization for geoengineering: A review of effectiveness, environmental impacts and emerging governance. Process Saf Environ Prot 90:475–488. https://doi.org/10.1016/j.psep.2012.10.007

Wilson DS, Wilson EO (2007) Rethinking the Theoretical Foundation of Sociobiology. Q Rev Biol 82:327–348. https://doi.org/10.1086/522809

Wilson EO (1975) Sociobiology The New Synthesis, Belknap Pr. Oxford

Wilson EO (1984) Biophilia. Harvard University Press

Wilson EO (2012) The Social Conquest of Earth. Liveright Publishing Corporation

Wilson ML, Boesch C, Fruth B, et al (2014) Lethal aggression in Pan is better explained by adaptive strategies than human impacts. Nature 513:414–417. https://doi.org/10.1038/nature13727

Wiseman R (2007) Quirkology: The Curious Science of Everyday Lives. Pan Macmillan

Witzel EJM (2012) The origins of the world's mythologies. Oxford University Press, Oxford and New York

WM (2020) The future of energy after COVID-19: three scenarios. In: Wood Mackenzie. www.woodmac.com/news/feature/the-future-of-energy-after-covid-19-three-scenarios/

WMA (2009) WMA Position Statement on The Environmental Impact of Using Silver Iodide As A Cloud Seeding Agent

WMO (2017) Greenhouse Gas Bulletin. The State of Greenhouse Gases in the Atmosphere Based on Global Observations through 2016

WMO (2018) Greenhouse Gas Bulletin. The State of Greenhouse Gases in the Atmosphere Based on Global Observations through 2018

WMO (2019) High-level synthesis report of latest climate science information convened by the Science Advisory Group of the UN Climate Action Summit 2019

Woetzel J, Madgavkar A, Rifai K, et al (2016) People on the move: Global migration's impact opportunity

Wolfers A (1962) Discord And Collaboration Essays On International Politics. In: Discord And Collaboration Essays On International Politics. John Hopkins University Press, Baltimore, pp 147–165

Woolhouse MEJ, Gowtage-Sequeria S (2005) Host Range and Emerging and Reemerging Pathogens. Emerg Infect Dis 11:1842–1847. https://doi.org/10.3201/eid1112.050997

Worstall T (2016) Gabriel Zucman Shows How Irrelevant Offshore Tax Evasion Is, How Trivial. In: Forbes. https://www.forbes.com/sites/timworstall/2016/04/10/gabriel-zucman-shows-how-irrelevant-offshore-tax-evasion-is-how-trivial/#6230af836764

WRI (2014) Historical Emissions Tool. In: World Resour. Inst. http://cait.wri.org/

Wright TP (1936) Factors Affecting the Cost of Airplanes. J Aeronaut Sci 3:122–128. https://doi.org/10.2514/8.155

WTTC (2013) Tourism for tomorrow: Position paper

WTTC (2018) World Travel and Tourism Council, Press release of 22 March. In: World Travel Tour. Counc. https://www.wttc.org/about/media-centre/press-releases/press-releases/2018/one-in-five-of-all-new-jobs-created-globally-in-2017-are-attributable-to-travel-and-tourism/

WWF (2016) Living Planet Report 2016. Risk and resilience in a new era

WWF (2018a) Living Planet Report - 2018: Aiming Higher. World Wide Fund For Nature (WWF)

WWF (2018b) Agricultural water file: Farming for a drier future. In: World Wide Fund Nat. https://www.wwf.org.za/water/?25441/Agricultural-water-file-Farming-for-a-drier-future

Wynn T (2009) Hafted spears and the archaeology of mind. Proc Natl Acad Sci 106:9544–9545. https://doi.org/10.1073/pnas.0904369106

Xia L, Robock A, Tilmes S, Neely III RR (2016) Stratospheric sulfate geoengineering could enhance the terrestrial photosynthesis rate. Atmos Chem Phys 16:1479–1489. https://doi.org/10.5194/acp-16-1479-2016

Yaari ME (1965) Uncertain Lifetime, Life Insurance, and the Theory of the Consumer. Rev Econ Stud 32:137. https://doi.org/10.2307/2296058

Yamada T, Kremer RJ, de Camargo e Castro PR, Wood BW (2009) Glyphosate interactions with physiology, nutrition, and diseases of plants: Threat to agricultural sustainability? Eur J Agron 31:111–113. https://doi.org/10.1016/j.eja.2009.07.004

Yan W (2019) The nuclear sins of the Soviet Union live on in Kazakhstan. Nature 568:22–24. https://doi.org/10.1038/d41586-019-01034-8

Yoon J-E, Yoo K-C, Macdonald AM, et al (2018) Reviews and syntheses: Ocean iron fertilization experiments – past, present, and future looking to a future Korean Iron Fertilization Experiment in the Southern Ocean (KIFES) project. Biogeosciences 15:5847–5889. https://doi.org/10.5194/bg-15-5847-2018

Zachos J (2001) Trends, Rhythms, and Aberrations in Global Climate 65 Ma to Present. Science 292:686–693. https://doi.org/10.1126/science.1059412

Zachos JC, Dickens GR, Zeebe RE (2008) An early Cenozoic perspective on greenhouse warming and carbon-cycle dynamics. Nature 451:279–283. https://doi.org/10.1038/nature06588

Zagorevski DV, Loughmiller-Newman JA (2012) The detection of nicotine in a Late Mayan period flask by gas chromatography and liquid chromatography mass spectrometry methods. Rapid Commun Mass Spectrom 26:403–411. https://doi.org/10.1002/rcm.5339

Zalasiewicz J, Kryza R, Williams M (2014) The mineral signature of the Anthropocene in its deep-time context. Geol Soc London, Spec Publ 395:109–117. https://doi.org/10.1144/SP395.2

Zalasiewicz J, Williams M, Waters CN, et al (2017) Scale and diversity of the physical technosphere: A geological perspective. Anthr Rev 4:9–22. https://doi.org/10.1177/2053019616677743

Zaman MQ (2012) Modern Islamic thought in a radical age. Religious Authority and Internal Criticism. Cambridge University Press

Zamyatin Y (1924) We. E.P. Dutton (1924), New York

Zanello SB, Jackson DM, Holick MF (2000) Expression of the Circadian Clock Genes clock and period1 in Human Skin. J Invest Dermatol 115:757–760. https://doi.org/10.1046/j.1523-1747.2000.00121.x

Zeebe RE, Ridgwell A, Zachos JC (2016) Anthropogenic carbon release rate unprecedented during the past 66 million years. Nat Geosci 9:325–329. https://doi.org/10.1038/ngeo2681

Zettler ER, Mincer TJ, Amaral-Zettler LA (2013) Life in the "Plastisphere": Microbial Communities on Plastic Marine Debris. Environ Sci Technol 47:7137–7146. https://doi.org/10.1021/es401288x

Zhang L, Rashad I (2008) Obesity and time preference: the health consequences of discounting the future. J Biosoc Sci 40:97–113. https://doi.org/10.1017/S0021932007002039

Zhang YG, Pagani M, Liu Z, et al (2013) A 40-million-year history of atmospheric CO2. Philos Trans R Soc A Math Phys Eng Sci 371:20130096. https://doi.org/10.1098/rsta.2013.0096

Zhou H, Chen X, Hu T, et al (2020a) A Novel Bat Coronavirus Closely Related to SARS-CoV-2 Contains Natural Insertions at the S1/S2 Cleavage Site of the Spike Protein. Curr Biol. https://doi.org/10.1016/j.cub.2020.05.023

Zhou P, Yang X-L, Wang X-G, et al (2020b) A pneumonia outbreak associated with a new coronavirus of probable bat origin. Nature 579:270–273. https://doi.org/10.1038/s41586-020-2012-7

Zilhão J, Angelucci DE, Badal-Garcia E, et al (2010) Symbolic use of marine shells and mineral pigments by Iberian Neandertals. Proc Natl Acad Sci 107:1023–1028. https://doi.org/10.1073/pnas.0914088107

Zook HA (2001) Spacecraft Measurements of the Cosmic Dust Flux. In: Accretion of Extraterrestrial Matter Throughout Earth's History. Springer US, Boston, MA, pp 75–92

Zuboff S (2019) The Age of Surveillance Capitalism: The Fight for a Human Future at the New Frontier of Power. Public Affairs

Zucman G (2013) The Missing Wealth of Nations: Are Europe and the U.S. net Debtors or net Creditors?. Q J Econ 128:1321–1364. https://doi.org/10.1093/qje/qjt012

Index

© Springer Nature Switzerland AG 2021

F. Duarte Santos, *Time, Progress, Growth and Technology*, The Frontiers Collection,
https://doi.org/10.1007/978-3-030-55334-0